Robert L. Boylestad

QUEENSBOROUGH COMMUNITY COLLEGE
AND
THAYER SCHOOL OF ENGINEERING

Introductory Circuit Analysis

4th Edition

CHARLES E. MERRILL PUBLISHING COMPANY
A BELL & HOWELL COMPANY
COLUMBUS • TORONTO • LONDON • SYDNEY

To Else Marie, Eric, Alison, and Stacey Jo

Published by
Charles E. Merrill Publishing Company
A Bell & Howell Company
Columbus, Ohio 43216

This book was set in Times Roman
Cover photo by Larry Hamill
Cover coordination: Will Chenoweth
Copy Editor: Margaret Shaffer

International Standard Book Number:
  0-675-09938-2
Library of Congress Catalog Card Number:
  81-82530
       6 7 8 9–87 86 85 84
Printed in the United States of America

A new edition of a text, particularly a fourth edition, invariably raises questions as to whether the changes are necessary and actually improve the quality of the presentation. I believe, however, that this is the edition you will be most satisfied with. It should now have all the qualities that one strives for when a project of this type is undertaken: It should be complete, accurate, up-to-date, and properly prepared and produced. Incorporating all of these attributes in a single new edition is very difficult, and it seems a particularly elusive goal in the early editions of a text.

The impetus for improving this text has largely been the numerous comments, literally hundreds from instructors, students, and users in industry, that I have received over the years. I've tried to respond to all who have called or written and I apologize to anyone I've missed. Thank you all. Naturally, I'm looking forward to hearing your reactions to this new fourth edition so that the quality of this evolving work can be sustained.

Those familiar with the third edition will find that the essential material has been retained. Certain important subjects have been expanded, and a few topics have been added. For example, the content has been updated with new methods and devices (such as IC's); SI and IEEE notation has been used throughout; and a totally new, more useful computer chapter has been written. The lab manual has been completely reworked so that lower power and less equipment are required for lab work. The study guide now includes a summary of formulae and a glossary of terms for each chapter.

I have tried to ensure that the text is as accurate as possible. There is certainly nothing more frustrating to an instructor than to recommend a text, probably the only source of reference for the student, and then discover numerous errors. To prevent this, I have personally reworked every example in the text and every problem at the end of each chapter; several other people have done the same as a further check. The handwritten instructor's manual was completely redone with extreme care.

The primary goal of this text remains to establish a firm understanding of the basic laws of electric circuits and to develop a working knowledge of the methods of analysis used most frequently in the industrial and research communities. The text can be used at a variety of levels, determined by the detail with which each section is examined; that is, there are topics investigated that do not have to appear on a course prospectus for a student to progress through the entire text in the sequence presented.

The text is designed for a two-semester course in the basics of dc and ac networks. The first half presents the fundamental concepts of circuit analysis solely from a dc viewpoint. The second half is devoted to sinusoidal ac circuits and to selected topics such as resonance, nonsinusoi-

Preface

dal circuits, transformers, two-port system parameters, and computer use in circuit analysis.

The current source continues to be examined in detail in this edition as an aid in the student's studies of the transistor, a current-controlled device. It appears throughout the analysis of dc and sinusoidal ac circuits under such headings as Superposition, Thevenin's Theorem, and Resonance.

Figures and examples are used extensively throughout the text. In teaching through the years, I have found that such illustrations permit a level of understanding otherwise unattainable. Problems provided at the end of each chapter progress from the simple to the more complex—the most difficult denoted by an asterisk. As an aid to both the instructor and the student, each section of the text has a corresponding set of problems. Answers to selected problems are provided following the appendix material. Many of the illustrations and charts are new for this edition, chosen to interest the reader by introducing some of the advances made in recent years.

Controlled sources continue to be examined in detail to ensure that the operation involved with these active devices is clearly understood. Too frequently students are adept at applying the important theorems to dc and ac networks but are lost when confronted with a controlled source network in their subsequent electronics courses. Each major theorem has a separate section examining the effect of controlled sources in the analysis.

The Hewlett-Packard desktop scientific computer is introduced in this edition to illustrate computer techniques in network analysis. To maintain continuity, the computer chapter is the last chapter in the book; however, the student would be wise to read this chapter at the earliest opportunity and apply some of the techniques while progressing through the dc and ac analysis. Anyone intimately involved with electronics today must become familiar with the computer and its analysis and design capabilities.

Through the years many individuals have contributed to the success of this text, including the reviewers for the current edition: Robert L. Reid, Arthur J. Sweat, and Louis G. Gross. There is one individual, however, whose support and guidance resulted in the actual writing of the first edition: Professor Joseph Aidala. Through his foresight and wisdom, he created an atmosphere in our department at Queensborough that has resulted in the publication of a number of successful texts.

There is no question that a large measure of the success of the text is also due to the efforts of the people at Charles E. Merrill Publishing Company. They always make sure that the text is well designed with the content properly presented to its intended audience.

There are many hours involved in producing any text. The process requires that those around you make many adjustments to afford you sufficient time to prepare and review the publication through all its production phases. For the above I thank my wife, Else, and our children, Eric, Alison, and Stacey Jo, for all their support and love.

Robert L. Boylestad

1

UNITS AND NOTATION, 1

2

CURRENT AND VOLTAGE, 17

3

RESISTANCE, 41

4

OHM'S LAW, POWER, AND ENERGY, 65

5

SERIES AND PARALLEL CIRCUITS, 81

Contents

6

SERIES-PARALLEL NETWORKS, 113

7

METHODS OF ANALYSIS AND SELECTED TOPICS (dc), 133

8

NETWORK THEOREMS, 185

9

CAPACITORS, 223

10

MAGNETIC CIRCUITS, 263

11

INDUCTORS, 303

12

dc INSTRUMENTS, 325

13

SINUSOIDAL ALTERNATING CURRENT, 347

14

(j)

PHASORS, 397

15

SERIES AND PARALLEL CIRCUITS, 419

16

SERIES—PARALLEL ac NETWORKS, 461

17

METHODS OF ANALYSIS AND SELECTED TOPICS (ac), 477

18

NETWORK THEOREMS (ac), 511

19

P_S^q

POWER (ac), 543

20

RESONANCE, 567

21

POLYPHASE SYSTEMS, 609

22

ac METERS, 641

UNITS AND NOTATION

1

Σ SI

1.1 INTRODUCTION

It is vital that the importance of units of measurement be understood and appreciated early in the development of a technically oriented background. Too frequently their effect on the most basic substitution is ignored. Consider, for example, the following very fundamental physics equation:

$$\boxed{F = ma}$$

$$F = \text{force}$$
$$m = \text{mass} \qquad \textbf{(1.1)}$$
$$a = \text{acceleration}$$

Assume, for the moment, that the following data are obtained:

$$m = 10 \text{ kilograms}$$
$$a = 50 \text{ ft/s}^2$$

and F is desired in newtons.

Often, without a second thought or consideration, the numerical values are simply substituted into the equation with the result

$$F = ma = 10 \cdot 50 = \cancel{500 \text{ newtons}}$$

As indicated above, the solution is quite incorrect. On occasion, since *kilo* indicates a multiplier of 1000, then 10,000 is substituted for m, and not simply 10. This will result in a solution of 500,000 newtons—an answer of ridiculous proportions for the data obtained. The point to be made, and made quite strongly, is that *due consideration must be given to the units of measurement of each term of an equation before substituting numerical values*. Returning to Eq. (1.1), we should first recognize that the kilogram *is* a unit of mass and acceleration *is* measured in distance/time^2 to insure that the data are in the form necessary for substitution. It must then be noted, or at least understood, that *kilo* is a prefix referring to a multiplying factor of 10^3 (prefixes of this type will be examined in detail in this chapter). The next important consideration is to *note whether each quantity is in the same system of units*. In the next section we will find that if the force is desired in the MKS system of units (newtons), the mass must be in kilograms, as provided, but acceleration must be in m/s^2. Another problem associated with units is now evident: How do we convert from ft/s^2 to m/s^2? One method will be presented in this chapter. For now it is given that

$$50 \text{ ft/s}^2 \cong 15.2 \text{ m/s}^2$$

Another question may now be asked: What are the relative sizes of a meter and a foot? Or, when the result is obtained: How much force can I relate to one (1) newton? These questions will also be considered in the sections to follow.

Substituting the proper values of $m = 10$ kilograms and $a = 15.2$ m/s^2, we have

$$F = ma = (10)(15.2) = \textbf{152 newtons}$$

which is certainly considerably smaller than the 500 or 500,000 newtons obtained earlier. Another important point can now be made: *If a unit of measurement is applicable to a result or piece of data, then it must be applied to the numerical value*. To conclude that $F = 152$ without including the unit of measurement, newtons, is *completely* meaningless.

A final important consideration following any mathematical calculation should be an examination of the result to see whether it falls within expected limits. Certainly, a result of 500,000 newtons should generate some questions as to its validity when consideration is given to the size of the mass and its acceleration. But some feeling for the newton as a unit of measurement, the mass of a kilogram, and an acceleration in m/s^2 must be established before a result can be properly examined.

Equation (1.1) is not a difficult one. A simple algebraic manipulation will result in the solution for any one of the three variables. However, in light of the number of questions arising from this equation, the reader may wonder if the difficulty associated with an equation will increase at the same rate as the number of terms in the equation. In the broad sense, this will not be the case. There is, of course, more room for a mathematical error with a more complex equation, but once the proper system of units is chosen and each term properly found in that system, there should be very little added difficulty associated with an equation requiring an increased number of mathematical calculations.

In review, before substituting numerical values into an equation, be absolutely sure of the following:

1. Each quantity has the proper unit of measurement as defined by the equation.
2. The proper magnitude of each quantity as determined by the defining equation is substituted.
3. Each quantity is in the same system of units (or as defined by the equation).
4. The magnitude of the result is of a reasonable nature when compared to the level of the substituted quantities.
5. The proper unit of measurement is applied to the result.

1.2 SYSTEMS OF UNITS

In the past, the *systems of units* most commonly employed were the English and metric, as outlined by Table 1.1. Note that while the English system is based on a single standard, the metric is subdivided into two interrelated standards: the MKS and CGS. Fundamental quantities of these systems are compared in Table 1.1 along with their abbreviations. The MKS and CGS systems draw their names from the unit of measurements used with each system; the MKS system uses *M*eters, *K*ilograms, and *S*econds, while the CGS system uses *C*entimeters, *G*rams, and *S*econds.

Understandably, the use of more than one system of units in a world that finds itself continually shrinking in size, due to advanced technical developments in communications and transportation, would introduce unnecessary complications to the basic understanding of any technical data. The need for a standard set of units to be adopted by all nations has become increasingly obvious. The International Bureau of Weights and Measures located at Sèvres, France, has been the host for the General Conference of Weights and Measures, attended by representatives from all nations of the world. In 1960, the General Conference adopted a system called Le Système International d'Unités (International System of Units) which has the international abbreviation SI. Since then, it has been adopted by the Institute of Electrical and Electronic Engineers (IEEE) in

TABLE 1.1 *Comparison of the English and metric systems of units.*

English	Metric		SI
	MKS	**CGS**	
Length: Yard (yd) (0.914 m)	Meter (m) (39.37 in.) (100 cm)	Centimeter (cm) (2.54 cm = 1 in.)	**Meter (m)**
Mass: Slug (14.6 kg)	Kilogram (kg) (1000 g)	Gram (g)	**Kilogram (kg)**
Force: Pound (lb) (4.45 N)	Newton (N) (100,000 dynes)	Dyne	**Newton (N)**
Temperature: Fahrenheit (°F) $\left(=\frac{9}{5}\,°C + 32\right)$	Celsius or Centigrade (°C) $\left(=\frac{5}{9}\,(°F - 32)\right)$	Centigrade (°C)	**Kelvin (K)** **K = 273.15 + °C**
Energy: Foot-Pound (ft-lb) (1.356 joules)	Newton-Meter (Nm) or Joule (J) (0.7378 ft-lb)	Dyne-Centimeter or Erg (1 joule = 10^7 ergs)	**Joule (J)**
Time: Second (s)	Second (s)	Second (s)	**Second (s)**

1965 and by the USASI in 1967 as a standard for all scientific and engineering literature.

The inevitable changeover to the metric system has already resulted in the use of *both* miles/hour (mi/h) and kilometers/hour (km/h) on some new road signs and the distribution of English-to-metric conversion charts as advertising literature by some firms. In fact, calculators are now available that are designed specifically to convert from one system to the other.

For comparison, the SI units of measurement and their abbreviations appear in Table 1.1. These abbreviations are those usually applied to each unit of measurement, and they were carefully chosen to be the most effective. Therefore, it is important that they be used whenever applicable to insure universal understanding. Note the similarities of the SI system to the MKS system. This text will employ, whenever possible and practical, all of the major units and abbreviations of the SI system in an effort to support the need for a universal system. For those requiring further information on the SI system, a complete kit has been assembled for general distribution by the American Society for Engineering Education (ASEE)*.

What is the true significance of a system of units? How will it affect our analysis? The example of Section 1.1 clearly demonstrated the necessity to stay within the boundaries of the chosen system of units. In the example, since the MKS system was chosen (F = newtons), *m must* be measured in kilograms (10), and not grams (10×10^3), and acceleration in m/s^2 and not ft/s^2. *Unless otherwise stated, the unit of measurement of each term of an equation will be within the structure of a system of units;* that is, whenever the MKS system is indicated, only those units of measurement in the second column of Table 1.1 can be employed. If a unit of measurement is not in the proper system, it must be found through a conversion process (Section 1.4).

Figure 1.1 should help the reader develop some feeling for the relative magnitudes of the units of measurement of each system of units. It is important that dimensions such as 0.6 m and 15.2 N have some physical meaning, rather than simply appearing as magnitudes of some mysterious quantity required by an equation. Figure 1.1 should also help in achieving the very useful ability to think in more than one system of units. Note in the figure the relatively small magnitude of the units of measurement for the CGS system.

A standard exists for each unit of measurement of each system. Some units are quite interesting. The *meter* was defined in 1960 as 1,650,763.73 wavelengths of the orange-red light of krypton 86. It was originally defined in 1790 to be 1/10,000,000 the distance between the equator and either pole at sea level. This length is preserved on a platinum-iridium bar at the International Bureau of Weights and Measures at Sèvres, France. The *kilogram* is defined as a mass equal to 1000 times the mass of one cubic centimeter of pure water at 4°C. This standard is preserved in the form of a platinum-iridium cylinder in Sèvres. The *second* was originally defined as 1/86,400 of the mean solar day. It was redefined in 1960 as 1/31,556,925.9747 of the tropical year 1900.

1.3 SCIENTIFIC NOTATION

It should be apparent from the relative magnitude of the various units of measurement that very large and very small numbers will frequently be encountered in the study

---

*American Society for Engineering Education (ASEE), One Dupont Circle, Suite 400, Washington, D.C. 20036.

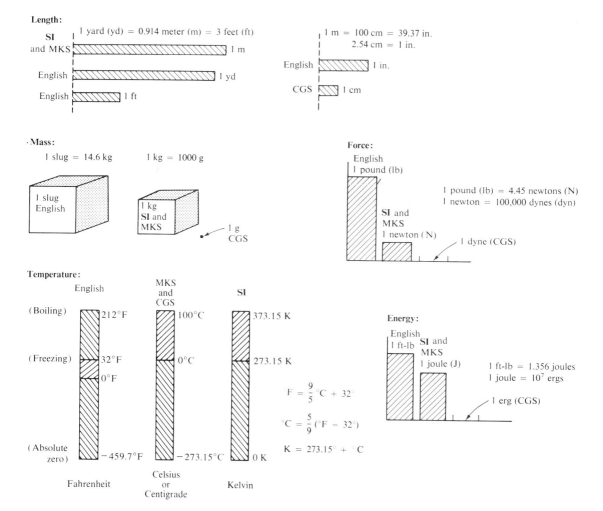

FIG. 1.1 *Comparison of units of the various systems of units.*

of the sciences. To ease the difficulty of mathematical operations with numbers of extreme size, *scientific notation* is usually employed. This notation takes full advantage of the mathematical properties of powers of 10. The notation used to represent numbers that are integer powers of 10 is as follows:

$$
\begin{array}{ll}
1 = 10^0 & 1/10 = \quad 0.1 = 10^{-1} \\
10 = 10^1 & 1/100 = \quad 0.01 = 10^{-2} \\
100 = 10^2 & 1/1000 = \quad 0.001 = 10^{-3} \\
1000 = 10^3 & 1/10,000 = 0.0001 = 10^{-4}
\end{array}
$$

A quick method of determining the proper power of 10 is to place a caret mark to the right of the numeral 1 wherever it may occur; then count from this point the number of places to the right or left before arriving at the decimal point. Moving to the right indicates a positive power of 10, while moving to the left indicates a negative power. For example,

$$10{,}000.0 = 1\underbrace{0\,.\,0\,0\,0}_{1\ \ 2\ \ 3\ \ 4}\,. = 10^{+4}$$

$$0.00001 = 0.\underbrace{0\,0\,0\,0\,1}_{5\ \ 4\ \ 3\ \ 2\ \ 1} = 10^{-5}$$

Since some of these powers of 10 appear quite frequently, a written form of abbreviation has been adopted (as indicated in Table 1.2), which when written in conjunction with the unit of measurement eliminates the need to include the power of 10 in numerical form.

TABLE 1.2

Power of 10	Prefix	Abbreviation
10^{12}	Tera	T
10^{9}	Giga	G
10^{6}	Mega	M
10^{3}	Kilo	k
10^{-3}	Milli	m
10^{-6}	Micro	μ
10^{-9}	Nano	n
10^{-12}	Pico	p

EXAMPLES

$$1{,}000{,}000 \text{ ohms} = 1 \times 10^6 \text{ ohms}$$
$$= 1 \text{ megohm (M}\Omega)$$

$$100{,}000 \text{ meters} = 100 \times 10^3 \text{ meters}$$
$$= 100 \text{ kilometers (km)}$$

$$0.0001 \text{ second} = 0.1 \times 10^{-3} \text{ second}$$
$$= 0.1 \text{ millisecond (ms)}$$

$$0.000001 \text{ farad} = 1 \times 10^{-6} \text{ farad}$$
$$= 1 \text{ microfarad } (\mu\text{F})$$

Some important mathematical relationships applying to powers of 10 are listed below with a few examples. In each case, n and m can be any positive or negative real number. However, the positive and negative sign must be included when substituting into the equation.

$$(10^n)(10^m) = 10^{(n+m)} \tag{1.2}$$

EXAMPLES

$$(1000)(10{,}000) = (10^3)(10^4) = 10^{(3+4)} = 10^7$$
$$(0.00001)(100) = (10^{-5})(10^2) = 10^{(-5+2)} = 10^{-3}$$

$$\frac{10^n}{10^m} = 10^{(n-m)} \tag{1.3}$$

EXAMPLES

$$\frac{100{,}000}{100} = \frac{10^5}{10^2} = 10^{(5-2)} = 10^3$$

$$\frac{1000}{0.0001} = \frac{10^3}{10^{-4}} = 10^{(3-(-4))} = 10^{(3+4)} = 10^7$$

$$\boxed{(10^n)^m = 10^{(nm)}} \qquad (1.4)$$

EXAMPLES

$$(100)^4 = (10^2)^4 = 10^{(2)(4)} = 10^8$$
$$(1000)^{-2} = (10^3)^{-2} = 10^{(3)(-2)} = 10^{-6}$$
$$(0.01)^{-3} = (10^{-2})^{-3} = 10^{(-2)(-3)} = 10^6$$

Let us now consider a few examples demonstrating the use of powers of 10 with arbitrary numbers. When working with arbitrary numbers, we can separate the operations involving powers of 10 from those of the whole numbers.

Addition:

$$\begin{aligned}
6300 + 75{,}000 &= (6.3)(1000) + (75.0)(1000) \\
&= 6.3 \times 10^3 + 75.0 \times 10^3 \\
&= (6.3 + 75.0) \times 10^3 \\
&= \mathbf{81.30 \times 10^3}
\end{aligned}$$

Subtraction:

$$\begin{aligned}
960{,}000 - 40{,}000 &= (96.0)(10{,}000) - (4.0)(10{,}000) \\
&= 96.0 \times 10^4 - 4.0 \times 10^4 \\
&= (96.0 - 4.0) \times 10^4 \\
&= \mathbf{92.0 \times 10^4}
\end{aligned}$$

Multiplication:

$$\begin{aligned}
(0.0002)(0.000007) &= [(2)(0.0001)][(7)(0.000001)] \\
&= (2 \times 10^{-4})(7 \times 10^{-6}) \\
&= (2)(7) \times (10^{-4})(10^{-6}) \\
&= \mathbf{14.0 \times 10^{-10}}
\end{aligned}$$

$$\begin{aligned}
(340{,}000)(0.00061) &= (3.4 \times 10^5)(61 \times 10^{-5}) \\
&= (3.4)(61) \times (10^5)(10^{-5}) \\
&= \mathbf{207.40}
\end{aligned}$$

Division:

$$\begin{aligned}
\frac{0.00047}{0.002} &= \frac{47.0 \times 10^{-5}}{2 \times 10^{-3}} = \left(\frac{47.0}{2}\right) \times \left(\frac{10^{-5}}{10^{-3}}\right) \\
&= \mathbf{23.50 \times 10^{-2}}
\end{aligned}$$

$$\begin{aligned}
\frac{690{,}000}{0.00000013} &= \frac{69 \times 10^4}{13 \times 10^{-8}} = \left(\frac{69}{13}\right) \times \left(\frac{10^4}{10^{-8}}\right) \\
&= \mathbf{5.310 \times 10^{12}}
\end{aligned}$$

Powers:

$$(0.00003)^3 = (3 \times 10^{-5})^3 = (3)^3 \times (10^{-5})^3$$
$$= \mathbf{27.0 \times 10^{-15}}$$
$$(90{,}800{,}000)^2 = (9.08 \times 10^7)^2 = (9.08)^2 \times (10^7)^2$$
$$= \mathbf{82.4464 \times 10^{14}}$$

The following examples include units of measurement.

EXAMPLES

a. 41,200 m is equivalent to 41.2×10^3 m = **41.20 km.**

b. 0.00956 g is equivalent to 9.56×10^{-3} g = **9.56 mg.**

c. 0.000768 s is equivalent to 768×10^{-6} s = **768.0 μs.**

d. $\dfrac{8400 \text{ m}}{0.06} = \dfrac{8.4 \times 10^3 \text{ m}}{6 \times 10^{-2}} = \left(\dfrac{8.4}{6}\right) \times \left(\dfrac{10^3}{10^{-2}}\right) \text{m}$

$\qquad = 1.4 \times 10^5 \text{ m} = 140 \times 10^3 \text{ m} = \mathbf{140.0\ km}$

e. $(0.0003)^4 \text{ s} = (3 \times 10^{-4})^4 \text{ s} = 81 \times 10^{-16} \text{ s}$

$\qquad = 0.0081 \times 10^{-12} \text{ s} = \mathbf{0.0081\ ps}$

To demonstrate the amount of work saved and the reduced possibility of error that results by using powers of 10, consider finding the solution to the last example in the following manner:

$$
\begin{array}{r}
0.0003 \\
\times\ 0.0003 \\
\hline
0.00000009 \\
\times\ 0.0003 \\
\hline
0.000000000027 \\
\times\ 0.0003 \\
\hline
0.0000000000000081 = 81 \times 10^{-16} \text{ s} = \mathbf{0.0081\ ps}
\end{array}
$$

1.4 CONVERSION WITHIN AND BETWEEN SYSTEMS OF UNITS

Conversions within and between systems of units usually present some difficulty for most students, but conversions are very simple operations if approached correctly. Consider the process of converting 10,000 cm to meters. That is,

$$10{,}000 \text{ cm} = (?) \text{ m}$$

Certainly, multiplying 10,000 cm by one (1) will not change the magnitude. Therefore,

$$(10{,}000 \text{ cm})(1) = (?) \text{ m}$$

Let us now consult our tables to determine the conversion factor between centimeters and meters. The result is

$$1 \text{ m} = 100 \text{ cm}$$

Dividing both sides by 100 cm will produce the following:

$$\frac{1 \text{ m}}{100 \text{ cm}} = \frac{100 \text{ cm}}{100 \text{ cm}} = (1)$$

Substituting this one (1) into the above equation will result in

$$(10,000 \cancel{\text{ cm}}) \left(\frac{1 \text{ m}}{100 \cancel{\text{ cm}}} \right) = (?) \text{ m}$$

Note that the centimeter units cancel, leaving the *desired* unit of measurement. Dividing through, we find that

$$\frac{10,000 \text{ m}}{100} = \textbf{100 m}$$

Let us now briefly review the mathematical manipulations applied above. We set up the factor one (1) or the conversion units such that the original unit of measurement was canceled, leaving the desired unit of measurement. In other words, the one (1) factor should be initiated by placing the units to be canceled in the denominator and then establishing a proper factor of one (1) through the conversion relationship. The above is a technique that, when properly learned, will eliminate the impossible answers that often occur when converting units. It would be time well spent to master the above technique to be used if another approach is unavailable.

Using scientific notation, we obtain

$$(10,000 \text{ m})(1) = (10^4 \cancel{\text{ m}}) \left(\frac{10^2 \text{ cm}}{1 \cancel{\text{ m}}} \right)$$

$$= (10^4)(10^2) \text{ cm} = 10^6 \text{ cm}$$

The above procedure can be carried on continuously until the desired unit of measurement remains in the numerator of the last factor (1) employed.

The following examples will develop some familiarity with the procedure just described.

EXAMPLES

a. Convert 6000 m to centimeters.

$$6000 \text{ m} = (6 \times 10^3 \text{ m})(1) = (6 \times 10^3 \cancel{\text{ m}}) \left(\frac{10^2 \text{ cm}}{1 \cancel{\text{ m}}} \right)$$

$$= \textbf{6.0} \times \textbf{10}^5 \textbf{ cm}$$

b. Convert 1.8 g to kilograms.

$$1.8 \text{ g} = (1.8 \text{ g})(1) = (1.8 \cancel{\text{ g}}) \left(\frac{1 \text{ kg}}{10^3 \cancel{\text{ g}}} \right) = \textbf{1.80} \times \textbf{10}^{-3} \textbf{ kg}$$

c. Convert 3 cm to millimeters. In this case, instead of finding the conversion factor between centimeters and millimeters, do the following:

$$3 \text{ cm} = (3 \text{ cm})(1)(1) = (3 \text{ cm})\left(\frac{1 \text{ m}}{10^2 \text{ cm}}\right)\left(\frac{1 \text{ mm}}{10^{-3} \text{ m}}\right)$$

$$= 3 \times 10^{-2} \times 10^3 \text{ mm} = \textbf{30.0 mm}$$

d. Convert 4.6 mm to kilometers.

$$4.6 \text{ mm} = (4.6 \text{ mm})\left(\frac{10^{-3} \text{ m}}{1 \text{ mm}}\right)\left(\frac{1 \text{ km}}{10^3 \text{ m}}\right)$$

$$= 4.6 \times 10^{-3} \times 10^{-3} \text{ km} = \textbf{4.60} \times \textbf{10}^{-6} \textbf{ km}$$

e. Convert 0.0432 h to milliseconds.

$$0.0432 \text{ h} = (0.0432 \text{ h})(1)(1)(1)$$

$$= (0.0432 \text{ h})\left(\frac{60 \text{ min}}{1 \text{ h}}\right)\left(\frac{60 \text{ s}}{1 \text{ min}}\right)\left(\frac{1 \text{ ms}}{10^{-3} \text{ s}}\right)$$

$$= \textbf{155.52} \times \textbf{10}^3 \textbf{ ms}$$

Other methods for converting from one system to another will become apparent after acquiring some confidence in the use of the conversion factors. This is one method, however, that can help correct some of the impossible results that occur from lack of a systematic approach.

Always examine the result of any conversion to determine whether it makes sense; that is, compare it to the magnitude of that quantity in the original system of units.

1.5 LAWS OF UNITS

A law of units seldom applied to its fullest potential is the following: *The units of measurement of each distinct term of an equation must be the same.*

In the following series, the units of measurement for the last two terms are dimensionally incorrect. The calculations involved in obtaining these incorrect results should therefore be checked over carefully before proceeding with the problem.

$$\text{Time (h)} = 4 \text{ h} + \underbrace{\frac{30 \text{ min}}{0.5 \text{ h}}}$$

$$+ (2 \text{ h})(6 \text{ m})\left(\frac{5}{\text{m}}\right) + \frac{10}{\text{h}} + (16 \text{ h})^2$$

The units of some newly defined terms can be determined through the equality that must exist on both sides of an equal sign. For example, the units of acceleration can be found from the following equation since those of velocity and time are known:

$$\text{Acceleration (?)} = \frac{\text{velocity (m/s)}}{\text{time (s)}}$$

$$= \frac{\text{m/s}}{\text{s}} = \textbf{m/s}^2$$

It should also be pointed out that only one system of units should be used with any one equation. That is, if the SI system of units is used on one side of an equation, it should also be used on the other side—not the CGS or English systems. For example, the force between two charged bodies is given by

$$F = \frac{kQ_1Q_2}{r^2}$$

where F = force in newtons (SI), $Q_{1,2}$ = charge on each in coulombs (SI), k = constant = 9.0×10^9 (SI), r = meters (SI).

Since we are dealing with the SI system of units, r must be measured in meters, not centimeters or inches; that is, if $r = 1000$ cm $= 10$ m, the number 10 is substituted into the equation, not 1000.

For another example, if the speed of an object is required in meters per second (SI) and the distance is given in inches and the time in minutes, the distance must be converted to meters and the time to seconds if the proper result is to be obtained. That is,

$$\frac{d(\text{in.})}{t(\text{min})(60 \text{ s/min})(39.37 \text{ in./m})}$$

1.6 SYMBOLS

Throughout the text, various symbols will be employed that the reader may not have had occasion to use. Some are defined in Table 1.3, and others will be defined in the text as the need arises.

1.7 CONVERSION TABLES

Conversion tables such as appearing in Appendix A can be very useful when time does not permit the application of methods described in this chapter. However, even though such tables appear easy to use, frequent errors occur because the operations appearing at the head of the table are not properly performed. In any case, when using such tables, try to establish mentally some order of magnitude for the quantity to be determined as compared to the magnitude of the quantity in its original set of units. This simple operation should prevent a number of the impossible results that may occur if the conversion operation is improperly applied.

TABLE 1.3

Symbol	Meaning
$\neq$	Not equal to $6.12 \neq 6.13$
$>$	Greater than $4.78 > 4.20$
$\gg$	Much greater than $840 \gg 16$
$<$	Less than $430 < 540$
$\ll$	Much less than $0.002 \ll 46$
$\geq$	Greater than or equal to $x \geq y$ is satisfied for $y = 3$ and $x > 3$ or $x = 3$
$\leq$	Less than or equal to $x \leq y$ is satisfied for $y = 3$ and $x < 3$ or $x = 3$
$\cong$	Approximately equal to $3.14159 \cong 3.14$
Σ	Sum of $\Sigma (4 + 6 + 8) = 18$
$\vert\ \vert$	Absolute magnitude of $\vert a \vert = 4$, where $a = -4$ or $+4$
$\therefore$	Therefore $x = \sqrt{4} \quad \therefore x = \pm 2$

For example, consider the following from such a conversion table:

To convert from	To	Multiply by
Miles	Meters	1.609×10^3

A conversion of 2.5 miles to meters would require that we multiply 2.5 by the conversion factor. That is,

$$2.5 \text{ mi } (1.609 \times 10^3) = 4.0225 \times 10^3 \text{ m}$$

A conversion from 4000 meters to miles would require a division process:

$$\frac{4000 \text{ m}}{1.609 \times 10^3} = 2486.02 \times 10^{-3} = 2.48602 \text{ mi}$$

In each of the above there should have been little difficulty realizing that 2.5 miles would convert to a few thousand meters, and 4000 meters would be only a few miles. As indicated above, this kind of prior thinking will eliminate the possibility of ridiculous conversion results.

PROBLEMS

Note: More difficult problems are indicated with an asterisk (*) throughout the text.

Section 1.3

1. Express the following numbers as powers of 10:
   a. 10,000   **b.** 0.0001   **c.** 1000
   d. 1,000,000   **e.** 0.0000001   **f.** 0.00001

2. Using only those powers of 10 listed in Table 1.2, express the following numbers in what seems to you the most logical form for future calculations:
   a. 15,000   **b.** 0.03000   **c.** 7,400,000
   d. 0.0000068   **e.** 0.00040200   **f.** 0.0000000002

Perform each of the following operations and express the result as a power of 10:

3. **a.** $(100)(100)$ **b.** $(0.01)(1000)$ **c.** $(10^3)(10^6)$
   d. $(1000)(0.00001)$ **e.** $(10^{-6})(10,000,000)$
   f. $(10,000)(10^{-8})(10^{35})$

4. **a.** $\dfrac{100}{1000}$ **b.** $\dfrac{0.01}{100}$ **c.** $\dfrac{10,000}{0.00001}$
   d. $\dfrac{0.0000001}{100}$ **e.** $\dfrac{10^{38}}{0.000100}$ **f.** $\dfrac{(100)^{1/2}}{0.01}$

5. **a.** $(100)^3$ **b.** $(0.0001)^{1/2}$ **c.** $(10,000)^8$
   d. $(0.00000010)^9$.

6. **a.** $(-0.001)^2$ **b.** $\dfrac{(100)(10^{-4})}{10}$ **c.** $\dfrac{(0.01)^2(100)}{10.000}$
   d. $\dfrac{(10^2)(10,000)}{0.001}$ **e.** $\dfrac{(0.0001)^3(100)}{1,000,000}$ ***f.** $\dfrac{[(100)(0.01)]^{-3}}{[(100)^2][0.001]}$

Perform the following operations:

*7. **a.** $\dfrac{(300)^2(100)}{10^4}$ **b.** $[(40,000)^2][(20)^{-3}]$

   c. $\dfrac{(60,000)^2}{(0.02)^2}$ **d.** $\dfrac{(0.000027)^{1/3}}{210,000}$

   e. $\dfrac{[(4000)^2][300]}{0.02}$ **f.** $[(0.000016)^{1/2}][(100,000)^5][0.02]$

   g. $\dfrac{[(0.003)^3][(0.00007)^2][(800)^2]}{[(100)(0.0009)]^{1/2}}$ (a challenge)

Section 1.4

Convert the following:

8. **a.** 1.5 min to seconds
   b. 0.04 h to seconds
   c. 0.05 s to microseconds
   d. 0.16 H to millihenries
   e. 0.00000012 s to nanoseconds
   f. 3,620,000 ohms (Ω) to megohms
   g. 1020 mm to meters

9. a. 0.1 μF (microfarad) to picofarads

   b. 0.467 kΩ to ohms

   c. 63.9 mH to henries

   d. 69 cm to kilometers

   e. 3.2 h to milliseconds

   f. 0.016 mH to microhenries

   g. 60 sq cm (cm²) to square meters (m²)

*10. a. 100 in. to meters

   b. 4 ft to meters

   c. 6 lb to newtons

   d. 60,000 dyn to pounds

   e. 150,000 cm to feet

   f. 0.002 mi to meters (5280 ft = 1 mi)

   g. 7800 m to yards

Section 1.5

11. Which terms of the following expressions are dimensionally incorrect?

   a. Length = 4 m + (6 m/s)(40 s) + (18 min)(7 ft/s)

   + (10 ft)(2 s) + (72 in.)(6 cm/s)

   + (80 ft²)(6 ft⁻¹)

   b. Time = 4 h + 6 min + $\dfrac{40 \text{ h}}{200 \text{ min}}$ + (50 s)(2 min)

   + $\dfrac{40 \text{ m}}{6 \text{ m/s}}$

12. Find the velocity in miles per hour (mi/h) of a mass that travels 50 ft in 2 minutes.

13. How long in minutes will it take a car traveling at 100 mi/h to travel the length of a football field (100 yd)?

*14. Find the distance in meters that a mass traveling at 600 cm/s will cover in 0.016 hour.

15. Convert 6 mi/h to meters per second.

16. Eventually, all speed limit signs in the U.S. will be in km/h. How fast in miles per hour would you be traveling if you were going the limit of 100 km/h? What is the conversion factor between the two?

Section 1.7

17. Using Appendix A, determine the number of:

   a. B.t.u. in 5 joules of energy.

   b. cubic meters in 24 ounces of a liquid.

   c. seconds in 1.4 days.

   d. pints in 1 cubic meter of a liquid.

GLOSSARY

CGS system The system of units employing the *Centimeter*, *Gram*, and *Second* as its fundamental units of measure.

English system The system of units that employs a yard, slug, and second as its fundamental units of measure.

Joule (J)   A unit of measurement for energy in the SI or MKS system. Equal to 0.7378 foot-pound in the English system and 10^7 ergs in the CGS system.

Kelvin (K)   A unit of measurement for temperature in the SI system. Equal to $273.15° + °C$ in the MKS and CGS systems.

Kilogram (kg)   A unit of measure for mass in the SI and MKS systems. Equal to 1000 grams in the CGS system.

Meter (m)   A unit of measure for length in the SI and MKS systems. Equal to 1.094 yards in the English system and 100 centimeters in the CGS system.

MKS system   The system of units employing the *Meter*, *Kilo-gram*, and *Second* as its fundamental units of measure.

Newton (N)   A unit of measurement for force in the SI and MKS systems. Equal to 100,000 dynes in the CGS system.

Pound (lb)   A unit of measurement for force in the English system. Equal to 4.45 newtons in the SI or MKS system.

Second (s)   A unit of measurement for time in the SI, MKS, English, and CGS systems.

Scientific notation   A method for describing very large and very small numbers through the use of powers of 10.

SI system   The system of units adopted by the IEEE in 1965 and the USASI in 1967 as the International System of Units (*Système International* d'Unités).

Slug   A unit of measure for mass in the English system. Equal to 14.6 kilograms in the SI or MKS system.

2.1 ATOMS AND THEIR STRUCTURE

A basic understanding of the fundamental concepts of current and voltage requires a degree of familiarity with the atom and its structure. The hydrogen atom is made up of two basic particles, the *proton* and the *electron*, in the relative positions shown in Fig. 2.1(a). The *nucleus* of the hydrogen atom is the proton, a positively charged particle. *The orbiting electron carries a negative charge that is equal in magnitude to the positive charge of the proton.* In all other elements, the nucleus also contains *neutrons,* which are slightly heavier than protons and have no electrical charge. The helium atom, for example, has two neutrons in addition to two electrons and two protons as shown in Fig. 2.1(b). *In all neutral atoms the number of electrons is equal to the number of protons,* with the number of neutrons determined by the difference between the mass number and number of protons. The mass of the electron is 9.11×10^{-28} g, and that of the proton is 1.672×10^{-24} g. The mass of the neutron is 1.672×10^{-24} g. The mass of the proton (or neutron) is therefore approximately 1836 times that of the electron. The radii of the proton, neutron, and electron are all of the order of magnitude of 2×10^{-15} m.

For the hydrogen atom, the radius of the smallest orbit followed by the electron is about 5×10^{-11} m. The radius

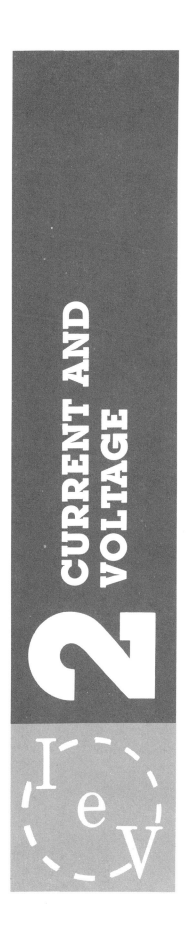

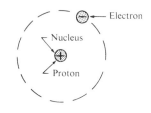

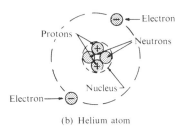

(a) Hydrogen atom

(b) Helium atom

FIG. 2.1 *The hydrogen and helium atoms.*

of this orbit is approximately 25,000 times that of the basic constituents of the atom. This is equivalent to a sphere the size of a dime rotating about another sphere of the same size more than a quarter of a mile away.

Different atoms will have various numbers of electrons in the concentric shells about the nucleus. The first shell, which is closest to the nucleus, can contain only two electrons. If an atom should have three electrons, the third must go to the next shell. The second shell can contain a maximum of eight electrons, the third 18, and the fourth 32, as determined by the equation $2n^2$, where n is the shell number. These shells are usually denoted by a number ($n = 1, 2, 3, \ldots$) or letter ($n = k, l, m, \ldots$).

Each shell is then broken down into subshells, where the first subshell can contain a maximum of 2 electrons, the second subshell 6 electrons, the third 10 electrons, and the fourth 14, as shown in Fig. 2.2. The subshells are usually denoted by the letters s, p, d, and f, in that order, outward from the nucleus.

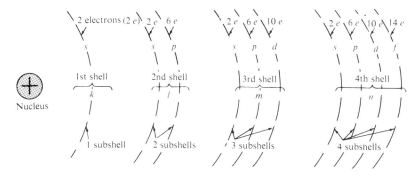

FIG. 2.2 *Shells and subshells of the atomic structure.*

It has been determined by experimentation that *unlike charges attract, and like charges repel.* The force of attraction or repulsion can be determined by Coulomb's law:

$$F \text{ (attraction or repulsion)} = \frac{kQ_1Q_2}{r^2} \qquad (2.1)$$

where F is in newtons, $k = $ constant $= 9.0 \times 10^9$, Q_1 and Q_2 are the charges in coulombs, and r is the distance in meters between the two charges.

In the atom, therefore, electrons will repel each other, and protons and electrons will attract each other. Since the nucleus consists of many positive charges (protons), a strong attractive force exists for the electrons in orbits close to the nucleus [note the effects of a large charge Q and a small distance r in Eq. (2.1)]. As the distance between the nucleus and the orbital electrons increases, the binding force diminishes until it becomes minimal at the outermost subshell (largest r). Due to the weaker binding forces, less energy must be expended to remove an electron from an outer subshell than from an inner subshell. Also, it is generally true that electrons are more readily removed from atoms having outer subshells that are incomplete *and*, in addition, possess few electrons. These properties of the atom that permit the removal of electrons under certain conditions are essential if motion of charge is to be created. Without this motion, this text could venture no further, since our basic quantities rely on it.

Copper is the most commonly used metal in the electrical field today. An examination of its atomic structure will help explain why it has such widespread applications. The copper atom (Fig. 2.3) has one more electron than needed to complete the first three shells. This incomplete outermost subshell, possessing only one electron, and the distance between this electron and the nucleus, indicate that the twenty-ninth electron is loosely bound to the copper atom. If this twenty-ninth electron gains sufficient energy from

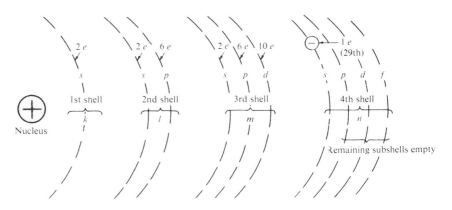

the surrounding media to leave its parent atom, it is called a *free electron*. In one cubic inch of copper at room temperature there are approximately $1.4 \times 10^{+24}$ free electrons. Copper also has the advantage of being able to be drawn into long thin wires (ductility) or worked into many different shapes (malleability). Other metals that exhibit the same properties as copper, but to a different degree, are silver, gold, platinum, and aluminum. Due to the cost factor, only aluminum has appeared as a competitor for copper for commercial use.

FIG. 2.3 *The copper atom.*

2.2 THE AMPERE

Consider a short length of copper wire cut with an imaginary perpendicular plane, producing the circular cross section shown in Fig. 2.4. At room temperature with no external forces applied, there exists within the copper wire the random motion of free electrons created by the thermal energy that the electrons gain from the surrounding media. When an atom loses its free electron, it acquires a net positive charge and is referred to as a *positive ion*. The free electron is able to move within these positive ions and leave the general area of the parent atom, while the positive ions only oscillate in a mean fixed position. For this reason, *the free electron is the charge carrier in the copper wire or in any other solid conductor of electricity.*

An array of positive ions and free electrons is depicted in Fig. 2.5. Within this array, the free electrons find themselves continually gaining or losing energy by virtue of their changing direction and velocity. Some of the factors responsible for this random motion include (1) the collisions with positive ions and other electrons, (2) the attractive forces for the positive ions, and (3) the force of repulsion that exists between electrons. This random motion of free electrons is such that over a period of time, the number of electrons moving to the right across the circular cross section of Fig. 2.4 is exactly equal to the number passing over to the left. *With no external forces applied,* the net flow of charge in any one direction is zero.

Let us now connect this copper wire between two battery terminals as shown in Fig. 2.6. The battery, at the expense of chemical energy, places a net positive charge on one terminal and a net negative charge on the other. The instant the wire is connected between these two terminals, the free electrons of the copper wire will drift toward the positive terminal, while the remaining positive ions will continue to oscillate in their mean fixed positions. The negative terminal is a supply of electrons to be drawn from when the electrons of the copper wire closest to the negative terminal drift toward the positive terminal. Since the electrons are supplied by the negative terminal at the same rate at which they are accepted by the positive, any section of the wire remains electrically neutral, as depicted by section *ab* of Fig. 2.6.

If the terminals are such that they can cause 6.242×10^{18} *electrons* to drift at uniform velocity to the right through the imaginary circular cross section of Fig. 2.6 in 1

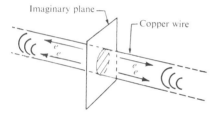

FIG. 2.4

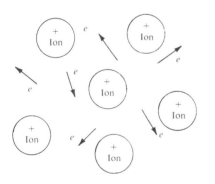

FIG. 2.5 *Random motion of free electrons in an atomic structure.*

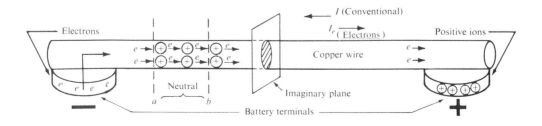

FIG. 2.6

second, the flow of charge, or *current,* is said to be 1 *ampere* (A). By definition, the charge associated with 6.242×10^{18} electrons is *one coulomb (C). The charge associated with each electron is therefore*

$$\frac{1 \text{ coulomb}}{6.242 \times 10^{18} \text{ electrons}} = 1.6 \times 10^{-19} \text{ C/electron}$$

The current in amperes can then be determined using the following equation:

$$\boxed{I = \frac{Q}{t}} \qquad \begin{array}{l} I = \text{amperes (A)} \\ Q = \text{coulombs (C)} \\ t = \text{seconds (s)} \end{array} \qquad \textbf{(2.2)}$$

The capital letter I was chosen from the French word for current: *intensité.* A second glance at Fig. 2.6 will reveal that two directions of charge flow have been indicated. One is called *conventional flow* while the other is called *electron flow.* This text will deal only with conventional flow. The reason for the choice of conventional flow is discussed in the introduction to Chapter 5.

EXAMPLE 2.1. The charge flowing through the imaginary surface of Fig. 2.6 is 0.16 C every 64 ms. Determine the current in amperes.

Solution:

Eq. (2.1):

$$I = \frac{Q}{t} = \frac{0.16}{64 \times 10^{-3}} = \frac{160 \times 10^{-3}}{64 \times 10^{-3}} = \textbf{2.50 A}$$

EXAMPLE 2.2. Determine the time required for 4×10^{10} electrons to pass through the imaginary surface of Fig. 2.6 if the current is 4 mA.

Solution:

$$4 \times 10^{10} \text{ electrons} \left[\frac{1 \text{ coulomb}}{6.242 \times 10^{18} \text{ electrons}} \right]$$
$$= 0.641 \times 10^{-8} \text{ C}$$

(Eq. 2.2):

$$t = \frac{Q}{I} = \frac{0.641 \times 10^{-8}}{4 \times 10^{-3}} = 0.160 \times 10^{-5} \text{ s}$$
$$= \textbf{1.60 } \mu\textbf{s}$$

2.3 VOLTAGE

The flow of charge described in the previous section is established by an external "pressure" derived from the energy that a mass has by virtue of its position: *potential energy.*

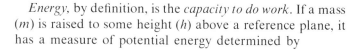

Energy, by definition, is the *capacity to do work*. If a mass (*m*) is raised to some height (*h*) above a reference plane, it has a measure of potential energy determined by

$$\boxed{\text{Potential energy (PE)} = mgh} \qquad \text{(joules)} \quad \textbf{(2.3)}$$

where g is the gravitational acceleration (9.754 m/s^2). This mass now has the ability to do work such as crush an object placed on the reference plane. If the weight is raised further, it has an increased measure of potential energy and can do additional work. There is an obvious *difference in potential* between the two heights above the reference plane.

In the battery of Fig. 2.6, the internal chemical action will establish (through an expenditure of energy) an accumulation of negative charges (electrons) on one terminal (the negative terminal) and positive charges (positive ions) on the positive terminal. A "positioning" of the charges has been established that will result in a *potential difference* between the terminals. If a conductor is connected between the terminals of the battery, the electrons at the negative terminal have sufficient potential energy to do the work necessary to overcome collisions with other particles in the conductor and the repulsion from similar charges to reach the positive terminal to which they are attracted. The difference in potential between the terminals is measured in *volts,* and for sources of potential difference such as the battery it is called an *electromotive force (emf)*. The greater the work (or expenditure of energy) required to establish the same accumulation of charge on the terminals, the greater the *potential difference* or *voltage* across the terminals. *A difference of 1 volt (V) will exist between two points if 1 joule (J) of energy is expended moving 1 coulomb (C) of charge between the two points.* In equation form, the potential difference between two points *a* and *b* is given by

$$\boxed{V_{ab} = \frac{W_{ab}}{Q}} \qquad \begin{array}{l} V_{ab} = \text{volts (V)} \\ W_{ab} = \text{joules (J)} \\ Q = \text{coulombs (C)} \end{array} \qquad \textbf{(2.4)}$$

where W_{ab} is the energy expended on the charge Q.

For a conductor, Eq. (2.4) will also determine the difference in potential between two points 1 and 2 as shown in Fig. 2.7 where W is the energy expended by the charge Q in moving from position 1 to position 2.

EXAMPLE 2.3 Find the potential difference between two points in an electrical system if 60 J of energy are expended by a charge of 20 C between these two points.

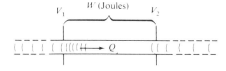

FIG. 2.7

Solution:

$$V = \frac{W}{Q} = \frac{60}{20} = 3 \text{ V}$$

EXAMPLE 2.4 Determine the energy expended moving a charge of 50 μC through a potential difference of 6 V.

Solution:

$$W = VQ = (50 \times 10^{-6})(6) = 300 \times 10^{-6} \text{ joules}$$
$$= 300 \text{ }\mu\text{J}$$

In summary, the applied emf in an electric circuit is the "pressure" to set the system in motion and "cause" the flow of charge or current through the electrical system. A mechanical analogy to the applied emf is the pressure applied to the water in a main. The resulting flow of water through the system is likened to the flow of charge through an electric circuit.

2.4 FIXED (dc) SUPPLIES

The terminology dc employed in the heading of this section is an abbreviation for *direct current,* which encompasses the various electrical systems in which there is a *unidirectional* flow of charge. A great deal more will be said about this terminology in the chapters to follow. For now, we will consider only those supplies that provide a fixed voltage or current.

dc Voltage Sources

Since the dc voltage, or current, source is the more familiar of the two types of supplies, it will be examined first. The symbol used for all dc voltage supplies in this text is indicated in Fig. 2.8. Sign convention for high and low potentials is determined by the positive charge. For dc voltage sources, therefore, the positive terminal is at a higher potential than the negative and is assigned the positive sign, while the negative terminal is assigned the negative sign. The relative lengths of the bars indicate the terminals they represent.

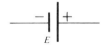

FIG. 2.8

Batteries DC voltage sources can be divided into the battery, generator, or rectification types. The battery is the most common. By definition, a battery (derived from the expression "battery of cells") consists of a series or parallel combination of two or more similar *cells,* a cell being the fundamental source of electrical energy developed through the conversion of chemical or solar energy. All cells can be divided into the *primary* or *secondary* types. The secondary is rechargeable, whereas the primary is not. That is, the chemical reaction of the secondary cell can be reversed to restore its capacity. The two most common rechargeable batteries include the lead-acid unit (used primarily in automobiles) and the nickel-cadmium battery which is used in calculators, tools, photoflash units, shavers, and so on. The obvious advantage of the rechargeable unit is the reduced costs associated with not having to continually replace discharged primary cells.

All of the cells appearing in this chapter except the *solar cell,* which absorbs energy from incident light in the form of photons, produce a voltage through the conversion of chemical energy. In addition, each has a positive and a negative *electrode* and an *electrolyte* to complete the circuit between electrodes within the battery. The electrolyte is the contact element and the source of ions for conduction between the terminals.

The popular carbon-zinc primary battery uses a zinc can as its negative electrode, a manganese dioxide mix and carbon rod as its positive electrode, and an electrolyte that is a mix of ammonium and zinc chlorides, flour, and starch, as shown in Fig. 2.9. Figure 2.10 shows a number of other types of primary units with an area of application and a rating to be considered later in this section.

For the secondary lead-acid unit appearing in Fig. 2.11, the electrolyte is sulfuric acid and the electrodes are spongy lead (Pb) and lead peroxide (PbO_2). When a load is applied

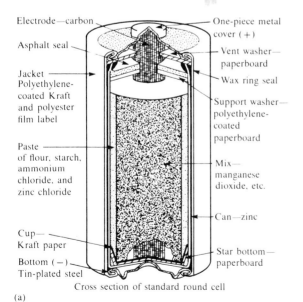

(a)

FIG. 2.9 *Carbon-zinc primary battery.* (*a*) *Construction;* (*b*) *appearance and ratings.*

(b)

"C" cell
1.5 V
0–80 mA

"D" cell
1.5 V
0–150 mA

"AA" cell
1.5 V
0–25 mA

"AAA" cell
1.5 V
0–20 mA

Courtesy of Union Carbide Corp.

Courtesy of Catalyst Research Corp.

Lithiode™ lithium-iodine cell
Long-life power sources with printed circuit
board mounting capability
2.8 V, 870 mAh
(a)

Courtesy of Catalyst Research Corp.

Lithium-iodine pacemaker cell
2.8 V, 2.0 Ah
(b)

Courtesy of Union Carbide Corp.

Eveready transistor battery 9 V, 450 mAh
(c)

to the battery terminals, there is a transfer of electrons from the spongy lead electrode to the lead peroxide electrode through the load. This transfer of electrons will continue until the battery is completely discharged. The discharge time is determined by how diluted the acid has become and how heavy the coating of lead sulfate is on each plate. The state of discharge of a lead storage cell can be determined by measuring the specific gravity of the electrolyte with a hydrometer. The specific gravity of a substance is defined to be the ratio of the weight of a given volume of the substance to the weight of an equal volume of water at 4°C. For fully charged batteries, the specific gravity should be somewhere between 1.28 and 1.30. When the specific gravity drops to about 1.1, the battery should be recharged.

FIG. 2.10 *Primary cells.*

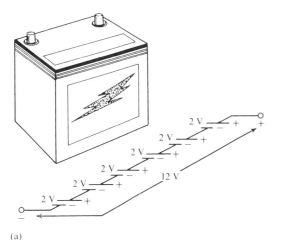

(a)

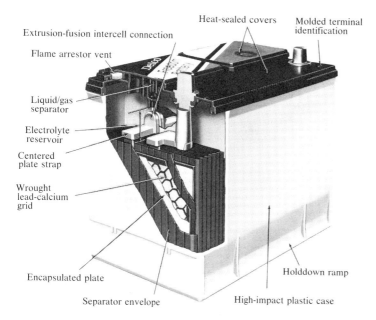

Heat-sealed covers

Molded terminal
identification

Extrusion-fusion intercell connection

Flame arrestor vent

Liquid/gas
separator

Electrolyte
reservoir

Centered
plate strap

Wrought
lead-calcium
grid

Encapsulated plate

Separator envelope

Holddown ramp

High-impact plastic case

*Courtesy of Delco-Remy, a division of General
Motors Corp.*

FIG. 2.11 (a) *Standard 12-V lead-
acid car battery; (b) maintenance-free
Delco-Remy battery.*

Since the lead storage cell is a secondary cell, it can be
recharged at any point during the discharge phase simply
by applying an external dc source across the cell that will
pass current through the cell in a direction opposite to that
in which the cell supplied current to the load. This will
remove the lead sulfate from the plates and restore the
concentration of sulfuric acid.

The output of a lead storage cell over most of the discharge phase is about 2 V. In the commercial lead storage
batteries used in the automobile, the 12-V can be produced
by six cells in series, as shown in Fig. 2.11(a). The use of a
new grid made from wrought lead-calcium alloy strip

rather than the lead-antimony cast grid commonly used has resulted in maintenance-free batteries such as appearing in Fig. 2.11(b). The lead-antimony structure was susceptible to corrosion, overcharge, gassing, water usage, and self-discharge. Improved design with the lead-calcium grid has either eliminated or substantially reduced most of these problems. Note in the figure, however, that an electrolyte is still employed and individual cells are connected in series to establish the desired 12 volts.

The nickel-cadmium battery is a rechargeable battery that has been receiving enormous interest and develop-

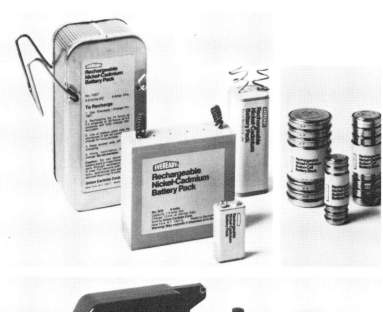

Courtesy of Union Carbide Corp.

FIG. 2.12 *Rechargeable nickel-cadmium batteries.*

ment in recent years. A number of such batteries manufactured by the Union Carbide Corporation appear in Fig. 2.12. The internal construction of the cylindrical type cell appears in Fig. 2.13. In the fully charged condition the positive electrode is nickel hydroxide ($NiOOH$); the negative electrode, metallic cadmium (Cd); and the electrolyte, potassium hydroxide (KOH). The oxidation (increased oxygen content) of the negative electrode occurring simultaneously with the reduction of the positive electrode provides the required electrical energy. The separator is required to isolate the two electrodes and maintain the location of the electrolyte. The advantage of such cells is that the active materials go through a change in oxidation state necessary to establish the required ion level without a change in the physical state. This establishes an excellent recovery mechanism for the recharging phase.

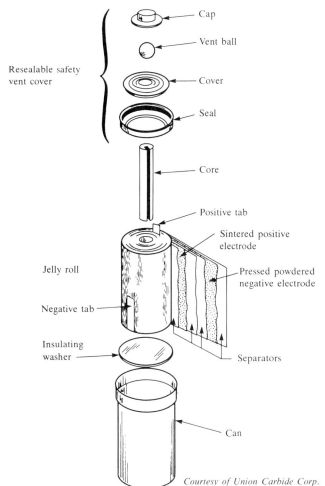

Courtesy of Union Carbide Corp.

FIG. 2.13 *Internal structure of the cylindrical-type nickel-cadmium rechargeable cell.*

Eveready® BH 500 cell
1.2 V, 500 mAh
Applications where vertical height is severe limitation

(a)   *Courtesy of Union Carbide Corp.*

Printed circuit board mountable 2.4 V battery (70 mAh)

(b)   *Courtesy of General Electric Co.*

Additional types of nickel cadmium cells appear in Fig. 2.14 with their areas of application and their ratings.

A high-density, 40-W solar cell appears in Fig. 2.15 with some of its associated data and areas of application. Since the maximum available wattage in an average bright sunlit day is 100 mW/cm² and conversion efficiencies are currently between 10% and 14%, the maximum available power per square centimeter from most commercial units is between 10 mW and 14 mW. For a square meter, however, the return would be 100 W to 140 W. A more detailed description of the solar cell will appear in your electronics courses. For now it is important to realize that a fixed illumination of the solar cell will provide a fairly steady dc voltage for driving various loads, from watches to automobiles.

Batteries have a capacity rating given in ampere-hours (Ah) or milliampere-hours (mAh). Some of these ratings are included in the above figures. A battery with an ampere-hour rating of 100 will theoretically provide a steady current of 1 A for 100 h, 2 A for 50 h, 10 A for 10 h, and so on, as determined by the following equation:

$$\text{Life (hours)} = \frac{\text{ampere-hour rating (Ah)}}{\text{amperes drawn (A)}} \quad \textbf{(2.5)}$$

Two factors that affect this rating, however, are the temperature and the rate of discharge. The disc-type

FIG. 2.14 *Nickel-cadmium batteries.*

40 watt, high density solar module
100 mm x 100 mm (4″ x 4″) square cells are used
to provide maximum power in a minimum of
space. The 33 series cell module provides a
strong 12-volt battery charging current for a wide
range of temperatures (− 40°C to 60°C)

Courtesy of Motorola Semiconductor Products

FIG. 2.15 *Solar module.*

EVEREADY® BH 500 cell appearing in Fig. 2.14(a) has the terminal characteristics appearing in Fig. 2.16. Note that for the 1-V unit the rating is above 500 mAh at a discharge current of 100 mA [Fig. 2.16(a)] but drops to 300 mAh at about 1 A. For a unit that is less than $1\frac{1}{2}$ inches in diameter and less than 1/2 inch in thickness, however, these are excellent terminal characteristics. Figure 2.16(b) reveals that the maximum mAh rating (at a current drain of 50 mA) occurs at about 75°F ($\cong 24$°C) or just above average room temperature. Note how the curve drops to the right and left of this maximum value. We are all aware of the reduced "strength" of a battery at low temperatures. Note that it has dropped to almost 300 mAh at −20°C.

Another curve of interest appears in Fig. 2.17. It provides the expected cell voltage at a particular drain over a period of hours of use. It is noteworthy that the loss in hours between 50 mA and 100 mA is much greater than between 100 mA and 150 mA, even though the increase in current is the same between levels.

EXAMPLE 2.5

a. Determine the capacity in milliampere-hours for the 0.9-V BH 500 cell of Fig. 2.16(a) if the discharge current is 600 mAh.

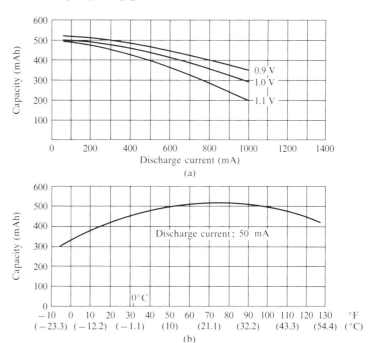

Courtesy of Union Carbide Corp.

FIG. 2.16 *EVEREADY® BH 500 cell characteristics. (a) Capacity vs. discharge current; (b) capacity vs. temperature.*

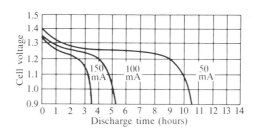

Courtesy of Union Carbide Corp.

b. At what temperature will the mAh rating of the cell of Fig. 2.16(b) be 90% of its maximum value if the discharge current is 50 mA?

FIG. 2.17 *EVEREADY® BH 500 cell discharge curves.*

Solution:

a. From Fig. 2.16(a), the capacity at 600 mA is about 450 mAh. Thus, from Eq. (2.5),

$$\text{Life} = \frac{450 \text{ mAh}}{600 \text{ mA}} = 0.75 \text{ h} = \textbf{45 min}$$

b. From Fig. 4.16(b), the maximum is approximately 520 mAh. The 90% level is therefore 468 mAh, which occurs just above freezing or **1°C** and at the higher temperature of **45°C**.

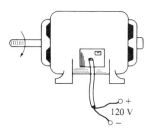

FIG. 2.18 *dc generator.*

Generators The dc generator is quite different, both in construction (Fig. 2.18) and in mode of operation, from the battery. When the shaft of the generator is rotating at the nameplate speed due to the applied torque of some external source of mechanical power, a voltage of rated value will appear across the external terminals. The terminal voltage and power-handling capabilities of the dc generator are typically higher than those of most batteries, and its lifetime is determined only by its construction. Commercially used dc generators are typically of the 120-V or 240-V variety. As pointed out earlier in this section, for the purposes of this text no distinction will be made between the symbol for a battery and a generator.

Rectification The dc supply encountered most frequently in the laboratory employs the rectification and filtering processes as its means toward obtaining a steady dc voltage. By this process, a time-varying voltage (such as ac voltage available from a home outlet) is converted to one of a fixed magnitude. This process will be covered in detail in the basic electronics courses. Two dc laboratory supplies of this type appear in Fig. 2.19.

(a)

(b)

Courtesy of Lambda Electronics Corp.

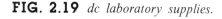

FIG. 2.19 *dc laboratory supplies.*

Most dc laboratory supplies have a regulated, adjustable voltage output with three available terminals, as indicated in Figs. 2.19 and 2.20(a). The symbol for ground or zero potential (the reference) is also shown in Fig. 2.20(a). If 10 volts above ground potential are required, then the connections are made as shown in Fig. 2.20(b). If 15 volts below ground potential are required, then the connections are made as shown in Fig. 2.20(c). If connections are as shown in Fig. 2.20(d), we say we have a "floating" voltage of 5 volts since the reference level is not included. Seldom is the

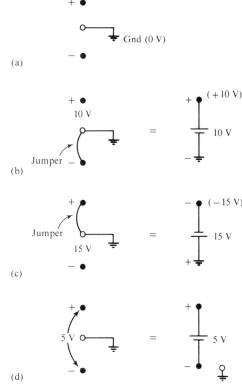

FIG. 2.20 *Possible output connections of a dc laboratory supply.*

configuration of Fig. 2.20(d) employed since the operator may (undesirably) serve as the path to ground potential for the floating supply.

dc Current Sources

The wide variety of types of and applications for the dc voltage source have resulted in its becoming a rather familiar device, the characteristics of which are understood, at least basically, by the layperson. For example, it is common knowledge that a 12-V car battery has a terminal voltage (at least approximately) of 12 V even though the current drain by the automobile may vary under different operating conditions. In other words, *a dc voltage source ideally will provide a fixed terminal voltage even though the current drain may vary.* A dc current source is just the reverse. *The current source, will, ideally, supply a fixed current to a load even though there will be variations in the terminal voltage as determined by the load.* (Do not become alarmed if the concept of a current source is strange and somewhat confusing at this point. It will be covered in great detail in later chapters).

The introduction of semiconductor devices such as the transistor has accounted in large measure for the increasing interest in current sources. A representative commercially available dc current source appears in Fig. 2.21.

Courtesy of Lambda Electronics Corp.

FIG. 2.21 *dc current source.*

2.5 CONDUCTORS AND INSULATORS

Different wires placed across the same two battery terminals will allow different amounts of charge to flow between the terminals. Many factors, such as the stability, density, and mobility of the material, account for these variations in charge flow. *Conductors are those materials that permit a generous flow of electrons with very little electromotive force applied.* Since copper is used most frequently, it serves as the standard of comparison for the relative conductivity in Table 2.1. Note that aluminum, which has lately seen some commercial use, has only 61% of the conductivity level of copper, but keep in mind that this must be weighed against the cost and weight factors.

Materials that have very few free electrons, high stability and density, and low mobility are called *insulators, since a very high electromotive force is required to produce any siza-*

TABLE 2.1 *Relative conductivity of various materials.*

Metal	Relative Conductivity (%)
Silver	105
Copper	**100**
Gold	70.5
Aluminum	61
Tungsten	31.2
Nickel	22.1
Iron	14
Constantan	3.52
Nichrome	1.73
Calorite	1.44

ble current flow through such materials. A common use of insulating material is for covering current-carrying wire which, if uninsulated, could cause dangerous side effects. For example, workers on high-voltage power lines wear rubber gloves as an additional safety measure. A number of different types of insulators and their applications appear in Fig. 2.22.

It must be pointed out, however, that even the best insulator will break down (permit charge to flow around and through it) if a sufficiently large potential is applied across it. The breakdown strengths of some common insulators are listed in Table 2.2.

(a)

TABLE 2.2 *Breakdown strength of some common insulators.*

Material	Average Breakdown Strength (kV/cm)
Air	30
Porcelain	70
Oils	140
Bakelite	150
Rubber	270
Paper (Paraffin-coated)	500
Teflon	600
Glass	900
Mica	2000

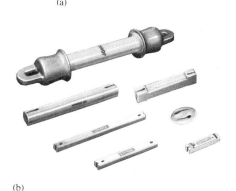

(b)

According to this table, for insulators with the same geometric shape, it would require 270/30 = 9 times as much potential to pass current through rubber as through air and approximately 67 times as much voltage to pass current through mica as through air.

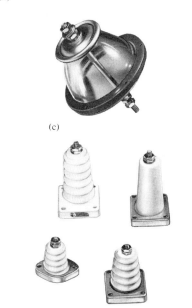

(c)

2.6 SEMICONDUCTORS

Between the class of elements called *insulators* and those exhibiting conductor properties, there exists a group of elements of significant importance called *semiconductors*. The entire electronic industry is dependent on these materials. The diodes, transistors, and integrated circuits (ICs) that we hear so much about are constructed of semiconductor materials. Although *silicon* is the most extensively employed, *germanium* is also used in a number of devices. Both of these materials will be examined in some detail in your electronics courses. It will then be demonstrated why they are so appropriate for the applications noted above.

(d)

Courtesy of Herman H. Smith, Inc.

FIG. 2.22 *Insulators.* (a) *Insulated thru-panel bushings;* (b) *antenna strain insulators;* (c) *feed-thru bowl assemblies;* (d) *porcelain stand-off insulators.*

PROBLEMS

Section 2.1

6×10^{-5} C 5×10^{-4} C

Q_1 (+) |——— r ———| (−) Q_2

FIG. 2.23

8μC 4000μC

Q_1 (+) |——— r ———| (+) Q_2

FIG. 2.24

1. The number of orbiting electrons in aluminum and silver is 13 and 47, respectively. Draw the electronic configuration, including all the shells and subshells, and discuss briefly why each is a good conductor.

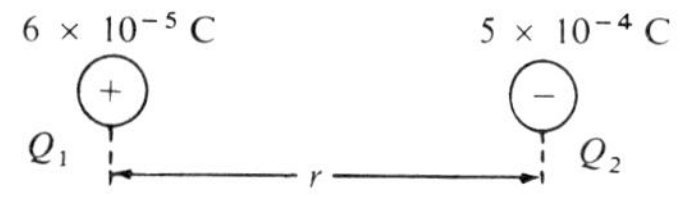

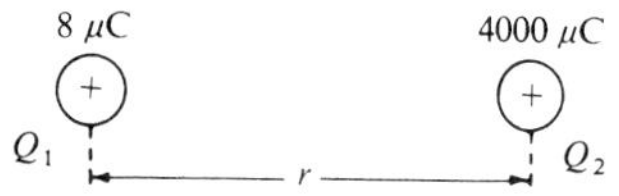

2. Find the force of attraction in newtons between the charges Q_1 and Q_2 in Fig. 2.23 when
   a. $r = 1$ m
   b. $r = 3$ m
   c. $r = 10$ m
   (Note how quickly the force drops with increase in r.)

*3. Find the force of repulsion in newtons between Q_1 and Q_2 in Fig. 2.24 when
   a. $r = 1$ mi
   b. $r = 0.01$ m
   c. $r = 1/16$ in.

4. Two charged bodies, Q_1 and Q_2, when separated by a distance of 2 m, experience a force of repulsion equal to 1.8 N.
   a. What will the force of repulsion be when they are 10 m apart?
   b. If the ratio of $Q_1/Q_2 = 1/2$, find Q_1 and Q_2.

Section 2.2

5. Find the current in amperes if 650 C of charge pass through a wire in 50 s.

6. If 465 C of charge pass through a wire in 2.5 min, find the current in amperes.

7. If a current of 40 A exists for 1 min, how many coulombs of charge have passed through the wire?

8. How many coulombs of charge pass through a lamp in 2 min if the current is constant at 750 mA?

9. If the current in a conductor is constant at 2 mA, how much time is required for 4600×10^{-6} C to pass through the conductor?

10. If $21.847 \times 10^{+18}$ electrons pass through a wire in 7 s, find the current.

11. How many electrons pass through a conductor in 1 min if the current is 1 A?

*12. If $0.784 \times 10^{+18}$ electrons pass through a wire in 643 ms, find the current.

13. Will a fuse rated at 1 A "blow" if 86 C pass through it in 1.2 min?

Section 2.3

14. In your own words, describe in a few sentences the quantities voltage, potential difference, and emf.

15. a. If 72 J of energy are required to move 8 C of charge from infinity to position x, what is the potential difference between position x and infinity?

    b. If 24 additional joules are required to move the same 8 C from position x to position y, what is the potential difference between x and y?

    c. What is the potential difference between position y and infinity?

16. If the potential difference between two points is 42 V, how much work is required to bring 6 C from one point to the other?

17. Find the charge Q that requires 96 J of energy to be moved through a potential difference of 16 V.

18. The electromotive force of a battery is 22.5 V. How much charge moves if the energy used is 90 J?

19. If a conductor with a current of 500 mA passing through it converts 40 J of electrical energy into heat in 30 s, what is the potential drop across the conductor?

20. Charge is flowing through a conductor at the rate of 420 C/min. If 742 J of electrical energy are converted to heat in 1 min, what is the potential drop across the conductor?

Section 2.4

21. What current will a battery with an Ah rating of 200 theoretically provide for 40 h?

22. What is the Ah rating of a battery that can provide 0.8 A for 76 h?

23. For how many hours will a battery with an Ah rating of 32 theoretically provide a current of 1.28 A?

24. Find the mAh rating of the EVEREADY® BH 500 battery at 100°F and 0°C at a discharge current of 50 mA using Fig. 2.16(b).

25. Find the mAh rating of the 1.0 V EVEREADY® BH 500 battery if the current drain is 550 mA using Fig. 2.16(a). How long will it supply this current?

26. For how long can 50 mA be drawn from the battery in Fig. 2.17 before its terminal voltage drops below 1 V? Determine the number of hours at a drain current of 150 mA, and compare the ratio of drain current to the resulting ratio of hours of availability.

27. Discuss briefly the difference between the three types of dc voltage supplies (batteries, rectification, and generators).

28. Indicate in a few sentences your concept of a current source. Employ its characteristics in your description.

Section 2.5

29. Discuss two properties of the atomic structure of copper that make it a good conductor.

30. Name two materials not listed in Table 2.1 that are good conductors of electricity.

31. Explain the terms *insulator* and *breakdown strength*.

32. List three uses of insulators not mentioned in Section 2.5.

Section 2.6

33. What is a semiconductor? How does it compare with a conductor and insulator?

34. Consult a semiconductor electronics text and note the extensive use of germanium and silicon semiconductor materials. Review the characteristics of each material.

GLOSSARY

Ampere (A)   The SI unit of measurement applied to the flow of charge through a conductor.

Ampere-hour rating   The rating applied to a source of energy that will reveal how long a particular level of current can be drawn from that source.

Cell   A fundamental source of electrical energy developed through the conversion of chemical or solar energy.

Conductors   Materials that permit a generous flow of electrons with very little emf applied.

Copper   A material possessing physical properties that make it particularly useful as a conductor of electricity.

Coulomb (C)   The fundamental SI unit of measure for charge. It is equal to the charge carried by 6.242×10^{18} electrons.

Coulomb's law   An equation defining the force of attraction or repulsion between two charges.

dc current source   A source that will provide a fixed current level even though the load to which it is applied may cause its terminal voltage to change.

dc generator   A source of dc voltage available through the turning of the shaft of the device by some external means.

Direct current   Current in which the magnitude does not change over a period of time.

Ductility   The property of a material that allows it to be drawn into long thin wires.

Electrolytes   The contact element and the source of ions between the electrodes of the battery.

Electromotive force (emf)   Force that causes current flow, equivalent to the potential difference between terminals.

Electron   The particle with negative polarity that orbits the nucleus of an atom.

Free electron   An electron unassociated with any particular atom, relatively free to move through a crystal lattice structure under the influence of external forces.

Insulators Materials in which a very high emf must be applied to produce any measurable current flow.

Malleability The property of a material that allows it to be worked into many different shapes.

Neutron The particle having no electrical charge, found in the nucleus of the atom.

Nucleus The structural center of an atom that contains both protons and neutrons.

Positive ion An atom having a net positive charge due to the loss of one of its negatively charged electrons.

Potential difference The difference in potential between two points in an electrical system.

Potential energy The energy that a mass possesses by virtue of its position.

Primary cell Sources of electromotive force that cannot be recharged.

Proton The particle of positive polarity found in the nucleus of the atom.

Secondary cell Sources of electromotive force that can be recharged.

Semiconductor A material having a conductance value between that of an insulator and that of a conductor. Of significant importance in the manufacture of semiconductor electronic devices.

Solar cell Sources of emf available through the conversion of light energy (photons) into electrical energy.

Specific gravity The ratio of the weight of a given volume of a substance to the weight of an equal volume of water at 4°C.

Rectification The process by which an ac signal is converted to one which has an average dc level.

Volt (V) The unit of measurement applied to the difference in potential between two points. If one joule of energy is required to move one coulomb of charge between two points, the difference in potential is said to be one volt.

3.1 INTRODUCTION

The flow of charge through any material encounters an opposing force similar in many respects to mechanical friction. This opposition, due to the collisions between electrons and between electrons and other atoms in the material, *which converts electrical energy into heat,* is called the *resistance* of the material. The unit of measurement of resistance is the *ohm,* for which the symbol is Ω, the capital Greek letter omega. The circuit symbol for resistance appears in Fig. 3.1 with the graphic abbreviation for resistance (*R*).

The resistance of any material with a uniform cross-sectional area is determined by the following four factors (depicted in Fig. 3.2 for a wire):

1. Material
2. Length
3. Cross-sectional area
4. Temperature

For two wires of the same physical size at the same temperature, as shown in Fig. 3.3(a), the relative resistances will be determined solely by the material. As indicated in Fig. 3.3(b), an increase in length will result in an increased resistance for similar areas, material, and temperature. Increased area, as in Fig. 3.3(c), for remaining similar determining variables will result in a decrease in resistance. Finally, increased temperature [Fig. 3.3(d)] for metallic wires

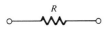

FIG. 3.1 *Resistance symbol and notation.*

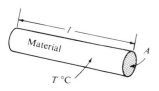

FIG. 3.2

of identical construction and material will result in an increased resistance.

The preceding discussion reveals that resistance is directly proportional to the material employed and its length, and inversely proportional to its area. In equation form, the resistance of a conductor at 20°C (room temperature) is expressed as follows:

$$R = \rho\frac{l}{A}$$

(3.1)

R = ohms

l = length in feet

A = cross-sectional area in circular mils (CM)

ρ = resistivity (characteristic of material at 20°C)

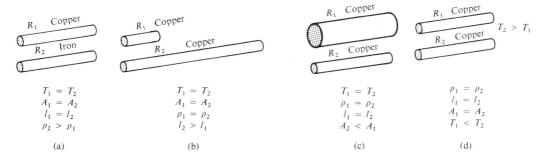

FIG. 3.3 *Cases in which $R_2 > R_1$.*

Note that the area of the conductor is measured in *circular mils* and not in square meters, inches, and so on, as determined by the equation

$$\text{Area (circle)} = \pi r^2 = \frac{\pi d^2}{4}$$

(3.2)

r = radius

d = diameter

Recall from Chapter 1 that

$$1 \text{ mil} = \frac{1}{1000} \text{ in.} = 0.001 \text{ in.} = 10^{-3} \text{ in.}$$

or

$$1000 \text{ mils} = 1 \text{ in.}$$

A square mil will appear as shown in Fig. 3.4(a). By definition, *a wire that has a diameter of 1 mil, as shown in Fig. 3.4(b), has an area of 1 circular mil (CM)*. One square mil was superimposed on the 1 CM area of Fig. 3.4(b) to show clearly that the square mil has a larger surface area than the circular mil.

Applying the above definition to a wire having a diameter of 1 mil, we have

1 square mil

1 mil

├─1 mil─┤

(a)

1 circular mil (CM)

1 mil

(b)

FIG. 3.4

by definition

$$A = \frac{\pi d^2}{4} = \frac{\pi}{4}(1)^2 = \frac{\pi}{4} \text{ sq mils} \equiv 1 \text{ CM}$$

Therefore,

$$1 \text{ CM} = \frac{\pi}{4} \text{ sq mils}$$

or

$$1 \text{ sq mil} = \frac{4}{\pi} \text{CM}$$

For conversion purposes,

$$\text{no. of CM} = \frac{4}{\pi} \times (\text{no. of sq mils})$$

$$\text{no. of sq mils} = \frac{\pi}{4} \times (\text{no. of CM})$$

(3.3)

For a wire with a diameter of N mils (where N can be any positive number),

$$A = \frac{\pi d^2}{4} = \frac{\pi N^2}{4} \text{ sq mils}$$

Substituting the fact that $4/\pi$ CM = 1 sq mil, we have

$$A = \frac{\pi N^2}{4} (\text{sq mils}) = \left(\frac{\pi N^2}{4}\right)\left(\frac{4}{\pi}\text{CM}\right) = N^2 \text{ CM}$$

Since $d = N$, the area in circular mils is simply equal to the diameter in mils squared; that is,

$$A_{\text{CM}} = (d_{\text{mils}})^2$$

(3.4)

Therefore, in order to find the area in circular mils, the diameter must first be converted to mils. Since 1 mil = 0.001 in., if the diameter is given in inches, simply move the decimal point three places to the right. For example,

$$0.123 \text{ in.} = 123.0 \text{ mils}$$

The constant ρ (resistivity) is different for every material. Its value is the resistance of a length of wire 1 ft by 1 mil in diameter, measured at 20°C (Fig. 3.5). The unit of measurement for ρ can be determined from Eq. (3.1) as follows:

$$R = \rho \frac{l}{A}$$

$$\text{Ohms} = \rho \frac{\text{ft}}{\text{CM}}$$

$$\text{Units of } \rho = \frac{\text{CM-ohms}}{\text{ft}}$$

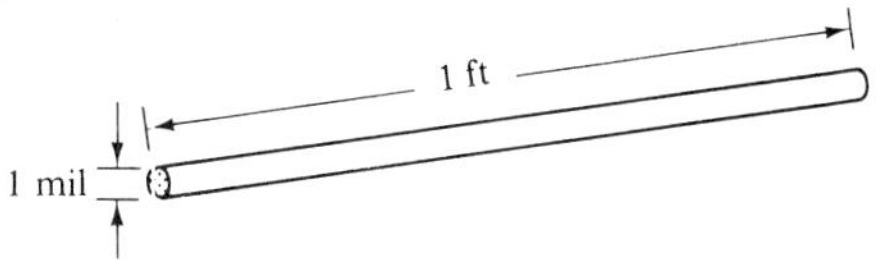

FIG. 3.5

The resistivity ρ is also measured in ohms per mil-ft as determined by Fig. 3.5, or *ohm-meters* in the SI system of units.

Some typical values of ρ are listed in Table 3.1.

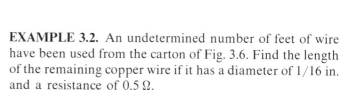

TABLE 3.1 *The resistivity of various materials.*

Material	$\rho \left(\dfrac{\text{CM-ohms}}{\text{ft}} \right)$ @ 20°C
Silver	9.9
Copper	10.37
Gold	14.7
Aluminum	17.0
Tungsten	33.0
Nickel	47.0
Iron	74.0
Constantan	295.0
Nichrome	600.0
Calorite	720.0
Carbon	21,000.0

EXAMPLE 3.1. What is the resistance of a 100-ft length of copper wire with a diameter of 0.020 in. at 20°C?

Solution:

$$\rho = 10.37 \qquad 0.020 \text{ in.} = 20 \text{ mils}$$
$$A_{\text{CM}} = (d_{\text{mils}})^2 = (20)^2 = 400$$
$$R = \rho \frac{l}{A} = \frac{(10.37)100}{400}$$
$$R = \textbf{2.59 } \boldsymbol{\Omega}$$

FIG. 3.6

EXAMPLE 3.2. An undetermined number of feet of wire have been used from the carton of Fig. 3.6. Find the length of the remaining copper wire if it has a diameter of 1/16 in. and a resistance of 0.5 Ω.

Solution:

$$\rho = 10.37 \qquad \frac{1}{16} \text{ in.} = 0.0625 \text{ in.} = 62.5 \text{ mils}$$
$$A_{\text{CM}} = (d_{\text{mils}})^2 = (62.5)^2 = 3906.25 \text{ CM}$$
$$R = \rho \frac{l}{A} \Rightarrow l = \frac{RA}{\rho} = \frac{(0.5)(3906.25)}{10.37} = \frac{1953.125}{10.37}$$
$$l = \textbf{188.34 ft}$$

EXAMPLE 3.3. What is the resistance of a copper bus-bar as used in the power distribution panel of a high-rise office building with the dimensions indicated in Fig. 3.7?

Solution:

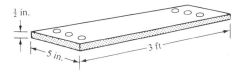

FIG. 3.7

$$A_{\text{CM}} \begin{cases} 5.0 \text{ in.} = 5000 \text{ mils} \\ \dfrac{1}{2} \text{ in.} = 500 \text{ mils} \\ A = (5000)(500) = 2.5 \times 10^6 \text{ sq mils} \\ \quad = 2.5 \times 10^6 \text{ sq mils} \left(\dfrac{4/\pi \text{ CM}}{1 \text{ sq mil}} \right) \\ A = 3.185 \times 10^6 \text{ CM} \end{cases}$$

$$R = \rho \frac{l}{A} = \frac{(10.37)(3)}{3.185 \times 10^6} = \frac{31.110}{3.185 \times 10^6}$$

$$R = \mathbf{9.768 \times 10^{-6}\ \Omega}$$
(quite small, 0.000009768 Ω)

3.2 TEMPERATURE EFFECTS

For most conductors, the resistance increases with increase in temperature, due to the increased molecular movement within the conductor which hinders the flow of charge. Figure 3.8 indicates that for copper (and most other metallic conductors), the resistance increases almost linearly (in a straight-line relationship) with increase in temperature. For the range of semiconductor materials such as employed in transistors, diodes, and so on, the resistance decreases with increase in temperature.

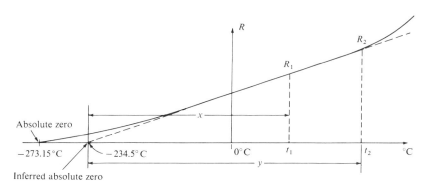

FIG. 3.8 *Effect of temperature on the resistance of copper.*

Since temperature can have such a pronounced effect on the resistance of a conductor, it is important that we have some method of determining the resistance at any temperature within operating limits. An equation for this purpose can be obtained by approximating the curve of Fig. 3.8 by the straight dashed line that intersects the temperature scale at $-234.5°C$. Although the actual curve extends to *absolute zero* ($-273.15°C$), the straight-line approximation is quite accurate for the normal operating temperature range. At two different temperatures, t_1 and t_2, the resistance of copper is R_1 and R_2, as indicated on the curve. Using a property of similar triangles, we may develop a mathematical relationship between these values of resistances at different temperatures. Let x equal the distance from $-234.5°C$ to t_1 and y the distance from $-234.5°C$ to t_2, as shown in Fig. 3.8. From similar triangles,

$$\frac{x}{R_1} = \frac{y}{R_2}$$

or

$$\boxed{\frac{234.5 + t_1}{R_1} = \frac{234.5 + t_2}{R_2}} \tag{3.5}$$

The temperature of $-234.5°C$ is called the *inferred absolute temperature* of copper. For different conducting materials, the intersection of the straight-line approximation will occur at different temperatures. A few typical values are listed in Table 3.2.

TABLE 3.2 *Inferred absolute temperatures.*

Material	Temperature (°C)
Silver	−243
Copper	−234.5
Gold	−274
Aluminum	−236
Tungsten	−204
Nickel	−147
Iron	−162
Constantan	−125,000
Nichrome	−2250

Equation (3.5) can easily be adapted to any material by inserting the proper inferred absolute temperature. It may therefore be written as follows:

$$\boxed{\frac{|T| + t_1}{R_1} = \frac{|T| + t_2}{R_2}} \tag{3.6}$$

where $|T|$ indicates that the inferred absolute temperature of the material involved is inserted as a positive value in the equation.

EXAMPLE 3.4. If the resistance of a copper wire is 50 Ω at 20°C, what is its resistance at 100°C (boiling point of water)?

Solution: Eq. (3.5):

$$\frac{234.5 + 20}{50} = \frac{234.5 + 100}{R_2}$$

$$R_2 = \frac{50(334.5)}{254.5} = \mathbf{65.72\ \Omega}$$

EXAMPLE 3.5. If the resistance of a copper wire at freezing (0°C) is 30 Ω, what is its resistance at −40°C?

Solution: Eq. (3.5):

$$\frac{234.5 + 0}{30} = \frac{234.5 - 40}{R_2}$$

$$R_2 = \frac{30(194.5)}{234.5} = \mathbf{24.88\ \Omega}$$

There is a second popular equation for calculating the resistance of a conductor at different temperatures. Defining

$$\alpha_1 = \frac{1}{|T| + t_1}$$

as the *temperature coefficient of resistance* at a temperature t_1, we have

$$\boxed{R_2 = R_1[1 + \alpha_1(t_2 - t_1)]} \qquad (3.7)$$

The values of α_1 for different materials at a temperature of 20°C have been evaluated, and a few are listed in Table 3.3. As indicated in the table, carbon and the remaining family of *semiconductor materials have negative temperature coefficients*. In other words, the resistance of the material will drop with increase in temperature and vice versa.

TABLE 3.3 *Temperature coefficient of resistance for various materials at 20°C.*

Material	Temperature Coefficient (α_1)
Silver	0.0038
Copper	0.00393
Gold	0.0034
Aluminum	0.00391
Tungsten	0.005
Nickel	0.006
Iron	0.0055
Constantan	0.000008
Nichrome	0.00044
Carbon	−0.0005

Resistance

Equation (3.7) can be rewritten in the following form:

$$m = \text{slope of the curve} = \frac{\Delta y}{\Delta x}$$

$$\alpha_1 = \frac{1}{R_1}\left[\overbrace{\frac{R_2 - R_1}{t_2 - t_1}}\right]$$

Referring to Fig. 3.8, we find that the temperature coefficient is directly proportional to the slope of the curve, so the greater the slope of the curve, the greater the value of α_1. We can then conclude that *the higher the value of α_1, the greater the rate of change of resistance with temperature.* Referring to Table 3.3, we find that copper is more sensitive to temperature variations than silver, gold, or aluminum, although the differences are quite small.

3.3 WIRE TABLES

The wire table was designed primarily to standardize the size of wire produced by manufacturers throughout the United States. As a result, the manufacturer has a larger market and the consumer knows that standard wire sizes will always be available. The table was designed to assist the user in every way possible; it usually includes such data as the cross-sectional area in circular mils, diameter in mils, ohms per 1000 ft at 20°C, and weight per 1000 ft.

The American Wire Gage (AWG) sizes are given in Table 3.4 for solid round copper wire. A column indicating the maximum allowable current in amperes, as determined by the National Fire Protection Association, has also been included.

These sizes are determined by their cross-sectional areas. For a drop in three gage numbers, the area increases approximately 1.99 the original, or a nearly 100% increase in area. In other words, *the area is doubled for every drop in three gage numbers.* Further analysis will indicate *an increase in area equivalent to 10 times the original for every drop in 10 gage numbers.*

Examining Eq. (3.1), we note also that *doubling the area cuts the resistance in half, and increasing the area by a factor of 10 decreases the resistance to 1/10 the original,* everything else kept constant.

The actual sizes of some of the gage wires listed in Table 3.4 are shown in Fig. 3.9 with a few of their areas of application. A few examples using Table 3.4 follow.

EXAMPLE 3.6. Find the resistance of 650 ft of #8 copper wire ($T = 20°C$).

Solution: For #8 copper wire (solid), $\Omega/1000$ ft at 20°C = 0.6282 Ω, and

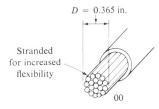

$D = 0.365$ in.

Stranded for increased flexibility

00

Power distribution

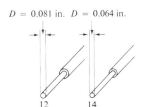

$D = 0.081$ in. $D = 0.064$ in.

12   14

Lighting, outlets, general home use

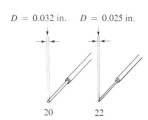

$D = 0.032$ in. $D = 0.025$ in.

20   22

Radio, television

$D = 0.013$ in.

28

Telephone, instruments

FIG. 3.9

TABLE 3.4 *American Wire Gage (AWG) sizes.*

	AWG #	Area (CM)	Ω/1000 ft at 20°C	Maximum Allowable Current for RHW Insulation (A)*
(4/0)	0000	211,600	0.0490	230
(3/0)	000	167,810	0.0618	200
(2/0)	00	133,080	0.0780	175
(1/0)	0	105,530	0.0983	150
	1	83,694	0.1240	130
	2	66,373	0.1563	115
	3	52,634	0.1970	100
	4	41,742	0.2485	85
	5	33,102	0.3133	—
	6	26,250	0.3951	65
	7	20,816	0.4982	—
	8	16,509	0.6282	45
	9	13,094	0.7921	—
	10	10,381	0.9989	30
	11	8,234.0	1.260	—
	12	6,529.0	1.588	20
	13	5,178.4	2.003	—
	14	4,106.8	2.525	15
	15	3,256.7	3.184	
	16	2,582.9	4.016	
	17	2,048.2	5.064	
	18	1,624.3	6.385	
	19	1,288.1	8.051	
	20	1,021.5	10.15	
	21	810.10	12.80	
	22	642.40	16.14	
	23	509.45	20.36	
	24	404.01	25.67	
	25	320.40	32.37	
	26	254.10	40.81	
	27	201.50	51.47	
	28	159.79	64.90	
	29	126.72	81.83	
	30	100.50	103.2	
	31	79.70	130.1	
	32	63.21	164.1	
	33	50.13	206.9	
	34	39.75	260.9	
	35	31.52	329.0	
	36	25.00	414.8	
	37	19.83	523.1	
	38	15.72	659.6	
	39	12.47	831.8	
	40	9.89	1049.0	

Left margin label (vertical): INCREASING AWG #; middle label (vertical): DECREASING DIAMETER (AND AREA)

*Not more than three conductors in raceway, cable, or direct burial.

Resistance

$$650 \, ft \left[\frac{0.6282 \, \Omega}{1000 \, ft} \right] = \mathbf{0.408 \, \Omega}$$

EXAMPLE 3.7. What is the diameter, in inches, of a #12 copper wire?

Solution: For #12 copper wire (solid), area = 6529.9 CM, and

$$d_{\text{mils}} = \sqrt{A_{\text{CM}}} = \sqrt{6529.9} \cong 80.81 \text{ mils}$$
$$d = \mathbf{0.0808 \text{ in.}} \text{ (or close to } 1/12 \text{ in.)}$$

EXAMPLE 3.8. For the system of Fig. 3.10, the total resistance of *each* power line cannot exceed 0.025 Ω, and the maximum current to be drawn by the load is 95 A. What gage wire should be used?

Solution:

$$R = \rho \frac{l}{A} \Rightarrow A = \rho \frac{l}{R} = \frac{(10.37)(100)}{0.025} = 41{,}480 \text{ CM}$$

Using the wire table, we choose the wire with the next largest area, which is #4, to satisfy the resistance requirement. We note, however, that 95 A must flow through the line. This specification requires that #3 wire be used, since the #4 wire can carry a maximum current of only 85 A.

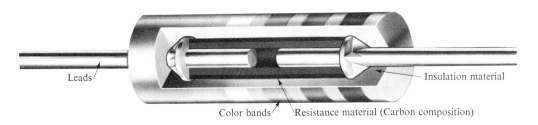

Input — Solid round copper wire — Load
— 100 ft —

FIG. 3.10

3.4 TYPES OF RESISTORS

Resistors are made in many forms, but all belong in either of two groups: fixed or variable. The most common of the low-wattage, fixed-type resistors is the molded-carbon composition resistor. The basic construction is shown in Fig. 3.11.

Leads — Color bands — Resistance material (Carbon composition) — Insulation material

FIG. 3.11 *Fixed composition resistor.*

The relative sizes of all fixed and variable resistors change with the wattage (power) rating, increasing in size for increased wattage ratings in order to withstand the higher currents and dissipation losses. The relative sizes of the molded-composition resistors for different wattage ratings are shown in Fig. 3.12. Resistors of this type are readily available in values ranging from 2.7 Ω to 22 MΩ.

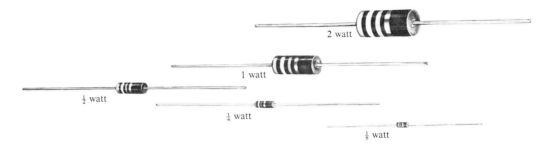

Courtesy of Ohmite Manufacturing Co.

FIG. 3.12 *Fixed composition resistors of different wattage ratings.*

The temperature-versus-resistance curve for a 10,000-Ω and 0.5-MΩ composition-type resistor is shown in Fig. 3.13.

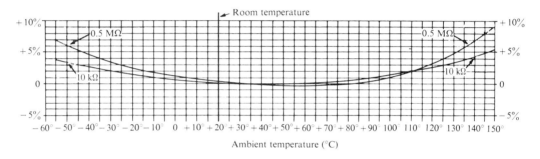

Ambient temperature (°C)

Courtesy of Allen-Bradley Co.

FIG. 3.13 *Curves showing percent temporary resistance changes from* +*25°C values.*

Note the small percent resistance change in the normal temperature operating range. Several other types of fixed resistors are shown in Fig. 3.14.

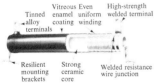

(a) Vitreous-enameled resistor
   App: All types of equipment

(b) "Corrib" corrugated ribbon resistor
   App: Where low resistances are
   required to   handle high currents

(c) Molded vitreous-enameled wirewound axial lead resistor
   App: For low-wattage applications
   in   electronic and similar circuits

(d) Thin resistors
   App: Where height above mounting surface is
   limited or where resistors must be closely
   stacked

(e) Metal-film precision resistors
   App: Where high stability, low temperature
   coefficient and low noise level are desired

Courtesy of Ohmite Manufacturing Co.

FIG. 3.14 *Fixed resistors.*

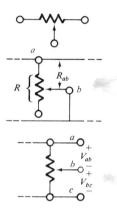

FIG. 3.15 *Rheostat.*

FIG. 3.16 *Potentiometer.*

(a) Tapered winding

(b) Linear winding

Courtesy of Ohmite Manufacturing Co.

FIG. 3.17 *Wirewound vitreous enameled rheostats.*

Variable resistors come in many forms, but basically they can be separated into the linear or nonlinear types. The symbol for a two-point linear or nonlinear *rheostat* is shown in Fig. 3.15. The three-point variable resistor may be called a *rheostat* or *potentiometer,* depending on how it is used. The symbol for a three-point variable resistor is shown in Fig. 3.16, along with the connections for its use as a rheostat or potentiometer. The arrow in the symbol of Fig. 3.16 is a contact that is movable on a continuous resistive element. As shown in Fig. 3.16, if the lug connected to the moving contact and a stationary lug are the only terminals used, the variable resistor is being used as a rheostat. The moving contact will determine whether R_{ab} is a minimum (zero ohms) or maximum value (R). If all three lugs are connected in the circuit, it is being employed as a potentiometer. The terminology *potentiometer* refers to the fact that the moving contact (wiper arm) will control by its position the *potential* differences V_{ab} and V_{bc} of Fig. 3.16.

Figure 3.17 shows both a linear and a tapered type of rheostat. In this case and those to follow, rather than calling each a variable resistor, we call it a rheostat or potentiometer on the basis of its most frequent application. In the linear type of Fig. 3.17, the number of turns of the high-resistance wire per unit length of the core is uniform; therefore, the resistance will vary linearly with the position of the rotating contact. One-half of a turn will result in half the total resistance between either stationary lug and the moving contact. Three-quarters of a turn will establish three-quarters of the total across two terminals and one-quarter between the other stationary lug and the moving contact. If the number of turns is not uniform as in the tapered rheostat, the resistance will vary nonlinearly with the position of the rotating contact. That is, a quarter turn may result in less or more than one-quarter the total resistance between a stationary lug and the moving contact. Rheostats of both types in Fig. 3.17 are made in all sizes, with a range of maximum values from 200 to 50,000 Ω.

The molded composition potentiometer shown in Fig. 3.18 is the type used in circuits with smaller power demands than the one previously described. It is smaller in size but has maximum values ranging from 20 Ω to 22 MΩ.

Two other types of variable resistors (Fig. 3.19) are the screw-drive rheostat and the carbon-pile rheostat, the screw-drive being linear and the carbon-pile nonlinear.

The resistance of the screw-drive type is determined by the position of the contact arm, which can be moved by using the handwheel. The stationary terminal used with the movable contact determines whether the resistance increases or decreases with movement of the contact arm.

The operation of the carbon-pile rheostat is different from any we have discussed to this point. To vary the resist-

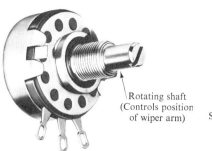

(a) External view

(b) Internal view

Rotating shaft
(Controls position
of wiper arm)

Sliding contact

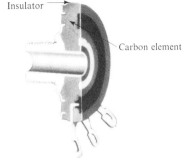

Insulator

Carbon element

(c) Carbon element
Courtesy of Allen-Bradley Co.

ance, pressure is applied to a column of graphite discs by means of a compression screw operated by the handwheel.

FIG. 3.18 *Molded composition type potentiometer.*

(a) Screw-drive rheostat   *Courtesy of James G. Biddle Co.*

Complete assembly
(b) Carbon-pile rheostat

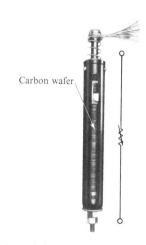

Carbon wafer

Internal view

Courtesy of Allen-Bradley Co.

FIG. 3.19 *Screw-drive and carbon-pile rheostats.*

Increasing the pressure will decrease the resistance, as shown in Fig. 3.20. The decrease in resistance is due to the better contact being made between the graphite discs.

The miniaturization of parts—used quite extensively in computers—requires that resistances of different values be placed in very small packages. Four steps leading to the packaging of three resistors in a single module are shown in Fig. 3.21.

For use with printed circuit boards, resistor networks in a variety of configurations are available in miniature packages such as shown in Fig. 3.22 with a photograph of the casing and pins. The LDP is a coding for the production series, while the second number, 14, is the number of pins. The last two digits indicate the internal circuit configuration. The resistance range for the discrete elements in each chip is $10 \, \Omega$ to $10 \, M\Omega$.

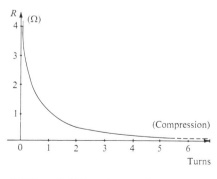

$R \, (\Omega)$

(Compression)

Turns

FIG. 3.20 *Terminal resistance characteristics of the carbon-pile rheostat.*

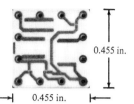

(a) Electrodes placed on module

(b) Resistance applied and adjusted to desired
value by air-abrasion techniques

(c) Module completely encased

Courtesy of International Business Machines Corp.

FIG. 3.21 *Placement of resistors on a module.*

(a)

Courtesy of Dale Electronics, Inc.

FIG. 3.22 *Resistor configuration microcircuit.*

As indicated in Fig. 3.23, trimmers (potentiometers employed for fine-tuning applications) and fixed resistors are now available in the same dual-in-line package.

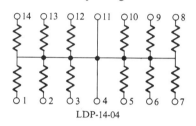

(b)   LDP-14-01

(c)   LDP-14-04

3.5 THERMISTORS

The *thermistor* is a two-terminal semiconductor device whose resistance, as the name suggests, is temperature-sensitive. A representative characteristic appears in Fig. 3.24 with the graphic symbol for the device. Note the non-

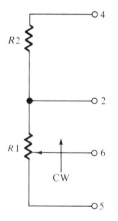

Courtesy of Bourns® Inc.

FIG. 3.23

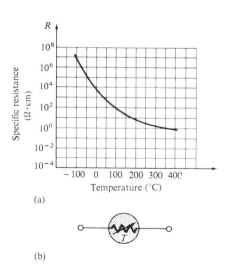

(a)

(b)

FIG. 3.24 *Thermistor. (a) Characteristics; (b) symbol.*

linearity of the curve and the drop in resistance from 5000 Ω to 100 Ω for an increase in temperature from 20°C to 100°C. The decrease in resistance with increase in temperature indicates a negative temperature coefficient.

The temperature of the device can be changed internally or externally. An increase in current through the device will raise its temperature, causing a drop in its terminal resistance. Any externally applied heat source will result in an increase in its body temperature and a drop in resistance. This type of action (internal or external) lends itself well to control mechanisms. A number of different types of thermistors are shown in Fig. 3.25. Materials employed in the

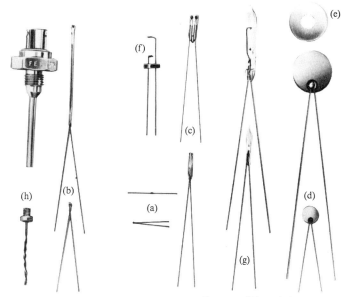

Courtesy of Fenwal Electronics, Inc.

manufacture of thermistors include oxides of cobalt, nickel, strontium, and manganese.

Note the use of a log scale in Fig. 3.24 for the vertical axis. The log scale permits the display of a wider range of specific resistance levels than a linear scale such as the horizontal axis. Note that it extends from 0.0001 Ω-cm to 100,000,000 Ω-cm over a very short interval. The log scale is used for both the vertical and the horizontal axis of Fig. 3.26, which appears in the next section.

FIG. 3.25 *Thermistors. (a) Beads; (b) glass probes; (c) iso-curve interchangeable probes and beads; (d) discs; (e) washers; (f) specially mounted beads; (g) vacuum and gas-filled probes; (h) special probe assemblies.*

3.6 PHOTOCONDUCTIVE CELL

The *photoconductive cell* is a two-terminal semiconductor device whose terminal resistance is determined by the intensity of the incident light on its exposed surface. As the applied illumination increases in intensity, the energy state of the surface electrons and atoms increases, with a result-

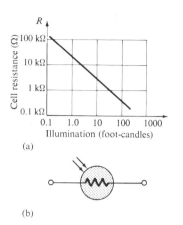

(b)

FIG. 3.26 *Photoconductive cell.*
(a) Characteristics; (b) symbol.

ant increase in the number of "free carriers" and a corresponding drop in resistance. A typical set of characteristics and its graphic symbol appear in Fig. 3.26. Note (as for the thermistor, which was also a semiconductor device) the negative temperature coefficient. A number of cadmium sulfide photoconductive cells appear in Fig. 3.27.

3.7 VARISTORS

Varistors are voltage-dependent, nonlinear resistors used to suppress high-voltage transients. That is, their characteristics are such as to limit the voltage that can appear across the terminals of a sensitive device or system. A typical set of characteristics appear in Fig. 3.28(a) along with a linear resistance characteristic for comparison purposes. Note

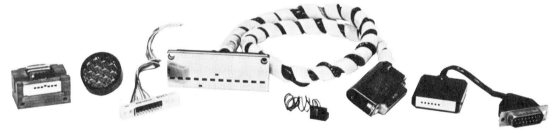

Courtesy of International Rectifier.

FIG. 3.27 *Photoconductive cells.*

that at a particular "firing voltage," the current rises rapidly but the voltage is limited to a level just above this

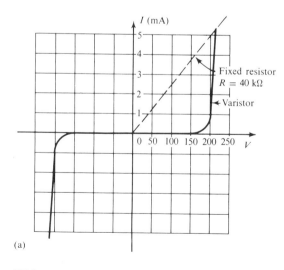

(a)

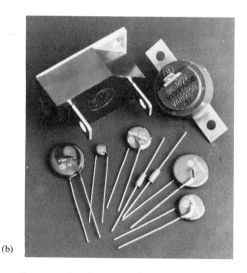

(b)

FIG. 3.28 *Varistors. (a) Characteristics; (b) photograph.*

firing potential. In other words, the magnitude of the voltage that can appear across this device cannot exceed that level defined by its characteristics. Through proper design techniques this device can therefore limit the voltage appearing across sensitive regions of a network. The current is simply limited by the network to which it is connected. A photograph of a number of commercial units appear in Fig. 3.28(b).

3.8 COLOR CODING AND STANDARD RESISTOR VALUES

A wide variety of resistors, fixed or variable, are large enough to have their resistance in ohms printed on the casing. There are some, however, that are too small to have numbers printed on them, so a system of color coding is used. For the fixed molded composition resistor, four color bands are printed on one end of the outer casing as shown in Fig. 3.29(a). Each color has the numerical value indicated in Table 3.5. The color bands are always read left to right from the end that has the band closest to it, as shown in Fig. 3.29(a). The first and second bands represent the first and second digits, respectively. The third band is the number of zeros that follow the second digit, or a multiplying factor determined by the gold and silver bands. The fourth band is the manufacturer's tolerance, which is a

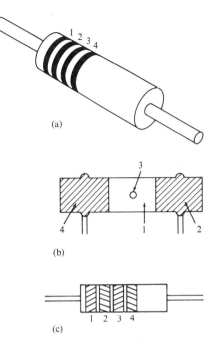

(a)

(b)

(c)

FIG. 3.29 *Color coding. (a) Fixed molded composition resistor; (b) fixed molded composition resistor with radial leads; (c) fixed wirewound resistor.*

TABLE 3.5 *Color coding.*

0	Black	7	Violet
1	Brown	8	Gray
2	Red	9	White
3	Orange	0.1	Gold
4	Yellow	0.01	Silver
5	Green	5%	Gold ⎫
6	Blue	10%	Silver ⎭ Tolerance

measure of the precision by which the resistor was made. If the fourth band is omitted, the tolerance is assumed to be ±20%.

EXAMPLE 3.9. Find the range in which a resistor having the following color bands must exist to satisfy the manufacturer's tolerance:

a. 1st band	2nd band	3rd band	4th band	
Gray	Red	Black	Gold	
8	2	0	±5%	= **82 Ω,**
				±5%

Since 5% of 82 = 4.10, the resistor should be within the range $82 \, \Omega \pm 4.10 \, \Omega$, or *between 77.90 and 86.10 Ω*.

b.	1st band	2nd band	3rd band	4th band	
	Orange	White	Gold	Silver	**= 39(0.1),**
	3	9	0.1	±10%	**±10%**
					= 3.9 Ω,
					±10%

The resistor should lie somewhere *between 3.51 and 4.29 Ω*.

The color coding for fixed-composition resistors with radial leads and the fixed wirewound resistor is provided in Fig. 3.29(b) and (c). The numerical value associated with each color is the same for all three methods of color coding.

Throughout the text, resistor values in the networks will be chosen to reduce the mathematical complexity of finding the solution. It was felt that the procedure or analysis technique was of primary importance and the mathematical exercise secondary. Many of the values appearing in the text are not *standard values*. That is, they are available only through special request. A list of readily available standard values appears in Table 3.6.

If a designer requires other than a standard value, such as 38 kΩ, then the 39 kΩ resistor will probably be employed. If we have a 39 kΩ resistor with a 10% tolerance, the 38 Ω value will certainly fall within its limits of

$$39 \text{ k}\Omega \pm (0.1)(39 \text{ k}\Omega) = 39 \text{ k}\Omega \pm 3.9 \text{ k}\Omega$$
$$= \textbf{35.1 k}\Omega \textbf{ to 42.9 k}\Omega$$

For readily available 10% resistors, all possible values in the range 0.1 Ω to 22 MΩ are included. That is, if we extend the 10% tolerance to each side of each resistor as shown in Fig. 3.30 for the 16-, 18-, and 20-Ω values, all possible intermediate values are included as shown in the figure.

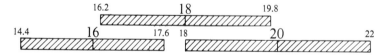

FIG. 3.30 *Range of coverage for 10% 16-Ω, 18-Ω, and 20-Ω resistors.*

3.9 CONDUCTANCE

By finding the reciprocal of the resistance of a material, we have a measure of how well the material will conduct electricity. This quantity is called *conductance,* has the symbol *G,* and is measured in *siemens* (S). For years it was measured in *mhos* (derived from ohm spelled backwards), for which the symbol was the inverted capital letter omega, ℧. However, because the SI system defines siemen as the correct unit of measure, it will be employed in this text.

In equation form, conductance is

TABLE 3.6 *Standard values of commercially available resistors.*

Ohms (Ω)					Kilohms (kΩ)		Megohms (MΩ)	
0.10	**1.0**	**10**	**100**	**1000**	**10**	**100**	**1.0**	**10.0**
0.11	1.1	11	110	1100	11	110	1.1	11.0
0.12	**1.2**	**12**	**120**	**1200**	**12**	**120**	**1.2**	**12.0**
0.13	1.3	13	130	1300	13	130	1.3	13.0
0.15	**1.5**	**15**	**150**	**1500**	**15**	**150**	**1.5**	**15.0**
0.16	1.6	16	160	1600	16	160	1.6	16.0
0.18	**1.8**	**18**	**180**	**1800**	**18**	**180**	**1.8**	**18.0**
0.20	2.0	20	200	2000	20	200	2.0	20.0
0.22	**2.2**	**22**	**220**	**2200**	**22**	**220**	**2.2**	**22.0**
0.24	2.4	24	240	2400	24	240	2.4	
0.27	**2.7**	**27**	**270**	**2700**	**27**	**270**	**2.7**	
0.30	3.0	30	300	3000	30	300	3.0	
0.33	**3.3**	**33**	**330**	**3300**	**33**	**330**	**3.3**	
0.36	3.6	36	360	3600	36	360	3.6	
0.39	**3.9**	**39**	**390**	**3900**	**39**	**390**	**3.9**	
0.43	4.3	43	430	4300	43	430	4.3	
0.47	**4.7**	**47**	**470**	**4700**	**47**	**470**	**4.7**	
0.51	5.1	51	510	5100	51	510	5.1	
0.56	**5.6**	**56**	**560**	**5600**	**56**	**560**	**5.6**	
0.62	6.2	62	620	6200	62	620	6.2	
0.68	**6.8**	**68**	**680**	**6800**	**68**	**680**	**6.8**	
0.75	7.5	75	750	7500	75	750	7.5	
0.82	**8.2**	**82**	**820**	**8200**	**82**	**820**	**8.2**	
0.91	9.1	91	910	9100	91	910	9.1	

NOTE: **Boldface** figures are 10% values.

$$G = \frac{1}{R} \qquad \text{(siemens, S)} \qquad (3.8)$$

A resistance of 1 MΩ is equivalent to a conductance of 10^{-6} S, and a resistance of 10 Ω is equivalent to a conductance of 10^{-1} S. The larger the conductance, therefore, the less the resistance and the greater the conductivity.

In equation form, the conductance is determined by

$$G = \frac{A}{\rho l} \qquad \text{(S)} \qquad (3.9)$$

indicating that increasing the area or decreasing either the length or the resistivity will increase the conductance.

EXAMPLE 3.10. What is the relative increase or decrease in conductivity of a conductor if the area is reduced by 80% and the length is increased by 40%? The resistivity is fixed.

Solution:

$$G_1 = \frac{\dfrac{A_1}{\rho_1 l_1}}{\dfrac{A_2}{\rho_2 l_2}} \quad G_2 = \frac{A_2}{\rho_2 l_2}$$

and for $\rho_1 = \rho_2$,

$$\frac{G_1}{G_2} = \frac{A_1 l_2}{A_2 l_1}$$

with $A_2 = \dfrac{1}{5}A_1$ and $l_2 = 1.4 l_1$, resulting in

$$\frac{G_1}{G_2} = \frac{A_1(1.4 l_1)}{(0.2 A_1)(l_1)} = \frac{1.4}{0.2} = 7$$

and

$$G_2 = \frac{1}{7}G_1$$

PROBLEMS

Section 3.1

1. Convert the following to mils:
   a. 0.5 in.   **b.** 0.01 in.   **c.** 0.004 in.
   d. 1 in.   **e.** 0.05 ft   **f.** 0.01 cm

2. Calculate the area in circular mils (CM) of wires having the following diameters:
   a. 0.050 in.   **b.** 0.016 in.   **c.** 0.41 in.
   d. 0.1 cm   **e.** 0.003 ft   **f.** 0.0042 m

3. The area in circular mils is
   a. 1600 CM   **b.** 900 CM   **c.** 60,000 CM
   d. 625 CM   **e.** 9.89 CM   **f.** 81 CM
   What is the diameter of each wire in inches?

4. What is the resistance of a copper wire 200 ft long and 0.01 inch in diameter ($T = 20°C$)?

5. Find the resistance of a silver wire 35 yd long and 0.0045 inch in diameter ($T = 20°C$).

6. **a.** What is the area in circular mils of an aluminum conductor that is 78 ft long with a resistance of 2.5 Ω?
   b. What is its diameter in inches?

7. **a.** What is the length of a copper wire with a diameter of 1/32 in. and a resistance of 0.005 kΩ ($T = 20°C$)?
   b. Repeat (a) for a silver wire and compare the results.

*8. **a.** What is the resistance of a copper bus-bar with the dimensions shown ($T = 20°C$) in Fig. 3.31?
   b. Repeat (a) for aluminum and compare the results.
   c. Without working out the numerical solution, determine

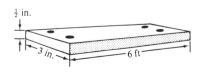

FIG. 3.31

whether the resistance of the bar (aluminum or copper) will increase or decrease with increase in length. Explain your answer.

   d. Repeat (c) for increase in cross-sectional area.

9. a. What is the area in CM of a copper wire that has a resistance of 2.5 Ω and is 300 ft long ($T = 20°C$)?

   b. Without working out the numerical solution, determine whether the area of an aluminum wire will be smaller or larger than that of the copper wire. Explain.

   c. Repeat (b) for a silver wire.

10. In Fig. 3.32 three conductors of different materials are presented.

   a. Without working out the numerical solution, guess at which section would appear to have the most resistance. Explain.

   b. Find the resistance of each section and compare with the result of (a) ($T = 20°C$).

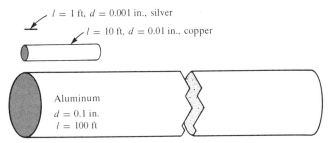

$l = 1$ ft, $d = 0.001$ in., silver

$l = 10$ ft, $d = 0.01$ in., copper

Aluminum
$d = 0.1$ in.
$l = 100$ ft

FIG. 3.32

11. A wire 1000 ft long has a resistance of 0.5 kΩ and an area of 94 CM. Of what material is the wire made ($T = 20°C$)?

*12. Determine the increase in resistance of a copper conductor if the area is reduced by a factor of 4 and the length doubled. The original resistance was 0.2 Ω. The temperature remains fixed.

Section 3.2

13. The resistance of a copper wire is 2 Ω at 10°C. What is its resistance at 60°C?

14. The resistance of an aluminum bus-bar is 0.03 Ω at 0°C. What is its resistance at 60°C?

15. The resistance of a copper wire is 4 Ω at 66°C. What is its resistance at 16°C?

16. The resistance of a copper wire is 0.76 Ω at 30°C. What is its resistance at −40°C?

17. a. The resistance of a copper wire is 0.002 Ω at room temperature (68°F). What is its resistance at 32°F (freezing) and 212°F (boiling)?

   b. For (a), determine the change in resistance for each 10° change in temperature between room temperature and 212°F.

18. If the resistance of a silver wire is 0.04 Ω at −30°C, what is its resistance at −2°C?

*19. **a.** The resistance of a copper wire is 0.92 Ω at 4°C. At what temperature (°C) will it be 1.06 Ω?
   b. At what temperature will it be 0.15 Ω?

20. Find the values of α_1 for copper and aluminum at 20°C, and compare them with those given in Table 3.3.

21. Using Eq. (3.7), find the resistance of a copper wire at 16°C if its resistance at 20°C is 40 Ω.

22. **a.** Find the value of α_1 at $t_1 = 40$°C for copper.
   b. Using the result of (a), find the resistance of a copper wire at 75°C if its resistance is 30 Ω at 40°C.

Section 3.3

23. **a.** Using Table 3.4, find the resistance of 450 ft of #11 and #14 AWG wires.
   b. Compare the resistances of the two wires.
   c. Compare the areas of the two wires.

24. **a.** Using Table 3.4, find the resistance of 1800 ft of #8 and #18 AWG wires.
   b. Compare the resistances of the two wires.
   c. Compare the areas of the two wires.

25. **a.** For the system of Fig. 3.33, the resistance of each line cannot exceed 0.006 Ω, and the maximum current drawn by the load is 110 A. What gage wire should be used?
   b. Repeat (a) for a maximum resistance of 0.003 Ω, $d = 30$ ft, and a maximum current of 110 A.

26. **a.** From Table 3.4 determine the maximum permissible current density (A/CM) for an AWG #0000 wire.
   b. Convert the result of (a) to A/in².
   c. Using the result of (b), determine the cross-sectional area required to carry a current of 5000 A.

Section 3.4

27. Give a brief description of the following resistors:
   a. Carbon-pile rheostat
   b. Tapered-type rheostat
   c. "Corrib" resistor

Section 3.5

28. Find the resistance of the thermistor having the characteristics of Fig. 3.24 at −50°C, 50°C, and 200°C. Note that it is a log scale. If necessary, consult a reference with an expanded log scale.

Section 3.6

29. Using the characteristics of Fig. 3.26, determine the resistance of the photoconductive cell at 10 and 100 foot-candle illumination. As in Problem 28, note that it is a log scale.

Section 3.7

30. **a.** Referring to Fig. 3.28(a), find the terminal voltage of the device at 0.5, 1, 3, and 5 mA.
   b. What is the total change in voltage for the indicated range of current levels?

$d = 30$ ft

E

Load

Solid round copper wire

FIG. 3.33

c. Compare the ratio of maximum to minimum current levels above to the corresponding ratio of voltage levels.

Section 3.8

31. Find the range in which a resistor having the following color bands must exist to satisfy the manufacturer's tolerance:

	1st band	2nd band	3rd band	4th band
a.	green	yellow	orange	gold
b.	red	red	brown	silver
c.	brown	black	black	—

Section 3.9

32. Find the conductance of each of the following resistances:
   a. $0.086\ \Omega$
   b. $4000\ \Omega$
   c. $0.05\ M\Omega$
Compare the three results.

33. Find the conductance of 1000 ft of #18 AWG wire made of
   a. copper
   b. aluminum
   c. iron

34. The conductance of a wire is 100 S. If the area of the wire is increased by a factor of 2/3 and the length is reduced by the same factor, find the new conductance of the wire if the temperature remains fixed.

GLOSSARY

Absolute zero The temperature at which all motion ceases in the atomic structure; ($-273.15°C$).

Circular mil (CM) The cross-sectional area of a wire having a diameter of one mil.

Color coding A technique employing bands of color to indicate the resistance levels and tolerance of resistors.

Conductance (G) An indication of the relative ease with which current can be established in a material. It is measured in siemens (S) or mhos (℧).

Inferred absolute temperature The temperature through which a straight-line approximation for the actual resistance vs. temperature curve will intersect the temperature axis.

Negative temperature coefficient of resistance The value which reveals that the resistance of a material will decrease with increase in temperature.

Ohm (Ω) The unit of measurement applied to resistance.

Photoconductive cell A two-terminal semiconductor device whose terminal resistance is determined by intensity of the incident light on its exposed surface.

Positive temperature coefficient of resistance The value which reveals that the resistance of a material will increase with increase in temperature.

Potentiometer A three-terminal device through which potential levels can be varied in a linear or nonlinear manner.

Resistance A measure of the opposition to the flow of charge through a material.

Resistivity (ρ) A constant of proportionality between the resistance of a material and its physical dimensions.

Rheostat An element whose terminal resistance can be varied in a linear or nonlinear manner.

Thermistor A two-terminal semiconductor device whose resistance is temperature-sensitive.

Varistor A voltage-dependent, nonlinear resistor used to suppress high-voltage transients.

4.1 OHM'S LAW

Consider the following expression:

$$\text{Effect} = \frac{\text{cause}}{\text{opposition}} \qquad \textbf{(4.1)}$$

Every conversion of energy from one form to another can be related to this equation. In electric circuits, the *effect* we are trying to establish is the flow of charge, or *current*. The *potential difference* between two points is the *cause* ("pressure") of this flow of charge, and the opposition to charge flow is the *resistance* encountered.

Substituting these terms into Eq. (4.1) gives

$$\text{Current} = \frac{\text{potential difference}}{\text{resistance}}$$

and

$$I = \frac{E}{R} \qquad \left. \begin{array}{l} I = \text{amperes (A)} \\ E = \text{volts (V)} \\ R = \text{ohms } (\Omega) \end{array} \right\} \text{Ohm's law} \quad \textbf{(4.2)}$$

By simple mathematical manipulations, the voltage and resistance can be found in terms of the other two quantities:

$$E = IR \qquad \text{(volts)} \qquad \textbf{(4.3)}$$

$$\boxed{R = \frac{E}{I}} \qquad \text{(ohms)} \qquad \textbf{(4.4)}$$

For voltage, the symbol E will represent all sources of emf, such as the battery, and the symbol V will represent the potential drop across a resistor or any other energy-converting device.

If we relate Ohm's law to the straight-line equation in the following manner:

$$V = R \cdot I + 0$$
$$\downarrow \quad \downarrow \quad \downarrow \quad \downarrow$$
$$y = m \cdot x + b$$

we find that the slope (m) of the line is determined directly by the resistance, as shown in Fig. 4.1. Note that for a voltage (ordinate) versus current (abscissa) plot, *the closer the line to the vertical axis, the greater the resistance.* The curve clearly indicates that for a fixed current (I_1), the greater the resistance, the greater the voltage as determined by Ohm's law.

For semiconductor devices, such as the transistor and diode to be introduced in the electronics courses, the plot is usually of current (ordinate) versus voltage (abscissa). In this case,

$$I = \frac{1}{R} \cdot V + 0 = G \cdot V + 0$$
$$\downarrow \qquad\qquad\qquad \downarrow \quad \downarrow \quad \downarrow$$
$$y \qquad\qquad\qquad = m \cdot x + b$$

and we find that the conductance will determine the slope of the straight line as indicated for two values in Fig. 4.2. In this case, the greater the conductance (or less the resistance), the closer the plot to the vertical axis. For a fixed voltage V_1, the greater the conductance (or less the resistance), the greater the resulting current.

Before considering a few examples, let us first investigate the characteristics of a very important semiconductor device called the diode, which will be examined in detail in the basic electronics courses. Its total behavior is much like that of a simple switch; that is, it will pass current in only one direction and not the other when operating within specified limits. A typical set of characteristics appears in Fig. 4.3. Without any mathematical calculations, the closeness of the characteristic to the voltage axis for negative values of applied voltage indicates that this is the low conductance (high resistance, switch opened) region. Note that this region extends to approximately 0.5 V positive. However, for values of applied voltage greater than 0.5 V, the vertical rise in the characteristics indicates a high conduc-

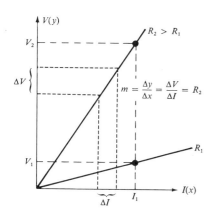

FIG. 4.1 *A plot of voltage versus current for two resistor values.*

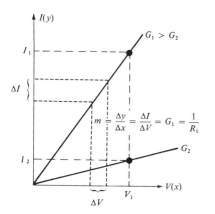

FIG. 4.2 *A plot of current versus voltage for two conductance values.*

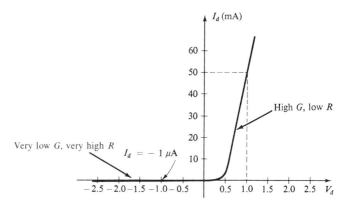

tivity (low resistance, switch closed) region. Application of Ohm's law will now verify the above conclusions.

At $V = +1$ V,

$$R_{\text{diode}} = \frac{V}{I} = \frac{1 \text{ V}}{50 \text{ mA}} = \frac{1}{50 \times 10^{-3}}$$
$$= 20 \ \Omega$$

(relatively low value for most applications)

At $V = -1$ V,

$$R_{\text{diode}} = \frac{V}{I} = \frac{1}{1 \ \mu\text{A}}$$
$$= 1 \text{ M}\Omega$$

FIG. 4.3 *Semiconductor diode characteristic.*

EXAMPLE 4.1. What is the current through a 2-Ω resistor (Fig. 4.4) that has a potential drop of 16 V across it?

Solution:

$$I = \frac{V}{R} = \frac{16}{2} = \textbf{8 A}$$

Note the polarities of the potential drop across the resistor as determined by the direction of the current I.

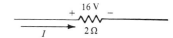

FIG. 4.4

EXAMPLE 4.2. Calculate the voltage that must be applied across the soldering iron of Fig. 4.5 to establish a current of 1.5 A through the iron if its internal resistance is 80 Ω.

Solution:

$$V = IR = 1.5 \cdot 80 = \textbf{120 V}$$

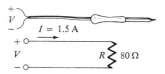

FIG. 4.5

EXAMPLE 4.3. Find the resistance of the bulb of Fig. 4.6 that has a current of 0.4 A through it and a potential drop of 120 V across it.

Solution:

$$R = \frac{V}{I} = \frac{120}{0.4} = \textbf{300 }\Omega$$

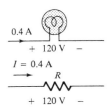

FIG. 4.6

4.2 POWER

Power is an indication of how much work (the conversion of energy from one form to another) can be accomplished in a specified amount of time, that is, a *rate* of doing work. Since work is measured in joules (J), a unit of measurement introduced in Chapter 1, and time in seconds, power (*P*) is measured in joules per second. The electrical unit of measurement for power is the *watt* (W), which is equivalent to 1 J/s. In equation form,

$$P = \frac{W \text{ (joules)}}{t \text{ (s)}} = \frac{W}{t} \qquad \text{(watts, W)} \qquad \textbf{(4.5)}$$

The unit of measurement, the watt, is derived from the surname of James Watt, who was instrumental in establishing the standards for power measurements. He introduced the *horsepower* (hp) as a measure of the average power of a strong dray horse over a full working day. It is approximately 50% more than can be expected from the average horse. The horsepower and watt are related in the following manner:

$$\boxed{1 \text{ horsepower} \cong 746 \text{ watts}}$$

The power delivered to, or absorbed by, an electrical device or system can be found in terms of the current and voltage by first substituting Eq. (2.3) into Eq. (4.5):

$$P = \frac{W}{t} = \frac{QV}{t}$$

But

$$I = \frac{Q}{t}$$

so that

$$\boxed{P = VI} \qquad \text{(watts)} \qquad \textbf{(4.6)}$$

 The power delivered by an energy source is given by

$$\boxed{P = EI} \qquad \text{(watts)} \qquad \textbf{(4.7)}$$

where *E* is the emf of the source and *I* is the current drain from the source.

By direct substitution of Ohm's law, the equation for power can be obtained in two other forms:

$$P = VI = V\left(\frac{V}{R}\right)$$

and

$$P = \frac{V^2}{R} \qquad \text{(watts)} \qquad \textbf{(4.8)}$$

or

$$P = VI = (IR)I$$

and

$$P = I^2R \qquad \text{(watts)} \qquad \textbf{(4.9)}$$

EXAMPLE 4.4. Find the power delivered to the dc motor of Fig. 4.7.

Solution:

$$P = VI = 120(5) = 600 \text{ W} = \textbf{0.6 kW}$$

EXAMPLE 4.5. What is the power dissipated in a 5-Ω resistor if the current through it is 4 A?

Solution:

$$P = I^2R = (4)^2 \cdot 5 = \textbf{80 W}$$

EXAMPLE 4.6. The *V-I* characteristics of a light bulb are provided in Fig. 4.8. Note the nonlinearity of the curve, indicating a wide range in resistance of the bulb with applied voltage. If the rated voltage is 120 V, find the wattage rating of the bulb. Also calculate the resistance of the bulb under rated conditions.

Solution: At 120 V,

$$I = 0.625 \text{ A}$$

and

$$P = VI = 120(0.625) = \textbf{75 W}$$

At 120 V,

$$R = \frac{V}{I} = \frac{120}{0.625} = \textbf{192 } \Omega$$

Since the curve is approaching (closer to) the current axis, the resistance is less for lesser values of applied voltage.

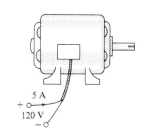

FIG. 4.7

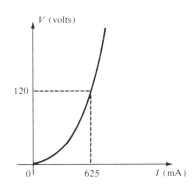

FIG. 4.8

4.3 EFFICIENCY

Any motor or similar device that converts energy from one form to another can be represented by a "black box" with an energy input and output terminal, as in Fig. 4.9.

Conservation of energy requires that

Energy input = (energy output) + (energy lost or stored inside the black box)

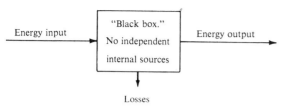

Losses

FIG. 4.9

Dividing both sides of the relationship by t gives

$$\frac{W_{\text{in}}}{t} = \frac{W_{\text{out}}}{t} + \frac{W_{\text{lost or stored inside the black box}}}{t}$$

Since $P = W/t$, we have the following:

$$\boxed{P_i = P_o + P_{\text{lost or stored}}} \tag{4.10}$$

The efficiency (η) of the device inside the black box is then given by the following equation:

$$\text{Efficiency} = \frac{\text{power output}}{\text{power input}}$$

and

$$\boxed{\eta = \frac{P_o}{P_i}} \tag{4.11}$$

Expressed as a percentage,

$$\boxed{\eta = \frac{P_o}{P_i} \times 100\%} \tag{4.12}$$

In terms of the input and output energy, the efficiency in percent is given by

$$\boxed{\eta = \frac{W_o}{W_i} \times 100\%} \tag{4.13}$$

The maximum possible efficiency is 100%, which occurs when $P_o = P_i$, or when the power lost or stored in the system is zero. Obviously, the greater the internal losses of the system in generating the necessary output power or energy, the lower the net efficiency.

EXAMPLE 4.7. A 2-hp motor operates at an efficiency of 75%. What is the power input in watts? If the input current is 9.05 A, what is the input voltage?

Solution:

$$\eta = \frac{P_o}{P_i} \times 100\%$$

$$0.75 = \frac{(2)(746)}{P_i}$$

and

$$P_i = \frac{1492}{0.75} = \textbf{1989.33 W}$$

$$P = EI \quad \text{or} \quad E = \frac{P}{I} = \frac{1990}{9.05} = 219.82 \text{ V} \cong \textbf{220 V}$$

EXAMPLE 4.8. What is the output in horsepower of a motor with an efficiency of 80% and an input current of 8 A at 120 V?

Solution:

$$\eta = \frac{P_o}{P_i} \times 100\%$$

$$0.80 = \frac{P_o}{(120)(8)}$$

and

$$P_o = (0.80)(120)(8) = 768 \text{ W}$$

$$768 \text{ W} \left(\frac{1 \text{ hp}}{746 \text{ W}} \right) = \textbf{1.029 hp}$$

EXAMPLE 4.9. What is the efficiency in percent of a system in which the input energy is 50 J and the output energy is 42.5 J?

Solution:

$$\eta = \frac{W_o}{W_i} \times 100\% = \frac{42.5}{50} \times 100\% = \textbf{85\%}$$

The very basic components of a generating (voltage) system are depicted in Fig. 4.10. The source of mechanical power is a structure such as a paddlewheel that is turned by the water rushing over the dam. The gear train will then insure that the rotating member of the generator is turning at rated speed. The output voltage must then be fed through a transmission system to the load. For each component of the system, an input and output power have been indicated. The efficiency of each system is given by

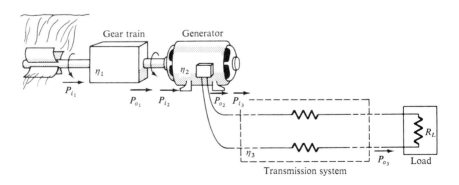

FIG. 4.10 *Basic components of a generating system.*

$$\eta_1 = \frac{P_{o_1}}{P_{i_1}} \qquad \eta_2 = \frac{P_{o_2}}{P_{i_2}} \qquad \eta_3 = \frac{P_{o_3}}{P_{i_3}}$$

If we form the product of these three efficiencies,

$$\eta_1 \cdot \eta_2 \cdot \eta_3 = \frac{P_{o_1}}{P_{i_1}} \cdot \frac{P_{o_2}}{P_{i_2}} \cdot \frac{P_{o_3}}{P_{i_3}}$$

and substitute the fact that $P_{i_2} = P_{o_1}$ and $P_{i_3} = P_{o_2}$, we find that the quantities indicated above will cancel, resulting in P_{o_3}/P_{i_1}, which is a measure of the efficiency of the entire system. In general, for the representative cascaded system of Fig. 4.11,

$$\boxed{\eta_{\text{total}} = \eta_1 \cdot \eta_2 \cdot \eta_3 \cdots \eta_n} \qquad \textbf{(4.14)}$$

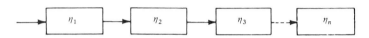

$$\eta_T = \eta_1 \cdot \eta_2 \cdot \eta_3 \cdot \eta_n$$

FIG. 4.11 *Cascaded system.*

EXAMPLE 4.10. Find the overall efficiency of the system of Fig. 4.10 if $\eta_1 = 90\%$, $\eta_2 = 85\%$, and $\eta_3 = 95\%$.

Solution:

$$\eta_T = \eta_1 \cdot \eta_2 \cdot \eta_3 = (0.90)(0.85)(0.95) = 0.727 \text{ or } \textbf{72.7\%}$$

EXAMPLE 4.11. If the efficiency η_1 drops to 60%, find the new overall efficiency and compare the result with that obtained in Example 4.10.

Solution:

$$\eta_T = \eta_1 \cdot \eta_2 \cdot \eta_3 = (0.60)(0.85)(0.95) = 0.485 \text{ or } \textbf{48.5\%}$$

Certainly 48.5% is noticeably less than 75%. The total efficiency of a cascaded system is therefore determined primarily by the lowest efficiency (weakest link) and is less than (or equal to if the remaining efficiencies are 100%) the least efficient link of the system.

4.4 ENERGY

In order for power, which is the rate of doing work, to produce an energy conversion of any form, it must be *used over a period of time*. For example, a motor may have the horsepower to run a heavy load, but unless the motor is *used* over a period of time, there will be no energy conversion. In addition, the longer the motor is used to drive the load, the greater will be the energy expended.

The energy lost or gained by any system is determined by

$$\boxed{W = P \cdot t}$$

W = energy in wattseconds (Ws)
    or joules (J)
P = power in watts (W)   **(4.15)**
t = time in seconds (s)

Since power is measured in watts (or joules per second) and time in seconds, the unit of energy is the *wattsecond* or *joules* as indicated above. The wattsecond, however, is too small a quantity for most practical purposes, so the *watthour* (Wh) or *kilowatthour* (kWh) was defined.

$$\boxed{\text{Energy (Wh)} = \text{power (W)} \times \text{time (h)}} \quad \textbf{(4.16)}$$

$$\boxed{\text{Energy (kWh)} = \frac{\text{power (W)} \times \text{time (h)}}{1000}} \quad \textbf{(4.17)}$$

The watthour meter is an instrument for measuring the energy supplied to the residential or commercial user of electricity. It is normally connected directly to the lines at a point just prior to entering the power distribution panel of the building. A typical set of dials is shown in Fig. 4.12 with a photograph of a watthour meter. As indicated, each power of 10 below a dial is in kilowatthours. The more rapidly the aluminum disc rotates, the greater the energy demand. The dials are connected through a set of gears to the rotation of this disc.

EXAMPLE 4.12. For the dial positions of Fig. 4.12, calculate the electricity bill if the previous reading was 4650 kWh and the average cost is 4.4¢ per kWh.

Solution:

$$5360 - 4650 = 710 \text{ kWh used}$$

$$710 \text{ kWh}\left(\frac{4.4\text{¢}}{\text{kWh}}\right) = \textbf{\$31.24}$$

EXAMPLE 4.13. How much energy (in kilowatthours) is required to light a 60-W bulb continuously for 1 year (365 days)?

Solution:

$$W = \frac{P \cdot t}{1000} = \frac{(60)(24)(365)}{1000} = \frac{525,600}{1000} = \textbf{525.60 kWh}$$

EXAMPLE 4.14. How long can a 205-W television set be on before using more than 4 kWh of energy?

Solution:

$$W = \frac{P \cdot t}{1000} = \frac{(205)(t)}{1000} \Rightarrow t(\text{hours}) = \frac{(4)(1000)}{205} = \textbf{19.51 h}$$

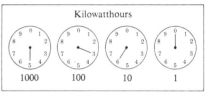

Kilowatthours

1000   100   10   1

Courtesy of Westinghouse Electric Corp.

FIG. 4.12 *Watthour meter.*

EXAMPLE 4.15. What is the cost of using a 5-hp motor for 3 h if the cost is 4.4¢ per kilowatthour?

Solution:

$$W \text{ (kilowatthours)} = \frac{(5)(746 \times 3)}{1000} = 11.2 \text{ kWh}$$

$$\text{Cost} = 11.2(4.4) = \textbf{49.28¢}$$

EXAMPLE 4.16. What is the total cost of using the following at 4.4¢ per kilowatthour?
a. a 1200-W toaster for 30 min
b. six 50-W bulbs for 4 h
c. a 400-W washing machine for 45 min
d. a 5600-W electric clothes dryer for 20 min

Solution:

$$W = \frac{(1200)(\frac{1}{2}) + (6)(50)(4) + (400)(\frac{3}{4}) + (5600)(\frac{1}{3})}{1000}$$

$$= \frac{600 + 1200 + 300 + 1867}{1000}$$

$$= \frac{3967}{1000}$$

$$W = \textbf{3.967 kWh}$$

$$\text{Cost} = (3.967)(4.4) = \textbf{17.45¢}$$

The chart in Fig. 4.13 shows the average cost per kilowatthour as compared to the kilowatthours used per customer.

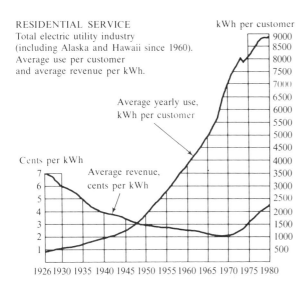

RESIDENTIAL SERVICE
Total electric utility industry
(including Alaska and Hawaii since 1960).
Average use per customer
and average revenue per kWh.

*Courtesy of Edison
Electric Institute*

FIG. 4.13

TABLE 4.1 *Typical wattage ratings of some common household appliances.*

Appliance	Wattage Rating	Appliance	Wattage Rating
Air conditioner	Up to 2,080	Stereo equipment	75
Clock	2	Iron, dry or steam	1,000
Clothes dryer:		Phonograph	75
Electric,	5,600	Projector	Up to 1,280
conventional		Clock radio	4
Electric,	8,400	Range:	
high-speed		Free-standing	Up to 16,000
Gas	400	Wall ovens	Up to 8,000
Clothes washer	400	Refrigerator	320
Coffee maker	Up to 1,000	Shaver	11
Dishwasher	1,400	Cassette player/	5
Fan:		recorder	
Portable	160	Toaster	1,200
Window	210	TV (color)	160
Heater	Up to 1,650		
Heating equipment:			
Furnace fan	320		
Oil-burner motor	230		

Courtesy of Con Edison.

Table 4.1 lists some common household items with their typical wattage ratings. It might prove interesting for the reader to calculate the cost of operating some of these appliances over a period of time using the preceding chart to find the cost per kilowatthour.

4.5 CIRCUIT BREAKERS AND FUSES

The incoming power to any large industrial plant, heavy equipment, simple circuit in the home, or meters used in the laboratory must be controlled to insure that the current through the lines is not above the rated value. Otherwise, the electric or electronic equipment may be damaged or dangerous side effects such as fire or smoke may result. To limit the current level, fuses or circuit breakers are installed at the point where the power enters the installation, such as in the panel in the basement of most homes at the point where the outside feeder lines enter the dwelling. The fuse (depicted in Fig. 4.14) has an internal bimetallic conductor through which the current will pass; it will begin to melt if the current through the system exceeds the rated value printed on the casing. Of course, if it melts through, the current path is broken and the load in its path protected.

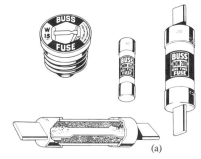

(a)

Courtesy of Bussman Manufacturing Co.

600 V series available from 5 to 800 A
Courtesy of International Rectifier Corp.

FIG. 4.14 *Bimetallic fuses.*

In homes built in recent years, the fuse has been replaced by the circuit breaker appearing in Fig. 4.15. When the current level exceeds rated conditions, an electromagnet in the device will have sufficient strength to draw a connecting metallic link in the breaker out of the circuit and open the current path. When the heavy current drawing element has been removed from the system, the breaker can be reset and used again, unlike the fuse that has to be replaced.

Mini-mag series magnetic hydraulic circuit breakers
Courtesy of Potter and Brumfield Division, AMF, Inc.

FIG. 4.15

PROBLEMS

Section 4.1

1. What is the potential drop across a 6-Ω resistor with a current of 2.5 A through it?

2. What is the current through a 72-Ω-resistor if the voltage drop across it is 12 V?

3. How much resistance is required to limit the current to 1.5 mA if the potential drop across the resistor is 6 V?

4. Find the current through a 3.4-MΩ resistance placed across a 125-V source.

5. If the current through a 0.02-Ω resistor is 3.6 μA, what is the voltage drop across the resistor?

6. If a voltmeter has an internal resistance of 15 kΩ, find the current through the meter when it reads 62 V.

7. If a refrigerator draws 2.2 A at 120 V, what is its resistance?

8. If a clock has an internal resistance of 7.5 kΩ, find the current through the clock if it is plugged into a 120-V outlet.

9. What electromotive force is required to pass 42 mA through a resistance of 0.04 MΩ?

10. If a soldering iron draws 0.76 A at 120 V, what is its resistance?

11. A heating element has a resistance of 20 Ω. Find the current through the element if 120 V are applied across the element.

12. The internal resistance of a dc generator is 0.5 Ω. Determine the loss in terminal voltage across this internal resistance if the current is 15 A.

Section 4.2

13. If 420 J of energy are absorbed by a resistor in 7 min, what is the power to the resistor?

14. The power to a device is 40 joules per second (J/s). How long will it take to deliver 640 J?

15. **a.** How many joules of energy does a 2-W nightlight dissipate in 8 h?
    b. How many kilowatthours does it dissipate?

16. A resistor of 10 Ω has charge flowing through it at the rate of 300 coulombs per minute (C/min). How much power is dissipated?

17. How long will a steady current of 2 A have to exist in a resistor that has 3 V across it to dissipate 12 J of energy?

18. What is the power delivered by a 6-V battery if the charge flows at the rate of 48 C/min?

19. The current through a 4-Ω resistor is 7 mA. What is the power delivered to the resistor?

20. The voltage drop across a 3-Ω resistor is 9 mV. What is the power input to the resistor?

21. If the power input to a 4-Ω resistor is 64 W, what is the current through the resistor?

22. A 1/2-W resistor has a resistance of 1000 Ω. What is the maximum current that it can safely handle?

23. If the power input to a 7.2-kΩ resistor is 88 W, what is the potential drop across the resistor?

24. A power supply can deliver 100 mA at 400 V. What is the power rating?

25. What are the resistance and current ratings of a 120-V, 100-W bulb?

26. What are the resistance and voltage ratings of a 450-W automatic washer that draws 3.75 A?

27. A 20-kΩ resistor has a rating of 100 W. What are the maximum current and the maximum voltage that can be applied to the resistor?

28. **a.** If a home is supplied with a 120-V, 100-A service, find the maximum power capability.
    b. Can the homeowner safely operate the following loads at the same time?
    1. a 5-hp motor
    2. a 3000-W clothes dryer
    3. a 2400-W electric range
    4. a 1000-W steam iron

Section 4.3

29. What is the efficiency of a device that has an output of 0.4 hp with an input of 373 W?

30. What is the input power in watts of a system with an efficiency of 95% and a power output of 4.2 hp?

31. What is the efficiency of a motor that delivers 1 hp when the input current and voltage are 4 A and 220 V, respectively?

32. If an electric motor having an efficiency of 87% and operating off a 220-V line delivers 3.6 hp, what input current does the motor draw?

33. A motor is rated to deliver 2 hp.
    a. If it runs on 110 V and is 90% efficient, how many watts does it draw from the power line?
    b. What is the input current?
    c. What is the input current if the motor is only 70% efficient?

34. An electric motor has an efficiency of 90%. If the input voltage is 220 V, what is the input current when the motor is delivering 4 hp?

35. If two systems in cascade each have an efficiency of 80% and the input energy is 60 J, what is the output energy?

36. The overall efficiency of two systems in cascade is 72%. If the efficiency of one is 0.9, what is the efficiency in percent of the other?

37. If the total input and output power of two systems in cascade is 400 W and 128 W, respectively, what is the efficiency of each system if one has twice the efficiency of the other?

38. **a.** What is the total efficiency of three systems in cascade with efficiencies of 0.98, 0.87, and 0.21?
    b. If the system with the least efficiency (0.21) were removed and replaced by one with an efficiency of 0.90, what would be the percent increase in total efficiency?

Section 4.4

39. A 10-Ω resistor is connected across a 15-V battery.
    a. How many joules of energy will it use in one min?
    b. If the resistor is left connected for two min instead of one, will the energy used increase? Will the power increase?

40. How much energy in kilowatthours is required to keep a 230-W oil-burner motor running continuously for 4 months?

41. How long can a 1500-W heater be on before using more than 10 kWh of energy?

42. How much does it cost to use a 30-W radio for 3 h at 4.4¢ per kilowatthour?

43. What is the total cost of using the following at 4.4¢ per kilowatthour?
    a. a 2000-W air conditioner for 24 h
    b. an 8000-W clothes dryer for 30 min
    c. a 400-W washing machine for 1 h
    d. a 1400-W dishwasher for 45 min

44. What is the total cost of using the following at 4.4¢ per kilowatthour?

a. a 200-W stereo set for 4 h
b. a 1200-W projector for 3 h
c. a 60-W tape recorder for 2 h
d. a 420-W color television set for 6 h

45. Repeat Problem 44 if the cost per kilowatthour is 7¢ (as it was in 1926).

GLOSSARY

Circuit breaker A two-terminal device designed to insure that current levels do not exceed safe levels. If "tripped," it can be reset with a switch or a reset button.

Diode A semiconductor device whose behavior is much like that of a simple switch; that is, it will pass current ideally in only one direction when operating within specified limits.

Efficiency (η) A ratio of output to input power that provides immediate information about the energy-converting characteristics of a system.

Energy (W) A quantity whose change in state is determined by the product of the rate of conversion (P) and the period involved (t). It is measured in joules (J) or wattseconds (Ws).

Fuse A two-terminal device whose sole purpose is to insure that current levels in a circuit do not exceed safe levels.

Horsepower (hp) Equivalent to 746 watts in the electrical system.

Ohm's law An equation that establishes a relationship between the current, voltage, and resistance of an electrical system.

Power An indication of how much work can be done in a specified amount of time; a *rate* of doing work. It is measured in joules/second (J/s) or watts (W).

Watthour meter An instrument for measuring kilowatthours of energy supplied to a residential or commercial user of electricity.

SERIES AND
PARALLEL CIRCUITS

5

5.1 INTRODUCTION

Two types of current are readily available to the consumer today. One is *direct current* (dc), in which ideally the flow of charge (current) does not change in magnitude or direction. *The other is sinusoidal alternating current* (ac), in which the flow of charge is continually changing in magnitude and direction. The next few chapters are an introduction to circuit analysis purely from a dc approach. The methods and concepts will be discussed in detail for direct current; and thus, when possible, a short discussion will suffice to cover any variations we might encounter when we consider ac.

The battery in Fig. 5.1, by virtue of the potential difference between its two terminals, has the ability to cause (or pressure) charge to flow through the simple circuit. The positive terminal attracts the electrons through the wire at the same rate at which electrons are supplied by the negative terminal. As long as the battery is connected in the circuit and maintains its terminal characteristics, the current (dc) through the circuit will not change in magnitude or direction.

If we consider the wire to be an ideal conductor (that is, having no opposition to flow), the potential difference V across the resistor will be equal in magnitude to the emf of the battery: V (volts) = E (volts).

The current in this circuit is limited only by the resistor R. The higher the resistance, the less the current, and conversely, as determined by Ohm's law.

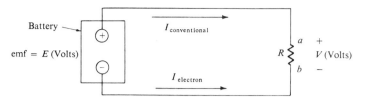

FIG. 5.1

By convention, the direction of I as shown in Fig. 5.1 is opposite to that of electron flow. The reason for defining conventional flow in the opposite direction stems from an assumption made at the time electricity was discovered: that the positive charge is the moving particle. Also, the uniform flow of charge dictates that the direct current I is the same everywhere in the circuit. By following the direction of conventional flow, we notice that there is a rise in potential across the battery ($-$ to $+$), and a drop in potential across the resistor ($+$ to $-$). For single-voltage-source dc circuits, conventional flow always passes from a low potential to a high potential when passing through a voltage source. However, conventional flow always passes from a high to low potential when passing through a resistor for any number of voltage sources in the same circuit (Fig. 5.2).

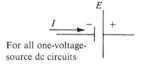

For all one-voltage-source dc circuits

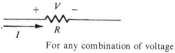

For any combination of voltage sources in the same dc circuit

FIG. 5.2

5.2 SERIES CIRCUITS

A *circuit* consists of any number of elements joined at terminal points, providing at least one closed path through which charge can flow. Two elements are in *series* if they have *only one* point in common that is not connected to a third element. A *branch* is a portion of a circuit consisting of one or more elements in series.

The resistors R_1 and R_2 in Fig. 5.3(a) are in series since they have only point b in common, with no other branches connected to this point. In Fig. 5.3(b), however, the resistors R_1 and R_2 are no longer in series since their common point b is also the junction for a third branch. A further examination will show that R_2 in Fig. 5.3(a) is in series with the voltage source E (point c in common) and that E is in series with R_1. This circuit is therefore referred to as a *series circuit*.

To find the total resistance of a series circuit, simply add the values of the various resistors. In Fig. 5.3(a), for example, the total resistance (R_T) is equal to $R_1 + R_2$. In general, to find the total resistance offered by N resistors in series, simply find the sum of the N resistors:

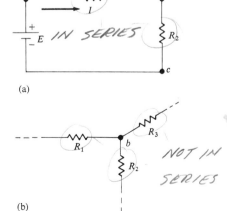

(a)

(b)

FIG. 5.3

$$\boxed{R_T = R_1 + R_2 + R_3 + \cdots + R_N} \qquad (5.1)$$

EXAMPLE 5.1. Find the total resistance of the series circuit of Fig. 5.4.

Solution:

$$R_T = R_1 + R_2 = 2 + 3 = 5\Omega$$

EXAMPLE 5.2. Find R_T for the circuit of Fig. 5.5.

Solution:

$$\begin{aligned} R_T &= R_1 + R_2 + R_3 + R_4 \\ &= 4 + 8 + 7 + 10 \\ &= 29 \ \Omega \end{aligned}$$

To find the total resistance of N resistors of the same value in series, simply multiply the value of one resistor by N:

$$\boxed{R_T = N \cdot R} \tag{5.2}$$

EXAMPLE 5.3. Calculate R_T for the series circuit of Fig. 5.6.

Solution:

$$\begin{aligned} R_T &= R_1 + R_3 + NR_2 \\ &= 1 + 9 + (3)(7) \\ &= 31 \ \Omega \end{aligned}$$

In a series circuit the current is the same through each series element. For the circuit of Fig. 5.3, therefore, the current I through each resistor is the same as that through the battery.

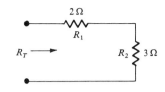

FIG. 5.4

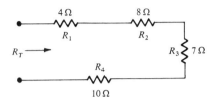

FIG. 5.5

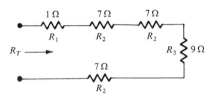

FIG. 5.6

5.3 KIRCHHOFF'S VOLTAGE LAW

Kirchhoff's voltage law (KVL) states that *the algebraic sum of the potential rises and drops around a closed loop (or path) is zero.*

A *closed loop* is any continuous connection of branches that allows us to trace a path which leaves a point in one direction and returns to that same point from another direction without leaving the circuit. In Fig. 5.7, by following the current, we can trace a continuous path that leaves point a through R_1 and returns through E without leaving the circuit. Therefore, *abca* is a closed loop. In order for us to be able to apply Kirchhoff's voltage law, the summation of potential rises and drops must be made in one direction around the closed loop. A plus sign is assigned to a potential rise and a minus sign to a potential drop. If we follow current flow in Fig. 5.7 from point a, we first encounter a potential drop V_1 (+ to −) across R_1 and then another

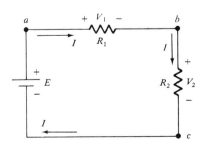

FIG. 5.7

potential drop V_2 across R_2. Continuing through the voltage source, we have a potential rise E ($-$ to $+$) before returning to point a. In symbolic form, where Σ represents summation, $\circlearrowright$ the closed loop, and V the potential drops and rises, we have

$$\boxed{\Sigma_\circlearrowright V = 0}$$ (Kirchhoff's voltage law in symbolic form)   **(5.3)**

which for the circuit of Fig. 5.7 yields (following current flow)

$$-V_1 - V_2 + E = 0$$

or

$$E = V_1 + V_2$$

implying that the potential impressed on the circuit by the battery is equal to the potential drops within the circuit. To go a step further, the sum of the potential rises will equal the sum of the potential drops around a closed path. Therefore, another way of stating Kirchhoff's voltage law is the following:

$$\boxed{\Sigma_\circlearrowright V_{\text{rises}} = \Sigma_\circlearrowright V_{\text{drops}}}$$ **(5.4)**

The text will emphasize the use of Eq. (5.3), however.

There is no requirement that the loop be taken in the direction of current flow. If the loop were taken opposite to that of current flow in Fig. 5.7, the following would result:

$$\Sigma_\circlearrowright V = 0$$
$$-E + V_2 + V_1 = 0$$

or, as before,

$$E = V_1 + V_2$$

Three different closed paths have been indicated on the network of Fig. 5.8. Applying Kirchhoff's voltage law to path 1, we have

$$+E - V_1 - V_2 = 0$$

and

$$E = V_1 + V_2$$

For path 2,

$$+V_2 - V_3 - V_4 = 0$$

and

$$V_2 = V_3 + V_4$$

For path 3,

$$E - V_1 - V_3 - V_4 = 0$$

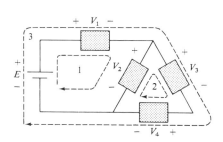

FIG. 5.8

and

$$E = V_1 + V_3 + V_4$$

To apply the basic concepts just introduced, consider the following examples.

EXAMPLE 5.4. Find V_1 and V_2 for the circuits of Figs. 5.9 and 5.10.

Solution (Fig. 5.9): For path 1,

$$10 - 6 - V_1 = 0$$

and

$$V_1 = \textbf{4 V}$$

For path 2,

$$10 - V_2 = 0$$

and

$$V_2 = \textbf{10 V}$$

Solution (Fig. 5.10): For path 1,

$$+25 - V_1 + 15 = 0$$

and

$$V_1 = \textbf{40 V}$$

For path 2,

$$+20 + V_2 = 0$$

and

$$V_2 = \textbf{-20 V}$$

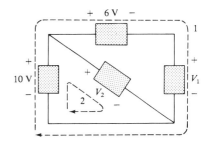

FIG. 5.9

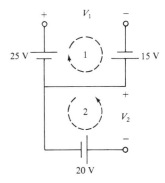

FIG. 5.10

The minus sign simply indicates that the actual polarities of the potential difference are opposite from the polarity indicated in Fig. 5.10.

EXAMPLE 5.5 For the circuit of Fig. 5.11:
a. Find R_T.
b. Find I.
c. Find V_1 and V_2.
d. Find the power input to the 4-Ω and 6-Ω resistors.

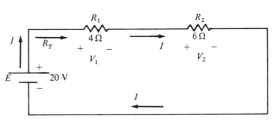

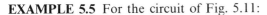

FIG. 5.11

e. Find the power output of the battery, and compare it with that dissipated by the 4-Ω and 6-Ω resistors combined.

f. Verify Kirchhoff's voltage law (following current flow).

Solutions:

a. $R_T = R_1 + R_2 = 4 + 6 = \mathbf{10\ \Omega}$

b. $I = \dfrac{E}{R_T} = \dfrac{20}{10} = \mathbf{2\ A}$

c. $V_1 = IR_1 = (2)(4) = \mathbf{8\ V}$
   $V_2 = IR_2 = (2)(6) = \mathbf{12\ V}$

d. $P_{4\Omega} = \dfrac{V_1^2}{R_1} = \dfrac{8^2}{4} = \dfrac{64}{4} = \mathbf{16\ W}$
   $P_{6\Omega} = I^2 R_2 = (2)^2 \cdot 6 = 4 \cdot 6 = \mathbf{24\ W}$

e. $P_E = EI = (20)(2) = \mathbf{40\ W}$
   $P_E = P_{4\Omega} + P_{6\Omega}$
   $40 = 16 + 24$
   $\underline{40 = 40 \qquad \text{(checks)}}$

f. $\Sigma_{\circlearrowleft} V = +E - V_1 - V_2 = 0$
   $E = V_1 + V_2$
   $20 = 8 + 12$
   $\underline{20 = 20 \qquad \text{(checks)}}$

EXAMPLE 5.6. For the circuit of Fig. 5.12:

a. Find R_T.
b. Find I.
c. Find V_1, V_2, and V_3.
d. Verify Kirchhoff's voltage law (following current flow).

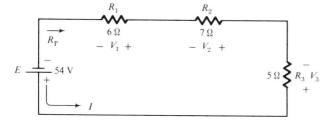

FIG. 5.12

Solutions:

a. $R_T = R_1 + R_2 + R_3 = 6 + 7 + 5 = \mathbf{18\ \Omega}$

b. $I = \dfrac{E}{R_T} = \dfrac{54}{18} = \mathbf{3\ A}$

c. $V_1 = IR_1 = (3)(6) = \mathbf{18\ V}$
   $V_2 = IR_2 = (3)(7) = \mathbf{21\ V}$
   $V_3 = IR_3 = (3)(5) = \mathbf{15\ V}$

d. $\Sigma_{\circlearrowleft} V = +E - V_3 - V_2 - V_1 = 0$
   $E = V_1 + V_2 + V_3$
   $54 = 18 + 21 + 15$
   $\underline{54 = 54 \qquad \text{(checks)}}$

5.4 VOLTAGE DIVIDER RULE

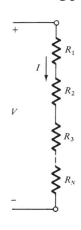

Evaluating the voltage across any resistor or combination of series resistors in a series circuit can be reduced to one step by using the *voltage divider rule*. The proof, which is quite short and direct, will be developed using the circuit of Fig. 5.13:

1. Total resistance: $R_T = R_1 + R_2 + R_3 + \cdots + R_N$
2. Current: $I = V/R_T$
3. Voltage across resistor R_x (where x can be any number from 1 to N): $V_x = IR_x$
4. Voltage across two or more resistors in series, having a total resistance equal to R'_T: $V'_T = IR'_T$
5. Substitute I from part (2) in the equations in parts (3) and (4).

FIG. 5.13

The voltage divider rule is stated in equation form as follows:

$$V_x = \frac{R_x V}{R_T}$$ [for any one resistor R_x **(5.5)** $(0 < x \le N)$]

$$V'_T = \frac{R'_T V}{R_T}$$ (for two or more resistors in series having a total   **(5.6)** resistance equal to R'_T)

In words, the rule states that for a series circuit, the voltage across any resistor (or combination of series resistors) is equal to the value of that resistor (or *sum* of two or more resistors in series), multiplied by the potential difference across the series circuit and divided by the total resistance of the circuit. Note that there is no necessity that V be a source of emf.

EXAMPLE 5.7. Determine the voltage V_2 and V' in Fig. 5.14.

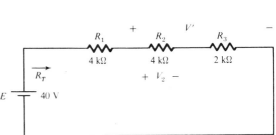

FIG. 5.14

Solution:

$$V_2 = \frac{R_2 E}{R_T} = \frac{R_2 E}{R_1 + R_2 + R_3}$$

$$= \frac{(4 \times 10^3)(40)}{(4 \times 10^3) + (4 \times 10^3) + (2 \times 10^3)}$$

$$= \frac{(4)(40)}{10}$$

$$= \mathbf{16\ V}$$

$$V' = \frac{R'_T E}{R_T} = \frac{(R_2 + R_3)E}{R_1 + R_2 + R_3}$$

$$= \frac{[(4 \times 10^3) + (2 \times 10^3)](40)}{10 \times 10^3}$$

$$= \frac{(6)(40)}{10}$$

$$= \mathbf{24\ V}$$

EXAMPLE 5.8. Determine the voltage drop across resistors R_1 and R_2 in Fig. 5.15.

Solution: The notation in Fig. 5.15 for the 24 volts is equivalent to saying that that point is 24 volts positive with respect to the ground or reference plane (the symbol for the ground or reference plane is shown in the figure). An equivalent circuit appears in Fig. 5.16. On large schematics where it is cumbersome to indicate all supplies, the notation of Fig. 5.15 is used quite frequently. Applying the voltage divider rule, we have

$$V_1 = \frac{4(24)}{4 + 2} = \mathbf{16\ V}$$

and

$$V_2 = \frac{2(24)}{4 + 2} = \mathbf{8\ V}$$

Note that the larger resistor has the greater voltage drop across it. In fact, twice the resistance will result in twice the potential drop, as shown in the example.

EXAMPLE 5.9. For the circuit of Fig. 5.17:

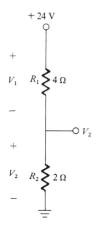

FIG. 5.15

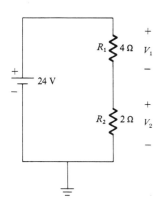

FIG. 5.16

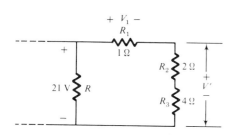

FIG. 5.17

a. Find V_1.
b. Find V'.

Solutions:

a. $V_1 = \dfrac{R_1 V}{R_T} = \dfrac{(1)(21)}{1 + 2 + 4} = \dfrac{21}{7} = \textbf{3 V}$

b. $V' = \dfrac{(R_2 + R_3)V}{R_T} = \dfrac{(6)(21)}{7} = \dfrac{126}{7} = \textbf{18 V}$

EXAMPLE 5.10. For the circuit of Fig. 5.18:
a. Find V_1.
b. Find V_2.
c. Find V.
d. Verify Kirchhoff's voltage law around the closed loop.

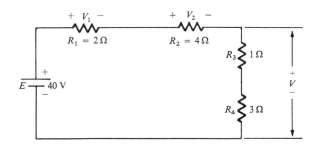

FIG. 5.18

Solutions:

a. $V_1 = \dfrac{R_1 E}{R_T} = \dfrac{(2)(40)}{2 + 4 + 1 + 3} = \dfrac{80}{10} = \textbf{8 V}$

b. $V_2 = \dfrac{R_2 E}{R_T} = \dfrac{(4)(40)}{10} = \textbf{16 V}$

c. $V = \dfrac{(R_3 + R_4)E}{R_T} = \dfrac{(4)(40)}{10} = \textbf{16 V}$

d. $\Sigma_{\circlearrowleft} V = E - V_1 - V_2 - V = 0$
$E = V_1 + V_2 + V$
$40 = 8 + 16 + 16$
$40 = 40 \quad$ (checks)

5.5 PARALLEL CIRCUITS

Two elements or branches are in *parallel* if they have *two points* in common. In Fig. 5.19, elements A and B are in parallel with each other and with element C. Since each element is in parallel with every other element, it is called a *parallel circuit*. In Fig. 5.20(a), A and B are still in parallel, but element C is now in series with the parallel elements A and B. In Fig. 5.20(b), neither A nor B is in parallel with C, but the series combination of A and B is in parallel with element C.

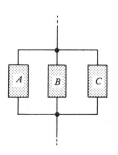

FIG. 5.19

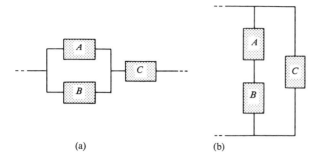

(a)   (b)

FIG. 5.20

Returning to actual elements, resistors R_1 and R_2 of Fig. 5.21 are in parallel since they both have points a and b in common. For the same reason, the impressed voltage E is in parallel with R_1 and R_2, and we note that *the voltage is always the same across parallel elements*. Applying Eq. (3.8), we find that resistors R_1 and R_2 have conductances $G_1 = 1/R_1$ and $G_2 = 1/R_2$, respectively. The total conductance G_T of a parallel circuit is found in a manner similar to that used to find the total resistance of a series circuit; that is, it is the sum of the conductances. The total conductance in Fig. 5.21 is

$$G_T = G_1 + G_2$$

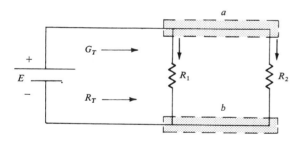

FIG. 5.21

In general, the total conductance of a parallel circuit is equal to the sum of the conductances of the individual branches:

$$G_T = G_1 + G_2 + G_3 + \cdots + G_N \qquad (5.7)$$

or

$$\frac{1}{R_T} = \frac{1}{R_1} + \frac{1}{R_2} + \frac{1}{R_3} + \cdots + \frac{1}{R_N} \qquad (5.8)$$

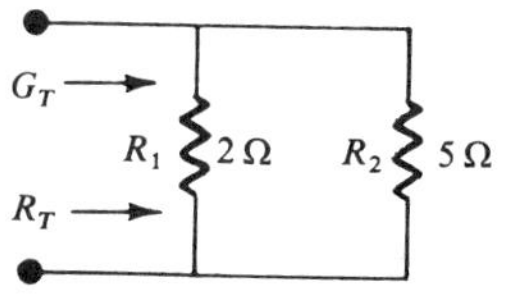

FIG. 5.22

EXAMPLE 5.11. Calculate the total conductance and resistance of the parallel network of Fig. 5.22.

Solution:

$$G_T = G_1 + G_2$$

$$G_1 = \frac{1}{R_1} = \frac{1}{2} = 0.5 \text{ S}$$

$$G_2 = \frac{1}{R_2} = \frac{1}{5} = 0.2 \text{ S}$$

$$G_T = 0.5 + 0.2 = \mathbf{0.7\,S}$$

$$R_T = \frac{1}{G_T} = \frac{1}{0.7} = \mathbf{1.429\ \Omega}$$

EXAMPLE 5.12. Evaluate G_T and R_T for the network of Fig. 5.23.

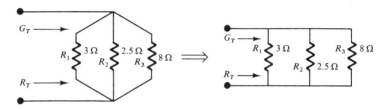

FIG. 5.23

Solution:

$$G_T = G_1 + G_2 + G_3$$

$$G_1 = \frac{1}{R_1} = \frac{1}{3} = 0.333 \text{ S}$$

$$G_2 = \frac{1}{R_2} = \frac{1}{2.5} = 0.40 \text{ S}$$

$$G_3 = \frac{1}{R_3} = \frac{1}{8} = 0.125 \text{ S}$$

$$G_T = 0.333 + 0.40 + 0.125 = \mathbf{0.858\,S}$$

$$R_T = \frac{1}{G_T} = \frac{1}{0.858} = \mathbf{1.165\ \Omega}$$

The conductance of a parallel circuit and the resistance of a series circuit are often called the *duals* of each other, since it is possible to develop some of the relationships for the parallel circuit directly from those of a series circuit (and vice versa) by simply interchanging R and G. For example, replacing G in Eq. (5.7) by R and keeping the same subscripts, we can derive the expression for the total resistance of a series circuit:

$$R_T = R_1 + R_2 + R_3 + \cdots + R_N \qquad (5.1)$$

Similarly, replacing R with G in Eq. (5.2), we obtain, using duality, the expression for the total conductance of a parallel circuit with N equal parallel conductances:

$$G_T = N \cdot G \qquad (5.9)$$

For the total resistance R_T,

$$\frac{1}{R_T} = N\frac{1}{R}$$

and

$$R_T = \frac{R}{N} \qquad (5.10)$$

In a parallel circuit, therefore, with N equal resistors in parallel, the total resistance is the value of *one* resistance divided by the total number (N) of resistors in parallel.

EXAMPLE 5.13. Compute G_T and R_T for the circuit of Fig. 5.24.

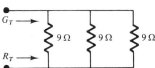

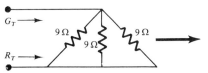

FIG. 5.24

Solution:

$$G_T = NG$$

$$G = \frac{1}{9} = 0.111\,\text{S}$$

$$G_T = (3)(0.111) = \mathbf{0.333\,S}$$

and

$$R_T = \frac{1}{G_T} = \frac{1}{0.333} = \mathbf{3\,\Omega}$$

or

$$R_T = \frac{R}{N} = \frac{9}{3} = \mathbf{3\,\Omega}$$

EXAMPLE 5.14. Calculate R_T and G_T for the circuit of Fig. 5.25.

Solution:

$$R_T = \frac{R}{N} = \frac{2}{4} = \mathbf{0.5\,\Omega}$$

$$G_T = \frac{1}{R_T} = \frac{1}{0.5} = \mathbf{2\,S}$$

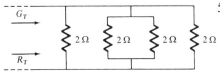

FIG. 5.25

It is often helpful to have at hand a formula for finding the total resistance of two parallel resistors when analyzing circuits. For example, consider the circuit of Fig. 5.26, where

$$G_T = \frac{1}{R_T} = \frac{1}{R_1} + \frac{1}{R_2}$$

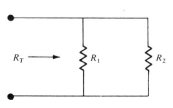

FIG. 5.26

$\overline{}^{S}\overline{}$
$\overline{\mathsf{I}}\mathsf{P}\overline{\mathsf{I}}$
               Parallel Circuits   **93**

and

$$R_T = \frac{1}{(1/R_1) + (1/R_2)} = \frac{1}{(R_1 + R_2)/(R_1 R_2)}$$

so that

$$\boxed{R_T = \frac{R_1 R_2}{R_1 + R_2}} \qquad (5.11)$$

For three parallel resistors, a derivation very similar to that for two parallel resistors will result in Eq. (5.12):

$$\boxed{R_T = \frac{R_1 R_2 R_3}{R_1 R_2 + R_1 R_3 + R_2 R_3}} \qquad (5.12)$$

In the examples to follow, note that *the total resistance of two parallel resistors is always less than either of the resistors but greater than one-half the smaller.*

EXAMPLE 5.15. Find the total resistance of the parallel network of Fig. 5.27.

Solution:

$$R_T = \frac{(3)(6)}{3 + 6} = \frac{18}{9} = \mathbf{2\ \Omega}$$

EXAMPLE 5.16. Calculate R_T for the network of Fig. 5.28.

Solution:

$$R_T = \frac{R_1 R_2 R_3}{R_1 R_2 + R_1 R_3 + R_2 R_3} = \frac{(4)(4)(3)}{(4)(4) + (4)(3) + (4)(3)}$$

$$= \frac{48}{16 + 12 + 12} = \frac{48}{40} = \mathbf{1.2\ \Omega}$$

or

$$R'_T = \frac{R}{N} = \frac{4}{2} = 2\ \Omega$$

and

$$R_T = \frac{(2)(3)}{2 + 3} = \frac{6}{5} = \mathbf{1.2\ \Omega}$$

EXAMPLE 5.17. Find R_T for the network of Fig. 5.29.

Solution:

$$R'_T = \frac{R}{N} = \frac{4}{3} = 1.333\ \Omega$$

$$R''_T = \frac{(3)(8)}{3 + 8} = \frac{24}{11} = 2.182\ \Omega$$

$$R_T = \frac{(1.333)(2.182)}{1.333 + 2.182} = \frac{2.909}{3.515} = \mathbf{0.8276\ \Omega}$$

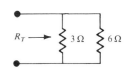

FIG. 5.27

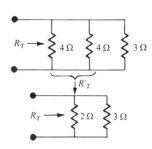

FIG. 5.28

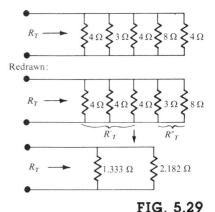

FIG. 5.29

5.6 KIRCHHOFF'S CURRENT LAW

Kirchhoff's current law (KCL) states that *the algebraic sum of the currents entering and leaving a node is zero.* (A *node* is a junction of two or more branches.) In other words, *the sum of the currents entering a node must equal the sum of the currents leaving a node.* In equation form,

$$\Sigma I_{\text{entering}} = \Sigma I_{\text{leaving}} \qquad (5.13)$$

In Fig. 5.30, the currents I_1, I_3, and I_4 enter the junction while I_2 and I_5 leave. Applying Eq. (5.13), we obtain

$$I_1 + I_3 + I_4 = I_2 + I_5$$

Substituting values yields

$$2 + 4 + 3 = 8 + 1$$
$$9 = 9$$

The shaded region of Fig. 5.30 is not limited to a terminal point for the branches. As demonstrated by Fig. 5.31, it may include a number of series and parallel elements or a very complex network. Applying Kirchhoff's current law to the inner region yields

$$I_A + I_B + I_C = I_D$$

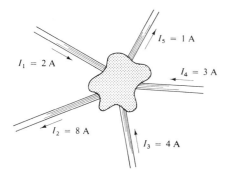

FIG. 5.30

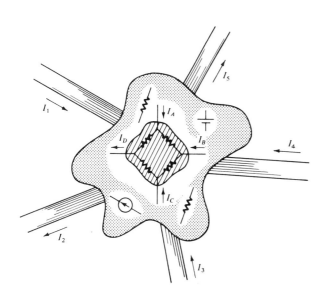

FIG. 5.31

We will now consider a few examples in which we determine the unknown currents by applying this law at the various nodes.

EXAMPLE 5.18. Determine the currents I_3 and I_5 of Fig. 5.32 through applications of Kirchhoff's current law.

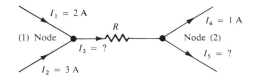

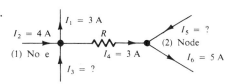

FIG. 5.32

Solution: Note that since node 2 has two unknown quantities and node 1 only has one, we must first apply KCL to node 1. The result can then be applied to node 2.

(1)	(2)
$I_1 + I_2 = I_3$	$I_3 = I_4 + I_5$
$2 + 3 = I_3$	or $I_5 = I_3 - I_4 = 5 - 1$
and $I_3 = 5\,A$	and $I_5 = 4\,A$

EXAMPLE 5.19. Find the currents I_3 and I_5 of Fig. 5.33.

Solution:

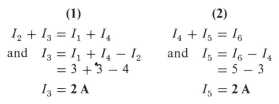

(1)	(2)
$I_2 + I_3 = I_1 + I_4$	$I_4 + I_5 = I_6$
and $I_3 = I_1 + I_4 - I_2$	and $I_5 = I_6 - I_4$
$= 3 + 3 - 4$	$= 5 - 3$
$I_3 = 2\,A$	$I_5 = 2\,A$

FIG. 5.33

EXAMPLE 5.20. For the circuit of Fig. 5.34:
a. Find the total conductance and total resistance.
b. Find the current I_T.
c. Find the current in each branch.
d. Verify Kirchhoff's current law at node a.
e. Find the power dissipated by each resistor and note whether the power delivered is equal to that dissipated.

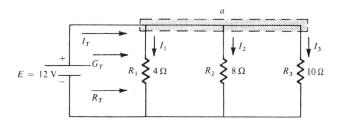

Solutions:

FIG. 5.34

a. $G_T = G_1 + G_2 + G_3 = \dfrac{1}{4} + \dfrac{1}{8} + \dfrac{1}{10}$
   $= 0.250 + 0.125 + 0.100$
   $G_T = \mathbf{0.475\,S}$
   $R_T = \dfrac{1}{G_T} = \dfrac{1}{0.475} = \mathbf{2.105\,\Omega}$

b. $I_T = EG_T = (12)(0.475) = \mathbf{5.70\,A}$
   or

$$I_T = \frac{E}{R_T} = \frac{12}{2.105} = \textbf{5.70 A}$$

c. $I_1 = \dfrac{V_1}{R_1} = \dfrac{E}{R_1} = \dfrac{12}{4} = \textbf{3 A}$

　　$I_2 = \dfrac{V_2}{R_2} = \dfrac{E}{R_2} = \dfrac{12}{8} = \textbf{1.5 A}$

　　$I_3 = \dfrac{V_3}{R_3} = \dfrac{E}{R_3} = \dfrac{12}{10} = \textbf{1.2 A}$

d. $+I_T - I_1 + I_2 - I_3 = 0$

　　or

　　$I_T = I_1 + I_2 + I_3$

　　$5.7 = 3 + 1.5 + 1.2$

　　$\underline{5.7 = 5.7 \text{ (checks)}}$

e. $P_{\text{deliv}} = EI_T = (12)(5.7) = \textbf{68.4 W}$

　　$P_{R_1} = EI_1 = (12)(3) = \textbf{36 W}$

　　$P_{R_2} = EI_2 = (12)(1.5) = \textbf{18 W}$

　　$P_{R_3} = EI_3 = (12)(1.2) = \textbf{14.4 W}$

　　$P_{\text{deliv}} = P_{R_1} + P_{R_2} + P_{R_3}$

　　$68.4 = 36 + 18 + 14.4$

　　$\underline{68.4 = 68.4 \qquad \text{(checks)}}$

5.7 CURRENT DIVIDER RULE

The *current divider rule* (CDR) will be derived through the use of the representative network of Fig. 5.35. The input

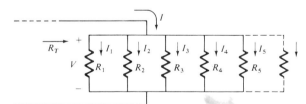

FIG. 5.35

current I equals V/R_T, where R_T is the total resistance of the parallel branches. Substituting $V = I_x R_x$ into the above equation, where I_x refers to the current through a parallel branch of resistance R_x, we have

$$I = \frac{V}{R_T} = \frac{I_x R_x}{R_T}$$

and

$$\boxed{I_x = \frac{R_T}{R_x} I} \qquad\qquad \textbf{(5.14)}$$

which is the general form for the current divider rule. In words, the current through any parallel branch is equal to the product of the *total* resistance of the parallel branches and the input current divided by the resistance of the branch through which the current is to be determined.

For the current I_1,

$$I_1 = \frac{R_T}{R_1} I$$

or for I_2,

$$I_2 = \frac{R_T}{R_2} I$$

For the particular case of *two parallel resistors* as shown in Fig. 5.36,

$$R_T = \frac{R_1 R_2}{R_1 + R_2}$$

and

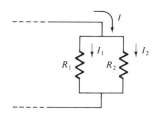

FIG. 5.36

$$I_1 = \frac{R_T}{R_1} I = \frac{\dfrac{R_1 R_2}{R_1 + R_2}}{R_1} I$$

and

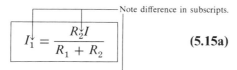

Note difference in subscripts.

$$I_1 = \frac{R_2 I}{R_1 + R_2} \tag{5.15a}$$

Similarly for I_2,

$$I_2 = \frac{R_1 I}{R_1 + R_2} \tag{5.15b}$$

In words, for two parallel branches, the current through either branch is equal to the product of the *other* parallel resistor and the input current divided by the *sum* (not total parallel resistance) of the two parallel resistances.

EXAMPLE 5.21. Find the current I_1 for the network of Fig. 5.37.

Solution: By Eq. (5.14),

$$R_T = 6 \,\|\, 24 \,\|\, 24 = 6 \,\|\, 12 = 4 \,\Omega$$

$$I_1 = \frac{4(42 \times 10^{-3})}{6} = \textbf{28 mA}$$

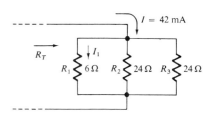

FIG. 5.37

EXAMPLE 5.22. Determine the magnitudes of the currents I_1 and I_2 of Fig. 5.38.

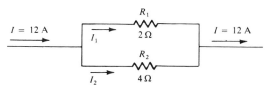

FIG. 5.38

Solution: By Eq. (5.15), the current divider rule,

$$I_1 = \frac{(4)(12)}{4 + 2} = \textbf{8 A}$$

By Kirchhoff's current law,

$$I = I_1 + I_2$$
$$I_2 = I - I_1 = 12 - 8$$
$$I_2 = \textbf{4 A}$$

or, using the current divider rule again,

$$I_2 = \frac{(2)(12)}{4 + 2} = \textbf{4 A}$$

EXAMPLE 5.23. Find I_1 and I_2 of Fig. 5.39.

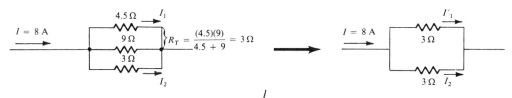

FIG. 5.39

Solution:

$$I'_1 = I_2 = \frac{8}{2} = \textbf{4 A}$$

Current divides equally between two equal resistors.

By the current divider rule (Fig. 5.40),

$$I_1 = \frac{(9)(4)}{9 + 4.5} = \frac{36}{13.5}$$
$$I_1 = \textbf{2.667 A}$$

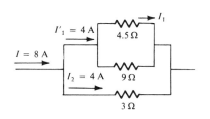

FIG. 5.40

From the examples just discussed, note the following:

1. Redrawing the circuit can often be helpful in analyzing circuits.
2. The current entering any number of parallel resistors divides into these resistors as the inverse ratio of their ohmic values. This is depicted in Fig. 5.41.
3. More current passes through the smaller of two parallel resistors.

A mechanical analogy often used to describe this division of current flow is the flow of water through pipes. The

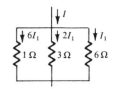

FIG. 5.41

water represents the flow of charge, and the tubes or pipes represent conductors. In this analogy, the greater the resistance of the corresponding electrical element, the smaller the diameter of the tubing.

The total current I in Fig. 5.42(a) divides equally between the two equal resistors. The analogy just described is shown to the right. Obviously, for two pipes of equal diameter, the water will divide equally.

In Fig. 5.42(b), one resistor is three times the other, resulting in the current dividing as shown. Its mechanical analogy is shown in the adjoining figure. Again, it should be obvious that three times as much water (current) will pass through one pipe as through the other. In both Figs. 5.42(a) and (b), the total water (current) entering the parallel systems from the left will equal that leaving to the right.

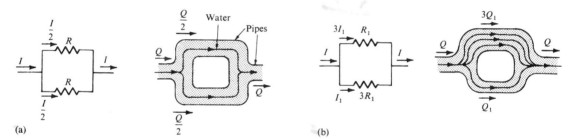

(a)   (b)

FIG. 5.42

5.8 SHORT CIRCUITS

An element is *short-circuited* if its effect is altered by a direct connection of low resistive value across its terminals. For example, the wire of effectively zero resistance in Fig. 5.43 is "shorting out" the 2-Ω resistor. The current through the short can be determined by the current divider rule:

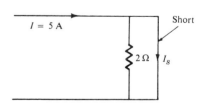

FIG. 5.43

$$I_s = \frac{(2)(5)}{2 + 0} = 5 = I$$

The total current entering the parallel branches is through the short. The resistor has no current through it; consequently there is no voltage drop across it. Its effect has been completely eliminated by the short.

This shorting effect can be dangerous if the current increases to a very high value after the load is shorted. In the system of Fig. 5.44(a), the current is limited by the load (99 Ω) and the resistance of the conductors (1 Ω) to

$$I = \frac{120}{99 + 1} = \frac{120}{100} = 1.2 \text{ A}$$

If the 99-Ω resistor is shorted out by foreign debris or system breakdown [Fig. 5.44(b)], the total resistance of the circuit is only that of the conductors (1 Ω). The current in the conductors and the short will then be

$$I = \frac{120}{1} = 120 \text{ A}$$

which may produce enough heat (I^2R) in the wire to melt the insulation and cause dangerous effects such as sparking and fire.

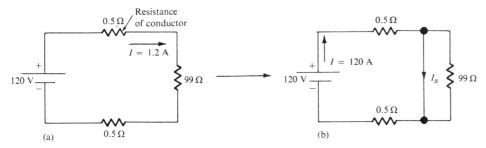

FIG. 5.44

5.9 VOLTAGE SOURCES IN SERIES

Two or more voltage sources in series may be replaced by one voltage source having the magnitude and polarity of the *resultant*. The resultant is found by summing all of the voltage sources having corresponding polarities in one direction and subtracting the sum of those with the opposite polarities. For example, for each of the circuits in Fig. 5.45, the resultant source is shown to the right.

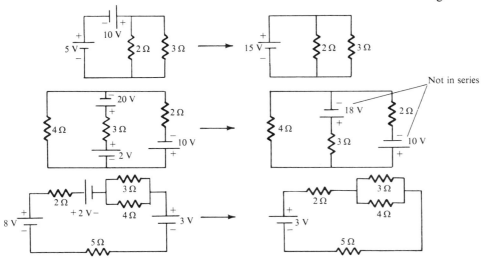

FIG. 5.45

5.10 INTERNAL RESISTANCE OF VOLTAGE SOURCES

Every source of emf, whether it be a generator, battery, or laboratory supply as shown in Fig. 5.46(a), will have some internal resistance. The equivalent circuit of any source of emf will therefore appear as shown in Fig. 5.46(b). In this section we will examine the effect of the internal resistance on the output voltage so that any difficulties with a particular source of emf can be explained.

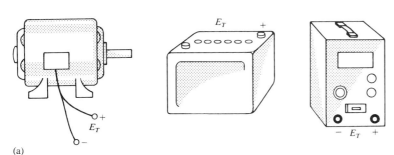

(a)

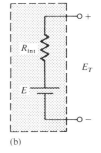

(b)

FIG. 5.46

In all circuit analyses to this point, the ideal voltage source (no internal resistance) was used [see Fig. 5.47(a)]. This will continue to be the case in all the analyses that follow, unless stated otherwise. On occasion, we will explain how to include the internal resistance in a particular analysis so that its effects can be considered.

The ideal voltage source of Fig. 5.47(a) has no internal resistance and an output voltage of E volts with no load or full load. In the practical case [Fig. 5.47(b)], where we consider the effects of the internal resistance, the output voltage will be E volts only when no-load ($I_L = 0$) conditions exist. When a load is connected [Fig. 5.47(c)], the output voltage of the voltage source will drop to E_T volts due to the voltage drop across the internal resistance.

By applying Kirchhoff's voltage law around the indicated loop of Fig. 5.47(c), we obtain

$$E - IR_{\text{int}} - E_T = 0$$

or, since

$$E = E_{NL}$$
$$E_{NL} - IR_{\text{int}} - E_T = 0$$

or

$$\boxed{E_T = E_{NL} - IR_{\text{int}}} \qquad (5.16)$$

If the value of R_{int} is not available, it can be found by first solving for R_{int} in the equation just derived for E_T. That is,

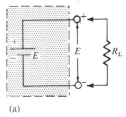

(a)

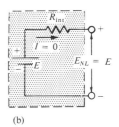

(b)

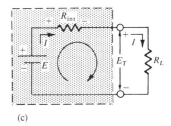

(c)

FIG. 5.47

$$R_{\text{int}} = \frac{E_{NL} - E_T}{I} = \frac{E_{NL}}{I} - \frac{IR_L}{I}$$

and

$$\boxed{R_{\text{int}} = \frac{E_{NL}}{I} - R_L} \qquad \textbf{(5.17)}$$

If we then measure the no-load output voltage and the current for a load R_L, we can find the value of R_{int} by substitution into Eq. (5.17).

A plot of the output voltage versus current demands is shown in Fig. 5.48(b) for the circuit in Fig. 5.48(a). Note that an increase in load demand (I_L) increases the current through, and thereby the voltage drop across, the internal

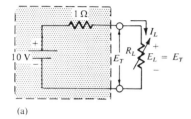

(a)

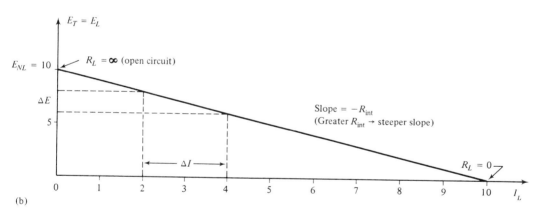

(b)

FIG. 5.48

resistance of the source, resulting in a decrease in terminal voltage. Eventually, as the load resistance approaches zero ohms, all the generated voltage will appear across the internal resistance and none at the output terminals. The steeper the slope of the curve of Fig. 5.48(b), the greater the internal resistance. In fact, for any chosen interval of voltage or current, the magnitude of the internal resistance is given by

$$\boxed{R_{\text{int}} = \frac{\Delta E}{\Delta I}} \qquad \text{where } \Delta \text{ signifies finite change} \quad \textbf{(5.18)}$$

For the chosen interval of $2 \rightarrow 4$ A on Fig. 5.48(b), ΔE is 2 V, so that $R_{\text{int}} = 2/2 = 1 \, \Omega$, as indicated in Fig. 5.48(a).

A direct consequence of the loss in output voltage is a loss in power delivered to the load. Multiplying both sides of Eq. (5.16) by the current I in the circuit, we obtain

$$
\underset{\substack{\text{Power}\\\text{to load}}}{IE_T} = \underset{\substack{\text{Power output}\\\text{by battery}}}{IE_{NL}} - \underset{\substack{\text{Power loss in}\\\text{the form of heat}}}{I^2 R_{\text{int}}}
\qquad\text{(5.19)}
$$

EXAMPLE 5.24. Before a load is applied, the terminal voltage of the power supply of Fig. 5.49 is set to 40 V. When a load of 500 Ω is attached, as shown in Fig. 5.49(b), the terminal voltage drops to 36 V. What happened to the remainder of the no-load voltage, and what is the internal resistance of the source?

Solution: The $40 - 36 = 4$ V now appear across the internal resistance of the source. The load current is $36/0.5$ kΩ $= 72$ mA. Applying Eq. (5.17), we have

$$
R_{\text{int}} = \frac{E_{NL}}{I} - R_L = \frac{40}{72 \text{ mA}} - 0.5 \text{ k}\Omega
$$

$$
= 555.55 - 500 = \textbf{55.55 } \boldsymbol{\Omega}
$$

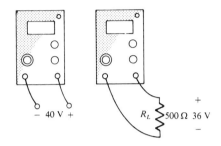

FIG. 5.49

EXAMPLE 5.25. The battery of Fig. 5.50 has an internal resistance of 2 Ω. Find the voltage E_T and the power lost to the internal resistance if the applied load is a 13-Ω resistor.

Solution:

$$
I = \frac{30}{2 + 13} = \frac{30}{15} = 2 \text{ A}
$$

$$
E_T = E_{NL} - IR_{\text{int}}
$$
$$
= 30 - (2)(2)
$$
$$
E_T = \textbf{26 V}
$$
$$
P_{\text{lost}} = I^2 R_{\text{int}} = (2)^2(2) = (4)(2)
$$
$$
= \textbf{8 W}
$$

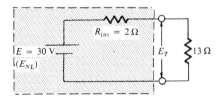

FIG. 5.50

EXAMPLE 5.26. The terminal characteristics of a dc generator appear in Fig. 5.51. Rated (full-load) conditions are indicated at 120 V, 8 A.
a. Calculate the average internal resistance of the supply.
b. At what load current will the terminal voltage drop to 100 V?

Solutions:

a. $R_{\text{int}} = \dfrac{\Delta E}{\Delta I} = \dfrac{4}{8} = \textbf{0.5 } \boldsymbol{\Omega}$

b. $\Delta I = \dfrac{\Delta E}{R_{\text{int}}} = \dfrac{24}{0.5} = I_L = \textbf{48 A}$

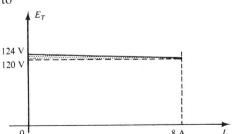

FIG. 5.51

5.11 VOLTAGE REGULATION

For any supply, ideal conditions dictate that for the range of load demand (I_L), the terminal voltage remain fixed in magnitude. In other words, if a supply is set for 12 volts, it is desirable that it maintain this terminal voltage even though the current demand on the supply may vary. A measure of how close a supply will come to ideal conditions is given by the voltage regulation characteristic. By definition, the voltage regulation of a supply between the limits of full-load and no-load conditions (Fig. 5.52) is given by the following:

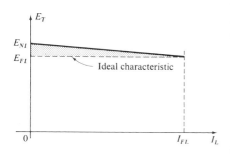

FIG. 5.52

$$\text{Voltage regulation } (VR)\% = \frac{E_{NL} - E_{FL}}{E_{FL}} \times 100\%$$

(5.20)

For ideal conditions, $E_{FL} = E_{NL}$ and $VR\% = 0$. Therefore, *the smaller the voltage regulation, the less the variation in terminal voltage with change in load.*

It can be shown with a short derivation that the voltage regulation is also given by

$$VR\% = \frac{R_{\text{int}}}{R_L} \times 100\%$$

(5.21)

In other words, the smaller the internal resistance for the same load, the smaller the regulation and more ideal the output.

EXAMPLE 5.27. Calculate the voltage regulation of a supply having the characteristics of Fig. 5.51.

Solution:

$$VR\% = \frac{E_{NL} - E_{FL}}{E_{FL}} \times 100\%$$

$$= \frac{124 - 120}{120} \times 100\% = \frac{4}{120} \times 100\%$$

and

$$VR = \mathbf{3.33\%}$$

EXAMPLE 5.28. Determine the voltage regulation of the supply of Fig. 5.50.

Solution:

$$VR\% = \frac{R_{\text{int}}}{R_L} \times 100\% = \frac{2}{13} \times 100\%$$

$$\cong \mathbf{15.38\%}$$

5.12 VOLTAGE SOURCES IN PARALLEL

For situations involving continuous operation, voltage sources should never be placed in parallel unless they have the same output voltage and similar characteristics (same internal resistance, ampere-hour rating, and so on). Two storage batteries with similar internal resistances but different voltage ratings are shown in parallel in Fig. 5.53. Note that with no load attached, the 12-V battery will be discharging. Eventually, the no-load voltage of the 12-V battery may drop to about 11 V, and the internal resistance may increase to perhaps 10 Ω (200 times the rated value). At this point, the potential drop across the internal resistance of the 12-V battery will be about 5 V [$V = 10(11 - 6)/(10 + 0.05) \cong 5$V], leaving only 6 V across the output terminals of the 12-V battery. The ampere-hour rating of the 12-V battery has also dropped to the point of no practical use. If a load is now connected, it will draw its current almost exclusively from the 6-V battery at its rated 6 V. Occasionally, batteries of different emfs are placed in parallel, but this is usually for very short periods of time.

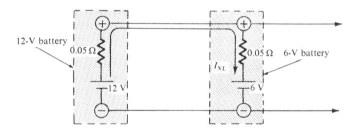

FIG. 5.53

Similar batteries are sometimes placed in parallel if a load requires an ampere-hour rating (at the same voltage) that is greater than that of one battery. For example, in Fig. 5.54, two similar 12-V batteries with ampere-hour ratings of 50 at 2 A have been placed in parallel. Ideally, since the batteries are similar, no discharging or charging will take place with no load. When a load is attached, the current I_T will be 4 A ($I_T = I_{B_1} + I_{B_2}$) if each battery is supplying its rated 2 A. The ampere-hour rating of the combined system has therefore changed to 100 Ah at 4 A. The supply voltage, however, is still 12 V.

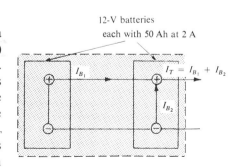

FIG. 5.54

PROBLEMS

Section 5.2

1. Find the total resistance and current I for each circuit of Fig. 5.55.

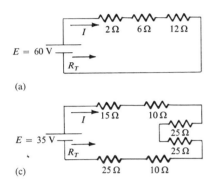

(a)

(b)

(c)

(d)

FIG. 5.55

2. For the circuits of Fig. 5.56, the total resistance is specified. Find the unknown resistances and the current I for each circuit.

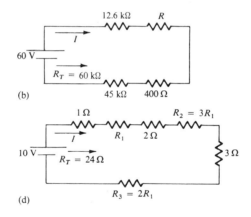

(a)

(b)

(c)

(d)

FIG. 5.56

3. Find the applied voltage E necessary to develop the current specified in each network of Fig. 5.57.

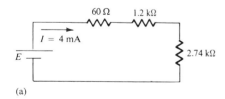

(a)

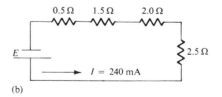

(b)

FIG. 5.57

Section 5.3

4. Find V_{ab} with polarity for the circuits of Fig. 5.58. Each box can contain a load or a power supply, or a combination of both.

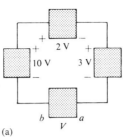

(a)

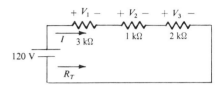

(b)

FIG. 5.58

5. For the circuit of Fig. 5.59:
   a. Find the total resistance, current, and unknown voltage drops.
   b. Verify Kirchhoff's voltage law around the closed loop.
   c. Find the power dissipated by each resistor, and note whether the power delivered is equal to the power dissipated.
   d. If the resistors are available with wattage ratings of 1/2, 1, and 2 W, what minimum wattage rating can be used for each resistor in this circuit?

6. Repeat Problem 5 for the circuit of Fig. 5.60.

FIG. 5.59

FIG. 5.60

*7. Find the unknown quantities in the circuits of Fig. 5.61 using the information provided.

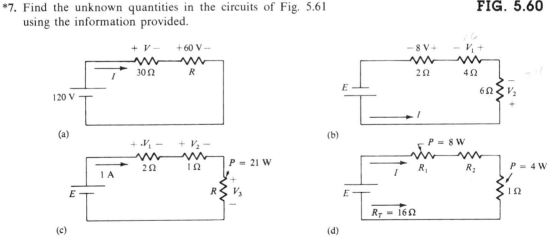

(a)

(b)

(c)

(d)

FIG. 5.61

8. There are 8 Christmas tree lights connected in series as shown in Fig. 5.62.
   a. If the set is connected to a 120-V source, what is the current through the bulbs if each bulb has an internal resistance of $28\frac{1}{8}\,\Omega$?
   b. Determine the power delivered to each bulb.
   c. Calculate the voltage drop across each bulb.
   d. If one bulb burns out (that is, the filament opens), what is the effect on the remaining bulbs?

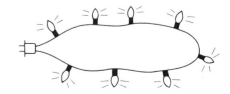

FIG. 5.62

Section 5.4

9. Using the voltage divider rule, find V_{ab} (with polarity) for the circuits of Fig. 5.63.

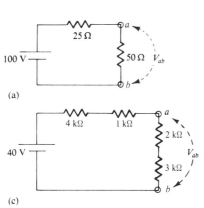

(a)

(c)

FIG. 5.63

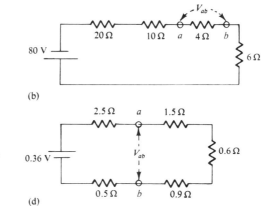

(b)

(d)

10. Find the unknown quantities using the information provided for the circuits of Fig. 5.64.

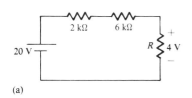

(a)

FIG. 5.64

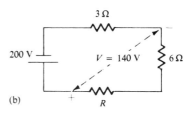

(b)

Section 5.5

11. Find the total conductance, total resistance, and current I for the networks of Fig. 5.65.

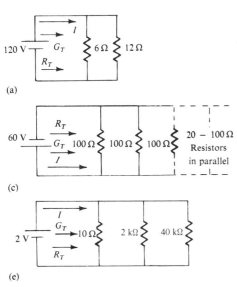

(a)

(c)

(e)

FIG. 5.65

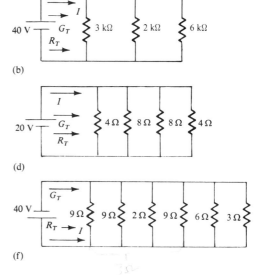

(b)

(d)

(f)

12. The total conductance of the networks of Fig. 5.66 is specified. Find the value in ohms of the unknown resistances.

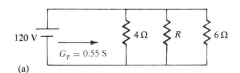

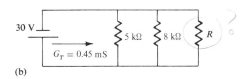

FIG. 5.66

13. The total resistance of the circuits of Fig. 5.67 is specified. Find the value in ohms of the unknown resistances.

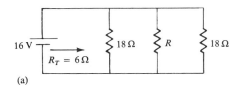

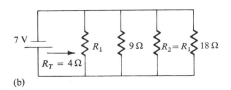

Section 5.6

14. Find all unknown currents and their directions in the circuits of Fig. 5.68.

FIG. 5.67

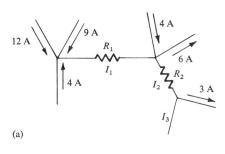

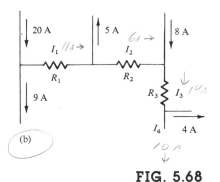

FIG. 5.68

15. For the network of Fig. 5.69:
   a. Find the total conductance and total resistance.
   b. Find I_T and the current through each parallel branch.
   c. Verify Kirchhoff's current law at one node.
   d. Find the power dissipated by each resistor, and note whether the power delivered is equal to the power dissipated.
   e. If the resistors are available with wattage ratings of 1/2, 1, 2, and 50 W, what minimum wattage rating can be used by each resistor in the circuit?

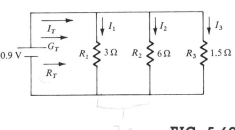

*16. Repeat Problem 15 for the network of Fig. 5.70.

FIG. 5.69

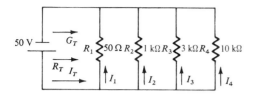

FIG. 5.70

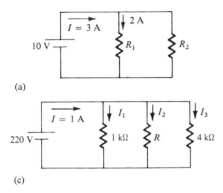

(a)

(c)

FIG. 5.71

*17. Find the unknown quantities for the circuits of Fig. 5.71 using the information provided.

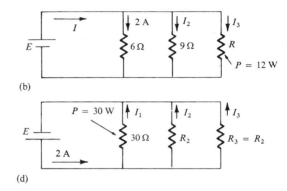

(b)

(d)

18. There are 8 Christmas tree lights connected in parallel as shown in Fig. 5.72.

FIG. 5.72

a. If the set is connected to a 120-V source, what is the current through each bulb if each bulb has an internal resistance of 1.8 kΩ?
b. Determine the total resistance of the network.
c. Find the power delivered to each bulb.
d. If one bulb burns out (that is, the filament opens) what is the effect on the remaining bulbs?
e. Compare the parallel arrangement of Fig. 5.72 to the series arrangement of Fig. 5.62. What are the relative advantages and disadvantages of the parallel system as compared to the series arrangement?

19. A portion of a residential service to a home is depicted in Fig. 5.73.

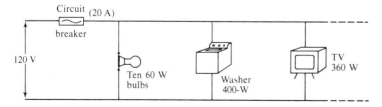

FIG. 5.73

a. Determine the current through each parallel branch of the network.
b. Calculate the current drawn from the 120-V source. Will the 20-A circuit breaker trip?

c. What is the total resistance of the network?

d. Determine the power supplied by the 120-V source. How does it compare to the total power of the load?

Section 5.7

20. Using the current divider rule, find the unknown currents for the networks of Fig. 5.74.

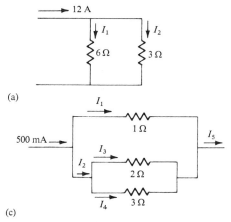

(a)

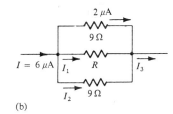

(b)

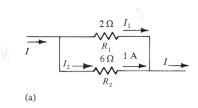

(c)

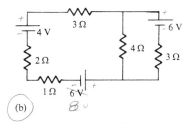

(d)

FIG. 5.74

21. Find the unknown quantities using the information provided for the networks of Fig. 5.75.

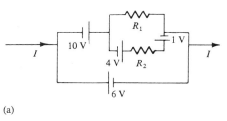

(a)

(b)

FIG. 5.75

Section 5.9

22. Reduce the number of voltage sources in the circuits of Fig. 5.76 to a minimum.

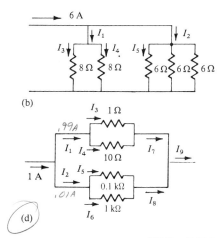

(a)

(b)

FIG. 5.76

Section 5.10

23. Find the internal resistance of a battery that has a no-load output voltage of 60 V and supplies a current of 2 A to a load of 28 Ω.

24. Find the voltage E_T and the power loss in the internal resistance for the configuration of Fig. 5.77.

25. Find the internal resistance of a battery that has a no-load output voltage of 6 V and supplies a current of 10 mA to a load of 1/2 kΩ.

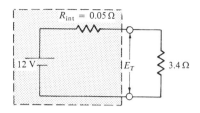

FIG. 5.77

Section 5.11

26. Determine the voltage regulation for the battery of Problem 23.

27. Calculate the voltage regulation for the supply of Fig. 5.77.

GLOSSARY

Branch   The portion of a circuit consisting of one or more elements in series.

Circuit   A combination of a number of elements joined at terminal points providing at least one closed path through which charge can flow.

Closed loop   Any continuous connection of branches that allows us to trace a path which leaves a point in one direction and returns to that same point from another direction without leaving the circuit.

Conventional current flow   A defined direction for the flow of charge in an electrical system that is opposite to that of the motion of electrons.

Current divider rule   A method by which the current through parallel elements can be determined without first finding the voltage across those parallel elements.

Electron flow   The flow of charge in an electrical system having the same direction as the motion of electrons.

Internal resistance   The inherent resistance found internal to any source of energy.

Kirchhoff's current law   Law which states that the algebraic sum of the currents entering and leaving a node is zero.

Kirchhoff's voltage law   Law which states that the algebraic sum of the potential rises and drops around a closed loop (or path) is zero.

Node   A junction of two or more branches.

Parallel circuit   A circuit configuration in which the elements have two points in common.

Series circuit   A circuit configuration in which the elements have only one point in common and each terminal is not connected to a third element.

Short circuit   A direct connection of low resistive value that can significantly alter the behavior of an element or system.

Voltage divider rule   A method by which a voltage in a series circuit can be determined without first calculating the current in the circuit.

Voltage regulation (*VR*)   A value, given in percent, which provides an indication of the change in terminal voltage of a supply with change in load demand.

6.1 ANALYSIS OF SERIES-PARALLEL NETWORKS

A firm understanding of the basic principles associated with series and parallel circuits is a sufficient background to approach most complicated series-parallel networks (a network being a combination of any number of series and parallel elements) with *one* source of emf. Multisource networks will be considered in Chapters 7 and 8. The variation introduced by series-parallel networks is that series and parallel circuit configurations may exist within the same network.

In general, when you are working with series-parallel dc networks, the following is a natural sequence:

1. Study the problem and make a brief mental "sketch" of the overall approach you plan to use. Doing this may result in time- and energy-saving shortcuts.
2. After the overall approach has been determined, consider each branch involved in your method independently before tying them together in series-parallel combinations. This will eliminate many of the errors that might develop due to the lack of a systematic approach.
3. When you have arrived at a solution, check to see that it is reasonable by considering the magnitudes

of the energy source and the elements in the circuit. If it does not seem reasonable, either solve the circuit using another approach, or check over your work very carefully.

In Fig. 6.1, black boxes B and C are in parallel (points b and c in common), and the voltage source E is in series with

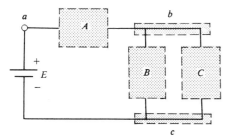

FIG. 6.1

black box A (point a in common). The parallel combination of boxes B and C is also in series with A and the voltage source E, due to the common points b and c, respectively. If each black box in Fig. 6.1 contained a single resistive element, the network of Fig. 6.2 would result.

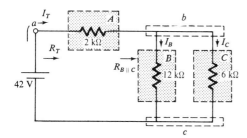

FIG. 6.2

In order to find the source current and then the resulting branch currents, it is first necessary to calculate the input resistance "seen" by the source of emf. The most direct method of finding the total resistance of a network is to consider first each branch (black box) independently and then the series and parallel combinations of branches.

EXAMPLE 6.1. In this case, a single resistor appears in each black box (Fig. 6.2), so that

$$R_A = 2 \text{ k}\Omega \qquad R_B = 12 \text{ k}\Omega \qquad R_C = 6 \text{ k}\Omega$$

In an effort to insure that the analysis to follow is as clear and uncluttered as possible. the following notation will be employed:

$$R_{x\|y} = R_x \| R_y = \frac{R_x R_y}{R_x + R_y}$$

$$R_{x,y} = R_x + R_y$$

The total resistance R_T is

$$R_T = R_A + R_{B\|C} = 2 \text{ k}\Omega + \frac{(12 \text{ k}\Omega)(6 \text{ k}\Omega)}{12 \text{ k}\Omega + 6 \text{ k}\Omega}$$
$$= 2 \text{ k}\Omega + 4 \text{ k}\Omega$$
$$= 6 \text{ k}\Omega$$

and the total current is

$$I_T = \frac{E}{R_T} = \frac{42}{6 \text{ k}\Omega} = 7 \text{ mA}$$
$$I_A = I_T = 7 \text{ mA}$$

The current divider rule must be employed to determine I_B and I_C:

$$I_B = \frac{6 \text{ k}\Omega(I_T)}{6 \text{ k}\Omega + 12 \text{ k}\Omega} = \frac{6}{18} I_T = \frac{1}{3}(7) = 2\tfrac{1}{3} \text{ mA}$$

$$I_C = \frac{12 \text{ k}\Omega(I_T)}{12 \text{ k}\Omega + 6 \text{ k}\Omega} = \frac{12}{18} I_T = \frac{2}{3}(7) = 4\tfrac{2}{3} \text{ mA}$$

or

$$I_C = I_T - I_B = 7 - 2\tfrac{1}{3} = \frac{21}{3} - \frac{7}{3} = \frac{14}{3} = 4\tfrac{2}{3} \text{ mA}$$

EXAMPLE 6.2. It is also possible that the black boxes A, B, and C may contain the elements and configurations of Fig. 6.3.

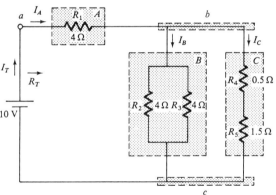

FIG. 6.3

$$R_A = 4 \, \Omega$$
$$R_B = R_2 \| R_3 = R_{2\|3} = \frac{R}{N} = \frac{4}{2} = 2 \, \Omega$$
$$R_C = R_4 + R_5 = R_{4,5} = 0.5 + 1.5 = 2 \, \Omega$$

so that

$$R_{B\|C} = \frac{R}{N} = \frac{2}{2} = 1 \, \Omega$$

with

$$R_T = R_A + R_{B\|C} \quad \text{(Note similarity between this equation and that obtained for Example 6.1.)}$$
$$= 4 + 1 = 5 \, \Omega$$

and

$$I_T = \frac{E}{R_T} = \frac{10}{5} = \mathbf{2\ A}$$

We can find the currents I_A, I_B, and I_C using the reduction of the network of Fig. 6.3 as found in Fig. 6.4:

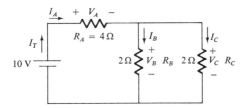

FIG. 6.4

$$I_A = I_T = \mathbf{2\ A}$$

and

$$I_B = I_C = \frac{I_A}{2} = \frac{I_T}{2} = \frac{2}{2} = \mathbf{1\ A}$$

Returning to the network of Fig. 6.3, we have

$$I_{R_2} = I_{R_3} = \frac{I_B}{2} = \mathbf{0.5\ A}$$

The voltages V_A, V_B, and V_C from either figure are

$$V_A = I_A R_A = (2)(4) = \mathbf{8\ V}$$
$$V_B = I_B R_B = (1)(2) = \mathbf{2\ V}$$
$$V_C = V_B = \mathbf{2\ V}$$

Applying Kirchhoff's voltage law for the loop indicated in Fig. 6.4, we obtain

$$\Sigma_{\circlearrowleft} V = E - V_A - V_B = 0$$

or

$$E = V_A + V_B = 8 + 2$$
$$\underline{10 = 10 \qquad \text{(checks)}}$$

EXAMPLE 6.3. Another possible variation of Fig. 6.1 is found in Fig. 6.5.

$$R_A = R_{1\|2} = \frac{(9)(6)}{9 + 6} = \frac{54}{15} = 3.6\ \Omega$$

$$R_B = R_3 + R_{4\|5} = 4 + \frac{(6)(3)}{6 + 3} = 4 + 2 = 6\ \Omega$$

$$R_C = 3\ \Omega$$

The network of Fig. 6.5 can then be redrawn in reduced form as shown in Fig. 6.6. Note the similarities between this circuit and those of Figs. 6.2 and 6.4.

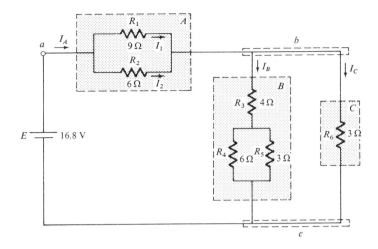

FIG. 6.5

$$R_T = R_A + R_{B\|C}$$
$$= 3.6 + \frac{(6)(3)}{6 + 3} = 3.6 + 2 = 5.6 \ \Omega$$

$$I_T = \frac{E}{R_T} = \frac{16.8}{5.6} = 3 \text{ A}$$

$$I_A = I_T = 3 \text{ A}$$

Applying the current divider rule yields

$$I_B = \frac{R_C(I_A)}{R_C + R_B} = \frac{3(3)}{3 + 6} = \frac{9}{9} = 1 \text{ A}$$

By Kirchhoff's current law,

$$I_C = I_A - I_B = 3 - 1 = 2 \text{ A}$$

By Ohm's law,

$$V_A = I_A R_A = (3)(3.6) = 10.8 \text{ V}$$
$$V_B = I_B R_B = V_C = I_C R_C = (2)(3) = 6 \text{ V}$$

Returning to the original network (Fig. 6.5), we have

$$I_1 = \frac{R_2(I_A)}{R_2 + R_1} = \frac{6(3)}{6 + 9} = \frac{18}{15} = 1.2 \text{ A}$$

By Kirchhoff's current law,

$$I_2 = I_A - I_1 = 3 - 1.2 = 1.8 \text{ A}$$

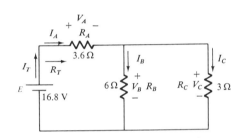

FIG. 6.6

Kirchhoff's voltage law for the loop indicated in the reduced network (Fig. 6.6) is

$$\Sigma_\circlearrowleft V = E - V_A - V_B = 0$$
$$E = V_A + V_B$$
$$16.8 = 10.8 + 6$$
$$\underline{16.8 = 16.8} \quad \text{(checks)}$$

Figure 6.1 is only one of an infinite variety of configurations that the black boxes can assume within a given network. There is no restriction on their number, location, or internal circuitry. They were included in our discussion of series-parallel networks to indicate the numerous configurations possible from the same basic setup and to emphasize the importance of considering each branch independently before finding the solution for the network as a whole. Many methods can be used to find a solution of the examples just considered. Proficiency in picking the best method, however, can be developed only by exposure to numerous examples and problems. A few are now considered.

6.2 DESCRIPTIVE EXAMPLES

EXAMPLE 6.4. Find the voltage V_2 and the current I_4 in the network of Fig. 6.7.

Solution: Since the three branches are in parallel (points a and b in common in Fig. 6.8), the voltage across each is the impressed emf E:

$$R'_T = R_2 \parallel R_3 = 3 \parallel 4 = \frac{3(4)}{3+4} = \frac{12}{7} = 1.71\ \Omega$$

By Ohm's law,

$$I_4 = \frac{E}{R_4} = \frac{12}{8} = 1.5\ \text{A}$$

By the voltage divider rule,

$$V_2 = \frac{(1.71)(12)}{1.71+4} = \frac{20.52}{5.71} = \textbf{3.6 V}$$

EXAMPLE 6.5. Find the indicated currents and voltages of the network of Fig. 6.9.

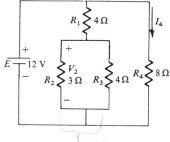

FIG. 6.7

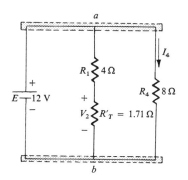

FIG. 6.8

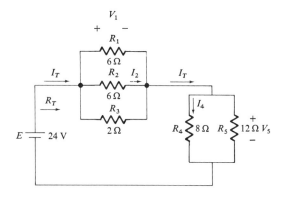

FIG. 6.9

Solution:

$$R_{1\|2} = \frac{R}{N} = \frac{6}{2} = 3\,\Omega$$

$$R_{1\|2\|3} = \frac{(3)(2)}{3 + 2} = \frac{6}{5} = 1.2\,\Omega$$

$$R_{4\|5} = \frac{(8)(12)}{8 + 12} = \frac{96}{20} = 4.8\,\Omega$$

The reduced form of Fig. 6.9 will then appear as shown in Fig. 6.10, and

$$R_T = R_{1\|2\|3} + R_{4\|5} = 1.2 + 4.8 = 6\,\Omega$$

$$I_T = \frac{E}{R_T} = \frac{24}{6} = 4\text{ A}$$

with

$$V_1 = I_T \cdot R_{1\|2\|3} = (4)(1.2) = 4.8\text{ V}$$
$$V_5 = I_T \cdot R_{4\|5} = (4)(4.8) = 19.2\text{ V}$$

Applying the current divider rule to the original network (Fig. 6.9) yields

$$I_4 = \frac{R_5(I_T)}{R_5 + R_4} = \frac{12(4)}{12 + 8} = \frac{48}{20} = 2.4\text{ A}$$

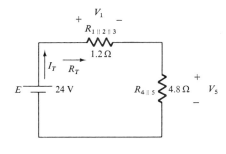

FIG. 6.10

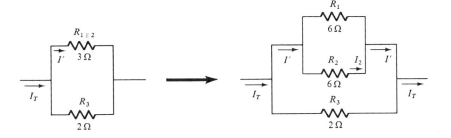

FIG. 6.11

For I_2, consider Fig. 6.11, where

$$I' = \frac{R_3(I_T)}{R_3 + R_{1\|2}} = \frac{2(4)}{2 + 3} = \frac{8}{5} = 1.6\text{ A}$$

and

$$I_2 = \frac{I'}{2} = \frac{1.6}{2} = 0.8\text{ A}$$

EXAMPLE 6.6. Find the voltages V_1 and V_{ab} for the network of Fig. 6.12.

Solution:

$$R_{1,2} = R_1 + R_2 = 5 + 3 = 8\,\Omega$$
$$R_{3,4} = R_3 + R_4 = 6 + 2 = 8\,\Omega$$

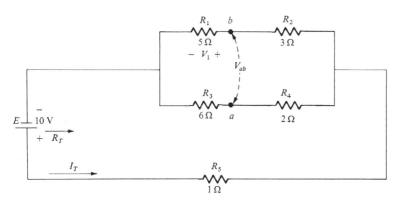

FIG. 6.12

The network can then be redrawn as shown in Fig. 6.13, and

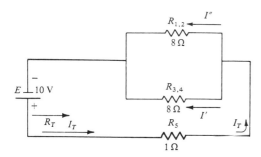

FIG. 6.13

$$R_T = R_6 + R_{3,4} \parallel R_{1,2} = 1 + \frac{8}{2} = 1 + 4 = \mathbf{5\,\Omega}$$

with

$$I_T = \frac{E}{R_T} = \frac{10}{5} = \mathbf{2\,A}$$

$$I' = I'' = \frac{I_T}{2} = \frac{2}{2} = \mathbf{1\,A}$$

For V_1 as shown in Fig. 6.14,

$$V_1 = I'' \cdot R_1 = (1)(5) = \mathbf{5\,V}$$

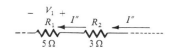

FIG. 6.14

We can find the voltage V_{ab} by applying Kirchhoff's voltage law around the closed loop indicated in Fig. 6.15:

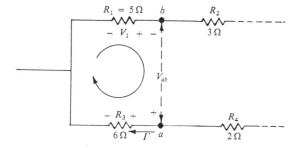

FIG. 6.15

$$+V_{ab} - I'(6) + V_1 = 0$$

or

$$V_{ab} = I'(6) - V_1$$
$$= (1)(6) - 5$$
$$V_{ab} = 1\,\text{V}$$

EXAMPLE 6.7. For the network of Fig. 6.16 determine the voltages V_1, V_2 and the current I.

Solution: The network is redrawn in Fig. 6.17. Note the common connection of the grounds and the replacing of the terminal voltage by an actual supply.

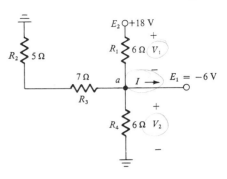

FIG. 6.16

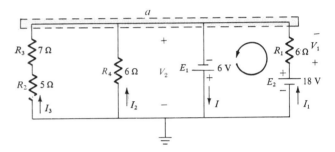

FIG. 6.17

From Fig. 6.17,

$$V_2 = -6\,\text{V}$$

The minus sign simply indicates that the chosen polarity for V_2 in Fig. 6.16 is opposite to the actual voltage. Applying Kirchhoff's voltage law to the loop indicated, we obtain

$$+18 - V_1 + 6 = 0$$
$$V_1 = 6 + 18 = 24\,\text{V}$$

Applying Kirchhoff's current law to node a yields

$$I = I_1 + I_2 + I_3$$
$$= \frac{V_1}{R_1} + \frac{E_1}{R_4} + \frac{E_1}{(R_2 + R_3)}$$
$$= \frac{24}{6} + \frac{6}{6} + \frac{6}{12}$$
$$= 4 + 1 + 0.5$$
$$I = 5.5\,\text{A}$$

EXAMPLE 6.8. For the transistor configuration of Fig. 6.18:

a. Determine the voltage V_E and the current I_E.
b. Calculate V_1.
c. Determine V_{BC} using the fact that the approximation $I_C = I_E$ is often applied to transistor networks.
d. Calculate V_{CE} using the information obtained in parts (a) through (c).

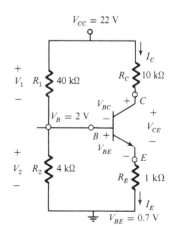

FIG. 6.18

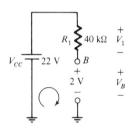

FIG. 6.19

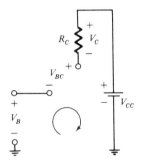

FIG. 6.20

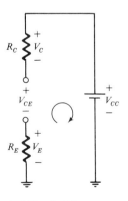

FIG. 6.21

FIG. 6.22

Solutions:

a. From Fig. 6.18 we find

$$V_2 = V_B = 2 \text{ V}$$

Writing Kirchhoff's voltage law around the lower loop yields

$$V_2 - V_{BE} - V_E = 0$$

or

$$V_E = V_2 - V_{BE} = 2 - 0.7 = \textbf{1.3 V}$$

and

$$I_E = \frac{V_E}{R_E} = \frac{1.3}{1000} = \textbf{1.3 mA}$$

b. The 22-V dc source can be split as shown in Fig. 6.19 as an aid in the required analysis. The section of interest can then be sketched as shown in Fig. 6.20, and Kirchhoff's law will result in

$$V_{CC} - V_1 - V_B = 0$$

or

$$V_1 = V_{CC} - V_B = 22 - 2 = \textbf{20 V}$$

c. Redrawing the section of the network of immediate interest will result in Fig. 6.21, where Kirchhoff's voltage law yields

$$V_B + V_{BC} + V_C - V_{CC} = 0$$

or

$$V_{BC} = V_{CC} - V_C - V_B$$

with

$$V_C = I_C R_C = (1.3 \text{ mA})(10 \text{ k}\Omega) = 13 \text{ V}$$

and

$$V_{BC} = 22 - 13 - 2 = \textbf{7 V}$$

d. The appropriate section appears in Fig. 6.22. Application of Kirchhoff's voltage law will result in

$$V_E + V_{CE} + V_C = V_{CC}$$

or

$$V_{CE} = V_{CC} - V_C - V_E$$

$$= 22 - 13 - 1.3 = \textbf{7.7 V}$$

Note in the above analysis that there was no requirement to know any of the details regarding transistor behavior (except $I_C = I_E$). Be assured that there will be a great deal of coverage of the type of analysis described above when electronic systems are examined.

EXAMPLE 6.9. Calculate the indicated currents and voltages of Fig. 6.23.

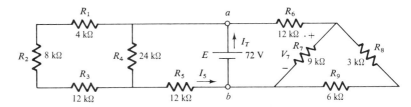

FIG. 6.23

Solution: Redrawing the network after combining series elements yields Fig. 6.24, and

$$I_5 = \frac{E}{R_{(1,2,3)\|4} + R_5} = \frac{72}{12 \text{ k}\Omega + 12 \text{ k}\Omega} = \frac{72}{24 \text{ k}\Omega} = 3 \text{ mA}$$

with

$$V_7 = \frac{(R_{7\|8,9})E}{R_{7\|(8,9)} + R_6} = \frac{4.5 \text{ k}\Omega(72)}{4.5 \text{ k}\Omega + 12 \text{ k}\Omega} = \frac{324}{16.5} = 19.6 \text{ V}$$

$$I_6 = \frac{V_7}{R_{7\|(8,9)}} = \frac{19.6}{4.5 \text{ k}\Omega} = 4.35 \text{ mA}$$

and

$$I_T = I_5 + I_6 = 3 + 4.35 = 7.35 \text{ mA}$$

Since the potential difference between points *a* and *b* of Fig. 6.23 is fixed at *E* volts, the circuit to the right or left is unaffected if the network is reconstructed as shown in Fig. 6.25.

FIG. 6.24

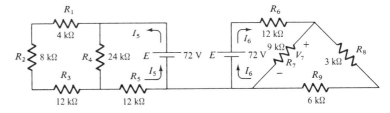

FIG. 6.25

We can find each quantity required, except I_T, by analyzing each circuit independently. To find I_T, we must find the source current for each circuit and add it as in the above solution; that is, $I_T = I_5 + I_6$.

6.3 LADDER NETWORKS

A *ladder* network is shown in Fig. 6.26. The reason for the terminology is quite obvious. There are basically two methods for solving networks of this type.

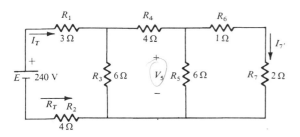

FIG. 6.26

Method One

Calculate the total resistance and resulting source current, and then work back through the ladder until the desired current or voltage is obtained. This method is now employed to determine V_5 and I_7 in Fig. 6.26.

Combining parallel and series elements as shown in Fig. 6.27 will result in the reduced network of Fig. 6.28, and

$$R_T = 3 + 3 + 4 = 10\,\Omega$$

$$I_T = \frac{E}{R_T} = \frac{240}{10} = 24\,\text{A}$$

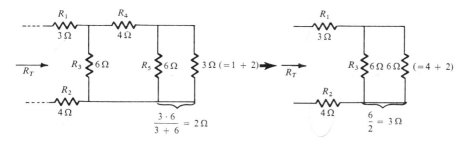

FIG. 6.27

Working our way back to V_5 and I_7 (Fig. 6.29), we have

$$I_4 = \frac{I_T}{2} = \frac{24}{2} = 12\,\text{A}$$

and finally (Fig. 6.30),

$$I_5 = \frac{3I_4}{3 + 6} = \frac{1}{3}(12) = 4\,\text{A}$$

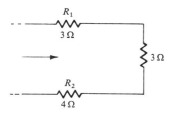

FIG. 6.28

and

$$V_5 = I_5 R_5 = 4 \cdot 6 = \textbf{24 V}$$

$$I_7 = I_4 - I_5 = 12 - 4 = \textbf{8 A}$$

Method Two

Assign a letter symbol to the last branch current and work back through the network to the source, maintaining this assigned current or other current of interest. The desired current can then be found directly. This method can best

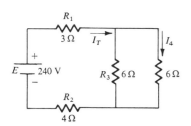

FIG. 6.29

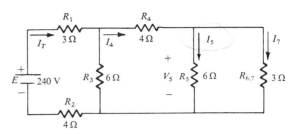

FIG. 6.30

be described through the analysis of the same network considered above, redrawn in Fig. 6.31.

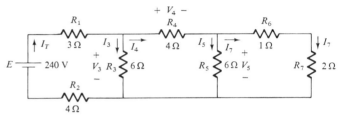

FIG. 6.31

The assigned notation for the current through the final branch is I_7:

$$I_7 = \frac{V_5}{R_6 + R_7} = \frac{V_5}{(1 + 2)} = \frac{V_5}{3}$$

or

$$V_5 = 3I_7$$

so that

$$I_5 = \frac{V_5}{R_5} = \frac{3I_7}{6} = 0.5I_7$$

and

$$I_4 = I_5 + I_7 = 0.5I_7 + I_7 = 1.5I_7$$
$$V_4 = I_4R_4 = (1.5I_7)4 = 6I_7$$

Also,

$$V_3 = V_4 + V_5 = 6I_7 + 3I_7 = 9I_7$$

so that

$$I_3 = \frac{V_3}{R_3} = \frac{9I_7}{6} = 1.5I_7$$

and

$$I_T = I_3 + I_4 = 1.5I_7 + 1.5I_7 = 3I_7$$

with

$$V_1 = I_1R_1 = I_TR_1 = 3I_T$$

so that

$$E = V_1 + V_3 + V_2 = 3I_T + 9I_7 + 4I_T$$
$$= 7I_T + 9I_7 = 7(3I_7) + 9I_7$$
$$E = 30I_7$$

and

$$I_7 = \frac{E}{30} = \frac{240}{30} = \textbf{8 A}$$

PROBLEMS

Section 6.2

1. For the network of Fig. 6.32:

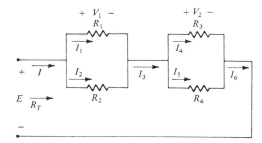

FIG. 6.32

   a. Does $I = I_3 = I_6$? Explain.
   b. If $I = 5$ A and $I_1 = 2$ A, find I_2.
   c. Does $I_1 + I_2 = I_4 + I_5$? Explain.
   d. If $V_1 = 6$ V and $E = 10$ V, find V_2.
   e. If $R_1 = 3\,\Omega$, $R_2 = 2\,\Omega$, $R_3 = 4\,\Omega$, and $R_4 = 1\,\Omega$, what is R_T?
   f. If the resistors have the values given in part (e) and $E = 10$ V, what is the value of I in amperes?
   g. Using values given in (e) and (f), find the power delivered by the battery E and dissipated by the resistors R_1 and R_2.

2. For the network of Fig. 6.33:
   a. Calculate R_T.
   b. Determine I and I_1.
   c. Find V_3.

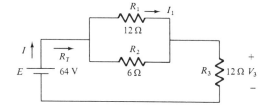

FIG. 6.33

3. For the network of Fig. 6.34:
   a. What are the currents I and I_1?
   b. Find the polarity and the voltage drop across each resistor.
   c. What are the currents I_3 and I_4?

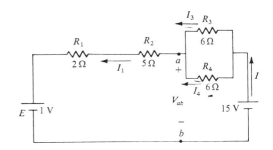

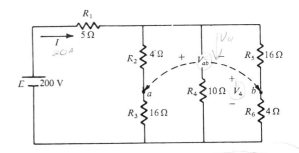

d. What is the power delivered to the 5-Ω resistor?

FIG. 6.34

e. Find the voltage V_{ab}.

4. For the network of Fig. 6.35:
  a. Calculate I.
  b. Find the voltage V_4.
  c. Find the voltage V_{ab}.

5. For the network of Fig. 6.36:
  a. Determine the current I_1.
  b. Calculate the currents I_2 and I_3.
  c. Determine the voltage levels V_A and V_B.

FIG. 6.35

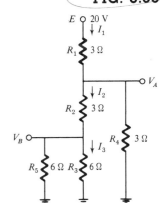

6. For the network of Fig. 6.37:
  a. Find V_{ab}, V_{bc}, V_{cd} using the voltage divider rule.
  b. Find the currents I_3 and I_6.
  c. Find the power to each 12-kΩ resistor.

FIG. 6.36

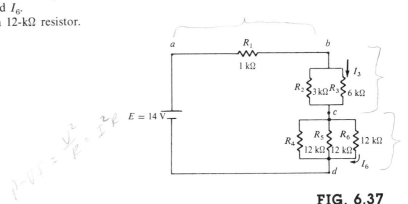

FIG. 6.37

7. For the network of Fig. 6.38:
   a. Find the currents I and I_6.
   b. Find the voltages V_1 and V_5.
   c. Find the power delivered to the 6-kΩ resistor.

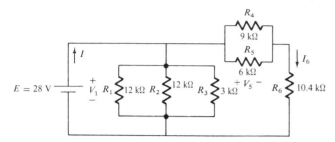

FIG. 6.38

8. For the series-parallel network of Fig. 6.39:
   a. Find the current I.
   b. Find the currents I_3 and I_9.
   c. Find the current I_8.
   d. Find the voltage V_{ab}.

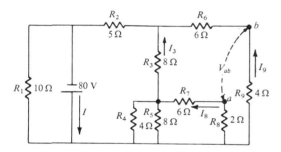

FIG. 6.39

*9. The network of Fig. 6.40 is the basic biasing arrangement for the *field effect transistor* (FET), a device of increasing importance in electronic design. (*Biasing* simply means the application of dc levels to establish a particular set of operating conditions.) Even though you may be unfamiliar with the FET, you can perform the following analysis using only the basic laws introduced in this chapter and the information provided on the diagram.

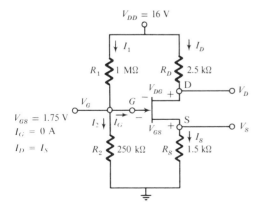

FIG. 6.40

a. Calculate the voltages V_G and V_S.

b. Determine V_{DG}.

c. Determine V_{DS}.

*10. The *difference amplifier* of Fig. 6.41 is a compound configuration that will establish an output (in the ac domain) that is the difference between the two input signals. Using the concepts introduced in this chapter and previous chapters, determine the following dc levels:

a. V_E (given that $I_1 = 2$ mA).

b. I_2 (magnitude and direction).

c. I_{E_1} and I_{E_2} using the fact that $I_C = I_E$ (an approximate relationship introduced in Example 6.8) and $I_{E_1} = I_{E_2}$ (balanced system).

d. V_C, V_{E_1} and V_{E_2} if $V_{CE} = 10.7$ V.

e. V_{B_1} and V_{B_2} using information obtained above.

f. V_{CE_1} and V_{CE_2} if $I_{C_1} = I_{E_1}$ and $I_{C_2} = I_{E_2}$.

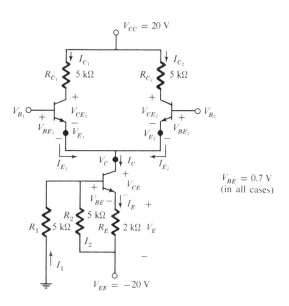

FIG. 6.41

11. For the series-parallel configuration of Fig. 6.42:

a. Find the current I.

b. Find the currents I_1, I_3, and I_8.

c. Find the power delivered to the 21-Ω resistor.

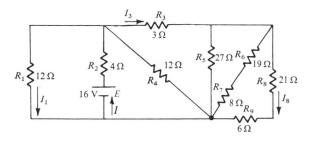

FIG. 6.42

12. For the network of Fig. 6.43:
    a. Determine the current I.
    b. Find V.

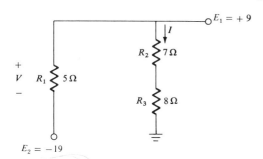

FIG. 6.43

13. For the configuration of Fig. 6.44:
    a. Find the currents I_2, I_6, and I_8.
    b. Find the voltages V_4 and V_8.

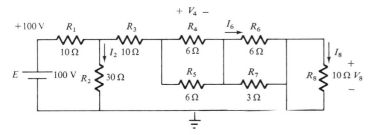

FIG. 6.44

*14. For the network of Fig. 6.45, find the resistance R_3 if the current through it is 2 A.

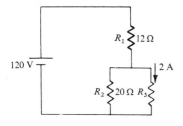

FIG. 6.45

Section 6.3

15. For the ladder network of Fig. 6.46:
    a. Find the current I.
    b. Find the current I_7.
    c. Determine the voltages V_3, V_5, and V_7.
    d. Calculate the power delivered to R_7 and compare it to the power delivered by the 240-V supply.

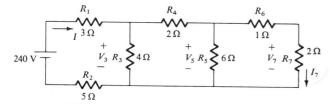

FIG. 6.46

16. For the ladder network of Fig. 6.47:
    a. Determine R_T.
    b. Calculate I.
    c. Find I_8.

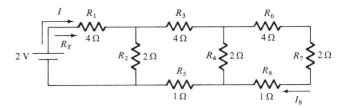

FIG. 6.47

*17. For the multiple ladder configuration of Fig. 6.48:
  a. Determine I.
  b. Calculate I_4.
  c. Find I_6.
  d. Find I_{10}.

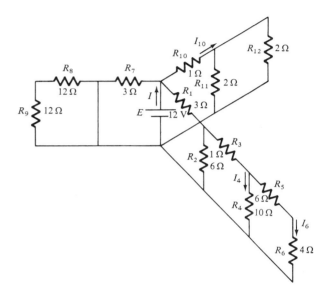

FIG. 6.48

GLOSSARY

Ladder network   A network that consists of a cascaded set of series-parallel combinations and has the appearance of a ladder.

Series-parallel network   A network consisting of a combination of both series and parallel branches.

Transistor   A three-terminal semiconductor electronic device that can be used for amplification and switching purposes.

7.1 INTRODUCTION

The circuits described in Chapters 5 and 6 had only one source or two or more sources in series present. The step-by-step procedure outlined in those chapters cannot be applied if two or more sources in the same network are not in series. There will be an interaction of sources that will not permit the reduction technique used in Chapter 6 to find such quantities as the total resistance and current.

Methods of analysis have been developed that allow us to approach a network with any number of sources in a systematic manner. These methods can also be applied to networks with one source. The methods to be discussed in detail in this chapter include *branch-current analysis, mesh analysis, and nodal analysis.* Each can be applied to the same network. The "best" method cannot be defined by a set of rules but can be determined only by acquiring a firm understanding of the relative advantages of each. All of the methods will be described for *linear bilateral* networks only. The term *linear* indicates that the characteristics of the network elements (such as the resistors) are independent of the voltage across or current through them. The second term, *bilateral,* refers to the fact that there is no change in the behavior or characteristics of an element if the current through or voltage across the element is reversed. Before discussing these methods, we shall consider the current source and the use of determinants. At the end of the chapter we shall consider bridge networks and Δ-Y and Y-Δ conversions.

Chapter 8 will present the important theorems of network analysis that can also be employed to solve networks with one or more sources.

7.2 CURRENT SOURCES

The concept of the current source was introduced in Section 2.4 with the photograph of a commercially available unit. We must now investigate its characteristics in greater detail so that we can properly determine its effect on the network to be considered in this chapter.

The current source is often referred to as the *dual* of the voltage source. That is, while a battery supplies a *fixed* voltage and the source current can vary, the current source supplies a *fixed* current to the branch in which it is located, while its terminal voltage may vary as determined by the network to which it is applied. Note from the above that *duality* simply implies an interchange of current and voltage to distinguish the characteristics of one source from the other.

The increasing interest in the current source is due fundamentally to semiconductor devices such as the transistor. In the basic electronics courses, you will find that the transistor is a current-controlled device. In the physical model (equivalent circuit) of a transistor used in the analysis of transistor networks, there appears a current source as indicated in Fig. 7.1. The symbol for a current source appears in Fig. 7.1. The direction of the arrow within the circle indicates the direction in which current is being supplied.

For further comparison, the terminal characteristics of a *dc* voltage and current source are presented in Fig. 7.2. Note that for the voltage source the terminal voltage is fixed at E volts for the range of current values. For the

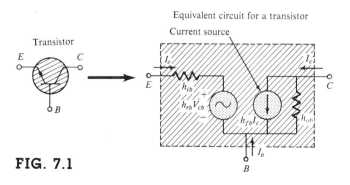

FIG. 7.1

region to the right of the voltage axis, the current will have one direction through the source, while to the left of the same axis it is reversed. In other words, as indicated in the associated figure of Fig. 7.2(a), the terminal voltage of a source is unaffected by the direction of current through the source. The characteristics of the current source, shown in Fig. 7.2(b), indicate that the current source will supply a fixed current even though the voltage across the source may vary in magnitude or reverse its polarity. This is indicated in the associated figure of Fig. 7.2(b). For the voltage source, the current direction will be determined by the remaining elements of the network. For all one-voltage-source networks it will have the direction indicated to the right of the battery in Fig. 7.2(a). For the current source, the network to which it is connected will also determine the magnitude and polarity of the voltage across the source. For all single-current-source networks, it will have the polarity indicated just to the right of the current source in Fig. 7.2(b).

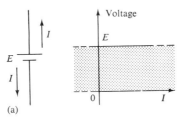

(a)   (b)

FIG. 7.2

It is not difficult to build a current source for a known load variation using a voltage source and series resistor. In Fig. 7.3 the designed current source appears within the shaded area. As indicated on the figure, it will supply a fixed current of 1 mA for a load variation from 0 to 500 Ω. The current

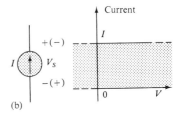

$$I_S = \frac{E}{R_S + R_L} \cong \frac{E}{R_S} = \frac{100}{100 \text{ k}\Omega} = 1 \text{ mA}$$

since R_S is much greater than any value R_L may assume in the range $0 \rightarrow 500$ Ω.

FIG. 7.3 *1 mA-current source.*

For changing values of R_L between 0 and 500 Ω, the current will remain essentially fixed at 1 mA, and the voltage across the load will be determined by

$$V_L = I_L \cdot R_L = I_S \cdot R_L = (1 \text{ mA})(R_L)$$

EXAMPLE 7.1. Find the voltage V_S and the current I_1 for the circuit of Fig. 7.4.

Solution:

$$I_1 = I = \mathbf{10 \text{ mA}}$$
$$V_S = V_1 = I_1 R_1 = (10 \text{ mA})(20 \text{ k}\Omega) = \mathbf{200 \text{ V}}$$

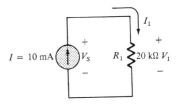

FIG. 7.4

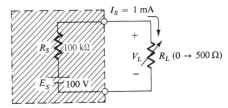

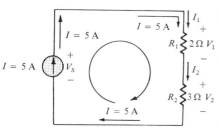

FIG. 7.5

EXAMPLE 7.2. Calculate the voltages V_1, V_2, and V_S for the circuit of Fig. 7.5.

Solution:

$$V_1 = I_1 R_1 = IR_1 = (5)(2) = \textbf{10 V}$$
$$V_2 = I_2 R_2 = IR_2 = (5)(3) = \textbf{15 V}$$

Applying Kirchhoff's law of voltages, we obtain

$$\Sigma_{\circlearrowright} V = +V_S - V_1 - V_2 = 0$$

or

$$V_S = V_1 + V_2 = 10 + 15$$

and

$$V_S = \textbf{25 V}$$

Note the polarity of V_S for the single-source circuit.

EXAMPLE 7.3. Consider the series-parallel circuit of Fig. 7.6.

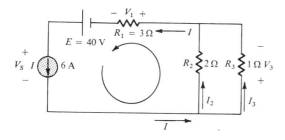

FIG. 7.6

Solution: By the current divider rule,

$$I_2 = \frac{(1)I}{1 + 2} = \frac{1}{3}(6) = \textbf{2 A}$$

By Ohm's law,

$$V_3 = I_3 R_3 = I_2 R_2 = (2)(2) = \textbf{4 V}$$
$$V_1 = I_1 R_1 = (6)(3) = \textbf{18 V}$$

By Kirchhoff's voltage law,

$$\Sigma_{\circlearrowright} V = -V_S - V_3 - V_1 + E = 0$$

or

$$V_S = -V_3 - V_1 + E = -4 - 18 + 40$$
$$V_S = \textbf{18 V}$$

Note the polarity of V_S as determined by the multisource network.

7.3 SOURCE CONVERSIONS

It is often necessary or convenient to have a voltage source rather than a current source or a current source rather than a voltage source. If we consider the basic voltage source with its internal resistance as shown in Fig. 7.7, we find that

$$I_L = \frac{E}{R_S + R_L}$$

Or by multiplying the numerator of the equation by a factor of one which we choose to be R_S/R_S, we obtain

$$I_L = \frac{(1)(E)}{R_S + R_L} = \frac{(R_S/R_S)E}{R_S + R_L} = \frac{R_S(E/R_S)}{R_S + R_L}$$

Since E/R_S relates to a current I, the above equation is an application of the current divider rule to the network of Fig. 7.8.

For the load resistor R_L of Fig. 7.7 or 7.8, it is immaterial which source is applied as long as each element has the corresponding value. That is, the voltage across or current through R_L will be the same for each network. For clarity, the equivalent sources are repeated in Fig. 7.9 with the equations necessary for the conversion. Note that the resistor R_S is unchanged in magnitude and is simply brought

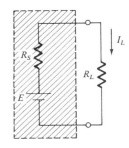

FIG. 7.7

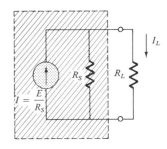

FIG. 7.8

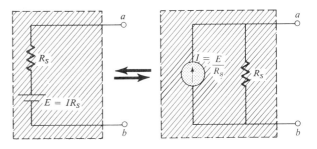

FIG. 7.9 *Source conversion.*

from a series position for the voltage source to the parallel arrangement for the current source. It was pointed out in some detail in Chapter 5 that every source of emf has some internal series resistance. *For the current source, some internal parallel resistance will always exist in the practical world.* However, in many cases, it is an excellent approximation to drop the internal resistance of a source due to the magnitude of the elements of the network to which it is applied. For this reason, in the analyses to follow, voltage sources may appear without a series resistor, and current sources may appear without a parallel resistance. Realize, however, that in order to perform a conversion from one type of source to another, a voltage source must have a resistor in series with it, and a current source must have a resistor in parallel.

EXAMPLE 7.4. Convert the voltage source of Fig. 7.10 to a current source and calculate the current through the load for each source.

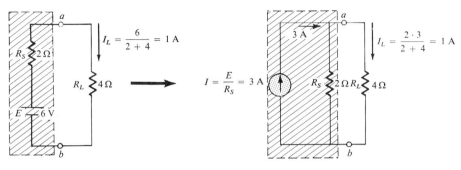

FIG. 7.10

Solution: See the right side of Fig. 7.10.

EXAMPLE 7.5. Convert the current source of Fig. 7.11 to a voltage source and find the current through the load for each source.

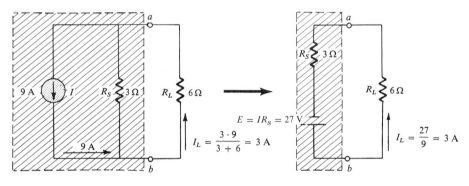

FIG. 7.11

Solution: See the right side of Fig. 7.11.

7.4 CURRENT SOURCES IN PARALLEL

If two or more current sources are in parallel, they may all be replaced by one current source having the magnitude and direction of the resultant, which can be found by summing the currents in one direction and subtracting the sum of the currents in the opposite direction. The new parallel resistance is determined by methods described in the discussion of parallel resistors in Section 5.5. Consider the following examples.

EXAMPLE 7.6. Reduce the left sides of Figs. 7.12 and 7.13 to a minimum number of elements.

Solution: See the right sides of the figures.

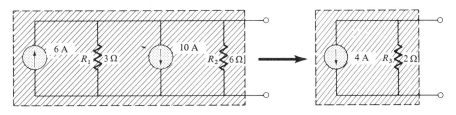

FIG. 7.12

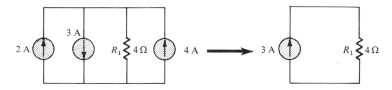

FIG. 7.13

7.5 CURRENT SOURCES IN SERIES

The current through any branch of a network can be only single-valued. For the situation indicated at point *a* in Fig. 7.14, we find by application of Kirchhoff's current law that the current leaving that point is greater than that entering—an impossible situation. Therefore, *current sources of different current ratings are not connected in series,* just as voltage sources of different voltage ratings are not connected in parallel.

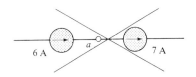

FIG. 7.14

7.6 DETERMINANTS

In the following analyses we will find it necessary to solve for the unknown variables of two or perhaps three simultaneous equations. Consider the following equations:

Col. 1	Col. 2	Col. 3	
$a_1 x$ +	$b_1 y$ =	c_1	**(7.1a)**
$a_2 x$ +	$b_2 y$ =	c_2	**(7.1b)**

It is certainly possible to solve for one variable in Eq. (7.1a) and substitute into Eq. (7.1b). That is, for x,

$$x = \frac{c_1 - b_1 y}{a_1}$$

and

$$a_2 \left(\frac{c_1 - b_1 y}{a_1} \right) + b_2 y = c_2$$

It is now possible to solve for y and then substitute into either equation for x. This is acceptable for two equations, but it becomes a very tedious and lengthy process for three or more simultaneous equations.

A mathematical procedure involving the use of *determinants* is most frequently applied to equations of this type. There is a set of basic steps for forming the determinant and finding its solution. These steps save time and effort and prevent the errors which frequently occur from the procedure discussed above. In Eqs. (7.1a) and (7.1b), x and y are the unknown variables and a_1, a_2, b_1, b_2, c_1, and c_2 are constants. To find the values of x and y that satisfy the two simultaneous equations by the method of determinants requires that the following format be used:

$$x = \frac{\begin{vmatrix} c_1 & b_1 \\ c_2 & b_2 \end{vmatrix}}{\begin{vmatrix} a_1 & b_1 \\ a_2 & b_2 \end{vmatrix}} \qquad y = \frac{\begin{vmatrix} a_1 & c_1 \\ a_2 & c_2 \end{vmatrix}}{\begin{vmatrix} a_1 & b_1 \\ a_2 & b_2 \end{vmatrix}} \qquad (7.2)$$

(Col. 1, Col. 2 labels over each determinant)

Each configuration in the numerator and denominator is referred to as a determinant (D), which can be expanded in the following manner.

Denominator

$$\text{Determinant} = D = \begin{vmatrix} a_1 & b_1 \\ a_2 & b_2 \end{vmatrix} = a_1b_2 - a_2b_1 \qquad (7.3)$$

(Col. 1, Col. 2 labels over the determinant)

The expanded value is obtained by first multiplying the top left element by the bottom right and then subtracting the product of the lower left and upper right elements. This particular determinant is referred to as a *second-order* determinant, since it contains two rows and two columns.

A closer examination of Eq. (7.2) will reveal that the denominator of each expression is the same—it is the determinant of the coefficients of x and y in Eqs. (7.1a) and (7.1b). It is important to remember when using determinants that the columns of the equations, as indicated in Eqs. (7.1a) and (7.1b), be placed in the same order within the determinant configuration. That is, since a_1 and a_2 are in column 1 of Eqs. (7.1a) and (7.1b), they must be in column 1 of the determinant. (The same is true for b_1 and b_2.)

Numerator

The numerator of each expression is determined by the variable to be evaluated. The numerator for the variable x

is formed by replacing the column in which x is located in the original equations by the elements to the right of the equal sign, and leaving the second column undisturbed. For y, the reverse is done.

Expanding the entire expression for x and y, we have the following:

$$x = \frac{\begin{vmatrix} c_1 & b_1 \\ c_2 & b_2 \end{vmatrix}}{\begin{vmatrix} a_1 & b_1 \\ a_2 & b_2 \end{vmatrix}} = \frac{c_1 b_2 - c_2 b_1}{a_1 b_2 - a_2 b_1} \qquad \text{(7.4a)}$$

$$y = \frac{\begin{vmatrix} a_1 & c_1 \\ a_2 & c_2 \end{vmatrix}}{\begin{vmatrix} a_1 & b_1 \\ a_2 & b_2 \end{vmatrix}} = \frac{a_1 c_2 - a_2 c_1}{a_1 b_2 - a_2 b_1} \qquad \text{(7.4b)}$$

EXAMPLE 7.7. Evaluate the following determinants:

a. $\begin{vmatrix} 2 & 2 \\ 3 & 4 \end{vmatrix} = (2)(4) - (3)(2) = 8 - 6 = \mathbf{2}$

b. $\begin{vmatrix} 4 & -1 \\ 6 & 2 \end{vmatrix} = (4)(2) - (6)(-1) = 8 + 6 = \mathbf{14}$

c. $\begin{vmatrix} 0 & -2 \\ -2 & 4 \end{vmatrix} = (0)(4) - (-2)(-2) = 0 - 4 = \mathbf{-4}$

d. $\begin{vmatrix} 0 & 0 \\ 3 & 10 \end{vmatrix} = (0)(10) - (3)(0) = \mathbf{0}$

EXAMPLE 7.8. Solve for x and y:

a. $2x + y = 3$
$\quad 3x + 4y = 2$

b. $-x + 2y = 3$
$\quad 3x - 2y = -2$

c. $\quad x = 3 - 4y$
$\quad 20y = -1 + 3x$

Solutions:

a. $x = \dfrac{\begin{vmatrix} 3 & 1 \\ 2 & 4 \end{vmatrix}}{\begin{vmatrix} 2 & 1 \\ 3 & 4 \end{vmatrix}} = \dfrac{(3)(4) - (2)(1)}{(2)(4) - (3)(1)} = \dfrac{12 - 2}{8 - 3} = \dfrac{10}{5} = 2$

$y = \dfrac{\begin{vmatrix} 2 & 3 \\ 3 & 2 \end{vmatrix}}{5} = \dfrac{(2)(2) - (3)(3)}{5} = \dfrac{4 - 9}{5} = \dfrac{-5}{5} = -1$

Check:

$$2x + y = (2)(2) + (-1)$$
$$= 4 - 1 = 3 \quad \text{(checks)}$$

$$3x + 4y = (3)(2) + (4)(-1)$$
$$= 6 - 4 = 2 \quad \text{(checks)}$$

b. $x = \dfrac{\begin{vmatrix} 3 & 2 \\ -2 & -2 \end{vmatrix}}{\begin{vmatrix} -1 & 2 \\ 3 & -2 \end{vmatrix}} = \dfrac{(3)(-2) - (-2)(2)}{(-1)(-2) - (3)(2)}$

$$= \frac{-6 + 4}{2 - 6} = \frac{-2}{-4} = \frac{1}{2}$$

$$y = \dfrac{\begin{vmatrix} -1 & 3 \\ 3 & -2 \end{vmatrix}}{-4} = \dfrac{(-1)(-2) - (3)(3)}{-4}$$

$$= \frac{2 - 9}{-4} = \frac{-7}{-4} = \frac{7}{4}$$

c. Equation (c) can be rewritten as

$$x + 4y = 3$$
$$-3x + 20y = -1$$

$$x = \dfrac{\begin{vmatrix} 3 & 4 \\ -1 & 20 \end{vmatrix}}{\begin{vmatrix} 1 & 4 \\ -3 & 20 \end{vmatrix}} = \dfrac{(3)(20) - (-1)(4)}{(1)(20) - (-3)(4)}$$

$$= \frac{60 + 4}{20 + 12} = \frac{64}{32} = 2$$

$$y = \dfrac{\begin{vmatrix} 1 & 3 \\ -3 & -1 \end{vmatrix}}{32} = \dfrac{(1)(-1) - (-3)(3)}{32}$$

$$= \frac{-1 + 9}{32} = \frac{8}{32} = \frac{1}{4}$$

The use of determinants is not limited to the solution of two simultaneous equations; determinants can be applied to any number of simultaneous linear equations. For our purposes, however, a shorthand method for solving the third-order (three-simultaneous-equation) determinant will be sufficient. Third- and higher-order determinants are considered in Appendix A.

Consider the three following simultaneous equations:

Col. 1		Col. 2		Col. 3		Col. 4
$a_1 x$	$+$	$b_1 y$	$+$	$c_1 z$	$=$	d_1
$a_2 x$	$+$	$b_2 y$	$+$	$c_2 z$	$=$	d_2
$a_3 x$	$+$	$b_3 y$	$+$	$c_3 z$	$=$	d_3

in which x, y, and z are the variables, and $a_{1,2,3}$, $b_{1,2,3}$, $c_{1,2,3}$, and $d_{1,2,3}$ are constants.

The determinant configuration for x, y, and z can be found in a manner similar to that for two simultaneous equations. That is, to solve for x, find the determinant in the numerator by replacing column 1 with the elements to the right of the equal sign. The denominator is the determinant of the coefficients of the variables (the same applies to y and z). Again, the denominator is the same for each variable.

$$x = \frac{\begin{vmatrix} d_1 & b_1 & c_1 \\ d_2 & b_2 & c_2 \\ d_3 & b_3 & c_3 \end{vmatrix}}{\begin{vmatrix} a_1 & b_1 & c_1 \\ a_2 & b_2 & c_2 \\ a_3 & b_3 & c_3 \end{vmatrix}}, \quad y = \frac{\begin{vmatrix} a_1 & d_1 & c_1 \\ a_2 & d_2 & c_2 \\ a_3 & d_3 & c_3 \end{vmatrix}}{D}, \quad z = \frac{\begin{vmatrix} a_1 & b_1 & d_1 \\ a_2 & b_2 & d_2 \\ a_3 & b_3 & d_3 \end{vmatrix}}{D}$$

$$D = \begin{vmatrix} a_1 & b_1 & c_1 \\ a_2 & b_2 & c_2 \\ a_3 & b_3 & c_3 \end{vmatrix}$$

A shorthand method for evaluating the third-order determinant consists simply of repeating the first two columns of the determinant to the right of the determinant and then summing the products along specific diagonals as shown below:

$$D = \begin{vmatrix} a_1 & b_1 & c_1 \\ a_2 & b_2 & c_2 \\ a_3 & b_3 & c_3 \end{vmatrix} \begin{matrix} a_1 & b_1 \\ a_2 & b_2 \\ a_3 & b_3 \end{matrix}$$

The products of the diagonals 1, 2, and 3 are positive and have the following magnitudes:

$$+a_1 b_2 c_3 + b_1 c_2 a_3 + c_1 a_2 b_3$$

The products of the diagonals 4, 5, and 6 are negative and have the following magnitudes:

$$-a_3 b_2 c_1 - b_3 c_2 a_1 - c_3 a_2 b_1$$

The total solution is the sum of the diagonals 1, 2, and 3 minus the sum of the diagonals 4, 5, and 6:

$$+(a_1 b_2 c_3 + b_1 c_2 a_3 + c_1 a_2 b_3) \\ - (a_3 b_2 c_1 + b_3 c_2 a_1 + c_3 a_2 b_1) \quad \text{(7.5)}$$

Warning: **This method of expansion is good only for third-order determinants!** It cannot be applied to fourth- and higher-order systems.

EXAMPLE 7.9. Evaluate the following determinant:

$$\begin{vmatrix} 1 & 2 & 3 \\ -2 & 1 & 0 \\ 0 & 4 & 2 \end{vmatrix} \Rightarrow \begin{matrix} 1 & 2 & 3 \\ -2 & 1 & 0 \\ 0 & 4 & 2 \end{matrix} \begin{matrix} 1 & 2 \\ -2 & 1 \\ 0 & 4 \end{matrix}$$

Solution:

$$[(1)(1)(2) + (2)(0)(0) + (3)(-2)(4)]$$
$$- [(0)(1)(3) + (4)(0)(1) + (2)(-2)(2)]$$
$$= (2 + 0 - 24) - (0 + 0 - 8) = (-22) - (-8)$$
$$= -22 + 8 = -14$$

EXAMPLE 7.10. Solve for x, y, and z:

$$1x + 0y - 2z = -1$$
$$0x + 3y + 1z = +2$$
$$1x + 2y + 3z = 0$$

Solution:

$$x = \frac{\begin{vmatrix} -1 & 0 & -2 \\ 2 & 3 & 1 \\ 0 & 2 & 3 \end{vmatrix}}{\begin{vmatrix} 1 & 0 & -2 \\ 0 & 3 & 1 \\ 1 & 2 & 3 \end{vmatrix}}$$

$$= \frac{[(-1)(3)(3) + (0)(1)(0) + (-2)(2)(2)]}{[(1)(3)(3) + (0)(1)(1) + (-2)(0)(2)]}$$
$$\frac{- [(0)(3)(-2) + (2)(1)(-1) + (3)(2)(0)]}{- [(1)(3)(-2) + (2)(1)(1) + (3)(0)(0)]}$$

$$= \frac{(-9 + 0 - 8) - (0 - 2 + 0)}{(9 + 0 + 0) - (-6 + 2 + 0)}$$

$$= \frac{-17 + 2}{9 + 4} = -\frac{15}{13}$$

$$y = \frac{\begin{vmatrix} 1 & -1 & -2 \\ 0 & 2 & 1 \\ 1 & 0 & 3 \end{vmatrix}}{13}$$

$$= \frac{[(1)(2)(3) + (-1)(1)(1) + (-2)(0)(0)]}{- [(1)(2)(-2) + (0)(1)(1) + (3)(0)(-1)]}$$

$$= \frac{(6 - 1 + 0) - (-4 + 0 + 0)}{13}$$

$$= \frac{5 + 4}{13} = \frac{9}{13}$$

$$z = \frac{\begin{vmatrix} 1 & 0 & -1 \\ 0 & 3 & 2 \\ 1 & 2 & 0 \end{vmatrix}}{13}$$

$$= \frac{[(1)(3)(0) + (0)(2)(1) + (-1)(0)(2)]}{- [(1)(3)(-1) + (2)(2)(1) + (0)(0)(0)]}$$
$$\frac{}{13}$$

$$= \frac{(0 + 0 + 0) - (-3 + 4 + 0)}{13}$$

$$= \frac{0 - 1}{13} = -\frac{1}{13}$$

or from

$$0x + 3y + 1z = +2$$
$$z = 2 - 3y$$
$$= 2 - 3\left(\frac{9}{13}\right)$$
$$= \frac{26}{13} - \frac{27}{13}$$
$$= -\frac{1}{13}$$

Check:

$$
\begin{array}{l}
\left. \begin{array}{l} 1x + 0y - 2z \\ = -1 \end{array} \right| \quad -\frac{15}{13} + 0 + \frac{2}{13} = -1 \left| \quad -\frac{13}{13} = -1 \checkmark \right. \\[2ex]
\left. \begin{array}{l} 0x + 3y + 1z \\ = +2 \end{array} \right| 0 + \frac{27}{13} + \frac{-1}{13} = +2 \left\} \quad \frac{26}{13} = +2 \checkmark \right. \\[2ex]
\left. \begin{array}{l} 1x + 2y + 3z \\ = 0 \end{array} \right| \quad -\frac{15}{13} + \frac{18}{13} + \frac{-3}{13} = 0 \left| \quad -\frac{18}{13} + \frac{18}{13} = 0 \checkmark \right.
\end{array}
$$

7.7 BRANCH-CURRENT METHOD

We will now consider the first in a series of methods for solving networks with two or more sources. Once this method is mastered, there is no linear bilateral dc network for which a solution cannot be found. Keep in mind that networks with two isolated voltage sources cannot be solved using the approach of Chapter 6. For further evidence of this fact, try solving for the unknown elements of Example 7.11 using the technique introduced in Chapter 6. The most direct introduction for a method of this type is to list the series of steps required for its application. There are four steps, as indicated below. Before continuing, understand that this method will produce the current through each branch of the network, the *branch current*. Once this is known, all other quantities, such as voltage or power, can be determined.

1. Assign a distinct current of *arbitrary* direction to each branch of the network.
2. Indicate the polarities for each resistor *as determined by the assumed direction of current flow*.
3. Apply Kirchhoff's voltage law around *each* closed loop and Kirchhoff's current law *at* the minimum

number of nodes that will include *all* of the branch currents of the network.

4. Solve the resulting simultaneous linear equations for assumed branch currents using determinants.

EXAMPLE 7.11. Apply the branch-current method to the network of Fig. 7.15.

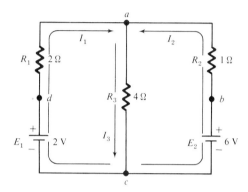

FIG. 7.15

Solution:

Step 1: Since there are three distinct branches (*cda, cba, ca*), three currents of arbitrary directions (I_1, I_2, I_3) are chosen as indicated in Fig. 7.15.

Step 2: Polarities for each resistor are drawn to agree with assumed current directions as indicated in Fig. 7.16.

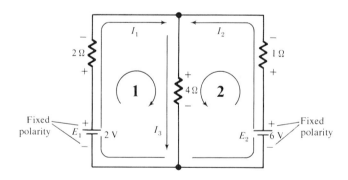

FIG. 7.16

Step 3: Kirchhoff's voltage law is applied around each closed loop (1 and 2):

loop 1: $\Sigma_0 V = +2 - 2I_1 - 4I_3 = 0$

(Rise in potential: +2 Battery potential; Drop in potential: $2I_1$ Voltage drop across 2-Ω resistor, $4I_3$ Voltage drop across 4-Ω resistor)

loop 2: $\Sigma_0 V = +6 - 1I_2 - 4I_3 = 0$

Kirchhoff's current law at node *a* is given by

$$I_1 + I_2 = I_3$$

Step 4: There are three equations and three unknowns:

$$2 - 2I_1 - 4I_3 = 0 \qquad \text{Rewritten: } -2I_1 + 0 - 4I_3 = -2$$
$$6 - I_2 - 4I_3 = 0 \qquad\qquad\qquad 0 - I_2 - 4I_3 = -6$$
$$I_1 + I_2 = I_3 \qquad\qquad\qquad I_1 + I_2 - I_3 = 0$$

Using third-order determinants, we have

$$I_1 = \frac{\begin{vmatrix} -2 & 0 & -4 \\ -6 & -1 & -4 \\ 0 & 1 & -1 \end{vmatrix}}{D = \begin{vmatrix} -2 & 0 & -4 \\ 0 & -1 & -4 \\ 1 & 1 & -1 \end{vmatrix}} = -1\,A$$

A negative sign in front of a branch current indicates only that the actual current flows in the direction opposite to that assumed.

$$I_2 = \frac{\begin{vmatrix} -2 & -2 & -4 \\ 0 & -6 & -4 \\ 1 & 0 & -1 \end{vmatrix}}{D} = 2\,A$$

$$I_3 = \frac{\begin{vmatrix} -2 & 0 & -2 \\ 0 & -1 & -6 \\ 1 & 1 & 0 \end{vmatrix}}{D} = 1\,A$$

Instead of using third-order determinants, we could reduce the three equations to two by substituting the third equation in the first and second equations:

$$\left.\begin{array}{l} 2 - 2I_1 - 4(\overbrace{I_1 + I_2}^{I_3}) = 0 \\[2mm] 6 - I_2 - 4(\overbrace{I_1 + I_2}^{I_3}) = 0 \end{array}\right\} \quad \begin{array}{l} 2 - 2I_1 - 4I_1 - 4I_2 = 0 \\[2mm] 6 - I_2 - 4I_1 - 4I_2 = 0 \end{array}$$

or

$$-6I_1 - 4I_2 = -2$$
$$-4I_1 - 5I_2 = -6$$

Multiplying through by -1 in each equation yields

$$6I_1 + 4I_2 = +2$$
$$4I_1 + 5I_2 = +6$$

and

$$I_1 = \frac{\begin{vmatrix} 2 & 4 \\ 6 & 5 \end{vmatrix}}{\begin{vmatrix} 6 & 4 \\ 4 & 5 \end{vmatrix}} = \frac{10 - 24}{30 - 16} = \frac{-14}{14} = -1\,A$$

$$I_2 = \frac{\begin{vmatrix} 6 & 2 \\ 4 & 6 \end{vmatrix}}{14} = \frac{36 - 8}{14} = \frac{28}{14} = 2\,A$$

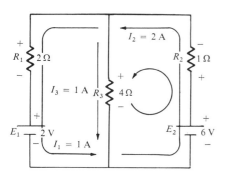

FIG. 7.17

$I_3 = I_1 + I_2$
$I_3 = -1 + 2$
$I_3 = \mathbf{1\,A}$

Returning to the original network (Fig. 7.17), we apply Kirchhoff's voltage law around the loop indicated:

$$\Sigma_\circlearrowright V = 6 - 1I_2 - 4I_3 = 0$$

or

$$6 = 1I_2 + 4I_3 = (1)(2) + (4)(1)$$
$$6 = 6 \quad \text{(checks)}$$

EXAMPLE 7.12. Apply branch-current analysis to the network of Fig. 7.18.

Solution: By Kirchhoff's voltage law,

loop 1: $+15 - 4I_1 + 10I_3 - 20 = 0$
loop 2: $+20 - 10I_3 - 5I_2 + 40 = 0$
node a: $I_1 + I_3 = I_2$

Substituting the third equation into the other two yields

$$15 - 4I_1 + 10I_3 - 20 = 0 \,\Big\} \text{ \footnotesize Substituting for } I_2 \text{ (since it occurs}$$
$$20 - 10I_3 - 5(I_1 + I_3) + 40 = 0 \,\Big\rfloor \text{ \footnotesize only once in the two equations)}$$

or

$$-4I_1 + 10I_3 = 5$$
$$-5I_1 - 15I_3 = -60$$

Multiplying the lower equation by -1, we have

$$-4I_1 + 10I_3 = 5$$
$$5I_1 + 15I_3 = 60$$

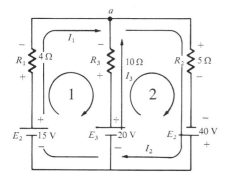

FIG. 7.18

$$I_1 = \frac{\begin{vmatrix} 5 & 10 \\ 60 & 15 \end{vmatrix}}{\begin{vmatrix} -4 & 10 \\ 5 & 15 \end{vmatrix}} = \frac{75 - 600}{-60 - 50} = \frac{-525}{-110} = \mathbf{4.773\,A}$$

$$I_3 = \frac{\begin{vmatrix} -4 & 5 \\ 5 & 60 \end{vmatrix}}{-110} = \frac{-240 - 25}{-110} = \frac{-265}{-110} = \mathbf{2.409\,A}$$

$$I_2 = I_1 + I_3 = 4.773 + 2.409$$
$$I_2 = \mathbf{7.182\,A}$$

7.8 MESH ANALYSIS (GENERAL APPROACH)

The second method of analysis to be described is called *mesh analysis*. The term *mesh* is derived from the similarities in appearance between the closed loops of a network and a wire mesh fence. Although this approach is on a more sophisticated plane than the branch-current method, it incorporates many of the ideas just developed. Of the two methods, mesh analysis is the one more frequently applied today. Branch-current analysis is introduced as a stepping stone to mesh analysis because branch currents are initially more "real" to the student than the loop currents employed in mesh analysis. Essentially, the mesh-analysis approach simply eliminates the need to substitute the results of Kirchhoff's current law into the equations derived from Kirchhoff's voltage law. It is now accomplished in the initial writing of the equations. The systematic approach outlined below should be followed when applying this method.

1. Assign a distinct current in the clockwise direction to each independent closed loop of the network. It is not absolutely necessary to choose the clockwise direction for each loop current. In fact, any direction can be chosen for each loop current with no loss in accuracy as long as the remaining steps are followed properly. However, by choosing the clockwise direction as a standard, we can develop a shorthand method for writing the required equations that will save time and eliminate many errors (Section 7.9).

This first step is most effectively accomplished by placing a loop current within each "window" of the network as demonstrated in the examples of this section. A window is simply any bounded area within the network. This insures that they are all independent. There are a variety of other loop currents that can be assigned. In each case, however, be sure that the information carried by any one loop equation is not included in a combination of the other network equations. This is the crux of the terminology: *independent*. No matter how you choose your loop currents, the number of loop currents required is always equal to the number of windows of a planar (no-crossovers) network. On occasion a network may appear to be nonplanar. However, a redrawing of the network may reveal that it is, in fact, planar. Such may be the case in one or two problems at the end of the chapter.

Before continuing to the next step, let us insure that the concept of a loop current is clear. For the network of Fig. 7.19, the loop current I_1 is the branch current of the branch containing the 2-Ω resistor and 2-V battery. The current through the 4-Ω resistor is not I_1, however, since there is also a loop current I_2 through it. Since they have opposite

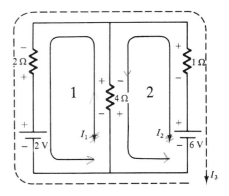

FIG. 7.19

directions, $I_{4\Omega}$ equals $I_1 - I_2$, as pointed out in the example to follow. In other words, a loop current is a branch current only when it is the only loop current assigned to that branch.

2. Indicate the polarities within each loop for each resistor as determined by the assumed direction of loop current for that loop.
3. Apply Kirchhoff's voltage law around each closed loop.

   a. If a resistor has two or more assumed currents flowing through it, the total current through the resistor is the assumed current of the loop in which Kirchhoff's voltage law is being applied, plus the assumed currents of the other loops passing through in the same direction, minus the assumed currents through in the opposite direction.
   b. The polarity of a voltage source is unaffected by the loop currents passing through it.
4. Solve the resulting simultaneous linear equations for the assumed loop currents using determinants.

EXAMPLE 7.13. Consider the same basic network as in Example 7.11 of the preceding section (Fig. 7.19).

Solution:

Step 1: Two loop currents (I_1 and I_2) are assigned in the clockwise direction. A third loop (I_3) could have been included, but the information carried by this loop is already included in the other two.

Step 2: Polarities are drawn to agree with assumed current directions. Note that for this case the polarities across the 4-Ω resistor are the opposite for each loop current.

Step 3: Kirchhoff's voltage law is applied around each loop:

Voltage drop across
4-Ω resistor

loop 1: $+2 - 2I_1 - \overbrace{4(I_1 - I_2)} = 0$

Total current
through
4-Ω resistor

Subtracted since both in opposite directions through resistor

loop 2: $-6 - 1I_2 - 4(I_2 - I_1) = 0$

Total current
through
4-Ω resistor

Step 4: The equations are rewritten as follows:

loop 1: $+2 - 2I_1 - 4I_1 + 4I_2 = 0$
loop 2: $-6 - 1I_2 - 4I_2 + 4I_1 = 0$

and

loop 1: $+2 - 6I_1 + 4I_2 = 0$
loop 2: $-6 - 5I_2 + 4I_1 = 0$

or

loop 1: $-6I_1 + 4I_2 = -2$
loop 2: $+4I_1 - 5I_2 = +6$

Applying determinants will result in

$$I_1 = -1 \text{ A} \quad \text{and} \quad I_2 = -2 \text{ A}$$

The minus signs indicate that the current in the network flows in the direction opposite to that indicated by the assumed loop current.

The current through the 2-V source and 2-Ω resistor is therefore 1 A in the other direction, and the current through the 6-V source and 1-Ω resistor is 2 A in the opposite direction indicated on the circuit. The current through the 4-Ω resistor is

$$I_1 - I_2 \quad \text{(from original network: loop 1)}$$

or

$$-1 - (-2) = -1 + 2 = 1 \text{ A} \quad \text{(in the direction of } I_1\text{)}$$

The outer loop (I_3) and *one* inner loop (either I_1 or I_2) would also have produced the correct results. This approach, however, will often lead to errors since the loop equations will usually be more difficult to write. The best method of picking these loop currents is to use the window approach.

EXAMPLE 7.14. Find the current through each branch of the network of Fig. 7.20.

Solution:
Steps 1 and 2 are as indicated in the circuit. Note that the polarities of the 6-Ω resistor are different for each loop current.
Step 3: Kirchhoff's voltage law is applied around each closed loop:

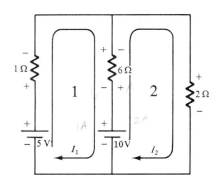

FIG. 7.20

Drop in potential

loop 1: $5 - 1I_1 \overset{\downarrow}{-} 6(I_1 - I_2) - 10 = 0$

I_2 flows through the 6-Ω resistor in the direction opposite to I_1.

loop 2: $10 - 6(I_2 - I_1) - 2I_2 = 0$

The equations are rewritten as

$$5 - I_1 - 6I_1 + 6I_2 - 10 = 0 \Big\} \quad -7I_1 + 6I_2 = 5$$
$$10 - 6I_2 + 6I_1 - 2I_2 = 0 \Big\} \quad +6I_1 - 8I_2 = -10$$

$$I_1 = \frac{\begin{vmatrix} 5 & 6 \\ -10 & -8 \end{vmatrix}}{\begin{vmatrix} -7 & 6 \\ 6 & -8 \end{vmatrix}} = \frac{-40 + 60}{56 - 36} = \frac{20}{20} = \textbf{1 A}$$

$$I_2 = \frac{\begin{vmatrix} -7 & 5 \\ 6 & -10 \end{vmatrix}}{20} = \frac{70 - 30}{20} = \frac{40}{20} = \textbf{2 A}$$

Since I_1 and I_2 are positive and flow in opposite directions through the 6-Ω resistor and 10-V source, the total current in this branch is equal to the difference of the two currents in the direction of the larger one:

$$I_2 > I_1 \quad (2 > 1)$$

Therefore, **1 A** $(2 - 1)$ flows in this branch in the direction of I_2.

It is sometimes impractical to draw all the branches of a circuit at right angles to one another. The next example demonstrates how a portion of a network may appear due to various constraints. The method of analysis does not change with this change in configuration.

EXAMPLE 7.15. Find the branch currents of the network of Fig. 7.21.

Solution:
Steps 1 and 2 are as indicated in the circuit.

Step 3: Kirchhoff's voltage law is applied around each closed loop:

loop 1: $-6 - 2I_1 - 4 - 4(I_1 - I_2) = 0$
loop 2: $-4(I_2 - I_1) + 4 - 6I_2 - 3 = 0$

which are rewritten as

$$\begin{aligned} -10 - 4I_1 - 2I_1 + 4I_2 &= 0 \\ +1 + 4I_1 - 4I_2 - 6I_2 &= 0 \end{aligned} \right\} \quad \begin{aligned} -6I_1 + 4I_2 &= +10 \\ +4I_1 - 10I_2 &= -1 \end{aligned}$$

or, by multiplying the top equation by -1, we obtain

$$6I_1 - 4I_2 = -10$$
$$4I_1 - 10I_2 = -1$$

and

$$I_1 = \frac{\begin{vmatrix} -10 & -4 \\ -1 & -10 \end{vmatrix}}{\begin{vmatrix} 6 & -4 \\ 4 & -10 \end{vmatrix}} = \frac{100 - 4}{-60 + 16} = \frac{96}{-44} = \textbf{-2.182 A}$$

$$I_2 = \frac{\begin{vmatrix} 6 & -10 \\ 4 & -1 \end{vmatrix}}{-44} = \frac{-6 + 40}{-44} = \frac{34}{-44} = \textbf{-0.773 A}$$

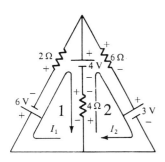

FIG. 7.21

The current in the 4-Ω resistor and 4-V source for loop 1 is

$$I_1 - I_2 = -2.182 - (-0.773)$$
$$= -2.182 + 0.773$$
$$= -1.409 \text{ A}$$

indicating that it is 1.409 A in a direction opposite to I_1 in loop 1.

It may happen that current sources will be present in the network to which we wish to apply mesh analysis. The first step will then be to convert all current sources to voltage sources as in the network of Fig. 7.22.

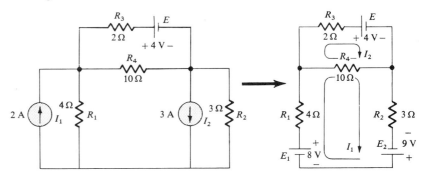

FIG. 7.22

If the current source has no resistance in parallel with it, as in the network of Fig. 7.23, there is no need to solve for I_1 since it is simply equal to that of the current source.

Applying Kirchhoff's voltage law around loop 2 yields

$$+4 - 4I_2 - 3I_2 - 2(I_1 + I_2) = 0$$

For loop 1, $I_1 = 1.5 \text{ A}$. I_2 can then be found by substituting $I_1 = 1.5$ A in the preceding equation:

$$+4 - 4I_2 - 3I_2 - 2(1.5 + I_2) = 0$$

which is rewritten as

$$-7I_2 - 2I_2 + 4 - 3 = 0$$

or

$$-9I_2 = -1$$

and

$$I_2 = 0.111 \text{ A}$$

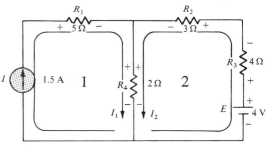

FIG. 7.23

7.9 MESH ANALYSIS (FORMAT APPROACH)

Now that the basis for the mesh-analysis approach has been established, in this section we will consider a technique for writing the mesh equations more rapidly and usually with fewer errors. As an aid in introducing the pro-

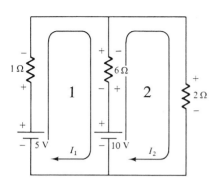

FIG. 7.24

cedure, the network of Example 7.14 (Fig. 7.20) has been redrawn in Fig. 7.24 with the assigned loop currents. (Note that each loop current has a clockwise direction.)

The equations obtained are

$$-7I_1 + 6I_2 = 5$$
$$6I_1 - 8I_2 = -10$$

which can also be written as

$$7I_1 - 6I_2 = -5$$
$$8I_2 - 6I_1 = 10$$

and expanded as

Col. 1		Col. 2		Col. 3
$(1 + 6)I_1$	$-$	$6I_2$	$=$	$(5 - 10)$
$(2 + 6)I_2$	$-$	$6I_1$	$=$	10

Note in the above equations that column 1 is composed of a loop current times the sum of the resistors through which that loop current passes. Column 2 is the product of the resistors common to another loop current times that other loop current. Note that in each equation this column is subtracted from column 1. Column 3 is the *algebraic* sum of the voltage sources through which the loop current of interest passes. A source of emf is assigned a positive sign if the loop current passes from the negative to the positive terminal, and a negative value if the polarities are reversed. The comments above are correct only for a standard direction of loop current in each window, the chosen being the clockwise direction.

Each statement above can be extended to develop the following *format approach* to mesh analysis:

1. Assign a loop current to each independent closed loop (as in the previous section) in a *clockwise* direction.
2. The number of required equations is equal to the number of chosen independent closed loops. Column 1 of each equation is formed by summing the resistance values of those resistors through which the loop current of interest passes and multiplying the result by that loop current.
3. We must now consider the mutual terms which, as noted in the examples above, are always subtracted from the first column. It is possible to have more than one mutual term if the loop current of interest has an element in common with more than one other loop current. This will be demonstrated in an example to follow. Each term is the product of the mutual resistor and the other loop current passing through the same element.

4. The column to the right of the equality sign is the algebraic sum of the voltage sources through which the loop current of interest passes. Positive signs are assigned to those sources of emf having a polarity such that the loop current passes from the negative to positive terminal. A negative sign is assigned to those potentials for which the reverse is true.

5. Solve the resulting simultaneous equations for the desired loop currents.

Let us now consider a few examples.

EXAMPLE 7.16. Write the mesh equations for the network of Fig. 7.25 and find the current through the 7-Ω resistor.

Solution:
Step 1: As indicated in Fig. 7.25, each assigned loop current has a clockwise direction.
Steps 2 to 4:

$$I_1: (8 + 6 + 2)I_1 - (2)I_2 = 4$$
$$I_2: (7 + 2)I_2 - (2)I_1 \quad\quad = -9$$

or

$$16I_1 - 2I_2 = 4$$
$$9I_2 - 2I_1 = -9$$

which for determinants are

$$16I_1 - 2I_2 = 4$$
$$-2I_1 + 9I_2 = -9$$

and

$$I_2 = I_{7\Omega} = \frac{\begin{vmatrix} 16 & 4 \\ -2 & -9 \end{vmatrix}}{\begin{vmatrix} 16 & -2 \\ -2 & 9 \end{vmatrix}} = \frac{-144 + 8}{144 - 4} = \frac{-136}{140}$$

$$= -0.971 \text{ A}$$

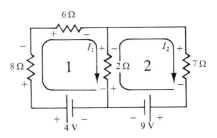

FIG. 7.25

EXAMPLE 7.17. Write the mesh equations for the network of Fig. 7.26.

Solution:
Step 1: Each window is assigned a loop current in the clockwise direction:

I_1 does not pass through an element mutual with I_3.
↓

$$I_1: \quad\quad (1 + 1)I_1 - (1)I_2 + 0 = 2 - 4$$
$$I_2: (1 + 2 + 3)I_2 - (1)I_1 - (3)I_3 = 4$$
$$I_3: \quad\quad (3 + 4)I_3 - (3)I_2 + 0 = 2$$
↑

I_3 does not pass through an element mutual with I_1.

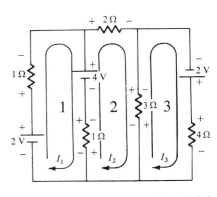

FIG. 7.26

Summing terms yields

$$2I_1 - I_2 + 0 = -2$$
$$6I_2 - I_1 - 3I_3 = 4$$
$$7I_3 - 3I_2 + 0 = 2$$

which are rewritten for determinants as

$$2I_1 - I_2 + 0 = -2$$
$$-I_1 + 6I_2 - 3I_3 = 4$$
$$0 - 3I_2 + 7I_3 = 2$$

which compares directly with the equations obtained in that example.

Note that the coefficients of the a and b diagonals are equal. This *symmetry* about the c axis will always be true for equations written using the format approach. It is a check on whether the equations were obtained correctly. Note the symmetry of the equations of Example 7.16.

We will now consider a network with only one source of emf to point out that mesh analysis can be used to advantage in other than multisource networks.

EXAMPLE 7.18. Find the current through the 10-Ω resistor of the network of Fig. 7.27.

Solution:

$$I_1: \quad (8 + 3)I_1 - (8)I_3 - (3)I_2 = 15$$
$$I_2: (3 + 5 + 2)I_2 - (3)I_1 - (5)I_3 = 0$$
$$I_3: (8 + 10 + 5)I_3 - (8)I_1 - (5)I_2 = 0$$

$$11I_1 - 8I_3 - 3I_2 = 15$$
$$10I_2 - 3I_1 - 5I_3 = 0$$
$$23I_3 - 8I_1 - 5I_2 = 0$$

or

$$11I_1 - 3I_2 - 8I_3 = 15$$
$$-3I_1 + 10I_2 - 5I_3 = 0$$
$$-8I_1 - 5I_2 + 23I_3 = 0$$

and

$$I_3 = I_{10\Omega} = \frac{\begin{vmatrix} 11 & -3 & 15 \\ -3 & 10 & 0 \\ -8 & -5 & 0 \end{vmatrix}}{\begin{vmatrix} 11 & -3 & -8 \\ -3 & 10 & -5 \\ -8 & -5 & 23 \end{vmatrix}} = \mathbf{1.220\ A}$$

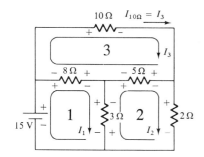

FIG. 7.27

The natural sequence of steps makes this method very useful for computer techniques to be examined briefly in Chapter 26. For students studying computer programming, this type of problem is an excellent exercise.

7.10 NODAL ANALYSIS (GENERAL APPROACH)

Recall from the development of loop analysis that the general network equations were obtained by applying Kirchhoff's voltage law around each closed loop. We will now employ Kirchhoff's current law to develop a method referred to as *nodal analysis.*

A *node* is defined as a junction of two or more branches. If we now define one node of any network as a reference (that is, a point of zero potential or ground), the remaining nodes of the network will all have a fixed potential relative to this reference. For a network of N nodes, therefore, there will exist $(N - 1)$ nodes with a fixed potential relative to the assigned reference node. Equations relating these nodal voltages can be written by applying Kirchhoff's current law at each of the $(N - 1)$ nodes. To obtain the complete solution of a network, these nodal voltages are then evaluated in the same manner in which loop currents were found in loop analysis.

To facilitate the writing of the network equations, all voltage sources within the network will be converted to current sources before Kirchhoff's current law is applied. The nodal analysis method is applied as follows:

1. Convert all voltage sources to current sources.
2. Determine the number of nodes within the network.
3. Pick a reference node and label each remaining node with a subscripted value of voltage: V_1, V_2, and so on.
4. Write Kirchhoff's current law at each node except the reference.
5. Solve the resulting equations for nodal voltages.

EXAMPLE 7.19. Apply nodal analysis to the network of Fig. 7.28.

Solution:
Step 1: Convert voltage sources to current sources (Fig. 7.29).

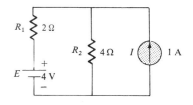

FIG. 7.28

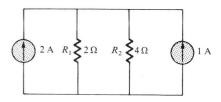

FIG. 7.29

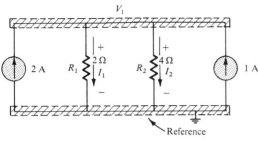

FIG. 7.30

Step 4 (One nodal equation at V_1): V_1 is positive with respect to the reference node. Therefore, current flows away from V_1 through the 2-Ω and 4-Ω resistors at a rate equal to

$$I_1 = \frac{V_1}{2} \quad \text{and} \quad I_2 = \frac{V_1}{4}$$

respectively. Applying Kirchhoff's current law yields

$$\underbrace{2 + 1}_{\text{Entering}} - \underbrace{\left(\frac{V_1}{2} + \frac{V_1}{4}\right)}_{\text{Leaving}} = 0$$

Step 5:

$$V_1\left(\frac{1}{2} + \frac{1}{4}\right) = 3 \quad \text{or} \quad V_1\left(\frac{3}{4}\right) = 3$$

$$V_1 = \frac{12}{3} = \mathbf{4\ V}$$

The potential across each current source and resistor is therefore 4 V, and

$$I_1 = \frac{V_1}{2} = \frac{4}{2} = \mathbf{2\ A}$$

$$I_2 = \frac{V_1}{4} = \frac{4}{4} = \mathbf{1\ A}$$

EXAMPLE 7.20. Determine the nodal voltages for the network of Fig. 7.31.

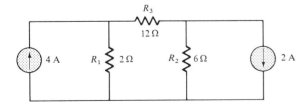

FIG. 7.31

Solution:
Steps 1 to 3: See Fig. 7.32.

Before attempting step 4, note that the current I_3 is determined by the difference in potential across the 12-Ω re-

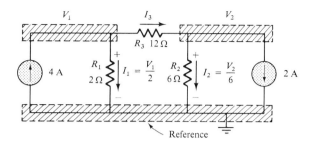

FIG. 7.32

sistor. Either V_1 or V_2 can be assumed the larger so that some direction of flow is assigned to I_3, but this same assumption must be used when the equations are written at each node. Even though the assumption may be incorrect (the current I_3 may, in actuality, be flowing in the opposite direction), the values of V_1 and V_2 determined by the resulting equations will be correct.

Step 4: Assuming $V_1 > V_2$, I_3 will flow as indicated through the 12-Ω resistor. The resulting nodal equations are

node 1: $4 - I_1 - I_3 = 0$
node 2: $I_3 - I_2 - 2 = 0$

Expanding in terms of V_1 and V_2, we have

node 1: $4 - \dfrac{V_1}{2} - \dfrac{(V_1 - V_2)}{12} = 0$

node 2: $\dfrac{(V_1 - V_2)}{12} - \dfrac{V_2}{6} - 2 = 0$

or

$$\boxed{\begin{array}{l} V_1\left(\dfrac{1}{2} + \dfrac{1}{12}\right) - V_2\left(\dfrac{1}{12}\right) = +4 \\[2mm] V_2\left(\dfrac{1}{12} + \dfrac{1}{6}\right) - V_1\left(\dfrac{1}{12}\right) = -2 \end{array}}$$

(7.6)

producing

$$\dfrac{7}{12}V_1 - \dfrac{1}{12}V_2 = +4 \qquad 7V_1 - V_2 = 48$$

$$-\dfrac{1}{12}V_1 + \dfrac{3}{12}V_2 = -2 \qquad -1V_1 + 3V_2 = -24$$

and

$$V_1 = \dfrac{\begin{vmatrix} 48 & -1 \\ -24 & 3 \end{vmatrix}}{\begin{vmatrix} 7 & -1 \\ -1 & 3 \end{vmatrix}} = \dfrac{120}{20} = +6\ \mathbf{V}$$

$$V_2 = \dfrac{\begin{vmatrix} 7 & 48 \\ -1 & -24 \end{vmatrix}}{2} = \dfrac{-120}{20} = -6\ \mathbf{V}$$

Since V_1 is greater than V_2, the assumed direction of I_3 was correct. Its value is

$$I_3 = \frac{V_1 - V_2}{12} = \frac{6 - (-6)}{12} = \frac{12}{12} = \mathbf{1\ A}$$

$$I_1 = \frac{V_1}{2} = \frac{6}{2} = \mathbf{3\ A}$$

$$I_2 = \frac{V_2}{6} = \frac{-6}{6} = \mathbf{-1\ A}$$

A negative sign indicates that the current in the original network has the opposite direction.

7.11 NODAL ANALYSIS (FORMAT APPROACH)

A close examination of Eq. (7.6) appearing in Example 7.20 reveals that the subscripted voltage at the node in which Kirchhoff's current law is applied is multiplied by the sum of the conductances attached to that node. Note also that the other nodal voltages within the same equation are multiplied by the negative of the conductance between the two nodes. The current sources are represented to the right of the equal sign with a positive sign if they supply current to the node and with a negative sign if they draw current from the node.

These conclusions can be expanded to include networks with any number of nodes. This will allow us to write nodal equations rapidly and in a form that is convenient for the use of determinants. Note the parallelism between the following four steps of application and those required for mesh analysis in Section 7.9:

1. Choose a reference node and assign a subscripted voltage label to $(N - 1)$ remaining nodes of the network.
2. The number of equations required for a complete solution is equal to the number of subscripted voltages $(N - 1)$. Column 1 of each equation is formed by summing the conductances tied to the node of interest and multiplying the result by that subscripted nodal voltage.
3. We must now consider the mutual terms which, as noted in the preceding example, are always subtracted from the first column. It is possible to have more than one mutual term if the nodal voltage of current interest has an element in common with more than one other nodal voltage. This will be demonstrated in an example to follow. Each mu-

tual term is the product of the mutual conductance and the other nodal voltage tied to that conductance.

4. The column to the right of the equality sign is the algebraic sum of the current sources tied to the node of interest. A current source is assigned a positive sign if it supplies current to a node and a negative sign if it draws current from the node.

5. Solve the resulting simultaneous equations for the desired voltages.

Let us now consider a few examples.

EXAMPLE 7.21. Write the nodal equations for the network of Fig. 7.33.

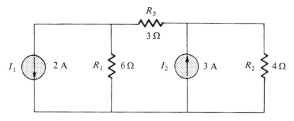

FIG. 7.33

Solution:

Step 1: The figure is redrawn with assigned subscripted voltages in Fig. 7.34.

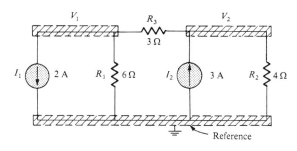

Steps 2 to 4:   **FIG. 7.34**

$$V_1: \left(\frac{1}{6} + \frac{1}{3}\right)V_1 - \left(\frac{1}{3}\right)V_2 = -2 \quad \text{Drawing current from node 1}$$

$$\underbrace{\phantom{\left(\frac{1}{6} + \frac{1}{3}\right)}}_{\substack{\text{Sum of} \\ \text{conductances} \\ \text{connected} \\ \text{to node 1}}} \quad \underbrace{\phantom{\left(\frac{1}{3}\right)}}_{\substack{\text{Mutual} \\ \text{conductance}}}$$

$$V_2: \left(\frac{1}{4} + \frac{1}{3}\right)V_2 - \left(\frac{1}{3}\right)V_1 = +3 \quad \text{Supplying current to node 2}$$

$$\underbrace{\phantom{\left(\frac{1}{4} + \frac{1}{3}\right)}}_{\substack{\text{Sum of} \\ \text{conductances} \\ \text{connected} \\ \text{to node 2}}} \quad \underbrace{\phantom{\left(\frac{1}{3}\right)}}_{\substack{\text{Mutual} \\ \text{conductance}}}$$

and

$$\frac{1}{2}V_1 - \frac{1}{3}V_2 = -2$$

$$-\frac{1}{3}V_1 + \frac{7}{12}V_2 = 3$$

EXAMPLE 7.22. Find the voltage across the 3-Ω resistor of Fig. 7.35 by nodal analysis.

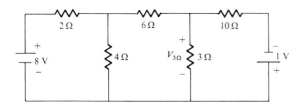

FIG. 7.35

Solution: Converting sources and choosing nodes (Fig. 7.36), we have

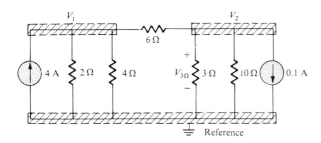

FIG. 7.36

$$\left(\frac{1}{2} + \frac{1}{4} + \frac{1}{6}\right)V_1 - \left(\frac{1}{6}\right)V_2 = +4$$

$$\left(\frac{1}{10} + \frac{1}{3} + \frac{1}{6}\right)V_2 - \left(\frac{1}{6}\right)V_1 = -0.1$$

$$\frac{11}{12}V_1 - \frac{1}{6}V_2 = 4$$

$$-\frac{1}{6}V_1 + \frac{3}{5}V_2 = -0.1$$

which is equivalent to

$$\frac{11}{12}V_1 - \frac{2}{12}V_2 = 4 \qquad 11V_1 - 2V_2 = +48$$

$$-\frac{5}{30}V_1 + \frac{18}{30}V_2 = -0.1 \qquad -5V_1 + 18V_2 = -3$$

$$11V_1 - 2V_2 = +48$$
$$-5V_1 + 18V_2 = -3$$

$$V_2 = V_{3\Omega} = \frac{\begin{vmatrix} 11 & 48 \\ -5 & -3 \end{vmatrix}}{\begin{vmatrix} 11 & -2 \\ -5 & 18 \end{vmatrix}} = \frac{-33 + 240}{198 - 10} = \frac{207}{188} = \mathbf{1.101\ V}$$

As demonstrated for mesh analysis, nodal analysis can also be a very useful technique for solving networks with only one source.

EXAMPLE 7.23. Using nodal analysis, determine the potential across the 4-Ω resistor in Fig. 7.37.

Solution: The reference and four subscripted voltage levels were chosen as shown in Fig. 7.38. A moment of reflection should reveal that for any difference in potential between V_1 and V_3, the current through and the potential drop across each 5-Ω resistor will be the same. Therefore, V_4 is simply a mid-voltage level between V_1 and V_3 and is known if V_1 and V_3 are available. We will therefore not include it in a nodal voltage and will redraw the network as shown in Fig. 7.39. Understand, however, that V_4 can be included if desired, although four nodal voltages will result rather than the three to be obtained in the solution of this problem.

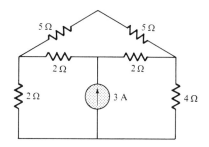

FIG. 7.37

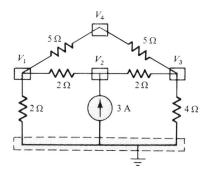

FIG. 7.38

$$V_1: \quad \left(\frac{1}{2} + \frac{1}{2} + \frac{1}{10}\right)V_1 - \left(\frac{1}{2}\right)V_2 - \left(\frac{1}{10}\right)V_3 = 0$$

$$V_2: \quad \left(\frac{1}{2} + \frac{1}{2}\right)V_2 - \left(\frac{1}{2}\right)V_1 - \left(\frac{1}{2}\right)V_3 = 3$$

$$V_3: \quad \left(\frac{1}{10} + \frac{1}{2} + \frac{1}{4}\right)V_3 - \left(\frac{1}{2}\right)V_2 - \left(\frac{1}{10}\right)V_1 = 0$$

which are rewritten as

$$1.1V_1 - 0.5V_2 - 0.1V_3 = 0$$
$$V_2 - 0.5V_1 - 0.5V_3 = 3$$
$$0.85V_3 - 0.5V_2 - 0.1V_1 = 0$$

For determinants,

$$\underset{c}{1.1}V_1 - \underset{b}{0.5}V_2 - \underset{a}{0.1}V_3 = 0$$

$$\underset{b}{-0.5}V_1 + \underset{}{1}V_2 - \underset{}{0.5}V_3 = 3$$

$$\underset{a}{-0.1}V_1 - \underset{}{0.5}V_2 + \underset{}{0.85}V_3 = 0$$

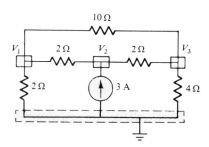

FIG. 7.39

Before continuing, note the symmetry about the major diagonal in the equation above. Recall a similar result for mesh analysis. Examples 7.21 and 7.22 also exhibit this property in the resulting equations. Keep this thought in mind as a check on future applications of nodal analysis.

$$V_3 = V_{4\Omega} = \frac{\begin{vmatrix} 1.1 & -0.5 & 0 \\ -0.5 & +1 & 3 \\ -0.1 & -0.5 & 0 \end{vmatrix}}{\begin{vmatrix} 1.1 & -0.5 & -0.1 \\ -0.5 & +1 & -0.5 \\ -0.1 & -0.5 & +0.85 \end{vmatrix}} = \textbf{4.645 V}$$

Another example with only one source involves ladder network.

EXAMPLE 7.24. Write the nodal equations and find the voltage across the 2-Ω resistor for the network of Fig. 7.40.

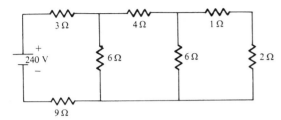

FIG. 7.40

Solution: The nodal voltages are chosen as shown in Fig. 7.41.

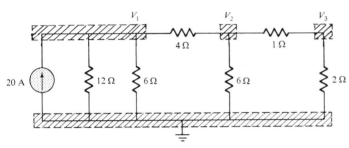

FIG. 7.41

$$V_1: \left(\frac{1}{12} + \frac{1}{6} + \frac{1}{4}\right)V_1 - \left(\frac{1}{4}\right)V_2 + \quad 0 = 20$$

$$V_2: \left(\frac{1}{4} + \frac{1}{6} + \frac{1}{1}\right)V_2 - \left(\frac{1}{4}\right)V_1 - \left(\frac{1}{1}\right)V_3 = 0$$

$$V_3: \left(\frac{1}{1} + \frac{1}{2}\right)V_3 - \left(\frac{1}{1}\right)V_2 + \quad 0 = 0$$

and

$$0.5V_1 - 0.25V_2 + \quad 0 = 20$$

$$-0.25V_1 + \frac{17}{12}V_2 - 1V_3 = 0$$

$$0 \quad - \quad 1V_2 + 1.5V_3 = 0$$

Note the symmetry present about the major axis. Application of determinants reveals that

$$V_3 = V_{2\Omega} = \mathbf{10.667\ V}$$

Another example of general interest is the bridge network. This will be set aside for the section on bridge networks to follow in this chapter.

There are various situations in which it may appear impossible to apply nodal analysis. However, by eliminating components or introducing new components, we can often resolve this problem. For example, in the network of Fig.

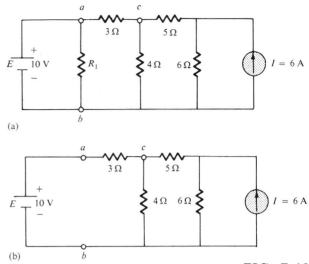

FIG. 7.42

7.42(a), the voltage source cannot be converted to a current source since it does not have a resistance in series with it.

Note that the voltage V_{ab} across the resistor R_1 is always the source voltage $E = 10$ V, no matter what value the resistance R_1 may be. The resistor R_1, therefore, does not affect the voltage V_{ab}, V_{ac}, or any other voltage within the network. Since we are solving only for voltage levels when we apply nodal analysis, the resistor R_1 can be removed without affecting our solution. Once the nodal analysis is complete (all nodal voltages known), the resistor R_1 must be considered if the current through it or through the battery is desired. After the resistor R_1 is removed, the circuit is as shown in the diagram of Fig. 7.42(b), and we can readily arrive at a solution by nodal analysis after we convert the source.

7.12 BRIDGE NETWORKS

This section will introduce the bridge network, a configuration that has a multitude of applications. In the chapters to follow, it will be employed in both dc and ac meters. In the electronics courses it will be encountered early in the discussion of rectifying circuits employed in converting a varying signal to one of a steady nature (such as dc). There are a number of other areas of application that require some knowledge of ac networks which will be discussed later.

The bridge network may appear in one of three forms as indicated in Fig. 7.43. The network of Fig. 7.43(c) is also called a symmetrical lattice network if $R_2 = R_3$ and $R_1 = R_4$. Figure 7.43(c) is an excellent example of how a planar network can be made to appear nonplanar. For the pur-

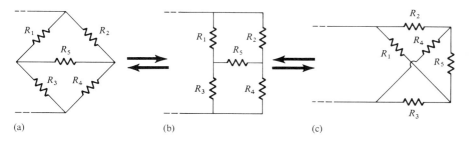

FIG. 7.43 *Bridge network.*

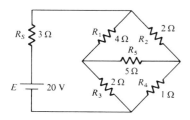

FIG. 7.44

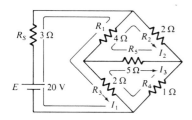

FIG. 7.45

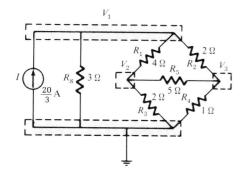

FIG. 7.46

poses of investigation, let us examine the network of Fig. 7.44 using mesh and nodal analysis.

Mesh analysis (Fig. 7.45) yields

$$(3 + 4 + 2)I_1 - (4)I_2 - (2)I_3 = 20$$
$$(4 + 5 + 2)I_2 - (4)I_1 - (5)I_3 = 0$$
$$(2 + 5 + 1)I_3 - (2)I_1 - (5)I_2 = 0$$

and

$$9I_1 - 4I_2 - 2I_3 = 20$$
$$-4I_1 + 11I_2 - 5I_3 = 0$$
$$-2I_1 - 5I_2 + 8I_3 = 0$$

with the result that

$$I_1 = \textbf{4 A}$$
$$I_2 = 2\tfrac{2}{3}\,\textbf{A}$$
$$I_3 = 2\tfrac{2}{3}\,\textbf{A}$$

The net current through the 5-Ω resistor is

$$I_{5\Omega} = I_2 - I_3 = 2\tfrac{2}{3} - 2\tfrac{2}{3} = \textbf{0 A}$$

Nodal analysis (Fig. 7.46) yields

$$\left(\frac{1}{3} + \frac{1}{4} + \frac{1}{2}\right)V_1 - \left(\frac{1}{4}\right)V_2 - \left(\frac{1}{2}\right)V_3 = \frac{20}{3}$$

$$\left(\frac{1}{4} + \frac{1}{2} + \frac{1}{5}\right)V_2 - \left(\frac{1}{4}\right)V_1 - \left(\frac{1}{5}\right)V_3 = 0$$

$$\left(\frac{1}{5} + \frac{1}{2} + \frac{1}{1}\right)V_3 - \left(\frac{1}{2}\right)V_1 - \left(\frac{1}{5}\right)V_2 = 0$$

and

$$\left(\frac{1}{3} + \frac{1}{4} + \frac{1}{2}\right)V_1 - \left(\frac{1}{4}\right)V_2 - \left(\frac{1}{2}\right)V_3 = \frac{20}{3}$$

$$-\left(\frac{1}{4}\right)V_1 + \left(\frac{1}{4} + \frac{1}{2} + \frac{1}{5}\right)V_2 - \left(\frac{1}{5}\right)V_3 = 0$$

$$-\left(\frac{1}{2}\right)V_1 - \left(\frac{1}{5}\right)V_2 + \left(\frac{1}{5} + \frac{1}{2} + \frac{1}{1}\right)V_3 = 0$$

Note the symmetry of the solution. The results are

$$V_1 = \mathbf{8\,V}$$
$$V_2 = \mathbf{2\tfrac{2}{3}\,V}$$
$$V_3 = \mathbf{2\tfrac{2}{3}\,V}$$

and the voltage across the 5-Ω resistor is

$$V_{5\Omega} = V_2 - V_3 = 2\tfrac{2}{3} - 2\tfrac{2}{3} = \mathbf{0\,V}$$

Since $V_{5\Omega} = 0$ V, we can insert a short in place of the bridge arm without affecting the network behavior. (Certainly $V = IR = I \cdot 0 = 0$ V.) In Fig. 7.47, a short circuit has replaced the resistor R_5, and the voltage across R_4 is to be determined. The network is redrawn in Fig. 7.48, and

$$V_{1\Omega} = \frac{(2 \parallel 1)20}{(2 \parallel 1) + (4 \parallel 2) + 3} \qquad \text{(voltage divider rule)}$$

$$= \frac{\dfrac{2}{3}(20)}{\dfrac{2}{3} + \dfrac{8}{6} + 3} = \frac{\dfrac{2}{3}(20)}{\dfrac{2}{3} + \dfrac{4}{3} + \dfrac{9}{3}}$$

$$= \frac{2(20)}{2 + 4 + 9} = \frac{40}{15} = \mathbf{2\tfrac{2}{3}\,V}$$

as obtained earlier.

We found through mesh analysis that $I_{5\Omega} = 0$ A, which has as its equivalent an open circuit as shown in Fig. 7.49. (Certainly $I = V/R = \dfrac{0}{\infty} = 0$ A.) The voltage across the resistor R_4 will again be determined and compared with the result above.

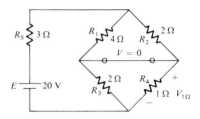

FIG. 7.47

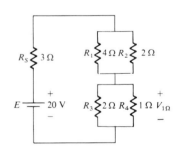

FIG. 7.48

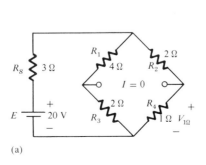

(a)

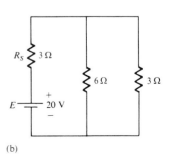

(b)

FIG. 7.49

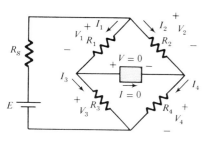

FIG. 7.50

The network is redrawn after combining series elements as shown in Fig. 7.49(b), and

$$V_{3\Omega} = \frac{(6 \parallel 3)(20)}{6 \parallel 3 + 3} = \frac{2(20)}{2 + 3} = 8 \text{ V}$$

and

$$V_{1\Omega} = \frac{1(8)}{1 + 2} = \frac{8}{3} = 2\tfrac{2}{3} \text{ V} \quad \text{as above}$$

The condition $V_{5\Omega} = 0$ V or $I_{5\Omega} = 0$ A exists only for a particular relationship between the resistors of the network. Let us now derive this relationship using the network of Fig. 7.50, in which it is indicated that $I = 0$ A and $V = 0$ V. Note that resistor R_S of the network of Fig. 7.49 will not appear in the following analysis.

The bridge network is said to be *balanced* when the condition of $I = 0$ A or $V = 0$ V exists. One application of this condition will be introduced in Chapter 12 in the discussion of dc instruments.

If $V = 0$ V (short circuit between a and b), then

$$V_1 = V_2$$

and

$$I_1 R_1 = I_2 R_2$$

or

$$I_1 = \frac{I_2 R_2}{R_1}$$

In addition, when $V = 0$ V,

$$V_3 = V_4$$

and

$$I_3 R_3 = I_4 R_4$$

If we set $I = 0$ A, then $I_3 = I_1$ and $I_4 = I_2$ with the result that the above equation becomes

$$I_1 R_3 = I_2 R_4$$

Substituting for I_1 from above yields

$$\left(\frac{I_2 R_2}{R_1}\right) R_3 = I_2 R_4$$

or, rearranging, we have

$$\boxed{\frac{R_1}{R_3} = \frac{R_2}{R_4}} \tag{7.7}$$

This conclusion states that if the ratio of R_1 to R_3 is equal to that of R_2 to R_4, the bridge will be balanced, and $I =$

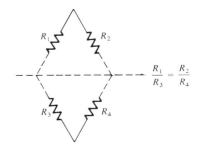

FIG. 7.51

0 A or $V = 0$ V. A method of memorizing this form is indicated in Fig. 7.51.

For the example above, $R_1 = 4\ \Omega$, $R_2 = 2\ \Omega$, $R_3 = 2\ \Omega$, $R_4 = 1\ \Omega$, and

$$\frac{R_1}{R_3} = \frac{R_2}{R_4} \Longrightarrow \frac{4}{2} = \frac{2}{1}$$

The emphasis in this section has been on the balanced situation. Understand that if the ratio is not satisfied, there will be a potential drop across the balance arm and a current through it. The methods just described (mesh and nodal analysis) will yield any and all potentials or currents desired, just as they were applied for the balanced situation.

7.13 Y-Δ (T-π) AND
Δ-Y (π-T) CONVERSIONS

Circuit configurations are often encountered in which the resistors do not appear to be in series or parallel. Under these conditions, it may be necessary to convert the circuit from one form to another in order to solve for any unknown quantities if mesh or nodal analysis is not applied. Two circuit configurations that often account for these difficulties are the wye (Y) and delta (Δ), depicted in Fig. 7.52(a). They are also referred to as the tee (T) and pi (π), respectively, as indicated in Fig. 7.52(b).

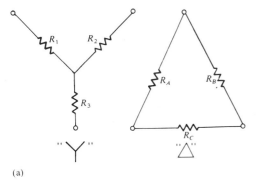

(a)

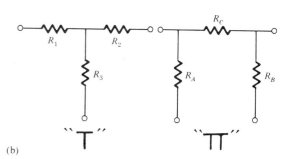

(b)

FIG. 7.52

The purpose of this section is to develop the equations for converting from Δ to Y or vice versa. This type of conversion will normally lead to a network that can be solved using techniques such as described in Chapter 6. In other words, in Fig. 7.52, with terminals *a, b,* and *c* held fast, if the wye (Y) configuration were desired *instead of* the delta (Δ) configuration, all that would be necessary is a direct application of the equations to be derived. The phrase *instead of* is emphasized to insure that it is understood that

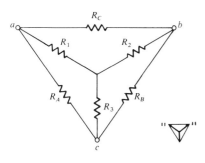

FIG. 7.53

only one of these configurations is to appear at one time between the indicated terminals.

It is our purpose (referring to Fig. 7.53) to find some expression for R_1, R_2, and R_3 in terms of R_A, R_B, and R_C and vice versa. If the two circuits are to be equivalent, the total resistance between any two terminals must be the same. Consider terminals a-c in the Δ-Y configurations of Fig. 7.54.

Let us first assume that we want to convert the Δ (R_A, R_B, R_C) to the Y (R_1, R_2, R_3). This requires that we have a relationship for R_1, R_2, and R_3 in terms of R_A, R_B, and R_C.

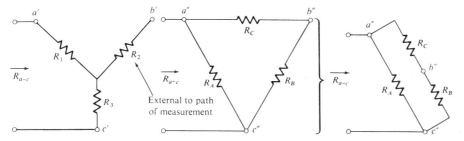

FIG. 7.54

If the resistance is to be the same between terminals a-c for both the Δ and the Y, the following must be true:

$$R_{a\text{-}c} = R_{a'\text{-}c'} = R_{a''\text{-}c''}$$

so that

$$\boxed{R_1 + R_3 = \frac{R_A(R_B + R_C)}{R_A + (R_B + R_C)}} \qquad \textbf{(7.8a)}$$

Using the same approach for a-b and b-c, we obtain the following relationships:

$$\boxed{R_{a\text{-}b} = R_1 + R_2 = \frac{R_C(R_A + R_B)}{R_C + (R_A + R_B)}} \qquad \textbf{(7.8b)}$$

and

$$\boxed{R_{b\text{-}c} = R_2 + R_3 = \frac{R_B(R_A + R_C)}{R_B + (R_A + R_C)}} \qquad \textbf{(7.8c)}$$

Subtracting Eq. (7.8a) from Eq. (7.8b), we have

$$(R_1 + R_2) - (R_1 + R_3) = \left(\frac{R_C R_A + R_C R_B}{R_A + R_B + R_C}\right) - \left(\frac{R_A R_B + R_A R_C}{R_A + R_B + R_C}\right)$$

so that

$$\boxed{R_2 - R_3 = \frac{R_B R_C - R_A R_B}{R_A + R_B + R_C}} \qquad \textbf{(7.8d)}$$

Subtracting Eq. (7.8d) from Eq. (7.8c) yields

$$(R_2 + R_3) - (R_2 - R_3) = \left(\frac{R_B R_A + R_B R_C}{R_A + R_B + R_C}\right)$$
$$- \left(\frac{R_B R_C - R_A R_B}{R_A + R_B + R_C}\right)$$

so that

$$2R_3 = \frac{2R_A R_B}{R_A + R_B + R_C}$$

resulting in the following expression for R_3 in terms of R_A, R_B, and R_C:

$$R_3 = \frac{R_A R_B}{R_A + R_B + R_C} \qquad \textbf{(7.9a)}$$

Following the same procedure for R_1 and R_2, we have

$$R_1 = \frac{R_A R_C}{R_A + R_B + R_C} \qquad \textbf{(7.9b)}$$

$$R_2 = \frac{R_B R_C}{R_A + R_B + R_C} \qquad \textbf{(7.9c)}$$

Note that each resistor of the Y is equal to the product of the resistors in the two closest branches of the Δ divided by the sum of the resistors in the Δ.

To obtain the relationships necessary to convert from a Y to a Δ, first divide Eq. (7.9a) by Eq. (7.9b):

$$\frac{R_3}{R_1} = \frac{(R_A R_B)/(R_A + R_B + R_C)}{(R_A R_C)/(R_A + R_B + R_C)} = \frac{R_B}{R_C}$$

or

$$R_B = \frac{R_C R_3}{R_1}$$

Then divide Eq. (7.9a) by Eq. (7.9c):

$$\frac{R_3}{R_2} = \frac{(R_A R_B)/(R_A + R_B + R_C)}{(R_B R_C)/(R_A + R_B + R_C)} = \frac{R_A}{R_C}$$

or

$$R_A = \frac{R_3 R_C}{R_2}$$

Substituting for R_A and R_B in Eq. (7.9c) yields

$$R_2 = \frac{(R_C R_3/R_1)R_C}{(R_3 R_C/R_2) + (R_C R_3/R_1) + R_C}$$

$$= \frac{(R_3/R_1)R_C}{(R_3/R_2) + (R_3/R_1) + 1}$$

Placing these over a common denominator, we obtain

$$R_2 = \frac{(R_3 R_C / R_1)}{(R_1 R_2 + R_1 R_3 + R_2 R_3)/(R_1 R_2)}$$

$$= \frac{R_2 R_3 R_C}{R_1 R_2 + R_1 R_3 + R_2 R_3}$$

and

$$R_C = \frac{R_1 R_2 + R_1 R_3 + R_2 R_3}{R_3} \qquad \textbf{(7.10a)}$$

We follow the same procedure for R_B and R_A:

$$R_B = \frac{R_1 R_2 + R_1 R_3 + R_2 R_3}{R_1} \qquad \textbf{(7.10b)}$$

and

$$R_A = \frac{R_1 R_2 + R_1 R_3 + R_2 R_3}{R_2} \qquad \textbf{(7.10c)}$$

Note that the value of each resistor of the Δ is equal to the sum of the possible product combinations of the resistances of the Y divided by the resistance of the Y farthest from the resistor to be determined.

Let us consider what would occur if all the values of a Δ or Y were the same. If $R_A = R_B = R_C$, Eq. (7.9a) would become (using R_A only)

$$R_3 = \frac{R_A R_B}{R_A + R_B + R_C} = \frac{R_A R_A}{R_A + R_A + R_A} = \frac{R_A^2}{3 R_A} = \frac{R_A}{3}$$

and, following the same procedure,

$$R_1 = \frac{R_A}{3} \qquad R_2 = \frac{R_A}{3}$$

In general, therefore,

$$R_Y = \frac{R_\Delta}{3} \qquad \textbf{(7.11a)}$$

or

$$R_\Delta = 3 R_Y \qquad \textbf{(7.11b)}$$

which indicates that *for a Y of three equal resistors, the value of each resistor of the Δ is equal to three times the value of any resistor of the Y.* If only two elements of a Y or a Δ are the same, the corresponding Δ or Y of each will also have two equal elements. The converting of equations will be left as an exercise for the reader.

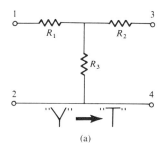

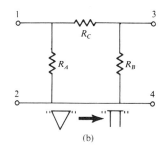

"Y" → "T"

(a)

"∇" → "∏"

(b)

FIG. 7.55

The Y and the Δ will often appear as shown in Fig. 7.55. They are then referred to as a *tee* (T) and *pi* (π) network. The equations used to convert from one form to the other are exactly the same as those developed for the Y and Δ transformation.

EXAMPLE 7.25. Convert the Δ of Fig. 7.56 to a Y.

Solution:

$$R_1 = \frac{R_A R_C}{R_A + R_B + R_C} = \frac{(20)(10)}{20 + 30 + 10} = \frac{200}{60} = 3\tfrac{1}{3}\ \Omega$$

$$R_2 = \frac{R_B R_C}{R_A + R_B + R_C} = \frac{(30)(10)}{60} = \frac{300}{60} = 5\ \Omega$$

$$R_3 = \frac{R_A R_B}{R_A + R_B + R_C} = \frac{(20)(30)}{60} = \frac{600}{60} = \mathbf{10\ \Omega}$$

The equivalent network is shown in Fig. 7.57.

EXAMPLE 7.26. Convert the Y of Fig. 7.58 to a Δ.

Solution:

$$R_A = \frac{R_1 R_2 + R_1 R_3 + R_2 R_3}{R_2}$$

$$= \frac{(60)(60) + (60)(60) + (60)(60)}{60}$$

$$= \frac{3600 + 3600 + 3600}{60} = \frac{10{,}800}{60}$$

$$R_A = \mathbf{180\ \Omega}$$

However, the three resistors for the Y are equal, permitting the use of Eq. (7.11) and yielding

$$R_\Delta = 3R_Y = 3(60) = 180\ \Omega$$

and

$$R_B = R_C = \mathbf{180\ \Omega}$$

The equivalent network is shown in Fig. 7.59.

EXAMPLE 7.27. Find the total resistance of the network of Fig. 7.60, where $R_A = 3\ \Omega$, $R_B = 3\ \Omega$, and $R_C = 6\ \Omega$.

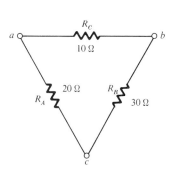

FIG. 7.56

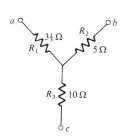

FIG. 7.57

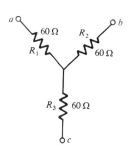

FIG. 7.58

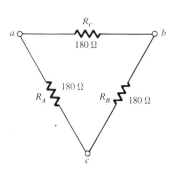

FIG. 7.59

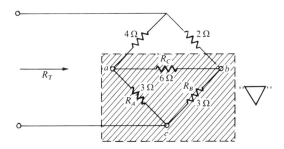

FIG. 7.60

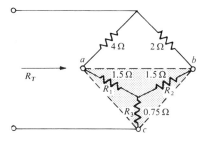

FIG. 7.61

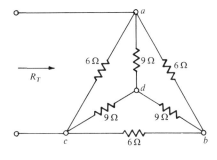

FIG. 7.62

Solution:

Two resistors of the Δ were equal; therefore, two resistors of the Y will be equal.

$$R_1 = \frac{R_A R_C}{R_A + R_B + R_C} = \frac{(3)(6)}{3 + 3 + 6} = \frac{18}{12} = \mathbf{1.5\ \Omega}$$

$$R_2 = \frac{R_B R_C}{R_A + R_B + R_C} = \frac{(3)(6)}{12} = \frac{18}{12} = \mathbf{1.5\ \Omega}$$

$$R_3 = \frac{R_A R_B}{R_A + R_B + R_C} = \frac{(3)(3)}{12} = \frac{9}{12} = \mathbf{0.75\ \Omega}$$

Replacing the Δ by the Y, as shown in Fig. 7.61, yields

$$R_T = 0.75 + \frac{(4 + 1.5)(2 + 1.5)}{(4 + 1.5) + (2 + 1.5)}$$

$$= 0.75 + \frac{(5.5)(3.5)}{5.5 + 3.5}$$

$$= 0.75 + \mathbf{2.139}$$

$$R_T = \mathbf{2.889\ \Omega}$$

EXAMPLE 7.28. Find the total resistance of the network of Fig. 7.62.

Solution: Since all the resistors of the Δ or Y are the same, Eqs. (7.11a) and (7.11b) can be used to convert either form to the other.

a. *Converting the Δ to a Y.* Note: When this is done, the resulting d' of the new Y will be the same as the point d shown in the original figure, only because both systems are "balanced." That is, the resistance in each branch of each system has the same value:

$$R_Y = \frac{R_\Delta}{3} = \frac{6}{3} = 2\ \Omega \qquad \text{(Fig. 7.63)}$$

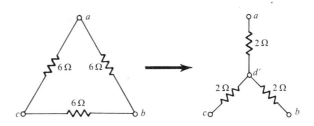

FIG. 7.63

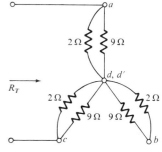

FIG. 7.64

The original circuit appears as shown in Fig. 7.64.

$$R_T = 2\left[\frac{(2)(9)}{2 + 9}\right] = \mathbf{3.2727\ \Omega}$$

b. *Converting the* Y *to a* Δ.

$$R_\Delta = 3R_Y$$
$$R_\Delta = (3)(9) = 27\ \Omega \qquad \text{(Fig. 7.65)}$$

$$R'_T = \frac{(6)(27)}{6 + 27} = \frac{162}{33} = 4.9091\ \Omega$$

$$R_T = \frac{R'_T(R'_T + R'_T)}{R'_T + (R'_T + R'_T)} = \frac{R'_T 2R'_T}{3R'_T} = \frac{2R'_T}{3} = \frac{2(4.9091)}{3}$$

and $R_T = \mathbf{3.2727\ \Omega}$, which checks with the previous solution.

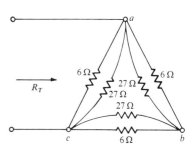

FIG. 7.65

PROBLEMS

Section 7.2

1. Find the voltage V_{ab} (with polarity) for the circuit of Fig. 7.66.

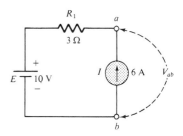

FIG. 7.66

2. For the network of Fig. 7.67:
  a. Find the voltages V_S and V_4.
  b. Find the current I_2.

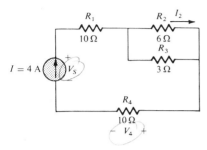

FIG. 7.67

*3. For the configuration of Fig. 7.68:
  a. Find the voltage V_{ab} and polarity of points a and b.
  b. Find the voltage V_{cd} and polarity of points c and d.

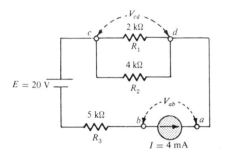

FIG. 7.68

4. Find the voltage V_3 and the current I_2 for the network of Fig. 7.69.

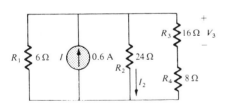

FIG. 7.69

Section 7.3

5. Convert the voltage sources of Fig. 7.70 to current sources.

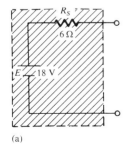

(a)

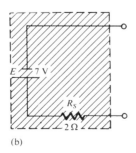

(b)

FIG. 7.70

6. Convert the current sources of Fig. 7.71 to voltage sources.

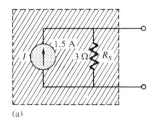

(a)

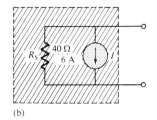

(b)

FIG. 7.71

7. For the network of Fig. 7.72:
   a. Find the current through the 2-Ω resistor.
   b. Convert the current source and 4-Ω resistor to a voltage source, and again solve for the current in the 2-Ω resistor. Compare the results.

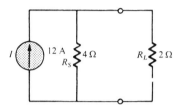

FIG. 7.72

8. For the configuration of Fig. 7.73:
   a. Convert the current source and 6-Ω resistor to a voltage source.
   b. Find the magnitude and direction of the current I_1.
   c. Find the voltage V_{ab} and the polarity of points a and b.

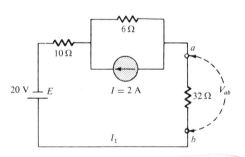

Section 7.4

FIG. 7.73

9. Find the voltage V_2 and the current I_1 for the network of Fig. 7.74.

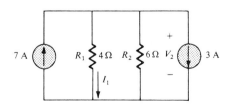

FIG. 7.74

10. **a.** Convert the voltage sources of Fig. 7.75 to current sources.
  b. Find the voltage V_{ab} and the polarity of points a and b.
  c. Find the magnitude and direction of the current I.

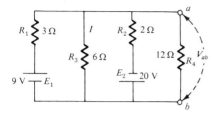

FIG. 7.75

*11. For the network of Fig. 7.76:
  a. Find the currents I_2 and I_5.
  b. Find the voltage V_1.
  c. Find the power delivered to R_2.

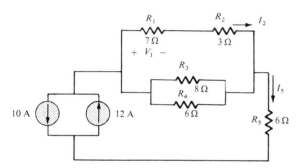

FIG. 7.76

Section 7.6

12. Evaluate the following determinants:

  a. $\begin{vmatrix} 2 & 4 \\ 1 & 2 \end{vmatrix}$ **b.** $\begin{vmatrix} 3 & 4 \\ 0 & -4 \end{vmatrix}$

  c. $\begin{vmatrix} 4 & 0 \\ 0 & 7 \end{vmatrix}$ **d.** $\begin{vmatrix} 6 & -9 \\ 5 & 0 \end{vmatrix}$

13. Evaluate the following determinants:

  a. $\begin{vmatrix} 3 & -4 & 0 \\ 2 & 7 & 0 \\ -1 & 0 & 1 \end{vmatrix}$ **b.** $\begin{vmatrix} 1 & 2 & 3 \\ 9 & 0 & 0 \\ 5 & 0 & -1 \end{vmatrix}$

14. Using determinants, solve for x and y.
  a. $4x + 2y = 8$
  $1x + 3y = 6$
  b. $3x - 6y = 1$
  $4x - 6y = 2$

  c. $3x + 7y = 0$
  $5y - 4 = 15x$
  d. $5x + 6y - 2x = -8$
  $5 + y - 6x = -4x$

*15. Solve for x, y, and z by determinants:
  a. $2x + 1y + 0 = 6$
  $0 + 4y + 3z = 12$
  $8x + 0 + 8z = 18$
  b. $1x + 0y + 2z = 1$
  $2x + 3y - 4z = 0$
  $-2x + 0y - 6z = 3$

Section 7.7

16. Using branch-current analysis, for each of the networks of Fig. 7.77, find the current through each resistor and the voltage V_{ab} with the polarity of points a and b.

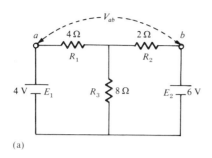

(a)

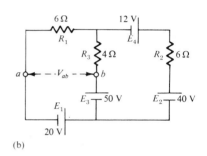

(b)

FIG. 7.77

17. Repeat Problem 16 for the networks of Fig. 7.78.

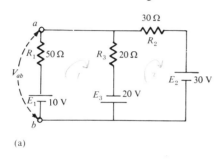

(a)

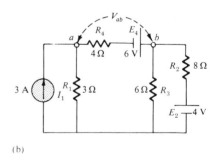

(b)

FIG. 7.78

*18. Repeat Problem 16 for the networks of Fig. 7.79.

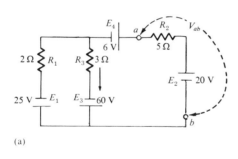

(a)

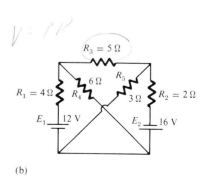

(b)

Section 7.8

FIG. 7.79

19. Repeat Problem 16 using mesh analysis.

20. Repeat Problem 17 using mesh analysis.

*21. Repeat Problem 18 using mesh analysis.

*22. Using mesh analysis, for each network of Fig. 7.80, determine the current through the 5-Ω resistor.

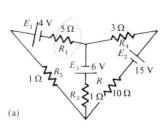

(a)

(b)

FIG. 7.80

*23. Write the mesh equations for each of the networks of Fig. 7.81, and, using determinants, solve for the loop currents in each circuit.

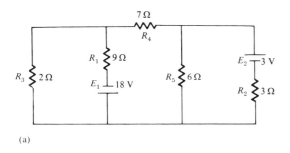

(a)

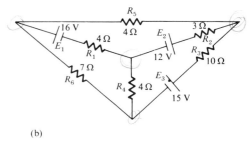

(b)

FIG. 7.81

*24. Repeat Problem 23 for the networks of Fig. 7.82.

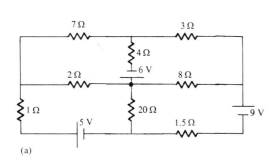

(a)

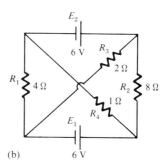

(b)

FIG. 7.82

Section 7.9

25. Using the format approach, write the mesh equations for the networks of Fig. 7.77. Is symmetry present? Using determinants, solve for the mesh currents.

26. Repeat Problem 25 for the networks of Fig. 7.78.

*27. Repeat Problem 25 for the networks of Fig. 7.79.

*28. Repeat Problem 25 for the networks of Fig. 7.80.

*29. Repeat Problem 25 for the networks of Fig. 7.81.

*30. Repeat Problem 25 for the networks of Fig. 7.82.

Section 7.11

31. Using the format approach, write the nodal equations for the networks of Fig. 7.83, and, using determinants, solve for the nodal voltages. Is symmetry present?

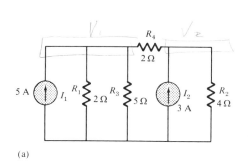

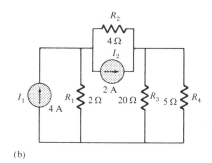

(a)

(b)

FIG. 7.83

*32. Repeat Problem 31 for the networks of Fig. 7.84.

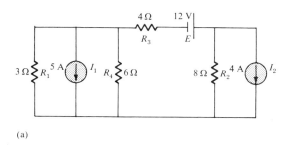

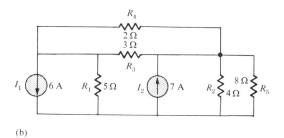

(a)

(b)

33. Repeat Problem 31 for the networks of Fig. 7.81.

FIG. 7.84

*34. For the networks of Fig. 7.85, using the format approach, write the nodal equations and solve for the nodal voltages.

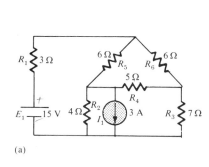

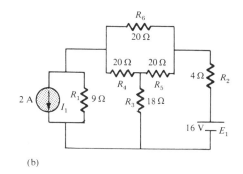

(a)

(b)

*35. Repeat Problem 34 for the networks of Fig. 7.86.

FIG. 7.85

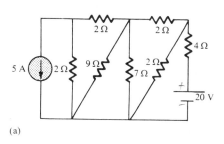

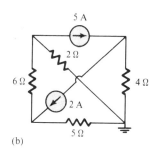

(a)

(b)

FIG. 7.86

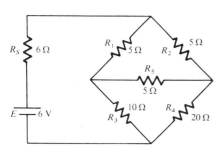

FIG. 7.87

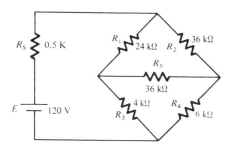

FIG. 7.88

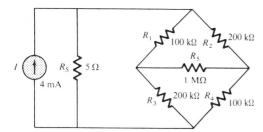

FIG. 7.89

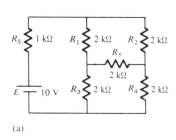

(a)

FIG. 7.90

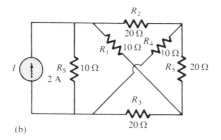

(b)

Section 7.12

*36. For the bridge network of Fig. 7.87:
  a. Using the format approach, write the mesh equations.
  b. Determine the current through R_5.
  c. Is the bridge balanced?
  d. Is Eq. (7.7) satisfied?

*37. For the network of Fig. 7.87:
  a. Using the format approach, write the nodal equations.
  b. Determine the voltage across R_5.
  c. Is the bridge balanced?
  d. Is Eq. (7.7) satisfied?

38. Repeat Problem 36 for the network of Fig. 7.88.

39. Repeat Problem 37 for the network of Fig. 7.88.

40. Write the mesh and nodal equations for the bridge configuration of Fig. 7.89. Use the format approach.

*41. a. Repeat Problem 40 for the networks of Fig. 7.90.
  b. Determine the current through the dc supply of each network.

Section 7.13

42. Using a Δ-Y or Y-Δ conversion, find the current I in each of the networks of Fig. 7.91.

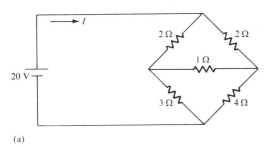

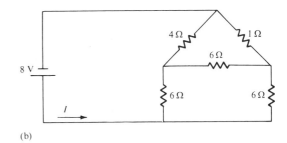

43. Repeat Problem 42 for the networks of Fig. 7.92.

FIG. 7.91

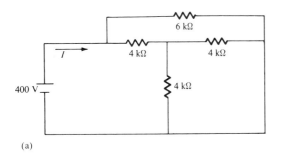

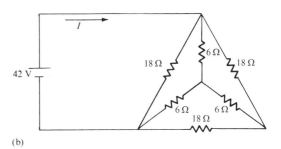

44. Determine the current *I* for the network of Fig. 7.93.

FIG. 7.92

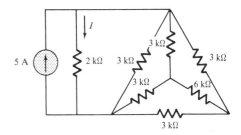

FIG. 7.93

GLOSSARY

Branch-current method A technique for determining the branch currents of a multiloop network.

Bridge network A network configuration typically having a diamond appearance in which no two elements are in series or parallel.

Current sources Sources that supply a fixed current to a network and have a terminal voltage dependent on the network to which they are applied.

Delta (Δ), pi (π) configuration A network structure that consists of three branches and has the appearance of the Greek letter delta (Δ) or pi (π).

Determinants method A mathematical technique for finding the unknown variables of two or more simultaneous linear equations.

Mesh analysis A technique for determining the mesh (loop) currents of a network that results in a reduced set of equations compared to the branch-current method.

Mesh (loop) current A labeled current assigned to each distinct closed loop of a network that can individually or, in combination with other mesh currents, define all of the branch currents of a network.

Nodal analysis A technique for determining the node voltages of a network.

Node A junction of two or more branches in a network.

Wye (Y) tee (T) configuration A network structure that consists of three branches and has the appearance of the capital letter Y or T.

8.1 INTRODUCTION

This chapter will introduce the important fundamental theorems of network analysis. Included are the *superposition, Thevenin's, Norton's, maximum power transfer, substitution, Millman's,* and *reciprocity* theorems. We will consider a number of areas of application for each. A thorough understanding of each theorem is important because each will be applied repeatedly in later courses.

8.2 SUPERPOSITION THEOREM

The *superposition theorem,* like the methods of the previous chapter, can be used to find the solution to networks with two or more sources that are not in series or parallel. The most obvious advantage of this method for networks of this type is that it does not require the use of a mathematical technique such as determinants to find the required voltages or currents. Instead, each source is treated independently, and the algebraic sum is found to determine a particular unknown quantity of the network. In other words, for a network with *n* sources, *n* independent series-parallel networks would have to be considered before a solution could be obtained.

The theorem states the following:

The current through, or voltage across, any element in a linear bilateral network is equal to the algebraic sum of the currents or voltages produced independently by each source.

To consider the effects of each source independently requires that sources be removed and replaced without affecting the final result. To remove a voltage source when applying this theorem, the difference in potential between the terminals of the voltage source must be set to zero (short-circuited); removing a current source requires that its terminals be opened (open circuit). Any internal resistance or conductance associated with the displaced sources is not eliminated but must still be considered.

The total current through any portion of the network is equal to the algebraic sum of the currents produced independently by each source. That is, for a two-source network, if the current produced by one source is in one direction, while that produced by the other is in the opposite direction through the same resistor, *the resulting current is the difference of the two and has the direction of the larger.* If the individual currents are in the same direction, *the resulting current is the sum of two in the direction of either current.* This rule holds true for the voltage across a portion of a network as determined by polarities, and it can be extended to networks with any number of sources.

The superposition principle is not applicable to power effects since the power loss in a resistor varies as the square (nonlinear) of the current or voltage. For this reason, the power to an element cannot be determined until the total current through (or voltage across) the element has been determined by superposition. This will be demonstrated in Example 8.4.

EXAMPLE 8.1. Determine V_1 for the network of Fig. 8.1 through superposition.

Solution: For the voltage source set to zero volts (short circuit), the circuit of Fig. 8.2(a) will result, and

$$V'_1 = I_1 R_1 = I R_1 = (2)(15) = 30 \text{ V}$$

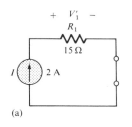

FIG. 8.1

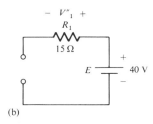

(a)   (b)

FIG. 8.2

The network of Fig. 8.2(b) will result for the current source set to zero (open circuit), and

$$V''_1 = 0 \text{ V}$$

The algebraic sum of the voltages is

$$V_1 = V'_1 + V''_1 = 30 + 0 = \mathbf{30\ V}$$

Note that the 40-V supply has no effect on V_1 since the current source determined the current through the 15-Ω resistor.

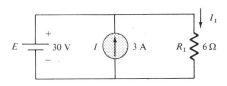

EXAMPLE 8.2. Determine I_1 for the network of Fig. 8.3.

Solution: Setting $E = 0$ for the network of Fig. 8.3 results in the network of Fig. 8.4(a), and

FIG. 8.3

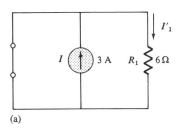 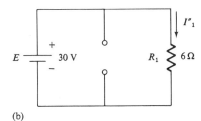

(a)   (b)

FIG. 8.4

$$I'_1 = 0\ \text{A}$$

For $I = 0$ (open circuit), the network of Fig. 8.4(b) will result, and

$$I''_1 = \frac{30}{6} = 5\ \text{A}$$

so that the algebraic sum of the currents results in

$$I_1 = I'_1 + I''_1 = 0 + 5 = \mathbf{5\ A}$$

Note in this case that the current source has no effect on the current through the 6-Ω resistor since the voltage across the resistor is fixed at $E = 30$ V.

EXAMPLE 8.3. Using superposition, determine the current through the 3-Ω resistor of Fig. 8.5. Note that this is a two-source network of the type considered in Chapter 7.

Solution: Considering the effects of a 72-V source (Fig. 8.6),

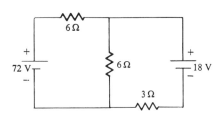

FIG. 8.5

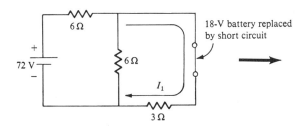

18-V battery replaced by short circuit

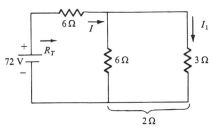

FIG. 8.6

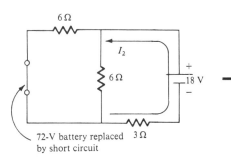

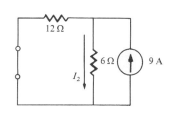

72-V battery replaced by short circuit

FIG. 8.7

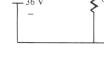

$I_1 = 6$ A

$3\,\Omega$

$I_2 = 3$ A

FIG. 8.8

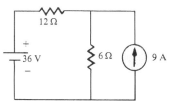

FIG. 8.9

Current source replaced by open circuit

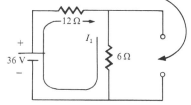

FIG. 8.10

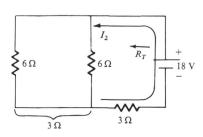

$$R_T = 6 + 2 = 8\,\Omega$$

$$I = \frac{E}{R_T} = \frac{72}{8} = 9 \text{ A}$$

$$I_1 = \frac{(6)(9)}{6 + 3} = 6 \text{ A}$$

Considering the effects of the 18-V source (Fig. 8.7),

$$R_T = 3 + 3 = 6\,\Omega$$

$$I_2 = \frac{E}{R_T} = \frac{18}{6} = 3 \text{ A}$$

The total current through the 3-Ω resistor (Fig. 8.8) is

$$I_{3\Omega} = \mathbf{3\ A} \qquad \text{(direction of } I_1)$$

EXAMPLE 8.4. Using superposition, find the current through the 6-Ω resistor of the network of Fig. 8.9.

Solution: Considering the effect of the 36-V source (Fig. 8.10),

$$I_1 = \frac{E}{R_T} = \frac{36}{12 + 6} = 2 \text{ A}$$

Considering the effect of the 9-A source (Fig. 8.11), by the current divider rule,

$$I_2 = \frac{(12)(9)}{12 + 6} = \frac{108}{18}$$

$$I_2 = 6 \text{ A}$$

The total current through the 6-Ω resistor (Fig. 8.12) is

$$I_{6\Omega} = \mathbf{8\ A}$$

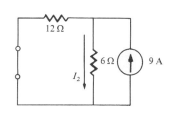

$12\,\Omega$

$6\,\Omega$ 9 A

I_2

FIG. 8.11

The actual power to the 6-Ω resistor is

$$\text{Power} = I^2R = (8^2)(6) = \textbf{384 W}$$

The calculated power to the 6-Ω resistor due to each source *misusing* the principle of superposition is

$$P_1 = I_1^2 R = (2)^2(6) = 24 \text{ W}$$
$$P_2 = I_2^2 R = (6)^2(6) = 216 \text{ W}$$
$$P_1 + P_2 = 240 \text{ W} \neq 384 \text{ W}$$

This results because $2 + 6 = 8$, but

$$(2)^2 + (6)^2 \neq (8)^2$$

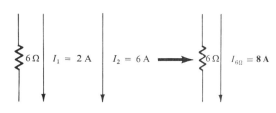

As mentioned previously, the superposition principle is not applicable to power effects, since power is proportional to the square of the current or voltage (I^2R or V^2/R).

Figure 8.13 is a plot of the power delivered to the 6-Ω resistor versus current.

FIG. 8.12

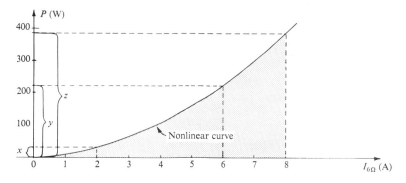

Obviously, $x + y \neq z$, or $24 + 216 \neq 384$, and superposition does not hold. However, for a linear relationship, such as that between the voltage and current of the fixed-type 6-Ω resistor, superposition can be applied, as demonstrated by the graph of Fig. 8.14, where $a + b = c$, or $2 + 6 = 8$.

FIG. 8.13

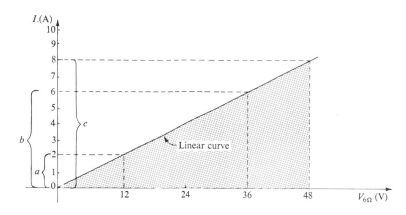

FIG. 8.14

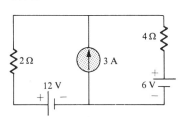

FIG. 8.15

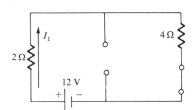

FIG. 8.16

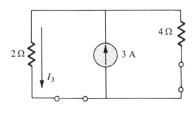

FIG. 8.17

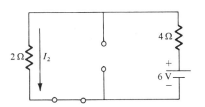

FIG. 8.18

FIG. 8.19

EXAMPLE 8.5. Find the current through the 2-Ω resistor of the network of Fig. 8.15.

Solution: Considering the effect of the 12-V source (Fig. 8.16),

$$I_1 = \frac{12}{2 + 4} = \frac{12}{6} = 2 \text{ A}$$

Considering the effect of the 6-V source (Fig. 8.17),

$$I_2 = \frac{E}{R_T} = \frac{6}{6} = 1 \text{ A}$$

Considering the effect of the 3-A source (Fig. 8.18), by the current divider rule,

$$I_3 = \frac{(4)(3)}{2 + 4} = \frac{12}{6} = 2 \text{ A}$$

The total current in the 2-Ω resistor is given in Fig. 8.19.

8.3 THEVENIN'S THEOREM

This theorem states the following:

*Any two-terminal linear bilateral dc network can be re-
placed by an equivalent circuit consisting of a voltage
source and a series resistor as shown in Fig. 8.20.*

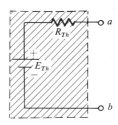

This theorem achieves two important objectives. First, as was true for all the methods previously described, it allows us to find any particular voltage or current in a linear network with one, two, or any other number of sources. Second, it allows us to concentrate on a specific portion of a network by replacing the remaining network by an equivalent circuit. In Fig. 8.21, for example, by finding the Thevenin equivalent circuit for the network in the shaded area, we can quickly calculate the change in current through or voltage across the variable resistor R for the various values that it may assume.

FIG. 8.20 *Thevenin equivalent circuit.*

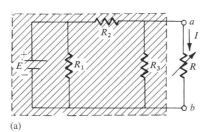

(a)

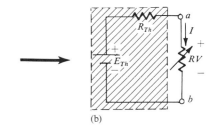

(b)

FIG. 8.21

Before we examine the steps involved in applying this theorem, it is important that an additional word be included here to insure that the implications of the Thevenin equivalent circuit are clear. In Fig. 8.21, the entire network, except R, is to be replaced by a single series resistor and battery as shown in Fig. 8.20. The values of these two elements of the Thevenin equivalent circuit must be chosen to insure that the resistor R will react to the network of Fig. 8.21(a) in the same manner as to the network of Fig. 8.21(b). In other words, the current through or voltage across R must be the same for either network for any value of R. One last note: The Thevenin equivalent circuit has only *one* form consisting of *only* two elements as shown in Fig. 8.20.

Now we present the steps leading to the *proper* value of R_{Th} and E_{Th}:

1. Remove that portion of the network across which the Thevenin equivalent circuit is to be found. In Fig. 8.21(a), this requires that the resistor R be removed from the network.
2. Mark the terminals of the remaining two-terminal network. (The importance of this step will become obvious as we progress through some complex networks.)
3. Calculate R_{Th} by first setting all sources to zero (voltage sources are replaced by short circuits and current sources by open circuits) and then finding the resultant resistance between the two marked terminals. (If the internal resistance of the voltage

and/or current sources is included in the original network, it must remain when the sources are set to zero.)

4. Calculate E_{Th} by first replacing the voltage and current sources and then finding the *open-circuit* voltage between the marked terminals. (This step is invariably the one that will lead to the most confusion and errors. In *all* cases, keep in mind that it is the *open-circuit* potential between the two terminals marked in step 2 above.)

5. Draw the Thevenin equivalent circuit with the portion of circuit previously removed replaced between the terminals of the equivalent circuit. This step is indicated by the placement of the resistor R between the terminals of the Thevenin equivalent circuit as shown in Fig. 8.21(b).

A few examples further demonstrate this very important and useful theorem.

EXAMPLE 8.6. Find the Thevenin equivalent circuit for the network in the shaded area of the network of Fig. 8.22. Then find the current through R_L for values of 2, 10, and 100 Ω.

Solution:
Steps 1 and 2 produce the network of Fig. 8.23.

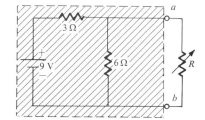

FIG. 8.22

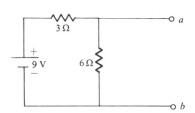

FIG. 8.23

Step 3: Shorting the voltage source (Fig. 8.24) yields

$$R_{Th} = \frac{(3)(6)}{3 + 6} = 2 \ \Omega$$

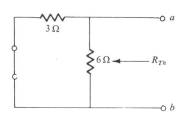

FIG. 8.24

Step 4: Replace the voltage source (Fig. 8.25). For this case only, E_{Th} is the same as the voltage drop across the 6-Ω resistor. By the voltage divider rule,

$$E_{Th} = \frac{(6)(9)}{6 + 3} = \frac{54}{9} = 6 \ \text{V}$$

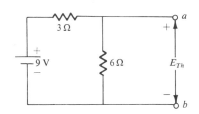

FIG. 8.25

Step 5 (Fig. 8.26):

$$R_L = 2\,\Omega: \quad I = \frac{6}{2 + 2} = \mathbf{1.5\ A}$$

$$R_L = 10\,\Omega: \quad I = \frac{6}{2 + 10} = \mathbf{0.5\ A}$$

$$R_L = 100\,\Omega: \quad I = \frac{6}{2 + 100} = \mathbf{0.059\ A}$$

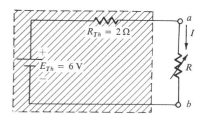

FIG. 8.26

EXAMPLE 8.7. Find the Thevenin equivalent circuit for the network in the shaded area of the network of Fig. 8.27.

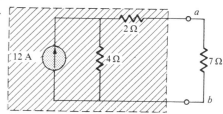

FIG. 8.27

Solution:
Steps 1 and 2 are shown in Fig. 8.28.

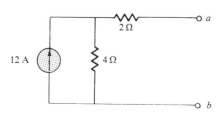

FIG. 8.28

Step 3 is shown in Fig. 8.29.

$$R_{Th} = 4 + 2 = \mathbf{6\ \Omega}$$

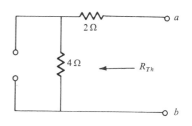

FIG. 8.29

Step 4 (Fig. 8.30): In this case, since there exists an open circuit between the two marked terminals, the current is zero between these terminals and through the 2-Ω resistor. There is, therefore, no voltage drop across the 2-Ω resistor, and E_{Th} is simply the voltage drop across the 4-Ω resistor:

$$E_{Th} = (4)(12) = \mathbf{48\ V}$$

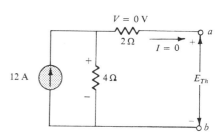

FIG. 8.30

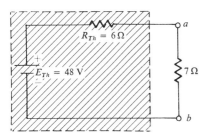

FIG. 8.31

Step 5 is shown in Fig. 8.31.

EXAMPLE 8.8. Find the Thevenin equivalent circuit for the network in the shaded area of the network of Fig. 8.32.

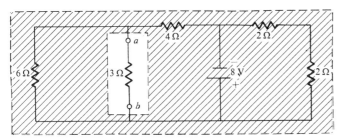

FIG. 8.32

Solution:
Steps 1 and 2: See Fig. 8.33.

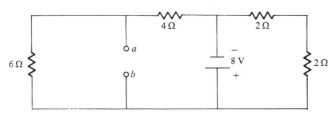

FIG. 8.33

Step 3: See Fig. 8.34.

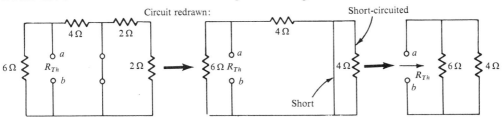

FIG. 8.34

$$R_{Th} = \frac{(6)(4)}{6 + 4} = \frac{24}{10}$$

$$R_{Th} = \mathbf{2.4 \ \Omega}$$

Step 4: See Fig. 8.35. By the voltage divider rule,

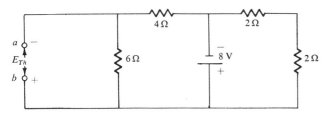

FIG. 8.35

$$E_{Th} = V_{6\Omega} = \frac{(6)(8)}{6+4} = \frac{48}{10} = \textbf{4.8 V}$$

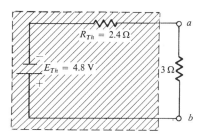

Step 5: See Fig. 8.36.

The importance of marking the terminals should be obvious from Example 8.8. Note that there is no requirement that the Thevenin voltage have the same polarity as the equivalent circuit originally introduced.

EXAMPLE 8.9. Find the Thevenin equivalent circuit for the network in the shaded area of the bridge network of Fig. 8.37.

FIG. 8.36

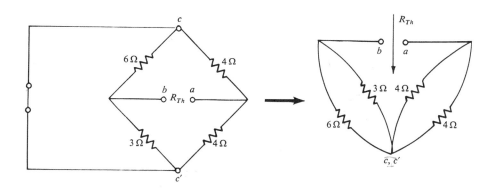

Solution:
Steps 1 and 2 are shown in Fig. 8.38.

FIG. 8.37

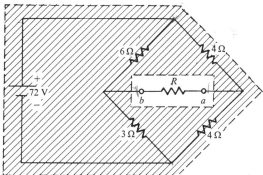

Step 3: See Fig. 8.39.

FIG. 8.38

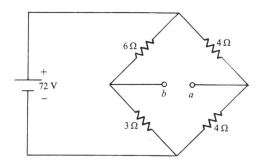

FIG. 8.39

$$R_{Th} = \frac{(3)(6)}{3 + 6} + \frac{4}{2} = 2 + 2$$

$$R_{Th} = \mathbf{4 \ \Omega}$$

Step 4: The circuit is redrawn in Fig. 8.40. By the voltage divider rule,

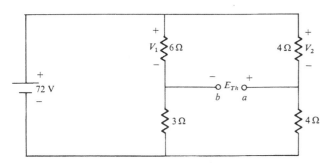

FIG. 8.40

$$V_1 = \frac{(6)(72)}{6 + 3} = \frac{432}{9} = 48 \text{ V}$$

$$V_2 = \frac{(4)(72)}{4 + 4} = \frac{72}{2} = 36 \text{ V}$$

Assuming the polarity shown for E_{Th} and applying Kirchhoff's voltage law, we have

$$\Sigma_{\circlearrowleft} V = +E_{Th} + V_2 - V_1 = 0$$

or

$$E_{Th} = V_1 - V_2$$
$$= 48 - 36$$

and

$$E_{Th} = +12 \text{ V}$$

_{Assumed correct polarity}

Step 5 is shown in Fig. 8.41.

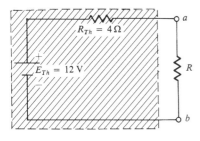

FIG. 8.41

Thevenin's theorem is not restricted to a single passive element, as shown in the preceding examples, but can be applied across sources, whole branches, portions of networks, or any circuit configuration, as shown in the following example. It is also possible that one of the methods previously described, such as mesh analysis or superposition, may have to be used to find the Thevenin equivalent circuit.

EXAMPLE 8.10. (Two sources) Find the Thevenin circuit for the network in the shaded area of Fig. 8.42.

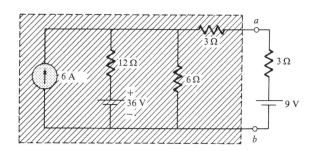

FIG. 8.42

Solution:

Steps 1 and 2 are shown in Fig. 8.43.

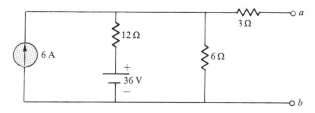

Step 3: See Fig. 8.44.

$$R_{Th} = 3 + \frac{(6)(12)}{6 + 12} = 3 + 4$$

$$R_{Th} = 7 \ \Omega$$

FIG. 8.43

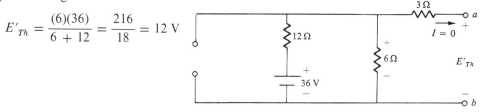

FIG. 8.44

Step 4: Use superposition. For the 36-V battery (Fig. 8.45) we can apply the voltage divider rule:

$$E'_{Th} = \frac{(6)(36)}{6 + 12} = \frac{216}{18} = 12 \ \text{V}$$

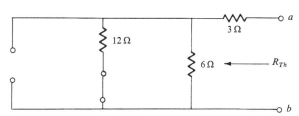

FIG. 8.45

For the 6-A source we have Fig. 8.46. By the current divider rule,

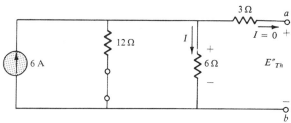

$$I_{6\Omega} = \frac{(12)(6)}{12 + 6} = \frac{72}{18} = 4 \ \text{A}$$

$$E''_{Th} = I_{6\Omega} \cdot 6 \ \Omega = (4)(6) = 24 \ \text{V}$$

$$E_{Th} = E'_{Th} + E''_{Th} = \textbf{36 V}$$

FIG. 8.46

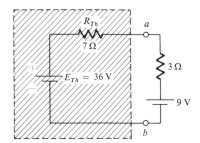

FIG. 8.47

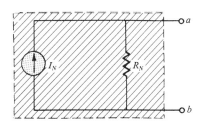

FIG. 8.48

Step 5: See Fig. 8.47.

8.4 NORTON'S THEOREM

It was demonstrated in Section 7.3 that every voltage source with a series internal resistance has a current source equivalent. The current source equivalent of the Thevenin network (which, you will note, satisfies the above conditions), as shown in Fig. 8.48, can be determined by Norton's theorem. It can also be found through the conversions of Section 7.4.

The theorem states the following:

Any two-terminal linear bilateral dc network can be replaced by an equivalent circuit consisting of a current source and a parallel resistor as shown in Fig. 8.48.

The discussion of Thevenin's theorem with respect to the equivalent circuit can also be applied to the Norton equivalent circuit.

The steps leading to the proper values of I_N and R_N are now listed:

1. Remove that portion of the network across which the Norton equivalent circuit is found.
2. Mark the terminals of the remaining two-terminal network.
3. Calculate R_N by first setting all sources to zero (voltage sources are replaced by short circuits and current sources by open circuits) and then finding the resultant resistance between the two marked terminals. (If the internal resistance of the voltage and/or current sources is included in the original network, it must remain when the sources are set to zero.)
4. Calculate I_N by first replacing the voltage and current sources and then finding the *short-circuit* current between the marked terminals.
5. Draw the Norton equivalent circuit with the portion of the circuit previously removed replaced between the terminals of the equivalent circuit.

The Norton and Thevenin equivalent circuits can also be found from each other by using the source transformation discussed earlier in this chapter and reproduced in Fig. 8.49.

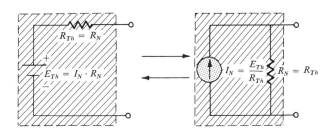

FIG. 8.49

EXAMPLE 8.11. Find the Norton equivalent circuit for the network in the shaded area of Fig. 8.50.

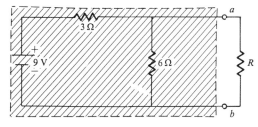

FIG. 8.50

Solution:
Steps 1 and 2 are shown in Fig. 8.51.

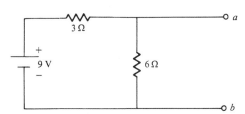

FIG. 8.51

Step 3 is shown in Fig. 8.52, and

$$R_N = \frac{(3)(6)}{3 + 6} = 2 \ \Omega$$

For R_N, NOT shorted out

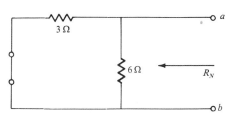

FIG. 8.52

Step 4 is shown in Fig. 8.53, and

$$I_N = \frac{9}{3} = 3 \ \mathbf{A}$$

For I_N, shorted out

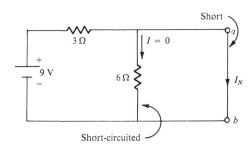

FIG. 8.53

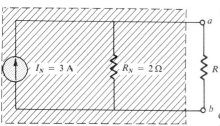

FIG. 8.54

This circuit is the same as the first one considered in the development of Thevenin's theorem. A simple conversion indicates that the Thevenin circuits are in fact the same (Fig. 8.55).

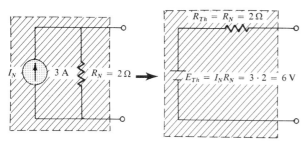

FIG. 8.55

EXAMPLE 8.12. Find the Norton equivalent circuit for the 9-Ω network in the shaded region of Fig. 8.56.

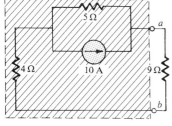

FIG. 8.56

Solution:
Steps 1 and 2: See Fig. 8.57.

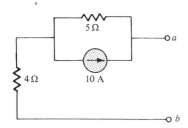

FIG. 8.57

Step 3: See Fig. 8.58, and

$$R_N = 5 + 4 = 9 \ \Omega$$

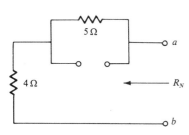

FIG. 8.58

Step 4 is shown in Fig. 8.59, and by the current divider rule,

$$I_N = \frac{(5)(10)}{4 + 5} = \frac{50}{9}$$

$$I_N = \textbf{5.556 A}$$

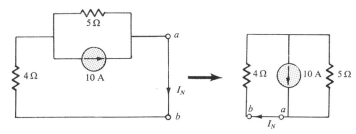

FIG. 8.59

Step 5: See Fig. 8.60.

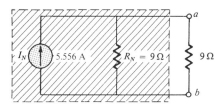

EXAMPLE 8.13. (Two sources) Find the Norton equivalent circuit for the portion of the network to the left of *a-b* in Fig. 8.61.

FIG. 8.60

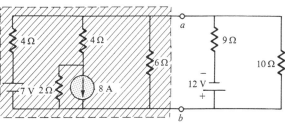

Solution:
Steps 1 and 2: See Fig. 8.62.

FIG. 8.61

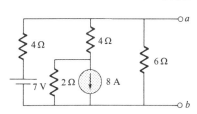

Step 3 is shown in Fig. 8.63, and

$$R_N = \frac{(6/2)(4)}{(6/2) + 4} = \frac{12}{7} = \textbf{1.714 } \Omega$$

FIG. 8.62

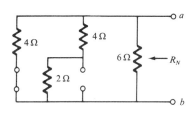

Step 4: (Using superposition) For the 7-V battery (Fig. 8.64),

$$I'_N = \frac{7}{4} = 1.75 \text{ A}$$

FIG. 8.63

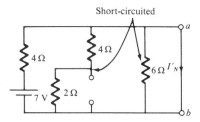

FIG. 8.64

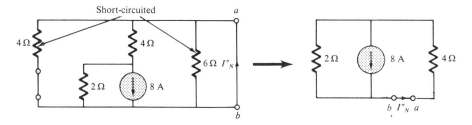

FIG. 8.65

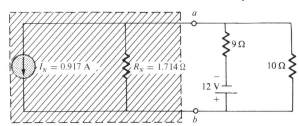

FIG. 8.66

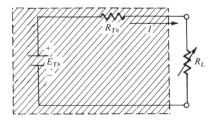

FIG. 8.67

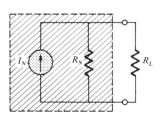

FIG. 8.68

For the 8-A source (Fig. 8.65),

$$I''_N = \frac{2(8)}{2 + 4} = \frac{16}{6} = 2.667 \text{ A}$$

and

$$I_N = I''_N - I'_N = 2.667 - 1.750$$

$$= \mathbf{0.917 \text{ A}} \qquad \text{(direction of } I''_N)$$

Step 5: See Fig. 8.66.

8.5 MAXIMUM POWER TRANSFER THEOREM

This theorem states the following:

A load will receive maximum power from a linear bilateral dc network when its total resistive value is exactly equal to the Thevenin resistance of the network as seen by the load.

In the network of Fig. 8.67, maximum power will be delivered to the load when

$$\boxed{R_L = R_{Th}} \tag{8.1}$$

From past discussions, we realize that a Thevenin equivalent circuit can be found across any element or group of elements in a linear bilateral dc network. Therefore, if we consider the case of the Thevenin equivalent circuit with respect to the maximum power transfer theorem, we are, in essence, considering the *total* effects of any network across a resistor R_L, such as in Fig. 8.67.

For the Norton equivalent circuit of Fig. 8.68, maximum power will be delivered to the load when

$$\boxed{R_L = R_N} \qquad\qquad (8.2)$$

This result [Eq. (8.2)] will be employed to its fullest advantage in the analysis of transistor networks where the most frequently applied transistor circuit model employs a current source rather than a voltage source.

For the network of Fig. 8.67,

$$I = \frac{E_{Th}}{R_{Th} + R_L}$$

and

$$P_L = I^2 R_L = \left(\frac{E_{Th}}{R_{Th} + R_L}\right)^2 R_L$$

so that

$$P_L = \frac{E_{Th}^2 R_L}{(R_{Th} + R_L)^2}$$

For $E_{Th} = 4$ V and $R_{Th} = 5\ \Omega$, the powers to R_L for different values of R_L are tabulated in Table 8.1.

TABLE 8.1

R_L (Ohms)	$P_L = \dfrac{16 R_L}{(5 + R_L)^2}$ (Watts)	
1	0.444	
2	0.653	
3	0.750	Increase
4	0.790	↓
5	0.800 ←	Maximum
6	0.793	
7	0.778	
8	0.757	Decrease
9	0.735	
10	0.711	↓

A plot of the above data (Fig. 8.69) clearly indicates that maximum power is delivered to R_L when it is exactly equal to R_{Th}.

For maximum power transfer conditions, the efficiency is now determined:

$$\eta\% = \frac{P_o}{P_i} \times 100\% = \frac{V_L I_L}{E_{Th} I_L} \times 100\%$$

$$= \frac{E_{Th}/2}{E_{Th}} \times 100\% = \frac{1}{2} \times 100\% = \mathbf{50\%}$$

It will always be 50% for maximum power transfer conditions.

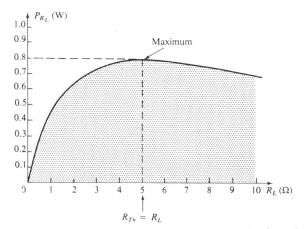

FIG. 8.69

A plot of the efficiency of the system versus load resistance appears in Fig. 8.70. Since this efficiency level is relatively low, this theorem is seldom applied in the field of power transmission, where 50% losses of energy could not be tolerated. However, it is widely used in electronics and communications engineering since the power levels are usually smaller and the efficiency criteria are of a less critical nature than in the field of power transmission.

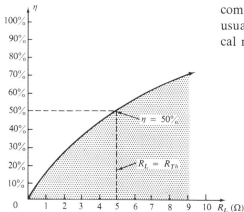

FIG. 8.70

A plot of load voltage versus R_L appears in Fig. 8.71, and of current versus R_L in Fig. 8.72. Note that the load voltage is one-half the Thevenin voltage at maximum power conditions while the current is one-half the value obtained with $R_L = 0\ \Omega$ (short circuit). For the condition, $R_L = 0\ \Omega$, and

$$I_L = I_{\max} = \frac{E_{Th}}{R_{Th}}$$

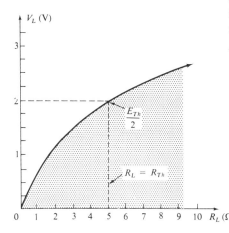

FIG. 8.71

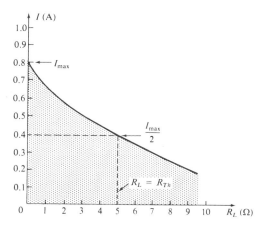

FIG. 8.72

It is interesting to note that it is not necessary to find E_{Th} in order to find the proper value of R for maximum power transfer. To find the maximum power delivered to R, however, requires either that E_{Th} be known or that the voltage across or current through the resistor R be found when $R = R_{Th}$ in the original network.

If P_L, V_L, and I_L are plotted on a log scale as shown in Fig. 8.73, the variation in levels for the wide range of resistor values becomes clear. Note that P_L reaches only one maximum (at $R_L = R_{Th}$), that V_L increases with increasing values of R_L as determined by the voltage divider rule, and that I_L drops with increasing levels of R_L as controlled by Ohm's law. One obvious advantage of log scales is that they permit a wide variation in the magnitude of a parameter (such as R_L in Fig. 8.73).

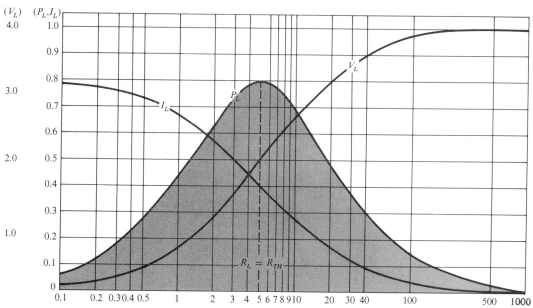

FIG. 8.73

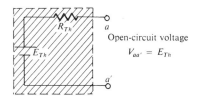

FIG. 8.74

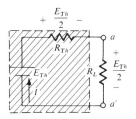

FIG. 8.75

For any physical network, the value of $R_L = R_{Th}$ can be determined by first measuring the open-circuit voltage across the load terminals as shown in Fig. 8.74; $V_{aa'} = E_{Th}$.

Then complete the circuit with R_L, and vary R_L until the voltage appearing across the load is one-half the open-circuit value, or $V_L = E_{Th}/2$ (Fig. 8.75). When this condition is established, the voltage across the Thevenin and the load resistances are the same. Since the current through each is also the same, the resistance values of each are equal, and

$$R_L = \frac{E_{Th}/2}{I} = \frac{E_{\text{open-circuit}}/2}{I(R_L = R_{Th})} = \frac{E_{oc}}{2I} = R_{Th} \quad \textbf{(8.3)}$$

Both E_{oc} and I can be determined by instruments properly connected to the load resistance branch. Meters will be considered in Chapter 12.

The power delivered to R_L under maximum power conditions ($R_L = R_{Th}$) is

$$I = \frac{E_{Th}}{R_{Th} + R_L} = \frac{E_{Th}}{2R_{Th}}$$

$$P_L = I^2 R_L = \left(\frac{E_{Th}}{2R_{Th}}\right)^2 R_{Th} = \frac{E_{Th}^2 R_{Th}}{4R_{Th}^2}$$

and

$$\boxed{P_{L_{\max}} = \frac{E_{Th}^2}{4R_{Th}}} \quad \text{(W)} \qquad \textbf{(8.4)}$$

For the Norton circuit of Fig. 8.69,

$$\boxed{P_{L_{\max}} = \frac{I_N^2 R_N}{4}} \quad \text{(W)} \qquad \textbf{(8.5)}$$

EXAMPLE 8.14. A dc generator, battery, and laboratory supply are connected to a resistive load R_L in Figs. 8.76(a), (b), and (c), respectively. For each, determine the value of R_L for maximum power transfer to R_L. (Review Fig. 5.46.)

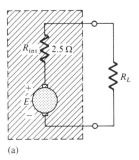

(a)

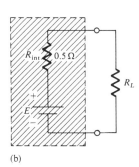

(b)

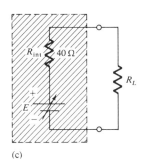

(c)

FIG. 8.76

Solution:

a. For the dc generator,

$$R_L = 2.5 \, \Omega$$

b. For the battery,

$$R_L = 0.5 \, \Omega$$

c. For the laboratory supply,

$$R_L = 40 \, \Omega$$

The results of the preceding example indicate that the following modified form of the statement of the maximum power theorem is correct:

For loads connected directly to a dc voltage supply, maximum power will be delivered to the load when the load resistance is equal to the internal resistance of the source; that is, when

$$\boxed{R_L = R_{\text{int}}} \qquad (8.6)$$

EXAMPLE 8.15. Analysis of a transistor network resulted in the reduced configuration of Fig. 8.77. Determine the R_L necessary to transfer maximum power to R_L, and calculate the power to R_L under these conditions.

Solution:

Eq. (8.2):

$$R_L = R_N = 40 \, \text{k}\Omega$$

Eq. (8.5):

$$P_{L_{\text{max}}} = \frac{I_N^2 R_N}{4} = \frac{(10 \times 10^{-3})^2 \, 40 \, \text{k}\Omega}{4} = 1 \, \text{W}$$

EXAMPLE 8.16. For the network of Fig. 8.78, determine the value of R for maximum power to R, and calculate the power delivered under these conditions.

Solution: See Fig. 8.79, and

$$R_{Th} = 8 + \frac{(3)(6)}{3 + 6} = 8 + 2$$

and

$$R = R_{Th} = 10 \, \Omega$$

See Fig. 8.80, and

$$E_{Th} = \frac{3(12)}{3 + 6} = \frac{36}{9} = 4 \, \text{V}$$

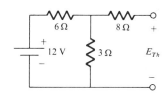

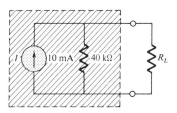

FIG. 8.77

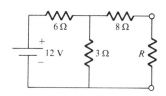

FIG. 8.78

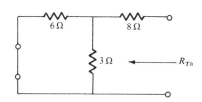

FIG. 8.79

FIG. 8.80

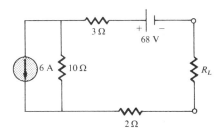

FIG. 8.81

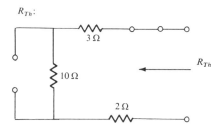

FIG. 8.82

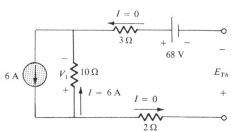

FIG. 8.83

and, by Eq. (8.4),

$$P_{L_{\max}} = \frac{E_{Th}^2}{4R_{Th}} = \frac{(4)^2}{4(10)} = 0.4 \text{ W}$$

EXAMPLE 8.17. Find the value of R_L in Fig. 8.81 for maximum power to R_L, and determine the maximum power.

Solution: See Fig. 8.82, and

$$R_{Th} = 3 + 10 + 2 = 15 \ \Omega$$

and

$$R_L = R_{Th} = 15 \ \Omega$$

See Fig. 8.83, and

$$V_1 = (6)(10) = 60 \text{ V}$$
$$E_{Th} = 68 + 60 = 128 \text{ V}$$
$$P_{L_{\max}} = \frac{E_{Th}^2}{4R_{Th}} = \frac{(128)^2}{(4)(15)} = 273.07 \text{ W}$$

8.6 MILLMAN'S THEOREM

Through the application of *Millman's theorem,* any number of parallel voltage sources can be reduced to one. In Fig. 8.84, for example, the three voltage sources can be reduced to one. This would permit finding the current through or voltage across R_L without having to apply a method such as mesh analysis, nodal analysis, superposition, and so on. The theorem can best be described by applying it to the network of Fig. 8.84. There are basically three steps included in its application.

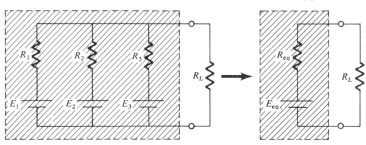

FIG. 8.84

Step 1: Convert all voltage sources to current sources as outlined in Section 7.3. This is performed in Fig. 8.85 for the network of Fig. 8.84.

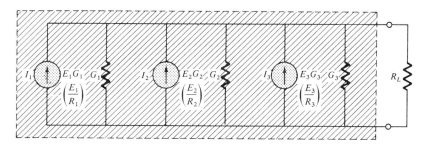

FIG. 8.85

Step 2: Combine parallel current sources as described in Section 7.4. The resulting network is shown in Fig. 8.86, where

$$I_T = I_1 + I_2 + I_3 \qquad G_T = G_1 + G_2 + G_3$$

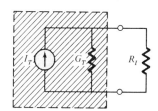

FIG. 8.86

Step 3: Convert the resulting current source to a voltage source, and the desired single-source network is obtained as shown in Fig. 8.87.

In general, Millman's theorem states that for any number of parallel voltage sources,

$$E_{eq} = \frac{I_T}{G_T} = \frac{\pm I_1 \pm I_2 \pm I_3 \pm \cdots \pm I_N}{G_1 + G_2 + G_3 + \cdots + G_N}$$

or

$$\boxed{E_{eq} = \frac{\pm E_1 G_1 \pm E_2 G_2 \pm E_3 G_3 \pm \cdots \pm E_N G_N}{G_1 + G_2 + G_3 + \cdots + G_N}}$$

(8.7)

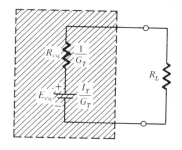

FIG. 8.87

The + and − signs are included in Eqs. (8.6) and (8.7) to include those cases where the sources may not be supplying energy in the same direction. (Note Example 8.18.)

$$\boxed{R_{eq} = \frac{1}{G_T} = \frac{1}{G_1 + G_2 + G_3 + \cdots + G_N}} \quad \textbf{(8.8)}$$

In terms of the resistance values,

$$\boxed{E_{eq} = \frac{\pm \dfrac{E_1}{R_1} \pm \dfrac{E_2}{R_2} + \dfrac{E_3}{R_3} \pm \cdots \pm \dfrac{E_N}{R_N}}{\dfrac{1}{R_1} + \dfrac{1}{R_2} + \dfrac{1}{R_3} + \cdots + \dfrac{1}{R_N}}} \quad \textbf{(8.9)}$$

and

$$\boxed{R_{eq} = \frac{1}{\dfrac{1}{R_1} + \dfrac{1}{R_2} + \dfrac{1}{R_3} + \cdots + \dfrac{1}{R_N}}} \quad \textbf{(8.10)}$$

The relatively few direct steps required may result in the student's applying each step rather than memorizing and employing Eqs. (8.7) through (8.10).

EXAMPLE 8.18. Using Millman's theorem, find the current through and voltage across the resistor R_L of Fig. 8.88.

Solution: By Eq. (8.9),

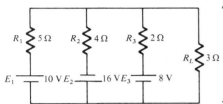

FIG. 8.88

$$E_{eq} = \frac{+\dfrac{E_1}{R_1} - \dfrac{E_2}{R_2} + \dfrac{E_3}{R_3}}{\dfrac{1}{R_1} + \dfrac{1}{R_2} + \dfrac{1}{R_3}}$$

The minus sign is used for E_2/R_2 because that supply has the opposite polarity of the other two. Our chosen reference direction is therefore that of E_1 and E_3. The total conductance is unaffected by the direction, and

$$E_{eq} = \frac{+\dfrac{10}{5} - \dfrac{16}{4} + \dfrac{8}{2}}{\dfrac{1}{5} + \dfrac{1}{4} + \dfrac{1}{2}} = \frac{2 - 4 + 4}{0.2 + 0.25 + 0.5}$$

$$= \frac{2}{0.95} = \textbf{2.105 V}$$

with

$$R_{eq} = \frac{1}{\dfrac{1}{5} + \dfrac{1}{4} + \dfrac{1}{2}} = \frac{1}{0.95} = \textbf{1.053 } \Omega$$

The resultant source is shown in Fig. 8.89, and

$$I_L = \frac{2.105}{1.053 + 3} = \frac{2.105}{4.053} = \textbf{0.519 A}$$

with

$$V_L = I_L R_L = (0.519)(3) = \textbf{1.557 V}$$

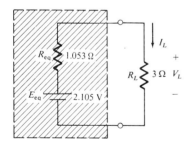

FIG. 8.89

EXAMPLE 8.19. Let us now consider the type of problem encountered in the introduction to mesh and nodal analysis in Chapter 7. Mesh analysis was applied to the network of Fig. 8.90 (Example 7.14). Let us now use Millman's theorem to find the current through the 2-Ω resistor and compare the results.

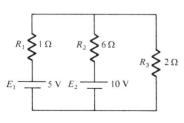

FIG. 8.90

Solution:
a. Let us first apply each step and, in the (b) solution, Eq. (8.9). Converting sources yields Fig. 8.91. Combining

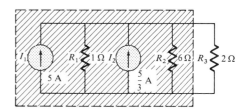

FIG. 8.91

sources and parallel conductance branches (Fig. 8.92) yields

$$I_T = I_1 + I_2 = 5 + \frac{5}{3} = \frac{15}{3} + \frac{5}{3} = \frac{20}{3} \text{ A}$$

$$G_T = G_1 + G_2 = 1 + \frac{1}{6} = \frac{6}{6} + \frac{1}{6} = \frac{7}{6} \text{ S}$$

Converting the current source to a voltage source (Fig. 8.93), we obtain

$$E_{eq} = \frac{I_T}{G_T} = \frac{\dfrac{20}{3}}{\dfrac{7}{6}} = \frac{6(20)}{3(7)} = \frac{40}{7} \text{ V}$$

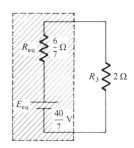

FIG. 8.92

and

$$R_{eq} = \frac{1}{G_T} = \frac{1}{\dfrac{7}{6}} = \frac{6}{7} \text{ } \Omega$$

so that

$$I_{2\Omega} = \frac{\dfrac{40}{7}}{\dfrac{6}{7} + 2} = \frac{\dfrac{40}{7}}{\dfrac{6}{7} + \dfrac{14}{7}} = \frac{40}{20} = \textbf{2 A}$$

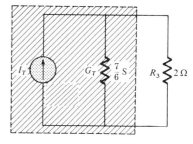

FIG. 8.93

which agrees with the result obtained in Example 7.14.
b. Let us now simply apply the proper equation, Eq. (8.9):

$$E_{eq} = \frac{+\dfrac{5}{1} + \dfrac{10}{6}}{\dfrac{1}{1} + \dfrac{1}{6}} = \frac{\dfrac{30}{6} + \dfrac{10}{6}}{\dfrac{6}{6} + \dfrac{1}{6}} = \frac{40}{7} \text{ V}$$

and

$$R_{eq} = \frac{1}{\dfrac{1}{1} + \dfrac{1}{6}} = \frac{1}{\dfrac{6}{6} + \dfrac{1}{6}} = \frac{1}{\dfrac{7}{6}} = \frac{6}{7} \text{ } \Omega$$

which are the same values obtained above.

The dual of Millman's theorem is the combining of series current sources. The dual of Fig. 8.84 is Fig. 8.94.

It can be shown that I_{eq} and R_{eq}, as shown in Fig. 8.94, are given by

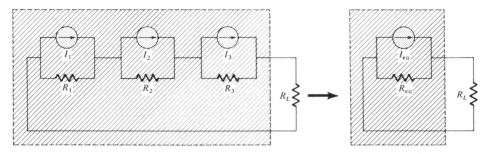

FIG. 8.94

$$I_{eq} = \frac{\pm I_1 R_1 \pm I_2 R_2 \pm I_3 R_3}{R_1 + R_2 + R_3} \qquad \textbf{(8.11)}$$

and

$$R_{eq} = R_1 + R_2 + R_3 \qquad \textbf{(8.12)}$$

The derivation will appear as a problem at the end of the chapter.

8.7 SUBSTITUTION THEOREM

This theorem states the following:

If the voltage across and current through any branch of a dc bilateral network are known, this branch can be replaced by any combination of elements that will maintain the same voltage across and current through the chosen branch.

More simply, the theorem states that for branch equivalence, the terminal voltage and current must be the same. Consider the circuit of Fig. 8.95 in which the voltage across and current through the branch a-b are determined. Through the use of the substitution theorem, a number of equivalent a-a' branches are shown in Fig. 8.96.

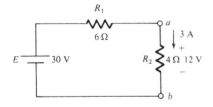

FIG. 8.95

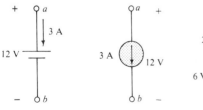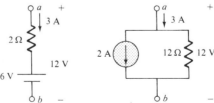

FIG. 8.96

Note that for each equivalent, the terminal voltage and current are the same. Also consider that the response of the remainder of the circuit of Fig. 8.95 is unchanged by substituting any one of the equivalent branches. As demonstrated by the single-source equivalents of Fig. 8.96, *a known potential difference or current in a network can be replaced by an ideal voltage source and current source, respectively.*

Understand that this theorem cannot be used to *solve* networks with two or more sources that are not in series or parallel. For it to be applied, a potential difference or current value must be known or found using one of the techniques discussed earlier. One application of the theorem is shown in Fig. 8.97. Note that in the figure the known potential difference V was replaced by a voltage source, permitting the isolation of the portion of the network including R_3, R_4, and R_5. Recall that this was basically the approach employed in the analysis of the ladder network as we worked our way back toward the terminal resistance R_5.

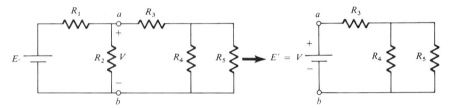

The current source equivalence of the above is shown in Fig. 8.98, where a known current is replaced by an ideal current source, permitting the isolation of R_4 and R_5.

FIG. 8.97

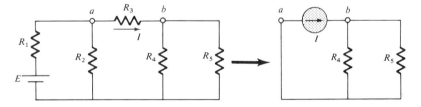

FIG. 8.98

You will also recall from the discussion of bridge networks that $V = 0$ and $I = 0$ were replaced by a short circuit and an open circuit, respectively. This substitution is a very specific application of the substitution theorem.

8.8 RECIPROCITY THEOREM

The reciprocity theorem is applicable only to single-source networks. It is, therefore, not a theorem employed in the analysis of multisource networks described thus far.

The theorem states the following:

The current I in any branch of a network, due to a single voltage source E anywhere else in the network, will equal the current through the branch in which the source was originally located if the source is placed in the branch in which the current I was originally measured.

In other words, the location of the voltage source and the resulting current may be interchanged without a change in current. The theorem requires that the polarity of the voltage source have the same correspondence with the direction of the branch current in each position.

In the representative network of Fig. 8.99(a), the current I due to the voltage source E was determined. If the position of each is interchanged as shown in Fig. 8.99(b), the

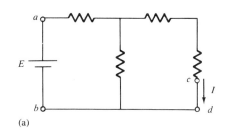

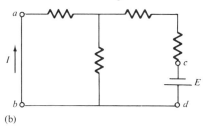

(a)   (b)

FIG. 8.99

current I will be the same value as indicated. To demonstrate the validity of this statement and the theorem, consider the network of Fig. 8.100, in which values for the elements of Fig. 8.99(a) have been assigned.

The total resistance is

$$R_T = 12 + 6 \| (2 + 4)$$
$$= 12 + 6 \| 6$$
$$= 12 + 3$$
$$R_T = 15 \ \Omega$$

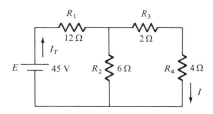

FIG. 8.100

and

$$I_T = \frac{E}{R_T} = \frac{45}{15} = 3 \ \text{A}$$

with

$$I = \frac{3}{2} = \textbf{1.5 A}$$

For the network of Fig. 8.101, which corresponds with that of Fig. 8.99(b), we find

$$R_T = 12 \| 6 + 2 + 4 = 4 + 6 = 10 \ \Omega$$

and

$$I_T = \frac{E}{R_T} = \frac{45}{10} = 4.5 \ \text{A}$$

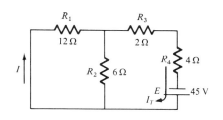

FIG. 8.101

so that

$$I = \frac{6(4.5)}{12 + 6} = \frac{4.5}{3} = \textbf{1.5 A}$$

which agrees with the above.

The uniqueness and power of such a theorem can best be demonstrated by considering a complex single-source network such as shown in Fig. 8.102.

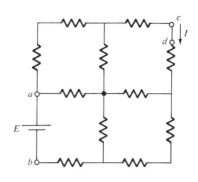

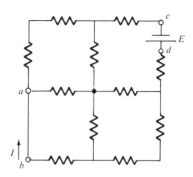

FIG. 8.102

PROBLEMS

Section 8.2

1. **a.** Using superposition, find the current through each resistor of the network of Fig. 8.103.
   b. Find the power delivered to R_1 for each source.
   c. Find the power delivered to R_1 using the total current through R_1.
   d. Does superposition apply to power effects? Explain.

2. Using superposition, find the current through the 10-Ω resistor for each of the networks of Fig. 8.104.

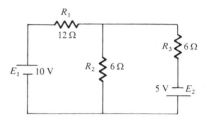

FIG. 8.103

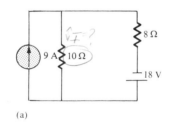

(a)

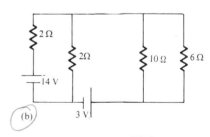

(b)

FIG. 8.104

*3. Using superposition, find the current through the 10-Ω resistor for each network of Fig. 8.105.

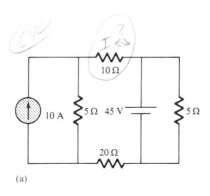

(a)

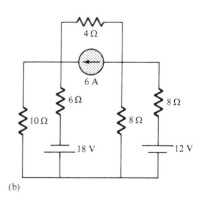

(b)

FIG. 8.105

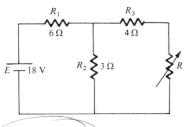

FIG. 8.106

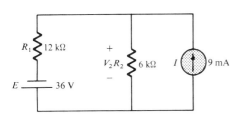

FIG. 8.107

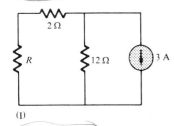

(I)

FIG. 8.108

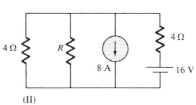

(1)   (II)

FIG. 8.109

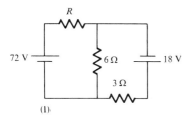

(I)

FIG. 8.110

4. Using superposition, find the voltage V_2 for the network of Fig. 8.106.

Section 8.3

5. a. Find the Thevenin equivalent circuit for the network external to the resistor R of Fig. 8.107.
   b. Find the current through R when R is 2, 30, and 100 Ω.

6. a. Find the Thevenin equivalent circuit for the network external to the resistor R in each of the networks of Fig. 8.108.
   b. Find the power delivered to R when R is 2 Ω and 100 Ω.

7. Repeat Problem 6 for the networks of Fig. 8.109.

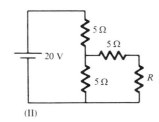

(II)

✓ ***8.** Repeat Problem 6 for the networks of Fig. 8.110.

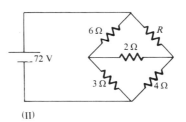

(II)

✓ **9.** Find the Thevenin equivalent circuit for the portions of the networks of Fig. 8.111 external to points a and b.

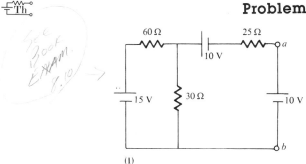

(I)

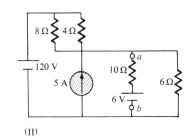

(II)

FIG. 8.111

*10. Repeat Problem 9 for the networks of Fig. 8.112.

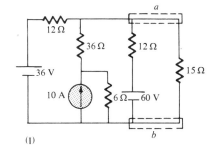

(I)

Section 8.4

11. Find the Norton equivalent circuit for the network external to the resistor R in each network of Fig. 8.108 by
    a. following the procedure outlined in the text.
    b. converting the Thevenin equivalent circuit to a Norton equivalent circuit if available from a previous assignment.

12. Repeat Problem 11 for the networks of Fig. 8.109.

*13. Repeat Problem 11 for the networks of Fig. 8.110.

14. Find the Norton equivalent circuit for the portions of the networks of Fig. 8.111 external to branch a-b by
    a. following the procedure outlined in the text.
    b. converting the Thevenin equivalent circuit to a Norton equivalent circuit if available from a previous assignment.

*15. Repeat Problem 14 for the networks of Fig. 8.112.

16. Find the Norton equivalent circuit for the portions of the networks of Fig. 8.113 external to branch a-b.

(II)

FIG. 8.112

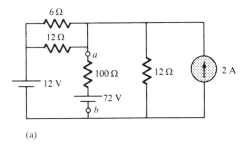

(a)

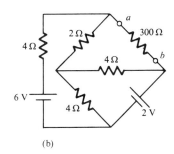

(b)

FIG. 8.113

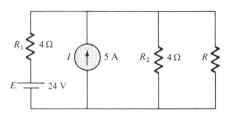

FIG. 8.114

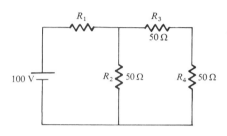

FIG. 8.115

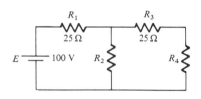

FIG. 8.116

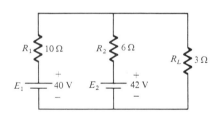

FIG. 8.117

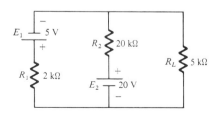

FIG. 8.118

Section 8.5

17. a. For each network of Fig. 8.108, find the value of R for maximum power to R.
  b. Determine the maximum power to R for each network.

18. Repeat Problem 17 for the networks of Fig. 8.109.

***19.** Repeat Problem 17 for the networks of Fig. 8.110.

20. a. For the network of Fig. 8.114, determine the value of R for maximum power to R.
  b. Determine the maximum power to R.
  c. Plot a curve of power to R vs. R for R equal to $\frac{1}{4}$, $\frac{1}{2}$, $\frac{3}{4}$, 1, $1\frac{1}{4}$, $1\frac{1}{2}$, $1\frac{3}{4}$, and 2 times the value obtained in part (a).

***21.** Find the resistance R_1 of Fig. 8.115 such that the resistor R_4 will receive maximum power. Think!

***22. a.** For the network of Fig. 8.116, determine the value R_2 for maximum power to R_4.
  b. Is there a general statement that can be made about situations such as presented in Problems 21 and 22?

Section 8.6

23. Using Millman's theorem, find the current through and voltage across the resistor R_L of Fig. 8.117.

24. Repeat Problem 23 for the network of Fig. 8.118.

25. Repeat Problem 23 for the network of Fig. 8.119.

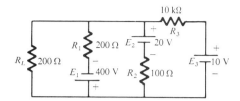

FIG. 8.119

26. Using the dual of Millman's theorem, find the current through and voltage across the resistor R_L of Fig. 8.120.

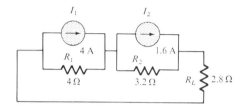

*27. Repeat Problem 26 for the network of Fig. 8.121.

FIG. 8.120

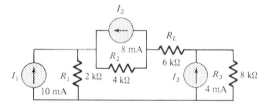

Section 8.7

FIG. 8.121

28. Using the substitution theorem, draw three equivalent branches for the branch a-b of the network of Fig. 8.122.

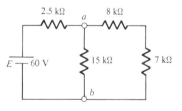

29. Repeat Problem 28 for the network of Fig. 8.123.

FIG. 8.122

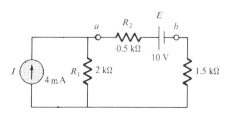

FIG. 8.123

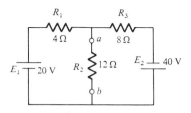

FIG. 8.124

*30. Repeat Problem 28 for the network of Fig. 8.124. Be careful!

Section 8.8

31. a. For the network of Fig. 8.125(a), determine the current I.
    b. Repeat part (a) for the network of Fig. 8.125(b).
    c. Is the reciprocity theorem satisfied?

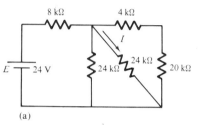

(a)

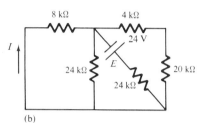

(b)

FIG. 8.125

32. Repeat Problem 31 for the networks of Fig. 8.126.

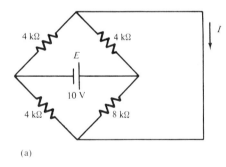

(a)

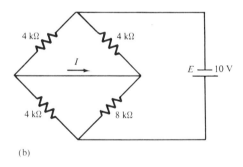

(b)

FIG. 8.126

33. a. Determine the voltage V for the network of Fig. 8.127(a).
    b. Repeat part (a) for the network of Fig. 8.127(b).
    c. Is the dual of the reciprocity theorem satisfied?

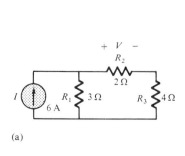

(a)

(b)

FIG. 8.127

GLOSSARY

Maximum power theorem A theorem used to determine the load resistance necessary to insure maximum power transfer to the load.

Millman's theorem A method employing source conversions that will permit the determination of unknown variables in a multiloop network.

Norton's theorem A theorem that permits the reduction of any two-terminal linear dc network to one having a single current source and parallel resistor.

Reciprocity theorem A theorem that states that for single-source networks, the current in any branch of a network, due to a single voltage source in the network, will equal the current through the branch in which the source was originally located if the source is placed in the branch in which the current was originally measured.

Substitution theorem A theorem that states that if the voltage across and current through any branch of a dc bilateral network are known, the branch can be replaced by any combination of elements that will maintain the same voltage across and current through the chosen branch.

Superposition theorem A network theorem that permits considering the effects of each source independently. The resulting current and/or voltage is the algebraic sum of the currents and/or voltages developed by each source independently.

Thevenin's theorem A theorem that permits the reduction of any two-terminal linear dc network to one having a single voltage source and series resistor.

CAPACITORS

9

9.1 INTRODUCTION

Thus far, the only passive device appearing in the text has been the resistor. We will now consider two additional passive devices called the *capacitor* and the *inductor,* which are quite different from the resistor in purpose, operation, and construction.

Unlike the resistor, these elements display their total characteristics only when a change in voltage or current is made in the circuit in which they exist. In addition, if we consider the *ideal* situation, they do not dissipate energy like the resistor but store it in a form that can be returned to the circuit whenever required by the circuit design.

Proper treatment of each requires that we devote this entire chapter to the capacitor and Chapter 11 to the inductor. Since electromagnetic effects are a major consideration in the design of inductors, Chapter 10 on magnetic circuits will appear first.

9.2 THE ELECTRIC FIELD

Recall from Chapter 2 that a force of attraction or repulsion exists between two charged bodies. We shall now examine this phenomenon in greater detail by considering the electric field that exists in the region around any charged body. This electric field is represented by electric flux lines, which are drawn to indicate the strength of the electric field at any point around the charged body; that is, the denser the lines of flux, the stronger the electric field. In Fig. 9.1, the electric field strength is stronger at position a than at position b because the flux lines are denser at a than at b. The symbol for electric flux is the Greek letter ψ (psi). The flux per unit area (flux density) is represented by the capital letter D and is determined by

$$D = \frac{\psi}{A}$$ (flux/unit area)   **(9.1)**

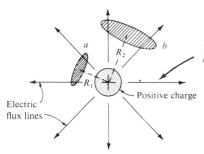

Flux lines radiate outward for positive charges and inward for negative charges.

FIG. 9.1

The larger the charge Q in coulombs, the greater the number of flux lines extending or terminating per unit area, independent of the surrounding medium. Twice the charge will produce twice the flux per unit area. The two can therefore be equated:

$$\psi \equiv Q$$ (coulombs, C)   **(9.2)**

By definition, the *electric field strength* at a point is the force acting on a unit positive charge at that point; that is,

$$\mathscr{E} = \frac{F}{Q}$$ (newtons/coulomb, N/C)   **(9.3)**

The force exerted on a unit positive charge ($Q_2 = 1$), by a charge Q_1, r meters away, as determined by Coulomb's law is

$$F = \frac{kQ_1Q_2}{r^2} = \frac{kQ_1(1)}{r^2} = \frac{kQ_1}{r^2}$$ $(k = 9 \times 10^9)$

Substituting this force F into Eq. (9.3) yields

$$\mathscr{E} = \frac{F}{Q_2} = \frac{kQ_1/r^2}{1}$$

$$\mathscr{E} = \frac{kQ_1}{r^2}$$ (N/C)   **(9.4)**

We can therefore conclude that the electric field strength at any point distance r from a point charge of Q coulombs is directly proportional to the magnitude of the charge and inversely proportional to the distance squared from the charge. The squared term in the denominator will result in a rapid decrease in the strength of the electric field with distance from the point charge. In Fig. 9.1, substituting distances R_1 and R_2 into Eq. (9.4) will verify our previous conclusion that the electric field strength is greater at a than at b.

Flux lines always extend from a positively charged to a negatively charged body, always extend or terminate perpendicular to the charged surfaces, and never intersect. For two charges of similar and opposite polarities, the flux distribution would appear as shown in Fig. 9.2.

The attraction and repulsion between charges can now be explained in terms of the electric field and its flux lines. In Fig. 9.2(a), the flux lines are not interlocked but tend to act as a buffer, preventing attraction and causing repulsion. Since the electric field strength is stronger (flux lines denser) for each charge the closer we are to the charge, the more we try to bring the two charges together, the stronger will be the force of repulsion between them. In Fig. 9.2(b), the flux lines extending from the positive charge are terminated at the negative charge. A basic law of physics states that electric flux lines always tend to be as short as possible. The two charges will therefore be drawn to each other. Again, the closer the two charges, the stronger the attraction between the two charges due to the increased field strengths.

9.3 CAPACITANCE

Up to this point we have considered only isolated positive and negative spherical charges, but the analysis can be extended to charged surfaces of any shape and size. In Fig. 9.3, for example, two parallel plates of a conducting material separated by an air gap have been connected through a switch and a resistor to a battery. If the parallel plates are initially uncharged and the switch is left open, no net positive or negative charge will exist on either plate. The instant the switch is closed, however, electrons are drawn from the upper plate through the resistor to the positive terminal of the battery. There will be a surge of current at first, limited in magnitude by the resistance present. The level of flow will then decline, as will be demonstrated in the sections to follow. This action creates a net positive charge on the top plate. Electrons are being repelled by the negative terminal through the lower conductor to the bottom plate at the same rate they are being drawn to the positive terminal. This transfer of electrons continues until

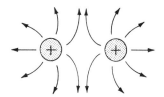

(a)

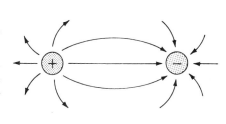

(b)

FIG. 9.2

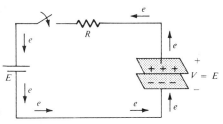

FIG. 9.3

the potential difference across the parallel plates is exactly equal to the battery emf. The final result is a net positive charge on the top plate and a negative charge on the bottom plate, very similar in many respects to the two isolated charges of Fig. 9.2(b).

This element, constructed simply of two parallel conducting plates separated by an insulating material (in this case, air), is called a *capacitor*. *Capacitance* is a measure of a capacitor's ability to store charge on its plates—in other words, its storage capacity. A capacitor has a capacitance of 1 *farad* if 1 coulomb of charge is deposited on the plates by a potential difference of 1 V across the plates. The farad is named after Michael Faraday, a nineteenth-century English chemist and physicist. The farad, however, is generally too large a measure of capacitance for most practical applications, so the microfarad (10^{-6}) or picofarad (10^{-12}) is more commonly used. Expressed as an equation, the capacitance is determined by

$$C = \frac{Q}{V}$$

$$\begin{aligned} C &= \text{farads (F)} \\ Q &= \text{coulombs (C)} \\ V &= \text{volts (V)} \end{aligned} \qquad \textbf{(9.5)}$$

Different capacitors for the same voltage across their plates will acquire greater or lesser amounts of charge on their plates. Hence the capacitors have a greater or lesser capacitance, respectively.

A cross-sectional view of the parallel plates is shown with the distribution of electric flux lines in Fig. 9.4(a). The number of flux lines per unit area (D) between the two plates is quite uniform. At the edges, the flux lines extend outside the common surface area of the plates, producing an effect known as *fringing*. This effect, which reduces the capacitance somewhat, can be neglected for most practical applications. For the analysis to follow, we will assume that all the flux lines leaving the positive plate will pass directly to the negative plate within the common surface area of the plates [Fig. 9.4(b)].

If a potential difference of V volts is applied across the two plates separated by a distance of d, the electric field strength between the plates is determined by

$$\mathcal{E} = \frac{V}{d} \qquad \text{(volts/meter, V/m)} \qquad \textbf{(9.6)}$$

The uniformity of the flux distribution in Fig. 9.4(b) also indicates that the electric field strength is the same at any point between the two plates.

Many values of capacitance can be obtained for the same set of parallel plates by the addition of certain insulating materials between the plates. In Fig. 9.5(a), an insu-

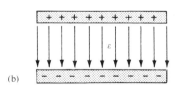
(b)

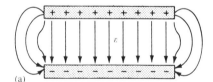
(a)

FIG. 9.4

lating material has been placed between a set of parallel plates having a potential difference of V volts across them.

Since the material is an insulator, the electrons within the insulator are unable to leave the parent atom and travel to the positive plate. The positive components (protons) and negative components (electrons) of each atom do shift, however [as shown in Fig. 9.5(a)], to form *dipoles*.

When all the atoms of the insulating material become dipoles and align themselves as shown in Fig. 9.5(a), the material is *polarized*. A close examination within this polarized material will indicate that the positive and negative components of adjoining dipoles are neutralizing the effects of each other [note the dashed area in Fig. 9.5(a)]. The layer of positive charge on one surface and the negative charge on the other are not neutralized, however, resulting in the establishment of an electric field within the insulator [$\mathscr{E}_{diel}$, Fig. 9.5(b)]. The net electric field between the plates ($\mathscr{E}_{resultant} = \mathscr{E}_{air} - \mathscr{E}_{diel}$) would therefore be reduced due to the insertion of the dielectric.

The purpose of the dielectric, therefore, is to create an electric field to oppose the electric field set up by free charges on the parallel plates. For this reason, the insulating material is referred to as a *dielectric, di* for *opposing* and *electric* for *electric field*.

In either case—with or without the dielectric—if the potential across the plates is kept constant and the distance between the plates is fixed, the net electric field within the plates must remain the same, as determined by the equation $\mathscr{E} = V/d$. We just ascertained, however, that the net electric field between the plates would decrease with insertion of the dielectric for a fixed amount of free charge on the plates. To compensate and keep the net electric field equal to the value determined by V and d, more charge must be deposited on the plates. [Look ahead to Eq. (9.11).] This additional charge for the same potential across the plates increases the capacitance, as determined by the following equation:

$$C\uparrow = \frac{Q\uparrow}{V}$$

For different dielectric materials between the same two parallel plates, different amounts of charge will be deposited on the plates. But $\psi \equiv Q$, so the dielectric is also determining the number of flux lines between the two plates and consequently the flux density ($D = \psi/A$) since A is fixed.

The ratio of the flux density to the electric field intensity in the dielectric is called the *permittivity* of the dielectric:

$$\epsilon = \frac{D}{\mathscr{E}} \qquad \text{(farads/meter, F/m)} \qquad \textbf{(9.7)}$$

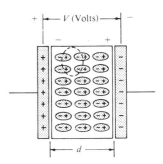

(a)

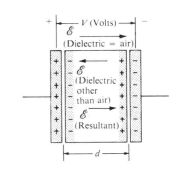

(b)

FIG. 9.5

It is a measure of how easily the dielectric will "permit" the establishment of flux lines within the dielectric. The greater its value, the greater the amount of charge deposited on the plates, and, consequently, the greater the flux density for a fixed area.

For a vacuum, the value of ϵ (denoted by ϵ_o) is 8.85×10^{-12} F/m. The ratio of the permittivity of any dielectric to that of a vacuum is called the *relative permittivity*, ϵ_r. In equation form,

$$\epsilon_r = \frac{\epsilon}{\epsilon_o} \qquad \textbf{(9.8)}$$

The value of ϵ for any material, therefore, is

$$\epsilon = \epsilon_r \epsilon_o$$

Note that ϵ_r is a dimensionless quantity. The relative permittivity or *dielectric constant*, as it is often called, for various dielectric materials is given in Table 9.1.

TABLE 9.1 *Relative permittivity* (*dielectric constant*) *of various dielectrics.*

Dielectric	ϵ_r (Average values)
Vacuum	1.0
Air	1.0006
Teflon®	2.0
Paper, paraffined	2.5
Rubber	3.0
Transformer oil	4.0
Mica	5.0
Porcelain	6.0
Bakelite	7.0
Glass	7.5
Water	80.0
Barium-strontium titanite (ceramic)	7500.0

Substituting for D and $\mathscr{E}$ in Eq. (9.7), we have

$$\epsilon = \frac{D}{\mathscr{E}} = \frac{\psi/A}{V/d} = \frac{Q/A}{V/d} = \frac{Qd}{VA}$$

But

$$C = \frac{Q}{V} \qquad \therefore \epsilon = \frac{Cd}{A}$$

and

$$C = \frac{\epsilon A}{d} \qquad \text{(farads, F)} \qquad \textbf{(9.9)}$$

or

$$C = \epsilon_o \epsilon_r \frac{A}{d} = 8.85 \times 10^{-12} \epsilon_r \frac{A}{d} \qquad \text{(F)}$$

(9.10)

where A is the area in square meters of the plates, d is the distance in meters between the plates, and ϵ_r is the relative permittivity. The capacitance, therefore, will be greater if the area of the plates is increased, or the distance between the plates is decreased, or the dielectric is changed so that ϵ_r is increased.

Solving for the distance d in Eq. (9.9), we have

$$d = \frac{\epsilon A}{C}$$

and substituting into Eq. (9.6) yields

$$\mathscr{E} = \frac{V}{d} = \frac{V}{\epsilon A/C} = \frac{CV}{\epsilon A}$$

But $Q = CV$, and therefore

$$\mathscr{E} = \frac{Q}{\epsilon A} \qquad \text{(V/m)} \qquad \textbf{(9.11)}$$

which gives the electric field intensity between the plates in terms of the permittivity ϵ, the charge Q, and the surface area A of the plates. The ratio

$$\frac{C = \epsilon A/d}{C_o = \epsilon_o A/d} = \frac{\epsilon}{\epsilon_o} = \epsilon_r$$

or

$$C = \epsilon_r C_o \qquad \textbf{(9.12)}$$

which, in words, states that for the same set of parallel plates, the capacitance using a dielectric of mica (or any other dielectric) is five (the relativity permittivity of the dielectric) times that obtained for a vacuum (or air, approximately) between the plates. This relationship between ϵ_r and the capacitances provides an excellent method for finding the value of ϵ_r for various dielectrics.

EXAMPLE 9.1. Determine the capacitance of each capacitor of Fig. 9.6.

Solutions:

a. $C = 3(5\,\mu\text{F}) = \textbf{15}\,\boldsymbol{\mu}\textbf{F}$

b. $C = \frac{1}{2}(0.1\,\mu\text{F}) = \textbf{0.05}\,\boldsymbol{\mu}\textbf{F}$

c. $C = 2.5(20\,\mu\text{F}) = \textbf{50}\,\boldsymbol{\mu}\textbf{F}$

d. $C = (5)\frac{4}{(1/8)}(1000\,\text{pF}) = (160)(1000\,\text{pF}) = \textbf{0.16}\,\boldsymbol{\mu}\textbf{F}$

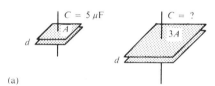

(a)

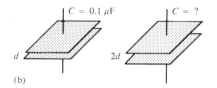

(b)

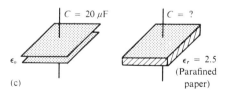

(c)

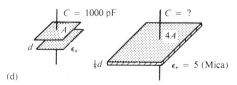

(d)

FIG. 9.6

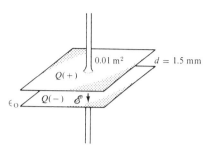

FIG. 9.7

EXAMPLE 9.2. For the capacitor of Fig. 9.7, determine
a. the capacitance.
b. the electric field strength between the plates.
c. the charge on each plate if 450 V are applied across the plates.

Solutions:

a. $C_o = \dfrac{\epsilon_o A}{d} = \dfrac{(8.85 \times 10^{-12})(0.01)}{1.5 \times 10^{-3}} = 59.0 \times 10^{-12}$

$\qquad = \textbf{59 pF}$

b. $\mathscr{E} = \dfrac{V}{d} = \dfrac{450}{1.5 \times 10^{-3}}$

$\qquad \cong \textbf{300} \times \textbf{10}^3 \textbf{ V/m}$

c. $C = \dfrac{Q}{V}$

or

$Q = CV = (59.0 \times 10^{-12})(450)$
$\qquad = 26.550 \times 10^{-9}$
$\qquad = \textbf{26.55 nC}$

EXAMPLE 9.3. A sheet of mica 1.5 mm thick having the same area as the plates is inserted between the plates of Example 9.2. Find
a. the electric field strength between the plates.
b. the charge on each plate.
c. the capacitance.

Solutions:
a. $\mathscr{E}$ is fixed by

$\mathscr{E} = \dfrac{V}{d} = \dfrac{450}{1.5 \times 10^{-3}}$

$\qquad \cong \textbf{300} \times \textbf{10}^3 \textbf{ V/m}$

b. $\mathscr{E} = \dfrac{Q}{\epsilon A}$

or

$Q = \epsilon \mathscr{E} A = \epsilon_r \epsilon_o \mathscr{E} A$
$\qquad = (5)(8.85 \times 10^{-12})(300 \times 10^3)(0.01)$
$\qquad = 132.75 \times 10^{-9} = \textbf{132.75 nC}$

(five times the amount for air between the plates)

c. $C = \epsilon_r C_o$
$\qquad = (5)(59 \times 10^{-12}) = \textbf{295 pF}$

9.4 DIELECTRIC STRENGTH

For every dielectric there is a potential that if applied across the dielectric will break the bonds within the dielectric and cause current to flow. The voltage required per unit length (electric field intensity) to establish conduction in a dielectric is an indication of its *dielectric strength* and is called the *breakdown voltage*. When breakdown occurs, the capacitor has characteristics very similar to those of a conductor. A typical example of breakdown is lightning, which occurs when the potential between the clouds and the earth is so high that charge can pass from one to the other through the atmosphere, which acts as the dielectric.

The average dielectric strengths for various dielectrics are tabulated in volts/mil in Table 9.2 (1 mil = 0.001 in.). The relative permittivity appears in parentheses to emphasize the importance of considering both factors in the design of capacitors. Take particular note of barium-strontium titanite and mica.

TABLE 9.2 *Dielectric strength of some dielectric materials.*

Dielectric	Dielectric Strength (Average Value), in Volts/Mil	(ϵ_r)
Air	75	(1.0006)
Barium-strontium titanite (ceramic)	75	(7500)
Porcelain	200	(6.0)
Transformer oil	400	(4.0)
Bakelite	400	(7.0)
Rubber	700	(3.0)
Paper, paraffined	1300	(2.5)
Teflon	1500	(2.0)
Glass	3000	(7.5)
Mica	5000	(5.0)

EXAMPLE 9.4. Find the maximum voltage that can be applied across a 0.2-μF capacitor having a plate area of 0.3 m². The dielectric is porcelain. Assume a linear relationship between the dielectric strength and the thick ,ss of the dielectric.

Solution:

$$C = \frac{8.85 \epsilon_r A}{10^{12} d}$$

or

$$d = \frac{8.85 \epsilon_r A}{10^{12} C} = \frac{(8.85)(6)(0.3)}{(10^{12})(0.2 \times 10^{-6})} = 7.965 \times 10^{-5}$$

$$\cong \textbf{79.65 } \mu\textbf{m}$$

Converting millimeters to mils, we have

$$79.65 \, \mu m \left(\frac{10^{-6} \, m}{\mu m} \right) \left(\frac{39.371 \, in.}{m} \right) \left(\frac{1000 \, mils}{1 \, in.} \right) = 3.136 \, mils$$

$$\text{Dielectric strength} = 200 \, \text{V/mil}$$

$$\therefore \left(\frac{200 \, \text{V}}{\text{mil}} \right) (3.136 \, \text{mils}) = \mathbf{627.20 \, V}$$

9.5 LEAKAGE CURRENT

Up to this point, we have assumed that the flow of electrons will occur in a dielectric only when the breakdown voltage is reached. This is the ideal case. In actuality, there are free electrons in every dielectric. These free electrons are due in part to impurities in the dielectric and forces within the material itself.

When a voltage is applied across the plates of a capacitor, a leakage current due to the free electrons flows from one plate to the other. The current is usually so small, however, that it can be neglected for most practical applications. This effect is represented by a resistor in parallel with the capacitor, as shown in Fig. 9.8(a). Its value is usually in the order of 1000 megohms (MΩ). There are some capacitors, however, such as the electrolytic type, that have high leakage currents. When charged and then disconnected from the charging circuit, these capacitors lose their charge in a matter of seconds because of the flow of charge (leakage current) from one plate to the other [Fig. 9.8(b)].

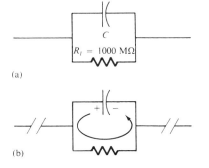

(a)

(b)

FIG. 9.8

9.6 TYPES OF CAPACITORS

Like resistors, all capacitors can be included under either of two general headings: fixed or variable. The symbol for a fixed capacitor is ⊥ and for a variable capacitor ⊀. The curved line represents the plate that is usually connected to the point of lower potential.

Many types of fixed capacitors are available today. Some of the most common are the mica, ceramic, electrolytic, tantalum, and polyester film capacitors. The typical *mica capacitor* consists basically of mica sheets separated by sheets of metal foil. The plates are connected to two electrodes, as shown in Fig. 9.9. The total area is the area of one sheet times the number of dielectric sheets. The entire system is encased in a plastic insulating material as shown in Fig. 9.10(a). The mica capacitor exhibits excellent characteristics under stress of temperature variations and high voltage applications (its dielectric strength is 5000 V/mil). Its leakage current is also very small (R_{leakage} is about 1000 MΩ).

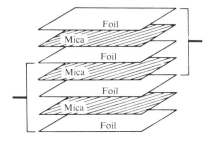

FIG. 9.9

Mica capacitors are typically between a few picofarads and 0.2 microfarad with voltages of 100 volts or more. The color code for the mica capacitors of Fig. 9.10(a) can be found in Appendix B.

A second type of mica capacitor appears in Fig. 9.10(b). Note in particular the cylindrical unit in the bottom left-hand corner of the figure. The ability to "roll" the mica to form the cylindrical shape is due to a process whereby the soluble contaminants in natural mica are removed, leaving a paperlike structure due to the cohesive forces in natural mica. It is commonly referred to as *reconstituted mica,* although the terminology does not mean "recycled" or "second-hand" mica. For some of the units in the photograph, different levels of capacitance are available between different sets of terminals.

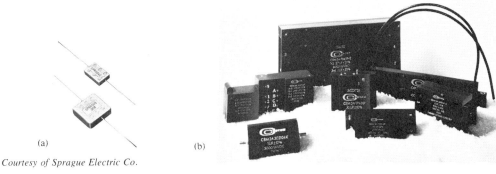

(a)

Courtesy of Sprague Electric Co.

(b)

Courtesy of Custom Electronics Inc.

FIG. 9.10 *Mica capacitors.*

The *ceramic capacitor* is made in many shapes and sizes, some of which are shown in Figs. 9.11 and 9.12. The basic construction, however, is about the same for each. A ceramic base is coated on two sides with a metal, such as copper or silver, to act as the two plates. The leads are then attached through electrodes to the plates. An insulating coating of ceramic or plastic is then applied over the plates and dielectric. Ceramic capacitors also have a very low leakage current (R_{leakage} about 1000 MΩ) and can be used in both dc and ac networks. They can be found in values ranging from a few picofarads to perhaps 2 microfarads, with very high working voltages such as 5000 volts or more.

Courtesy of Sprague Electric Co.

FIG. 9.11 *Ceramic disc capacitors.*

(a)

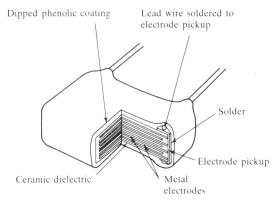

Dipped phenolic coating

Lead wire soldered to electrode pickup

Solder

Electrode pickup

Metal electrodes

Ceramic dielectric

(Alternately deposited layers of ceramic dielectric material and metal electrodes fired into a single homogeneous block)

(b)

Courtesy of Sprague Electric Co.

FIG. 9.12 *Multilayer, radial-lead ceramic capacitors.*

A molded ceramic capacitor employing the "stacking" method of Fig. 9.9 appears in Fig. 9.12.

In recent years there has been increasing interest in monolithic (single-structure) chip capacitors such as appearing in Fig. 9.13(a) due to their application on hybrid circuitry [networks using both discrete and integrated circuit (IC) components]. There has also been increasing use of microstrip (strip-line) circuitry such as appearing in Fig. 9.13(b).

(a)

Courtesy of Vitramon, Inc.

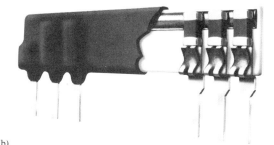

(b)

FIG. 9.13 *Monolithic chip capacitors.*

Note the small chips in this cutaway section. The *L* and *H* of Fig. 9.13(a) indicate the level of capacitance. For example, the letter *H* in black letters represents 16 units of capacitance (in picofarads), or 16 pF. If blue ink is used, a multiplier of 100 is applied, resulting in 1600 pF. Although the size is similar, the type of ceramic material controls the capacitance level.

The capacitance level of the chips is sensitive to temperature and applied dc voltage, as shown in Fig. 9.14. Note the sharp change in capacitance at temperature levels below freezing (0°C) and the linear (straight-line) decline in capacitance with increase in voltage. Consider, however, that for a temperature near 50°C and an applied voltage of 40 V, the percent *change* in capacitance for small variations in temperature and voltage is quite reasonable.

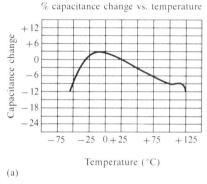

(a)

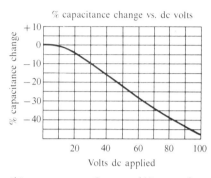

(b)   *Courtesy of Vitramon, Inc.*

FIG. 9.14 *Effects of temperature and applied voltage on the characteristics of a chip capacitor.*

The *electrolytic capacitor* is used most commonly in situations where capacitances of the order of one to several thousand microfarads are required. They are designed primarily for use in networks where only dc voltages will be applied across the capacitor. There are electrolytic capacitors available that can be used in ac circuits (for starting motors) and in cases where the polarity of the dc voltage will reverse across the capacitor for short periods of time.

The basic construction of the electrolytic capacitor consists of a roll of aluminum foil coated on one side with an aluminum oxide, the aluminum being the positive plate and the oxide the dielectric. A layer of paper or gauze saturated with an electrolyte is placed over the aluminum oxide on the positive plate. Another layer of aluminum without the oxide coating is then placed over this layer to assume the role of the negative plate. In most cases the negative plate is connected directly to the aluminum container, which then serves as the negative terminal for external connections. Because of the size of the roll of aluminum foil, the overall area of this capacitor is large; and due to the use of an oxide as the dielectric, the distance between the plates is extremely small. The negative terminal of the electrolytic capacitor is usually the one with no visible identification on the casing. The positive is usually indicated by such designs as +, △, □, and so on. Due to the polarity requirement, the symbol for an electrolytic will normally appear as ⊥+.

Associated with each electrolytic capacitor are the dc working voltage and the surge voltage. The *working voltage* is the voltage that can be applied across the capacitor for long periods of time without breakdown. The *surge voltage* is the maximum dc voltage that can be applied for a short period of time. Electrolytic capacitors are characterized as having low breakdown voltages and high leakage currents ($R_{leakage}$ about 1 MΩ). Various types of electrolytic capacitors are shown in Fig. 9.15. They can be found in values extending from a few microfarads to several thousand microfarads and working voltages as high as 500 volts. However, increased levels of voltage are normally associated with lower values of available capacitance.

(a)

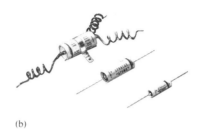

(b)

Courtesy of Sprague Electric Co.

FIG. 9.15 *Electrolytic capacitors.*

There are fundamentally two types of *tantalum* capacitors: the *solid* and the *wet-slug*. In each case, tantalum powder of high purity is pressed into a cylindrical or rectangular shape as shown in Fig. 9.16. The anode connection is then simply pressed into the resulting structures as shown in the figure. The resulting unit is then sintered (baked) in a vacuum at very high temperatures to establish a very porous material. The result is a structure with a very large surface area in a limited volume. Through immersion in an acid solution, a very thin tantalum pentoxide (Ta_2O_5) dielectric coating is established on the large, porous surface area. An electrolyte is then added to establish contact between the surface area and the cathode. If manganese nitrate is used to form a manganese dioxide (MnO_2) layer, the result is a solid tantalum capacitor. If an appropriate "wet" acid is introduced, it is called a *wet-slug* tantalum capacitor.

The last type of fixed capacitor to be introduced is the *polyester film capacitor,* the basic construction of which is shown in Fig. 9.17. It consists simply of two metal foils separated by a strip of polyester material such as Mylar®. The outside layer of polyester is applied to act as an insulating jacket. Each metal foil is connected to a lead which extends either axially or radially from the capacitor. The rolled construction results in a large surface area, and the use of the plastic dielectric results in a very thin layer between the conducting surfaces.

Data such as capacitance and working voltage are printed on the outer wrapping if the polyester capacitor is large enough. Color coding is used on smaller devices (see

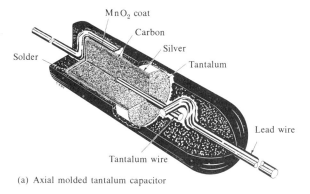

(a) Axial molded tantalum capacitor

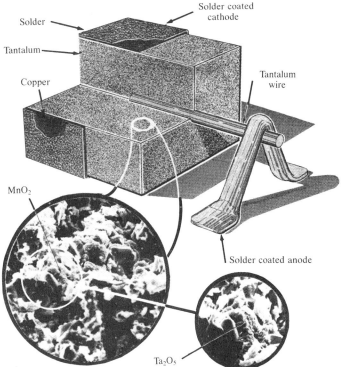

(b) High temperature chip capacitor

Courtesy Union Carbide Corporation

FIG. 9.16 *Tantalum capacitors.*

Appendix C). A band (usually black) is sometimes printed near the lead that is connected to the outer metal foil. The lead nearest this band should always be connected to the point of lower potential. This capacitor can be used for both dc and ac networks. Its leakage resistance is of the order of 100 MΩ. A typical polyester capacitor appears in Fig. 9.18. Polyester capacitors range in value from a few hundred picofarads to 10–20 microfarads, with working voltages as high as a few thousand volts.

The most common of the variable-type capacitors is shown in Fig. 9.19. The dielectric for each capacitor is air. The capacitance in Fig. 9.19(a) is changed by turning the shaft at one end to vary the common area of the movable and fixed plates. The greater the common area, the larger

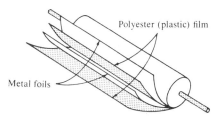

FIG. 9.17 *Polyester film capacitor.*

Capacitors

Courtesy of Sprague Electric Co.

FIG. 9.18 *Orange Drop® tubular capacitors.*

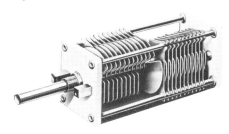

(a)

Courtesy of James Millen Manufacturing Co.

FIG. 9.19 *Variable air capacitors.*

(b)

Courtesy of Johnson Manufacturing Co.

Courtesy of Global Specialties Corp.

FIG. 9.20 *Digital reading capacitance meter.*

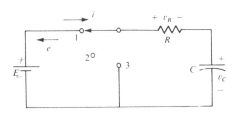

FIG. 9.21

the capacitance, as determined by Eq. (9.10). The capacitance of the trimmer capacitor in Fig. 9.19(b) is changed by turning the screw, which will vary the distance between the plates and thereby the capacitance. A digital reading capacitance meter appears in Fig. 9.20.

9.7 TRANSIENTS IN CAPACITIVE NETWORKS

We discussed how the capacitor acquires its charge in Section 9.3. Let us now extend this discussion to include the potentials and currents developed within the network. In any electrical system, the capacitor passes through two transient phases: *charging* and *discharging*.

The circuit in Fig. 9.21 was designed both to charge and to discharge the capacitor. During the charging phase (switch position 1), electrons are drawn from the top plate and deposited on the bottom plate by the battery, resulting in a net positive charge on the top plate and a negative charge on the bottom plate. The transfer of electrons continues until the *potential across the capacitor is exactly equal to the emf (E) applied* (note Fig. 9.23).

The current in the circuit at any instant of time is determined by the amount of charge that passes through a perpendicular cross section of the wire in a vanishingly small period of time. This is also equivalent to the amount of charge acquired by either plate in the same period of time.

The instantaneous current is described by the following equation:

$$i = \frac{dq}{dt} \qquad (9.13)$$

where d signifies a vanishingly small change in charge and time. The expression dq/dt is called the *derivative* of the quantity q with respect to time. For this situation, if charge is not deposited on the plates or passing through the conductor, at a particular instant, $dq = 0$, and $i = dq/dt = 0$. From this point on, lower-case letters will be used to represent quantities that vary with time (called *variables*) such as the current (i), charge (q), and time (t).

An equation relating the current of the capacitor to the voltage across the capacitor can be found by substituting for q in Eq. (9.13). That is,

$$q = Cv \text{ [Eq. (9.5)] } \text{ and } i = \frac{dq}{dt} = \frac{d}{dt}Cv$$

But $C = $ constant (time independent), and therefore

$$\boxed{i_C = C\frac{dv_C}{dt}} \tag{9.14}$$

The impact of this equation is discussed in greater detail in Section 9.9.

The mathematical expression for the voltage across the capacitor can be found by first applying Kirchhoff's voltage law around the closed loop:

$$E - v_R - v_C = 0 \quad \text{or} \quad v_R + v_C = E$$

Substituting $v_R = iR$, we have

$$iR + v_C = E$$

Since the current i is the same for the resistor and capacitor,

$$i = C\frac{dv_C}{dt} \quad \text{and} \quad C\frac{dv_C}{dt}R + v_C = E$$

or

$$RC\frac{dv_C}{dt} + v_C = E$$

The voltage v_C can then be determined using methods of *calculus:*

$$\boxed{v_C(t) = E(1 - e^{-t/RC})} \tag{9.15}$$

The units of measurement for both R and C in Eq. (9.15) are

$$R = \frac{E}{I} = \frac{\text{volts}}{\text{amperes}} = \frac{\text{volts}}{\text{coulombs/second}}$$

and

$$C = \frac{Q}{V} = \frac{\text{coulombs}}{\text{volts}}$$

with their product resulting in

$$RC = \left(\frac{\text{volts}}{\text{coulombs/second}}\right)\left(\frac{\text{coulombs}}{\text{volts}}\right) = \textbf{seconds}$$

The product RC is called the *time constant* of the system and has the units of time (seconds). Its symbol is the Greek letter τ (tau); that is,

$$\boxed{\tau = RC} \qquad \text{(s)} \qquad\qquad \textbf{(9.16)}$$

Equation (9.15) can then be written as

$$\boxed{v_C(t) = E(1 - e^{-t/\tau})} \quad {}_{charging} \qquad \textbf{(9.17)}$$

For further analysis, the exponential functions $e^{-t/\tau}$ and $1 - e^{-t/\tau}$ have been plotted in Fig. 9.22. Note the presence of $(1 - e^{-t/\tau})$ in Eqs. (9.15) and (9.17). The plot for $v_C(t)$ is therefore as shown in Fig. 9.23. In Appendix G, values of e^{-x} are tabulated. They are also directly available on most modern calculators.

The rate of change of v_C in percent during each time interval RC is shown in Table 9.3. Note that the capacitor charges to 63.2% of its final value in only one time constant and to 86.5% in only two time constants. Consider also that the change in $v_C(t)$ from the fifth to sixth time constant is only 0.4% and quite a bit less for an increasing number of time constants. For all practical purposes, therefore, a capacitor will charge to its final voltage in *five time constants*. Since C is usually found in microfarads or picofarads, the time constant τ will never be greater than a few seconds unless R is very large.

TABLE 9.3

$(0 \rightarrow 1)\tau$	63.2%
$(1 \rightarrow 2)\tau$	23.3%
$(2 \rightarrow 3)\tau$	8.6%
$(3 \rightarrow 4)\tau$	3.0%
$(4 \rightarrow 5)\tau$	1.2%
$(5 \rightarrow 6)\tau$	0.4%

If we keep R constant and reduce C, the product RC will decrease, and the rise time of five time constants will decrease. The change in transient behavior of the voltage v_C is plotted in Fig. 9.24 for various values of C. The product RC will always have some numerical value, even though it may be very small in some cases. For this reason, *the voltage across a capacitor cannot change instantaneously*. In fact, the capacitance of a network is also a measure of how much it will oppose a change in voltage across the network. The larger the capacitance, the larger the time constant and the longer it takes to charge up to its final value (curve of C_3 in Fig. 9.24). A lesser capacitance would permit the voltage to build up more quickly since the time constant is less (curve of C_1 in Fig. 9.24).

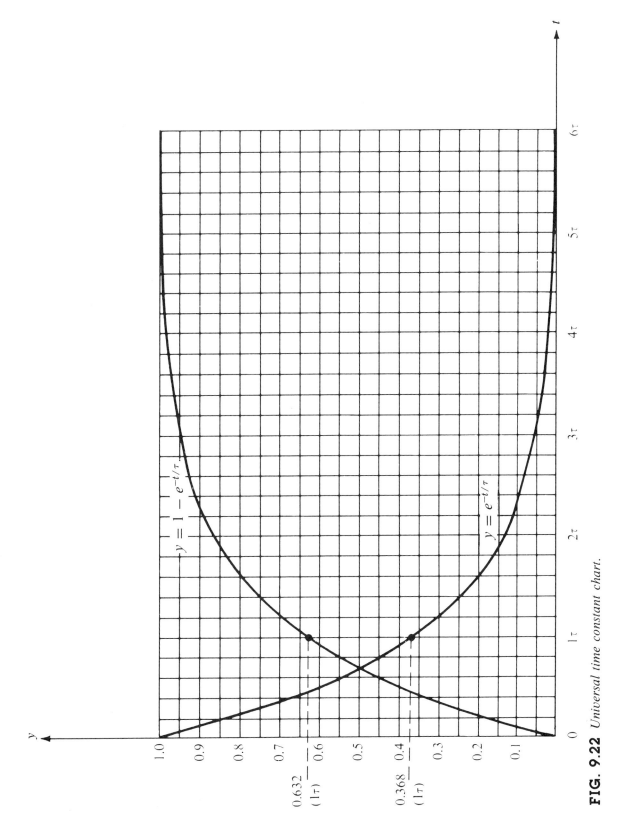

FIG. 9.22 *Universal time constant chart.*

241

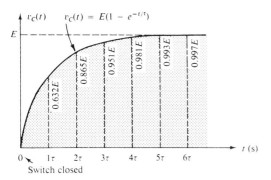

FIG. 9.23

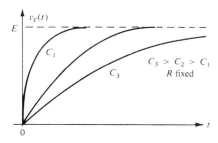

FIG. 9.24

The rate at which charge is deposited on the plates during the charging phase can be found by substituting the following for $v_C(t)$ in Eq. (9.17):

$$v_C(t) = \frac{q(t)}{C}$$

and

$$\boxed{q(t) = Cv_C(t) = CE(1 - e^{-t/\tau})} \quad \text{charging} \quad \textbf{(9.18)}$$

indicating that the charging rate is very high during the first few time constants and less than 1% after five time constants.

Let us now examine the current in the circuit for this charging phase. If we substitute $t = 0$ into Eq. (9.17), we find that $v_C(t) = 0$ volts when the switch is closed. *The capacitor, therefore, is effectively a short circuit when the switch is first closed* ($t = 0$). The current in the network is then determined by the remaining elements of the circuit. Applying Ohm's law, we find $I = E/R$ at $t = 0$. Since the charging rate (depositing of charge on the plates) determined by Eq. (9.18) will drop with an increasing number of time constants, the current $i_C(t)$ will also begin to decay, eventually falling to zero when $v_C(t) = E$ volts. A mathematical expression for $i_C(t)$ can be found by *differentiating* (calculus) $q(t)$ with respect to time; that is,

$$i_C(t) = \frac{d}{dt} q(t) = \frac{d}{dt} CE(1 - e^{-t/\tau})$$

and

$$\boxed{i_C(t) = \frac{E}{R} e^{-t/\tau}} \quad \text{charging} \quad \textbf{(9.19)}$$

Note the presence of the factor $e^{-t/\tau}$ which appears in Fig. 9.22 and results in the plot of Fig. 9.25.

The voltage across the resistor is determined by Ohm's law:

$$v_R(t) = Ri_C(t) = R \frac{E}{R} e^{-t/\tau}$$

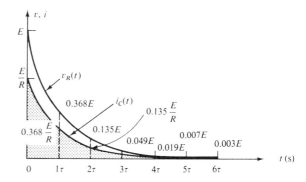

FIG. 9.25

or

$$v_R(t) = Ee^{-t/\tau}$$ *charging* **(9.20)**

A plot of $v_R(t)$ appears in Fig. 9.25.

Due to the short-circuit representation for the capacitor at $t = 0$, the voltage v_R jumped to the input voltage E and then declined toward zero at the same rate at which v_C approached E (Fig. 9.25). Eventually, therefore, the voltage v_R and current i_C will reach zero at the same instant at which v_C reaches E. These terminal characteristics, as shown in Fig. 9.26, indicate that when a capacitor has charged to its final value in a dc network, it can, for all practical purposes, be replaced by an open circuit. For most applications, a capacitor in a dc circuit may be replaced by an open circuit after a period of time equal to five time constants has passed. The capacitor is, of course, quite different from a simple open circuit since it stores a measurable amount of charge on the two plates at opposite ends of the open circuit.

Once the voltage across the capacitor has reached the input voltage E, the capacitor is fully charged and will remain in this state if no further changes are made in the circuit.

If the switch is moved to position 2, as shown in Fig. 9.27(a), the capacitor will retain its charge for a period of time determined by its leakage current. For capacitors such

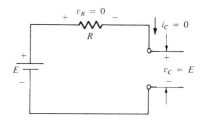

FIG. 9.26

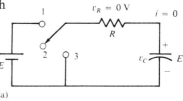

(a)

as the mica and ceramic, the leakage current is very small, so the capacitor will retain its charge, and hence the potential difference across its plates, for a long period of time. For electrolytic capacitors, which have very high leakage currents, the capacitor will discharge more rapidly, as shown in Fig. 9.27(b). In any event, to insure that they are completely discharged, capacitors should be shorted by a lead or a screwdriver before they are handled.

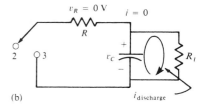

(b)

FIG. 9.27

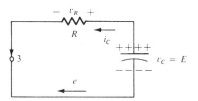

FIG. 9.28

If the switch is then moved to position 3, the capacitor will enter the discharge phase as shown in Fig. 9.28. The electrons on the negative plate will travel toward the positive plate the instant the switch is closed. This reduces the charge on the plates and hence the potential across the plates ($V = Q/C$). The capacitor is acting as a source of emf that is decreasing toward zero with time. The mathematical expressions for the current i_C and potentials v_C and v_R can again be found by using calculus. Note that each is an exponentially decaying function. The functions are

$$\boxed{v_C(t) = Ee^{-t/\tau}} \quad discharging \qquad \textbf{(9.21)}$$

$$\boxed{i_C(t) = \frac{E}{R}e^{-t/\tau}} \quad discharging \qquad \textbf{(9.22)}$$

$$\boxed{v_R(t) = Ee^{-t/\tau} = v_C(t)} \quad discharging \qquad \textbf{(9.23)}$$

The complete discharge will occur, for all practical purposes, in five time constants. Note that the current i_C has reversed direction and the voltage v_R has reversed polarity.

If the switch of Fig. 9.21 is moved between the various positions every five time constants, the wave shapes will be as shown in Fig. 9.29 for the current i_C and voltages v_C and v_R.

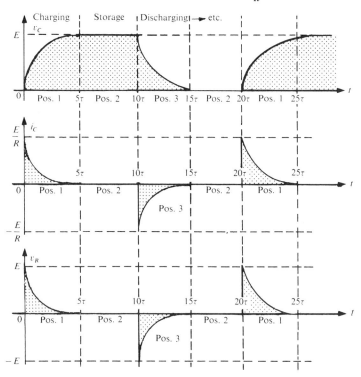

FIG. 9.29

EXAMPLE 9.5. Find the mathematical expressions for the transient behavior of v_C, i_C, and v_R for the circuit of Fig.

9.30 after the closing of the switch. Plot the curves of v_C, i_C and v_R.

Solution:

$$\tau = RC = (8 \times 10^3)(4 \times 10^{-6}) = 32 \times 10^{-3} = \mathbf{32\ ms}$$

By Eq. (9.17),

$$v_C = E(1 - e^{-t/\tau}) = \mathbf{40(1 - e^{-t/32 \times 10^{-3}})}$$

By Eq. (9.19),

$$i_C = \frac{E}{R}e^{-t/\tau} = \frac{40}{8\ k\Omega}e^{-t/32 \times 10^{-3}}$$
$$= \mathbf{5 \times 10^{-3}e^{-t/32 \times 10^{-3}}}$$

By Eq. (9.20),

$$v_R = Ee^{-t/\tau} = \mathbf{40e^{-t/32 \times 10^{-3}}}$$

The curves appear in Fig. 9.31.

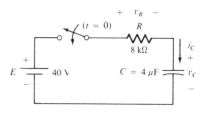

FIG. 9.30

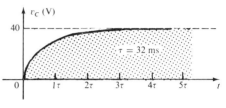

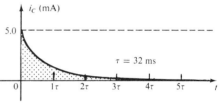

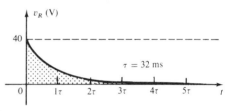

FIG. 9.31

EXAMPLE 9.6. After v_C has reached its final value of 40 V in Fig. 9.30, a second switch is thrown, as shown in Fig. 9.32. Find the mathematical expressions for the transient behavior of v_C, i_C, and v_R after the closing of the switch. Plot the curves for v_C, i_C, and v_R.

Solution:

$$\tau = 32\ ms$$

By Eq. (9.21),

$$v_C = Ee^{-t/\tau} = \mathbf{40e^{-t/32 \times 10^{-3}}}$$

By Eq. (9.22),

$$i_C = \frac{E}{R}e^{-t/\tau} = \mathbf{5 \times 10^{-3}e^{-t/32 \times 10^{-3}}}$$

By Eq. (9.23),

$$v_R = Ee^{-t/\tau} = \mathbf{40e^{-t/32 \times 10^{-3}}}$$

The curves appear in Fig. 9.33.

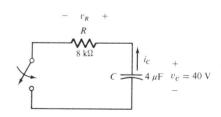

FIG. 9.32

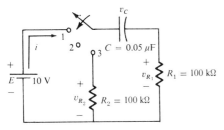

FIG. 9.33

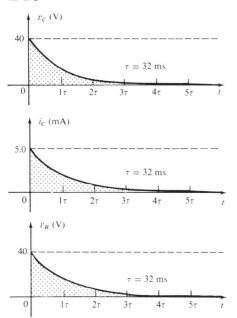

FIG. 9.34

EXAMPLE 9.7.

a. Find the mathematical expression for the transient behavior of the voltage across the capacitor of Fig. 9.34 after the switch is thrown into position 1 (at $t = 0$ s). Calculate the voltage v_C at $t = 10$ ms.

b. Repeat part (a) for i_C and v_{R_1}.

c. Find the mathematical expression for the transient behavior of the voltages v_C and v_{R_1} and the current i_C if the switch is then thrown into position 2 at $t = 30$ ms. (Assume that the leakage resistance of capacitor is equal to ∞ ohms.)

d. Find the mathematical expression for the transient behavior of the voltages v_C and $v_{R_1} + v_{R_2}$ and the current i_C if the switch is thrown into position 3 at $t = 48$ ms.

e. Plot the waveforms obtained in parts (a) and (d) on the same time axis for each voltage and the current.

Solutions:

a. $v_C(t) = E(1 - e^{-t/\tau})$

       $\tau = RC = (100 \times 10^3)(0.05 \times 10^{-6}) = 5 \times 10^{-3}$
                               $= 5$ ms

       $\therefore v_C(t) = \mathbf{10(1 - e^{-t/5 \times 10^{-3}})}$

       at $t = 10$ ms.

       Note that

            $\tau = RC = 5$ ms
        $\therefore t = 10$ ms $= 2\tau$

Fig. 9.22 indicates that

$$v_C = 0.865E \text{ in } 2\tau$$
$$\therefore v_C = (0.865)(10) = \mathbf{8.65 \ V}$$

or

$$v_C(t) = 10[1 - e^{-(10\times10^{-3})/(5\times10^{-3})}]$$
$$= 10[1 - e^{-2}] = 10(1 - 0.135) = 10(0.865)$$
$$= \mathbf{8.65 \ V}$$

b. $i_C(t) = \dfrac{E}{R}e^{-t/\tau}$

$$= \left(\frac{10}{100 \times 10^3}\right)e^{-t/5\times10^{-3}}$$
$$= \mathbf{0.1 \times 10^{-3}}e^{-t/5\times10^{-3}}$$

At $t = 10$ ms,

$$i_C(t) = 10^{-4}e^{-10\times10^{-3}/5\times10^{-3}}$$
$$= (10^{-4})(e^{-2}) = (10^{-4})(0.135)$$
$$= \mathbf{13.5 \ \mu A}$$
$$v_{R_1}(t) = Ee^{-t/RC}$$
$$= 10e^{-t/5\times10^{-3}}$$

At $t = 10$ ms,

$$v_{R_1}(t) = 10(e^{-2}) = 10(0.135)$$
$$= \mathbf{1.35 \ V}$$

c. $v_C(t) = E = \mathbf{10 \ V}$

$\quad i_C(t) = \mathbf{0}$

$\quad v_R(t) = \mathbf{0}$

d. $\qquad\qquad v_C(t) = Ee^{-t/\tau'}$

$$\tau' = (R_1 + R_2)C = R_T C$$
$$= (100 \ \text{k}\Omega + 100 \ \text{k}\Omega)(0.005 \times 10^{-6})$$
$$= (200 \times 10^3)(0.05 \times 10^{-6})$$
$$= (2 \times 10^5)(5 \times 10^{-8})$$
$$= 10 \times 10^{-3}$$
$$= 10 \text{ ms (twice the original } \tau)$$
$$v_C(t) = \mathbf{10}e^{-t/10\times10^{-3}}$$
$$i_C(t) = \frac{E}{R}e^{-t/\tau'}$$
$$= \frac{10}{200 \ \text{k}\Omega}e^{-t/10\times10^{-3}}$$
$$i_C(t) = \mathbf{0.05 \times 10^{-3}}e^{-t/10\times10^{-3}}$$
$$v_{R_T}(t) = Ee^{-t/\tau'}$$
$$= \mathbf{10}e^{-t/10\times10^{-3}}$$

e. Fig. 9.35 (Note the reversal in i_C and v_R but not in v_C.)

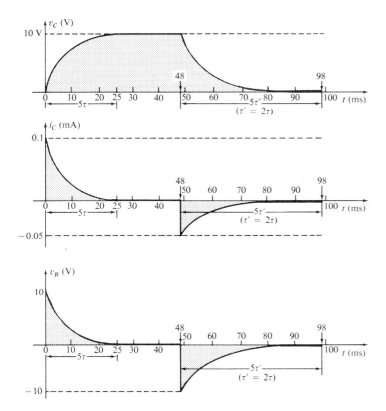

FIG. 9.35

9.8 $\tau = R_{Th}C$

Occasions will arise in which the network does not have the simple series form of Fig. 9.21. It will then be necessary to first find the Thevenin equivalent circuit for the network external to the capacitive element. E_{Th} will then be the source voltage E of Eqs. (9.15) through (9.23) and R_{Th} the resistance R. The time constant is then $\tau = R_{Th}C$.

EXAMPLE 9.8. For the network of Fig. 9.36:

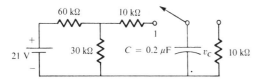

FIG. 9.36

    a. Find the mathematical expression for the transient behavior of the voltage v_C after the closing of the switch (position 1 at $t = 0$ s).

    b. Find the mathematical expression for the voltage v_C as a function of time if the switch is thrown into position 2 at $t = 15$ ms.

    c. Draw the resultant waveform of parts (a) and (b) on the same time axis.

Solutions:

a. Applying Thevenin's theorem to the 0.2-μF capacitor, we obtain Fig. 9.37:

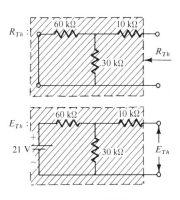

$$R_{Th} = \frac{(60 \text{ k}\Omega)(30 \text{ k}\Omega)}{90 \text{ k}\Omega} + 10 \text{ k}\Omega$$
$$= 20 \text{ k}\Omega + 10 \text{ k}\Omega$$
$$= 30 \text{ k}\Omega$$

$$E_{Th} = \frac{(30 \text{ k}\Omega)(21)}{30 \text{ k}\Omega + 60 \text{ k}\Omega} = \frac{1}{3}(21) = 7 \text{ V}$$

The resultant Thevenin equivalent circuit with the capacitor replaced is shown in Fig. 9.38:

FIG. 9.37

$$v_C(t) = E(1 - e^{-t/\tau})$$
$$\tau = RC = (30 \text{ k}\Omega)(0.2 \text{ }\mu\text{F})$$
$$= (30 \times 10^3)(0.2 \times 10^{-6}) = 6 \times 10^{-3}$$
$$= 6 \text{ ms}$$
$$v_C(t) = \mathbf{7(1 - e^{-t/6\times10^{-3}})}$$

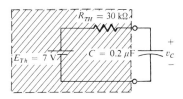

b. At $t = 15$ ms,

$$v_C(t) = 7[1 - e^{-15\times10^{-3}/6\times10^{-3}}]$$
$$= 7(1 - e^{-2.5}) = 7(1 - 0.082) = 7(0.918)$$
$$= \mathbf{6.426 \text{ V}}$$

$$v_C(t) = Ee^{-t/\tau'}$$
$$\tau' = RC = (10 \text{ k}\Omega)(0.2 \text{ }\mu\text{F})$$
$$= (10 \times 10^3)(0.2 \times 10^{-6}) = 2 \times 10^{-3}$$
$$= 2 \text{ ms}$$
$$v_C(t) = \mathbf{6.426}e^{-t/2\times10^{-3}}$$

FIG. 9.38

c. See Fig. 9.39.

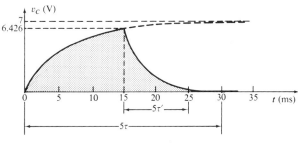

FIG. 9.39

EXAMPLE 9.9. The capacitor of Fig. 9.40 is initially charged to 80 V. Find the mathematical expression for v_C after the closing of the switch.

Solution: The network is redrawn in Fig. 9.41:

$$R_{Th} = 5 \text{ k}\Omega + 18 \text{ k}\Omega \parallel (7 \text{ k}\Omega + 2 \text{ k}\Omega)$$
$$= 5 \text{ k}\Omega + 18 \text{ k}\Omega \parallel 9 \text{ k}\Omega = 5 \text{ k}\Omega + 6 \text{ k}\Omega = 11 \text{ k}\Omega$$
$$\tau = R_{Th}C = (11 \times 10^3)(40 \times 10^{-6})$$
$$= 440 \times 10^{-3} = 0.44 \text{ s}$$

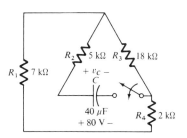

FIG. 9.40

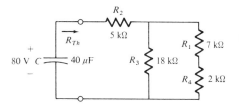

FIG. 9.41

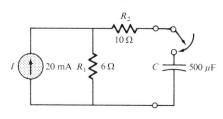

FIG. 9.42

and

$$v_C(t) = Ee^{-t/\tau} = \mathbf{80}e^{-t/0.44}$$

EXAMPLE 9.10. For the network of Fig. 9.42, find the mathematical expression for the voltage v_C after the closing of the switch (at $t = 0$).

Solution:

$$R_{Th} = 6 + 10 = 16\ \Omega$$
$$E_{Th} = V_{6\Omega} + V_{10\Omega} = IR_1 + 0$$
$$= (20 \times 10^{-3})(6) = 120 \times 10^{-3} = 0.12\ \text{V}$$

and

$$\tau = R_{Th}C = (16)(500 \times 10^{-6}) = 8\ \text{ms}$$

so that

$$v_C(t) = \mathbf{0.12(1 - }e^{-t/8\times10^{-3}}\mathbf{)}$$

9.9 THE CURRENT i_C

In the previous section the current i_C associated with a capacitance C was found to be related to the voltage across the capacitor by

$$i_C = C\frac{dv_C}{dt}$$

which, in words, states that the current associated with a capacitor at any instant of time is determined by the capacitance C and the instantaneous change in voltage across the plates. If the voltage fails to change at a particular instant,

$$dv_C = 0 \quad \text{and} \quad i_C = C\frac{dv_C}{dt} = 0$$

In an effort to develop a clearer understanding of Eq. (9.14), let us calculate the average current associated with a capacitor for various voltages impressed across the capacitor.

The average current is defined by the equation

$$i_{Cav} = \frac{\Delta q}{\Delta t} = C\frac{\Delta v_C}{\Delta t} \tag{9.24}$$

where Δ indicates a finite change in charge, voltage, or time. The instantaneous current can be derived from Eq. (9.24) by letting Δt become vanishingly small; that is,

$$i_{Cinst} = \lim_{\Delta t \to 0} C\frac{\Delta v_C}{\Delta t} = C\frac{dv_C}{dt}$$

In the following example, the change in voltage Δv_C will be considered for each slope of the voltage waveform. If the voltage increases with time, the average current is the change in voltage over the change in time, with a positive sign. If the voltage decreases with time, the average current is again the change in voltage over the change in time, but with a negative sign.

EXAMPLE 9.11. Find the waveform for the average current if the voltage across a 2-μF capacitor is as shown in Fig. 9.43.

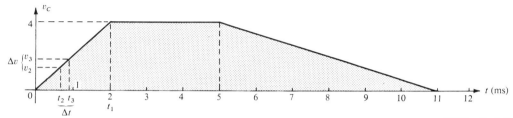

FIG. 9.43

Solutions:

a. From 0 to 2 ms, the voltage increases linearly from 0 to 4 V, the change in voltage $\Delta v = 4 - 0 = 4$ (with a positive sign, since the voltage increases with time). The change in time $\Delta t = 2 - 0 = 2$ ms, and

$$i_C = C\frac{\Delta v_C}{\Delta t} = (2 \times 10^{-6})\left(\frac{4}{2 \times 10^{-3}}\right)$$

$$= 4 \times 10^{-3} = 4 \text{ mA}$$

b. From 2 to 5 ms, the voltage remains constant at 4 V; the change in voltage $\Delta v = 0$. The change in time $\Delta t = 3$ ms, and

$$i_C = C\frac{\Delta v_C}{\Delta t} = C\frac{0}{\Delta t} = 0$$

c. From 5 to 11 ms, the voltage decreases from 4 to 0 V. The change in voltage Δv is therefore $4 - 0 = 4$ V (with a negative sign, since the voltage is decreasing with time). The change in time $\Delta t = 11 - 5 = 6$ ms, and

$$i_C = C\frac{\Delta v_C}{\Delta t} = -(2 \times 10^{-6})\left(\frac{4}{6 \times 10^{-3}}\right)$$

$$= -1.33 \times 10^{-3} = -1.33 \text{ mA}$$

d. From 11 ms on, the voltage remains constant at 0 and $\Delta v = 0$, so $i_C = 0$. The waveform for the average current for the impressed voltage is as shown in Fig. 9.44.

The average value for the preceding example is actually the same as the instantaneous value at any point along the slope over which the average value was found. For example, if the interval Δt is reduced from $0 \rightarrow t_1$ to $t_2 \rightarrow t_3$ as

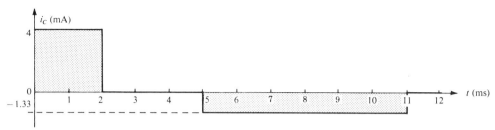

FIG. 9.44

noted in Fig. 9.33, $\Delta v/\Delta t$ is still the same. In fact, no matter how small the interval Δt, the slope will be the same, and therefore the current $i_{C_{av}}$ will be the same. If we consider the limit as $\Delta t \rightarrow 0$, the slope will still remain the same, and therefore $i_{C_{av}} = i_{C_{inst}}$ at any instant of time between 0 and t_1. The same can be said about any portion of the voltage waveform that has a constant slope.

The important point to be gained by this discussion is that the current associated with a capacitor is zero if the voltage across the capacitor fails to change with time. It should also be noted in the example just completed that *the greater the rate of change of voltage* across the capacitor, *the greater the current* associated with the capacitor.

The method described above is only for waveforms with straight line (linear) segments. For nonlinear (curved) waveforms, a method of calculus (differentiation) must be employed.

9.10 CAPACITORS IN SERIES AND PARALLEL

Capacitors, like resistors, can be placed in series and parallel. For capacitors in series, the charge is the same on each capacitor (Fig. 9.45).

FIG. 9.45

$$\boxed{Q_T = Q_1 = Q_2 = Q_3} \qquad (9.25)$$

Applying Kirchhoff's voltage law around the closed loop gives

$$E = V_1 + V_2 + V_3$$

However,

$$V = \frac{Q}{C}$$

so that

$$\frac{Q_T}{C_T} = \frac{Q_1}{C_1} + \frac{Q_2}{C_2} + \frac{Q_3}{C_3}$$

Using Eq. (9.25) and dividing both sides by Q yields

$$\frac{1}{C_T} = \frac{1}{C_1} + \frac{1}{C_2} + \frac{1}{C_3} \qquad \textbf{(9.26)}$$

which is similar to the manner in which we found the total resistance of a parallel resistive circuit. The total capacitance of two capacitors in series is

$$C_T = \frac{C_1 C_2}{C_1 + C_2} \qquad \textbf{(9.27)}$$

For capacitors in parallel as shown in Fig. 9.46, the voltage is the same across each capacitor, and the total charge is the sum of that on each capacitor:

$$Q_T = Q_1 + Q_2 + Q_3 \qquad \textbf{(9.28)}$$

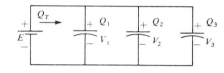

FIG. 9.46

However,

$$Q = CV \qquad \therefore C_T E = C_1 V_1 + C_2 V_2 + C_3 V_3$$

and

$$E = V_1 = V_2 = V_3$$

Thus,

$$C_T = C_1 + C_2 + C_3 \qquad \textbf{(9.29)}$$

which is similar to the manner in which the total resistance of a series circuit is found.

EXAMPLE 9.12. For the circuit of Fig. 9.47:
a. Find the total capacitance.
b. Determine the charge on each plate.
c. Find the voltage across each capacitor.

Solutions:

a. $\dfrac{1}{C_T} = \dfrac{1}{C_1} + \dfrac{1}{C_2} + \dfrac{1}{C_3} = \dfrac{1}{200 \times 10^{-6}}$

$$+ \frac{1}{50 \times 10^{-6}} + \frac{1}{10 \times 10^{-6}}$$
$$= 0.005 \times 10^6 + 0.02 \times 10^6$$
$$+ 0.1 \times 10^6$$
$$= 0.125 \times 10^6$$

and

$$C_T = \frac{1}{0.125 \times 10^6} = \textbf{8 } \boldsymbol{\mu}\textbf{F}$$

FIG. 9.47

b. $Q_T = Q_1 = Q_2 = Q_3$
$Q_T = C_T E = (8 \times 10^{-6})(60) = \textbf{480 } \boldsymbol{\mu}\textbf{C}$

c. $V_1 = \dfrac{Q_1}{C_1} = \dfrac{480 \times 10^{-6}}{200 \times 10^{-6}} = \textbf{2.4 V}$

$$V_2 = \frac{Q_2}{C_2} = \frac{480 \times 10^{-6}}{50 \times 10^{-6}} = \mathbf{9.6\ V}$$

$$V_3 = \frac{Q_3}{C_3} = \frac{480 \times 10^{-6}}{10 \times 10^{-6}} = \mathbf{48.0\ V}$$

and

$$E = V_1 + V_2 + V_3 = 2.4 + 9.6 + 48$$
$$= \mathbf{60\ V} \quad \text{(checks)}$$

EXAMPLE 9.13. For the network of Fig. 9.48:
a. Find the total capacitance.
b. Determine the charge on each plate.
c. Find the total charge.

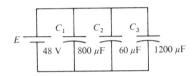

FIG. 9.48

Solutions:

a. $C_T = C_1 + C_2 + C_3 = 800\ \mu F + 60\ \mu F + 1200\ \mu F$
$$= \mathbf{2060\ \mu F}$$

b. $Q_1 = C_1 E = (800 \times 10^{-6})(48) = \mathbf{0.038\ C}$

$Q_2 = C_2 E = (60 \times 10^{-6})(48) = \mathbf{0.00288\ C}$

$Q_3 = C_3 E = (1200 \times 10^{-6})(48) = \mathbf{0.0576\ C}$

c. $Q_T = Q_1 + Q_2 + Q_3 = 0.038 + 0.00288 + 0.0576$
$$= \mathbf{0.06428\ C}$$

EXAMPLE 9.14. Find the voltage across and charge on each capacitor for the network of Fig. 9.49.

Solution:

$$C'_T = C_2 + C_3 = 4 + 2 = 6\ \mu F$$

$$C_T = \frac{C_1 C'_T}{C_1 + C'_T} = \frac{3 \times 6}{3 + 6} = 2\ \mu F$$

$$Q_T = C_T E = (2 \times 10^{-6})(120)$$

$$Q_T = \mathbf{240\ \mu C}$$

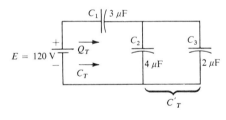

FIG. 9.49

An equivalent circuit (Fig. 9.50) has

$$Q_T = Q_1 = Q'_T$$
$$\therefore Q_1 = \mathbf{240\ \mu C}$$

and

$$V_1 = \frac{Q_1}{C_1} = \frac{240 \times 10^{-6}}{3 \times 10^{-6}} = \mathbf{80\ V}$$

$$Q'_T = 240\ \mu C$$

$$\therefore V'_T = \frac{Q'_T}{C'_T} = \frac{240 \times 10^{-6}}{6 \times 10^{-6}} = \mathbf{40\ V}$$

and

$$Q_2 = C_2 V'_T = (4 \times 10^{-6})(40) = \mathbf{160\ \mu C}$$
$$Q_3 = C_3 V'_T = (2 \times 10^{-6})(40) = \mathbf{80\ \mu C}$$

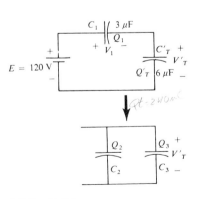

FIG. 9.50

EXAMPLE 9.15. Find the voltage across and charge on capacitor C_1 of Fig. 9.51 after it has charged up to its final value.

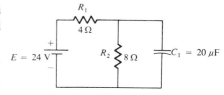

FIG. 9.51

Solution: As previously discussed, the capacitor is effectively an open circuit for dc after charging up to its final value (Fig. 9.52). Therefore,

$$V_C = \frac{(8)(24)}{4 + 8} = \mathbf{16\ V}$$

$$Q_1 = C_1 V_C = (20 \times 10^{-6})(16)$$

$$Q_1 = \mathbf{320\ \mu C}$$

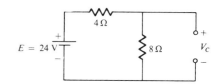

FIG. 9.52

EXAMPLE 9.16. Find the voltage across and charge on each capacitor of the network of Fig. 9.53 after each has charged up to its final value.

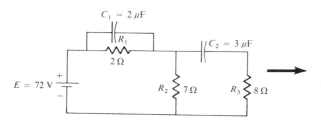

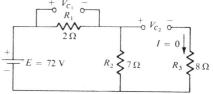

FIG. 9.53

Solution:

$$V_{C_2} = \frac{(7)(72)}{7 + 2} = \mathbf{56\ V}$$

$$V_{C_1} = \frac{(2)(72)}{2 + 7} = \mathbf{16\ V}$$

$$Q_1 = C_1 V_{C_1} = (2 \times 10^{-6})(16) = \mathbf{32\ \mu C}$$

$$Q_2 = C_2 V_{C_2} = (3 \times 10^{-6})(56) = \mathbf{168\ \mu C}$$

9.11 ENERGY STORED BY A CAPACITOR

The ideal capacitor does not dissipate any of the energy supplied to it. It stores the energy in the form of an electric field. A plot of the voltage, current, and power to a capacitor is shown during the charging phase in Fig. 9.54. The power curve can be obtained by finding the product of the voltage and current at selected instants in time and connecting the points obtained (Note p_1 on the curve of Fig. 9.54). The energy stored is represented by the shaded area under the power curve. Using calculus, we can determine the area under the curve:

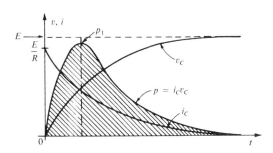

FIG. 9.54

$$\boxed{W_C = \frac{1}{2} CE^2} \qquad \text{(joules)} \qquad \textbf{(9.30)}$$

In terms of Q and C,

$$W_C = \frac{1}{2} C \left(\frac{Q}{C}\right)^2$$

$$\boxed{W_C = \frac{Q^2}{2C}} \qquad \text{(joules)} \qquad \textbf{(9.31)}$$

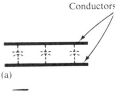

(a)

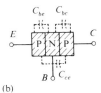

(b)

(c)

FIG. 9.55

9.12 STRAY CAPACITANCES

In addition to the capacitors discussed so far in this chapter, there are stray capacitances that exist not through design but simply because two conducting surfaces are relatively close to each other. Two conducting wires in the same network will have a capacitive effect between them, as shown in Fig. 9.55(a). In electronic circuits, capacitance levels exist between conducting surfaces of the transistor as shown in Fig. 9.55(b). In Chapter 10 we will discuss another element called the *inductor* which will have capacitive effects between the windings [Fig. 9.55(c)]. Stray capacitances can often lead to serious errors in system design if not considered carefully.

PROBLEMS

Section 9.2

1. Find the electric field strength at a point 2 m from a charge of 4 μC.

2. The electric field strength is 36 newtons/coulomb (N/C) at a point r meters from a charge of 0.064 μC. Find the distance r.

Section 9.3

3. Find the capacitance of a parallel plate capacitor if 1400 μC of charge are deposited on its plates when 20 V are applied across the plates.

4. How much charge is deposited on the plates of a 0.05-μF capacitor if 45 V are applied across the capacitor?

5. Find the electric field strength between the plates of a parallel plate capacitor if 100 mV are applied across the plates and the plates are 2 mm apart.

6. Repeat Problem 5 if the plates are separated by 4 mils.

7. A 4-μF parallel plate capacitor has 160 μC of charge on its plates. If the plates are 5 mm apart, find the electric field strength between the plates.

8. Find the capacitance of a parallel plate capacitor if the area of each plate is 0.075 m^2 and the distance between the plates is 1.77 mm. The dielectric is air.

9. Repeat Problem 8 if the dielectric is paraffin-coated paper.

10. Find the distance in mils between the plates of a 2-μF capacitor if the area of each plate is 0.09 m^2 and the dielectric is transformer oil.

11. The capacitance of a capacitor with a dielectric of air is 1200 pF. When a dielectric is inserted between the plates, the capacitance increases to 0.006 μF. Of what material is the dielectric made?

12. The plates of a parallel plate air capacitor are 0.2 mm apart, have an area of 0.08 m^2, and 200V are applied across the plates. Find
   a. the capacitance.
   b. the electric field intensity between the plates.
   c. the charge on each plate if the dielectric is air.

13. A sheet of bakelite 0.2 mm thick having an area of 0.08 m^2 is inserted between the plates of Problem 12. Find
   a. the electric field strength between the plates.
   b. the charge on each plate.
   c. the capacitance.

Section 9.4

14. Find the maximum voltage ratings of the capacitors of Problems 12 and 13 assuming a linear relationship between the breakdown voltage and the thickness of the dielectric.

15. Find the maximum voltage that can be applied across a parallel plate capacitor of 0.006 μF. The area of one plate is 0.02 m^2 and the dielectric is mica. Assume a linear relationship between the dielectric strength and the thickness of the dielectric.

16. Find the distance in millimeters between the plates of a parallel plate capacitor if the maximum voltage that can be applied across the capacitor is 1250 V. The dielectric is mica. Assume a linear relationship between the breakdown strength and the thickness of the dielectric.

Section 9.7

17. For the circuit of Fig. 9.56:

Capacitors

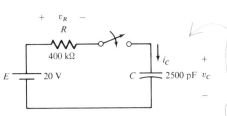

FIG. 9.56

a. Find the mathematical expression for the voltage across the capacitor following the closing of the switch.
b. Determine the voltage v_C after 1, 3, and 5 time constants.
c. Write the equations for the current i_C and voltage v_R.
d. Repeat part (b) for i_C and v_R.
e. Sketch the waveforms of v_C, i_C, and v_R.

18. For the circuit of Fig. 9.57:
   a. Find the mathematical expression for the voltage across the capacitor after the switch is thrown into position 1.
   b. Repeat part (a) for i_C and v_{R_2}.
   c. Find the voltages v_C and v_{R_2} and the current i_C if the switch is thrown into position 2 at $t = 500$ ms. (Assume that the leakage resistance of the capacitor equals ∞ ohms.)
   d. Find the mathematical expressions for the voltages v_C and v_{R_2} and the current i_C after the switch is thrown into position 3 at $t = 1000$ ms (1 s).
   e. Plot the waveforms obtained in parts (a) through (d) on the same time axis for each voltage and the current.

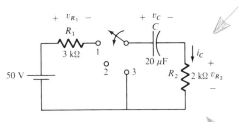

FIG. 9.57

19. Repeat Problem 18 for $R_2 = 20$ kΩ and compare the results.

*20. For the circuit of Fig. 9.58:
   a. Find the mathematical expression for the voltage across the capacitor after the switch is thrown into position 1.
   b. Repeat part (a) for the voltage v_{R_2} and current i_C.
   c. Find the mathematical expressions for the voltages v_C and v_{R_2} and the current i_C if the switch is thrown into position 2 at a time equal to 6 time constants of the charging circuit.
   d. Plot the waveforms obtained in parts (a) through (c) on the same time axis for each voltage and the current.

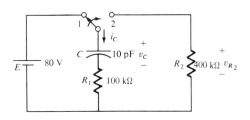

FIG. 9.58

21. Repeat Problem 20 for the circuit of Fig. 9.59. Replace v_{R_2} by v_{R_1} in the statement of the problem.

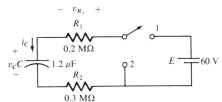

FIG. 9.59

Section 9.8

22. For the circuit of Fig. 9.60:
   a. Find the mathematical expressions for the transient behavior of the voltage v_C and the current i_C following the closing of the switch.
   b. Sketch the waveforms of v_C and i_C.

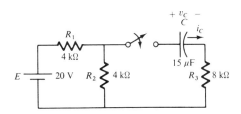

FIG. 9.60

*23. Repeat Problem 22 for the circuit of Fig. 9.61.

Section 9.9

24. Find the waveform for the average current if the voltage across a 0.06-μF capacitor is as shown in Fig. 9.62.

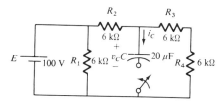

FIG. 9.61

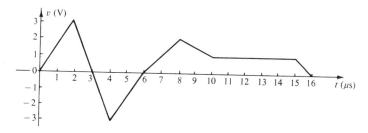

25. Repeat Problem 24 for the waveform of Fig. 9.63.

FIG. 9.62

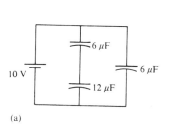

Section 9.10

26. Find the total capacitance C_T between points a and b of the circuits of Fig. 9.64.

FIG. 9.63

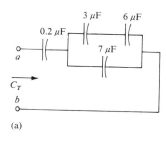

(a)

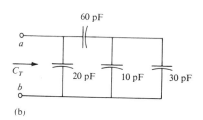

(b)

27. Find the voltage across and charge on each capacitor for the circuits of Fig. 9.65.

FIG. 9.64

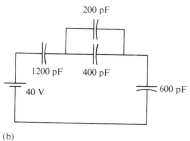

(a)

(b)

FIG. 9.65

28. For the circuits of Fig. 9.66, find the voltage across and charge on each capacitor after each capacitor has charged to its final value.

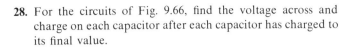

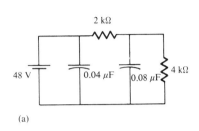

(a)

FIG. 9.66

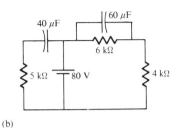

(b)

Section 9.11

29. Find the energy stored by a 120-pF capacitor with 12 V across its plates.

30. If the energy stored by a 6-μF capacitor is 1200 J, find the charge Q on each plate of the capacitor.

*31. An electronic flashgun has a 1000-μF capacitor which is charged to 100 V.
   a. How much energy is stored by the capacitor?
   b. What is the charge on the capacitor?
   c. When the photographer takes a picture, the flash fires for 1/2000 s. What is the average current through the flash-tube?
   d. Find the power delivered to the flashtube.
   e. After a picture is taken, the capacitor has to be recharged by a power supply which delivers a maximum current of 10 mA. How long will it take to charge the capacitor?

GLOSSARY

Breakdown voltage   Another term for *dielectric strength*.

Capacitance   A measure of a capacitor's ability to store charge; measured in farads (F).

Capacitive time constant   The product of resistance and capacitance that establishes the required time for the charging and discharging phases of a capacitive transient.

Capacitive transient   The waveforms for the voltage and current of a capacitor that result during the charging and discharging phases.

Capacitor   A fundamental electrical element having two conducting surfaces separated by an insulating material and having the capacity to store charge on its plates.

Coulomb's law   An equation relating the force between two like or unlike charges.

Dielectric   The insulating material between the plates of a capacitor that can have a pronounced effect on the charge stored on the plates of a capacitor.

Dielectric constant Another term for *relative permittivity*.

Dielectric strength An indication of the voltage required for unit length to establish conduction in a dielectric.

Electric field strength The force acting on a unit positive charge in the region of interest.

Electric flux lines Lines drawn to indicate the strength and direction of an electric field in a particular region.

Fringing An effect established by flux lines that do not pass directly from one conducting surface to another.

Leakage current The current that will result in the total discharge of a capacitor if the capacitor is disconnected from the charging network for a sufficient length of time.

Permittivity A measure of how well a dielectric will *permit* the establishment of flux lines within the dielectric.

Relative permittivity The permittivity of a material compared to that of air.

Stray capacitance Capacitances that exist not through design but simply because two conducting surfaces are relatively close to one another.

Surge voltage The maximum voltage that can be applied across the capacitor for very short periods of time.

Working voltage The voltage that can be applied across a capacitor for long periods of time without concern for dielectric breakdown.

10.1 INTRODUCTION

Magnetism plays an integral part in almost every electrical device used today in industry, research, or the home. Generators, motors, transformers, circuit breakers, televisions, radios, and telephones all employ magnetic effects induced by the flow of charge or current, *electromagnetism*. Many computers use magnetic tapes, disks, or "bubbles" to store the numerous bits of data.

The compass, used by Chinese sailors as early as the second century A.D., relies on a *permanent magnet* for indicating direction. The permanent magnet is made of a material, such as steel, that will remain magnetized for long periods of time without the aid of external means, such as the current required for electromagnetism.

In 1820, the Danish physicist Hans Christian Oersted discovered that the needle of a compass would deflect if brought near a current-carrying conductor. For the first time it was demonstrated that electricity and magnetism were related, and in the same year the French physicist André Marie Ampère performed experiments in this area and developed what is presently known as *Ampère's circuital law*. In subsequent years, men such as Michael Faraday, Karl Friedrich Gauss, and James Clerk Maxwell continued to experiment in this area and developed many of the basic concepts of electromagnetism.

There is a great deal of similarity between the analyses of electric circuits and magnetic circuits. This will be demonstrated later in this chapter when we compare the basic equations and methods used to solve magnetic circuits with those used for electric circuits.

Difficulty in understanding methods used with magnetic circuits will often arise in simply learning to use the proper set of units, and not because of the equations themselves. The problem exists because three different systems of units are still being used with magnetic circuits. To the extent practical, the SI will be used throughout this chapter. For the CGS and English systems, a conversion table is provided in Appendix G.

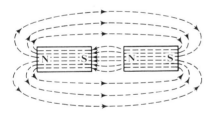

FIG. 10.1

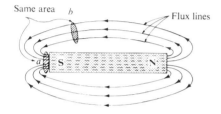

FIG. 10.2

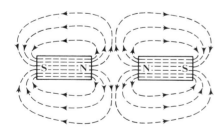

FIG. 10.3

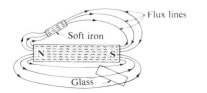

FIG. 10.4

10.2 MAGNETIC FIELDS

In the region surrounding a permanent magnet there exists a magnetic field, which can be represented by magnetic flux lines similar to electric flux lines. Magnetic flux lines, however, do not have origins or terminating points like the electric flux lines but exist in continuous loops, as shown in Fig. 10.1. The symbol for magnetic flux is the Greek letter Φ (phi).

The magnetic flux lines radiate from the north pole to the south pole, returning to the north pole through the metallic bar. Note the equal spacing between the flux lines within the core and the symmetric distribution outside the magnetic material. These are additional properties of magnetic flux lines in homogeneous materials (that is, materials having uniform structure or composition throughout). It is also important to realize that the continuous magnetic flux line will strive to occupy as small an area as possible. This will result in magnetic flux lines of minimum length between the like poles, as shown in Fig. 10.2. The strength of a magnetic field in a particular region is directly related to the density of flux lines in that region. In Fig. 10.1, for example, the magnetic field strength at *a* is twice that at *b* since there are twice as many magnetic flux lines associated with the perpendicular plane at *a* than at *b*.

If unlike poles of two permanent magnets are brought together, the magnets will attract, and the flux distribution will be as shown in Fig. 10.2. If like poles are brought together, the magnets will repel, and the flux distribution will be as shown in Fig. 10.3.

If a nonmagnetic material, such as glass or copper, is placed in the flux paths surrounding a permanent magnet, there will be an almost unnoticeable change in the flux distribution (Fig. 10.4). However, if a magnetic material, such as soft iron, is placed in the flux path, the flux lines will pass through the soft iron rather than the surrounding air because flux lines pass with greater ease through magnetic materials than through air. This principle is put to use in the shielding of sensitive electrical elements and instruments that can be affected by stray magnetic fields (Fig. 10.5).

As indicated in the introduction, a magnetic field (represented by concentric magnetic flux lines, as in Fig. 10.6) is also present around every wire that carries an electric current.

The direction of the magnetic flux lines can be found by placing the thumb of the right hand in the direction of conventional current flow and noting the direction of the fingers. (This method is commonly called the *right-hand rule*.) If the conductor is wound in a single-turn coil (Fig. 10.7), the resulting flux will flow in a common direction through the center of the coil.

A coil of more than one turn would produce a magnetic field that would exist in a continuous path through and around the coil (Fig. 10.8).

The flux distribution of the coil is quite similar to that of the permanent magnet. The flux lines leaving the coil from the left and entering to the right simulate a north and south pole, respectively. The principal difference between the two flux distributions is that the flux lines are more concentrated for the permanent magnet than for the coil. Also, since the strength of a magnetic field is determined by the density of the flux lines, the coil has a weaker field strength. The field strength of the coil can be effectively increased by placing certain materials, such as iron, steel, or cobalt, within the coil that allows flux to pass through it more easily than through air, and thus all the flux lines will tend to pass through the material, thereby increasing the flux density and the field strength. By increasing the field strength with the addition of the core, we have devised an *electromagnet* (Fig. 10.9) which, in addition to having all the properties of a permanent magnet, also has a field strength that can be varied by changing one of the component values (current, turns, and so on). Of course, current must pass through the coil of the electromagnet in order for magnetic flux to be developed, whereas there is no need for the coil or current in the permanent magnet. The direction of flux lines can be determined for the electromagnet (or in any core with a wrapping of turns) by placing the fingers of the right hand in the direction of current flow around the core. The thumb will then point in the direction of the north pole of the induced magnetic flux. This is demonstrated in Fig. 10.10. A cross section of the same electromagnet was included in the figure to introduce the convention for directions perpendicular to the page. Quite obviously, the cross and dot refer to the tail and head of the arrow, respectively.

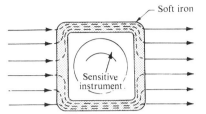

FIG. 10.5

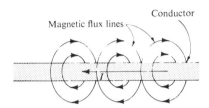

FIG. 10.6

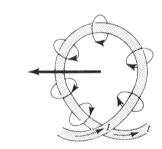

FIG. 10.7

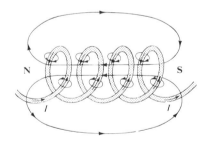

FIG. 10.8

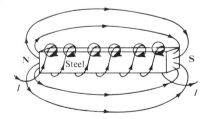

FIG. 10.9

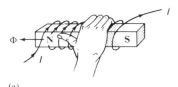

(a)

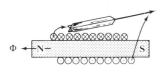

(b)

FIG. 10.10

Other areas of application for electromagnetic effects are shown in Fig. 10.11. The flux path for each is indicated in each figure.

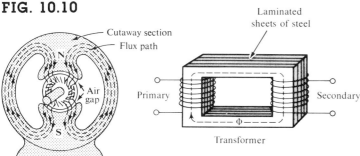

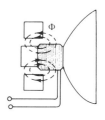

Loudspeaker

Generator

Laminated sheets of steel

Transformer

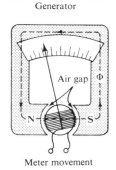

Meter movement

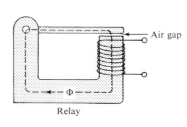

Relay

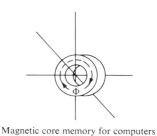

Magnetic core memory for computers

FIG. 10.11

10.3 FLUX DENSITY

In the SI system of units, magnetic flux is measured in webers (Wb) and has the symbol Φ. The number of flux lines per unit area is called the *flux density* and is denoted by the capital letter B. Its magnitude is determined by the following equation:

$$B = \frac{\Phi}{A} \qquad \begin{array}{l} B = \text{teslas (T)} \\ \Phi = \text{Wb} \\ A = \text{m}^2 \end{array} \qquad \textbf{(10.1)}$$

FIG. 10.12

where Φ is the number of flux lines passing through the area A (Fig. 10.12). The flux density at position a in Fig. 10.1 is twice that at b because twice as many flux lines are passing through the same area.

As noted in Eq. (10.1), magnetic flux density in the SI system of units is measured in *teslas,* for which the symbol is T. By definition,

$$1 \text{ tesla} = 1 \text{ Wb/m}^2$$

EXAMPLE 10.1. For the core of Fig. 10.13, determine the flux density B in teslas.

Solution:

$$B = \frac{\Phi}{A} = \frac{6 \times 10^{-5}}{1.2} = \mathbf{5 \times 10^{-5} \ T}$$

EXAMPLE 10.2. In Fig. 10.13, if the flux density is 1.2 Wb/m^2 and the area is 0.25 in.^{2}, determine the flux through the core.

Solution: By Eq. (10.1),

$$\Phi = BA$$

However,

$$0.25 \ \text{in.}^2 \left[\frac{1 \text{ m}}{39.37 \text{ in.}} \right] \left[\frac{1 \text{ m}}{39.37 \text{ in.}} \right] = 1.613 \times 10^{-4} \text{ m}^2$$

and

$$\Phi = (1.2)(1.613 \times 10^{-4})$$
$$\Phi = \mathbf{1.936 \times 10^{-4} \ Wb}$$

An instrument designed to measure flux density in gauss (CGS system) appears in Fig. 10.14. Appendix G reveals that 1 Wb/m^2 = 10^4 gauss. The magnitude of the meter reading in Wb/m^2 or in teslas is therefore

$$1.964 \text{ gauss} \left[\frac{1 \text{ Wb/m}^2}{10^4 \text{ gauss}} \right] = 1.964 \times 10^{-4} \text{ Wb/m}^2$$
$$= 1.964 \times 10^{-4} \text{ T}$$

This digital reading instrument can measure the static (dc) flux density or the *rms* of sinusoidal magnetic field from dc to 1 kHz. (The latter terminology will be defined in Chapter 13.)

10.4 PERMEABILITY

If cores of different materials with the same physical dimensions are used in the electromagnet described in Section 10.2, the strength of the magnet will vary in accordance with the core used. This variation in strength is due to the greater or lesser number of flux lines passing through the core. Materials in which flux lines can readily be set up are said to be *magnetic* and to have *high permeability.* The permeability (μ) of a material, therefore, is a measure of the ease with which magnetic flux lines can be established

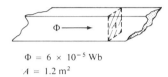

$\Phi = 6 \times 10^{-5}$ Wb
$A = 1.2$ m^2

FIG. 10.13

Courtesy of LDJ Electronics, Inc.

FIG. 10.14 *Digital display gauss-meter.*

in the material. It is similar in many respects to conductivity in electric circuits. The permeability of free space μ_o (vacuum) is

$$\mu_o = 4\pi \times 10^{-7} \text{ Wb/Am}$$

As indicated above, μ has the units of Wb/Am. Practically speaking, the permeability of all nonmagnetic materials, such as copper, aluminum, wood, glass, and air, is the same as that for free space. Materials that have permeabilities slightly less than that of free space are said to be *diamagnetic,* and those with permeabilities slightly greater than that of free space are said to be *paramagnetic.* Magnetic materials, such as iron, nickel, steel, cobalt, and alloys of these metals, have permeabilities hundreds and even thousands of times that of free space. Materials with these very high permeabilities are referred to as *ferromagnetic.*

The ratio of the permeability of a material to that of free space is called its *relative permeability;* that is,

$$\mu_r = \frac{\mu}{\mu_o} \qquad \textbf{(10.2)}$$

In general, for ferromagnetic materials, $\mu_r \geq 100$, and for nonmagnetic materials, $\mu_r = 1$.

Since μ_r is a variable, dependent on other quantities of the magnetic circuit, values of μ_r are not tabulated. Methods of calculating μ_r from the data supplied by manufacturers will be considered in a later section.

10.5 RELUCTANCE

The resistance of a conductor to the flow of charge (current) is determined for electric circuits by the equation

$$R = \rho \frac{l}{A} \qquad (\Omega)$$

The *reluctance* of a material to the setting up of magnetic flux lines in the material is determined by the following equation:

$$\mathscr{R} = \frac{l}{\mu A} \qquad \text{(rels or At/Wb)}$$

$$\textbf{(10.3)}$$

where $\mathscr{R}$ is the reluctance, l is the length of the magnetic path, and A is its cross-sectional area. Note that the resistance and reluctance are inversely proportional to the area, indicating that an increase in area will result in a reduction in each and an *increase* in the desired result: current and flux. For an increase in length the opposite is true, and the desired effect is reduced. The reluctance, however, is inversely proportional to the permeability, while the resistance is directly proportional to the resistivity. The larger the μ or smaller the ρ, the smaller the reluctance and resistance, respectively. Obviously, therefore, materials with high permeability, such as the ferromagnetics, have very small reluctances and will result in an increased measure of flux through the core. There is no widely accepted unit for reluctance, although the *rel* and the At/Wb are usually applied. The latter unit of measure will be derived in a later section.

10.6 OHM'S LAW FOR MAGNETIC CIRCUITS

Recall the equation

$$\text{Effect} = \frac{\text{cause}}{\text{opposition}}$$

appearing in Chapter 4 to introduce Ohm's law for electric circuits. For magnetic circuits, the effect desired is the flux Φ. The cause is the *magnetomotive force* (mmf) $\mathscr{F}$, which is the external force (or "pressure") required to set up the magnetic flux lines within the magnetic material. The opposition to the setting up of the flux Φ is the reluctance $\mathscr{R}$.

Substituting, we have

$$\boxed{\Phi = \frac{\mathscr{F}}{\mathscr{R}}} \qquad \textbf{(10.4)}$$

The magnetomotive force $\mathscr{F}$ is proportional to the product of the number of turns around the core (in which the flux is to be established) and the current through the turns of wire (Fig. 10.15). In equation form:

$$\boxed{\mathscr{F} = NI} \qquad \text{(ampere-turns, At)} \qquad \textbf{(10.5)}$$

The equation clearly indicates that an increase in the number of turns or the current through the wire will result in an increased "pressure" on the system to establish flux lines through the core.

Although there is a great deal of similarity between electric and magnetic circuits, one must continue to realize that the flux Φ is not a "flow" variable such as current in an electric circuit. Magnetic flux is established in the core through the alteration of the atomic structure of the core due to external pressure and is not a measure of the flow of some charged particles through the core.

FIG. 10.15

10.7 MAGNETIZING FORCE

The magnetomotive force per unit length is called the *magnetizing force* (*H*). In equation form,

$$H = \frac{\mathscr{F}}{l} \qquad \text{(At/m)} \qquad \textbf{(10.6)}$$

Substituting for the magnetomotive force will result in

$$H = \frac{NI}{l} \qquad \text{(At/m)} \qquad \textbf{(10.7)}$$

For the magnetic circuit of Fig. 10.16, if $NI = 40$ At and $l = 0.2$ m, then

$$H = \frac{NI}{l} = \frac{40}{0.2} = 200 \text{ At/m}$$

In words, the result indicates that there are 200 ampere-turns of "pressure" per meter to establish flux in the core.

Note in Fig. 10.16 that the direction of the flux Φ can be determined by placing the fingers of the right hand in the direction of current flow around the core and noting the direction of the thumb. It is interesting to realize that *the magnetizing force is independent of the type of core material*—it is determined solely by the number of turns, the current, and the length of the core.

The applied magnetizing force has a pronounced effect on the resulting permeability of a magnetic material. As the magnetizing force increases, the permeability rises to a maximum and then drops to a minimum, as shown in Fig. 10.17 for three commonly employed magnetic materials.

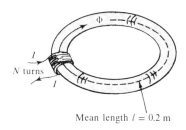

Mean length $l = 0.2$ m

FIG. 10.16

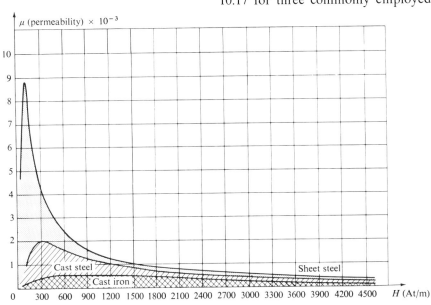

FIG. 10.17

An equation relating the flux density B to the magnetizing force can be derived if we first substitute for the reluctance $\mathscr{R}$ in the following equation:

$$\Phi = \frac{\mathscr{F}}{\mathscr{R}} = \frac{\mathscr{F}}{l/\mu A} \quad \text{or} \quad \Phi = \frac{\mathscr{F}\mu A}{l}$$

But

$$B = \frac{\Phi}{A} = \frac{\mathscr{F}\mu A}{lA} = \frac{\mu\mathscr{F}}{l} \quad \text{or} \quad l = \frac{\mu\mathscr{F}}{B}$$

and

$$H = \frac{\mathscr{F}}{l} = \frac{\mathscr{F}}{\mu\mathscr{F}/B}$$

so that

$$H = \frac{B}{\mu}$$

or

$$\boxed{B = \mu H} \qquad\qquad \textbf{(10.8)}$$

This equation indicates that for a particular magnetizing force, the greater the permeability, the greater will be the induced flux density. It was this equation that was employed to determine the units of μ in Section 10.4. That is,

$$\mu = \frac{B}{H} = \frac{B}{NI/l}$$
$$= \frac{\text{Wb/m}^2}{\text{A/m}} \text{ (for } N = 1)$$
$$= \frac{\textbf{Wb}}{\textbf{A} \cdot \textbf{m}}$$

10.8 HYSTERESIS

A curve of the flux density B versus the magnetizing force H of a material is of particular importance to the engineer. Curves of this type can usually be found in manuals and descriptive pamphlets and brochures published by manufacturers of magnetic materials. A typical B-H curve for a ferromagnetic material such as steel can be derived using the setup of Fig. 10.18.

The core is initially unmagnetized and the current $I = 0$. If the current I is increased to some value above zero, the magnetizing force H will increase to a value determined by

$$H{\uparrow} = \frac{NI{\uparrow}}{l}$$

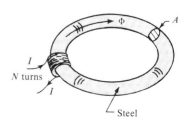

FIG. 10.18

The flux Φ and the flux density B $(B = \Phi/A)$ will also increase with the current I (or H). If the magnetizing force H is increased to some value H_a, the B-H curve will follow the path shown in Fig. 10.19 between o and a. If the magnetizing force H is increased until saturation (H_s) occurs, the curve will continue as shown in the figure to point b. When saturation occurs, the flux density has, *for all practical purposes,* reached its maximum value. Any further increase in current through the coil increasing $H = NI/l$ will result in a very small increase in flux density B.

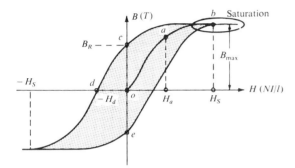

FIG. 10.19

If the magnetizing force is reduced to zero by letting I decrease to zero, the curve will follow the path of the curve between b and c. The flux density B_R, which remains when the magnetizing force is zero, is called the *residual flux density.* It is this residual flux density that makes it possible to create permanent magnets. If the coil is now removed from the core of Fig. 10.18, the core will still have the magnetic properties determined by the residual flux density, a measure of its "retentivity." If the current I is reversed, developing a magnetizing force, $-H$, the flux density B will decrease with increase in I. Eventually, the flux density will be zero when $-H_d$ (the portion of curve from c to d) is reached. The magnetizing force H_d required to "coerce" the flux density to reduce its level to zero is called the *coercive force,* a measure of the coercivity of the magnetic sample. As the force $-H$ is increased until saturation again occurs and is then reversed and brought back to zero, the path shown from d to e will result. If the magnetizing force is increased in the positive direction $(+H)$, the curve will trace the path shown from e to b. The entire curve represented by *bcded* is called the *hysteresis curve* for the ferromagnetic material, from the Greek *hysterein,* meaning "to lag behind." The flux density B *lagged* behind the magnetizing force H during the entire plotting of the curve. When H was zero at c, B was not zero but had only begun to decline. Long after H had passed through zero and had become equal to $-H_d$ did the flux density B finally become equal to zero.

If the entire cycle is repeated, the curve obtained for the same core will be determined by the maximum H applied. Three hysteresis loops for the same material for maximum values of H less than the saturation value are shown in Fig. 10.20. In addition, the saturation curve is repeated for comparison purposes.

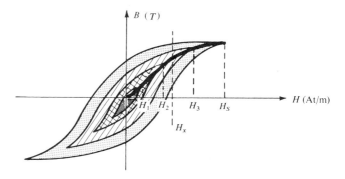

FIG. 10.20

Note from the various curves that for a particular value of H, say, H_x, the value of B can vary widely, as determined by the history of the core. In an effort to assign a particular value of B to each value of H, we compromise by connecting the tips of the hysteresis loops. The resulting curve, shown by the dashed line in Fig. 10.20 and for various materials in Fig. 10.21, is called the *normal magnetization curve*. An expanded view of one region appears in Fig. 10.22.

A comparison of Figs. 10.17 and 10.21 shows that for the same value of H, the value of B is higher in Fig. 10.21 for the materials with the higher μ in Fig. 10.17. This is particularly obvious for low values of H. This correspondence between the two figures must exist, since $B = \mu H$. In fact, if in Fig. 10.21 we find μ for each value of H using the equation $\mu = B/H$, we will obtain the curves of Fig. 10.17.

An instrument that will provide a plot of the B-H curve for a magnetic sample appears in Fig. 10.23.

It is interesting to note that the hysteresis curves of Fig. 10.20 have a "point symmetry" about the origin. That is, the inverted pattern to the left of the vertical axis is the same as that appearing to the right of the vertical axis. In addition, you will find that a further application of the same magnetizing forces to the sample will result in the same plot. For a current I in $H = NI/l$ that will move between positive and negative maximums at a fixed rate, the same B-H curve will result during each cycle. Such will be the case when we examine ac (sinusoidal) networks in the later chapters. The reversal of the field (Φ) due to the changing current direction will result in a loss of energy that can best be described by first introducing the *domain theory of magnetism*.

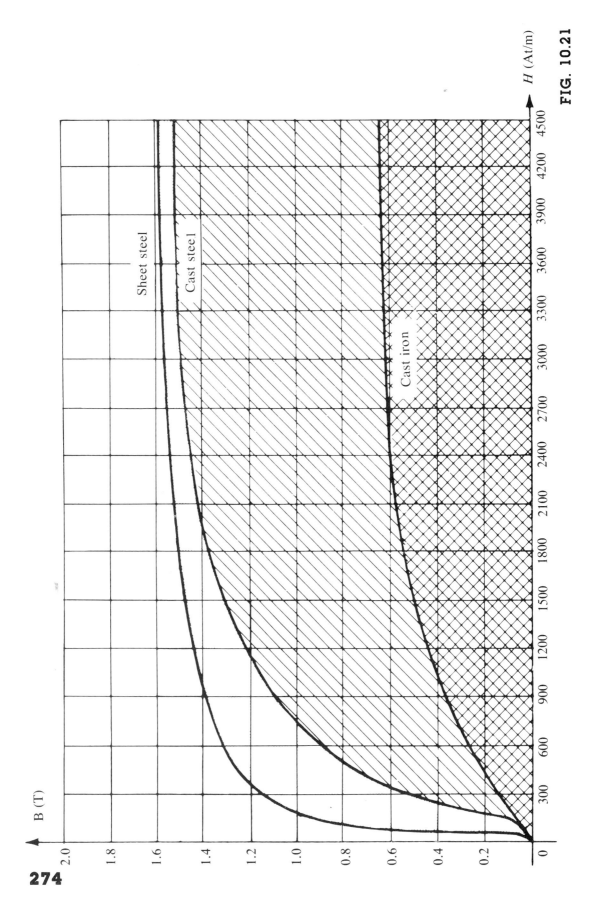

FIG. 10.21

274

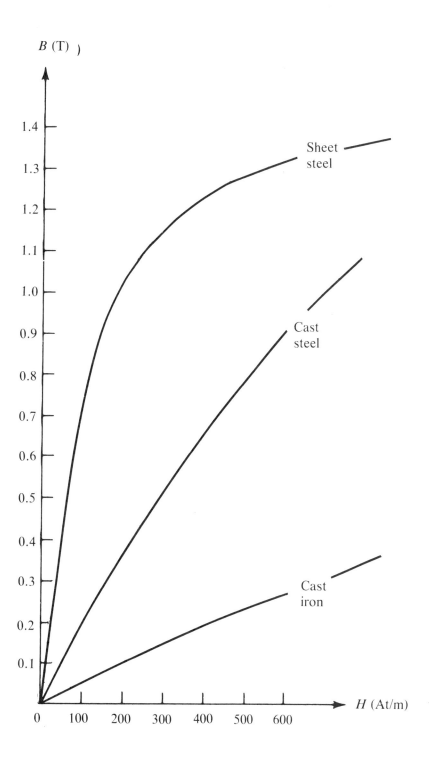

FIG. 10.22

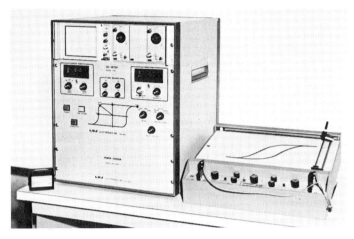

FIG. 10.23

Within each atom, the orbiting electrons (described in Chapter 2) are also spinning as they revolve around the nucleus. The atom, due to its spinning electrons, has a magnetic field associated with it. In nonmagnetic materials, the net magnetic field is effectively zero, since the magnetic fields due to the atoms of the material oppose each other. In magnetic materials such as iron and steel, however, the magnetic fields of groups of atoms numbering in the order of 10^{12} are aligned, forming very small bar magnets. This group of magnetically aligned atoms is called a *domain*. Each domain is a separate entity; that is, each domain is independent of the surrounding domains. For an unmagnetized sample of magnetic material, these domains appear in a random manner, such as shown in Fig. 10.24(a). The net magnetic field in any one direction is zero.

When an external magnetizing force is applied, the domains that are nearly aligned with the applied field will grow at the expense of the less favorably oriented domains, such as shown in Fig. 10.24(b). Eventually, if a sufficiently strong field is applied, all of the domains will have the orientation of the applied magnetizing force, and any further increase in external field will not increase the strength

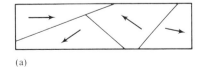

(a)

(b)

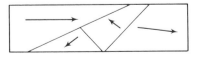

FIG. 10.24

of the magnetic flux through the core—a condition referred to as *saturation*. The elasticity of the above is evidenced by the fact when the magnetizing force is removed, the alignment will be lost to some measure and the flux density will drop to B_R. In other words, the removal of the magnetizing force will result in the return of a number of misaligned domains within the core. The continued alignment of a number of the domains, however, accounts for our ability to create permanent magnets.

At a point just before saturation, the opposing unaligned domains are reduced to small cylinders of various shapes referred to as *bubbles*. These bubbles can be moved within the magnetic sample through the application of *controlling* magnetic field. It is these magnetic bubbles that form the basis of the recently designed bubble memory system for computers.

If an alternating signal as described above is applied to the magnetizing coil, these domains will have to pass through a 180° reversal during each cycle of the input. The natural opposition (friction) of the domains to this reversal will result in a transfer of heat to the core. The hysteresis curves for three different ferromagnetic materials appear in Fig. 10.25. It has been determined through mathematical

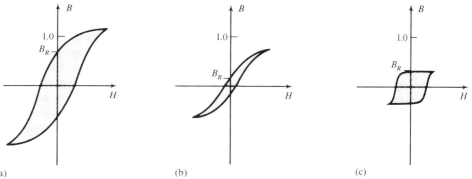

and experimental techniques that the energy transfer to the core is directly proportional to the area enclosed by the hysteresis loop. Note that it is a maximum for materials having a high retentivity [Fig. 10.25(a)], such as those used for permanent magnet applications (motors, generators, meters, and so on). The energy transferred, however, is also directly proportional to the frequency (the rate at which the current oscillates between the positive and negative peak values) of the applied current. The energy and frequency are proportional because the same energy transaction will occur during each cycle of the reversing current, and the more cycles per second, the more energy is absorbed by the core. Since permanent magnets are not asked to realign these domains on a continuing basis, this cycle factor is not a real concern. However, for applications involving a high switching rate, such as in magnetic control devices, relays,

FIG. 10.25 *Hysteresis curves.* (*a*) *Sheet steel;* (*b*) *soft iron;* (*c*) *ferrites.*

and so on, the area should be minimal, and therefore materials such as soft iron [Fig. 10.25(b)] are employed. Fortunately, there is little need for high levels of retentivity in such applications.

The characteristics of Fig. 10.25(c) were obtained using a core constructed of a mixture of metallic oxides. A number of such *ferrite* cores appear in Fig. 10.26. The range of applications for these cores include inductors and transformers (to be described in this text), antennas, and magnetic shields. The ferrite core of Fig. 10.11 is employed as a memory unit for computers. Its outside diameter is as small as 20/1000 inch, with a hole barely visible by the naked eye. Note that the ferrite core will saturate at levels lower than the other materials and will exhibit a *B-H* curve very sensitive to the magnetizing force *H*. That is, the flux density *B* (and therefore Φ) remain essentially constant in magnitude until a particular value of *H* is reached. The level of *B* and the direction of Φ will then change abruptly from one state to the other. In effect, the magnetic flux density *B* and flux Φ are in either one state or the other. These characteristics make the ferrite core excellent for computer memory applications in which the direction of flux will determine whether a 0 or a 1 is being stored. The area within the hysteresis loop is small, making this core suitable for such high-speed applications.

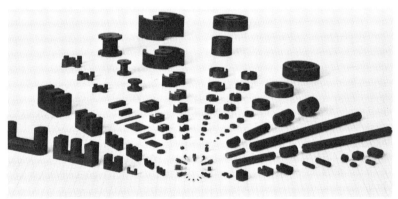

Courtesy of Siemens Corp.

FIG. 10.26 *Ferrite cores.*

The hysteresis loss is determined essentially by the following equation:

$$P_h = k \cdot \text{Vol.} \; f \cdot B_{\text{max}}^n \quad \text{(W)} \qquad \textbf{(10.9)}$$

As indicated by the equation, hysteresis loss is also directly related to the volume of material exposed to this alternating flux. In other words, the larger the core, the greater the energy transfer. The B_{max}^n factor is directly related to the area of the loop, with *n* determined by the material. The B_{max}^n factor normally has a magnitude of 1.4 to 2.6. Note the appearance of the frequency factor *f* which reflects the number of cycles to which the core will be exposed in one second. The factor k is determined by the core material.

10.9 INCREMENTAL AND AVERAGE PERMEABILITY

If written in the following form, Eq. (10.8) can be used to determine μ at a particular point on the normal magnetization curve:

$$\mu_{dc} = \frac{B}{H} \qquad \textbf{(10.10)}$$

This value of μ is often called the *static* or *dc* value since it is defined by a fixed level of B and H, such as shown for μ_1 in Fig. 10.27(a).

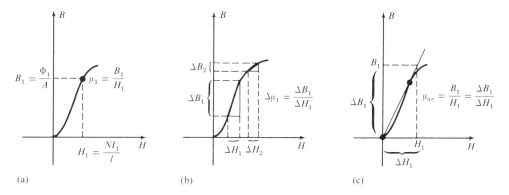

(a)   (b)   (c)

FIG. 10.27 *Permeabilities.* (*a*) *Static;* (*b*) *incremental;* (*c*) *average.*

In some instances the applied signal may result in a range of values for μ such as indicated in Fig. 10.27(b). In such situations an *incremental* permeability is defined by

$$\mu_\Delta = \frac{\Delta B}{\Delta H} \qquad \textbf{(10.11)}$$

Note in Fig. 10.27(b) that the value of μ is a maximum (compare the change in B to the same change in H for each region) in the mid-range, indicating that this is the region that should be used when large changes in Φ ($B = \Phi/A$) are required due to limited changes in current I ($H = NI/l$), such as in communication equipment (loudspeakers, telephone receivers, and so on).

Since μ will vary from point to point and is not fixed, such as ρ for resistive elements and ϵ for capacitive elements, the analysis of magnetic circuits normally requires the use of the normal magnetization curves to find the proper value of B or H. However, an approximate solution is possible if we define an average value of μ as defined by the following equation and Fig. 10.27(c):

$$\mu_{av} = \frac{\Delta B}{\Delta H} = \frac{B_1}{H_1}$$ **(10.12)**

This value will provide an approximate solution to magnetic circuit problems of the type to be described shortly for levels of B below saturation. Note that μ is defined by the origin and the knee of the curve. Any two values of B_1 and H_1 along the straight line segment will result in the same value.

EXAMPLE 10.3. For the cast steel sample of Fig. 10.21, determine

a. μ_{dc} at $B = 1.0$ T.
b. μ_Δ in the region of $B = 1.0$ T.
c. μ_{av}.

Solutions:

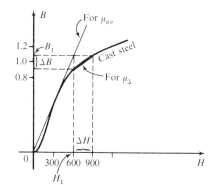

FIG. 10.28

a. $\mu_{dc} = \dfrac{B}{H} = \dfrac{1.0}{750} \dfrac{\text{Wb/m}^2}{\text{At/m}} = \mathbf{1.33 \times 10^{-3}\,Wb/Am}$

b. As indicated in Fig. 10.28,

$$\mu_\Delta = \frac{\Delta B}{\Delta H} = \frac{1.1 - 0.9}{900 - 600} = \frac{0.2}{300}$$
$$= \mathbf{0.667 \times 10^{-3}\,Wb/Am}$$

c. As indicated in Fig. 10.28,

$$\mu_{av} = \frac{B_1}{H_1} = \frac{1.0}{600} = \mathbf{1.67 \times 10^{-3}\,Wb/Am}$$

10.10 AMPÈRE'S CIRCUITAL LAW

It was mentioned in the introduction to this chapter that there is a great similarity between electric and magnetic circuits. This has already been demonstrated to some extent for the quantities in Table 10.1.

TABLE 10.1

	Electric Circuits	Magnetic Circuits
Cause	E	$\mathscr{F}$
Effect	I	Φ
Opposition	R	$\mathscr{R}$

If we apply the cause analogy to Kirchhoff's voltage law $(\Sigma_{\circlearrowleft} V = 0)$, we obtain the following:

$$\Sigma_{\circlearrowleft} \mathscr{F} = 0$$ (for magnetic circuits) **(10.13)**

which, in words, states that the algebraic sum of the rises and drops of the mmf around a closed loop of a magnetic circuit is equal to zero; that is, the sum of the mmf rises equals the sum of the mmf drops around a closed loop.

Equation (10.13) is referred to as *Ampère's circuital law*. When it is applied to magnetic circuits, sources of mmf are expressed by the equation

$$\boxed{\mathscr{F} = NI} \qquad \text{(At)} \qquad \textbf{(10.14)}$$

The equation for mmf drop across a portion of a magnetic circuit can be found by applying the relationships listed in the Table 10.1. That is, for electric circuits,

$$V = IR$$

resulting in the following for magnetic circuits:

$$\boxed{\mathscr{F} = \Phi\mathscr{R}} \qquad \text{(At)} \qquad \textbf{(10.15)}$$

where Φ is the flux passing through a section of the magnetic circuit and $\mathscr{R}$ is the reluctance of that section. The reluctance, however, is seldom calculated in the analysis of magnetic circuits. A more practical equation for the mmf drop is

$$\boxed{\mathscr{F} = Hl} \qquad \text{(At)} \qquad \textbf{(10.16)}$$

as derived from Eq. (10.6), where H is the magnetizing force on a section of a magnetic circuit and l is the length of the section. As an example of Eq. (10.13), consider the magnetic circuit appearing in Fig. 10.29 constructed of three different ferromagnetic materials.

Applying Ampère's circuital law, we have

$$\Sigma_\circlearrowright \mathscr{F} = 0$$

$$\underbrace{+ NI}_{\text{rise}} - \underbrace{H_{ab}l_{ab}}_{\text{drop}} - \underbrace{H_{bc}l_{bc}}_{\text{drop}} - \underbrace{H_{ca}l_{ca}}_{\text{drop}} = 0$$

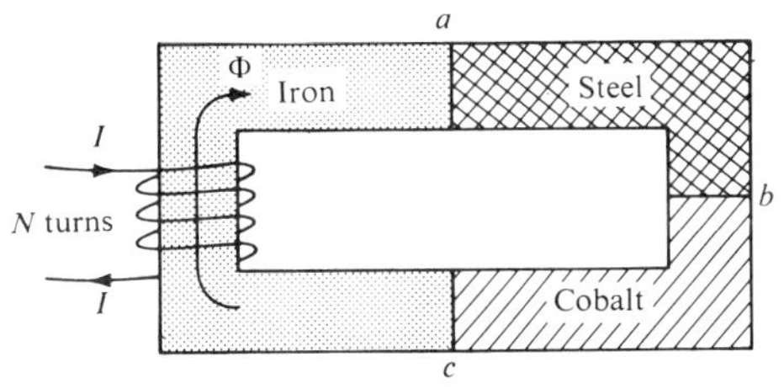

or

$$\underbrace{NI}_{\substack{\text{impressed}\\\text{mmf}}} = \underbrace{H_{ab}l_{ab} + H_{bc}l_{bc} + H_{ca}l_{ca}}_{\text{mmf drops}}$$

FIG. 10.29

All of the terms of the equation are known except the magnetizing force for each portion of the magnetic circuit, which can be found by using the *B-H* curve if the flux density *B* is known.

10.11 THE FLUX Φ

If we continue to apply the relationships described in the previous section to Kirchhoff's current law, we will find

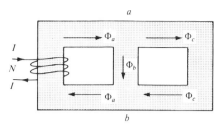

FIG. 10.30

that the sum of the fluxes entering a function is equal to the sum of the fluxes leaving a junction; that is, for the circuit of Fig. 10.30,

$$\Phi_a = \Phi_b + \Phi_c \qquad \text{at junction } a$$

or

$$\Phi_b + \Phi_c = \Phi_a \qquad \text{at junction } b$$

both of which are equivalent.

10.12 SERIES MAGNETIC CIRCUITS: DETERMINING NI

We are now in a position to solve a few magnetic circuit problems, which are basically of two types. In one type, Φ is given, and the impressed mmf NI must be computed. This is the type of problem encountered in the design of motors, generators, and transformers. In the other type, NI is given, and the flux Φ of the magnetic circuit must be found. This type of problem is encountered primarily in the design of magnetic amplifiers and is more difficult since the approach is "hit or miss."

As indicated in earlier discussions, the value of μ will vary from point to point along the magnetization curve. This eliminates the possibility of finding the reluctance of each "branch" or the "total reluctance" of a network as was done for electric circuits where ρ had a fixed value for any applied current of voltage. If the total reluctance could be determined, Φ could then be determined using the Ohm's law analogy for magnetic circuits.

For magnetic circuits, the level of B or H is determined from the other using the B-H curve, and μ is seldom calculated unless asked for. Of course, if an approximate solution is acceptable, μ_{av} could be determined and quantities such as the "total reluctance" could be determined and applied toward a solution. However, since this latter approach is applied only on a very limited basis, the emphasis in this chapter will be on the use of the normal magnetization curves.

An approach frequently employed in the analysis of magnetic circuits is the *table* method. Before a problem is analyzed in detail, a table is prepared listing in the extreme left-hand column the various sections of the magnetic circuit. The columns on the right are reserved for the quantities to be found for each section. In this way, the individual doing the problem can keep track of what is required to complete the problem and also of what his next step should be. After a few examples, the usefulness of this method should become clear.

This section will consider only *series* magnetic circuits in which the flux Φ is the same throughout. In each example, the magnitude of the magnetomotive force is to be determined.

EXAMPLE 10.4. Determine the current *I* required to establish a flux of 2.5×10^{-4} Wb in the core of the transformer of Fig. 10.31. The open secondary has no effect on the flux in the core.

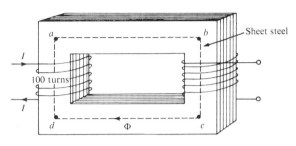

Area (throughout) $= 0.2 \times 10^{-3}$ m^2
$a - b = c - d = 0.05$ m
$b - c = d - a = 0.03$ m

FIG. 10.31

Solution: The data are provided in Table 10.2. Note that it is now only necessary to fill in all the blanks until the value of *Hl* is determined for each section. Then Ampère's circuital law can be applied to obtain the solution.

TABLE 10.2

Section	Φ	A	B	H	l	Hl
$a-b = c-d$	2.5×10^{-4} Wb	0.2×10^{-3} m^2			0.05 m	
$b-c = d-a$	2.5×10^{-4} Wb	0.2×10^{-3} m^2			0.03 m	

$$B \text{ (throughout)} = \frac{\Phi}{A} = \frac{2.5 \times 10^{-4}}{0.2 \times 10^{-3}} = 1.25 \text{ T}$$

$$H \text{ (throughout)} \cong 420 \text{ At/m}$$
$$\text{(as obtained from Fig. 10.21)}$$

Applying Ampère's circuital law results in

$$\Sigma_{\circlearrowleft} \mathscr{F} = 0$$
$$NI = Hl$$

and

$$NI = 2Hl_{a-b} + 2Hl_{b-c}$$
$$= 2H(l_{a-b} + l_{b-c})$$
$$= 2(420)(0.05 + 0.03)$$
$$100I = 67.20$$
$$I = \textbf{0.672 A}$$

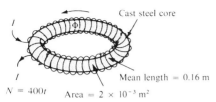

Cast steel core

I

Φ

Mean length $= 0.16$ m

$N = 400t$ Area $= 2 \times 10^{-3}$ m^2

FIG. 10.32 *Toroid for Example 10.5.*

EXAMPLE 10.5. For the series-magnetic circuit of Fig. 10.32:

a. Find the value of I required to develop a magnetic flux of $\Phi = 4 \times 10^{-4}$ Wb.
b. Determine the permeability of the material under these conditions.

Solutions: The magnetic circuit can be represented by the system shown in Fig. 10.33(a). The electric circuit analogy is shown in Fig. 10.33(b). Analogies of this type can be very helpful in the solution of magnetic circuits. Table 10.3 is for part (a) of this problem.

TABLE 10.3

Section	Φ (Wb)	A (m^2)	B (T)	H (At/m)	l (m)	Hl (At)
One continuous section	4×10^{-4}	2×10^{-3}			0.16	

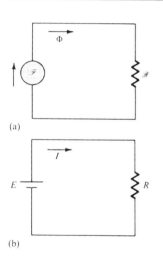

(a)

(b)

FIG. 10.33 (a) *Magnetic circuit equivalent and* (b) *electric circuit analogy.*

a. The flux density B is

$$B = \frac{\Phi}{A} = \frac{4 \times 10^{-4}}{2 \times 10^{-3}} = 2 \times 10^{-1} = 0.2 \text{ T}$$

Using the B-H curves of Fig. 10.22, we can determine the magnetizing force H:

$$H \text{ (cast steel)} = 110 \text{ At/m}$$

Applying Ampère's circuital law yields

$$NI = Hl$$

and

$$I = \frac{Hl}{N} = \frac{(110)(0.16)}{400} = 0.044 = \textbf{44 mA}$$

b. The permeability of the material can be found using Eq. (10.10):

$$\mu = \frac{B}{H} = \frac{0.2}{110} = \textbf{1.818} \times \textbf{10}^{-3} \textbf{ Wb/Am}$$

EXAMPLE 10.6. The electromagnet in Fig. 10.34 has picked up a section of cast iron. Determine the current I to establish the flux indicated.

Solution: All of the dimensions must first be converted to the metric system to use Figs. 10.21 and 10.22:

$$3.5 \text{ in.} \left[\frac{1 \text{ m}}{39.37 \text{ in.}}\right] = 88.90 \times 10^{-3} \text{ m}$$

$$4 \text{ in.} \left[\frac{1 \text{ m}}{39.37 \text{ in.}}\right] = 101.6 \times 10^{-3} \text{ m}$$

$$0.5 \text{ in.} \left[\frac{1 \text{ m}}{39.37 \text{ in.}} \right] = 12.70 \times 10^{-3} \text{ m}$$

$$1 \text{ in.}^2 \left[\frac{1 \text{ m}}{39.37 \text{ in.}} \right] \left[\frac{1 \text{ m}}{39.37 \text{ in.}} \right] = 6.452 \times 10^{-4} \text{ m}^2$$

The magnetic circuit and its electric circuit analogy appear in Fig. 10.35. Note that even though different elements appear in the system, it is still a series configuration.

The data are provided in Table 10.4.

The information available from the specifications of the problem has already been included. When the problem has been completed, each space will contain some data. Sufficient data to complete the problem can be found if we fill in each column from left to right. As the various quantities are calculated, they will be placed in a similar table found at the end of the problem.

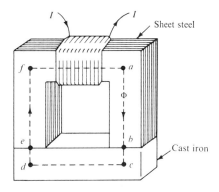

$l_{a-b} = l_{e-f} = 3.5$ in.
$l_{b-c} = l_{d-e} = 0.5$ in.
$l_{c-d} = l_{f-a} = 4$ in.
Area (throughout) $= 1$ in.^{2}
$\Phi = 3.5 \times 10^{-4}$ Wb

FIG. 10.34 *Electromagnetic and section for Example 10.6.*

TABLE 10.4

Section	Φ (Wb)	A (m²)	B (T)	H (At/m)	l (m)	Hl (At)
$a\text{-}b = e\text{-}f$	3.5×10^{-4}	6.452×10^{-4}			88.90×10^{-3}	
$c\text{-}d$	3.5×10^{-4}	6.452×10^{-4}			101.6×10^{-3}	
$b\text{-}c = d\text{-}e$	3.5×10^{-4}	6.452×10^{-4}			12.70×10^{-3}	
$f\text{-}a$	3.5×10^{-4}	6.452×10^{-4}			101.6×10^{-3}	

The flux density for each section is

$$B = \frac{\Phi}{A} = \frac{3.5 \times 10^{-4}}{6.45 \times 10^{-4}} = 0.5425 \text{ T}$$

The magnetizing force is

$$H \text{ (sheet steel, Fig. 10.22)} = 60 \text{ At/m}$$

$$H \text{ (cast iron, Fig. 10.21)} = 1600 \text{ At/m}$$

Note the extreme difference in magnetizing force for each material for the required flux density. In fact, when we apply Ampère's circuital law, we will find that the sheet steel section could be ignored with a minimal error in the solution.

Determining *Hl* for each section and inserting into the table will result in the following (see also Table 10.5):

$$H_{a-b}l_{a-b} = (60)(88.90 \times 10^{-3})$$
$$= 5334 \times 10^{-3} = 5.334 \text{ At} = H_{e-f}l_{e-f}$$
$$H_{c-d}l_{c-d} = (1600)(101.6 \times 10^{-3}) = 162.56 \text{ At}$$
$$H_{b-c}l_{b-c} = (1600)(12.70 \times 10^{-3}) = 20.320 \text{ At} = H_{d-e}l_{d-e}$$
$$H_{f-a}l_{f-a} = (60)(101.6 \times 10^{-3}) = 6.096 \text{ At}$$

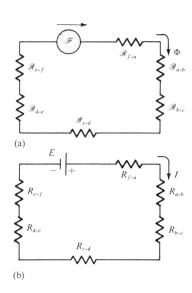

FIG. 10.35 *(a) Magnetic circuit equivalent and (b) electric circuit analogy for Example 10.6.*

TABLE 10.5

Section	Φ (Wb)	A (m^2)	B (T)	H (At/m)	l (m)	Hl (At)
$a\text{-}b = e\text{-}f$	3.5×10^{-4}	6.45×10^{-4}	0.5425	60	88.90×10^{-3}	5.334
$c\text{-}d$	3.5×10^{-4}	6.45×10^{-4}	0.5425	1600	101.6×10^{-3}	162.56
$b\text{-}c = d\text{-}e$	3.5×10^{-4}	6.45×10^{-4}	0.5425	1600	12.70×10^{-3}	20.320
$f\text{-}a$	3.5×10^{-4}	6.45×10^{-4}	0.5425	60	101.6×10^{-3}	6.096

Applying Ampère's circuital law, we obtain

$$NI = H_{a\text{-}b}l_{a\text{-}b} + H_{b\text{-}c}l_{b\text{-}c} + H_{c\text{-}d}l_{c\text{-}d} + H_{d\text{-}e}l_{d\text{-}e}$$
$$+ H_{e\text{-}f}l_{e\text{-}f} + H_{f\text{-}a}l_{f\text{-}a}$$

But

$$H_{a\text{-}b}l_{a\text{-}b} = H_{e\text{-}f}l_{e\text{-}f}$$

and

$$H_{b\text{-}c}l_{b\text{-}c} = H_{d\text{-}e}l_{d\text{-}e}$$

Therefore,

$$NI = 2(H_{a\text{-}b}l_{a\text{-}b}) + 2(H_{b\text{-}c}l_{b\text{-}c}) + H_{c\text{-}d}l_{c\text{-}d} + H_{f\text{-}a}l_{f\text{-}a}$$
$$= 2(5.334) + 2(20.320) + 162.56 + 6.096$$
$$= 10.668 + 40.64 + 162.56 + 6.096$$

$$50I = 219.964$$

$$I = \frac{219.964}{50} = \mathbf{4.399 \ A}$$

EXAMPLE 10.7. Determine the secondary current I_2 for the transformer of Fig. 10.36 if the flux in the core is 1.5×10^{-5} Wb.

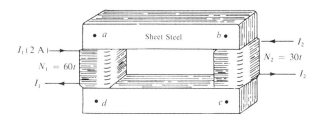

Area (throughout) $= 0.15 \times 10^{-3}$ m^2
$l_{a\text{-}b} = l_{c\text{-}d} = 0.05$ m
$l_{b\text{-}c} = l_{d\text{-}a} = 0.03$ m

FIG. 10.36 *Transformer for Example 10.7.*

Solution: This is the first example with two magnetizing forces to consider. In the analogies of Fig. 10.37 you will note that the resulting flux of each is opposing, just as the two sources of emf are opposing in the electric circuit analogy.

The data are given in Table 10.6.

The flux density throughout is

$$B = \frac{\Phi}{A} = \frac{1.5 \times 10^{-5}}{0.15 \times 10^{-3}} = 10 \times 10^{-2} = 0.10 \text{ T}$$

and

$$H \text{ (from Fig. 10.22)} \cong 7 \text{ At/m}$$

Applying Ampère's circuital law, we have

$$N_1 I_1 - N_2 I_2 = 2(H_{a\text{-}b} l_{a\text{-}b}) + 2(H_{b\text{-}c} l_{b\text{-}c})$$
$$60(2) - 30(I_2) = 2(7)(0.05) + 2(7)(0.03)$$
$$120 - 30 I_2 = 0.7 + 0.42$$

and

$$30 I_2 = 120 - 1.12$$

or

$$I_2 = \frac{118.88}{30} = \textbf{3.963 A}$$

For the analysis of most transformer systems, the equation $N_1 I_1 = N_2 I_2$ is employed. This would result in 4 A versus 3.963 A above. This difference is normally ignored, however, and the equation above considered exact.

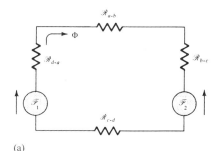

(a)

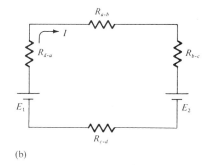

(b)

FIG. 10.37 (a) *Magnetic circuit equivalent and* (b) *electric circuit analogy for the transformer of Fig. 10.36.*

TABLE 10.6

Section	Φ (Wb)	A (m^2)	B (T)	H (At/m)	l (m)	Hl (At)
$a\text{-}b = c\text{-}d$	1.5×10^{-5}	0.15×10^{-3}			0.05	
$b\text{-}c = d\text{-}a$	1.5×10^{-5}	0.15×10^{-3}			0.03	

Because of the nonlinearity of the *B-H* curve, *it is not possible to apply superposition to magnetic circuits;* that is, in the previous example, we cannot consider the effects of each source independently and then find the total effects by using superposition.

10.13 AIR GAPS

Before continuing with the illustrative examples, let us consider the effects an air gap has on a magnetic circuit. Note the presence of air gaps in the magnetic circuits of the motor and meter of Fig. 10.11. The spreading of the flux lines outside the common area of the core for the air gap in Fig. 10.38(a) is known as *fringing*. For our purposes, we shall neglect this effect and assume the flux distribution to be as in Fig. 10.38(b).

The flux density of the air gap in Fig. 10.38(b) is given by

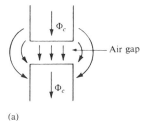

(a)

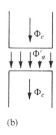

(b)

FIG. 10.38

$$B_g = \frac{\Phi_g}{A_g} \qquad \text{(10.17)}$$

where, for our purposes,

$$\Phi_g = \Phi_{\text{core}}$$

and

$$A_g = A_{\text{core}}$$

For most practical applications, the permeability of air is taken to be equal to that of free space. The magnetizing force of the air gap is then determined by

$$H_g = \frac{B_g}{\mu_o} \qquad \text{(10.18)}$$

and the mmf drop across the air gap is equal to $H_g l_g$. An equation for H_g is as follows:

$$H_g = \frac{B_g}{\mu_o} = \frac{B_g}{4\pi \times 10^{-7}}$$

and

$$H_g = 7.97 \times 10^5 B_g \qquad \text{(At/m)} \qquad \text{(10.19)}$$

EXAMPLE 10.8. Find the value of I required to establish a magnetic flux of $\Phi = 0.75 \times 10^{-4}$ Wb in the series magnetic circuit of Fig. 10.39.

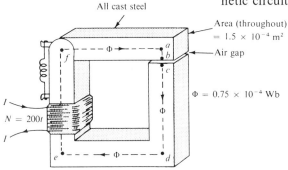

All cast steel

Area (throughout) = 1.5×10^{-4} m²

Air gap

$\Phi = 0.75 \times 10^{-4}$ Wb

$N = 200t$

$l_{a-b} = 3.4 \times 10^{-3}$ m $l_{d-e} = l_{e-f} = l_{f-a} = 50 \times 10^{-3}$ m
$l_{b-c} = 1.6 \times 10^{-3}$ m
$l_{c-d} = 45 \times 10^{-3}$ m

FIG. 10.39 *Relay for Example 10.8.*

Solution: An equivalent magnetic circuit and its electric circuit analogy are shown in Fig. 10.40. The table follows shortly.

The flux density for each section is

$$B = \frac{\Phi}{A} = \frac{0.75 \times 10^{-4}}{1.5 \times 10^{-4}} = 0.5 \text{ T}$$

From the *B-H* curves of Fig. 10.22;

$$H \text{ (cast steel)} \cong 280 \text{ At/m}$$

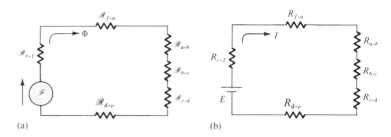

FIG. 10.40 (a) Magnetic circuit equivalent and (b) electric circuit analogy for the relay of Fig. 10.39.

Applying Eq. (10.19), we obtain

$$H_g = 7.97 \times 10^5 B_g$$
$$= 7.97 \times 10^5 (0.5)$$
$$= 3.985 \times 10^5 \text{ At/m}$$

The mmf drops are

$$H_{a-b}l_{a-b} = (280)(3.4 \times 10^{-3}) = 0.9520 \text{ At}$$
$$H_{b-c}l_{b-c} = (3.985 \times 10^5)(1.6 \times 10^{-3}) = 637.60 \text{ At}$$
$$H_{c-d}l_{c-d} = (280)(45 \times 10^{-3}) = 12.60 \text{ At}$$
$$H_{d-e}l_{d-e} = H_{f-a}l_{f-a} = H_{e-f}l_{e-f} = (280)(50 \times 10^{-3}) = 14.0 \text{ At}$$

The data are inserted in Table 10.7.

TABLE 10.7

Section	Φ (Wb)	A (m²)	B (T)	H (At/m)	l (m)	Hl (At)
a-b	0.75×10^{-4}	1.5×10^{-4}	0.5	280	3.4×10^{-3}	0.9520
b-c	0.75×10^{-4}	1.5×10^{-4}	0.5	3.985×10^5	1.6×10^{-3}	637.60
c-d	0.75×10^{-4}	1.5×10^{-4}	0.5	280	45×10^{-3}	12.60
d-e = e-f = f-a	0.75×10^{-4}	1.5×10^{-4}	0.5	280	50×10^{-3}	14.0

Applying Ampère's circuital law yields

$$NI = 3(H_{d-e}l_{d-e}) + H_{a-b}l_{a-b} + H_{b-c}l_{b-c} + H_{c-d}l_{c-d}$$
$$= 3(14.0) + 0.9520 + 637.60 + 12.60$$
$$200I = 693.152$$
$$I = \textbf{3.466 A}$$

Note from the table that the air gap requires the biggest share (by far) of the impressed NI due to the fact that air is nonmagnetic.

10.14 SERIES-PARALLEL MAGNETIC CIRCUITS

As one might expect, the close analogies between electric and magnetic circuits will eventually lead to series-parallel magnetic circuits similar in many respects to those encountered in Chapter 6. In this section, the table approach will be employed to its fullest advantage. Without it, it would be extremely difficult to keep track of the various quantities known and to be found.

EXAMPLE 10.9. Determine the current I required to establish a flux in the core indicated in Fig. 10.41 of 4×10^{-4} Wb.

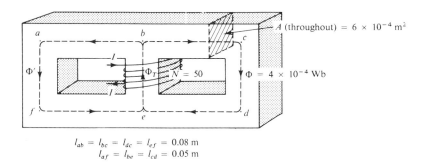

$$l_{ab} = l_{bc} = l_{dc} = l_{ef} = 0.08 \text{ m}$$
$$l_{af} = l_{be} = l_{cd} = 0.05 \text{ m}$$

FIG. 10.41

Solution: An equivalent magnetic circuit and its electrical analogy appear in Fig. 10.42.

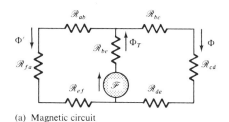

(a) Magnetic circuit

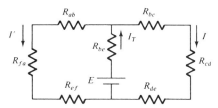

(b) Electric circuit analogy

FIG. 10.42

Due to symmetry,

$$\Phi' = \Phi = 4 \times 10^{-4} \text{ Wb}$$

and

$$\Phi_T = \Phi' + \Phi = 2\Phi \text{ (or } 2\Phi' \text{ if preferred)}$$
$$= 8 \times 10^{-4} \text{ Wb}$$

For each outside leg,

$$B = \frac{\Phi}{A} = \frac{4 \times 10^{-4}}{6 \times 10^{-4}} = 0.667 \text{ T}$$

and for the center leg,

$$B = \frac{\Phi}{A} = \frac{8 \times 10^{-4}}{6 \times 10^{-4}} = 1.333 \text{ T}$$

The data then appear as shown in Table 10.8.

TABLE 10.8

Section	Φ (Wb)	A (m²)	B (T)	H	l (m)	Hl
bcde or *efab*	4×10^{-4}	6×10^{-4}	0.667		0.21	
b-e	8×10^{-4}	6×10^{-4}	1.333		0.05	

From Fig. 10.21,

$$H_{bcde} \cong 100 \text{ At/m}$$
$$H_{be} \cong 650 \text{ At/m}$$

Before applying Ampère's circuital law, consider the electrical analog of Fig. 10.42, where Kirchhoff's voltage law around the right-hand loop would result in

$$E - I_T R_{be} - I(R_{bc} + R_{cd} + R_{de}) = 0$$

or

$$E - I_T R_{be} - I(R_{bcde}) = 0$$

Application of Ampère's circuital law to the magnetic circuit will result in a very similar equation; that is,

$$NI - H_{be}l_{be} - H_{bcde}l_{bcde} = 0$$

Therefore, if necessary, the electric circuit analogy can be employed to determine the form of the equation to be obtained by Ampère's circuital law.

Substituting yields

$$NI = +H_{be}l_{be} + H_{bcde}l_{bcde} = (650)(0.05) + (100)(0.21)$$

or

$$50I = 32.50 + 21.0 = 53.50$$

and

$$I = \frac{53.5}{50} = \textbf{1.070 A}$$

EXAMPLE 10.10. For the series-parallel magnetic circuit of Fig. 10.43, find the value of I required to establish a flux in the gap $\Phi_g = 2 \times 10^{-4}$ Wb.

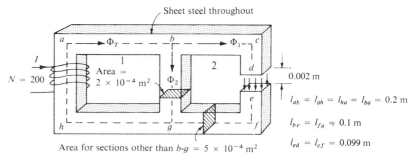

FIG. 10.43

Solution: An equivalent magnetic circuit and its electrical analogy are shown in Fig. 10.44. Table 10.9 is completed for this problem.

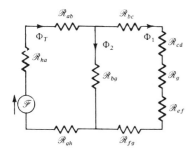

(a) Magnetic circuit

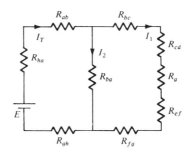

(b) Electric circuit analogy

FIG. 10.44

TABLE 10.9

Section	Φ (Wb)	A (m²)	B (T)	H	l (m)	Hl
a-b, g-h	√√	5×10^{-4}			0.2	
b-c, f-g	2×10^{-4}	5×10^{-4}	√	√	0.1	√
c-d, e-f	2×10^{-4}	5×10^{-4}	√	√	0.099	√
a-h	√√	5×10^{-4}			0.2	
b-g	√√	2×10^{-4}	√√	√√	0.2	√√
d-e	2×10^{-4}	5×10^{-4}	√	√	0.002	√

The flux densities are

$$B_{bc} = B_{cd} = B_g = B_{ef} = B_{fg} = \frac{\Phi}{A} = \frac{2 \times 10^{-4}}{5 \times 10^{-5}}$$

$$= 0.4 \text{ T}$$

For the air gap,

$$H_g = (7.97 \times 10^5)(0.4) \doteq 3.188 \times 10^5 \text{ At/m}$$

and

$$H_g l_g = (3.188 \times 10^5)(0.002) = 6.376 \times 10^2 = 637.60 \text{ At}$$

Using the *B-H* curves of Fig. 10.22, we find the magnetizing force *H* and mmf drops for the following sections:

$$H_{bc} = H_{cd} = H_{ef} = H_{fg} = 55 \text{ At/m}$$

and

$$H_{bc} l_{bc} = H_{fg} l_{fg} = (55)(0.1) = 5.50 \text{ At}$$
$$H_{cd} l_{cd} = H_{ef} l_{ef} = (55)(0.099) = 5.445 \text{ At}$$

With the results of the calculations just completed, we can fill in the squares indicated by single check marks in Table 10.9. From this table it is obvious that we now know all that is necessary about the section *bcdefg*. We must now consider the other portions of the magnetic circuit.

If we apply Ampère's circuital law around loop 2, as indicated in the original magnetic circuit, we can find the flux Φ_2; that is,

$$\Sigma_{\circlearrowleft} \mathcal{F} = 0$$

so that

$$H_{bc}l_{bc} + H_{cd}l_{cd} + H_g l_g + H_{ef}l_{ef} + H_{fg}l_{fg} - H_{gb}l_{gb} = 0$$

Note the difference in sign for the $H_{gb}l_{gb}$ term due to the direction of the flux Φ_2. (Compare this equation with that obtained for the same loop of the electric circuit analogy.)

Substituting for the known terms, we have

$$5.5 + 5.445 + 637.60 + 5.445 + 5.50 - H_{gb}l_{gb} = 0$$

and

$$H_{gb}l_{gb} = 659.49$$

or

$$H_{gb} = \frac{659.49}{0.2} = 3297.45 \text{ At/m}$$

From the *B-H* curves,

$$B_{gb} \cong 1.55 \text{ T}$$

and

$$\Phi_2 = B_{gb}A = 1.55(2 \times 10^{-4}) = 3.10 \times 10^{-4} \text{ Wb}$$

Substituting for Φ_1 and Φ_2 in the following equation yields

$$\Phi_T = \Phi_1 + \Phi_2$$
$$= 2 \times 10^{-4} + 3.10 \times 10^{-4}$$
$$\Phi_T = 5.10 \times 10^{-4} \text{ Wb} = \Phi_{ab} = \Phi_{ha} = \Phi_{gh}$$

With the results of the calculations just completed, we can now fill in the squares indicated by double check marks in the preceding table. The next few steps should be evident:

$$B_{ab} = B_{ha} = B_{gh} = \frac{\Phi_T}{A} = \frac{5.10 \times 10^{-4}}{5 \times 10^{-4}} = 1.020 \text{ T}$$

From the *B-H* curves,

$$H_{ab} = H_{ha} = H_{gh} \cong 160 \text{ At/m}$$

and the mmf drops:

$$H_{ab}l_{ab} = (160)(0.2) = 32.0 \text{ At}$$
$$H_{ha}l_{ha} = (160)(0.2) = 32.0 \text{ At}$$
$$H_{gh}l_{gh} = (160)(0.2) = 32.0 \text{ At}$$

which completes the table.

If we now apply Ampère's circuital law to loop 1, as indicated in the original circuit diagram, we have

$$\Sigma_{\circlearrowleft} \mathscr{F} = 0$$
$$NI = H_{ab}l_{ab} + H_{bg}l_{bg} + H_{gh}l_{gh} + H_{ah}l_{ah}$$

Substituting for the known values yields

$$200I = 32.0 + 659.49 + 32.0 + 32.0$$
$$200I = 755.49$$
$$I \cong \mathbf{3.780\ A}$$

The procedure for the series-parallel magnetic circuit of Fig. 10.45 is quite similar, requiring fewer steps due to the absence of the air gap.

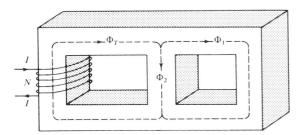

FIG. 10.45

10.15 DETERMINING Φ

The examples of this section are of the second type, where NI is given and the flux Φ must be found. This is a relatively straightforward problem if only one magnetic section, such as a toroid, is involved. Then

$$H = \frac{NI}{l} \qquad H \to B \text{ (B-H curve)}$$

and

$$\Phi = BA$$

For magnetic circuits with more than one section, there is no set order of steps that will lead to an exact solution for every problem on the first attempt. In general, however, we proceed as follows. We must find the impressed mmf for a *calculated guess* of the flux Φ and then compare this with the specified value of mmf. We can then make adjustments on our guess to bring it closer to the actual value. For most applications, a value within $\pm 5\%$ of the actual Φ or specified NI is acceptable.

We can make a reasonable guess at the value of Φ if we realize that the maximum mmf drop appears across the material with the smallest permeability if the length and area of each material are the same. As shown in Example 10.8, if there is an air gap in the magnetic circuit, there will be a considerable drop in mmf across the gap. As a starting point for problems of this type, therefore, we shall assume that the total mmf (NI) is across the section with the lowest μ or greatest $\mathcal{R}$ (if the other physical dimensions are relatively similar). This assumption gives a value of Φ that will produce a calculated NI greater than the specified value. Then, after considering the results of our original assumption very carefully, we shall *cut* Φ and NI by introducing the effects (reluctance) of the other portions of the magnetic circuit and *try* the new solution. For obvious reasons, this approach is frequently called the *cut and try* method.

EXAMPLE 10.11. Calculate the magnetic flux Φ for the magnetic circuit of Fig. 10.46.

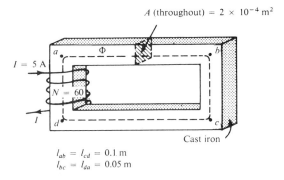

A (throughout) $= 2 \times 10^{-4}\,\text{m}^2$

$I = 5$ A

$N = 60$

Cast iron

$l_{ab} = l_{cd} = 0.1$ m
$l_{bc} = l_{da} = 0.05$ m

FIG. 10.46

Solution: By Ampère's circuital law,

$$NI = H_{abcda}l_{abcda}$$

or

$$H_{abcda} = \frac{NI}{l_{abcda}} = \frac{(60)(5)}{(0.1 + 0.05 + 0.1 + 0.05)}$$

$$= \frac{300}{0.3} = 1000\,\text{At/m}$$

and

$$B_{abcda}\ (\text{from Fig. 10.20}) \cong 0.38\,\text{T}$$

Since $B = \Phi/A$, we have

$$\Phi = BA = (0.38)(2 \times 10^{-4}) = \mathbf{0.760 \times 10^{-4}\,Wb}$$

EXAMPLE 10.12. Find the magnetic flux Φ for the series magnetic circuit of Fig. 10.47 for the specified impressed mmf.

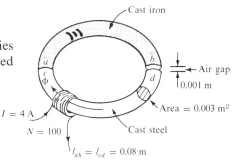

Cast iron

Air gap

0.001 m

Area $= 0.003$ m^2

$I = 4$ A

$N = 100$

Cast steel

$l_{ab} = l_{cd} = 0.08$ m

FIG. 10.47

Solution: Assuming that the total impressed mmf NI is across the air gap,

$$NI = H_g l_g$$

or

$$H_g = \frac{NI}{l_g} = \frac{400}{0.001} = 4 \times 10^5 \text{ At/m}$$

and

$$B_g = \mu_o H_g = (4\pi \times 10^{-7})(4 \times 10^5)$$
$$= 50.265 \times 10^{-2} \text{ T}$$

The flux

$$\Phi_g = \Phi_{\text{core}} = B_g A$$
$$= (50.265 \times 10^{-2})(0.003)$$
$$\Phi_{\text{core}} = 150.795 \times 10^{-5} \text{ Wb}$$

Using this value of Φ, we can find NI. The data are inserted in Table 10.10.

TABLE 10.10

Section	Φ (Wb)	A (m^2)	B (T)	H (At/m)	l (m)	Hl (At)
a-b	150.795×10^{-5}	0.003	50.265×10^{-2}	1500 (B-H curve)	0.08	
Gap	150.795×10^{-5}	0.003	50.265×10^{-2}	4×10^5	0.001	400
d-c	150.795×10^{-5}	0.003	50.265×10^{-2}	260 (B-H curve)	0.08	

$$H_{ab} l_{ab} = (1500)(0.08) = 120.0 \text{ At}$$
$$H_{dc} l_{dc} = (260)(0.08) = 20.8 \text{ At}$$

Applying Ampère's circuital law results in

$$NI = H_{ab} l_{ab} + H_g l_g + H_{dc} l_{dc}$$
$$= 120 + 400 + 20.8$$
$$NI = 540.8 > 400.0$$

Since we neglected the reluctance of all the magnetic paths but the air gap, the calculated value is greater than the specified value. We must therefore reduce this value by including the effect of these reluctances. Since approximately $(540.8 - 400.0)/540.8 = 140.8/540.8 \cong 26\%$ of our calculated value is above the desired value, let us reduce Φ by 26% and see how close we come to the impressed mmf of 400:

$$(0.26)(150.795 \times 10^{-5}) = 39.207 \times 10^{-5}$$

$$\begin{array}{r} 150.795 \times 10^{-5} \\ -39.207 \times 10^{-5} \\ \hline \Phi = 111.588 \times 10^{-5} \text{ Wb} \end{array}$$

See Table 10.11.

TABLE 10.11

Section	Φ (Wb)	A (m²)	B	H	l (m)	Hl
a-b	111.588×10^{-5}	0.003			0.08	
Gap	111.588×10^{-5}	0.003			0.001	
d-c	111.588×10^{-5}	0.003			0.08	

$$B = \frac{\Phi}{A} = \frac{111.588 \times 10^{-5}}{0.003} = 37.196 \times 10^{-2} \text{ T}$$

$$\begin{aligned}
H_g l_g &= 7.97 \times 10^5 B_g l_g \\
&= (7.97 \times 10^5)(37.196 \times 10^{-2})(0.001) \\
&\cong 297.45 \text{ At}
\end{aligned}$$

From B-H curves,

$$H_{ab} \cong 840 \text{ At/m}$$

$$H_{dc} \cong 210 \text{ At/m}$$

$$H_{ab} l_{ab} = (840)(0.08) = 67.20 \text{ At}$$

$$H_{dc} l_{dc} = (210)(0.08) = 16.80 \text{ At}$$

Applying Ampère's circuital law yields

$$NI = H_{ab} l_{ab} + H_g l_g + H_{dc} l_{dc}$$

$$= 67.20 + 297.45 + 16.80$$

$$NI = \textbf{381.45} < 400 \quad \text{(but within } \pm 5\%$$
$$\text{and therefore acceptable)}$$

PROBLEMS

Section 10.3

1. Using Appendix G, fill in the blanks in the following table. Indicate the units for each quantity.

	Φ	B
SI	5×10^{-4} Wb	8×10^{-4} T
CGS	——	——
English	——	——

2. Repeat Problem 1 for the following table if area = 2 in².

	Φ	B
SI	——	——
CGS	60,000 maxwells	——
English	——	——

3. For the electromagnet of Fig. 10.48:
  a. Find the flux density in the core.
  b. Sketch the magnetic flux lines and indicate their direction.

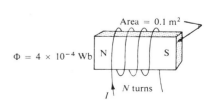

FIG. 10.48

c. Indicate the north and south poles of the magnet.

Section 10.5

4. Which section of Fig. 10.49 [(a), (b), or (c)] has the largest reluctance to the setting up of flux lines through its longest dimension?

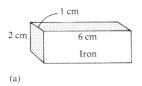

(a)

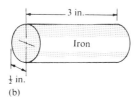

(b)

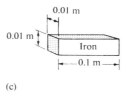

(c)

FIG. 10.49

Section 10.6

5. Find the reluctance of a magnetic circuit if a magnetic flux $\Phi = 4.2 \times 10^{-4}$ Wb is established by an impressed mmf of 400 At.

6. Repeat Problem 5 for $\Phi = 72,000$ maxwells and an impressed mmf of 120 gilberts.

Section 10.7

7. Find the magnetizing force H for Problem 5 in SI units if the magnetic circuit is 6 inches in length.

8. If a magnetizing force H of 600 At/m is applied to a magnetic circuit, a flux density B of 1200×10^{-4} Wb/m² is established. Find the permeability μ of a material that will produce twice the original flux density for the same magnetizing force.

Section 10.9

9. For the sheet steel sample of Figs. 10.21 and 10.22, determine
   a. μ_{dc} at $B = 0.8$ T.
   b. μ_Δ in the region of $B = 0.8$ T.
   c. μ_{av}.

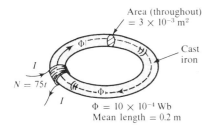

FIG. 10.50

10. For the cast iron sample of Figs. 10.21 and 10.22, determine
   a. μ_{dc} at $B = 0.2$T.
   b. μ_Δ in the region of $B = 0.2$ T.
   c. μ_{av}. (Although the knee is not well defined, choose a straight line segment that you feel best represents the curve below the saturation region.)

For the following problems, use the B-H curves of Figs. 10.21 and 10.22.

Section 10.12

11. For the series magnetic circuit of Fig. 10.50, determine the current I necessary to establish the indicated flux.

12. Find the current necessary to establish a flux of $\Phi = 3 \times 10^{-4}$ Wb in the series magnetic circuit of Fig. 10.51.

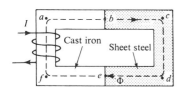

$l_{ab} = l_{bc} = l_{cd} = l_{de} = l_{ef} = l_{fa} = 0.1$ m
Area (throughout) $= 5 \times 10^{-4}$ m²
$N = 100$ turns

FIG. 10.51

13. For the series magnetic circuit of Fig. 10.52 with two impressed sources of magnetic pressure, determine the current I. Each applied mmf establishes a flux pattern in the clockwise direction.

$\phi = 0.8 \times 10^{-4}$ Wb

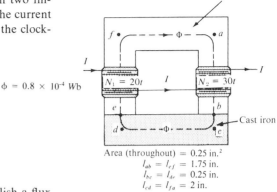

Area (throughout) = 0.25 in.²
$l_{ab} = l_{ef} = 1.75$ in.
$l_{bc} = l_{de} = 0.25$ in.
$l_{cd} = l_{fa} = 2$ in.

FIG. 10.52

14. a. Find the number of turns N required to establish a flux $\Phi = 12 \times 10^{-4}$ Wb in the magnetic circuit of Fig. 10.53.
   b. Find the permeability μ of the material.

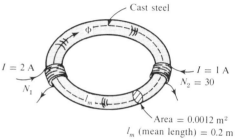

Area = 0.0012 m²
l_m (mean length) = 0.2 m

FIG. 10.53

15. a. Find the mmf (NI) required to establish a flux $\Phi = 80,000$ lines in the magnetic circuit of Fig. 10.54.
   b. Find the permeability of each material.

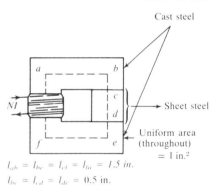

$l_{ab} = l_{bc} = l_{ef} = l_{fa} = 1.5$ in.
$l_{bc} = l_{cd} = l_{de} = 0.5$ in.

FIG. 10.54

Section 10.13

16. a. Find the current I required to establish a flux $\Phi = 2.4 \times 10^{-4}$ Wb in the magnetic circuit of Fig. 10.55.
   b. Compare the mmf drop across the air gap to that across the rest of the magnetic circuit. Discuss your results using the value of μ for each material.

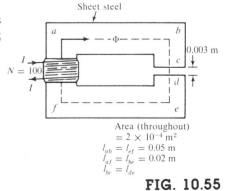

Area (throughout)
= 2×10^{-4} m²
$l_{ab} = l_{ef} = 0.05$ m
$l_{af} = l_{be} = 0.02$ m
$l_{bc} = l_{de}$

FIG. 10.55

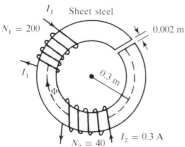

$N_1 = 200$ Sheet steel
I_1
0.002 m
0.3 m
Φ
$N_2 = 40$ $I_2 = 0.3$ A
Area (throughout) $= 1.3 \times 10^{-4}$ m²

FIG. 10.56

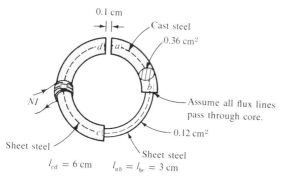

0.1 cm
Cast steel
0.36 cm²
d a
b
NI
Assume all flux lines pass through core.
0.12 cm²
c
Sheet steel
Sheet steel
$l_{cd} = 6$ cm $l_{ab} = l_{bc} = 3$ cm

FIG. 10.57

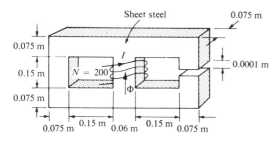

Spring
0.04 m
Cast steel
0.005 m
0.002 m
I
0.01 m
$N = 200$
0.01 m
Φ
0.023 m
0.01 m Cast steel
0.01 m

FIG. 10.58

17. Determine the current I_1 required to establish a flux of $\Phi = 2 \times 10^{-4}$ Wb in the magnetic circuit of Fig. 10.56.

*18. Find the mmf required to establish a flux $\Phi = 1200$ maxwells in the series-magnetic circuit of Fig. 10.57.

*19. a. If the Φ passing through the relay system of Fig. 10.58 is 0.3×10^{-4} Wb, find the current passing through the coil.
 b. The force exerted on the armature is determined by the equation

$$F \text{ (newtons)} = \frac{1}{2} \cdot \frac{B_g^2 A}{\mu_o}$$

where B_g is the flux density within the air gap and A is the common area of the air gap. Find the force exerted when the flux Φ specified in part (a) is established.

Section 10.14

*20. Find the current I required to establish a flux $\Phi_g = 20 \times 10^{-4}$ Wb in the air gap of the series-parallel magnetic circuit of Fig. 10.59.

Sheet steel
0.075 m
0.075 m
0.15 m $N = 200$
0.0001 m
0.075 m
0.075 m 0.15 m 0.06 m 0.15 m 0.075 m

FIG. 10.59

*21. Find the current I_1 required to establish a flux $\Phi_g = 7500$ lines in the air gap of the series-parallel magnetic circuit of Fig. 10.60. Carefully note the direction of current for each coil.

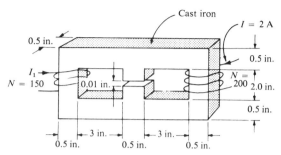

FIG. 10.60

Section 10.15

22. Find the magnetic flux Φ established in the series-magnetic circuit of Fig. 10.61.

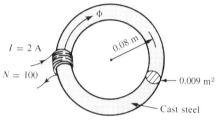

FIG. 10.61

23. Find the magnetic flux Φ established in the series-magnetic circuit of Fig. 10.62.

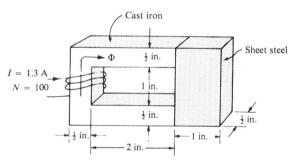

FIG. 10.62

*24. Determine the magnetic flux Φ established in the series-magnetic circuit of Fig. 10.63.

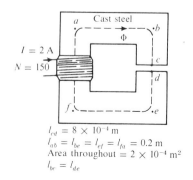

$l_{cd} = 8 \times 10^{-4}$ m
$l_{ab} = l_{bc} = l_{ef} = l_{fa} = 0.2$ m
Area throughout $= 2 \times 10^{-4}$ m^2
$l_{bc} = l_{de}$

FIG. 10.63

GLOSSARY

Ampère's circuital law   A law establishing the fact that the algebraic sum of the rises and drops of the mmf around a closed loop of a magnetic circuit is equal to zero.

Average permeability (μ_{av})   Permeability that provides an approximate value for the range of flux density from zero to (but not including) the saturation level.

Diamagnetic materials   Materials that have permeabilities slightly less than that of free space.

Domain   A group of magnetically aligned atoms.

Electromagnetism   Magnetic effects introduced by the flow of charge or current.

Ferrite   A ferromagnetic material constructed of a mixture of metallic oxides that exhibits a "square" hysteresis curve.

Ferromagnetic materials   Materials having permeabilities hundreds and thousands of times greater than that of free space.

Flux density (B)   A measure of the flux per unit area perpendicular to a magnetic flux path. It is measured in teslas (T) or weber/m^2 (Wb/m^2).

Hysteresis   The lagging effect between flux density of a material and the magnetizing force applied.

Incremental permeability (μ_{Δ})   Permeability sensitive to small changes in flux density through the core.

Magnetic flux lines   Lines of a continuous nature that reveal the strength and direction of magnetic field.

Magnetizing force (H)   A measure of the magnetomotive force per unit length of a magnetic circuit.

Magnetomotive force ($\mathscr{F}$)   The "pressure" required to establish magnetic flux in a ferromagnetic material. It is measured in ampere-turns (At).

Paramagnetic materials   Materials that have permeabilities slightly greater than that of free space.

Permanent magnet   A material such as steel that will remain magnetized for long periods of time without the aid of external means.

Permeability (μ)   A measure of the ease with which magnetic flux can be established in a material. It is measured in Wb/Am.

Relative permeability (μ_r)   The ratio of the permeability of a material to that of free space.

Reluctance ($\mathscr{R}$)   A quantity determined by the physical characteristics of a material that will provide an indication of the "reluctance" of that material to the setting up of magnetic flux lines in the material. It is measured in rels or At/Wb.

Static permeability (μ)   Permeability determined at a particular point on a normal magnetization curve.

INDUCTORS

11

11.1 INTRODUCTION

We have examined the resistor and the capacitor in detail. In this chapter we shall consider a third element, the *inductor,* which has a number of response characteristics similar in many respects to those of the capacitor. In fact, some sections of this chapter will proceed parallel to those for the capacitor to emphasize the similarity that exists between the two elements.

11.2 FARADAY'S LAW OF ELECTROMAGNETIC INDUCTION

If a conductor is moved through a magnetic field so that it cuts magnetic lines of flux, a voltage will be induced across the conductor, as shown in Fig. 11.1. The greater the number of flux lines cut per unit time (by increasing the speed with which it passes through the field), or the stronger the magnetic field strength (for the same traversing speed), the greater will be the induced voltage across the conductor. If the conductor is held fixed and the magnetic field is moved so that its flux lines cut the conductor, the same effect will be produced.

Inductors

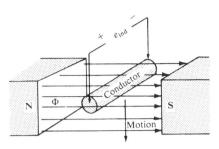

FIG. 11.1

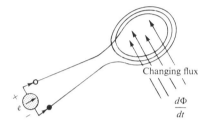

FIG. 11.2

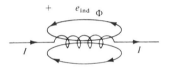

FIG. 11.3

If a coil of N turns is placed in the region of a changing flux, as in Fig. 11.2, a voltage will be induced across the coil as determined by *Faraday's law:*

$$e = N\frac{d\phi}{dt} \qquad \text{(volts)} \qquad \textbf{(11.1)}$$

where N = the number of turns of the coil and $d\phi/dt$ is the instantaneous change in flux (in webers) linking the coil. The term *linking* refers to that flux within the turns of wire. The term *changing* simply indicates that either the strength of the field linking the coil changes in magnitude or the coil is moved through the field in such a way that the number of flux lines through the coil changes with time.

If the flux linking the coil ceases to change, such as when the coil simply sits still in a magnetic field of fixed strength, $d\phi/dt = 0$, and the induced voltage $e = N(d\phi/dt) = N(0) = 0$.

11.3 LENZ'S LAW

In Section 10.2 it was shown that the magnetic flux linking a coil of N turns with a current I passing through it has the distribution of Fig. 11.3.

If the current passing through the coil increases in magnitude, the flux linking the coil also increases. It was shown in Section 11.2, however, that a changing flux linking a coil induces a voltage across the coil. For this coil, therefore, an induced emf is developed across the coil due to the change in current through the coil. The polarity of this induced emf tends to establish a current in the coil which produces a flux that will oppose any change in the original flux. In other words, the induced effect (e_{ind}) is a result of the increasing current through the coil. However, the resulting induced voltage will tend to establish a current that will oppose the increasing change in current through the coil. Keep in mind that this is all occurring simultaneously. The instant the current begins to increase in magnitude, there will be an opposing effect trying to limit the change. It is "choking" the change in current through the coil. Hence the term *choke* is often applied to the inductor or coil. In fact, we will find shortly that the current through a coil cannot change instantaneously. A period of time determined by the coil and the resistance of the circuit is required before the inductor discontinues its opposition to a momentary change in current. Recall a similar situation for the voltage across a capacitor in Chapter 9. The reaction above is true for increasing or decreasing levels of current through the coil. This effect is an example of a general principle known as *Lenz's law*, which states that *an induced effect is always such as to oppose the cause that produces it.*

11.4 SELF-INDUCTANCE

Since the induced emf opposes any change in the flux linking the coil and thereby any change in the current passing through the coil, it is called the *counter-emf*. The ability of a coil to oppose any change in current is a measure of the *self-inductance L* of the coil. For brevity, the prefix *self* is usually dropped. Inductance is measured in henries (H), after the American physicist Joseph Henry.

Inductors are coils of various dimensions designed to introduce specified amounts of inductance into a circuit. The inductance of a coil varies directly with the magnetic properties of the coil. Ferromagnetic materials, therefore, are frequently employed to increase the inductance by increasing the flux linking the coil.

A close approximation, in terms of physical dimensions, for the inductance of the coils of Fig. 11.4 can be found using the following equation:

$$L = \frac{N^2 \mu A}{l} \qquad \text{(henries, H)} \qquad \textbf{(11.2)}$$

where

N = number of turns

μ = permeability of core (recall that μ is not a constant but depends on the level of B and H since $\mu = B/H$)

A = area of core in m^2

l = mean length of core in meters

Equations for the inductance of coils different from those shown above can be found in reference handbooks. Most of the equations are more complex than those just described.

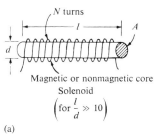

Magnetic or nonmagnetic core
Solenoid
$\left(\text{for } \dfrac{l}{d} \gg 10 \right)$

(a)

Magnetic or nonmagnetic core

Toroid

(b)

FIG. 11.4

EXAMPLE 11.1. Find the inductance of the coil of Fig. 11.5.

Solution:

$$
\begin{aligned}
L = \frac{N^2 \mu A}{l} &= \frac{[(100^2)(4\pi \times 10^{-7})][\pi(0.004^2)/4]}{0.08} \\
&= \frac{(10^4)(10^{-7})(4\pi)(\pi/4)(0.004^2)}{0.08} \\
&= \frac{(10^{-3})(9.87)(16 \times 10^{-6})}{0.08} \\
&= \frac{(10^{-9})(157.92)}{8 \times 10^{-2}} = \frac{(10^{-7})(157.92)}{8} \\
&= 19.74 \times 10^{-7} \\
L &= \textbf{1.974 } \boldsymbol{\mu}\textbf{H}
\end{aligned}
$$

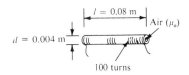

FIG. 11.5

EXAMPLE 11.2. Repeat Example 11.1, but with an iron core and conditions such that $\mu_r = 200$.

Solution:

$$L = \frac{N^2 \mu_r \mu_o A}{l} = \mu_r \left(\frac{N^2 \mu_o A}{l} \right)$$
$$= (200)(1.974 \; \mu\text{H})$$
$$L = \textbf{0.3948 mH}$$

11.5 TYPES OF INDUCTORS

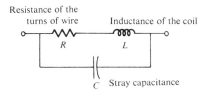

Resistance of the turns of wire — R
Inductance of the coil — L
C *Stray capacitance*

FIG. 11.6

Associated with every inductor are a resistance equal to the resistance of the turns and a stray capacitance due to the capacitance between the turns of the coil. To include these effects, the equivalent circuit for the inductor is as shown in Fig. 11.6.

The primary function of the inductor, however, is to introduce inductance—not resistance or capacitance—into the network. For this reason, the symbols employed for inductance are as shown in Fig. 11.7.

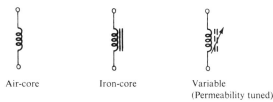

Air-core   Iron-core   Variable
                                 (Permeability tuned)

FIG. 11.7

All inductors, like capacitors, can be listed under two general headings: *fixed* and *variable*. The fixed air-core and iron-core inductors were described in the previous section. The permeability-tuned variable coil has a ferromagnetic shaft that can be moved within the coil to vary the flux linkages of the coil and thereby its inductance. Several fixed and variable inductors are shown in Fig. 11.8.

11.6 INDUCED VOLTAGE

The inductance of a coil is also a measure of the instantaneous change in flux linking a coil due to an instantaneous change in current through the coil; that is,

$$\boxed{L = N\frac{d\phi}{di}} \qquad \text{(henries, H)} \qquad \textbf{(11.3)}$$

where

N = number of turns

ϕ = flux in webers

i = current through the coil

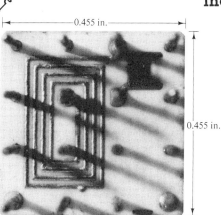

(a) Inductor and resistor on a module
Courtesy of International Business Machines Corp.

(b) 1.0 H at 8 A, 8 kV working voltage
Courtesy of Basler Electric Co.

(c) 0.025 to 0.11 H total inductance dependent on series or parallel connections ⁵⁄₁₆ in. dia. × ³⁄₁₆ in. high, weight ¹⁄₂₀ oz.
Courtesy of United Transformer Corp.

(f) molded inductors, 0.022–10,000 μH
Courtesy of Delevan, Division of American Precision Industries, Inc.

(d) Variable inductor 0.2 to 2 H
Courtesy of United Transformer Corp.

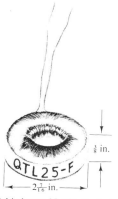

(e) Toroidal inductor 25 mH. Typical distributed capacitance 127 pF
Courtesy of Microtran Company, Inc.

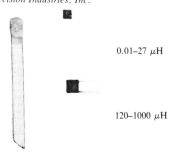

0.01–27 μH

120–1000 μH

12–100 μH

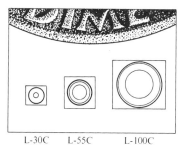

L-30C   L-55C   L-100C

Courtesy of Thinco Division, Hull Corp.

(g) μ Chip Inductors: L-30C (30-mil outside diameter, 2–56 μH), L-55C (55-mil outside diameter, up to 250 μH), and L-100C (100-mil outside diameter, up to 500 μH).

(h) Micro-i® chip inductors

FIG. 11.8 *Various types of inductors.*

The equation states that the larger the inductance of a coil (with N fixed), the larger will be the instantaneous change in flux linking the coil due to an instantaneous change in current through the coil.

If we write Eq. (11.1) as

$$e_L = N\frac{d\phi}{dt} = \left(N\frac{d\phi}{di}\right)\left(\frac{di}{dt}\right)$$

and substitute Eq. (11.3), we then have

$$\boxed{e_L = L\frac{di}{dt}} \qquad \text{(volts, V)} \qquad \textbf{(11.4)}$$

revealing that the magnitude of the voltage across an inductor is directly related to the inductance L and the instantaneous rate of change of current through the coil. Obviously, therefore, the greater the *rate* of change of current through the coil, the greater will be the induced voltage. This certainly agrees with our earlier discussion of Lenz's law. In addition, for the same change in current through the coil, the greater will be the inductance and the greater the counter-emf induced. If the current through the coil fails to change at a particular instant, the induced voltage across the coil will be zero. For dc applications, after the transient effect has passed, $di/dt = 0$, and the induced voltage

$$e_L = L\frac{di}{dt} = L(0) = 0 \text{ V}$$

Since the voltage induced across the coil opposes the source of emf that produced it, it is frequently called the *counter-emf,* abbreviated *cemf.* This opposition is frequently included with the addition of a negative sign as indicated by Eq. (11.5):

$$\boxed{e_{\text{cemf}} = -L\frac{di}{dt}} \qquad \text{(V)} \qquad \textbf{(11.5)}$$

We will simply use Eq. (11.4) to determine the magnitude and apply the proper sign as determined by Fig. 11.3.

Recall that the equation for the current of a capacitor is the following:

$$i_C = C\frac{dv_C}{dt}$$

Note the similarity between this equation and Eq. (11.4). In fact, if we apply the duality $e \rightleftarrows i$ (that is, interchange the two) and $L \rightleftarrows C$ for capacitance and inductance, each equation can be derived from the other.

The average voltage across the coil is defined by the equation

$$\boxed{e_{L_{av}} = L\frac{\Delta i}{\Delta t}} \qquad \text{(V)} \qquad \textbf{(11.6)}$$

where Δ signifies finite change (a measurable change). Compare this to $i_C = C(\Delta v/\Delta t)$, and the meaning of Δ and application of this equation should be clarified from Chapter 9. An example now follows.

EXAMPLE 11.3. Find the waveform for the average voltage across the coil if the current through a 4-mH coil is as shown in Fig. 11.9.

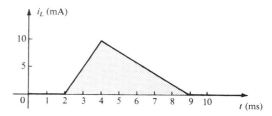

FIG. 11.9

Solutions:

a. 0–2 ms. Since there is no change in current through the coil, there is no voltage induced across the coil; that is,

$$e_L = L\frac{\Delta i}{\Delta t} = L\frac{(0)}{\Delta t} = 0$$

b. 2–4 ms

$$e_L = L\frac{\Delta i}{\Delta t} = 4 \times 10^{-3}\left(\frac{10 \times 10^{-3}}{2 \times 10^{-3}}\right) = 20 \times 10^{-3}$$
$$= 20 \text{ mV}$$

c. 4–9 ms

$$e_L = L\frac{\Delta i}{\Delta t} = -4 \times 10^{-3}\left(\frac{10 \times 10^{-3}}{5 \times 10^{-3}}\right)$$
$$= -8 \times 10^{-3}$$
$$= -8 \text{ mV}$$

d. 9 ms $\rightarrow \infty$

$$e_L = L\frac{\Delta i}{\Delta t} = L\frac{(0)}{\Delta t} = 0$$

The waveform for the average voltage across the coil is shown in Fig. 11.10. Note from the curve that *the voltage*

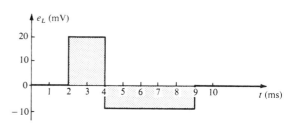

FIG. 11.10

across the coil is not determined by the magnitude of the change in current through the coil (Δi), but by the **rate** of change of current through the coil ($\Delta i/\Delta t$). A similar statement was made for the current of a capacitor due to a change in voltage across the capacitor.

11.7 *R-L* CIRCUITS

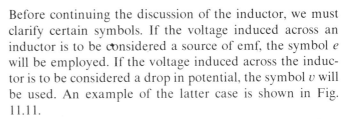

FIG. 11.11

Before continuing the discussion of the inductor, we must clarify certain symbols. If the voltage induced across an inductor is to be considered a source of emf, the symbol e will be employed. If the voltage induced across the inductor is to be considered a drop in potential, the symbol v will be used. An example of the latter case is shown in Fig. 11.11.

In order to describe fully the behavior of an inductor in a circuit with changing voltages and currents, let us consider the effects of moving a switch between the two positions indicated in Fig. 11.11. At the instant the switch is thrown into position 1, the inductance of the coil will prevent an instantaneous change in current through the circuit. The potential drop across the coil will then be equal to the impressed emf as determined by Kirchhoff's voltage law, since $v = iR = 0 \cdot R = 0$ volts. The current i will then build up from zero, resulting in a voltage drop across the resistor and a reduction in v_L. The voltage across the resistor then increases as i increases, while the voltage across the inductor decreases. Eventually, the voltage across the inductor will be zero, while that across the resistor will be the impressed emf E.

The mathematical expression for the current through the coil can be found by first applying Kirchhoff's voltage law around the closed loop:

$$E - v_R - v_L = 0$$

or

$$v_R + v_L = E$$

Substituting $v_R = iR$, we have

$$iR + v_L = E$$

Since the current i is the same for the resistor and inductor,

$$i_L R + L \frac{di_L}{dt} = E$$

The current i_L can then be found using calculus:

$$\boxed{i_L(t) = \frac{E}{R}(1 - e^{-t/(L/R)})} \qquad \textbf{(11.7)}$$

or substituting $I_m = E/R$,

$$\boxed{i_L(t) = I_m(1 - e^{-t/(L/R)})} \tag{11.8}$$

which is very similar to Eq. (9.15) for the voltage across a capacitor during the charging phase.

For capacitive circuits, the product RC was the time constant. For inductive circuits, the time constant (τ) is L/R, which also has the units of time:

$$\boxed{\tau = \frac{L}{R}} \qquad \text{(seconds, s)} \tag{11.9}$$

Equation (11.8) can then be written as

$$\boxed{i_L(t) = I_m(1 - e^{-t/\tau})} \tag{11.10}$$

For convenience, Fig. 9.22 is repeated here (Fig. 11.12) due to the presence of the factor $(1 - e^{-t/\tau})$ in Eq. (11.10) and $e^{-t/\tau}$ in equations to follow.

As described for the *R-C* circuit, in one time constant the exponential factor becomes $e^{-1} = 0.368$; for two time constants, 0.135; and so on. From Eq. (11.10), the current i_L will be $0.632 I_m$ in one time constant, and $0.865 I_m$ in two time constants. A plot of i_L versus time is shown in Fig. 11.13. Note that it has the same shape as $v_C(t)$ for the *R-C* circuit during the charging phase. At each time constant it has the same numerical multiplier. It was shown for the *R-C* circuit that the voltage v_C, for all practical purposes, reaches its final value in five time constants. The same is true for the current i_L in an *R-L* circuit.

If we keep R constant and reduce L, the ratio L/R decreases and the rise time of five time constants decreases. The change in transient behavior for the current i_L is plotted in Fig. 11.14 for various values of L. Note again the duality between these curves and those for the *R-C* circuit in Fig. 9.24.

The ratio L/R always has some numerical value, even though it may be very small in some cases. For this reason, *the current through an inductor, as determined by Eq. (11.10), cannot change instantaneously*. In fact, as mentioned earlier, the inductance of a network is a measure of how much it will oppose a change in current in the network. The larger the inductance, the larger will be the time constant, and the longer it will take i_L to reach its final value (curve of L_3 in Fig. 11.14).

The voltage across the coil in an *R-L* circuit can change instantaneously. When the switch is thrown into position 1, the voltage across the coil jumps to a value equal to the impressed voltage E. It then begins to decline, as determined by the following equation:

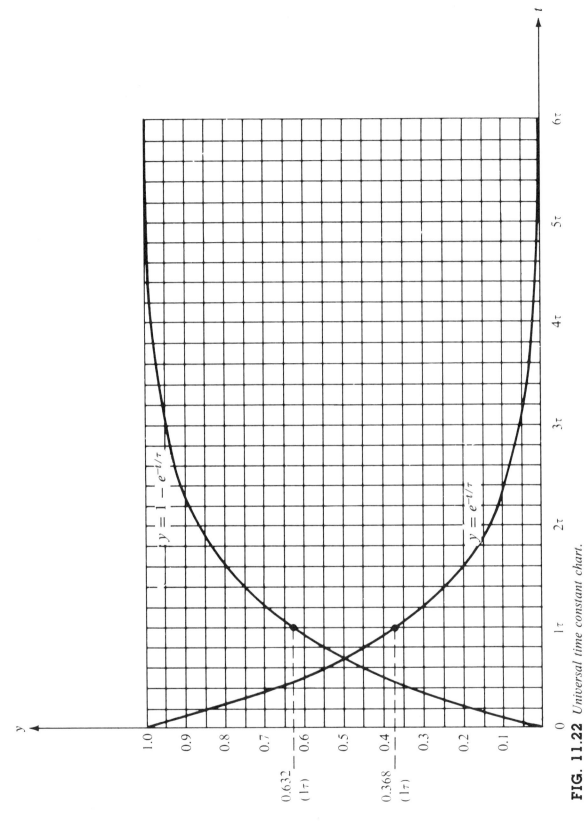

FIG. 11.22 *Universal time constant chart.*

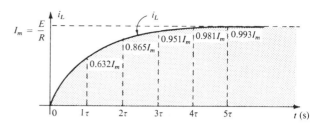

FIG. 11.13

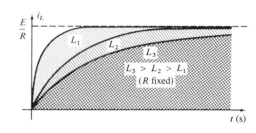

FIG. 11.14

$$\boxed{v_L(t) = Ee^{-t/\tau}}$$ **(11.11)**

A plot of this voltage appears in Fig. 11.15.

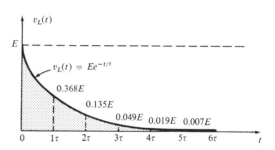

FIG. 11.15

The voltage across the resistor increases at a rate determined by the current through the circuit; that is,

$$v_R(t) = i(t)R$$

$$= \left[\frac{E}{R}(1 - e^{-t/\tau})\right]R$$

$$\boxed{v_R(t) = E(1 - e^{-t/\tau})}$$ **(11.12)**

Noting the presence of the same factor $e^{-t/\tau}$ in all of the transient equations, we can conclude that the voltage v_R and the current i_L increase toward their final values at the same rate at which v_L decreases in value. Eventually, the voltage v_R and the current i_L will reach their final values when the voltage v_L drops to zero. The inductor then has the characteristics of a short circuit, as shown in Fig. 11.16. For most practical applications, an inductor in a dc circuit may be replaced by a short circuit after a period of time equal to five time constants has passed.

At the instant the switch of Fig. 11.11 is thrown into position 2, the voltage across the resistor will remain at E volts, since the current through the coil cannot change in-

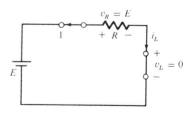

FIG. 11.16

stantaneously and $v_R = i_L R = E$. Applying Kirchhoff's voltage law around the closed loop, we find that $v_L = E$ volts. At this instant, therefore, the change in flux linking the coil is such that an induced voltage E is established across the coil. The polarity of the induced emf, as shown in Fig. 11.17, will cause a current to flow in the circuit in the same direction as when the switch was in position 1. This current will continue to flow until the induced voltage across the coil drops to zero. The voltage v_R also drops to zero, since its value is determined by the current i_L.

The mathematical expressions for v_L, and v_R during the decay phase, are as follows:

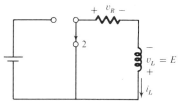

FIG. 11.17

$$v_L(t) = Ee^{-t/\tau} \qquad \text{(11.13)}$$

$$i_L(t) = \frac{E}{R} e^{-t/\tau} \quad DECAY \qquad \text{(11.14)}$$

$$v_R(t) = Ee^{-t/\tau} \qquad \text{(11.15)}$$

If the switch is moved instantaneously between positions 1 and 2 every five time constants, the wave shapes shown in Fig. 11.18 will be obtained for the voltage v_L, the current i_L, and the voltage v_R. Note the reversal in the polarity of the voltage across the coil due to the switching action of the circuit.

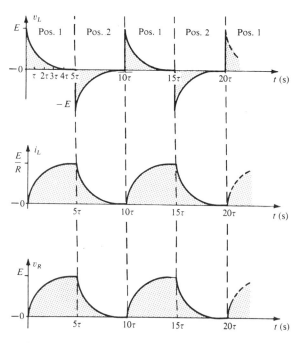

FIG. 11.18

EXAMPLE 11.4. Find the mathematical expressions for the transient behavior of i_L and v_L for the circuit of Fig. 11.19 after the closing of the switch. Sketch the resulting curves.

Solution:

$$\tau = \frac{L}{R} = \frac{4}{2 \text{ k}\Omega} = 2 \text{ ms}$$

By Eq. (11.10),

$$I_m = \frac{E}{R} = \frac{50}{2 \text{ k}\Omega} = 25 \times 10^{-3} \text{ A} = 25 \text{ mA}$$

and

$$i_L(t) = 25 \times 10^{-3}(1 - e^{-t/2 \times 10^{-3}})$$

By Eq. (11.11),

$$v_L(t) = 50e^{-t/2 \times 10^{-3}}$$

Both waveforms appear in Fig. 11.20.

EXAMPLE 11.5. For the network of Fig. 11.21:
a. Find the mathematical expression for the transient behavior of the current through the inductor after the switch is thrown into position 1. Calculate the current i_L at $t = 6$ ms.
b. Repeat part (a) for v_L and v_{R_1}.
c. Find the mathematical expression for the transient behavior of the voltages v_L and v_{R_1} and the current i_L if the switch is thrown instantaneously into position 2 at $t = 24$ ms.
d. Plot the waveforms obtained in parts (a) through (c) on the same time axis for each voltage and for the current.

Solutions:

a.
$$\tau = \frac{L}{R_1} = \frac{0.12}{30} = 4 \text{ ms}$$

By Eq. (11.10),

$$i_L(t) = I_m(1 - e^{-t/\tau}) = \frac{15}{30}(1 - e^{-t/4 \times 10^{-3}})$$

and

$$i_L(t) = 0.5(1 - e^{-t/4 \times 10^{-3}})$$

At $t = 6$ ms,

$$i_L(6 \text{ ms}) = 0.5\left[1 - \exp\left(-\frac{6 \times 10^{-3}}{4 \times 10^{-3}}\right)\right]$$
$$= 0.5(1 - e^{-1.5})$$
$$= (0.5)(1 - 0.223) = (0.5)(0.777)$$
$$= \mathbf{388.5 \text{ mA}}$$

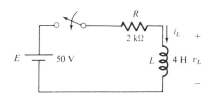

FIG. 11.19

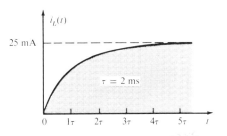

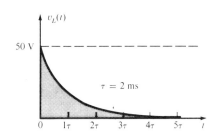

FIG. 11.20

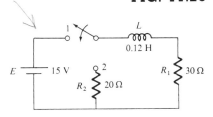

FIG. 11.21

b. By Eq. (11.11),

$$v_L(t) = Ee^{-t/\tau}$$
$$v_L(t) = \mathbf{15e^{-t/4\times 10^{-3}}}$$

At $t = 6$ ms,

$$v_L\,(6\text{ ms}) = 15\exp\left(-\frac{6\times 10^{-3}}{4\times 10^{-3}}\right) = (15)(0.223)$$
$$v_L = \mathbf{3.345\text{ V}}$$

By Eq. (11.12),

$$v_{R_1}(t) = E(1 - e^{-t/\tau})$$
$$v_{R_1}(t) = \mathbf{15(1 - e^{-t/4\times 10^{-3}})}$$

At $t = 6$ ms,

$$v_{R_1}\,(6\text{ ms}) = 15\left[1 - \exp\left(\frac{6\times 10^{-3}}{4\times 10^{-3}}\right)\right]$$
$$= (15)(1 - 0.223) = (15)(0.777)$$
$$v_{R_1}\,(6\text{ ms}) = \mathbf{11.655\text{ V}}$$

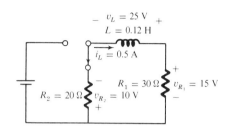

FIG. 11.22

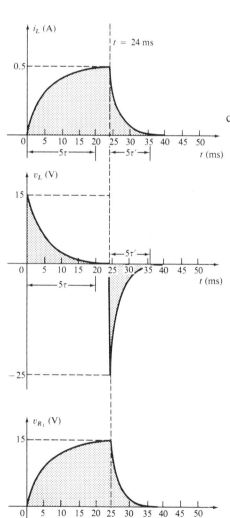

FIG. 11.23

c. Since $t = 24$ ms $> 5\tau$, we will assume that v_L, i_L, and v_R have reached their final values; that is, $i_L = 0.5$ A, $v_L = 0$ V, and $v_R = 15$ V. This circuit appears as shown in Fig. 11.22 the instant the switch is closed. The current cannot change instantaneously. Therefore

$$v_{R_1} = (0.5)(30) = 15\text{ V}$$
$$v_{R_2} = (0.5)(20) = 10\text{ V}$$

The voltage across the coil is then determined by Kirchhoff's voltage law:

$$v_{L_1} = v_{R_1} + v_{R_2} = 25\text{ V}$$

The time constant has changed to

$$\tau' = \frac{L}{R_1 + R_2} = \frac{0.12}{50} = 2.4\text{ ms}$$

Defining $t' = t - 24$ ms, so that

$$v_L(t) = \mathbf{25e^{-t'/2.4\times 10^{-3}}}$$

and

$$i_L(t) = \mathbf{0.5e^{-t'/2.4\times 10^{-3}}}$$

with

$$v_R(t) = \mathbf{15e^{-t'/2.4\times 10^{-3}}}$$

d. See Fig. 11.23.

11.8 $\tau = L/R_{Th}$

In Chapter 9 (capacitors) we found that there are occasions when the circuit does not have the basic form of Fig. 11.11. The same is true for inductive networks. Again, it is necessary to find the Thevenin equivalent circuit before proceeding in the manner described in this chapter. Consider the following example.

EXAMPLE 11.6. For the network of Fig. 11.24:

a. Find the mathematical expression for the transient behavior of the current i_L and the voltage v_L after the closing of the switch.

b. Draw the resultant waveform for each.

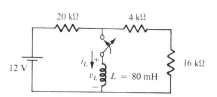

FIG. 11.24

Solutions:

a. Applying Thevenin's theorem to the 80-mH inductor (Fig. 11.25) yields

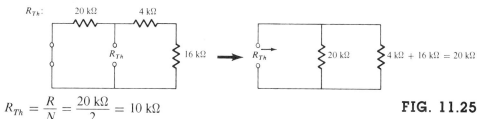

$$R_{Th} = \frac{R}{N} = \frac{20 \text{ k}\Omega}{2} = 10 \text{ k}\Omega$$

FIG. 11.25

By the voltage divider rule (Fig. 11.26),

$$E_{Th} = \frac{(4 \text{ k}\Omega + 16 \text{ k}\Omega)(12)}{(4 \text{ k}\Omega + 16 \text{ k}\Omega) + 20 \text{ k}\Omega}$$

$$= \frac{(20)(12)}{40}$$

$$E_{Th} = 6 \text{ V}$$

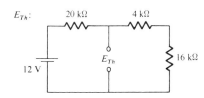

FIG. 11.26

The Thevenin equivalent circuit is shown in Fig. 11.27. By Eq. (11.10),

$$i_L(t) = \frac{E}{R}(1 - e^{-t/\tau})$$

$$\tau = \frac{L}{R} = \frac{80 \times 10^{-3}}{10 \times 10^{-3}} = 8 \times 10^{-6} \text{ s}$$

$$I_m = \frac{E}{R} = \frac{6}{10 \times 10^3} = 0.6 \times 10^{-3} \text{ A}$$

Thevenin equivalent circuit:

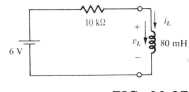

FIG. 11.27

and

$$i_L(t) = \mathbf{0.6 \times 10^{-3}(1 - e^{-t/8 \times 10^{-6}})}$$

By Eq. (11.11),

$$v_L(t) = Ee^{-t/\tau}$$

so that

$$v_L(t) = \mathbf{6e^{-t/8 \times 10^{-6}}}$$

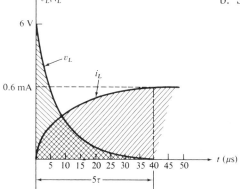

FIG. 11.28

b. See Fig. 11.28.

11.9 INDUCTORS IN SERIES AND PARALLEL

For inductors in series, the total inductance is found in the same manner as the total resistance of resistors in series (Fig. 11.29):

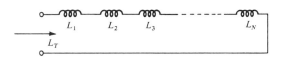

FIG. 11.29

$$L_T = L_1 + L_2 + L_3 + \cdots + L_N \qquad \textbf{(11.16)}$$

For inductors in parallel, the total inductance is found in the same manner as the total resistance of resistors in parallel (Fig. 11.30):

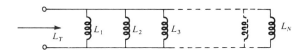

FIG. 11.30

$$\frac{1}{L_T} = \frac{1}{L_1} + \frac{1}{L_2} + \frac{1}{L_3} + \cdots + \frac{1}{L_N} \qquad \textbf{(11.17)}$$

For two inductors in parallel,

$$L_T = \frac{L_1 L_2}{L_1 + L_2} \qquad \textbf{(11.18)}$$

11.10 *R-L* AND *R-L-C* CIRCUITS WITH dc INPUTS

We found in Section 11.7 that for all practical purposes, an inductor can be replaced by a short circuit in a dc circuit after a period of time greater than five time constants has passed. If in the following circuits we assume that all of the currents and voltages have reached their final values, the current through each inductor can be found by replacing each inductor by a short circuit. For the circuit of Fig. 11.31, for example,

$$I_1 = \frac{10}{2} = 5 \text{ A}$$

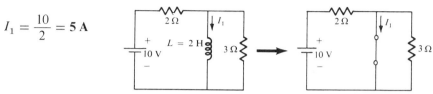

FIG. 11.31

For the circuit of Fig. 11.32,

$$I = \frac{21}{2} = 10.5 \text{ A}$$

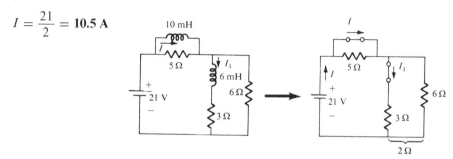

By the current divider rule,

FIG. 11.32

$$I_1 = \frac{(6)(10.5)}{3 + 6} = \frac{63}{9} = 7 \text{ A}$$

In the following examples we will assume that the voltage across the capacitors and the current through the inductors have reached their final values. Under these conditions, the inductors can be replaced by short circuits, and the capacitors by open circuits.

EXAMPLE 11.7. Find the current I_L and the voltage V_C for the network of Fig. 11.33.

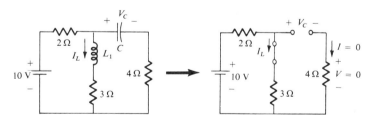

FIG. 11.33

Solution:

$$I_L = \frac{10}{5} = 2 \text{ A}$$

$$V_C = \frac{(3)(10)}{3 + 2} = 6 \text{ V}$$

EXAMPLE 11.8. Find the currents I_1 and I_2 and the voltages V_1 and V_2 for the network of Fig. 11.34.

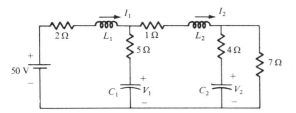

FIG. 11.34

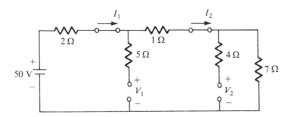

FIG. 11.35

Solution: Note Fig. 11.35:

$$I_1 = I_2$$

$$I_1 = \frac{50}{2 + 1 + 7} = \frac{50}{10} = 5 \text{ A}$$

$$V_2 = (7)(5) = 35 \text{ V}$$

By the voltage divider rule,

$$V_1 = \frac{(8)(50)}{2 + 8} = 40 \text{ V}$$

11.11 ENERGY STORED IN AN INDUCTOR

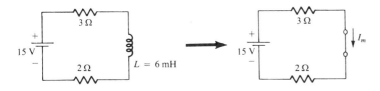

FIG. 11.36

The ideal inductor, like the ideal capacitor, does not dissipate the electrical energy supplied to it. It stores the energy in the form of a magnetic field. A plot of the voltage, current, and power to an inductor is shown in Fig. 11.36 during the buildup of the magnetic field surrounding the inductor. The energy stored is represented by the shaded area under the power curve. Using calculus, we can show that the evaluation of the area under the curve yields

$$\boxed{W_{\text{stored}} = \frac{1}{2} L I_m^2}$$ (joules, J) **(11.19)**

EXAMPLE 11.9. Find the energy stored by the inductor in the circuit of Fig. 11.37 when the current through it has reached its final value.

FIG. 11.37

Solution:

$$I_m = \frac{15}{5} = 3\ \text{A}$$

$$W_s = \frac{1}{2}LI_m^2$$

$$= \frac{1}{2}(6 \times 10^{-3})(3^2)$$

$$= \frac{54}{2} \times 10^{-3}$$

$$W_s = 27 \times 10^{-3}\ \textbf{J}$$

PROBLEMS

Section 11.2

1. If the flux linking a coil of 50 turns changes at a rate of 0.085 Wb/s, what is the induced emf across the coil?

2. Determine the rate of change of flux linking a coil if 20 V are induced across a coil of 40 turns.

3. How many turns does a coil have if 42 mV are induced across the coil by a change of flux of 0.003 Wb/s?

Section 11.4

4. Find the inductance L in henries of the inductor of Fig. 11.38.

5. Repeat Problem 4 with $l = 4$ in. and $d = 0.25$ in.

6. **a.** Find the inductance L in henries of the inductor of Fig. 11.39.

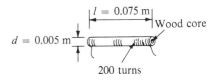

FIG. 11.38

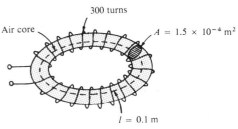

FIG. 11.39

    b. Repeat part (a) if a ferromagnetic core is added having a μ_r of 2000.

Section 11.6

7. Find the voltage induced across a coil of 5 H if the rate of change of current through the coil is
   a. 0.5 A/s.
   b. 60 mA/s.
   c. 0.04 A/ms.

8. Find the induced voltage across a 50-mH inductor if the current through the coil changes at a rate 0.1 mA/μs.

Inductors

9. Find the waveform for the voltage across a 200-mH coil if the current through the coil is as shown in Fig. 11.40.

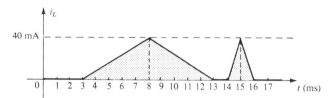

FIG. 11.40

10. Repeat Problem 9 for the waveform of Fig. 11.41.

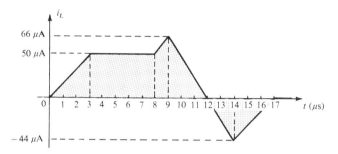

FIG. 11.41

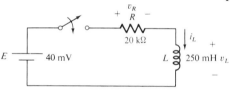

FIG. 11.42

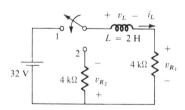

FIG. 11.43

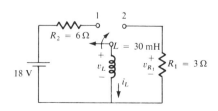

FIG. 11.44

Section 11.7

11. For the circuit of Fig. 11.42:
   a. Write the mathematical expression for the transient behavior of the current i_L directly after the switch is closed.
   b. Determine the current at 1, 2, 3, 4, and 5 time constants.
   c. Repeat part (a) for v_L and v_R.
   d. Repeat part (b) for v_L and v_R.
   e. Sketch the waveforms of i_L, v_L, and v_R.

12. For the circuit of Fig. 11.43:
   a. Find the mathematical expression for the transient behavior of the current through the inductor after the switch is thrown into position 1.
   b. Repeat part (a) for v_L and v_{R_1}.
   c. Find the mathematical expressions for the voltages v_L, v_{R_1} and v_{R_2} and the current i_L after the switch is thrown into position 2 at $t = 4$ ms.
   d. Plot the waveforms obtained in parts (a) through (c) on the same time axis for each voltage and the current.

13. Repeat Problem 12 for $R_2 = 60$ kΩ.

14. For the circuit of Fig. 11.44:
   a. Find the mathematical expression for the transient behavior of the current i_L after the switch is thrown into position 1.
   b. Repeat part (a) for the voltage v_L.
   c. Find the mathematical expressions for the voltages v_L and v_{R_1} and the current i_L if the switch is thrown instantaneously into position 2 at $t = 30$ ms.
   d. Plot the waveforms obtained in parts (a) through (c) on the same time axis for each voltage and the current.

15. Repeat Problem 14 for $L = 3$ H.

Section 11.8

16. For the circuit of Fig. 11.45:
  a. Find the mathematical expression for the transient behavior of the current i_L after the closing of the switch.
  b. Repeat part (a) for the voltage v_L.
  c. Plot the results of parts (a) and (b) for a period of time equal to five time constants.

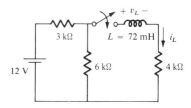

FIG. 11.45

*17. For the circuit of Fig. 11.46:
  a. Find the mathematical expression for the transient behavior of the current i_L after the closing of the switch.
  b. Repeat part (a) for the voltage v_L.
  c. Plot the results of parts (a) and (b) for a period of time equal to five time constants.

Section 11.9

18. Find the total inductance of the circuits of Fig. 11.47.

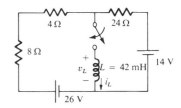

FIG. 11.46

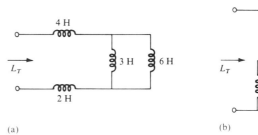

(a)   (b)

FIG. 11.47

Section 11.10

For Problems 19 through 22, assume that the voltage across each capacitor and the current through each inductor have reached their final values.

19. Find the voltages V_1 and V_2 and the current I_1 for the circuit of Fig. 11.48.

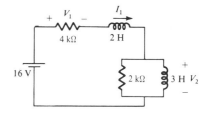

FIG. 11.48

20. Find the current I_1 and the voltage V_1 for the circuit of Fig. 11.49.

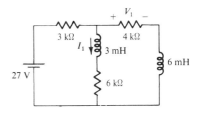

FIG. 11.49

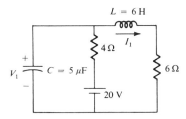

FIG. 11.50

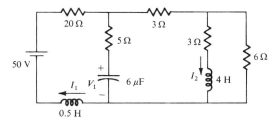

FIG. 11.51

21. Find the current I_1 and the voltage V_1 for the circuit of Fig. 11.50.

22. Find the voltage V_1 and the current through each inductor in the circuit of Fig. 11.51.

Section 11.11

23. Find the energy stored in each inductor of
    a. Problem 19.
    b. Problem 20.

24. Find the energy stored in the capacitor and inductor of Problem 21.

25. Find the energy stored in each inductor of Problem 22.

GLOSSARY

Choke A term often applied to an inductor, due to the ability of an inductor to resist a change in current through it.

Counter-emf The induced emf across the coil that opposes any change in flux linking the coil and thereby any change in current passing through the coil.

Faraday's law A law relating the voltage induced across a coil to the number of turns in the coil and the rate at which the flux linking the coil is changing.

Inductor A fundamental element of electrical systems constructed of numerous turns of wire around a ferromagnetic or air core.

Lenz's law A law stating that an induced effect is always such as to oppose the cause that produced it.

Self-inductance A measure of the ability of a coil to oppose any change in current through the coil and to store energy in the form of a magnetic field in the region surrounding the coil.

12.1 INTRODUCTION

The concepts and methods discussed in the previous chapters become more meaningful when applied to an actual dc network. This chapter is devoted to the instruments used to measure current, voltage, power, and component values (resistance, inductance, and capacitance). In recent years there has been a major shift from the analog (continuous-scale) reading instruments to the digital. As with the analog reading watch, however, there will be a sharing of responsibility between analog and digital in the years to come. Consider the tremendous teaching aid lost if the terms *clockwise* and *counterclockwise* lose their significance because only digital reading watches are available. This chapter will demonstrate that there are definite advantages associated with each.

The first series of instruments to be introduced employ an analog scale and use a *d'Arsonval* movement, developed by the French physicist Jacques Arsène d'Arsonval in 1881. It is suitable for use in a dc/ac ammeter, dc/ac voltmeter, ohmmeter, and Wheatstone bridge. It is, without question, the most common of the analog indicating movements. Other types will be introduced later in the chapter.

Courtesy of Weston Instruments, Inc.
FIG. 12.1 *d'Arsonval movement.*

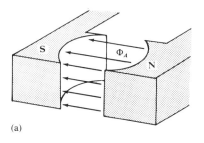

(a)

FIG. 12.2

(a)

(b)

Courtesy of Simpson Electric Co.

FIG. 12.3 *Meter scales.*

12.2 THE d'ARSONVAL MOVEMENT

This movement, shown in Fig. 12.1, consists basically of an iron-core coil mounted on bearings between a permanent magnet. The helical springs limit the turning motion of the coil and provide a path for the current to reach the coil. The magnetic fields of the magnet and the coil are as shown in Fig. 12.2.

The fluxes generated by the coil and magnet interact to develop a torque on the coil, which causes it to rotate on the bearings. Since the flux Φ of the magnet is the same at any instant, any change in torque is determined by a

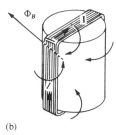

(b)

change in flux Φ_B of the coil. The flux Φ_B, however, is directly proportional to the current through the coil, so the torque is also directly proportional to the current in the coil. The greater the current through the coil, the greater the torque and deflection of the pointer connected to the coil. Consequently, the pointer indicates the magnitude of the current passing through the coil. The scales used to indicate the magnitude of the current are generally of two types: those with the zero to the far left, and those with the zero in the center (Fig. 12.3).

The movement is adjusted to indicate zero deflection when the current through the coil is zero. If a meter has the scale shown in Fig. 12.3(a), an up-scale deflection will occur only when the current passes in one direction through the coil. If the current is reversed, the pointer will deflect to the left, below zero. A meter with the scale shown in Fig. 12.3(b) can be used with the current flowing in either direction through the coil, since a reversal in current will simply require the use of the other scale. Most meters with the scale shown in Fig. 12.3(a) indicate by polarities the proper way of connecting the meter in a circuit.

D'Arsonval movements are usually rated by current and resistance. The specifications of a typical movement might be 1 mA, 50 Ω. The 1 mA is the *current sensitivity* (*CS*) of the movement, which is the current required for a full-scale deflection. It will be noted by the symbol I_{CS}. The 50-Ω resistance is the internal resistance (R_m) of the movement

itself. A common notation for the movement and its specifications are shown in Fig. 12.4.

12.3 THE AMMETER

As the name implies, an ammeter is an instrument designed to measure the magnitude of the current. The
proper measurement of the current through a branch of a
network requires that the branch be broken and the meter
inserted as indicated in Fig. 12.5. If the insertion of the
meter is to have the least effect possible, its internal resistance should be zero. We will find in this section and in the
discussions to follow that it is not exactly zero but is sufficiently small compared to that of the series elements to be
ignored for most applications.

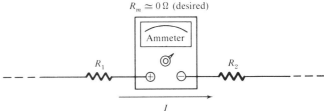

FIG. 12.4 *Movement notation.*

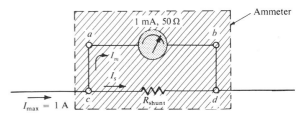

FIG. 12.5

The maximum current that the d'Arsonval movement
can read independently is equal to the current sensitivity of
the movement. However, higher currents can be measured
if additional circuitry is introduced. This additional circuitry, as shown in Fig. 12.6, results in the basic construction of an ammeter.

FIG. 12.6 *Basic ammeter.*

The resistance R_{shunt} is chosen for the ammeter of Fig.
12.6 to allow 1 mA to flow through the movement when a
maximum current of 1 A enters the ammeter. If less than
1 A should flow through the ammeter, the movement will
have less than 1 mA flowing through it and will indicate
less than full-scale deflection.

Since the voltage across parallel elements must be the
same, the potential drop across *a-b* in Fig. 12.6 must equal
that across *c-d;* that is,

$$(1 \text{ mA})(50 \, \Omega) = (R_{shunt})(I_s)$$

I_s must equal 1 A − 1 mA = 999 mA if the current is to be limited to 1 mA through the movement (Kirchhoff's current law). Therefore,

$$(1 \times 10^{-3})(50) = (999 \times 10^{-3})(R_{shunt})$$

$$R_{shunt} = \frac{(1 \times 10^{-3})(50)}{(999 \times 10^{-3})}$$

$$R_{shunt} \cong 0.05 \ \Omega$$

In general,

$$R_{shunt} = \frac{R_m \cdot I_{CS}}{I_{max} - I_{CS}} \qquad \textbf{(12.1)}$$

One method of constructing a multirange ammeter is shown in Fig. 12.7, where the rotary switch determines the

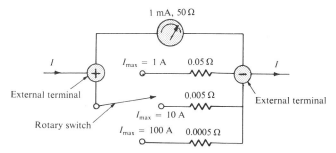

FIG. 12.7 *Multirange ammeter.*

R_{shunt} to be used for the maximum current indicated on the face of the meter. Most meters employ the same scale for various values of maximum current, as shown in Fig. 12.8.

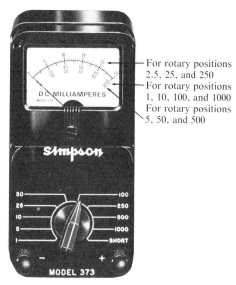

FIG. 12.8 *Direct current milliammeter.*

If you read 375 on the 0–50 mA scale with the switch on the 5 setting, the current is 3.75 mA. On the 50 setting, the current is 37.5 mA, and so on.

The polarity markings on the external terminals are the same as those that would exist for a potential drop across the internal resistance of the ammeter if the current being measured is to cause an up-scale deflection. To insure proper connection in a circuit, *the ammeter is always placed in series with the branch in which the current is to be determined, and with conventional current entering the positive terminal,* as shown in Fig. 12.5.

12.4 THE VOLTMETER

The voltmeter is an instrument designed to measure the magnitude of voltage levels in a network. It is placed in parallel with the element across which the voltage is to be measured, as indicated in Fig. 12.9. The insertion of the

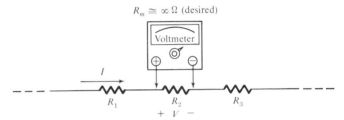

FIG. 12.9

meter for measurement purposes should not alter the behavior of the network. To insure that it does not, be certain that the internal resistance of the meter is very large compared to that of the element itself. Typical levels of magnitude for the resistance of a voltmeter will be introduced in this section and in the discussions to follow.

A variation in the additional circuitry will permit the use of the d'Arsonval movement in the design of a voltmeter. The 1-mA, 50-Ω movement can also be rated as a 50-mV (1 mA $\times$ 50 Ω), 50-Ω movement, indicating that the maximum voltage that the movement can measure independently is 50 mV. The millivolt rating is sometimes referred to as the *voltage sensitivity* (*VS*). The basic construction of the voltmeter is shown in Fig. 12.10.

The R_{series} is adjusted to limit the current through the movement to 1 mA when the maximum voltage is applied across the voltmeter. A lesser voltage would simply reduce the current in the circuit, and thereby the deflection of the movement.

Applying Kirchhoff's voltage law around the closed loop of Fig. 12.10, we obtain

$$[10 - (1 \times 10^{-3})R_{series}] - 50 \times 10^{-3} = 0$$

or

$$R_{series} = \frac{10 - (50 \times 10^{-3})}{1 \times 10^{-3}}$$

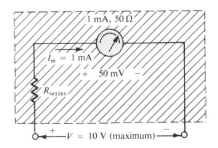

FIG. 12.10

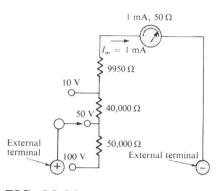

FIG. 12.11 *Multirange voltmeter.*

and

$$R_{series} = 9950 \ \Omega$$

In general,

$$R_{series} = \frac{V_{max} - V_{VS}}{I_{CS}} \qquad (12.2)$$

One method of constructing a multirange voltmeter is shown in Fig. 12.11. If the rotary switch is at 10 V, R_{series} = 9.950 kΩ; at 50 V, R_{series} = 40 kΩ + 9.950 kΩ = 49.950 kΩ; and at 100 V, R_{series} = 50 kΩ + 40 kΩ + 9.950 kΩ = 99.950 kΩ.

Most voltmeters employ the same scale for various values of maximum voltage as shown in Fig. 12.12.

FIG. 12.12 *Direct current voltmeter.*

For an up-scale deflection, the polarities of the potential difference being measured must correspond to the polarity markings on the external terminals of the voltmeter, as shown in Fig. 12.9.

12.5 OHM/VOLT RATING

Below the scales of most dc voltmeters there is an *ohm/volt rating,* which can be used to determine both the current sensitivity of the movement being used in the voltmeter and the resistance between the output terminals for any scale that might be used. It is related to the current sensitivity in the following manner:

$$I_{CS} = \frac{1}{\text{ohm/volt rating}} \qquad (12.3)$$

For a 1000 ohm/volt rating,

$$I_{CS} = \frac{1}{1000} = 1 \text{ mA}$$

For a 20,000 ohm/volt rating,

$$I_{CS} = \frac{1}{20,000} = 50 \text{ μA}$$

The resistance for each scale can be found simply by multiplying the maximum voltage of the scale being used by the ohm/volt rating. For the 100-V scale of a meter with a 1000 ohm/volt rating,

$$\text{Internal resistance} = (100)(1000) = 100 \text{ k}\Omega$$

For the 10-V scale of the same meter,

$$\text{Internal resistance} = (10)(1000) = 10 \text{ k}\Omega$$

In general,

$$R_{\text{scale}} = (V_{\text{max}})(\text{ohm/volt rating}) \qquad (12.4)$$

or

$$R_{\text{scale}} = \frac{V_{\text{max}}}{I_{CS}} \qquad (12.5)$$

Although we may measure only 2 V on the 10-V scale, the internal resistance is still that determined by the full-scale voltage (10 kΩ). This is true for every scale of the voltmeter.

12.6 OHMMETERS

In general, ohmmeters are designed to measure resistance in the low, mid-, or high range. The most common, and the first to be described, is applied to the midrange. It is referred to as the *series ohmmeter* due to the series network configuration inside the meter. It is different from the ammeter and voltmeter in that it will show a full-scale deflection for zero ohms and no deflection for infinite resistance. Its basic construction is shown in Fig. 12.13.

To determine R_s, the external terminals are shorted (a direct connection of zero ohms between the two) to simulate zero ohms, and the zero-adjust is set to half its maximum value. R_s is then adjusted to allow a current equal to the current sensitivity of the movement (1 mA) to flow in

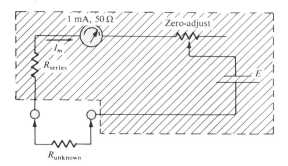

FIG. 12.13 *Series ohmmeter.*

the circuit. The zero-adjust is set to half its value so that any variation in the components of the meter that may produce a current more or less than the current sensitivity can be compensated for. The current

$$I_m \text{ (full-scale)} = I_{CS} = \cfrac{E}{R_s + R_m + \cfrac{\text{zero-adjust}}{2}}$$

(12.6)

and

$$R_s = \frac{E}{I_{CS}} - R_m - \frac{\text{zero-adjust}}{2} \qquad \textbf{(12.7)}$$

If an unknown resistance is then placed between the external terminals, the current will be reduced, causing a deflection less than full-scale. If the terminals are left open, simulating infinite resistance, the pointer will not deflect since the current through the circuit is zero.

An ohmmeter of any type is never applied to an energized circuit. The additional potential introduced by an energized circuit may be enough to establish a current through the movement much greater in magnitude than its current sensitivity, which may result in the permanent damage of the meter. In addition, the scale of the ohmmeter will be meaningless, since it was calibrated to the internal emf of the ohmmeter circuit.

The scale used by the multirange ohmmeter portion of the Simpson 260 is shown in Fig. 12.14. Note the nonlinearity of the scale. A multirange ohmmeter can be constructed by using a voltage divider arrangement similar to the one used in the multirange voltmeter.

An instrument designed to read very low values of resistance appears in Fig. 12.15. Because of its low range capability, the network design must be a great deal more sophisticated than described above. It employs electronic

Courtesy of Simpson Electric Co.

FIG. 12.14 *Series ohmmeter scale.*

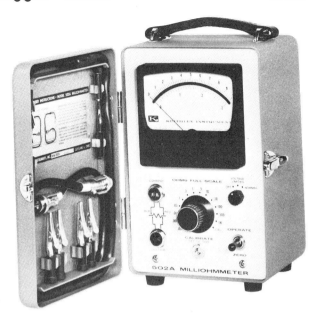

FIG. 12.15 *Milliohmmeter.*

components that eliminate the inaccuracies introduced by lean and contact resistances. It is similar to the above system in the sense that it is completely portable and does require a dc battery to establish measurement conditions. Note the special leads designed to limit any introduced resistance levels. Also note that the maximum scale setting can be set as low as 0.00352 (3 milliohms).

12.7 THE MEGGER® TESTER

The Megger tester is an instrument for measuring very high resistance values. The terminology *Megger* is derived from the fact that the device measures resistance values in the megohm range. Its primary function is to test the insulation found in power-transmission systems, electrical machinery, transformers, and so on. In order to measure the high-resistance values, a high dc voltage is established by a hand-driven generator. If the shaft is rotated above some set value, the output of the generator will be fixed at one selectable voltage, typically 250, 500, or 1000 volts. A photograph of the commercially available Megger tester is shown in Fig. 12.16. The unknown resistance is connected between the terminals marked *Line* and *Earth*. For this instrument, the range is zero to 2000 megohms.

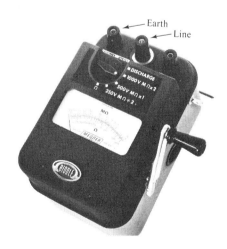

FIG. 12.16 *The Megger® tester.*

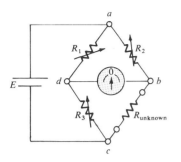

FIG. 12.17 *The Wheatstone bridge.*

12.8 THE WHEATSTONE BRIDGE

The Wheatstone bridge is an instrument that measures resistance with a much higher accuracy than that obtainable from the typical ohmmeter. The basic construction is shown in Fig. 12.17, where $R_{unknown}$ is the *unknown resistance*. The movement is a sensitive galvanometer that measures small amounts of current passing in either direction between b and d. When the standard resistors R_1, R_2, and R_3 are adjusted such that the galvanometer has a zero reading ("null" condition), $R_{unknown}$ can be found in terms of R_1, R_2, and R_3. The bridge is said to be *balanced* when $I_{gal} = 0$. The equation relating $R_{unknown}$ to the known resistors was derived in Chapter 7, Eq. (7.7), for a balanced bridge. It is repeated here in terms of the resistance values of Fig. 12.17:

$$R_{unknown} = \frac{R_2 R_3}{R_1} \qquad \textbf{(12.8)}$$

Note that the equation takes on the form of

$$R_{unknown} = \frac{\text{product of adjacent arms}}{\text{opposite resistor}} \qquad \textbf{(12.9)}$$

Wheatstone bridges are made with varying appearances, one of which is shown in Fig. 12.18. When the external dials are adjusted for zero deflection, the value of the unknown resistance can be read directly from the dials.

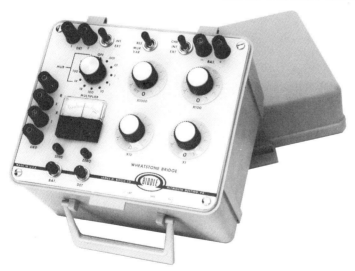

Courtesy of James G. Biddle Co.

FIG. 12.18 *Wheatstone bridge.*

12.9 MULTIMETERS

Since ammeters, voltmeters, and ohmmeters can use the same movement, they are often combined into a *volt-ohm-milliammeter* (VOM), which has standard dial settings for voltages from 2.5 to 500 V, currents from 1 to 500 mA dc, and resistances from 0 to ∞ Ω, with special connections extending the direct current range down to 50 μA and up to 10 A. These are the characteristics of the 260 Simpson VOM appearing in Fig. 12.19.

In a moment, digital display instruments will be introduced that perform all of the basic functions just described. As pointed out earlier, they are presently a very popular item (such as the digital watch) and are receiving a great deal of interest and development by manufacturers. There are, however, certain advantages associated with a continuous analog display that should be appreciated.

First, analog displays provide an immediate indication of continuity (short) or noncontinuity (open) in a network. Many digital units are now being modified to provide a more rapid indication of the above. For mild variations in level, the analog scale will more easily define the limits of the variations. Percent changes in magnitude are also more obvious when the same scale is used for two or more readings. Although some would argue that the analog scale displays all possible levels, there is certainly a limit to the accuracy with which an analog scale can be read. Digital meters are now available with a level of accuracy that certainly matches (or surpasses) the accuracy possible from a typical analog scale. However, the analog scale is one that the technologist will encounter on a continuing basis on graphs, scales, and displays and therefore one that the student should develop a high level of competence in reading. The early use of analog instruments to measure the basic quantities such as current, voltage, and resistance will contribute to this valuable form of development.

As will be noted in the descriptions to follow, there are certainly numerous advantages associated with the digital display. The discussion here was to simply point out those benefits associated with an analog display.

Courtesy of Simpson Electric Co.

FIG. 12.19 *260 Simpson VOM.*

12.10 METER SCHEMATIC

The overall schematic for the Simpson VOM 260 is shown in Fig. 12.20 to introduce the art of reading schematics. The circuitry included within the instrument when it is used as a dc voltmeter and dc milliammeter is shown by the two marked paths. The dc voltmeter is set on the 10-V dc scale, and the milliammeter on the 100-mA dc scale.

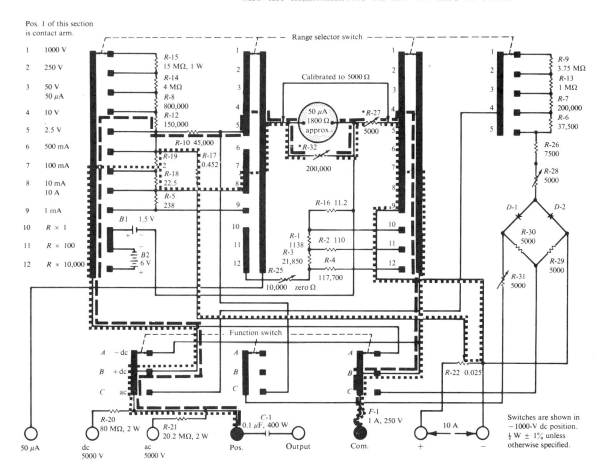

* *R*-32 is adjusted for a circuit current of 50 μA. Then *R*-27 is adjusted for a circuit resistance of 5000 Ω.

Courtesy of Simpson Electric Co.

FIG. 12.20 *Overall schematic of Simpson VOM 260, series 4 and series 4m.*

Before we find the resultant networks for each case in its simplest form, a few portions of the schematic must be explained. When the range-selector switch and function switch are moved, the corresponding four and three sets of contacts will all move to the same horizontal position on the schematic. The black bands are all continuous, conducting surfaces. The variable resistor *R*-32 is adjusted to insure a current of 50 μA through the movement. The variable resistor *R*-27 is adjusted until the resistance between the two points indicated by the arrows is 5000 Ω. By carefully following the path for the 10-V dc voltmeter portion of the multimeter, we obtain the circuit shown in Fig. 12.21.

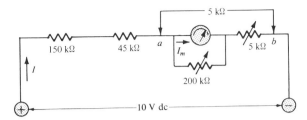

FIG. 12.21 *10-V dc voltmeter (Simpson 260).*

Since we have 5 kΩ between points *a* and *b*, the total resistance of the circuit is 200 kΩ. The current *I* in the circuit is therefore

$$I = \frac{E}{R} = \frac{10}{200 \times 10^3} = 50 \, \mu A$$

If all of the resistors are their indicated values (ideal case), such that the current is exactly 50 μA, the 200-kΩ resistor will be set at its maximum value to minimize its effects.

The internal circuitry for the 100-mA dc milliammeter portion of the multimeter is shown in Fig. 12.22. The current

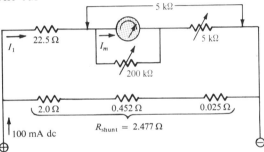

FIG. 12.22 *100-mA dc milliammeter (Simpson 260).*

$$I_1 \cong I_m = \frac{(2.477)(100 \times 10^{-3})}{(5022.5 + 2.477)} = 49.3 \, \mu A$$

which, for all practical purposes, is 50 μA.

12.11 LOADING EFFECTS AND TOLERANCE

When an ammeter or a voltmeter is used in a circuit, there is a loading effect introduced by the internal resistance of the meter. An ammeter places its internal resistance in series with the branch in which it is measuring the current (Fig. 12.23).

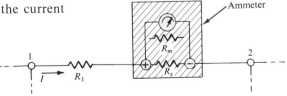

FIG. 12.23 *Ammeter loading.*

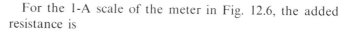

For the 1-A scale of the meter in Fig. 12.6, the added resistance is

$$\frac{(50)(0.05)}{50 + 0.05} \cong 0.05 \ \Omega$$

For most ammeters, the internal resistance is usually small enough to be neglected for each scale used. It must be considered, of course, if the resistors in the circuit external to the meter are of the same order of magnitude.

The voltmeter has a much higher internal resistance than the ammeter, but this resistance is always in parallel with the portion of the network across which it is reading the voltage (Fig. 12.24).

For the 10-V scale of the meter in Fig. 12.11, $R_s + R_m$ = 10 kΩ, and applying the current divider rule yields

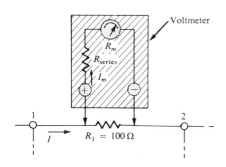

FIG. 12.24 *Voltmeter loading.*

$$I_m = \frac{100I}{100 + 10,000} \cong 0.01I$$

The current diverted from the circuit by the voltmeter is usually quite small due to the magnitude of the internal resistance. The larger the internal resistance of the voltmeter, the less the current drawn from the circuit, and the greater the accuracy of the reading. The typical VOM has an ohm/volt rating of 20,000, while the typical digital reading voltmeter has a constant input resistance of 10 MΩ for each dc voltage scale. This would introduce internal resistances of 2 MΩ and 10 MΩ for the 100-V scale of the respective meters.

The percent tolerance expresses the degree of accuracy that can be expected from readings taken with an instrument. A 2% tolerance indicates that between 10% and 100% of the maximum value of the scale being used, the readings of the meter are accurate to within ±2% of the full-scale value (the first 10% of most scales is considered inaccurate due to the meter construction). For a ±2% tolerance on the 100-V scale of a multimeter, the true value of a reading of 20 V will be somewhere between 18 and 22 V. On the 25-V scale of the same meter, the readings will be accurate to within ±0.5 V (±2% of 25 V); the true value of the 20-V reading will be somewhere between 19.5 and 20.5 V. The most accurate readings, therefore, are usually found on scales that have maximum values closest to the voltages or currents being measured.

On many occasions, the ohm/volt rating of a voltmeter must be carefully considered. For example, if a meter with a low ohm/volt rating of 1000 is used to measure the voltage across the 20-kΩ resistor in the network shown in Fig. 12.25, the reading will be off by a factor determined by the scale used.

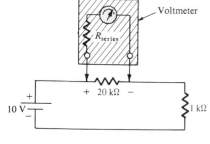

FIG. 12.25

For the 10-V scale, the internal resistance of the voltmeter is $(1000)(10) = 10$ kΩ. The parallel combination of R_{int} and 20 kΩ equals 6.7 kΩ, reducing the actual voltage across the resistor to 6.7 kΩ $(10)/(6.7$ kΩ $+ 1$ kΩ$) = 8.7$ V instead of the 20 kΩ $(10)/(20$ kΩ $+ 1$ kΩ$) = 9.52$ V that it should be. Using a meter with a higher ohm/volt rating, such as 20,000, $R_{int} = (20,000)(10) = 200$ kΩ, and the parallel combination of R_{int} and 20 kΩ $= 18.2$ kΩ, producing a voltage of 18.2 kΩ $(10)/(18.2$ kΩ $+ 1$ kΩ$) = 9.47$ V, which is much closer to 9.52 V. The higher the ohm/volt rating of the meter, the greater the accuracy of the reading.

For the ohmmeter, the greatest accuracy is usually found at midscale. The low end is inaccurate due to the meter construction; the high end is inaccurate due to the nonlinearity of the scale.

12.12 DIGITAL READING METERS

The hand-held *multimeter* of Fig. 12.26(a) has a number of different resistance settings that allow it to measure resistance levels with a maximum of 200 Ω and a resolution of 0.1 Ω to levels as high as 20 MΩ with a resolution of 10 kΩ. The resolution simply defines the variation around the maximum value of the scale as defined by the percent accuracy. It also has dc voltage scales with maximum readings of 200 mV to 1000 V and dc ranges with maximums of 2 mA to 2.0 A. Note the presence of nS units (10^{-9} siemen) for very high insulation resistance measurements. A reading of 20 nS would represent a resistance level of $1/(20 \times 10^{-9}) = 50$ MΩ. A positive test for continuity will result in an audible beep and a down arrow in the LCD (Liquid Crystal Display) readout. Recall that this was seen as one of the advantages of the analog display versus the digital output. When combined with a K-type thermocouple, this instrument can also indicate the temperature in °C from -20°C to $+300$°C

A bench-type portable unit appears in Fig. 12.26(b), manufactured by the same company. It can measure current ranges in the dc and ac mode with maximum values of 200 µA to 2.0 A. This unit will also permit the measurement of decibel (dB) levels with a choice of 16 reference impedances rather than the one (600-Ω) reference normally available with analog units. As indicated by its face, the meter has a wide range of capabilities not described in the paragraph above.

Thus far, the introduction to digital meters has not included a description of the type of movement or a similar device used to provide some relative magnitude for each measurement. Digital reading meters depend on integrated

(a)

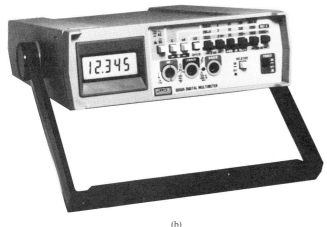

(b)

Courtesy of John Fluke Manufacturing Co., Inc.

FIG. 12.26 *Digital reading VOMs.*

circuits (ICs) or move, specifically, microprocessors to compare the measured quantity to some interval reference level. The result will then energize particular numbers on the display and set the position of the decimal point. A reference level requires an internal battery supply (not necessary for the analog voltmeters and ammeters), such as a single 9-V battery for the unit of Fig. 12.26(a) and internal rechargeable batteries for the unit of Fig. 12.26(b).

Two instruments designed to read very low levels of current and voltage appear in Fig. 12.27. The same company also manufacturers instruments with analog displays that can read similar levels of current and voltage.

An instrument capable of measuring resistance, inductance, and capacitance levels with an accuracy of 0.02% appears in Fig. 12.28. The instrument has now been designed to the point that almost no prior training in making such measurements is necessary.

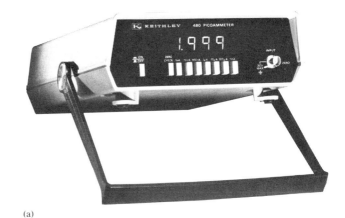

(a)

(b)   *Courtesy of Keithley Instruments, Inc.*

12.13 THE WATTMETER

FIG. 12.27 (*a*) *Picoammeter and* (*b*) *nanovoltmeter.*

The wattmeter is an instrument designed to read the power to an element or network. It employs an *electrodynamome-*

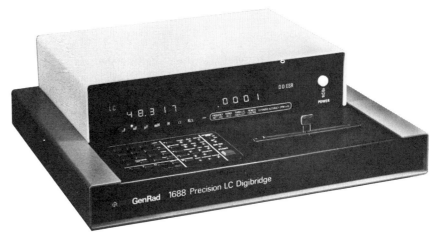

Courtesy of GenRad, Inc.

FIG. 12.28 *Precision LC Digibridge.*

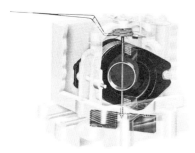

Courtesy of Weston Instruments, Inc.

FIG. 12.29 *Electrodynamometer movement.*

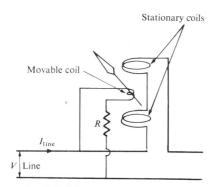

FIG. 12.30

Courtesy of Weston Instruments, Inc.

FIG. 12.31 *Wattmeter.*

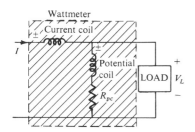

FIG. 12.32

ter-type movement such as appearing in Fig. 12.29 and not the d'Arsonval movement employed thus far. The electrodynamometer movement can also be used in ammeters or voltmeters, but it is, in general, more expensive than the d'Arsonval movement.

The coil in the d'Arsonval movement rotates within a fixed magnetic field created by the permanent magnet. In the electrodynamometer movement, the moving coil rotates in a magnetic field produced by current passing through a stationary coil. The fluxes of the stationary coils and movable coils interact to develop a torque on the pointer connected to the movable coil. When the electrodynamometer movement is used in an ammeter or a voltmeter, the torque developed is directly proportional to the product of the two fluxes. Since the fluxes are dependent on the current, the torque is also directly proportional to the product of the current in the moving coil and the current in the stationary coil. This product of similar quantities requires that a squared scale be used for the electrodynamometer voltmeter or ammeter. A reversal in current flow will not affect the position of the pointer of the meter since the fluxes of the stationary and movable coils will reverse at the same instant. For this reason, an ac meter using this movement can also measure dc quantities.

The sensitivity of the electrodynamometer movement requires that some additional circuitry be employed to limit the current through or voltage across the movement. A shunt and a multiplier are used in the ammeter and voltmeter, respectively. In the wattmeter configuration (Fig. 12.30), the current in the stationary coils is the line current, while the current in the moving coil is derived from the line voltage. This instrument then indicates power in watts on a linear scale. A typical wattmeter using the electrodynamometer movement is shown in Fig. 12.31.

For an up-scale deflection, the wattmeter is connected as shown in Fig. 12.32. Some wattmeters will always give a wattage reading that is higher than that actually delivered to the load. They are high by the amount of power consumed by the potential coil (V_{pc}^2/R_{pc}). This correction is important and should be considered with every set of data. Many wattmeters are designed to compensate for this correction, and therefore they eliminate the need for any other adjustment in the reading. The wattmeter is always connected with the potential coil in parallel, and the current coil in series, with the portion of the network to which the power is being measured.

The power delivered to R_1 in Fig. 12.33 can be found by connecting the wattmeter as shown in Fig. 12.33(a). To find the power delivered to the total network, it should be connected as shown in Fig. 12.33(b).

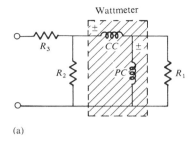 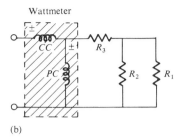

(a)   (b)

FIG. 12.33

PROBLEMS

Section 12.3

1. A d'Arsonval movement is rated 1 mA, 100 ohms.
   a. What is the current sensitivity?
   b. Design a 20-A ammeter using the above movement. Show the circuit and component values.

2. Using a 50-μA, 1000-Ω d'Arsonval movement, design a multirange milliammeter having scales of 25 mA, 50 mA, and 100 mA. Show the circuit and component values.

Section 12.4

3. A d'Arsonval movement is rated 50 μA, 1000 Ω.
   a. Design a 15-V dc voltmeter. Show the circuit and component values.
   b. What is the ohm/volt rating of the voltmeter?

4. Using a 1-mA, 100-Ω d'Arsonval movement, design a multirange voltmeter having scales of 5, 50, and 500 V. Show the circuit and component values.

5. A digital meter has an internal resistance of 10 MΩ on its 0.5-V range. If you had to build a voltmeter with a d'Arsonval movement, what current sensitivity would you need if the meter were to have the same internal resistance on the same voltage scale?

Section 12.6

*6. **a.** Design a series ohmmeter using a 100-μA, 1000-Ω movement, a zero-adjust with a maximum value of 2 kΩ, a battery of 3 V, and a series resistor whose value is to be determined.
   b. Find the resistance required for full-scale, $\frac{3}{4}$-scale, $\frac{1}{2}$-scale, and $\frac{1}{4}$-scale deflection.
   c. Using the results of part (b), draw the scale to be used with the ohmmeter.

Section 12.7

7. Describe the basic construction and operation of the Megger.

Section 12.8

8. Find the unknown resistance R_{unknown} for the balanced Wheatstone bridge of Fig. 12.34.

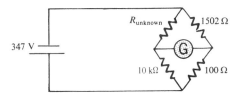

FIG. 12.34

9. Find the current I in the network of Fig. 12.35.

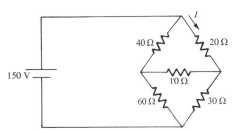

FIG. 12.35

Section 12.10

*10. **a.** Using the schematic of Fig. 12.20, find the internal circuitry for the 1000-V dc voltmeter portion of the 260 multimeter.

b. Calculate the current through the movement when 1000 volts dc are applied across the external terminals.

c. Repeat part (b) for 500 V across the external terminals.

*11. **a.** Using the schematic of Fig. 12.20, find the internal circuitry for the 1-mA dc milliammeter portion of the 260 multimeter.

b. Calculate the current through the movement when one milliampere dc is being measured by the milliammeter.

c. Repeat part (b) for 0.5 mA dc.

*12. **a.** Using the schematic of Fig. 12.20, find the internal circuitry for the $R \times$ 1-ohmmeter portion of the 260 multimeter.

b. Calculate the current through the movement when the external leads are shorted ($R_{\text{adjust}} = \frac{1}{2}$ maximum).

*13. Repeat Problem 12 for the $R \times$ 10,000-ohmmeter portion of the 260 multimeter.

Section 12.11

14. For the circuit of Fig. 12.36:

a. Determine the voltage across each resistor.

b. What voltage will a dc voltmeter with an ohm/volt rating of 1000 read when placed across the resistor R_1 if the 50-V scale is used?

c. Compare the results of parts (a) and (b).

d. Repeat parts (b) and (c) if the 100-V scale is used.

15. Repeat Problem 14 using a dc voltmeter with an ohm/volt rating of 20,000.

16. Repeat Problem 14 with a digital meter having an input resistance of 10 MΩ.

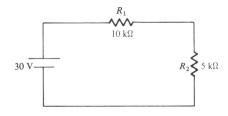

FIG. 12.36

Section 12.13

17. **a.** Hook up a wattmeter in the circuit of Fig. 12.37 to read the power delivered to the 40-Ω resistor. Indicate the positions of the current coil and the potential coil.
    b. Find the number of watts that the wattmeter will read over the actual value if the resistance of the potential coil is 10 kΩ.
    c. Repeat part (a) for the 26-Ω resistor.

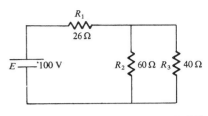

FIG. 12.37

GLOSSARY

Ammeter An instrument designed to read the current through elements in series with the meter.

d'Arsonval movement An iron-core coil mounted on bearings between a permanent magnet. A pointer connected to the movable core indicates the strength of the current passing through the coil.

Digital reading meter An instrument that employs a digital display to indicate the magnitude of the quantity measured.

Electrodynamometer movement A movement that employs a set of stationary coils and a movable coil and that can be used in the design of wattmeters, ammeters, and voltmeters.

Megger® tester An instrument for measuring very high resistance levels, such as in the megohm range.

Multimeter An instrument capable of measuring more than one level of current, voltage, and resistance.

Ohm/volt rating A rating used to determine both the current sensitivity of the movement and the internal resistance of the meter.

Series ohmmeter A resistance-measuring instrument in which the movement is placed in series with the unknown resistance.

Shunt ohmmeter A resistance-measuring instrument in which the movement is placed in parallel with the unknown resistance.

Voltmeter An instrument designed to read the voltage across elements in parallel with the meter.

Wattmeter An instrument capable of measuring the power delivered to an element by sensing both the voltage across the element and the current through the element.

Wheatstone bridge An instrument that employs a bridge configuration to establish a balance condition which can be used to determine an unknown resistance.

SINUSOIDAL ALTERNATING CURRENT

13

13.1 INTRODUCTION

The analysis thus far has been limited to dc networks, networks in which the currents or voltages are fixed in magnitude except for transient effects. We will now turn our attention to the analysis of networks in which the magnitude of the source of emf varies in a set manner. Of particular interest is the time-varying emf that is commercially available in large quantities and is commonly called the *ac voltage*. (The letters *ac* are an abbreviation for *alternating current*.) To be absolutely rigorous, the terminology *ac voltage* or *ac current* is not sufficient to describe the type of signal we will be analyzing. Each waveform of Fig. 13.1 is an alternating waveform available from commercial power supplies. The term *alternating* indicates only that the waveform alternates between two prescribed levels (Fig. 13.1). To be absolutely correct, the term *sinusoidal, square wave,* or *triangular* must also be applied. The pattern of particular interest is the *sinusoidal* ac voltage of Fig. 13.1. Since this type of signal is encountered in the vast majority of instances, the abbreviated phrases *ac voltage* or *ac current* are commonly applied without confusion. For the other patterns of Fig. 13.1, the descriptive term is always present, but frequently the ac abbreviation is dropped, resulting in the designation *square-wave* or *triangular* sources of emf.

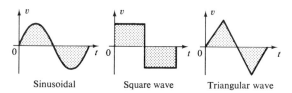

Sinusoidal Square wave Triangular wave

FIG. 13.1 *Alternating waveforms.*

13.2 SINUSOIDAL ac VOLTAGE GENERATION

The characteristics of the sinusoidal voltage and current and their effect on the basic R, L, C elements will be described in some detail in this chapter and those to follow. Of immediate interest is their generation.

The terminology *ac generator* or *alternator* should not be new to most technically oriented students. It is an electromechanical device capable of converting mechanical power to electrical power. As shown in the very basic ac generator of Fig. 13.2, it is constructed of two main components: the

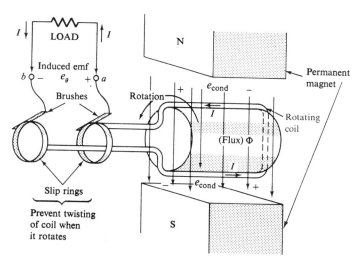

FIG. 13.2

rotor (or armature, in this case) and the *stator*. As implied by the terminology, the rotor rotates within the framework of the stator, which is stationary. When the rotor is caused to rotate by some mechanical power such as is available from the forces of rushing water (dams) or steam-turbine engines, the conductors on the rotor will cut magnetic lines of force established by the poles of the stator, as shown in Fig. 13.2. The poles may be those of a permanent magnet or may result from the turns of wire around the ferromagnetic core of the pole through which a dc current is passed to establish the necessary mmf for the required flux density.

Dictated by Eq. (11.1), the length of conductor passing through the magnetic field will have an emf induced across it as shown in Fig. 13.2. Note that the induced emf across each conductor is additive, so the generated terminal voltage is the sum of the two induced emf's. Since the armature of Fig. 13.2 is rotating, and the output terminals *a* and *b* are connected to some fixed external load, there is the necessity for the induced *slip rings*. The slip rings are circular conducting surfaces that provide a path of conduction from the generated voltage to the load and prevent a twisting of the coil at *a* and *b* when the coil rotates. The induced emf will have a polarity at terminals *a* and *b* and will develop a current *I* having the direction indicated in Fig. 13.2. Note that the direction of *I* is also the direction of increasing induced emf within the generator.

A method will now be described for determining the direction of the resulting current or of the increasing induced emf. For the generator, the thumb, forefinger, and middle finger are placed at right angles, as shown in Fig. 13.3. The

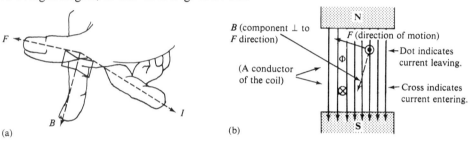

thumb is placed in the direction of force or motion of a conductor, the index finger in the direction of the magnetic flux lines, and the middle finger in the direction of current flow resulting in the conductor if a load is attached. If a load is not attached, the middle finger indicates the direction of increasing induced emf. The placement of the fingers is indicated in Fig. 13.3(a) for the top conductor of the rotor of Fig. 13.2 as it passes through the position indicated in Fig. 13.3. From this point on, we will assume that a load has been applied so that the current directions can be included using the dot (·)-cross (×) convention described in Chapter 10. Note that the resulting direction for the upper conductor is opposite that of the lower conductor. This is a necessary condition for the current *I* in the series configuration. The reversal in the direction of motion (the thumb) in this region will result in the opposite direction for *I*.

Let us now consider a few representative positions of the rotating coil and determine the relative magnitude and polarity of the generated voltage at these positions. At the instant the coil passes through position 1 in Fig. 13.4(a), there are no flux lines being cut, and the induced emf is zero. As the coil moves from position 1 to position 2, indicated in Fig. 13.4(b), the number of flux lines cut per unit time will increase, resulting in an increased induced emf

FIG. 13.3 *(a) Right-hand rule; (b) current directions as determined using the right-hand rule for the indicated position of the rotating coil.*

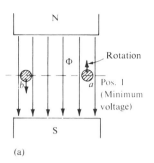
(a)

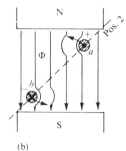
(b)

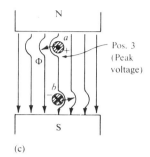
(c)

FIG. 13.4

across the coil. For position 2, the resulting current direction and polarity of terminals a and b are indicated as determined by the right-hand rule. At position 3, the number of flux lines being cut per unit time is a maximum, resulting in a maximum induced voltage. The polarities and current direction are the same as at position 2.

As the coil continues to rotate toward position 4, indicated in Fig. 13.5(a), the polarity of the induced emf and

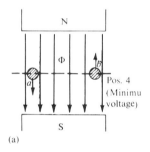
(a)

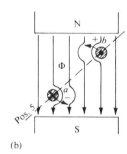
(b)

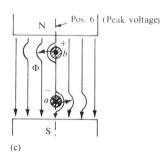
(c)

FIG. 13.5

the current direction remain the same, as shown in the figure, although the induced emf will drop due to the reduced number of flux lines cut per unit time. At position 4, the induced emf is again zero, since the number of flux lines cut per unit time has dropped to zero. As the coil now turns toward position 5, the magnitude of the induced emf will again increase, but note the change in polarity for terminals a and b and the reversal of current direction in each conductor. The similarities between the coil positions of positions 2 and 5 [Fig. 13.5(c)], and of 3 and 6 [Fig. 13.5(c)], indicate that the magnitude of the induced emf is the same although the polarity of a-b has reversed.

A continuous plot of the induced emf e_g appears in Fig. 13.6. The polarities of the induced emf are shown for terminals a and b to the left of the vertical axis.

Take a moment to relate the various positions to the resulting waveform of Fig. 13.6. This waveform will become very familiar in the discussions to follow. Note some of its obvious characteristics. As shown in the figure, if the coil is allowed to continue rotating, the generated emf will repeat itself in equal intervals of time. Note also that the pattern is exactly the same below the axis as it is above, and that it

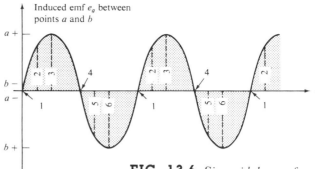

FIG. 13.6 *Sinusoidal waveform.*

continually changes with time (the horizontal axis). At the risk of being repetitious, let us again state that the waveform of Fig. 13.6 is the appearance of a *sinusoidal ac voltage*.

The alternator is only one source of sinusoidal voltages and currents. The function generator of Fig. 13.7, which employs semiconductor electronic components in its internal construction, will provide sinusoidal, square-wave, and triangular output waveforms as required.

Courtesy of Hewlett Packard Co.

FIG. 13.7 *Function generator.*

13.3 DEFINED POLARITIES AND DIRECTION

In the following analysis, we will find it necessary to establish a set of polarities for the sinusoidal ac voltage and a direction for the sinusoidal ac current. In each case, the polarity and current direction will be for an instant of time in the positive portion of the sinusoidal waveform. This is shown in Fig. 13.8 with the symbols for the sinusoidal ac voltage and current. A lower-case letter is employed for each to indicate that the quantity is time dependent; that is, its magnitude will change with time. The need for defining polarities and current direction will become quite obvious when we consider multisource ac networks. Note in the last sentence the absence of the term *sinusoidal* before the phrase *ac networks*. This will occur to an increasing degree as we progress; *sinusoidal* is to be understood unless otherwise indicated.

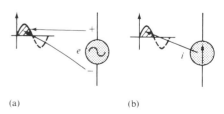

(a)   (b)

FIG. 13.8 (*a*) *Sinusoidal ac voltage sources;* (*b*) *sinusoidal ac current sources.*

Referring to Fig. 13.7, note that a "HI" and "LO" side have been indicated for each output set, with the ground or zero potential level associated with the LO terminal. A positive sign and a negative sign would then be appropriate for the HI and LO terminals, respectively.

13.4 DEFINITIONS

The sinusoidal waveform of Fig. 13.9 with its additional notation will now be used as a model in defining a few basic terms. These terms can, however, be applied to any waveform.

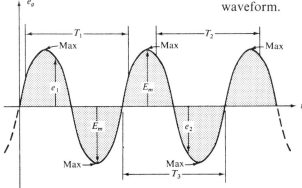

FIG. 13.9 *Sinusoidal voltage.*

Waveform: The path traced by a quantity, such as the emf in Fig. 13.9, plotted as a function of some variable such as time (as above), position, degrees, radians, temperature, and so on.

Instantaneous value: The magnitude of a waveform at any instant of time; denoted by lower-case letters (e_1, e_2).

Amplitude, or *peak value:* The maximum value of a waveform; denoted by upper-case letters (E_m).

Periodic waveform: A waveform that continually repeats itself after the same time interval. The waveform of Fig. 13.9 is a periodic waveform.

Period (T): The time interval between successive repetitions of a periodic waveform; the period $T_1 = T_2 = T_3$ in Fig. 13.9, so long as successive similar points of the periodic waveform are used in determining T.

Cycle: The portion of a waveform contained in *one period* of time. The cycles within T_1, T_2, and T_3 of Fig. 13.9 may appear different in Fig. 13.10, but they are all bounded by one period of time and therefore satisfy the definition of a cycle.

Frequency (f): The number of cycles that occur in 1 second. The frequency of the waveform of Fig. 13.11(a) is 1 cycle per second, and for Fig. 13.11(b), $2\frac{1}{2}$ cycles per second. If a waveform of similar shape had a period of 0.5 second [Fig. 13.11(c)], the frequency would be 2 cycles per second. For many years, the units for frequency were *cycles per*

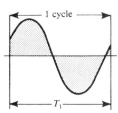

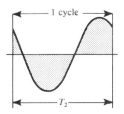

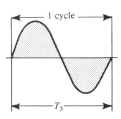

second (abbreviated cps). Recently, however, the emphasis
has been on the use of *hertz*, where

FIG. 13.10

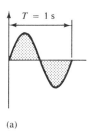

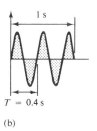

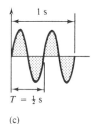

(a)   (b)   (c)

$$\boxed{1 \text{ hertz (Hz)} = 1 \text{ cycle per second (c/s)}} \quad \textbf{(13.1)}$$

FIG. 13.11

The unit hertz is derived from the surname of Heinrich
Rudolph Hertz, who did original research in the area of
alternating currents and voltages and their effect on the
basic R, L, and C elements. The frequency standard for this
country is 60 Hz.

   Since the frequency is inversely related to the period—
that is, as one increases the other decreases by an equal
amount—the two can be related by the following equation:

$$\boxed{f = \frac{1}{T}} \qquad \begin{array}{l} f = \text{Hz} \\ T = \text{seconds (s)} \end{array} \quad \textbf{(13.2)}$$

or

$$\boxed{T = \frac{1}{f}} \qquad\qquad \textbf{(13.3)}$$

EXAMPLE 13.1. Find the period of a periodic waveform
with a frequency of
a. 60 Hz.
b. 1000 Hz.

Solutions:

a. $T = \dfrac{1}{f} = \dfrac{1}{60} = 0.01667$ s or **16.67 ms**

   (a recurring value since 60 Hz is so prevalent)

b. $T = \dfrac{1}{f} = \dfrac{1}{1000} = 10^{-3}$ s $= $ **1 ms**

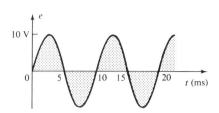

FIG. 13.12

EXAMPLE 13.2. Determine the frequency of the waveform of Fig. 13.12.

Solution: From the figure, $T = 10$ ms, and

$$f = \frac{1}{T} = \frac{1}{10 \times 10^{-3}} = \mathbf{100\ Hz}$$

EXAMPLE 13.3. The oscilloscope (appearing in Chapter 22) is an instrument that will display alternating waveforms such as described above. A sinusoidal pattern appears on the oscilloscope of Fig. 13.13 with the indicated scale settings. Determine the period, frequency, and peak value of the waveform.

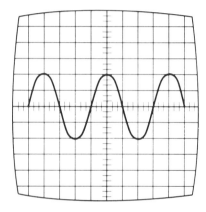

Vertical sensitivity = 0.1 V/cm

Horizontal sensitivity = 50 μs/cm

FIG. 13.13

Solution: One cycle spans 4 centimeters. The period is

$$T = 4(50\ \mu s) = \mathbf{200\ \mu s}$$

and the frequency is

$$f = \frac{1}{T} = \frac{1}{200 \times 10^{-6}} = \mathbf{5\ kHz}$$

The vertical height above the horizontal axis encompasses 2 centimeters. Therefore,

$$V_{\text{peak}} = 2(0.1) = \mathbf{0.2\ V}$$

13.5 THE SINE WAVE

The terms defined in the previous section can be applied to any type of periodic waveform, whether smooth or discontinuous. The sinusoidal waveform is of particular importance, however, since it lends itself readily to the mathematics and the physical phenomena associated with electric circuits. Consider the power of the following statement: The sine wave is the *only* waveform whose appearance is unaffected by the response characteristics of R, L, and C elements. In other words, if the voltage across a resistor, coil, or capacitor is sinusoidal in nature, the resulting current for each will also have sinusoidal characteristics. If a square wave or a triangular wave were applied, such would not be the case. It must be pointed out that the above statement is also applicable to the cosine wave since they differ only by a 90° shift on the horizontal axis, as shown in Fig. 13.14.

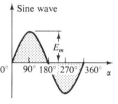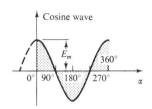

FIG. 13.14

The unit of measurement for the horizontal axis of Fig. 13.14 is the *degree*. A second unit of measurement frequently used is the *radian* (rad). It is defined by a quadrant of a circle such as in Fig. 13.15 where the distance subtended on the circumference equals the radius of the circle.

If we define x as the number of intervals of r (the radius) around the circumference of the circle, then

$$C = 2\pi r = x \cdot r$$

and we find

$$x = 2\pi$$

Therefore, there are 2π radians around a 360° circle, or

$$\boxed{2\pi \text{ radians} = 360°} \tag{13.4}$$

and

$$\boxed{1 \text{ radian} \cong 57.3°} \tag{13.5}$$

A number of electrical formulas contain a multiplier of π. This is one reason it is sometimes preferable to measure angles in radians rather than in degrees.

For 360°, the two units of measurement are related as shown in Fig. 13.16. The conversion equations between the two are the following:

$$\boxed{\text{Radians} = \left(\frac{\pi}{180°}\right) \times (\text{degrees})} \tag{13.6}$$

$$\boxed{\text{Degrees} = \left(\frac{180°}{\pi}\right) \times (\text{radians})} \tag{13.7}$$

Applying these equations, we have

$$30°: \text{Radians} = \frac{\pi}{180°}(30°) = \frac{\pi}{6}\text{rad} \cong 0.524 \text{ rad}$$

$$90°: \text{Radians} = \frac{\pi}{180°}(90°) = \frac{\pi}{2}\text{rad} \cong 1.571 \text{ rad}$$

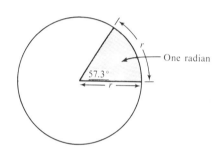

FIG. 13.15 *Defining the radian.*

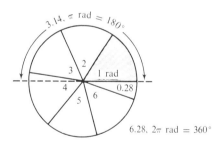

FIG. 13.16

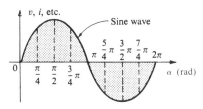

FIG. 13.17

$$\frac{\pi}{3}: \text{Degrees} = \frac{180°}{\pi}\left(\frac{\pi}{3}\right) = 60°$$

$$\frac{3\pi}{2}: \text{Degrees} = \frac{180°}{\pi}\left(\frac{3\pi}{2}\right) = 270°$$

Using the radian as the unit of measurement for the abscissa, we would obtain a sine wave as shown in Fig. 13.17.

It is of particular interest that the sinusoidal waveform can be derived from the length of the *vertical projection* of a radius vector rotating in a uniform circular motion about a fixed point. Starting as shown in Fig. 13.18(a) and plotting the amplitude (above and below zero) on the coordinates drawn to the right [Figs. 13.18(b)–(i)], we will trace a complete sinusoidal waveform after the radius vector has completed a 360° rotation about the center.

The velocity with which the radius vector rotates about the center, called the *angular velocity,* can be determined from the following equation:

$$\text{Angular velocity} = \frac{\text{distance (degrees or radians)}}{\text{time (seconds)}}$$

$$\textbf{(13.8)}$$

Substituting into Eq. (13.8) and assigning the Greek letter omega (ω) to the angular velocity, we have

$$\omega = \frac{\alpha}{t} \qquad \textbf{(13.9)}$$

In Fig. 13.18, the time required to complete one revolution is equal to the period (T) of the sinusoidal waveform of Fig. 13.18(i). The radians subtended in this time interval are 2π. Substituting, we have

$$\omega = \frac{2\pi}{T} \qquad \text{(rad/s)} \qquad \textbf{(13.10)}$$

In words, this equation states that the smaller the period of the sinusoidal waveform of Fig. 13.18(i), or the smaller the time interval before one complete cycle is generated, the greater must be the angular velocity of the rotating radius vector. Certainly this statement agrees with what we have learned thus far. We can now go one step further and apply the fact that the frequency of the generated waveform is inversely related to the period of the waveform; that is, $f = 1/T$. Thus,

$$\omega = 2\pi f \qquad \text{(rad/s)} \qquad \textbf{(13.11)}$$

This equation states that the higher the frequency of the generated sinusoidal waveform, the higher must be the angular velocity. Equations (13.10) and (13.11) are verified

(a)

(b)

(c)
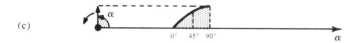

(d)
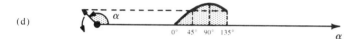

(e)

(f)
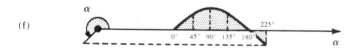

(g)
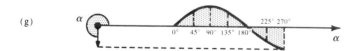

(h)
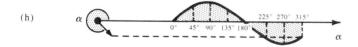

(i)
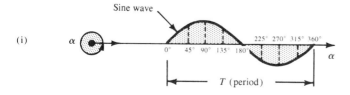

FIG. 13.18

somewhat by Fig. 13.19, where for the same radius vector, $\omega = 100$ rad/s and 500 rad/s.

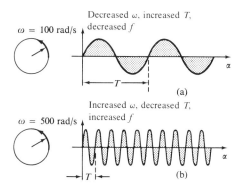

FIG. 13.19

EXAMPLE 13.4. Determine the angular velocity of a sine wave having a frequency of 60 Hz.

Solution:

$$\omega = 2\pi f = (6.28)(60)$$
$$\cong \mathbf{377\ rad/s}\ \text{(a recurring value due to}$$
$$\text{60-Hz predominance)}$$

EXAMPLE 13.5. Determine the frequency and period of the sine wave of Fig. 13.19(b).

Solution:

$$\omega = \frac{2\pi}{T} \qquad \therefore T = \frac{2\pi}{\omega} = \frac{2\pi}{500} = \frac{6.28}{500} = \mathbf{12.56\ ms}$$

$$f = \frac{1}{T} = \frac{1}{12.56 \times 10^{-3}} = \mathbf{79.62\ Hz}$$

13.6 GENERAL FORMAT FOR THE SINUSOIDAL VOLTAGE OR CURRENT

The basic mathematical format for the sinusoidal waveform is

$$\boxed{A_m \sin \alpha} \qquad \textbf{(13.12)}$$

where A_m is the peak value of the waveform and α is the unit of measure for the horizontal axis as shown in Fig. 13.20.
   Since

$$\omega = \frac{\alpha}{t} \qquad \text{[Eq. (13.9)]}$$

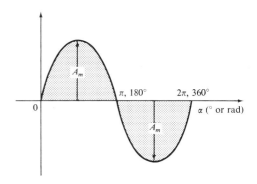

then

FIG. 13.20

$$\boxed{\alpha = \omega t} \qquad (13.13)$$

Equation (13.13) states that the angle α through which the rotating vector of Fig. 13.18 will pass is determined by angular velocity of the rotating vector and the length of time that it rotates. For example, for a particular angular velocity (fixed ω), the longer the radius vector is permitted to rotate (that is, the greater the value of t), the greater will be the number of degrees or radians through which the vector will pass. Relating this statement to the sinusoidal waveform, for a particular angular velocity, the longer the time, the greater the number of cycles shown. For a fixed time interval, the greater the angular velocity, the greater the number of cycles generated.

Due to Eq. (13.13), the general format of a sine wave can also be written

$$\boxed{A_m \sin \omega t} \qquad (13.14)$$

with ωt as the horizontal unit of measure.

For electrical quantities such as current and voltage, the general format is

$$i = I_m \sin \omega t = I_m \sin \alpha$$
$$e = E_m \sin \omega t = E_m \sin \alpha$$

where the capital letters with the subscript m represent the amplitude and the lower-case letters i and e represent the instantaneous value of current or voltage, respectively, at any time t.

The sine wave can also be plotted against *time* on the horizontal axis. For each degree or radian measure, the corresponding time can be determined from the frequency and then broken into segments corresponding to the degree or radian measure. The latter technique will be demonstrated in Example 13.6.

Before giving an example, we must note the relative simplicity of the mathematical equation that can represent a

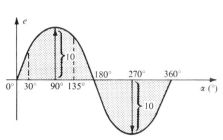

FIG. 13.21

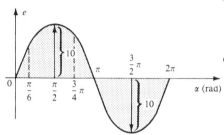

FIG. 13.22

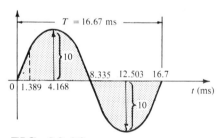

FIG. 13.23

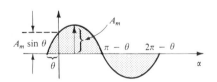

FIG. 13.24

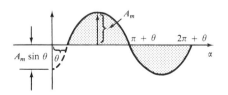

FIG. 13.25

sinusoidal waveform. Any alternating waveform whose characteristics differ from that of the sine wave cannot be represented by a single term, but may require two, four, six, or perhaps an infinite number of terms to be accurately represented. A further description of nonsinusoidal waveforms can be found in Chapter 22.

EXAMPLE 13.6. Sketch $e = 10 \sin 377t$ with the abscissa
a. angle in degrees (α).
b. angle in radians (α).
c. time in seconds (t).

Solutions:
a. See Fig. 13.21. (Note that no calculations are required.)
b. See Fig. 13.22. (Once the relationship between degrees and radians is understood, there is again no need for calculations.)

c. $\omega = \dfrac{2\pi}{T}$, or $T = \dfrac{2\pi}{\omega} = \dfrac{6.28}{377} = 16.67$ ms

$$\frac{T}{2} = \frac{16.67}{2} \times 10^{-3} = 8.335 \text{ ms}$$

$$\frac{T}{4} = \frac{16.67}{4} \times 10^{-3} = 4.168 \text{ ms}$$

$$\frac{T}{12} = \frac{16.67}{12} \times 10^{-3} = 1.389 \text{ ms}$$

See Fig. 13.23.

13.7 PHASE RELATIONS

So far, we have considered only sine waves that have maxima at $\pi/2$ and $3\pi/2$, with a zero value at 0, π, and 2π, as shown in Fig. 13.21. If the waveform is shifted to the right or left of $0°$, the expression becomes

$$\boxed{A_m \sin(\omega t \pm \theta)} \qquad \text{(13.15)}$$

where θ is the angle in degrees or radians that the waveform has been shifted.

If it passes through the horizontal axis with a *positive-going* (increasing with time) slope before $0°$, as shown in Fig. 13.24, the expression is

$$\boxed{A_m \sin(\omega t + \theta)} \qquad \text{(13.16)}$$

At $\omega t = \alpha = 0°$, the magnitude is determined by $A_m \sin \theta$. If it passes through the horizontal axis with a positive-going slope after $0°$, as shown in Fig. 13.25, the expression is

$$\boxed{A_m \sin(\omega t - \theta)} \qquad \text{(13.17)}$$

and at $\omega t = \alpha = 0°$, the magnitude is $A_m \sin(-\theta)$, which by a trigonometric identity is $-A_m \sin \theta$.

If the waveform crosses the horizontal axis with a positive-going slope 90° ($\pi/2$) sooner, as shown in Fig. 13.26, it is called a *cosine wave*. That is,

FIG. 13.26

$$\boxed{\sin(\omega t + 90°) = \sin\left(\omega t + \frac{\pi}{2}\right) = \cos \omega t}$$

(13.18)

or

$$\boxed{\sin \omega t = \cos(\omega t - 90°) = \cos\left(\omega t - \frac{\pi}{2}\right)}$$

(13.19)

The terms *lead* and *lag* are used to indicate the relationship between two sinusoidal waveforms of the *same frequency* plotted on the same set of axes. In Fig. 13.26, the cosine curve is said to *lead* the sine curve by 90°, and the sine is said to *lag* the cosine curve by 90°. The 90° is referred to as the phase angle between the two waveforms. In language commonly applied, they are *out of phase* by 90°. Note that the phase angle between the two waveforms is measured between those two points on the horizontal axis through which each passes with the same slope.

A few additional geometric relations that may prove useful in applications involving sines or cosines in phase relationships are the following:

$$\boxed{\begin{aligned} \sin(-\alpha) &= -\sin \alpha \\ \cos(-\alpha) &= \cos \alpha \\ -\sin(\alpha) &= \sin(\alpha \pm 180°) \\ -\cos(\alpha) &= \cos(\alpha \pm 180°) \end{aligned}}$$

(13.20)

If a sinusoidal expression should appear as

$$e = -E_m \sin \omega t$$

the negative sign is associated with the sine portion of the expression, not the peak value E_m. In other words, the expression, if not for convenience, would be written

$$e = E_m(-\sin \omega t)$$

Since

$$-\sin \omega t = \sin(\omega t \pm 180°)$$

the expression can also be written

$$e = E_m \sin(\omega t \pm 180°)$$

revealing that a negative sign can be replaced by a 180° change in phase angle (+ or −). That is,

$$e = -E_m \sin \omega t = E_m \sin(\omega t + 180°)$$
$$= E_m \sin(\omega t - 180°)$$

A plot of each will clearly show their equivalence. There are, therefore, two correct mathematical representations for the functions.

The phase relationship between two waveforms indicates which one leads or lags, and by how many degrees or radians.

EXAMPLE 13.7. What is the phase relationship between the following sinusoidal waveforms?

a. $v = 10 \sin(\omega t + 30°)$
   $i = 5 \sin(\omega t + 70°)$

b. $i = 15 \sin(\omega t + 60°)$
   $v = 10 \sin(\omega t - 20°)$

c. $i = 2 \cos(\omega t + 10°)$
   $v = 3 \sin(\omega t - 10°)$

d. $i = -\sin(\omega t + 30°)$
   $v = 2 \sin(\omega t + 10°)$

e. $i = -2 \cos(\omega t - 60°)$
   $v = 3 \sin(\omega t - 150°)$

Solutions:
a. See Fig. 13.27.
   ***i* leads *v* by 40° or *v* lags *i* by 40°.**

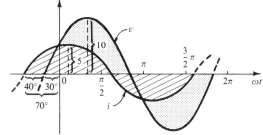

FIG. 13.27

b. See Fig. 13.28.
   ***i* leads *v* by 80° or *v* lags *i* by 80°.**

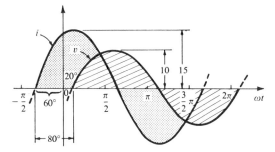

FIG. 13.28

c. See Fig. 13.29.

$$i = 2 \cos(\omega t + 10°) = 2 \sin(\omega t + 10° + 90°)$$
$$= 2 \sin(\omega t + 100°)$$

i leads v by 110° or v lags i by 110°.

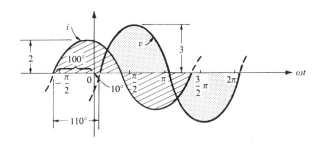

FIG. 13.29

d. See Fig. 13.30.

$$-\sin(\omega t + 30°) = \sin(\omega t + 30° \overset{\text{Note}}{-} 180°)$$
$$= \sin(\omega t - 150°)$$

v leads i by 160° or i lags v by 160°.

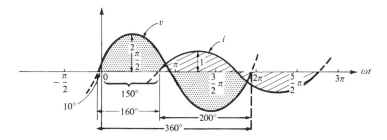

FIG. 13.30

Or using

$$-\sin(\omega t + 30°) = \sin(\omega t + 30° \overset{\text{Note}}{+} 180°)$$
$$= \sin(\omega t + 210°)$$

and **i leads v by 200° or v lags i by 200°.**

e. See Fig. 13.31.

$$i = -2 \cos(\omega t - 60°) = 2 \cos(\omega t - 60° \overset{\text{By choice}}{-} 180°)$$
$$= 2 \cos(\omega t - 240°)$$

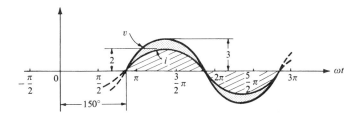

FIG. 13.31

However,

$$\cos \alpha = \sin(\alpha + 90°)$$

so that

$$2 \cos(\omega t - 240°) = 2 \sin(\omega t - 240° + 90°)$$
$$= 2 \sin(\omega t - 150°)$$

v and i are in phase.

13.8 AVERAGE VALUE

After traveling a considerable distance by car, some drivers like to calculate their average speed for the entire trip. This is usually done by dividing the miles traveled by the hours required to drive that distance. For example, if a person traveled 180 mi in 5 h, his average speed was 180/5 or 36 mi/h. This same distance may have been traveled at various speeds for various intervals of time, as shown in Fig. 13.32.

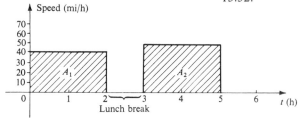

FIG. 13.32

By finding the total area under the curve for the 5 h and then dividing the area by 5 h (the total time for the trip), we obtain the same result of 36 mi/h; that is,

$$\text{Average speed} = \frac{\text{area under curve}}{\text{length of curve}} \quad \textbf{(13.21)}$$

$$= \frac{A_1 + A_2}{5}$$

$$= \frac{(40)(2) + (50)(2)}{5}$$

$$= \frac{180}{5}$$

$$= \textbf{36 mi/h}$$

Equation (13.21) can be extended to include any variable quantity, such as current or voltage, if we let G denote the average value, as follows:

$$G \text{ (average value)} = \frac{\text{algebraic sum of areas}}{\text{length of curve}}$$

(13.22)

The algebraic sum of the areas must be determined, since some area contributions will be from below the horizontal axis. Areas above the axis will be assigned a positive sign, and those below a negative sign. A positive average value will then be above the axis, and a negative value below.

The average value of *any* current or voltage is the value indicated on a dc meter. In other words, over a complete cycle, the average value is the equivalent dc value. In the analysis of electronic circuits to be considered in a later course, both dc and ac sources of emf will be applied to the same network. It will then be necessary to know or determine the dc (or average value) and ac components of the voltage or current in various parts of the system.

EXAMPLE 13.8. Find the average values of the following waveforms over one full cycle:

a. Fig. 13.33.

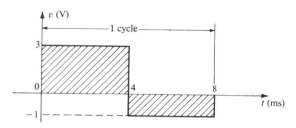

FIG. 13.33

b. Fig. 13.34.

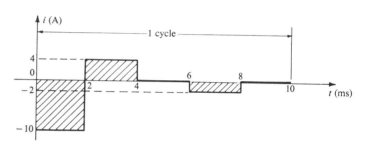

FIG. 13.34

Solutions:

a. $G = \dfrac{+(3)(4) - (1)(4)}{8} = \dfrac{12 - 4}{8} = \mathbf{1\ V}$

Note Fig. 13.35.

b. $G = \dfrac{-(10)(2) + (4)(2) - (2)(2)}{10}$

$\qquad = \dfrac{-20 + 8 - 4}{10}$

$\qquad = -\dfrac{16}{10}$

$\qquad = \mathbf{-1.6\ A}$

Note Fig. 13.36.

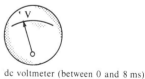

dc voltmeter (between 0 and 8 ms)

FIG. 13.35

dc ammeter (between 0 and 10 ms)

FIG. 13.36

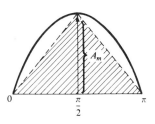

FIG. 13.37

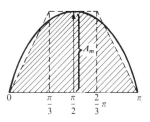

FIG. 13.38

We found the areas under the curves in the preceding example by using a simple geometric formula. If we should encounter a sine wave or any other unusual shape, however, we must find the area by some other means. We can obtain a good approximation of the area by attempting to reproduce the original wave shape using a number of small rectangles or other familiar shapes, the area of which we already know through simple geometric formulas. For example, *the actual area of the positive (or negative) pulse of a sine wave is* $2A_m$. Approximating this waveform by two triangles (Fig. 13.37), we obtain (using *area* $= \frac{1}{2}$ *base* $\times$ *height* for the area of a triangle) a rough idea of the actual area:

$$\text{Area (shaded)} = 2\left(\frac{1}{2}bh\right) = 2\left[\left(\frac{1}{2}\right)\overbrace{\left(\frac{\pi}{2}\right)}^{b}\overbrace{(A_m)}^{h}\right] = \frac{\pi}{2}A_m$$
$$\cong 1.58A_m$$

A closer approximation might be a rectangle with two similar triangles (Fig. 13.38):

$$\text{Area} = A_m\frac{\pi}{3} + 2\left(\frac{1}{2}bh\right) = A_m\frac{\pi}{3} + \frac{\pi}{3}A_m = \frac{2}{3}\pi A_m$$
$$= 2.094A_m$$

which is certainly close to the actual area. If an infinite number of forms were used, an exact answer of $2A_m$ could be obtained. For irregular waveforms, this method can be especially useful if data such as the average value are desired.

The procedure of calculus that gives the exact solution $2A_m$ is known as *integration*. Finding the area under the positive pulse of a sine wave using integration, we have

$$\text{Area} = \int_0^\pi A_m \sin \alpha \, d\alpha$$

where $\int$ is the sign of integration;
      π and 0 are the limits of integration;
      $A_m \sin \alpha$ is the function to be integrated; and
      $d\alpha$ indicates that we are integrating with respect to α.

Integrating, we obtain

$$\text{Area} = A_m[-\cos \alpha]_0^\pi$$
$$= -A_m(\cos \pi - \cos 0°)$$
$$= -A_m[-1 - (+1)] = -A_m(-2)$$

$$\boxed{\text{Area} = 2A_m}$$ **(13.23)**

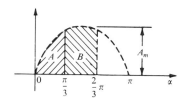

Integration is presented here only to make the method recognizable to the reader; it is not necessary to be proficient in its use to continue with this text. It is a useful mathematical tool, however, and should be learned.

Since we know the area under the positive (or negative) pulse, we can easily determine the average value of the positive (or negative region) of a sine wave pulse by applying Eq. (13.22):

$$G = \frac{2A_m}{\pi}$$

and

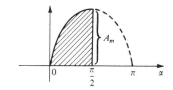

FIG. 13.39

$$\boxed{G = 0.637A_m}$$ 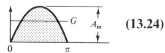 **(13.24)**

For the waveform of Fig. 13.39,

$$G = \frac{(2A_m/2)}{\pi/2} = \frac{2A_m}{\pi}$$ (average the same as for a full pulse)

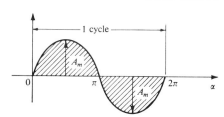

FIG. 13.40

The area of A or B in Fig. 13.40 is not $\frac{1}{3}(2A_m)$ and must be determined using methods of calculus.

EXAMPLE 13.9. Find the average value of the following waveforms over one full cycle:
a. Fig. 13.41.

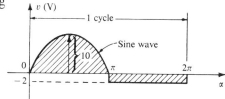

FIG. 13.41

b. Fig. 13.42.

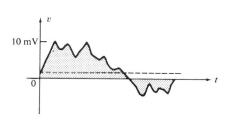

FIG. 13.42

c. Fig. 13.43. For this waveform, simply indicate whether the average value is positive and what its approximate value is.

FIG. 13.43

Solutions:

a. $G = \dfrac{\overbrace{(2)(10)}^{2Am} - (2)(\pi)}{2\pi} = \dfrac{20 - 2\pi}{2\pi} = \dfrac{10 - \pi}{\pi}$

    $= \textbf{2.183 V}$

b. $G = \dfrac{+2A_m - 2A_m}{2\pi} = \textbf{0}$

c. From the appearance of the waveform, the average value is positive and in the vicinity of 2 mV. Occasionally, judgments of this type will have to be made.

The average value of a sine wave (or cosine wave) is zero, which should be obvious from the appearance of the waveform over one full cycle.

13.9 EFFECTIVE VALUES

This section will begin to relate dc and ac quantities with respect to the power delivered to a load. It will help us determine the amplitude of a sinusoidal ac current required to deliver the same power as a particular dc current. The question frequently arises, How is it possible for a sinusoidal ac quantity to deliver a net power if, over a full cycle, the net current in any one direction is zero (average value = 0)? It would almost appear that the power delivered during the positive portion of the sinusoidal waveform is withdrawn during the negative portion, and since the two are equal in magnitude, the net power delivered is zero. However, understand that *irrespective of direction,* current of any magnitude through a resistor will deliver power *to that resistor.* In other words, during the positive or negative portions of a sinusoidal ac current, power is being delivered at *each instant of time* to the resistor. The power delivered at each instant will, of course, vary with the magnitude of the sinusoidal ac current, but there will be a net flow during either the positive or negative pulses with a net flow over the full cycle. The net power flow will equal twice that delivered by either the positive or negative regions of the sinusoidal quantity.

A fixed relationship between ac and dc voltages and currents can be derived from the experimental setup shown in Fig. 13.44. A resistor in a water bath is connected by switches to a dc and an ac supply. If switch 1 is closed, a dc current I, determined by the resistance R and battery voltage E, will flow through the resistor R. The temperature reached by the water is determined by the dc power dissipated in the form of heat by the resistor.

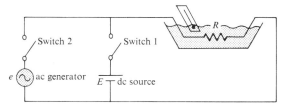

If switch 2 is closed and switch 1 left open, the ac current through the resistor will have a peak value of I_m. The temperature reached by the water is now determined by the ac power dissipated in the form of heat by the resistor. The ac input is varied until the temperature is the same as that reached with the dc input. When this is accomplished, the average electrical power delivered to the resistor R by the ac source is the same as that delivered by the dc source.

The power delivered by the ac supply at any instant of time is

$$P_{ac} = (i_{ac})^2 R = (I_m \sin \omega t)^2 R = (I_m^2 \sin^2 \omega t) R$$

but

$$\sin^2 \omega t = \tfrac{1}{2}(1 - \cos 2\omega t) \qquad \text{(trigonometric identity)}$$
$$\therefore P_{ac} = I_m^2 [\tfrac{1}{2}(1 - \cos 2\omega t)] R$$

and

$$P_{ac} = \frac{I_m^2 R}{2} - \frac{I_m^2 R}{2} \cos 2\omega t \qquad \textbf{(13.25)}$$

The *average power* delivered by the ac source is just the first term, since the average value of a cosine wave is zero even though it may have twice the frequency of the original input current waveform. Equating the average power delivered by the ac generator to that delivered by the dc source,

$$P_{av(ac)} = P_{dc}$$
$$\frac{I_m^2 R}{2} = I_{dc}^2 R$$

and

$$I_m = \sqrt{2} I_{dc}$$

or

$$I_{dc} = \frac{I_m}{\sqrt{2}} = 0.707 I_m$$

which, in words, states that *the equivalent dc value of a sinusoidal current or voltage is $1/\sqrt{2}$ or 0.707 of its maximum value. The equivalent dc value is called the effective value of the sinusoidal quantity.*

FIG. 13.44

In summary,

$$\boxed{I_{(equiv\ dc)} = I_{eff} = 0.707I_m} \qquad \textbf{(13.26)}$$

or

$$\boxed{I_m = \sqrt{2}I_{eff} = 1.414I_{eff}} \qquad \textbf{(13.27)}$$

and

$$\boxed{E_{eff} = 0.707E_m} \qquad \textbf{(13.28)}$$

or

$$\boxed{E_m = \sqrt{2}E_{eff} = 1.414E_{eff}} \qquad \textbf{(13.29)}$$

As a simple numerical example, it would require an ac current with a peak value $= \sqrt{2}(10) = 14.14$ A to deliver the same power to the resistor in Fig. 13.44 as a dc current of 10 A. The effective value of any quantity plotted as a function of time can be found by using the following equation derived from the experiment just described:

$$\boxed{I_{eff} = \sqrt{\dfrac{\displaystyle\int_0^T i(t)^2\, dt}{T}}} \qquad \textbf{(13.30)}$$

or

$$\boxed{I_{eff} = \sqrt{\dfrac{\text{area}\ [i(t)^2]}{T}}} \qquad \textbf{(13.31)}$$

which, in words, states that to find the effective value, the function $i(t)$ must first be squared. After $i(t)$ is squared, the area under the curve is found by integration. It is then divided by T, the length of the cycle or period of the waveform, to obtain the average or *mean* value of the squared waveform. The final step is to take the *square root* of the mean value. This procedure gives us another designation for the effective value, the *root-mean-square* (rms) value.

EXAMPLE 13.10. Find the effective values of the sinusoidal waveforms in each part of Fig. 13.45.

Solutions:

a. $I_{eff} = 0.707(12 \times 10^{-3}) = \textbf{8.484 mA}$

b. $I_{eff} = \textbf{8.484 mA}$

Note that frequency did not change the effective value in (b) above as compared with (a).

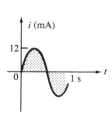

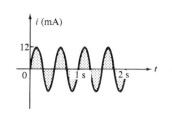

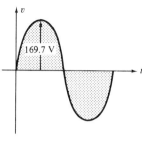

(a)   (b)   (c)

FIG. 13.45

c. $V_{\text{eff}} = 0.707(169.73) \cong$ **120 V,** as available from a home
outlet.

EXAMPLE 13.11. Find the effective or rms value of the
waveform of Fig. 13.46.

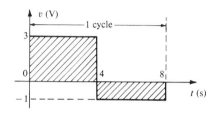

FIG. 13.46

Solution:

v^2 (Fig. 13.47):

$$V_{\text{eff}} = \sqrt{\frac{(9)(4) + (1)(4)}{8}} = \sqrt{\frac{40}{8}}$$

$$V_{\text{eff}} = \textbf{2.236 V}$$

EXAMPLE 13.12. Calculate the effective value of the voltage of Fig. 13.48.

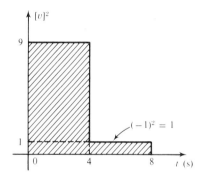

FIG. 13.47

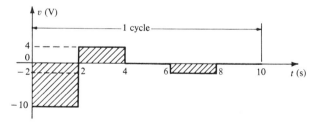

FIG. 13.48

Solution:

v^2 (Fig. 13.49):

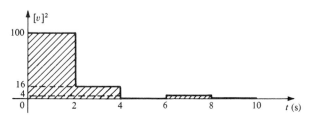

FIG. 13.49

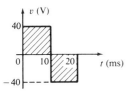

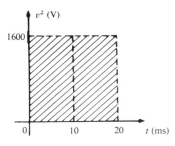

FIG. 13.50

FIG. 13.51

$$V_{\text{eff}} = \sqrt{\frac{(100)(2) + (16)(2) + (4)(2)}{10}} = \sqrt{\frac{240}{10}}$$

$$V_{\text{eff}} = \textbf{4.899 V}$$

EXAMPLE 13.13. Determine the average and effective values of the square wave of Fig. 13.50.

Solution:

By inspection, the average value = 0.

v^2 (Fig. 13.51):

$$V_{\text{eff}} = \sqrt{\frac{(1600)(10 \times 10^{-3}) + (1600)(10 \times 10^{-3})}{20 \times 10^{-3}}}$$

$$V_{\text{eff}} = \sqrt{\frac{32{,}000 \times 10^{-3}}{20 \times 10^{-3}}} = \sqrt{1600}$$

and $V_{\text{eff}} = \textbf{40 V}$ (the maximum value of the waveform of Fig. 13.50).

The waveforms appearing in these examples are the same as those used in the examples on the average value. It might prove interesting to compare the effective and average values of these waveforms.

The effective values of sinusoidal quantities such as voltage or current will be represented by E and I. These symbols are the same as those used for dc voltages and currents. To avoid confusion, the peak value of a waveform will always have a subscript m associated with it: $I_m \sin \omega t$. *Caution:* When finding the effective value of the positive pulse of a sine wave, note that the squared area is *not* simply $(2A_m)^2 = 4A_m^2$; it must be found by a completely new integration. This will always be the case for any waveform that is not rectangular.

13.10 THE DERIVATIVE

Before examining a few basic ac circuits, we must first be able to find the derivative of the sine wave or cosine wave, since it will appear in the initial description of these systems.

Recall from Section 9.7 that the derivative dx/dt is defined as the rate of change of x with respect to time. If x fails to change at a particular instant, $dx = 0$, and the derivative is zero. For the sinusoidal waveform, dx/dt is zero only at the positive and negative peaks ($\omega t = \pi/2$ and $\frac{3}{2}\pi$ in Fig. 13.52), since x fails to change at these instants of time.

A close examination of the sinusoidal waveform will also indicate that the greatest change in x will occur at the instants $\omega t = 0, \pi$, and 2π. The derivative is therefore a maxi-

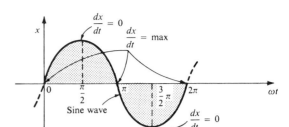

FIG. 13.52

mum at these points. For various values of ωt between these maxima and minima, the derivative will exist and have values from the minimum to the maximum inclusive. A plot of the derivative in Fig. 13.53 shows that the derivative of a sine wave is a cosine wave.

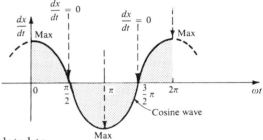

The peak value of the cosine wave is directly related to the frequency of the original waveform. The higher the frequency, the steeper the slope at the abscissa, and the greater the value of dx/dt, as shown in Fig. 13.54.

FIG. 13.53

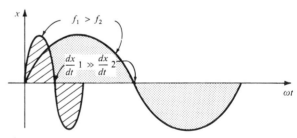

The derivative of a sine wave can be found directly by differentiation (calculus), resulting in the following:

FIG. 13.54

$$\frac{d}{dt}[E_m \sin(\omega t \pm \theta)] = \omega E_m \cos(\omega t \pm \theta) \tag{13.32}$$

By similar means, if

$$x(t) = e(t) = E_m \cos(\omega t \pm \theta)$$

then

$$\frac{d}{dt}[E_m \cos(\omega t \pm \theta)] = -\omega E_m \sin(\omega t \pm \theta) \tag{13.33}$$

For the sine wave in Fig. 13.52, we find $\theta = 0°$, and $x = X_m \sin \omega t$, so that

$$\frac{dx}{dt} = \omega X_m \cos \omega t$$

or

$$\frac{dx}{dt} = \overbrace{2\pi f X_m}^{\text{peak value}} \cos \omega t$$

Note the effect of the frequency on the peak value of the waveform. As mentioned earlier, the higher the frequency, the greater the peak value of the derivative.

13.11 RESPONSE OF BASIC R, L, AND C ELEMENTS TO SINUSOIDAL VOLTAGE OR CURRENT

Using Ohm's law and the basic equations for the capacitor and inductor, we can now apply the sinusoidal voltage or current to the basic R, L, and C elements.

Resistor

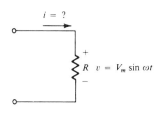

FIG. 13.55

For power-line frequencies and frequencies up to a few kilohertz, resistance is, for all practical purposes, unaffected by the frequency of the applied sinusoidal voltage or current. For this frequency region, the resistor R of Fig. 13.55 can be treated as a constant, and an application of Ohm's law will result in

$$i = \frac{v}{R} = \frac{V_m \sin \omega t}{R} = \frac{V_m}{R} \sin \omega t = I_m \sin \omega t$$

where

$$I_m = \frac{V_m}{R}$$

In addition, for a given i,

$$v = iR = (I_m \sin \omega t)R = I_m R \sin \omega t = V_m \sin \omega t$$

where

$$V_m = I_m R$$

A plot of v and i in Fig. 13.56 reveals that for a purely resistive element, the voltage across and the current through the element are *in phase*.

Inductor

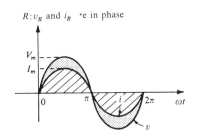

FIG. 13.56

For the series configuration of Fig. 13.57, the voltage appearing across the boxed-in element opposes the source of

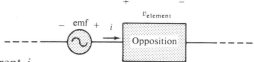

FIG. 13.57

emf and thereby reduces the magnitude of the current i. The magnitude of the voltage across the element is directly related to the opposition of the element to the flow of charge, or current i. For a resistive element, we have found that the opposition was its resistance and that $v_{element}$ and i were determined by $v_{element} = i \cdot R$.

For the inductor, we found in Chapter 11 that the counter-emf (or voltage appearing across an inductor) was directly related to the rate of change of current through the coil. Consequently, the higher the frequency, the greater will be the rate of change of current through the coil, and the greater the magnitude of the counter-emf. In addition, we found in the same chapter that the inductance of a coil will determine the rate of change of flux linking a coil for a particular change in current through the coil. The higher the inductance, the greater the rate of change of the flux linkages, and the greater the resulting voltage across the coil.

The counter-emf, therefore, is directly related to the frequency (or, more specifically, the angular velocity of the sinusoidal ac current through the coil) and the inductance of the coil. In Fig. 13.58, the counter-emf (cemf) is represented by v_L, and the current through the coil by i_L. For increased ω and inductance L, there will be an increased value of v_L. As shown in the figure, the product of these two determining quantities represents an opposition to the current i_L, since the greater the value of v_L in Fig. 13.57 for the same emf, the less will be the current i_L.

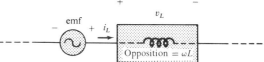

FIG. 13.58

We will now verify some of the preceding conclusions using a more mathematical approach and then define a few important quantities to be employed in the sections and chapters to follow.

For the inductor of Fig. 13.59, we know that

$$v_L = L\frac{di_L}{dt}$$

and, applying differentiation (calculus),

$$\frac{di_L}{dt} = \frac{d}{dt}(I_m \sin \omega t) = \omega I_m \cos \omega t$$

$$\therefore v_L = L\frac{di_L}{dt} = L(\omega I_m \cos \omega t) = \omega L I_m \cos \omega t$$

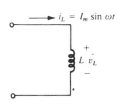

FIG. 13.59

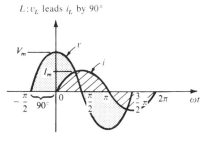

$L : v_L$ leads i_L by 90°

FIG. 13.60

or

$$v_L = V_m \sin(\omega t + 90°)$$

where

$$V_m = \omega L I_m$$

Note that the peak value of v_L is directly related to ω and L as predicted in the preceding discussion.

A plot of v_L and i_L in Fig. 13.60 reveals that v_L *leads* i_L by 90°, or i_L *lags* v_L by 90°.

For

$$i_L = I_m \sin(\omega t \pm \theta)$$
$$v_L = \omega L I_m \sin(\omega t \pm \theta + 90°)$$

The opposition to current developed by an inductor in a sinusoidal ac network can be found by applying Eq. (4.1):

$$\text{Effect} = \frac{\text{cause}}{\text{opposition}}$$

which for our purposes can be written

$$\text{Opposition} = \frac{\text{cause}}{\text{effect}}$$

Substituting values, we have

$$\text{Opposition} = \frac{V_m}{I_m} = \frac{\omega L I_m}{I_m} = \omega L$$

which agrees with the results obtained earlier.

The quantity ωL, called the *reactance* (from the word *reaction*) of an inductor, is symbolically represented by X_L and is measured in *ohms;* that is,

$$\boxed{X_L = \omega L} \qquad \text{(ohms)} \qquad \textbf{(13.34)}$$

Inductive reactance is the opposition to the flow of current, which results in the continual interchange of energy between the source and the magnetic field of the inductor. In other words, reactance, unlike resistance (which dissipates energy in the form of heat), does not dissipate electrical energy.

Capacitor

Let us now return to the series configuration of Fig. 13.58 and insert the capacitor as the element of interest. For the capacitor, however, we will determine i for a particular voltage across the element. When this approach reaches its conclusion, the relationship between the voltage and current will be known, and the opposing voltage (v_{element}) can be determined for any sinusoidal current i.

Our investigation of the inductor revealed that the cemf developed across a coil opposed the instantaneous change in current through the coil. For capacitive networks, the voltage across the capacitor is limited by the rate at which charge can be deposited on, or released by, the plates of the capacitor during the charging and discharging phases, respectively. In other words, an instantaneous change in voltage across a capacitor is opposed by the fact that there is an element of time required to deposit charge on (or release charge from) the plates of a capacitor, and $V = Q/C$.

Since capacitance is a measure of the rate at which a capacitor will store charge on its plates *for a particular change in voltage across the capacitor, the greater the value of capacitance, the greater will be the resulting capacitive current.* In addition, the fundamental equation relating the voltage across a capacitor to the current of a capacitor $[i = C(dv/dt)]$ indicates that for a particular capacitance, *the greater the rate of change of voltage across the capacitor, the greater the capacitive current. Certainly, an increase in frequency corresponds to an increase in the rate of change of voltage* across the capacitor and to an increase in the current of the capacitor.

The current of a capacitor is therefore directly related to the frequency (or, again more specifically, the angular velocity) and the capacitance of the capacitor. An increase in either quantity will result in an increase in the current of the capacitor. For the basic configuration of Fig. 13.61,

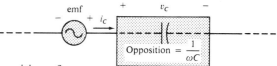

FIG. 13.61

however, we are interested in determining the opposition of the capacitor as related to the resistance of a resistor and ωL for the inductor. Since an increase in current corresponds to a decrease in opposition, and i_C is proportional to ω and C, the opposition of a capacitor is directly related to the reciprocal of ωC, or $1/\omega C$, as shown in Fig. 13.61. In other words, the higher the angular velocity (or frequency) and capacitance, the less the opposition to the current i_C or the lower the back emf of the capacitor (v_C), which limits the current i_C as indicated by Fig. 13.57.

We will now verify, as we did for the inductor, some of these conclusions using a more mathematical approach. Certain important quantities used repeatedly in the following analysis will then be defined.

For the capacitor of Fig. 13.62, we know that

$$i_C = C\frac{dv_C}{dt}$$

and that

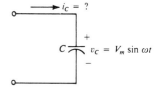

FIG. 13.62

$$\frac{dv_C}{dt} = \frac{d}{dt}(V_m \sin \omega t) = \omega V_m \cos \omega t$$

$$\therefore i_C = C\frac{dv_C}{dt} = C(\omega V_m \cos \omega t) = \omega C V_m \cos \omega t$$

or

$$i_C = I_m \sin(\omega t + 90°)$$

where

$$I_m = \omega C V_m$$

Note that the peak value of i_C is directly related to ω and C, as predicted in the discussion above.

A plot of v_C and i_C in Fig. 13.63 reveals that i_C *leads* v_C by 90°, or v_C *lags* i_C by 90°.

For

$$v_C = V_m \sin(\omega t \pm \theta)$$
$$i_C = \omega C V_m \sin(\omega t \pm \theta + 90°)$$

Applying

$$\text{Opposition} = \frac{\text{cause}}{\text{effect}}$$

and substituting values, we have

$$\text{Opposition} = \frac{V_m}{I_m} = \frac{V_m}{\omega C V_m} = \frac{1}{\omega C}$$

which agrees with the results obtained above.

The quantity $1/\omega C$, called the *reactance* of a capacitor, is symbolically represented by X_C and is measured in ohms; that is,

$$\boxed{X_C = \frac{1}{\omega C}} \qquad \text{(ohms)} \qquad \textbf{(13.35)}$$

Capacitive reactance is the opposition to the flow of charge, which results in the continual interchange of energy between source and the electric field of the capacitor. Like the inductor, the capacitor does *not* dissipate energy in any form (ignoring the effects of the leakage resistance).

In the circuits just considered, the current was given in the inductive circuit, and the voltage in the capacitive circuit. This was done to avoid the use of integration in finding the unknown quantities.

In the inductive circuit,

$$v_L = L\frac{di_L}{dt}$$

but

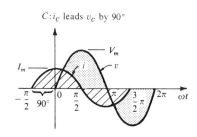

$C: i_C$ leads v_C by 90°

FIG. 13.63

$$i_L = \frac{1}{L} \int v_L \, dt \qquad \textbf{(13.36)}$$

In the capacitive circuit,

$$i_C = C \frac{dv_C}{dt}$$

but

$$v_C = \frac{1}{C} \int i_C \, dt \qquad \textbf{(13.37)}$$

Shortly, we shall consider a method of analyzing ac circuits that will permit us to solve for an unknown quantity with sinusoidal input without having to use direct integration or differentiation.*

It is possible to determine whether a circuit with one or more elements is predominantly capacitive or inductive by noting the phase relationship between the input voltage and current. *If the current leads the voltage, the circuit is predominantly capacitive, and if the voltage leads the current, it is predominantly inductive.*

Since we now have an equation for the reactance of an inductor or capacitor, we do not need to use derivatives or integration in the examples to be considered. Simply applying Ohm's law, $I_m = E_m/X_L$ (or X_C), and keeping in mind the phase relationship between the voltage and current for each element, will be sufficient to complete the examples.

EXAMPLE 13.14. The voltage across a resistor is indicated. Find the sinusoidal expression for the current if the resistor is 10 Ω. Sketch the v and i curves with the angle ωt as the abscissa.
a. $v = 100 \sin 377t$
b. $v = 25 \sin(377t + 60°)$

Solutions:

a. $i = \dfrac{v}{R} = \dfrac{100}{10} \sin 377t$

and

$i = \textbf{10 sin 377}t$

The curves are sketched in Fig. 13.64.

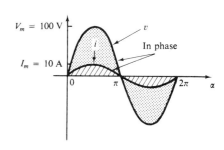

FIG. 13.64

---

*A mnemonic phrase sometimes used to remember the phase relationship between the voltage and current of a coil and capacitor is "*ELI* the *ICE* man." Note that the *L* (inductor) has the *E* before the *I* (*e* leads *i* by 90°), and the *C* (capacitor) has the *I* before the *E* (*i* leads *e* by 90°).

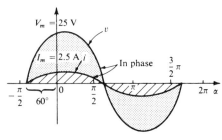

FIG. 13.65

b. $i = \dfrac{v}{R} = \dfrac{25}{10} \sin(377t + 60°)$

and

$$i = \mathbf{2.5 \sin(377t + 60°)}$$

The curves are sketched in Fig. 13.65.

EXAMPLE 13.15. The current through a 5-Ω resistor is given. Find the sinusoidal expression for the voltage across the resistor for $i = 40 \sin(377t + 30°)$.

Solution:

$$v = iR = (40)(5) \sin(377t + 30°)$$

and

$$v = \mathbf{200 \sin(377t + 30°)}$$

EXAMPLE 13.16. The current through a 0.1-H coil is given. Find the sinusoidal expression for the voltage across the coil. Sketch the v and i curves.
a. $i = 10 \sin 377t$
b. $i = 7 \sin(377t - 70°)$

Solutions:

a. $X_L = \omega L = 37.7 \, \Omega$
$V_m = I_m X_L = (10)(37.7) = 377 \, \text{V}$
and we know that for a coil v leads i by 90°. Therefore,
$$v = \mathbf{377 \sin(377t + 90°)}$$
The curves are sketched in Fig. 13.66.

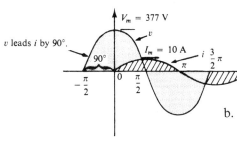

FIG. 13.66

b. $X_L = 37.7 \, \Omega$
$V_m = I_m X_L = (7)(37.7) = 263.9 \, \text{V}$
and we know that for a coil v leads i by 90°. Therefore,
$v = 263.9 \sin(377t - 70° + 90°)$
and
$$v = \mathbf{263.9 \sin(377t + 20°)}$$
The curves are sketched in Fig. 13.67.

EXAMPLE 13.17. The voltage across a 0.5-H coil is given. What is the sinusoidal expression for the current?

$$v = 100 \sin 20t$$

Solution:

$$X_L = \omega L = (20)(0.5) = 10 \, \Omega$$

$$I_m = \dfrac{V_m}{X_L} = \dfrac{100}{10} = 10 \, \text{A}$$

FIG. 13.67

and we know that i lags v by 90°. Therefore,

$$i = 10 \sin(20t - 90°)$$

EXAMPLE 13.18. The voltage across a 1-μF capacitor is given. What is the sinusoidal expression for the current? Sketch the v and i curves.

$$v = 30 \sin 400t$$

Solution:

$$X_C = \frac{1}{\omega C} = \frac{1}{(400)(1 \times 10^{-6})} = \frac{10^6}{400} = 2500 \ \Omega$$

$$I_m = \frac{V_m}{X_C} = \frac{30}{2500} = 0.0120 \ \text{A} = 12.0 \ \text{mA}$$

and we know that for a capacitor i leads v by 90°. Therefore,

$$i = 12 \times 10^{-3} \sin(400t + 90°)$$

The curves are sketched in Fig. 13.68.

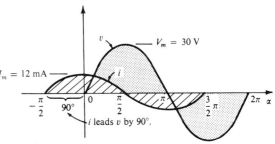

EXAMPLE 13.19. The current through a 100-μF capacitor is given. Find the sinusoidal expression for the voltage across the capacitor.

FIG. 13.68

$$i = 40 \sin(500t + 60°)$$

Solution:

$$X_C = \frac{1}{\omega C} = \frac{1}{(500)(100 \times 10^{-6})} = \frac{10^6}{5 \times 10^4} = \frac{10^2}{5} = 20 \ \Omega$$

$$V_m = I_m X_C = (40)(20) = 800 \ \text{V}$$

and we know for a capacitor that v lags i by 90°. Therefore,

$$v = 800 \sin(500t + 60° - 90°)$$

and

$$v = 800 \sin(500t - 30°)$$

EXAMPLE 13.20. For the following pairs of voltages and currents, indicate whether the element involved is a capacitor, inductor, or resistor, and determine the value of C, L, or R, if sufficient data are given (Fig. 13.69).

a. $v = 100 \sin(\omega t + 40°)$
   $i = 20 \sin(\omega t + 40°)$

b. $v = 1000 \sin(377t + 10°)$
   $i = 5 \sin(377t - 80°)$

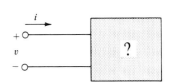

FIG. 13.69

c. $v = 500 \sin(157t + 30°)$
   $i = 1 \sin(157t + 120°)$

d. $v = 50 \cos(\omega t + 20°)$
   $i = 5 \sin(\omega t + 110°)$

Solutions:

a. Since v and i are *in phase*, the element is a *resistor*.

$$R = \frac{V_m}{I_m} = \frac{100}{20} = \mathbf{5\ \Omega}$$

b. Since v leads i by 90°, the element is an *inductor*.

$$X_L = \frac{V_m}{I_m} = \frac{1000}{5} = 200\ \Omega$$

so that

$$X_L = \omega L = 200\ \Omega \quad \text{or} \quad L = \frac{200}{\omega} = \frac{200}{377}$$

and

$$L = \mathbf{0.531\ H}$$

c. Since i *leads* v by 90°, the element is a *capacitor*.

$$X_C = \frac{V_m}{I_m} = \frac{500}{1} = 500\ \Omega$$

so that

$$X_C = \frac{1}{\omega C} = 500\ \Omega \quad \text{or} \quad C = \frac{1}{\omega 500} = \frac{1}{(157)(500)}$$

and

$$C = \mathbf{12.74\ \mu F}$$

d. $v = 50 \cos(\omega t + 20°) = 50 \sin(\omega t + 20° + 90°)$
   $= 50 \sin(\omega t + 110°)$

Since v and i are *in phase*, the element is a *resistor*.

$$R = \frac{V_m}{I_m} = \frac{50}{5} = \mathbf{10\ \Omega}$$

For dc circuits, the frequency is zero, since the currents and voltages have constant magnitudes. The reactance of the coil for dc is therefore

$$X_L = 2\pi f L = 2\pi 0 L = 0\ \Omega$$

hence the short-circuit representation for the inductor in dc circuits (Chapter 11). At very high frequencies, $X_L \uparrow = 2\pi f \uparrow L$ is very large, and for some practical applications, the inductor can be replaced by an open circuit.

The capacitor can be replaced by an open circuit in dc circuits since $f = 0$, and

$$X_C = \frac{1}{2\pi f C} = \frac{1}{2\pi 0 C} = \infty \ \Omega$$

once again substantiating our previous action (Chapter 9).
At very high frequencies, for finite capacitances,

$$X_C\!\downarrow = \frac{1}{2\pi f\!\uparrow C}$$

is very small, and for some practical applications, the capacitor can be replaced by a short circuit.

13.12 AVERAGE POWER AND POWER FACTOR

The instantaneous power to the load of Fig. 13.70 is

$$p = vi$$

If we consider the general case where

$$v = V_m \sin(\omega t + \beta) \quad \text{and} \quad i = I_m \sin(\omega t + \psi)$$

then

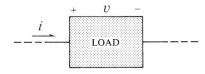

FIG. 13.70

$$\begin{aligned} p = vi &= V_m \sin(\omega t + \beta) I_m \sin(\omega t + \psi) \\ &= V_m I_m \sin(\omega t + \beta) \sin(\omega t + \psi) \end{aligned}$$

Using the trigonometric identity, we have

$$\sin A \sin B = \frac{\cos(A - B) - \cos(A + B)}{2}$$

$$\sin(\omega t + \beta) \sin(\omega t + \psi)$$

$$= \frac{\cos[(\omega t + \beta) - (\omega t + \psi)] - \cos[(\omega t + \beta) + (\omega t + \psi)]}{2}$$

$$= \frac{\cos(\beta - \psi) - \cos(2\omega t + \beta + \psi)}{2}$$

so that

$$p = \left[\frac{V_m I_m}{2} \cos(\beta - \psi) \right] - \left[\frac{V_m I_m}{2} \cos(2\omega t + \beta + \psi) \right]$$

The plots of the current, voltage, and power on the same set of axes are shown in Fig. 13.71.

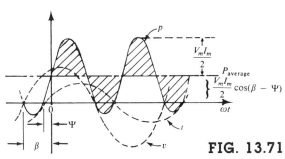

FIG. 13.71

Note that the second factor in the preceding equation is a cosine wave with an amplitude of $V_m I_m/2$, and a frequency twice that of the voltage or current. The average value of this term is zero, producing no net transfer of energy in any one direction.

The first term in the preceding equation, however, has a constant magnitude (no time dependence) and therefore provides some net transfer of energy. This term is referred to as the *average power,* the reason for which is obvious from Fig. 13.71. The average power, or *real* power as it is sometimes called, is the power delivered to and dissipated by the load. It corresponds to the power calculations performed for dc networks. The angle $(\beta - \psi)$ is the phase angle between v and i. Since $\cos(-\alpha) = \cos \alpha$, *the magnitude of average power is independent of whether v leads i or i leads v.* Defining θ as equal to $|\beta - \psi|$, where $|-|$ indicates that only the magnitude is important and the sign is immaterial, we have

$$\boxed{P = \frac{V_m I_m}{2} \cos \theta} \qquad \text{(watts, W)} \qquad \textbf{(13.38)}$$

where P is the average power in watts. This equation can also be written

$$P = \frac{V_m}{\sqrt{2}} \frac{I_m}{\sqrt{2}} \cos \theta$$

Or, since,

$$V_{\text{eff}} = \frac{V_m}{\sqrt{2}} \quad \text{and} \quad I_{\text{eff}} = \frac{I_m}{\sqrt{2}}$$

Eq. (13.38) becomes

$$\boxed{P = V_{\text{eff}} I_{\text{eff}} \cos \theta} \qquad \textbf{(13.39)}$$

Let us now apply Eqs. (13.38) and (13.39) to the basic R, L, C elements.

Resistor

In a purely resistive circuit, since v and i are in phase, $|\beta - \psi| = \theta = 0°$, and $\cos \theta = \cos 0° = 1$, so that

$$\boxed{P = \frac{V_m I_m}{2} = V_{\text{eff}} I_{\text{eff}}} \qquad \text{(W)} \qquad \textbf{(13.40)}$$

Or, since

$$I_{\text{eff}} = \frac{V_{\text{eff}}}{R}$$

$$\boxed{P = \frac{V_{\text{eff}}^2}{R} = I_{\text{eff}}^2 R} \quad \text{(W)} \quad \textbf{(13.41)}$$

Inductor

In a purely inductive circuit, since v leads i by $90°$, $|\beta - \psi| = \theta = 90°$. Therefore,

$$P = \frac{V_m I_m}{2} \cos 90° = \frac{V_m I_m}{2}(0) = \mathbf{0}$$

The average power or power dissipated by the ideal inductor (no associated resistance) is always zero.

Capacitor

In a purely capacitive circuit, since i leads v by $90°$, $|\beta - \psi| = \theta = |-90°| = 90°$. Therefore,

$$P = \frac{V_m I_m}{2} \cos(90°) = \frac{V_m I_m}{2}(0) = \mathbf{0}$$

The average power or power dissipated by the ideal capacitor (no associated resistance) is always zero.

EXAMPLE 13.21. Find the average power dissipated in a circuit whose input current and voltage are the following:

$$i = 5 \sin(\omega t + 40°)$$
$$v = 10 \sin(\omega t + 40°)$$

Solution: Since v and i are in phase, the circuit appears at the input terminals to be purely resistive. Therefore,

$$P = \frac{V_m I_m}{2} = \frac{(10)(5)}{2} = \mathbf{25 \ W}$$

or

$$R = \frac{V_m}{I_m} = \frac{10}{5} = 2 \ \Omega$$

and

$$P = \frac{V_{\text{eff}}^2}{R} = \frac{[(0.707)(10)]^2}{2} = \mathbf{25 \ W}$$

or

$$P = I_{\text{eff}}^2 R = [(0.707)(5)]^2(2) = \mathbf{25 \ W}$$

For the following examples, the circuit consists of a combination of resistances and reactances producing phase angles between the input current and voltage different from $0°$ or $90°$.

EXAMPLE 13.22. Determine the average power delivered to networks having the following input voltage and current:

a. $v = 100 \sin(\omega t + 40°)$
   $i = 20 \sin(\omega t + 70°)$

b. $v = 150 \sin(\omega t - 70°)$
   $i = 3 \sin(\omega t - 50°)$

Solutions:

a. $V_m = 100, \beta = 40°$
   $I_m = 20, \psi = 70°$
   $\theta = |\beta - \psi| = |40° - 70°| = |-30°| = 30°$

   and

   $$P = \frac{V_m I_m}{2} \cos \theta = \frac{(100)(20)}{2} \cos(30°) = 1000(0.866)$$
   $$= \textbf{866 W}$$

b. $V_m = 150, \beta = -70°$
   $I_m = 3, \psi = -50°$
   $\theta = |\beta - \psi| = |-70° - (-50°)|$
   $= |-70° + 50°| = |-20°| = 20°$

   and

   $$P = \frac{V_m I_m}{2} \cos \theta = \frac{(150)(3)}{2} \cos(20°) = (225)(0.9397)$$
   $$= \textbf{211.43 W}$$

Power Factor

The frequency and the elements of the parallel network of Fig. 13.72(a) were chosen such that

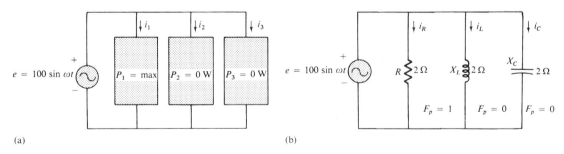

(a)

(b)

FIG. 13.72

$i_1 = 50 \sin \omega t$
$i_2 = 50 \sin(\omega t + 90°)$
$i_3 = 50 \sin(\omega t - 90°)$

Therefore, for each parallel branch, the peak value (or effective value) of the voltage and current are the same. Yet, as indicated in the figure, the power to two of the branches is zero and a maximum to the third. In our power equation, the only factor that accounts for this variation is $\cos\theta$, related to the phase relationship between v and i. This factor, $\cos\theta$, is called the *power factor* and is symbolically represented by F_P; that is,

$$\boxed{\text{Power factor} = F_p = \cos\theta} \qquad \textbf{(13.42)}$$

The more reactive a load, the lower the power factor, and the smaller the average power delivered. The more resistive the load, the higher the power factor, and the greater the real power delivered. The elements within each "black box" of Fig. 13.72(a) are indicated in Fig. 13.72(b). Note the power factor for each element in the figure. Low power factors are usually avoided, since a high current would be required to deliver any appreciable power. This higher current demand produces higher heating losses, and, consequently, the system operates at a lower efficiency.

In terms of the average power and the terminal voltage and current,

$$\boxed{F_p = \cos\theta = \frac{P}{V_{\text{eff}}I_{\text{eff}}}} \qquad \textbf{(13.43)}$$

The terms *leading* and *lagging* are often written in conjunction with the power factor. *They are defined by the current through the load.* If the current leads the voltage across a load, the load has a leading power factor. If the current lags the voltage across the load, the load has a lagging power factor. In other words, *capacitive networks have leading power factors, and inductive networks have lagging power factors.*

EXAMPLE 13.23. Determine the power factors of the following loads, and indicate whether they are leading or lagging:

a. Fig. 13.73

b. Fig. 13.74

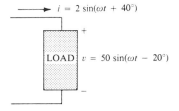

FIG. 13.73

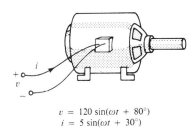

$v = 120\sin(\omega t + 80°)$
$i = 5\sin(\omega t + 30°)$

FIG. 13.74

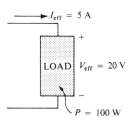

FIG. 13.75

c. Fig. 13.75

Solutions:

a. $F_p = \cos\theta = \cos 60° = \mathbf{0.5\ leading}$

b. $F_p = \cos\theta = \cos 50° = \mathbf{0.6428\ lagging}$

c. $F_p = \cos\theta = \dfrac{P}{V_{eff}I_{eff}} = \dfrac{100}{(20)(5)} = \dfrac{100}{100} = \mathbf{1}$

The load is resistive, and F_p is neither leading nor lagging.

PROBLEMS

Section 13.4

1. For the cycles of the periodic waveform shown in Fig. 13.76:
   a. Find the period T.
   b. How many cycles are shown?
   c. What is the frequency?
   d. What is the maximum amplitude?

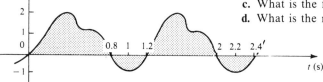

FIG. 13.76

2. Repeat Problem 1 for the periodic waveform of Fig. 13.77.

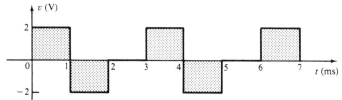

FIG. 13.77

3. Find the period of a periodic waveform whose frequency is
   a. 25 Hz   b. 35 MHz   c. 55 kHz   d. 1 Hz

4. Find the frequency of a repeating waveform whose period is
   a. 1/60 s   b. 0.01 s   c. 34 ms   d. 25 μs

5. Find the period of a sinusoidal waveform that completes 80 cycles in 24 ms.

6. If a periodic waveform has a frequency of 20 Hz, how long (in seconds) will it take to complete 5 cycles?

7. What is the frequency of a periodic waveform that completes 42 cycles in 6 s?

Section 13.5

8. Convert the following degrees to radians:
   a. 45°   b. 60°   c. 150°
   d. 270°   e. 178°   f. 221°

9. Convert the following radians to degrees:
   a. $\pi/4$
   b. $\pi/6$
   c. $\frac{5}{6}\pi$
   d. $\frac{7}{6}\pi$
   e. $\frac{4}{3}\pi$
   f. 0.55π

10. Find the angular velocity of a waveform with a period of
    a. 2 s
    b. 0.3 ms
    c. 0.5 s
    d. 1/25 s

11. Find the angular velocity of a waveform with a frequency of
    a. 50 Hz
    b. 600 Hz
    c. 0.1 Hz
    d. 0.004 MHz

12. Find the frequency and period of sine waves having an angular velocity of
    a. 754 rad/s
    b. 8.4 rad/s
    c. 6000 rad/s
    d. 1/16 rad/s

Section 13.6

13. Find the amplitude and frequency of the following waves:
    a. $20 \sin 377t$
    b. $5 \sin 754t$
    c. $10^6 \sin 10,000t$
    d. $0.001 \sin 942t$
    e. $-7.6 \sin 43.6t$
    f. $1/42 \sin 6.28t$

14. Sketch $5 \sin 754t$ with the abscissa
    a. angle in degrees
    b. angle in radians
    c. time in seconds

15. Sketch $10^6 \sin 10,000t$ with the abscissa
    a. angle in degrees
    b. angle in radians
    c. time in seconds

16. Sketch $-7.6 \sin 43.9t$ with the abscissa
    a. angle in degrees
    b. angle in radians
    c. time in seconds

17. If $e = 300 \sin 157t$, how long (in seconds) does it take this waveform to complete 1/2 cycle?

Section 13.7

18. Sketch $\sin(377t + 60°)$ with the abscissa
    a. angle in degrees
    b. angle in radians
    c. time in seconds

19. Sketch the following waveforms:
    a. $50 \sin(\omega t + 0°)$
    b. $-20 \sin(\omega t + 2°)$
    c. $5 \sin(\omega t + 60°)$
    d. $4 \cos \omega t$
    e. $2 \cos(\omega t + 10°)$
    f. $-5 \cos(\omega t + 20°)$

20. Find the phase relationship between the waveforms of each set:
    a. $v = 4 \sin(\omega t + 50°)$
       $i = 6 \sin(\omega t + 40°)$
    b. $v = 25 \sin(\omega t - 80°)$
       $i = 10 \sin(\omega t - 4°)$
    c. $v = 0.2 \sin(\omega t - 65°)$
       $i = 0.1 \sin(\omega t + 25°)$
    d. $v = 200 \sin(\omega t - 210°)$
       $i = 25 \sin(\omega t - 60°)$

*21. Repeat Problem 20 for the following sets:
    a. $v = 2 \cos(\omega t - 30°)$
       $i = 5 \sin(\omega t + 60°)$
    b. $v = -1 \sin(\omega t + 20°)$
       $i = 10 \sin(\omega t - 70°)$

c. $v = -4\cos(\omega t + 90°)$
       $i = -2\sin(\omega t + 10°)$

22. Write the analytical expression for the waveforms of Fig. 13.78 with the phase angle in degrees.

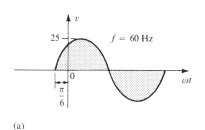

(a)

FIG. 13.78

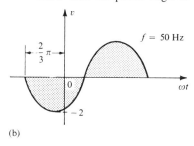

(b)

23. Repeat Problem 22 for the waveforms of Fig. 13.79.

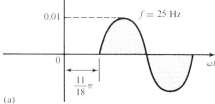

(a)

FIG. 13.79

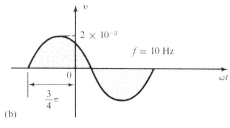

(b)

Section 13.8

24. Find the average value of the periodic waveforms of Fig. 13.80 over one full cycle.

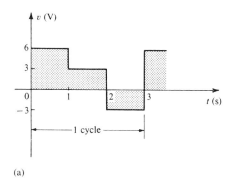

(a)

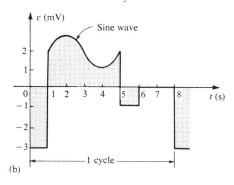

(b)

FIG. 13.80

25. Repeat Problem 24 for the waveforms of Fig. 13.81.

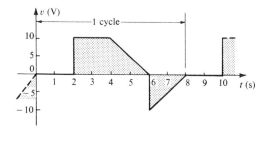

(a)

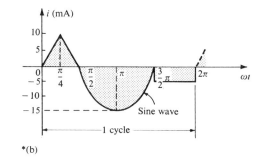

*(b)

FIG. 13.81

26. **a.** By the method of approximation, using familiear geometric shapes, find the area under the curve of Fig. 13.82 from 0 to 10 s. Compare your solution with the actual area of 5 volt-seconds (V · s).

    b. Find the average value of the waveform from 0 to 10 s.

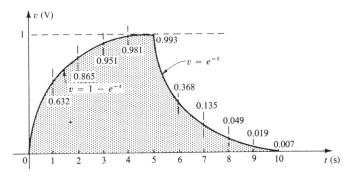

FIG. 13.82

Section 13.9

27. Find the effective values of the following sinusoidal waveforms:

    a. $v = 20 \sin 754t$

    b. $v = 7.07 \sin 377t$

    c. $i = 0.006 \sin(400t + 20°)$

    d. $i = 1.76 \cos(377t - 10°)$

28. Write the sinusoidal expressions for voltages and currents having the following effective values at a frequency of 60 Hz with zero phase shift:

    a. 1.414 V   **b.** 70.7 V   **c.** 0.06 A   **d.** 24 μA

29. Find the effective value of the periodic waveform of Fig. 13.83 over one full cycle.

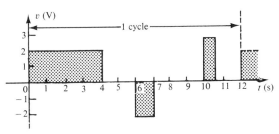

FIG. 13.83

30. Repeat Problem 29 for the waveform of Fig. 13.84.

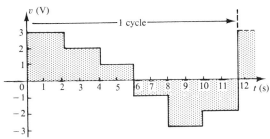

FIG. 13.84

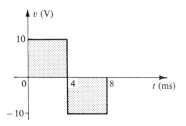

FIG. 13.85

31. What are the average and effective values of the square wave of Fig. 13.85?

Section 13.10

32. What is the derivative of each of the following sinusoidal expressions?
  a. 10 sin 377t
  b. 0.6 cos 754t
  c. 0.05 cos(157t − 10°)
  d. 25 cos(20t − 150°)

Section 13.11

33. The voltage across a 5-Ω resistor is as indicated. Find the sinusoidal expression for the current. In addition, sketch the v and i curves with the abscissa in radians.
  a. 150 sin 377t
  b. 30 sin(377t + 20°)
  c. 40 cos(ωt + 10°)
  d. −80 sin(ωt + 40°)

34. The current through a 7-kΩ resistor is as indicated. Find the sinusoidal expression for the voltage. In addition, sketch the v and i curves with the abscissa in radians.
  a. 0.03 sin 754t
  b. 12 × 10^{-3} sin(400t − 120°)
  c. 6 × 10^{-6} cos(ωt − 2°)
  d. −0.004 cos(ωt − 90°)

35. Determine the inductive reactance (in ohms) of a 2-H coil for
  a. dc
  and for the following frequencies:
  b. 25 Hz   **c.** 60 Hz   **d.** 2000 Hz   **e.** 100,000 Hz

36. Determine the inductance of a coil that has a reactance of
  a. 20 Ω at f = 2 Hz
  b. 1000 Ω at f = 60 Hz
  c. 5280 Ω at f = 1000 Hz

37. Determine the frequency at which a 10-H inductance has the following inductive reactances:
  a. 50 Ω
  b. 3770 Ω
  c. 15.7 kΩ
  d. 243 Ω

38. The current through a 20-Ω inductive reactance is given. What is the sinusoidal expression for the voltage? Sketch the v and i curves with the abscissa in radians.
  a. i = 5 sin ωt
  b. i = 0.4 sin(ωt + 60°)
  c. i = −6 sin(ωt − 30°)
  d. i = 3 cos(ωt + 10°)

39. The current through an 0.1-H coil is given. What is the sinusoidal expression for the voltage?
  a. 30 sin 30t
  b. 0.006 sin 377t
  c. 5 × 10^{-6} sin(400t + 20°)
  d. −4 cos(20t − 70°)

40. The voltage across a 50-Ω inductive reactance is given. What is the sinusoidal expression for the current? Sketch the v and i curves with the abscissa in radians.
  a. 50 sin ωt
  b. 30 sin(ωt + 20°)
  c. 40 cos(ωt + 10°)
  d. −80 sin(377t + 40°)

41. The voltage across an 0.2-H coil is given. What is the sinusoidal expression for the current?
  a. 1.5 sin 60t
  b. 0.016 sin(t + 4°)
  c. −4.8 sin(0.05t + 50°)
  d. 9 × 10^{-3} cos(377t + 360°)

42. Determine the capacitive reactance (in ohms) of a 5-μF capacitor for
  a. dc
  and for the following frequencies:
  b. 60 Hz   **c.** 120 Hz   **d.** 1800 Hz   **e.** 24,000 Hz

43. Determine the capacitance in microfarads if a capacitor has a reactance of
  a. 250 Ω at $f = 60$ Hz   **b.** 55 Ω at $f = 312$ Hz
  c. 10 Ω at $f = 25$ Hz

44. Determine the frequency at which a 50-μF capacitor has the following capacitive reactance:
  a. 342 Ω **b.** 684 Ω
  c. 171 Ω **d.** 2000 Ω

45. The voltage across a 2.5-Ω capacitive reactance is given. What is the sinusoidal expression for the current? Sketch the v and i curves with the abscissa in radians.
  a. $100 \sin \omega t$ **b.** $0.4 \sin(\omega t + 20°)$
  c. $8 \cos(\omega t + 10°)$ **d.** $-70 \sin(\omega t + 40°)$

46. The voltage across a 1-μF capacitor is given. What is the sinusoidal expression for the current?
  a. $30 \sin 200t$ **b.** $90 \sin 377t$
  c. $-120 \sin(374t + 30°)$ **d.** $70 \cos(800t - 20°)$

47. The current through a 10-Ω capacitive reactance is given. Write the sinusoidal expression for the voltage. Sketch the v and i curves with the abscissa in radians.
  a. $i = 50 \sin \omega t$ **b.** $i = 40 \sin(\omega t + 60°)$
  c. $i = -6 \sin(\omega t - 30°)$ **d.** $i = 3 \cos(\omega t + 10°)$

48. The current through a 0.5-μF capacitor is given. What is the sinusoidal expression for the voltage?
  a. $0.20 \sin 300t$ **b.** $0.007 \sin 377t$
  c. $0.048 \cos 754t$ **d.** $0.08 \sin(1600t - 80°)$

*49. For the following pairs of voltages and currents, indicate whether the element involved is a capacitor, inductor, or resistor, and the value of C, L, or R if sufficient data are given:
  a. $v = 550 \sin(377t + 40°)$ **b.** $v = 36 \sin(754t + 80°)$
      $i = 11 \sin(377t - 50°)$ $i = 4 \sin(754t + 170°)$
  c. $v = 10.5 \sin(\omega t + 13°)$
      $i = 1.5 \sin(\omega t + 13°)$

*50. Repeat Problem 49 for the following pairs of voltages and currents:
  a. $v = 2000 \sin \omega t$ **b.** $v = 80 \sin(157t + 150°)$
      $i = 5 \cos \omega t$ $i = 2 \sin(157t + 60°)$
  c. $v = 35 \sin(\omega t - 20°)$
      $i = 7 \cos(\omega t - 110°)$

Section 13.12

51. Find the average power loss in watts for each set of Problem 49.

52. Repeat Problem 51 for Problem 50.

*53. Find the average power loss and power factor for each of the circuits whose input current and voltage are as follows:

a. $v = 60 \sin(\omega t + 30°)$
   $i = 15 \sin(\omega t + 60°)$

b. $v = -50 \sin(\omega t - 20°)$
   $i = -2 \sin(\omega t + 40°)$

c. $v = 50 \sin(\omega t + 80°)$
   $i = 3 \cos(\omega t + 20°)$

d. $v = 75 \sin(\omega t - 5°)$
   $i = 0.08 \sin(\omega t - 35°)$

54. If the current through and voltage across an element are $i = 8 \sin(\omega t + 40°)$ and $v = 48 \sin(\omega t + 40°)$, compute the power by I^2R, $(V_m I_m/2)\cos\theta$, and $VI\cos\theta$, and compare answers.

55. A circuit dissipates 100 W (average power) at 150 V (effective input voltage) and 2 A (effective input current). What is the power factor? Repeat if the power is 0 W; 300 W.

*56. The power factor of a circuit is 0.5 lagging. The power delivered in watts is 500. If the input voltage is $50 \sin(\omega t + 10°)$, find the sinusoidal expression for the input current.

57. In Fig. 13.86, $e = 30 \sin(377t + 20°)$.
   a. What is the sinusoidal expression for the current?
   b. Find the power loss in the circuit.
   c. How long (in seconds) does it take the current to complete 6 cycles?

58. In Fig. 13.87, $e = 100 \sin(157t + 30°)$.
   a. Find the sinusoidal expression for i.
   b. Find the value of the inductance L.
   c. Find the average power loss by the inductor.

59. In Fig. 13.88, $i = 3 \sin(377t - 20°)$.
   a. Find the sinusoidal expression for e.
   b. Find the value of the capacitance C in μF.
   c. Find the average power loss in the capacitor.

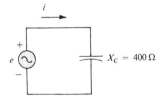

FIG. 13.86

FIG. 13.87

FIG. 13.88

GLOSSARY

Alternating waveform A waveform that oscillates above and below a defined reference level.

Angular velocity The velocity with which a radius vector projecting a sinusoidal function rotates about its center.

Average or real power The power delivered to and dissipated by the load over a full cycle.

Average value The level of a waveform defined by the condition that the area enclosed by the curve above this level is exactly equal to the area enclosed by the curve below this level.

Cycle A portion of a waveform contained in one period of time.

Derivative The instantaneous rate of change of a function with respect to time or another variable.

Effective value The equivalent dc value of any alternating voltage or current.

Frequency (f) The number of cycles of a periodic waveform that occur in one second.

Instantaneous value The magnitude of a waveform at any instant of time, denoted by lower-case letters.

Leading and lagging power factors An indication of whether a network is primarily capacitive or inductive in nature. Leading power factors are associated with capacitive networks, and lagging power factors with inductive networks.

Peak-to-peak value The magnitude of the total swing of a signal from positive to negative peaks. The sum of the absolute values of the positive and negative peak values.

Peak value The maximum value of a waveform, denoted by upper-case letters.

Period (T) The time interval between successive repetitions of a periodic waveform.

Periodic waveform A waveform that continually repeats itself after a defined time interval.

Phase relationship An indication of which of two waveforms leads or lags the other, and by how many degrees or radians.

Power factor (F_p) An indication of how reactive or resistive an electrical system is: The higher the power factor, the greater the resistive component.

Radian A unit of measure used to define a particular segment of a circle. One radian is approximately equal to 57.3°; 2π radians are equal to 360°.

Reactance The opposition of an inductor or capacitor to the flow of charge that results in the continual exchange of energy between the circuit and magnetic field of an inductor or the electric field of a capacitor.

RMS value The root-mean-square or effective value of a waveform.

Sinusoidal ac waveform An alternating waveform of unique characteristics that oscillates with equal amplitude above and below a given axis.

Waveform The path traced by a quantity, plotted as a function of some variable such as position, time, degrees, temperature, and so on.

14.1 INTRODUCTION

In our analysis of dc networks we found it necessary to find the algebraic sum of voltages and currents. Since the same will also be true for ac networks, the question arises: How do we determine the algebraic sum of two or more voltages (or currents) that are varying sinusoidally? Although one solution would be to find the algebraic sum on a point-to-point basis (as shown in Section 14.7), this would be a long and tedious process in which accuracy would be directly related to the scale employed.

It is the purpose of this chapter to introduce a system of *complex numbers* which, when related to the sinusoidal ac waveform, will result in a technique for finding the algebraic sum of sinusoidal waveforms that is quick, direct, and accurate. In the following chapters the technique will be extended to permit the analysis of sinusoidal ac networks in a manner very similar to that applied to dc networks. The methods and theorems as described for dc networks can then be applied to sinusoidal ac networks with little difficulty.

A *complex number* represents a point in a two-dimensional plane located with reference to two distinct axes. This point can also determine a radius vector drawn from the origin to the point. The horizontal axis is called the *real* axis, while the vertical axis is called the *imaginary* axis.

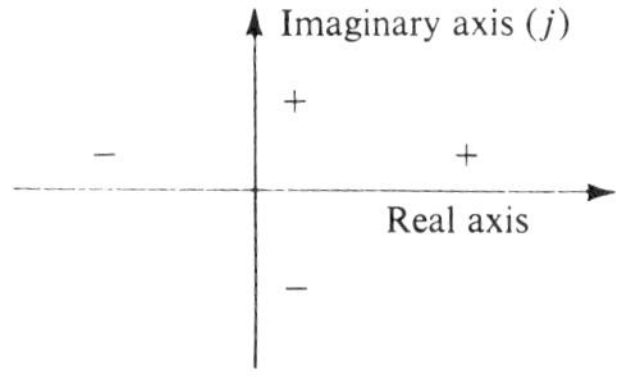

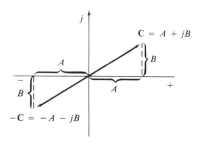

FIG. 14.1

Both are labeled in Fig. 14.1. For reasons that will be obvious later, the real axis is sometimes called the *resistance axis,* and the imaginary axis, the *reactance axis.* Every number from 0 to $\pm\infty$ can be represented by some point along the real axis. Prior to the development of this system of complex numbers, it was believed that any number not on the real axis would not exist—hence the term *imaginary* for the vertical axis.

In the complex plane, the horizontal or real axis represents all positive numbers to the right of the imaginary axis and all negative numbers to the left of the imaginary axis. All positive imaginary numbers are represented above the real axis, and all negative imaginary numbers, below the real axis. The symbol j (or sometimes i) is used to denote an imaginary number.

There are two forms used to represent a complex number: the *rectangular* and the *polar.* Each can represent a point in the plane or a radius vector drawn from the origin to that point.

14.2 RECTANGULAR FORM

The format for the rectangular form is

$$C = \pm A \pm jB \qquad (14.1)$$

The effect of a negative sign is shown in Fig. 14.2.

EXAMPLE 14.1. Sketch the following complex numbers in the complex plane:
a. $C = 3 + j4$
b. $C = 0 - j6$
c. $C = -10 - j20$

Solutions:
a. See Fig. 14.3.

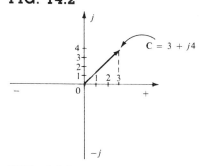

FIG. 14.2

b. See Fig. 14.4.

FIG. 14.3

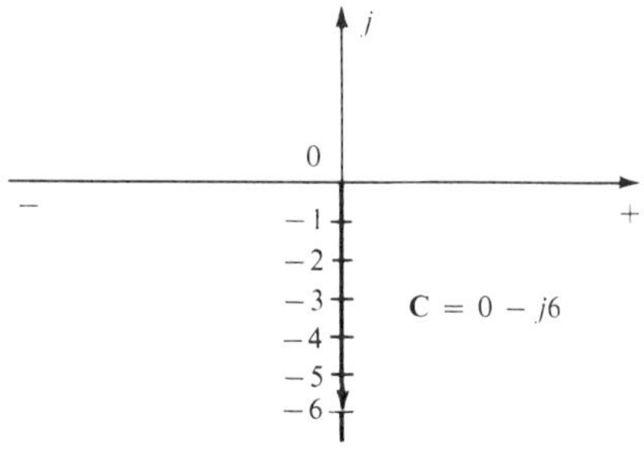

FIG. 14.4

c. See Fig. 14.5.

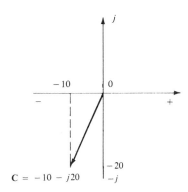

FIG. 14.5

14.3 POLAR FORM

The format for the polar form is

$$\boxed{\mathbf{C} = C \angle \theta} \qquad \textbf{(14.2)}$$

where C indicates magnitude only and θ is always meas-ured counterclockwise (CCW) from the *positive real axis*, as shown in Fig. 14.6.

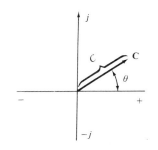

FIG. 14.6

A negative sign has the effect shown in Fig. 14.7:

$$\boxed{-\mathbf{C} = -C \angle \theta = C \angle \theta \pm \pi} \qquad \textbf{(14.3)}$$

EXAMPLE 14.2. Sketch the following complex numbers in the complex plane:
a. $\mathbf{C} = 5 \angle 30°$
b. $\mathbf{C} = 7 \angle 120°$
c. $\mathbf{C} = -4.2 \angle 60°$

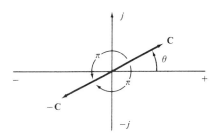

FIG. 14.7

Solutions:
a. See Fig. 14.8.

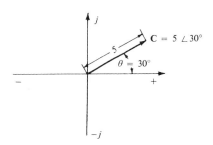

FIG. 14.8

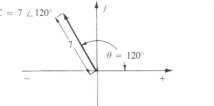

FIG. 14.9

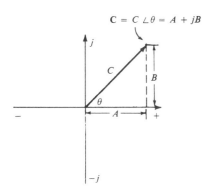

FIG. 14.10

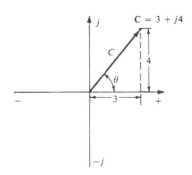

FIG. 14.11

b. See Fig. 14.9.

c. See Fig. 14.10.

14.4 CONVERSION BETWEEN FORMS

The two forms are related by the following equations.

Rectangular to Polar

$$C = \sqrt{A^2 + B^2} \qquad \textbf{(14.4a)}$$

$$\theta = \tan^{-1} \frac{B}{A} \qquad \textbf{(14.4b)}$$

Note Fig. 14.11.

Polar to Rectangular

$$A = C \cos \theta \qquad \textbf{(14.5a)}$$

$$B = C \sin \theta \qquad \textbf{(14.5b)}$$

EXAMPLE 14.3. Convert the following from rectangular to polar form:

$$\mathbf{C} = 3 + j4 \qquad \text{(Fig. 14.12)}$$

Solution:

$$C = \sqrt{3^2 + 4^2} = \sqrt{25} = 5$$
$$\theta = \tan^{-1}\left(\frac{4}{3}\right) = 53.13°$$

and

$$\mathbf{C} = 5 \angle 53.13°$$

EXAMPLE 14.4. Convert the following from polar to rectangular form:

$$\mathbf{C} = 10 \angle 45° \qquad \text{(Fig. 14.13)}$$

FIG. 14.12

Solution:

$$A = 10 \cos 45° = (10)(0.707) = 7.07$$
$$B = 10 \sin 45° = (10)(0.707) = 7.07$$

and

$$\mathbf{C} = 7.07 + j7.07$$

If the complex number should appear in the second, third, or fourth quadrant, simply convert it in that quadrant, and carefully determine the proper angle to be associated with the magnitude of the vector.

EXAMPLE 14.5. Convert the following from rectangular to polar form:

$$\mathbf{C} = -6 + j3 \qquad \text{(Fig. 14.14)}$$

Solution:

$$C = \sqrt{3^2 + 6^2} = \sqrt{45} = 6.71$$
$$\beta = \tan^{-1}\left(\frac{3}{6}\right) = 26.57°$$
$$\theta = 180 - 26.57° = 153.43°$$

and

$$\mathbf{C} = 6.71 \angle 153.43°$$

EXAMPLE 14.6. Convert the following from polar to rectangular form:

$$\mathbf{C} = 10 \angle 230° \qquad \text{(Fig. 14.15)}$$

Solution:

$$A = C \cos \beta = 10 \cos(230° - 180°) = 10 \cos 50°$$
$$= (10)(0.6428) = 6.428$$
$$B = C \sin \beta = 10 \sin 50° = (10)(0.7660) = 7.660$$
$$\mathbf{C} = -6.428 - j7.660$$

In conversions from rectangular to polar form, if the ratio of the magnitude of the imaginary to the real part, or the magnitude of the real to the imaginary part, is greater than ten (10), the magnitude of **C** in polar form is generally taken to be equal to the larger of the two. The angle θ is found in the same manner as in the previous examples.

EXAMPLE 14.7. Convert $1.2 + j14$ to polar form (Fig. 14.16).

Solution: The ratio is

$$\frac{14}{1.2} = 11.7 > 10$$
$$\therefore C = 14 \qquad \text{(the larger)}$$

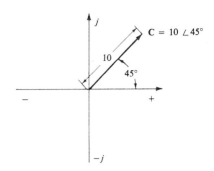

FIG. 14.13

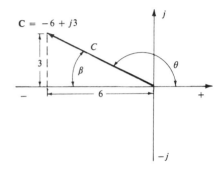

FIG. 14.14

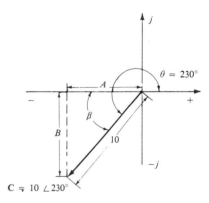

FIG. 14.15

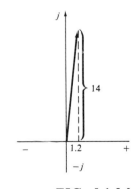

FIG. 14.16

with

$$\theta = \tan^{-1}\left(\frac{14}{1.2}\right) = \tan^{-1}(11.7) = 85.10°$$

and

$$1.2 + j14 = 14 \angle \mathbf{85.10°}$$

using

$$C = \sqrt{1.2^2 + 14^2} = \sqrt{1.44 + 196} \cong \sqrt{196} = 14$$

In conversions from polar to rectangular form, if the angle between the line drawn from the origin to the complex number and the real or imaginary axis is less than 5.7°, the magnitude of the larger component of the rectangular form is usually taken to be equal to the magnitude of **C.** The other component is found in the same manner as in the previous examples.

EXAMPLE 14.8. Convert $2 \angle 88°$ to rectangular form (Fig. 14.17).

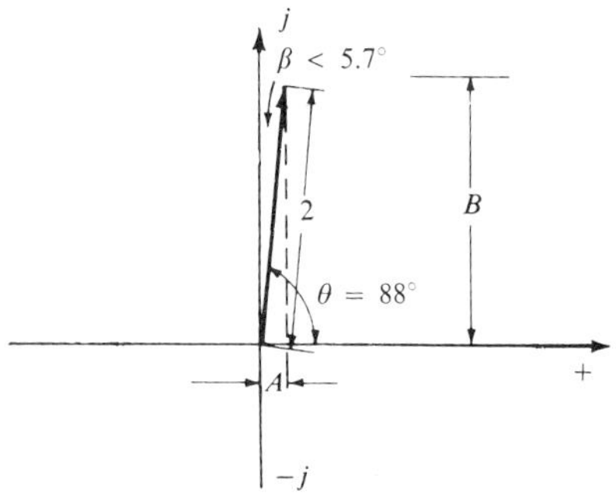

Solution:

$$\beta < 5.7°$$
$$\therefore B \cong C = 2$$

and

$$A = C \cos 88° = (2)(0.0349) = 0.0698$$

and

$$2 \angle 88° = \mathbf{0.0698 + j2}$$

using

$$B = 2 \sin 88° = (2)(0.9994) \cong 2$$

FIG. 14.17

The conversions from one form to another using the calculator are discussed in Appendix D. The time saved in making these conversions with the calculator will make it well worthwhile to examine this appendix thoroughly.

14.5 MATHEMATICAL OPERATIONS WITH COMPLEX NUMBERS

Complex numbers lend themselves readily to the basic mathematical operations of addition, subtraction, multiplication, and division. A few basic rules and definitions must be understood before considering these operations.

Let us first examine the symbol j associated with imaginary numbers. By definition,

$$\boxed{j = \sqrt{-1}} \qquad \qquad \textbf{(14.6)}$$

Thus

$$\boxed{j^2 = -1} \qquad (14.7)$$

and

$$j^3 = j^2j = -1j = -j$$

with

$$j^4 = j^2j^2 = (-1)(-1) = +1$$
$$j^5 = j$$

and so on.

The *reciprocal* of a complex number is 1 divided by the complex number. For example, the reciprocal of

$$\mathbf{C} = A + jB$$

is

$$\frac{1}{A + jB}$$

and of $C \angle \theta$,

$$\frac{1}{C \angle \theta}$$

Further,

$$\frac{1}{j} = (1)\left(\frac{1}{j}\right) = \left(\frac{j}{j}\right)\left(\frac{1}{j}\right) = \frac{j}{j^2} = \frac{j}{-1}$$

and

$$\boxed{\frac{1}{j} = -j} \qquad (14.8)$$

The *conjugate* or *complex conjugate* of a complex number can be found by simply changing the sign of the imaginary part in the rectangular form or by negating the angle of the polar form. For example, the conjugate of

$$\mathbf{C} = 2 + j3$$

is

$$2 - j3$$

as shown in Fig. 14.18. The conjugate of

$$\mathbf{C} = 2 \angle 30°$$

is

$$2 \angle -30°$$

as shown in Fig. 14.19.

We are now prepared to consider the four basic operations of *addition*, *subtraction*, *multiplication*, and *division* with complex numbers.

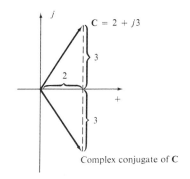

Complex conjugate of **C**

FIG. 14.18

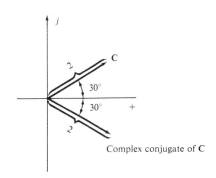

Complex conjugate of **C**

FIG. 14.19

Addition

To add two or more complex numbers, simply add the real and imaginary parts separately. For example, if

$$\mathbf{C}_1 = \pm A_1 \pm jB_1 \quad \text{and} \quad \mathbf{C}_2 = \pm A_2 \pm jB_2$$

then

$$\boxed{\mathbf{C}_1 + \mathbf{C}_2 = (\pm A_1 \pm A_2) + j(\pm B_1 \pm B_2)} \quad \textbf{(14.9)}$$

There is really no need to memorize the equation. Simply set one above the other and consider the real and imaginary parts separately, as shown in Solution (b) of Example 14.9.

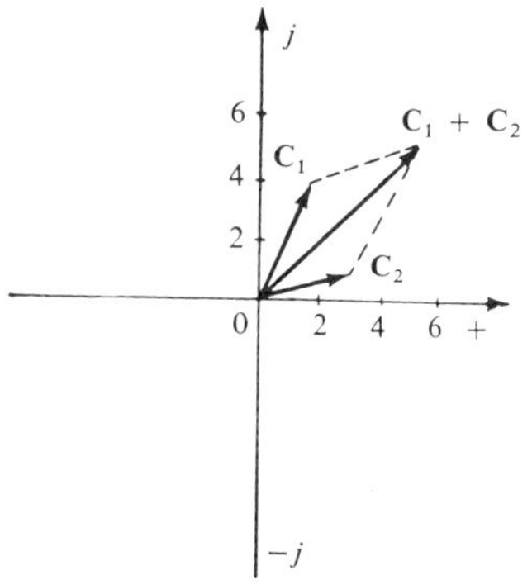

EXAMPLE 14.9.
a. Add $\mathbf{C}_1 = 2 + j4$ and $\mathbf{C}_2 = 3 + j1$.
b. Add $\mathbf{C}_1 = 3 + j6$ and $\mathbf{C}_2 = -6 + j3$.

Solutions:
a. By Eq. (14.9),

$$\mathbf{C}_1 + \mathbf{C}_2 = (2 + 3) + j(4 + 1) = \mathbf{5 + j5}$$

Note Fig. 14.20. An alternate method is

$$\begin{array}{c} 2 + j4 \\ \underline{3 + j1} \\ \downarrow \quad \downarrow \\ \mathbf{5 + j5} \end{array}$$

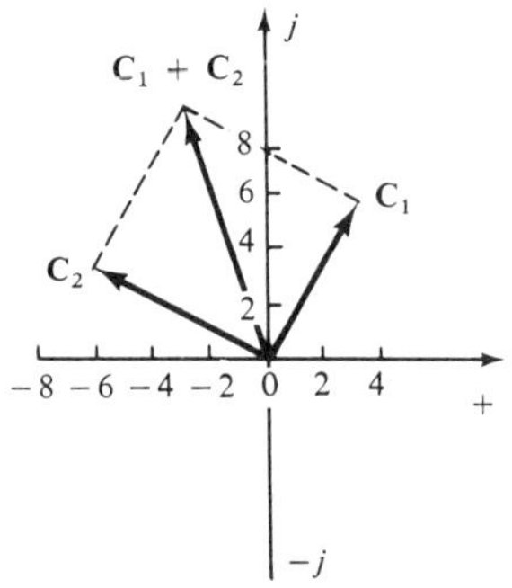

b. By Eq. (14.9),

$$\mathbf{C}_1 + \mathbf{C}_2 = (3 - 6) + j(6 + 3) = \mathbf{-3 + j9}$$

Note Fig. 14.21. An alternate method is

$$\begin{array}{c} 3 + j6 \\ \underline{-6 + j3} \\ \downarrow \quad \downarrow \\ \mathbf{-3 + j9} \end{array}$$

FIG. 14.20

FIG. 14.21

Subtraction

In subtraction, the real and imaginary parts are again considered separately. For example, if

$$\mathbf{C}_1 = \pm A_1 \pm jB_1 \quad \text{and} \quad \mathbf{C}_2 = \pm A_2 \pm jB_2$$

then

$$\boxed{\mathbf{C}_1 - \mathbf{C}_2 = [\pm A_1 - (\pm A_2)] + j[\pm B_1 - (\pm B_2)]}$$

$$\textbf{(14.10)}$$

Again, there is no need to memorize the equation if the alternate method of solution in Example 14.10 is employed.

EXAMPLE 14.10.

a. Subtract $C_2 = 1 + j4$ from $C_1 = 4 + j6$.
b. Subtract $C_2 = -2 + j5$ from $C_1 = +3 + j3$.

Solutions:

a. By Eq. (14.10),

$$C_1 - C_2 = (4 - 1) + j(6 - 4) = 3 + j2$$

Note Fig. 14.22. An alternate method is

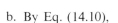

$$
\begin{array}{r}
4 + j6 \\
-(1 + j4) \\
\hline
 \downarrow \quad \downarrow \\
3 + j2
\end{array}
$$

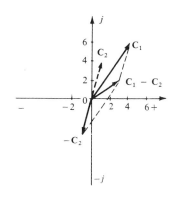

FIG. 14.22

b. By Eq. (14.10),

$$C_1 - C_2 = [3 - (-2)] + j(3 - 5) = 5 - j2$$

Note Fig. 14.23. An alternate method is

$$
\begin{array}{r}
3 + j3 \\
-(-2 + j5) \\
\hline
 \downarrow \quad \downarrow \\
5 - j2
\end{array}
$$

Addition or subtraction cannot be performed in polar form unless the complex numbers have the same angle θ or differ only by multiples of $180°$.

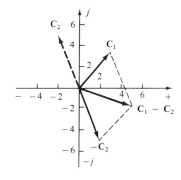

FIG. 14.23

EXAMPLE 14.11.

$$2 \angle 45° + 3 \angle 45° = 5 \angle 45°$$

Note Fig. 14.24. Or

$$2 \angle 0° - 4 \angle 180° = 6 \angle 0°$$

Note Fig. 14.25.

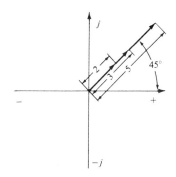

FIG. 14.24

Multiplication

To multiply two complex numbers in *rectangular* form, multiply the real and imaginary parts of one in turn by the real and imaginary parts of the other. For example, if

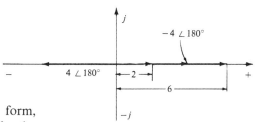

FIG. 14.25

$$\mathbf{C}_1 = A_1 + jB_1 \quad \text{and} \quad \mathbf{C}_2 = A_2 + jB_2$$

then

$$
\begin{array}{r}
\mathbf{C}_1 \cdot \mathbf{C}_2 = \quad A_1 + jB_1 \\
A_2 + jB_2 \\
\hline
A_1A_2 + jB_1A_2 \\
+ jA_1B_2 + j^2B_1B_2 \\
\hline
A_1A_2 + j(B_1A_2 + A_1B_2) + B_1B_2(-1)
\end{array}
$$

and

$$\boxed{\mathbf{C}_1 \cdot \mathbf{C}_2 = (A_1A_2 - B_1B_2) + j(B_1A_2 + A_1B_2)}$$

(14.11)

In Example 14.12(b), we obtain a solution without resorting to memorizing Eq. (14.11). Simply carry along the j factor when multiplying each part of one vector with the real and imaginary parts of the other.

EXAMPLE 14.12.
a. Find $\mathbf{C}_1 \cdot \mathbf{C}_2$ if

$$\mathbf{C}_1 = 2 + j3 \quad \text{and} \quad \mathbf{C}_2 = 5 + j10$$

b. Find $\mathbf{C}_1 \cdot \mathbf{C}_2$ if

$$\mathbf{C}_1 = -2 - j3 \quad \text{and} \quad \mathbf{C}_2 = +4 - j6$$

Solutions:
a. Using the format above, we have

$$
\begin{aligned}
\mathbf{C}_1 \cdot \mathbf{C}_2 &= [(2)(5) - (3)(10)] + j[(3)(5) + (2)(10)] \\
&= -20 + j35
\end{aligned}
$$

b. Without using the format, we obtain

$$
\begin{array}{r}
-2 - j3 \\
+4 - j6 \\
\hline
-8 - j12 \\
+ j12 + j^218 \\
\hline
-8 - [18 + j(-12 + 12)
\end{array}
$$

and

$$\mathbf{C}_1 \cdot \mathbf{C}_2 = -26 = 26 \angle 180°$$

In *polar* form, the magnitudes are multiplied and the angles added algebraically. For example, for

$$\mathbf{C}_1 = C_1 \angle \theta_1 \quad \text{and} \quad \mathbf{C}_2 = C_2 \angle \theta_2$$

$$\boxed{\mathbf{C}_1 \cdot \mathbf{C}_2 = C_1C_2 \,\underline{/\theta_1 + \theta_2}}$$ **(14.12)**

EXAMPLE 14.13.
a. Find $\mathbf{C}_1 \cdot \mathbf{C}_2$ if

$$\mathbf{C}_1 = 5 \angle 20° \quad \text{and} \quad \mathbf{C}_2 = 10 \angle 30°$$

b. Find $C_1 \cdot C_2$ if

$$C_1 = 2 \angle -40° \quad \text{and} \quad C_2 = 7 \angle +120°$$

Solutions:

a. $C_1 \cdot C_2 = (5)(10) \underline{/20° + 30°} = \mathbf{50} \angle \mathbf{50°}$

b. $C_1 \cdot C_2 = (2)(7) \underline{/-40° + 120°} = \mathbf{14} \angle \mathbf{+80°}$

To multiply a complex number in rectangular form by a real number requires that both the real part and the imaginary part be multiplied by the real number. For example,

$$(10)(2 + j3) = 20 + j30$$

and

$$50 \angle 0°(0 + j6) = j300 = 300 \angle 90°$$

Division

To divide two complex numbers in rectangular form, multiply the numerator and denominator by the conjugate of the denominator and the resulting real and imaginary parts collected. That is, for

$$C_1 = A_1 + jB_1 \quad \text{and} \quad C_2 = A_2 + jB_2,$$

$$\frac{C_1}{C_2} = \frac{(A_1 + jB_1)(A_2 - jB_2)}{(A_2 + jB_2)(A_2 - jB_2)}$$

$$= \frac{(A_1 A_2 + B_1 B_2) + j(A_2 B_1 - A_1 B_2)}{A_2^2 + B_2^2}$$

and

$$\boxed{\frac{C_1}{C_2} = \frac{A_1 A_2 + B_1 B_2}{A_2^2 + B_2^2} + j \frac{A_2 B_1 - A_1 B_2}{A_2^2 + B_2^2}} \quad \text{(14.13)}$$

The equation does not have to be memorized if the steps above used to obtain Eq. (14.13) are employed. That is, first multiply the numerator by the complex conjugate of the denominator and separate the real and imaginary terms. Then divide each term by the real number obtained by multiplying the denominator by its conjugate.

EXAMPLE 14.14.

a. Find C_1/C_2 if

$$C_1 = 1 + j4 \quad \text{and} \quad C_2 = 4 + j5$$

b. Find C_1/C_2 if

$$C_1 = -4 - j8 \quad \text{and} \quad C_2 = +6 - j1$$

Solutions:

a. By Eq. (14.13),

$$\frac{C_1}{C_2} = \frac{(1)(4) + (4)(5)}{4^2 + 5^2} + j \frac{(4)(4) - (1)(5)}{4^2 + 5^2}$$

$$= \frac{24}{41} + \frac{j11}{41} \cong 0.585 + j0.268$$

b. Using an alternate method, we obtain

$$
\begin{array}{r}
-4 - j8 \\
+6 + j1 \\
\hline
-24 - j48 \\
- j4 - j^2 8 \\
\hline
-24 - j52 + 8 = -16 - j52 \\
+6 \cdot j1 \\
+6 + j1 \\
\hline
36 + j6 \\
- j6 - j^2 1 \\
\hline
36 + 0 + 1 \quad = 37
\end{array}
$$

and

$$\frac{\mathbf{C}_1}{\mathbf{C}_2} = \frac{-16}{37} - \frac{j52}{37} = -0.432 - j1.405$$

To divide a complex number in rectangular form by a real number, both the real part and the imaginary part must be divided by the real number. For example,

$$\frac{8 + j10}{2} = 4 + j5$$

and

$$\frac{6.8 - j0}{2} = 3.4 - j0 = 3.4 \angle 0°$$

In *polar form,* division is accomplished by simply dividing the magnitude of the numerator by the magnitude of the denominator and subtracting the angle of the denominator from that of the numerator. That is, for

$$\mathbf{C}_1 = C_1 \angle \theta_1 \quad \text{and} \quad \mathbf{C}_2 = C_2 \angle \theta_2$$

$$\boxed{\frac{\mathbf{C}_1}{\mathbf{C}_2} = \frac{C_1}{C_2} \underline{/\theta_1 - \theta_2}} \qquad (14.14)$$

EXAMPLE 14.15.
a. Find $\mathbf{C}_1/\mathbf{C}_2$ if

$$\mathbf{C}_1 = 15 \angle 10° \quad \text{and} \quad \mathbf{C}_2 = 2 \angle 7°$$

b. Find $\mathbf{C}_1/\mathbf{C}_2$ if

$$\mathbf{C}_1 = 8 \angle 120° \quad \text{and} \quad \mathbf{C}_2 = 16 \angle -50°$$

Solutions:

a. $\dfrac{\mathbf{C}_1}{\mathbf{C}_2} = \dfrac{15}{2} \underline{/10° - 7°} = 7.5 \angle 3°$

b. $\dfrac{C_1}{C_2} = \dfrac{8}{16} \underline{/120° - (-50°)} = \mathbf{0.5 \angle 170°}$

We obtain the *reciprocal* in the rectangular form by multiplying the numerator and denominator by the complex conjugate of the denominator:

$$\frac{1}{A + jB} = \left(\frac{1}{A + jB}\right)\left(\frac{A - jB}{A - jB}\right) = \frac{A - jB}{A^2 + B^2}$$

and

$$\boxed{\frac{1}{A + jB} = \frac{A}{A^2 + B^2} - j\frac{B}{A^2 + B^2}} \qquad \textbf{(14.15)}$$

In the polar form the reciprocal is

$$\boxed{\frac{1}{C \angle \theta} = \frac{1}{C} \angle -\theta} \qquad \textbf{(14.16)}$$

Some concluding examples using the four basic operations follow.

EXAMPLE 14.16. Perform the following operations, leaving the answer in polar or rectangular form.

a. $\dfrac{(2 + j3) + (4 + j6)}{(7 + j7) - (3 - j3)}$

$$= \frac{(2 + 4) + j(3 + 6)}{(7 - 3) + j(7 + 3)}$$

$$= \frac{(6 + j9)(4 - j10)}{(4 + j10)(4 - j10)}$$

$$= \frac{[(6)(4) + (9)(10)] + j[(4)(9) - (6)(10)]}{4^2 + 10^2}$$

$$= \frac{114 - j24}{116} = \mathbf{0.983 - j0.207}$$

b. $\dfrac{(50 \angle 30°)(5 + j5)}{10 \angle -20°}$

$$= \frac{(50 \angle 30°)(7.07 \angle 45°)}{10 \angle -20°} = \frac{353.5 \angle 75°}{10 \angle -20°}$$

$$= 35.35 \underline{/75° - (-20°)} = \mathbf{35.35 \angle 95°}$$

c. $\dfrac{(2 \angle 20°)^2(3 + j4)}{8 - j6} = \dfrac{(2 \angle 20°)(2 \angle 20°)(5 \angle 53.13°)}{10 \angle -36.87°}$

$$= \frac{(4 \angle 40°)(5 \angle 53.13°)}{10 \angle -36.87°} = \frac{20 \angle 93.13°}{10 \angle -36.87°}$$

$$= 2 \underline{/93.13° - (-36.87°)} = \mathbf{2.0 \angle 130°}$$

d. $3 \angle 27° - 6 \angle -40° = (2.673 + j1.362)$
$$- (4.596 - j3.857)$$

FIG. 14.26 *Scientific calculator.*

$$= (2.673 - 4.596)$$
$$+ j(1.362 + 3.857)$$
$$= -1.923 + j5.219$$

14.6 TECHNIQUES OF CONVERSION

For many years the technologist was dependent on tables, charts, and the slide rule to perform conversions between complex numbers. Today, calculators such as shown in Fig. 14.26 are available that can perform the conversion to eight-place accuracy. In fact, this particular calculator is programmed to perform this particular operation by your pressing just a few buttons. The →R and →P refer to the rectangular and polar forms, respectively.

You will appreciate the speed and accuracy of calculator conversion after employing other techniques of conversion. There are inexpensive calculators that have the necessary functions to perform the conversions using Eqs. (14.4) and (14.5). For those who seriously expect to stay in this field, a calculator with the proper functions would be a wise investment.

14.7 PHASORS

As noted in the introduction to this chapter, the addition of sinusoidal voltages and currents will frequently be required in the analysis of ac circuits. One indicated method of performing this operation is to place both sinusoidal waveforms on the same set of axes and add algebraically the magnitudes of each at every point along the abscissa, as shown for $c = a + b$ in Fig. 14.27. This, however, can be a

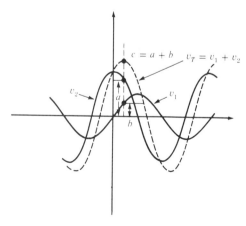

FIG. 14.27

long and tedious process with limited accuracy. A shorter method uses the rotating radius vector shown in Fig. 13.14. This *radius vector*, having a *constant magnitude* (length) with *one end fixed at the origin*, is called a *phasor* when applied to electric circuits. During its rotational development of the sine wave, the phasor will, at the instant $t = 0$, have the positions shown in Fig. 14.28(a) for each waveform in Fig. 14.28(b).

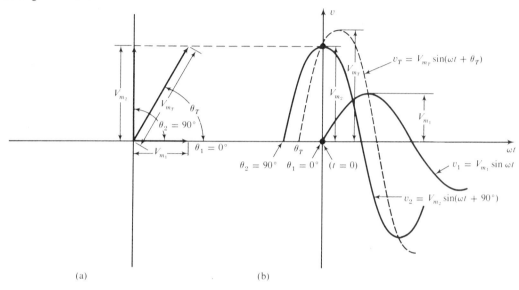

(a) (b)

FIG. 14.28

Note in Fig. 14.28(b) that v_1 passes through the horizontal axis at $t = 0$, requiring that the radius vector in Fig. 14.28(a) be on the horizontal axis. Its length in Fig. 14.28(a) is equal to the peak value of the sinusoid as required by the radius vector of Fig. 13.14. The other sinusoid (actually a cosine wave) has passed through 90° of its rotation by the time $t = 0$ is reached and therefore has its maximum vertical projection as shown in Fig. 14.28(a). (Recall that 90° separates the minimum and maximum values of any sinusoidal function.) Since its vertical projection is a maximum, the peak value of the sinusoid that it will generate is also attained at $t = 0$ in Fig. 14.28(b). Note also that $v_T = v_2$ at $t = 0$ since $v_1 = 0$ at this instant.

It can be shown [see Fig. 14.28(a)] using the vector algebra described in Section 14.5 that

$$V_{m_1} \angle 0° + V_{m_2} \angle 90° = V_{m_T} \angle \theta_T$$

In other words, if we convert v_1 and v_2 to the phasor form using

$$v = V_m \sin(\omega t \pm \theta) \Rightarrow V_m \angle \pm\theta$$

and add them using vector algebra, we can find the phasor form for v_T with very little difficulty. It can then be converted to the time domain and plotted on the same set of axes as shown in Fig. 14.28(b). Figure 14.28(a), showing the magnitudes and relative positions of the various phasors, is called a *phasor diagram*. It is actually a "snapshot" of the phasors representing the sinusoidal waveforms at $t = 0$.

In the future, therefore, if the addition of two sinusoids is required, they should first be converted to the phasor domain and the sum found using complex algebra. The result can then be converted to the time domain if required.

As an example, consider the case of

$$v_1 = 5 \sin \omega t$$
$$v_2 = 10 \sin(\omega t + 90°)$$

in Figs. 14.27 and 14.28. In the phasor domain,

$$\mathbf{V}_1 = 5 \angle 0°$$
$$\mathbf{V}_2 = 10 \angle 90°$$
$$\mathbf{V}_T = \mathbf{V}_1 + \mathbf{V}_2 = 5 + j10 = 11.180 \underline{/63.43°} = V_{m_T} \underline{/\theta_T}$$

In other words,

$$V_{m_T} = 11.180 \text{ V}$$
$$\theta_T = 63.43°$$
$$v_T = 11.180 \sin(\omega t + 63.43°)$$

as verified by Figs. 14.27 and 14.28.

The case of two sinusoidal functions having phase angles different from 0° and 90° appears in Fig. 14.29. Note again that the vertical height of the functions in Fig. 14.29(b) is determined by the rotational positions of the radius vectors in Fig. 14.29(a).

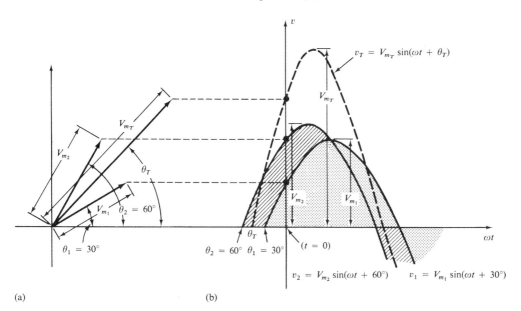

(a)   (b)

FIG. 14.29

Since the effective, rather than the peak, values are used almost exclusively in the analysis of ac circuits, the phasor will now be redefined for the purposes of practicality and uniformity as having a magnitude equal to the *effective value* of the sine wave it represents. The angle associated with the phasor will remain as previously described, the phase angle. The phasor diagram will therefore be as shown in Fig. 14.30, replacing that of Fig. 14.29(a).

In general, for all of the analyses to follow, the phasor form of a sinusoidal voltage or current will be

$$\mathbf{V} = V \angle \theta \quad \text{and} \quad \mathbf{I} = I \angle \theta$$

where V and I are effective values and θ is the phase angle. It should be pointed out that in phasor notation, the sine wave is always the reference, and the frequency is not represented. Phasor algebra for sinusoidal quantities is applicable only for waveforms having the *same frequency*, so that it can be carried along without any special notation.

EXAMPLE 14.17. Convert the following from the time to the phasor domain.

Time Domain	Phasor Domain
a. $\sqrt{2}(50) \sin \omega t$	$50 \angle 0°$
b. $69.6 \sin(\omega t + 72°)$	$(0.707)(69.6) \angle 72° = \mathbf{49.21} \angle \mathbf{72°}$
c. $45 \cos \omega t$	$(0.707)(45) \angle 90° = \mathbf{31.82} \angle \mathbf{90°}$

EXAMPLE 14.18. Write the sinusoidal expression for the following phasors if the frequency is 60 Hz.

Phasor Domain	Time Domain
a. $\mathbf{I} = 10 \angle 30°$	$i = \sqrt{2}(10) \sin(2\pi 60 t + 30°)$
	and $i = \mathbf{14.14} \sin(\mathbf{377}t + \mathbf{30°})$
b. $\mathbf{V} = 115 \angle -70°$	$v = \sqrt{2}(115) \sin(377t - 70°)$
	and $v = \mathbf{162.6} \sin(\mathbf{377}t - \mathbf{70°})$

EXAMPLE 14.19. Find the input voltage of the circuit of Fig. 14.31 if

$$\left. \begin{array}{l} v_a = 50 \sin(377t + 30°) \\ v_b = 30 \sin(377t + 60°) \end{array} \right\} f = 60 \text{ Hz}$$

Solution: Applying Kirchhoff's voltage law, we have

$$e_{\text{in}} = v_a + v_b$$

Converting from the time to the phasor domain yields

$$v_a = 50 \sin(377t + 30°) \Longrightarrow \mathbf{V}_a = 35.35 \angle 30°$$
$$v_b = 30 \sin(377t + 60°) \Longrightarrow \mathbf{V}_b = 21.21 \angle 60°$$

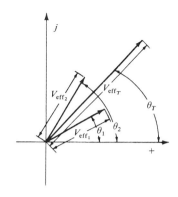

FIG. 14.30

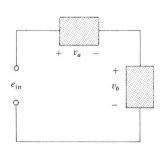

FIG. 14.31

Converting from polar to rectangular form for addition yields

$$\mathbf{V}_a = 35.35 \; \angle 30° = 30.61 + j17.68$$
$$\mathbf{V}_b = 21.21 \; \angle 60° = 10.61 + j18.37$$

Then

$$\mathbf{E}_{in} = \mathbf{V}_a + \mathbf{V}_b = (30.61 + j17.68) + (10.61 + j18.37)$$
$$\mathbf{E}_{in} = 41.22 + j36.05$$

Converting from rectangular to polar form, we have

$$\mathbf{E}_{in} = 41.22 + j36.05 = 54.76 \; \angle 41.17°$$

Converting from the phasor to the time domain, we obtain

$$\mathbf{E}_{in} = 54.76 \; \angle 41.17° \Rightarrow e_{in} = \sqrt{2}(54.76) \sin(377t + 41.17°)$$

and

$$e_{in} = \mathbf{77.43 \; sin(377t + 41.17°)}$$

A plot of the three waveforms is shown in Fig. 14.32. Note that at each instant of time, the sum of the two waveforms does in fact add up to e_{in}. At $t = 0$ ($\omega t = 0$), e_{in} is the sum of the two positive values, while at a value of ωt almost midway between $\pi/2$ and π, the sum of the positive value of v_a and the negative value of v_b results in $e_{in} = 0$.

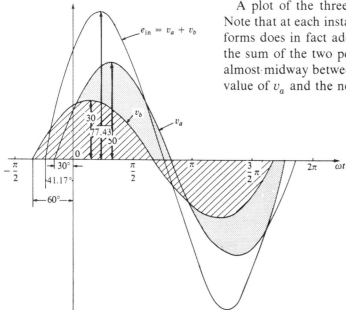

FIG. 14.32

EXAMPLE 14.20. Determine the current i_2 for the network of Fig. 14.33.

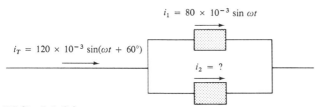

$i_1 = 80 \times 10^{-3} \sin \omega t$

$i_T = 120 \times 10^{-3} \sin(\omega t + 60°)$

$i_2 = \; ?$

FIG. 14.33

Solution: Applying Kirchhoff's current law, we obtain

$$i_T = i_1 + i_2$$
$$\text{or} \quad i_2 = i_T - i_1$$

Converting from the time to the phasor domain yields

$$i_T = 120 \times 10^{-3} \sin(\omega t + 60°) \Rightarrow 84.84 \times 10^{-3} \angle 60°$$
$$i_1 = 80 \times 10^{-3} \sin \omega t \Rightarrow 56.56 \times 10^{-3} \angle 0°$$

Converting from the polar to rectangular form for subtraction yields

$$\mathbf{I}_T = 84.84 \times 10^{-3} \angle 60°$$
$$= 42.42 \times 10^{-3} + j73.47 \times 10^{-3}$$
$$\mathbf{I}_1 = 56.56 \times 10^{-3} \angle 0°$$
$$= 56.56 \times 10^{-3} + j0$$

Then

$$\mathbf{I}_2 = \mathbf{I}_T - \mathbf{I}_1$$
$$= (42.42 \times 10^{-3} + j73.47 \times 10^{-3})$$
$$- (56.56 \times 10^{-3} + j0)$$

and

$$\mathbf{I}_2 = -14.14 \times 10^{-3} + j73.47 \times 10^{-3}$$

Converting from rectangular to polar form, we have

$$\mathbf{I}_2 = 74.82 \times 10^{-3} \angle 100.89°$$

Converting from the phasor to the time domain, we have

$$\mathbf{I}_2 = 74.82 \times 10^{-3} \angle 100.89° \Rightarrow$$
$$i_2 = \sqrt{2}(74.82 \times 10^{-3}) \sin(\omega t + 100.89°)$$

and

$$i_2 = \mathbf{105.8 \times 10^{-3} \sin(\omega t + 100.89°)}$$

A plot of the three waveforms appears in Fig. 14.34. The waveforms clearly indicate that $i_T = i_1 + i_2$.

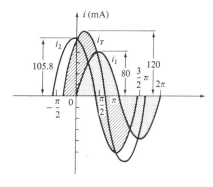

FIG. 14.34

PROBLEMS

Section 14.4

1. Convert the following from rectangular to polar form:

    a. $4 + j3$ **b.** $2 + j2$

    c. $3.5 + j16$ **d.** $100 + j650$

    e. $1000 + j750$ **f.** $0.001 + j0.0065$

    g. $7.6 - j9$ **h.** $-5.4 + j4$

    i. $-15 - j60$ **j.** $+78 - j65$

    k. $-2400 + j3600$ **l.** $5 \times 10^{-3} - j25 \times 10^{-3}$

2. Convert the following from polar to rectangular form:
   a. $6 \angle 30°$ b. $42 \angle 45°$
   c. $7400 \angle 70°$ d. $4 \times 10^{-4} \angle 8°$
   e. $8.49 \angle 80°$ f. $0.0093 \angle 23°$
   g. $65 \angle 150°$ h. $1.2 \angle 135°$
   i. $540 \angle 210°$ j. $6320 \angle -35°$
   k. $7.52 \angle -125°$ l. $0.008 \angle 310°$

3. Convert the following from rectangular to polar form:
   a. $1 + j15$ b. $60 + j5$
   c. $0.01 + j0.3$ d. $150 - j2500$
   e. $-5.6 + j86$ f. $-2.7 - j38.6$

4. Convert the following from polar to rectangular form:
   a. $13 \angle 5°$ b. $160 \angle 87°$
   c. $7 \times 10^{-6} \angle 2°$ d. $8.7 \angle 177°$
   e. $76 \angle -4°$ f. $396 \angle +265°$

Section 14.5

Perform the following operations.

5. Addition and subtraction (express your answers in rectangular form):
   a. $(4.2 + j6.8) + (7.6 + j0.02)$
   b. $(142 + j7) + (9.8 + j42) + (0.1 + j0.9)$
   c. $(4 \times 10^{-6} + j76) + (7.2 \times 10^{-7} - j0.9)$
   d. $(9.8 + j6.2) - (4.6 + j4.6)$
   e. $(167 + j243) - (-42.3 - j68)$
   f. $(-36.0 + j78) - (-4 - j6) + (10.8 - j72)$
   g. $6 \angle 20° + 10 \angle 30°$
   h. $42 \angle 45° + 62 \angle 60° - 70 \angle 120°$

6. Multiplication [express your answers in rectangular form for parts (i) to (l), and in polar form for parts (m) to (p)]:
   i. $(2 + j3)(6 + j8)$
   j. $(7.8 + j1)(4 + j2)(7 + j6)$
   k. $(0.002 + j0.006)(-2 + j2)$
   l. $(460 - j260)(-0.01 - j0.5)(-1 + j3)$
   m. $(2 \angle 60°)(4 \angle 22°)$
   n. $(6.9 \angle 8°)(7.2 \angle -72°)$
   o. $(0.002 \angle 120°)(0.5 \angle 200°)(40 \angle -60°)$
   p. $(540 \angle -20°)(-5 \angle 180°)(6.2 \angle 0°)$

7. Division (express your answers in polar form):
   q. $(42 \angle 10°)/(7 \angle 60°)$
   r. $(0.006 \angle 120°)/(50 \angle -20°)$
   s. $(4360 \angle -20°)/(40 \angle 210°)$
   t. $(650 \angle -80°)/(8.5 \angle 360°)$
   u. $(8 + j8)/(2 + j2)$
   v. $(8 + j42)/(-6 + j66)$
   w. $(0.05 + j0.25)/(8 - j60)$
   x. $(-4.5 - j6)/(0.1 - j0.4)$

Perform the following operations (express your answers in rectangular form):

8. a. $\dfrac{(4 + j3) + (6 - j8)}{(3 + j3) - (2 + j3)}$

b. $\dfrac{(1 + j5)(7 \angle 60°)}{(2 \angle 0°) + (100 + j100)}$

c. $\dfrac{(6 \angle 20°)(120 \angle -40°)(3 + j4)}{2 \angle -30°}$

d. $\dfrac{(0.4 \angle 60°)^2(300 \angle 40°)}{(3 + j6) - (27 + j6)}$

e. $\dfrac{(150 \angle 2°)(4 \times 10^{-6} \angle 88°)}{(1 \angle 10°)^3(4 \angle 30°)}$

Section 14.7

9. Express the following in phasor form:
   a. $\sqrt{2}(100) \sin(\omega t + 30°)$
   b. $\sqrt{2}(0.25) \sin(157t - 40°)$
   c. $100 \sin(\omega t - 90°)$
   d. $42 \sin(377t + 0°)$
   e. $6 \times 10^{-6} \cos \omega t$
   f. $3.6 \times 10^{-6} \cos(754t - 20°)$

10. Express the following phasor currents and voltages as sine waves if the frequency is 60 Hz:

a. $\mathbf{I} = 40 \angle 20°$	**b.** $\mathbf{V} = 120 \angle 0°$
c. $\mathbf{I} = 8 \times 10^{-3} \angle 120°$	**d.** $\mathbf{V} = 7.6 \angle 90°$
e. $\mathbf{I} = 1200 \angle -120°$	**f.** $\mathbf{V} = \dfrac{6000}{\sqrt{2}} \angle -180°$

11. For the system of Fig. 14.35, find the sinusoidal expression for the unknown voltage v_a if

    $$e_{in} = 60 \sin(377t + 20°)$$
    $$v_b = 20 \sin 377t$$

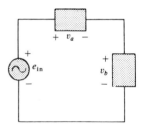

FIG. 14.35

12. For the system of Fig. 14.36, find the sinusoidal expression for the unknown current i_1 if

    $$i_T = 20 \times 10^{-6} \sin(\omega t + 90°)$$
    $$i_2 = 6 \times 10^{-6} \sin(\omega t - 60°)$$

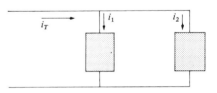

FIG. 14.36

13. Find the sinusoidal expression for the voltage v_c for the system of Fig. 14.37 if

    $$e_{in} = 120 \sin(\omega t - 30°)$$
    $$v_a = 60 \cos \omega t$$
    $$v_b = 30 \sin \omega t$$

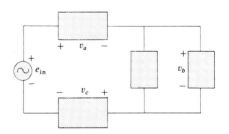

FIG. 14.37

14. Find the sinusoidal expression for the current i_T for the system of Fig. 14.38 if

    $$i_1 = 6 \times 10^{-3} \sin(377t + 180°)$$
    $$i_2 = 8 \times 10^{-3} \sin(377t - 20°)$$
    $$i_3 = 2i_2$$

GLOSSARY

Complex conjugate A complex number defined by simply changing the sign of an imaginary component of a complex number in the rectangular form.

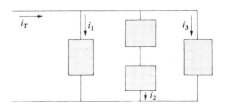

FIG. 14.38

Complex number A number that represents a point in a two-dimensional plane located with reference to two distinct axes. It defines a vector drawn from the origin to that point.

Phasor A radius vector that has a constant magnitude at a fixed angle from the positive real axis and that represents a sinusoidal voltage or current in the vector domain.

Phasor diagram A "snapshot" of the phasors that represent a number of sinusoidal waveforms at $t = 0$.

Polar form A method of defining a point in a complex plane that includes a single magnitude to represent the distance from the origin, and an angle to reflect the counterclockwise distance from the positive real axis.

Reciprocal A format defined by one (1) over the complex number.

Rectangular form A method of defining a point in a complex plane that includes the magnitude of the real component and the magnitude of the imaginary component, the latter component being defined by an associated letter j.

15.1 INTRODUCTION

In this chapter, phasor algebra will be used to develop a quick, direct method for solving both the series and the parallel ac circuits. The close relationship that exists between this method for solving for unknown quantities and the approach used for dc circuits will become apparent after a few simple examples are considered. Once this association is established, many of the rules (current divider rule, voltage divider rule, and so on) for dc circuits can be readily applied to ac circuits.

SERIES ac CIRCUITS

15.2 IMPEDANCE AND THE PHASOR DIAGRAM

In Chapter 13, we found, for the purely resistive circuit of Fig. 15.1, that v and i were in phase, and the magnitude

$$I_m = \frac{V_m}{R} \quad \text{or} \quad V_m = I_m R$$

In phasor form,

$$v = V_m \sin \omega t \Rightarrow \mathbf{V} = V \angle 0° \quad \text{where } V = 0.707 V_m$$

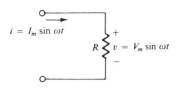

FIG. 15.1

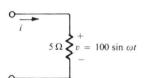

FIG. 15.2

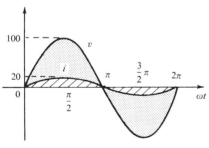

FIG. 15.3

FIG. 15.4

Applying Ohm's law and using phasor algebra, we have

$$\mathbf{I} = \frac{V \angle 0°}{R \angle \theta_R} = \frac{V}{R} \underline{/0° - \theta_R}$$

Since i and v are in phase, the angle associated with i also must be $0°$. To satisfy this condition, θ_R must equal $0°$. Substituting $\theta_R = 0°$, we find

$$\mathbf{I} = \frac{V \angle 0°}{R \angle 0°} = \frac{V}{R} \underline{/0° - 0°} = \frac{V}{R} \angle 0°$$

so that in the time domain,

$$i = \sqrt{2}\left(\frac{V}{R}\right) \sin \omega t$$

The complex number in the denominator of the above equation,

$$\boxed{\mathbf{R} = R \angle 0°} \qquad\qquad (15.1)$$

does not represent a sinusoidal function in the phasor domain even though it has the same format. It is a radius vector in the complex plane that has a fixed magnitude R at an angle of $0°$ (the positive real axis). The relative advantages of associating $\angle 0°$ with purely resistive elements is demonstrated in the following examples. You will find that it is no longer necessary to keep in mind that v and i are in phase. This fact was included when we associated an angle of $0°$ with R.

EXAMPLE 15.1. Using phasor algebra, find the current i for the circuit of Fig. 15.2. Sketch the waveforms of v and i.

Solution: Note Fig. 15.3:

$$v = 100 \sin \omega t \Rightarrow \text{phasor form } \mathbf{V} = 70.7 \angle 0°$$

$$\mathbf{I} = \frac{\mathbf{V}}{\mathbf{R}} = \frac{70.7 \angle 0°}{5 \angle 0°} = 14.14 \angle 0°$$

$$i = \sqrt{2}(14.14) \sin \omega t = \mathbf{20 \sin \omega t}$$

EXAMPLE 15.2. Using phasor algebra, find the voltage v for the circuit of Fig. 15.4. Sketch the waveforms of v and i.

Solution: Note Fig. 15.5:

$$i = 4 \sin(\omega t + 30°) \Rightarrow \text{phasor form } \mathbf{I} = 2.828 \angle 30°$$
$$\mathbf{V} = \mathbf{IR} = (2.828 \angle 30°)(2 \angle 0°) = 5.656 \angle 30°$$

and

$$v = \sqrt{2}(5.656) \sin(\omega t + 30°) = \mathbf{8.0 \sin(\omega t + 30°)}$$

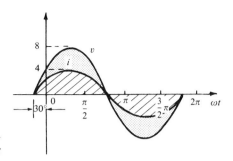

It is often helpful in the analysis of networks to have a *phasor diagram,* which shows at a glance the *magnitudes* and *phase relations* between the various quantities within the network. For example, the phasor diagrams of the circuits considered in the preceding examples would be as shown in Fig. 15.6. In both cases, it is immediately obvious that v and i are in phase, since they both have the same phase angle.

FIG. 15.5

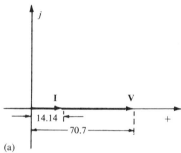

(a)

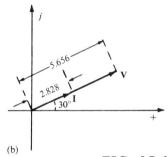

(b)

FIG. 15.6

For the pure inductor of Fig. 15.7, it was learned in Chapter 13 that the voltage leads the current by 90°, and that the reactance of the coil X_L is determined by ωL.

$$v = V_m \sin \omega t \Rightarrow \text{phasor form } \mathbf{V} = V \angle 0°$$

By Ohm's law,

$$\mathbf{I} = \frac{V \angle 0°}{X_L \angle \theta_L} = \frac{V}{X_L} \underline{/0° - \theta_L}$$

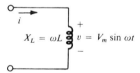

$$X_L = \omega L \quad v = V_m \sin \omega t$$

FIG. 15.7

Since v leads i by 90°, i must have an angle of $-90°$ associated with it. To satisfy this condition, θ_L must equal $+90°$. Substituting $\theta_L = 90°$, we obtain

$$\mathbf{I} = \frac{V \angle 0°}{X_L \angle 90°} = \frac{V}{X_L} \underline{/0° - 90°} = \frac{V}{X_L} \angle -90°$$

so that in the time domain,

$$i = \sqrt{2}\left(\frac{V}{X_L}\right) \sin(\omega t - 90°)$$

The complex number in the denominator of the preceding equation,

$$\boxed{\mathbf{X_L} = X_L \angle 90°} \qquad \textbf{(15.2)}$$

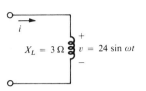

FIG. 15.8

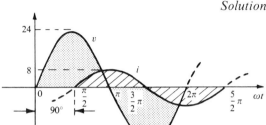

FIG. 15.9

does not represent a sinusoidal function in the phasor domain even though it has the same format. It is a radius vector in the complex plane that has a fixed magnitude X_L at an angle of $90°$.

EXAMPLE 15.3. Using phasor algebra, find the current i for the circuit of Fig. 15.8. Sketch the v and i curves.

Solution: Note Fig. 15.9:

$$v = 24 \sin \omega t \Rightarrow \text{phasor form } \mathbf{V} = 16.968 \angle 0°$$

$$\mathbf{I} = \frac{\mathbf{V}}{\mathbf{X}_L} = \frac{16.968 \angle 0°}{3 \angle 90°} = 5.656 \angle -90°$$

and

$$i = \sqrt{2}(5.656) \sin(\omega t - 90°) = \mathbf{8.0 \sin(\omega t - 90°)}$$

EXAMPLE 15.4. Using phasor algebra, find the voltage v for the circuit of Fig. 15.10. Sketch the v and i curves.

Solution: Note Fig. 15.11:

FIG. 15.10

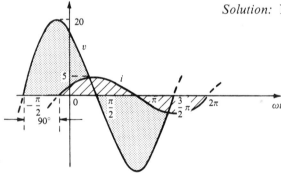

FIG. 15.11

$$i = 5 \sin(\omega t + 30°) \Rightarrow \text{phasor form } \mathbf{I} = 3.535 \angle 30°$$

$$\mathbf{V} = \mathbf{I}\mathbf{X}_L = (3.535 \angle 30°)(4 \angle +90°) = 14.140 \angle 120°$$

$$v = \sqrt{2}(14.140) \sin(\omega t + 120°) = \mathbf{20 \sin(\omega t + 120°)}$$

The phasor diagrams for the two circuits of the preceding examples are shown in Fig. 15.12. Both indicate quite clearly that the voltage leads the current by $90°$.

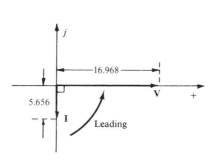

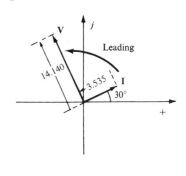

For the pure capacitor of Fig. 15.13, it was learned in Chapter 13 that the current leads the voltage by 90°, and that the reactance of the capacitor X_C is determined by $1/\omega C$.

$$v = V_m \sin \omega t \Rightarrow \text{phasor form } \mathbf{V} = V \angle 0°$$

FIG. 15.12

Applying Ohm's law and using phasor algebra, we find

$$\mathbf{I} = \frac{V \angle 0°}{X_C \angle \theta_C} = \frac{V}{X_C} \underline{/0° - \theta_C}$$

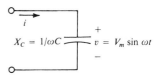

FIG. 15.13

Since we know i leads v by 90°, i must have an angle of $+90°$ associated with it. To satisfy this condition, θ_C must equal $-90°$. Substituting $\theta_C = -90°$ yields

$$\mathbf{I} = \frac{V \angle 0°}{X_C \angle -90°} = \frac{V}{X_C} \underline{/0° - (-90°)} = \frac{V}{X_C} \angle 90°$$

so, in the time domain,

$$i = \sqrt{2}\left(\frac{V}{X_C}\right) \sin(\omega t + 90°)$$

Once more, the complex number in the denominator of the above equation,

$$\boxed{\mathbf{X}_C = X_C \angle -90°} \qquad (15.3)$$

does not represent a sinusoidal function in the phasor domain even though it has the same format. It is a radius vector in the complex plane that has a fixed magnitude X_C at an angle of $-90°$.

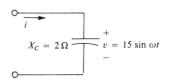

EXAMPLE 15.5. Using phasor algebra, find the current i in the circuit of Fig. 15.14. Sketch the v and i curves.

Solution: Note Fig. 15.15:

FIG. 15.14

$$v = 15 \sin \omega t \Rightarrow \text{phasor notation } \mathbf{V} = 10.605 \angle 0°$$

$$\mathbf{I} = \frac{\mathbf{V}}{\mathbf{X}_C} = \frac{10.605 \angle 0°}{2 \angle -90°} = 5.303 \angle 90°$$

and

$$i = \sqrt{2}(5.303) \sin(\omega t + 90°) = \mathbf{7.5 \sin(\omega t + 90°)}$$

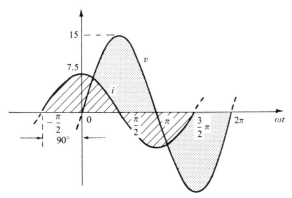

FIG. 15.15

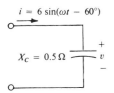

FIG. 15.16

EXAMPLE 15.6. Using phasor algebra, find the voltage v in the circuit of Fig. 15.16. Sketch the v and i curves.

Solution: Note Fig. 15.17:

$$i = 6\sin(\omega t - 60°) \Rightarrow \text{phasor notation } \mathbf{I} = 4.242 \angle -60°$$
$$\mathbf{V} = \mathbf{I}X_C = (4.242 \angle -60°)(0.5 \angle -90°) = 2.121 \angle -150°$$

and

$$v = \sqrt{2}(2.121)\sin(\omega t - 150°) = \mathbf{3.0\sin(\omega t - 150°)}$$

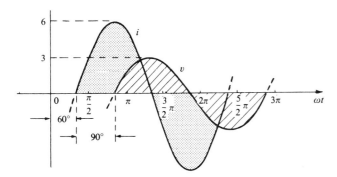

FIG. 15.17

The phasor diagrams for the two circuits of the preceding examples are shown in Fig. 15.18. Both indicate quite clearly that the current i leads the voltage by 90°.

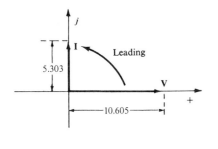

(a)

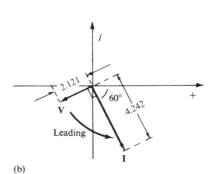

(b)

FIG. 15.18

A plot of resistance, inductive reactance, and capacitive reactance appears in Fig. 15.19. For any network, the resistance will *always* appear on the positive real axis, the inductive reactance on the positive imaginary axis, and capacitive reactance on the negative imaginary axis.

Any *one or combination* of these elements in an ac circuit is called the *impedance* of the circuit. It is a measure of how much the circuit will *impede* or hinder the flow of current through it. The diagram of Fig. 15.19 is referred to as an *impedance diagram*. The symbol for impedance is Z.

For the individual elements,

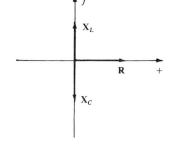

FIG. 15.19

$$\text{Resistance: } \mathbf{Z} = \mathbf{R} = R \angle 0° = R + j0 \quad \text{(15.4)}$$

$$\text{Inductive reactance: } \mathbf{Z} = \mathbf{X}_L$$
$$= X_L \angle 90° = 0 + jX_L \quad \text{(15.5)}$$

$$\text{Capacitive reactance: } \mathbf{Z} = \mathbf{X}_C$$
$$= X_C \angle -90° = 0 - jX_C \quad \text{(15.6)}$$

15.3 SERIES CONFIGURATION

The overall properties of series ac circuits (Fig. 15.20) are the same as those for dc circuits; that is, *the current is the*

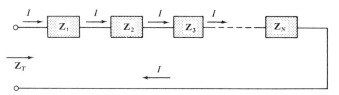

same through series elements, and the total impedance of a system is the sum of the individual impedances:

FIG. 15.20

$$\boxed{\mathbf{Z}_T = \mathbf{Z}_1 + \mathbf{Z}_2 + \mathbf{Z}_3 + \cdots + \mathbf{Z}_N} \quad \text{(15.7)}$$

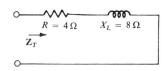

EXAMPLE 15.7. Draw the impedance diagram for the circuit of Fig. 15.21 and find the total impedance.

FIG. 15.21

Solution: As indicated by Fig. 15.22, the input impedance can be found graphically from the impedance diagram by properly scaling the real and imaginary axes and finding the length of the resultant vector Z_T and angle θ_T. Or, by using vector algebra, we obtain

$$\mathbf{Z}_T = \mathbf{Z}_1 + \mathbf{Z}_2$$
$$= R \angle 0° + X_L \angle 90°$$
$$= R + jX_L = 4 + j8$$
$$\mathbf{Z}_T = \mathbf{8.944} \angle \mathbf{63.43°}$$

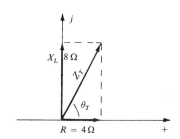

EXAMPLE 15.8. Determine the input impedance to the series network of Fig. 15.23. Draw the impedance diagram.

FIG. 15.22

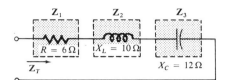

FIG. 15.23

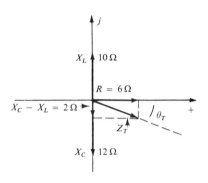

FIG. 15.24

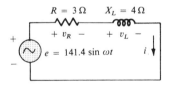

FIG. 15.25

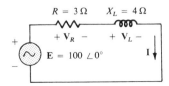

FIG. 15.26

Solution:

$$\mathbf{Z}_T = \mathbf{Z}_1 + \mathbf{Z}_2 + \mathbf{Z}_3$$
$$= R \angle 0° + X_L \angle 90° + X_C \angle -90°$$
$$= R + jX_L - jX_C$$
$$= R + j(X_L - X_C) = 6 + j(10 - 12) = 6 - j2$$
$$\mathbf{Z}_T = \mathbf{6.325} \angle -\mathbf{18.43°}$$

The impedance diagram appears in Fig. 15.24. Note that in this example series inductive and capacitive reactances are in direct opposition. For the circuit of Fig. 15.23, if the inductive reactance were equal to the capacitive reactance, the input impedance would be purely resistive. We will have more to say about this particular condition in a later chapter.

As mentioned in Chapter 13, it can be determined whether a circuit is predominantly inductive or capacitive by noting the phase relationship between the input current and voltage. The term to be applied can also be determined by noting the angle θ_T associated with the total impedance $\mathbf{Z}_T$ of a circuit. If θ_T is in the first quadrant, or $0° < \theta_T < 90°$, the circuit is predominantly inductive, and if θ_T is in the fourth quadrant, or $-90° < \theta_T < 0°$, the circuit is predominantly capacitive. If $\theta_T = 0°$, the circuit is resistive.

In many of the circuits to be considered, $3 + j4 = 5 \angle 53.13°$ and $4 + j3 = 5 \angle 36.87°$ will be used quite frequently to insure that the approach is as clear as possible and not lost in mathematical complexity.

Let us now examine the *R-L*, *R-C*, and *R-L-C* series networks. Their basic nature dictates that they be examined in some detail. Numerical values were assigned to make the description as informative as possible.

R-L (Fig. 15.25)

Phasor Notation:

$$e = 141.4 \sin \omega t \Rightarrow \mathbf{E} = 100 \angle 0°$$

Note Fig. 15.26.

$\mathbf{Z}_T$:

$$\mathbf{Z}_T = \mathbf{Z}_1 + \mathbf{Z}_2 = 3 \angle 0° + 4 \angle 90° = 3 + j4$$

and

$$\mathbf{Z}_T = 5 \angle 53.13°$$

Impedance diagram: As shown in Fig. 15.27.

I:

$$I = \frac{E}{Z_T} = \frac{100 \angle 0°}{5 \angle 53.13°} = 20 \angle -53.13°$$

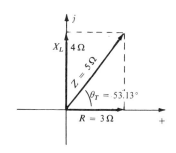

V_R, V_L:

Kirchhoff's voltage law:

$$\Sigma_\circ V = E - V_R - V_L = 0$$

or

$$E = V_R + V_L$$

In rectangular form,

$$V_R = 60 \angle -53.13° = 36.0 - j48.0$$
$$V_L = 80 \angle +36.87° = 64.0 + j48.0$$

and

$$E = V_R + V_L = (36 - j48) + (64 + j48) = 100 + j0$$
$$= 100 \angle 0° \quad \text{as applied}$$

FIG. 15.27

Phasor diagram: Note that for the phasor diagram of Fig. 15.28, **I** is in phase with the voltage across the resistor and lags the voltage across the inductor by 90°.

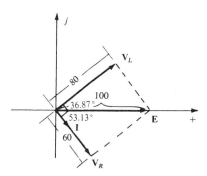

FIG. 15.28

Power: The total power in watts delivered to the circuit is

$$P_T = EI \cos\theta_T$$
$$= (100)(20) \cos 53.13° = 2000(0.6)$$
$$= \textbf{1200 W}$$

where E and I are effective values and θ_T is the phase angle between E and I, or

$$P_T = I^2 R$$
$$= (20^2)(3) = (400)(3)$$
$$= \textbf{1200 W}$$

where I is the effective value, or, finally,

$$P_T = P_R + P_L = V_R I \cos\theta_R + V_L I \cos\theta_L$$
$$= (60)(20) \cos 0° + (80)(20) \cos 90°$$
$$= 1200 + 0$$
$$= \textbf{1200 W}$$

where θ_R is the phase angle between V_R and I, and θ_L is the phase angle between V_L and I.

Power factor: The power factor F_p of the circuit is $\cos 53.13° = \textbf{0.6 lagging}$ where $53.13°$ is the phase angle between **E** and **I**.

If we write the basic power equation $P = EI \cos \theta$ in the following form:

$$\cos \theta = \frac{P}{EI}$$

where E and I are the input quantities and P is the power delivered to the network, and then perform the following substitutions from the basic series ac circuit:

$$\cos \theta = \frac{P}{EI} = \frac{I^2 R}{EI} = \frac{IR}{E} = \frac{R}{E/I} = \frac{R}{Z_T}$$

we find

$$\boxed{F_p = \cos \theta_T = \frac{R}{Z_T}} \qquad (15.8)$$

Reference to Fig. 15.27 also indicates that θ is the impedance angle θ_T as written in Eq. (15.8). In other words, *the impedance angle θ_T is also the phase angle between the input voltage and current for a series ac circuit.* To determine the power factor, it is necessary only to form the ratio of the total resistance to the magnitude of the input impedance.

For the case at hand,

$$F_p = \cos \theta = \frac{R}{Z_T} = \frac{3}{5} = 0.6 \text{ lagging} \quad \text{as found above}$$

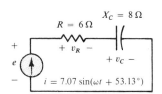

FIG. 15.29

R-C (Fig. 15.29)

Phasor notation:

$$i = 7.07 \sin(\omega t + 53.13°) \Rightarrow I = 5 \angle 53.13°$$

Note Fig. 15.30.

$\mathbf{Z}_T$:

$$\mathbf{Z}_T = \mathbf{Z}_1 + \mathbf{Z}_2 = 6 \angle 0° + 8 \angle -90° = 6 - j8$$

and

$$\mathbf{Z}_T = 10 \angle -53.13°$$

Impedance diagram: As shown in Fig. 15.31.

E:

$$\mathbf{E} = \mathbf{IZ}_T = (5 \angle 53.13°)(10 \angle -53.13°) = 50 \angle 0°$$

$\mathbf{V}_R$, $\mathbf{V}_C$:

$$\mathbf{V}_R = \mathbf{IR} = (5 \angle 53.13°)(6 \angle 0°) = 30 \angle 53.13°$$

$$\mathbf{V}_C = \mathbf{IX}_C = (5 \angle 53.13°)(8 \angle -90°) = 40 \angle -36.87°$$

Kirchhoff's voltage law:

$$\Sigma_\circlearrowleft \mathbf{V} = \mathbf{E} - \mathbf{V}_R - \mathbf{V}_C = 0$$

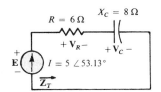

FIG. 15.30

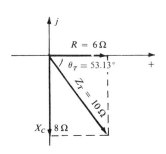

FIG. 15.31

or

$$\mathbf{E} = \mathbf{V}_R + \mathbf{V}_C$$

which can be verified by vector algebra as demonstrated for the *R-L* circuit.

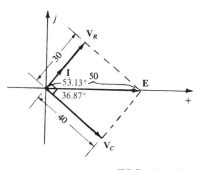

Phasor diagram: Note on the phasor diagram of Fig. 15.32 that the current **I** is in phase with the voltage across the resistor and leads the voltage across the capacitor by 90°.

Time domain: In the time domain,

$$e = \sqrt{2}(50) \sin \omega t = \mathbf{70.70 \sin \omega t}$$
$$v_R = \sqrt{2}(30) \sin(\omega t + 53.13°) = \mathbf{42.42 \sin(\omega t + 53.13°)}$$
$$v_C = \sqrt{2}(40) \sin(\omega t - 36.87°) = \mathbf{56.56 \sin(\omega t - 36.87°)}$$

FIG. 15.32

A plot of all of the voltages and the current of the circuit appears in Fig. 15.33. Note again that *i* and v_R are in phase and that v_C lags *i* by 90°.

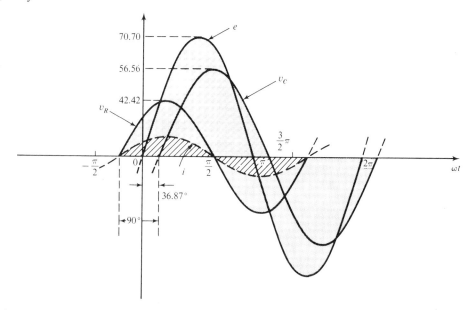

Power: The total power in watts delivered to the circuit is

FIG. 15.33

$$P_T = EI \cos \theta_T = (50)(5) \cos 53.13°$$
$$= (250)(0.6) = \mathbf{150\ W}$$

or

$$P_T = I^2 R = (5^2)(6) = (25)(6)$$
$$= \mathbf{150\ W}$$

or, finally,

$$P_T = P_R + P_C = V_R I \cos \theta_R + V_C I \cos \theta_C$$
$$= (30)(5) \cos 0° + (40)(5) \cos 90°$$
$$= 150 + 0 = \mathbf{150\ W}$$

Power factor: The power factor of the circuit is

$$F_p = \cos \theta = \cos 53.13° = \mathbf{0.6\ leading}$$

Using Eq. (15.8), we obtain

$$F_p = \cos \theta = \frac{R}{Z_T} = \frac{6}{10}$$

$$= \mathbf{0.6\ leading} \quad \text{as determined above}$$

R-L-C (Fig. 15.34)

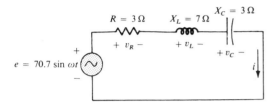

FIG. 15.34

Phasor notation: As shown in Fig. 15.35.

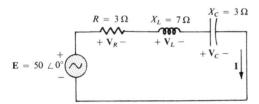

FIG. 15.35

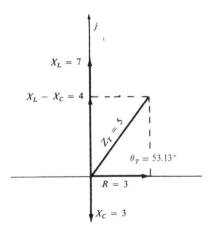

FIG. 15.36

$\mathbf{Z}_T$:

$$\mathbf{Z}_T = \mathbf{Z}_1 + \mathbf{Z}_2 + \mathbf{Z}_3 = R \angle 0° + X_L \angle 90° + X_C \angle -90°$$
$$= 3 + j7 - j3$$
$$= 3 + j4 = 5 \angle \mathbf{53.13°}$$

Impedance diagram: As shown in Fig. 15.36.

I:

$$\mathbf{I} = \frac{\mathbf{E}}{\mathbf{Z}_T} = \frac{50 \angle 0°}{5 \angle 53.13°} = \mathbf{10 \angle -53.13°}$$

$\mathbf{V}_R$, $\mathbf{V}_L$, $\mathbf{V}_C$:

$$\mathbf{V}_R = \mathbf{I}R = (10 \angle -53.13°)(3 \angle 0°) = \mathbf{30 \angle -53.13°}$$
$$\mathbf{V}_L = \mathbf{I}X_L = (10 \angle -53.13°)(7 \angle 90°) = \mathbf{70 \angle 36.87°}$$
$$\mathbf{V}_C = \mathbf{I}X_C = (10 \angle -53.13°)(3 \angle -90°) = \mathbf{30 \angle -143.13°}$$

Kirchhoff's voltage law:

$$\Sigma_{\bigcirc} \mathbf{V} = \mathbf{E} - \mathbf{V}_R - \mathbf{V}_L - \mathbf{V}_C = 0$$

or

$$\mathbf{E} = \mathbf{V}_R + \mathbf{V}_L + \mathbf{V}_C$$

which can also be verified through vector algebra.

Phasor diagram: The phasor diagram of Fig. 15.37 indicates that the current **I** is in phase with the voltage across the resistor, lags the voltage across the inductor by 90°, and leads the voltage across the capacitor by 90°.

Time domain:

$$i = \sqrt{2}(10)\sin(\omega t - 53.13°) = \mathbf{14.14\sin(\omega t - 53.13°)}$$
$$v_R = \sqrt{2}(30)\sin(\omega t - 53.13°) = \mathbf{42.42\sin(\omega t - 53.13°)}$$
$$v_L = \sqrt{2}(70)\sin(\omega t + 36.87°) = \mathbf{98.98\sin(\omega t + 36.87°)}$$
$$v_C = \sqrt{2}(30)\sin(\omega t - 143.13°) = \mathbf{42.42\sin(\omega t - 143.13°)}$$

A plot of all the voltages and the current of the circuit appears in Fig. 15.38.

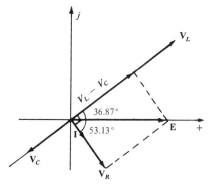

FIG. 15.37

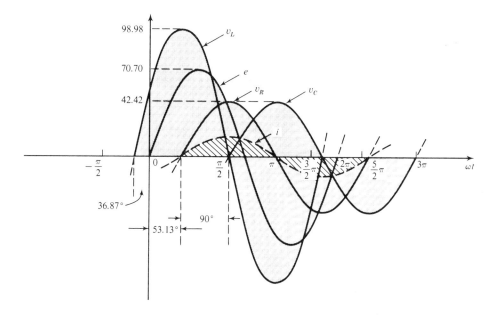

FIG. 15.38

Power: The total power in watts delivered to the circuit is

$$P_T = EI\cos\theta_T = (50)(10)\cos 53.13° = (500)(0.6) = \mathbf{300\ W}$$

or

$$P_T = I^2R = (10^2)(3) = (100)(3) = \mathbf{300\ W}$$

or

$$
\begin{aligned}
P_T &= P_R + P_L + P_C \\
&= V_R I \cos\theta_R + V_L I \cos\theta_L + V_C I \cos\theta_C \\
&= (30)(10)\cos 0° + (70)(10)\cos 90° + (30)(10)\cos 90° \\
&= (30)(10) + 0 + 0 = \mathbf{300\ W}
\end{aligned}
$$

Power factor: The power factor of the circuit is

$$F_p = \cos\theta_T = \cos 53.13° = \mathbf{0.6\ lagging}$$

Using Eq. (15.8), we obtain

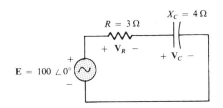

$$F_p = \cos\theta = \frac{R}{Z_T} = \frac{3}{5} = \textbf{0.6 lagging}$$

15.4 VOLTAGE DIVIDER RULE

The basic format for the voltage divider rule in ac circuits is exactly the same as that for dc circuits:

$$\boxed{\mathbf{V}_x = \frac{\mathbf{Z}_x\mathbf{E}}{\mathbf{Z}_T}} \qquad (15.9)$$

where $\mathbf{V}_x$ = voltage across one or more elements in series that have total impedance $\mathbf{Z}_x$

$\mathbf{E}$ = total voltage appearing across the series circuit

$\mathbf{Z}_T$ = total impedance of series circuit

EXAMPLE 15.9. Using the voltage divider rule, find the voltage across each element of the circuit of Fig. 15.39.

Solution:

$$\mathbf{V}_C = \frac{\mathbf{X}_C\mathbf{E}}{\mathbf{X}_C + \mathbf{R}} = \frac{(4 \angle -90°)(100 \angle 0°)}{4 \angle -90° + 3 \angle 0°} = \frac{400 \angle -90°}{3 - j4}$$
$$= \frac{400 \angle -90°}{5 \angle -53.13°} = \textbf{80} \angle \textbf{-36.87}°$$

$$\mathbf{V}_R = \frac{\mathbf{R}\mathbf{E}}{\mathbf{X}_C + \mathbf{R}} = \frac{(3 \angle 0°)(100 \angle 0°)}{5 \angle -53.13°} = \frac{300 \angle 0°}{5 \angle -53.13°}$$
$$= \textbf{60} \angle \textbf{+53.13}°$$

EXAMPLE 15.10. Using the voltage divider rule, find the unknown voltages $\mathbf{V}_R$, $\mathbf{V}_L$, $\mathbf{V}_C$, and $\mathbf{V}_1$ for the circuit of Fig. 15.40.

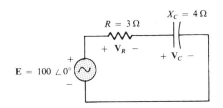

$X_C = 4\,\Omega$
$R = 3\,\Omega$
$+\ \mathbf{V}_R\ -$ $+\ \mathbf{V}_C\ -$
$\mathbf{E} = 100 \angle 0°$

FIG. 15.39

$X_C = 17\,\Omega$
$R = 6\,\Omega$ $X_L = 9\,\Omega$
$+\ \mathbf{V}_R\ -$ $+\ \mathbf{V}_L\ -$ $+\ \mathbf{V}_C\ -$
$\mathbf{E} = 50 \angle 30°$ $\mathbf{V}_1$

FIG. 15.40

Solution:

$$\mathbf{V}_R = \frac{\mathbf{R}\mathbf{E}}{\mathbf{R} + \mathbf{X}_L + \mathbf{X}_C} = \frac{(6 \angle 0°)(50 \angle 30°)}{6 \angle 0° + 9 \angle 90° + 17 \angle -90°}$$
$$= \frac{300 \angle 30°}{6 + j9 - j17} = \frac{300 \angle 30°}{6 - j8}$$
$$= \frac{300 \angle 30°}{10 \angle -53.13°} = \textbf{30} \angle \textbf{83.13}°$$

$$V_L = \frac{X_L E}{Z_T} = \frac{(9 \angle 90°)(50 \angle 30°)}{10 \angle -53.13°} = \frac{450 \angle 120°}{10 \angle -53.13°}$$
$$= 45 \angle \textbf{173.13}°$$

$$V_C = \frac{X_C E}{Z_T} = \frac{(17 \angle -90°)(50 \angle 30°)}{10 \angle -53°} = \frac{850 \angle -60°}{10 \angle -53°}$$
$$= 85 \angle \textbf{-6.87}°$$

$$V_1 = \frac{(X_L + X_C)E}{Z_T} = \frac{(9 \angle 90° + 17 \angle -90°)(50 \angle 30°)}{10 \angle -53.13°}$$
$$= \frac{(8 \angle -90°)(50 \angle 30°)}{10 \angle -53.13°}$$
$$= \frac{400 \angle -60°}{10 \angle -53.13°} = 40 \angle \textbf{-6.87}°$$

EXAMPLE 15.11. For the circuit of Fig. 15.41:
a. Calculate i, v_R, v_L, and v_C in phasor form.
b. Calculate the total power factor.
c. Calculate the average power delivered to the circuit.
d. Draw the phasor diagram.
e. Obtain the phasor sum of $\mathbf{V}_R$, $\mathbf{V}_L$, and $\mathbf{V}_C$, and show that it equals the input voltage $\mathbf{E}$.
f. Find $\mathbf{V}_R$ and $\mathbf{V}_C$ using the voltage divider rule.

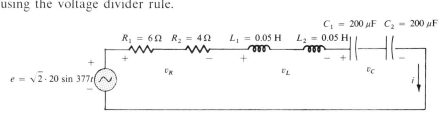

FIG. 15.41

Solutions:
a. Combining common elements and finding the reactance of the inductor and capacitor, we obtain

$$R_T = 6 + 4 = 10 \, \Omega$$
$$L_T = 0.05 + 0.05 = 0.1 \, H$$
$$C_T = \frac{200}{2} = 100 \, \mu F$$

$$X_L = \omega L = (377)(0.1) = 37.70 \, \Omega$$
$$X_C = \frac{1}{\omega C} = \frac{1}{(377)(100 \times 10^{-6})} = \frac{10^6}{37,700} = 26.53 \, \Omega$$

Redrawing the circuit using phasor notation results in Fig. 15.42.

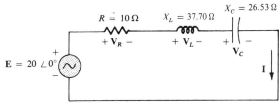

FIG. 15.42

For the circuit of Fig. 15.42,

$$\mathbf{Z}_T = R \angle 0° + X_L \angle 90° + X_C \angle -90°$$
$$= 10 + j37.70 - j26.53$$
$$= 10 + j11.17 = \mathbf{15} \angle \mathbf{48.16°}$$

The current **I** is

$$\mathbf{I} = \frac{\mathbf{E}}{\mathbf{Z}_T} = \frac{20 \angle 0°}{15 \angle 48.16°} = \mathbf{1.33} \angle \mathbf{-48.16°}$$

The voltage across the resistor, inductor, and capacitor can be found using Ohm's law:

$$\mathbf{V}_R = \mathbf{I}R = (1.33 \angle -48.16°)(10 \angle 0°)$$
$$= \mathbf{13.30} \angle \mathbf{-48.16°}$$
$$\mathbf{V}_L = \mathbf{I}X_L = (1.33 \angle -48.16°)(37.70 \angle 90°)$$
$$= \mathbf{50.14} \angle \mathbf{41.84°}$$
$$\mathbf{V}_C = \mathbf{I}X_C = (1.33 \angle -48.16°)(26.53 \angle -90°)$$
$$= \mathbf{35.28} \angle \mathbf{-138.16°}$$

b. The total power factor is the angle between the applied emf $\mathbf{E}_x$ and the resulting current **I** is 48.16°.

$$F_p = \cos \theta = \cos 48.16° = \mathbf{0.667 \ lagging}$$

or

$$F_p = \cos \theta = \frac{R}{Z_T} = \frac{10}{15} = \mathbf{0.667 \ lagging}$$

c. The total power in watts delivered to the circuit is

$$P_T = EI \cos \theta = (20)(1.33)(0.667) = \mathbf{17.74 \ W}$$

d. The phasor diagram appears in Fig. 15.43.

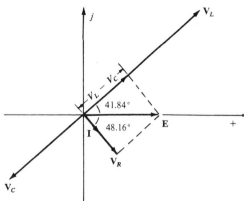

FIG. 15.43

e. The phasor sum of $\mathbf{V}_R$, $\mathbf{V}_L$, and $\mathbf{V}_C$ is

$$\mathbf{E} = \mathbf{V}_R + \mathbf{V}_L + \mathbf{V}_C$$
$$= 13.30 \angle -48.16° + 50.14 \angle 41.84° +$$
$$35.28 \angle -138.16°$$
$$\mathbf{E} = 13.30 \angle -48.16° + 14.86 \angle 41.84°$$

Therefore,

$$E = \sqrt{(13.30)^2 + (14.86)^2}$$
$$E = 20 \quad \text{and} \quad \theta_E = 0° \quad \text{(from phasor diagram)}$$

and

$$\mathbf{E} = 20 \angle 0°$$

f. $\mathbf{V}_R = \dfrac{\mathbf{RE}}{\mathbf{Z}_T} = \dfrac{(10 \angle 0°)(20 \angle 0°)}{15 \angle 48.16°} = \dfrac{200 \angle 0°}{15 \angle 48.16°}$

$$= 13.3 \angle -48.16°$$

$\mathbf{V}_C = \dfrac{\mathbf{X}_C\mathbf{E}}{\mathbf{Z}_T} = \dfrac{(26.5 \angle -90°)(20 \angle 0°)}{15 \angle 48.16°} = \dfrac{530 \angle -90°}{15 \angle 48.16°}$

$$= 35.28 \angle -138.16°$$

PARALLEL ac CIRCUITS

15.5 ADMITTANCE AND SUSCEPTANCE

The discussion for parallel ac circuits will be very similar to that for dc circuits. In dc circuits, *conductance* (G) was defined as equal to $1/R$. The total conductance of a parallel circuit was then found by adding the conductance of each branch. The total resistance R_T was then simply $1/G_T$.

In ac circuits, we define *admittance* (Y) as equal to $1/Z$. The unit of measure for admittance as defined by the SI system is *siemens,* which has the symbol S. For many years it was *mhos,* which had the inverted ohm symbol ℧. Admittance is a measure of how well an ac circuit will *admit* or allow current to flow in the circuit. The larger its value, therefore, the heavier the current flow for the same applied source of emf. The total admittance of a circuit can also be found by finding the sum of the parallel admittances. The total impedance Z_T of the circuit is then $1/Y_T$; that is, for the network of Fig. 15.44,

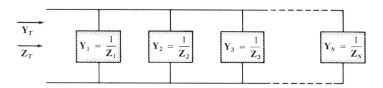

FIG. 15.44

$$\boxed{\mathbf{Y}_T = \mathbf{Y}_1 + \mathbf{Y}_2 + \mathbf{Y}_3 + \cdots + \mathbf{Y}_N} \qquad \textbf{(15.10)}$$

or, since $\mathbf{Z} = 1/\mathbf{Y}$,

$$\boxed{\frac{1}{\mathbf{Z}_T} = \frac{1}{\mathbf{Z}_1} + \frac{1}{\mathbf{Z}_2} + \frac{1}{\mathbf{Z}_3} + \cdots + \frac{1}{\mathbf{Z}_N}} \qquad \textbf{(15.11)}$$

For two impedances in parallel,

$$\frac{1}{\mathbf{Z}_T} = \frac{1}{\mathbf{Z}_1} + \frac{1}{\mathbf{Z}_2}$$

If the manipulations used in Chapter 5 to find the total resistance of two parallel resistors are now applied, the following similar equation will result:

$$\boxed{\mathbf{Z}_T = \frac{\mathbf{Z}_1 \mathbf{Z}_2}{\mathbf{Z}_1 + \mathbf{Z}_2}} \qquad \textbf{(15.12)}$$

For three parallel impedances,

$$\boxed{\mathbf{Z}_T = \frac{\mathbf{Z}_1 \mathbf{Z}_2 \mathbf{Z}_3}{\mathbf{Z}_1 \mathbf{Z}_2 + \mathbf{Z}_2 \mathbf{Z}_3 + \mathbf{Z}_1 \mathbf{Z}_3}} \qquad \textbf{(15.13)}$$

As pointed out in the introduction to this section, conductance is the reciprocal of resistance, and

$$\mathbf{Y} = \frac{1}{\mathbf{R}} = \frac{1}{R \angle 0^\circ} = G \angle 0^\circ$$

so that

$$\boxed{\mathbf{G} = G \angle 0^\circ} \qquad \textbf{(15.14)}$$

The reciprocal of reactance $(1/X)$ is called *susceptance* and is a measure of how *susceptible* an element is to the passage of current through it. Susceptance is also measured in *siemens* and is represented by the capital letter B.

For the inductor,

$$\mathbf{Y} = \frac{1}{\mathbf{X}_L} = \frac{1}{X_L \angle 90^\circ} = \frac{1}{\omega L \angle 90^\circ} = \frac{1}{\omega L} \angle -90^\circ$$

Defining

$$\boxed{B_L = \frac{1}{X_L}} \qquad \text{(siemens, S)} \qquad \textbf{(15.15)}$$

we have

$$\boxed{\mathbf{B}_L = B_L \angle -90^\circ} \qquad \textbf{(15.16)}$$

Note that for inductance an increase in frequency or inductance will result in a decrease in susceptance or, correspondingly, in admittance.

The capacitor,

$$\mathbf{Y} = \frac{1}{\mathbf{X}_C} = \frac{1}{X_C \angle -90°} = \frac{1}{1/\omega C \angle -90°} = \omega C \angle 90°$$

Defining

$$\boxed{B_C = \frac{1}{X_C}} \qquad (S) \qquad \textbf{(15.17)}$$

we have

$$\boxed{\mathbf{B}_C = B_C \angle 90°} \qquad \textbf{(15.18)}$$

For the capacitor, therefore, an increase in frequency or capacitance will result in an increase in its susceptibility.

In summary, for parallel circuits,

Resistance:

$$\mathbf{Y} = \frac{1}{\mathbf{R}} = \mathbf{G} = G \angle 0° = G + j0 \qquad \textbf{(15.19)}$$

Inductance:

$$\mathbf{Y} = \frac{1}{\mathbf{X}_L} = \mathbf{B}_L = B_L \angle -90° = 0 - jB_L \qquad \textbf{(15.20)}$$

Capacitance:

$$\mathbf{Y} = \frac{1}{\mathbf{X}_C} = \mathbf{B}_C = B_C \angle 90° = 0 + jB_C \qquad \textbf{(15.21)}$$

For parallel ac circuits, the *admittance diagram* is used with the three admittances represented as shown in Fig. 15.45.

Note in Fig. 15.45 that the conductance (like resistance) is on the positive real axis, while inductive and capacitive susceptance are still in direct opposition on the imaginary axis.

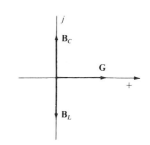

FIG. 15.45

EXAMPLE 15.12. For the network of Fig. 15.46:
a. Find the admittance of each parallel branch.
b. Determine the input admittance.
c. Calculate the input impedance.
d. Draw the admittance diagram.

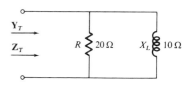

FIG. 15.46

Solutions:

a. $\mathbf{Y}_1 = \mathbf{G} = G \angle 0° = \frac{1}{R} \angle 0° = \frac{1}{20} \angle 0°$

$\qquad = 0.05 \angle 0° = 0.05 + j0$

$\mathbf{Y}_2 = \mathbf{B}_L = B_L \angle -90° = \frac{1}{X_L} \angle -90° = \frac{1}{10} \angle -90°$

$\qquad = 0.1 \angle -90° = 0 - j0.1$

b. $\mathbf{Y}_T = \mathbf{Y}_1 + \mathbf{Y}_2 = (0.05 + j0) + (0 - j0.1)$
$= \mathbf{0.05 - j0.1} = G - jB_L$

c. $\mathbf{Z}_T = \dfrac{1}{\mathbf{Y}_T} = \dfrac{1}{0.05 - j0.1} = \dfrac{1}{0.112 \angle -63.43°}$
$= \mathbf{8.93 \angle 63.43°}$

or Eq. (15.12):

$$\mathbf{Z}_T = \frac{\mathbf{Z}_1\mathbf{Z}_2}{\mathbf{Z}_1 + \mathbf{Z}_2} = \frac{(20 \angle 0°)(10 \angle 90°)}{20 + j10}$$
$$= \frac{200 \angle 90°}{22.36 \angle 26.57°} = \mathbf{8.93 \angle 63.43°}$$

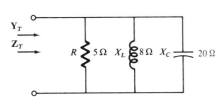

FIG. 15.47

d. The admittance diagram appears in Fig. 15.47.

EXAMPLE 15.13. Repeat Example 15.12 for the parallel network of Fig. 15.48.

Solutions:

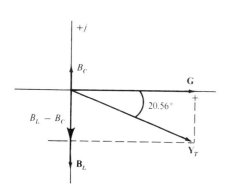

FIG. 15.48

a. $\mathbf{Y}_1 = \mathbf{G} = G \angle 0° = \dfrac{1}{R} \angle 0° = \dfrac{1}{5} \angle 0°$
$= \mathbf{0.2 \angle 0°} = 0.2 + j0$

$\mathbf{Y}_2 = \mathbf{B}_L = B_L \angle -90° = \dfrac{1}{X_L} \angle -90° = \dfrac{1}{8} \angle -90°$
$= \mathbf{0.125 \angle -90°} = 0 - j0.125$

$\mathbf{Y}_3 = \mathbf{B}_C = B_C \angle 90° = \dfrac{1}{X_C} \angle 90° = \dfrac{1}{20} \angle 90°$
$= \mathbf{0.050 \angle +90°} = 0 + j0.050$

b. $\mathbf{Y}_T = \mathbf{Y}_1 + \mathbf{Y}_2 + \mathbf{Y}_3$
$= (0.2 + j0) + (0 - j0.125) + (0 + j0.050)$
$= 0.2 - j0.075 = \mathbf{0.2136 \angle -20.56°}$

c. $\mathbf{Z}_T = \dfrac{1}{0.2136 \angle -20.56°} = \mathbf{4.68 \angle 20.56°}$

or

$$\mathbf{Z}_T = \frac{\mathbf{Z}_1\mathbf{Z}_2\mathbf{Z}_3}{\mathbf{Z}_1\mathbf{Z}_2 + \mathbf{Z}_2\mathbf{Z}_3 + \mathbf{Z}_1\mathbf{Z}_3}$$
$$= \frac{(5 \angle 0°)(8 \angle 90°)(20 \angle -90°)}{(5 \angle 0°)(8 \angle 90°) + (8 \angle 90°)(20 \angle -90°)}$$
$$\qquad\qquad\qquad\qquad + (5 \angle 0°)(20 \angle -90°)$$
$$= \frac{800 \angle 0°}{40 \angle 90° + 160 \angle 0° + 100 \angle -90°}$$
$$= \frac{800}{160 + j40 - j100} = \frac{800}{160 - j60}$$
$$= \frac{800}{170.88 \angle -20.56°}$$
$$= \mathbf{4.68 \angle 20.56°}$$

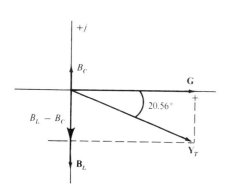

FIG. 15.49

d. The admittance diagram appears in Fig. 15.49.

On many occasions, the inverse relationship $\mathbf{Y}_T = 1/\mathbf{Z}_T$ or $\mathbf{Z}_T = 1/\mathbf{Y}_T$ will require that we divide the number 1 by a complex number having a real and an imaginary part. This division, if not performed in the polar form, requires that we multiply the numerator and denominator by the conjugate of the denominator, as follows:

$$\mathbf{Y}_T = \frac{1}{\mathbf{Z}_T} = \frac{1}{4 + j6} = \frac{1}{4 + j6}\frac{(4 - j6)}{(4 - j6)} = \frac{4 - j6}{4^2 + 6^2}$$

and

$$\mathbf{Y}_T = \frac{4}{52} - j\frac{6}{52}$$

To avoid this laborious task each time we want to find the reciprocal of a complex number in rectangular form, a format can be developed using the following complex number, which is symbolic of any impedance or admittance in the first or fourth quadrant:

$$\frac{1}{a_1 \pm jb_1} = \left(\frac{1}{a_1 \pm jb_1}\right)\left(\frac{a_1 \mp jb_1}{a_1 \mp jb_1}\right) = \frac{a_1 \mp jb_1}{a_1^2 + b_1^2}$$

or

$$\boxed{\frac{1}{a_1 \pm jb_1} = \frac{a_1}{a_1^2 + b_1^2} \mp j\frac{b_1}{a_1^2 + b_1^2}} \qquad \textbf{(15.22)}$$

Note that the denominator is simply the sum of the squares of each term. The sign is inverted between the real and imaginary parts. A few examples will develop some familiarity with the use of this equation.

EXAMPLE 15.14. Find the admittance of each set of series elements in Fig. 15.50.

Solutions:

a. $\mathbf{Z} = 6 - j8$

$$\mathbf{Y} = \frac{1}{6 - j8} = \frac{6}{6^2 + 8^2} + j\frac{8}{6^2 + 8^2} = \frac{6}{100} + j\frac{8}{100}$$

b. $\mathbf{Z} = 10 + j4 + (-j0.1) = 10 + j3.9$

$$\mathbf{Y} = \frac{1}{\mathbf{Z}} = \frac{1}{10 + j3.9} = \frac{10}{10^2 + 3.9^2} - j\frac{3.9}{10^2 + 3.9^2}$$

$$= \frac{10}{115.21} - j\frac{3.9}{115.21} = \mathbf{0.087 - j0.034}$$

To determine whether a series ac circuit is predominantly capacitive or inductive, it is necessary only to note whether the sign of the imaginary part of the total impedance is positive or negative. For parallel circuits, a negative sign in front of the imaginary part of the total admittance indicates that the circuit is predominantly inductive, and a positive sign indicates that it is predominantly capacitive.

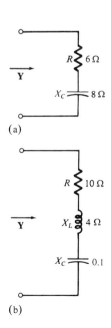

(a)

(b)

FIG. 15.50

15.6 *R-L, R-C,* AND *R-L-C* PARALLEL ac NETWORKS

R-L (Fig. 15.51)

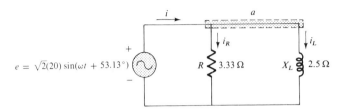

$e = \sqrt{2}(20)\sin(\omega t + 53.13°)$

R 3.33 Ω X_L 2.5 Ω

FIG. 15.51

Phasor notation: As shown in Fig. 15.52.

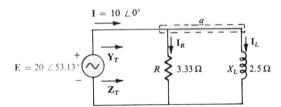

$\mathbf{I} = 10 \angle 0°$

$\mathbf{E} = 20 \angle 53.13°$ $\mathbf{Y}_T$ $\mathbf{Z}_T$ R 3.33 Ω X_L 2.5 Ω

FIG. 15.52

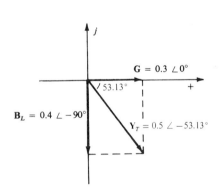

FIG. 15.53

$\mathbf{Y}_T(\mathbf{Z}_T)$:

$$\mathbf{Y}_T = \mathbf{Y}_1 + \mathbf{Y}_2 = \mathbf{G} + \mathbf{B}_L = \frac{1}{3.33} \angle 0° + \frac{1}{2.5} \angle -90°$$

$$= 0.3 \angle 0° + 0.4 \angle -90° = 0.3 - j0.4$$

$$= \mathbf{0.5 \angle -53.13°}$$

$$\mathbf{Z}_T = \frac{1}{\mathbf{Y}_T} = \frac{1}{0.5 \angle -53.13°} = \mathbf{2 \angle 53.13°}$$

Admittance diagram: As shown in Fig. 15.53.

I:

$$\mathbf{I} = \frac{\mathbf{E}}{\mathbf{Z}_T} = \mathbf{EY}_T = (20 \angle 53.13°)(0.5 \angle -53.13°) = \mathbf{10 \angle 0°}$$

$\mathbf{I}_R$, $\mathbf{I}_L$:

$$\mathbf{I}_R = \frac{\mathbf{E}}{\mathbf{R}} = \mathbf{EG} = (20 \angle 53.13°)(0.3 \angle 0°) = \mathbf{6 \angle 53.13°}$$

$$\mathbf{I}_L = \frac{\mathbf{E}}{\mathbf{X}_L} = \mathbf{EB}_L = (20 \angle 53.13°)(0.4 \angle -90°)$$

$$= \mathbf{8 \angle -36.87°}$$

Kirchhoff's current law: At node *a*,

$$\mathbf{I} - \mathbf{I}_R - \mathbf{I}_L = 0$$

or

$$\mathbf{I} = \mathbf{I}_R + \mathbf{I}_L$$

$$10 \angle 0° = 6 \angle 53.13° + 8 \angle -36.87°$$
$$10 \angle 0° = (3.60 + j4.80) + (6.40 - j4.80) = 10 + j0$$

and

$$10 \angle 0° = 10 \angle 0° \quad \text{(checks)}$$

Phasor diagram: The phasor diagram of Fig. 15.54 indicates that the applied voltage $\mathbf{E}$ is in phase with the current $\mathbf{I}_R$ and leads the current $\mathbf{I}_L$ by 90°.

Power: The total power in watts delivered to the circuit is

$$P_T = EI \cos \theta_T$$
$$= (20)(10) \cos 53.13° = (220)(0.6)$$
$$= \mathbf{120 \ W}$$

or

$$P_T = I^2 R = \frac{V_R^2}{R} = V_R^2 G = (20^2)(0.3) = \mathbf{120 \ W}$$

or, finally,

$$P_T = P_R + P_L = EI_R \cos \theta_R + EI_L \cos \theta_L$$
$$= (20)(6) \cos 0° + (20)(8) \cos 90° = 120 + 0$$
$$= \mathbf{120 \ W}$$

Power factor: The power factor of the circuit is

$$F_p = \cos \theta = \cos 53.13° = \mathbf{0.6 \ lagging}$$

or, through an analysis similar to that employed for a series ac circuit,

$$\cos \theta = \frac{P}{EI} = \frac{E^2/R}{EI} = \frac{EG}{I} = \frac{G}{I/V} = \frac{G}{Y_T}$$

and

$$\boxed{F_p = \cos \theta_T = \frac{G}{Y_T}} \qquad \textbf{(15.23)}$$

where G and Y_T are the magnitudes of the total conductance and admittance of the parallel network. For this case,

$$F_p = \cos \theta_T = \frac{0.3}{0.5} = \mathbf{0.6 \ lagging}$$

Impedance approach: The current $\mathbf{I}$ can also be found by first finding the total impedance of the network:

$$\mathbf{Z}_T = \frac{\mathbf{Z}_1 \mathbf{Z}_2}{\mathbf{Z}_1 + \mathbf{Z}_2} = \frac{(3.33 \angle 0°)(2.5 \angle 90°)}{3.33 \angle 0° + 2.5 \angle 90°}$$
$$= \frac{8.325 \angle 90°}{4.164 \angle 36.87°} = \mathbf{2 \angle 53.13°}$$

And then, using Ohm's law, we obtain

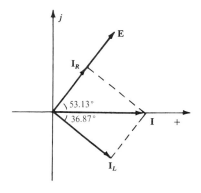

FIG. 15.54

$$I = \frac{E}{Z_T} = \frac{20 \angle 53.13°}{2 \angle 53.13°} = 10 \angle 0°$$

R-C (Fig. 15.55)

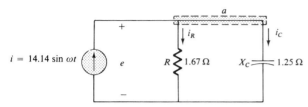

FIG. 15.55

Phasor notation: As shown in Fig. 15.56.

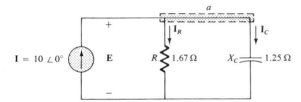

FIG. 15.56

$Y_T(Z_T)$:

$$Y_T = Y_1 + Y_2 = G + B_C = \frac{1}{1.67} \angle 0° + \frac{1}{1.25} \angle 90°$$
$$= 0.6 \angle 0° + 0.8 \angle 90° = 0.6 + j0.8 = \mathbf{1.0} \angle \mathbf{53.13°}$$

$$Z_T = \frac{1}{Y_T} = \frac{1}{1.0 \angle 53.13°} = 1 \angle -53.13°$$

Admittance diagram: As shown in Fig. 15.57.

E:

$$E = IZ_T = \frac{I}{Y_T} = \frac{10 \angle 0°}{1 \angle 53.13°} = 10 \angle -53.13°$$

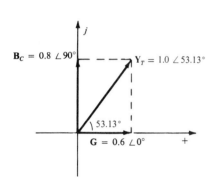

FIG. 15.57

I_R, I_C:

$$I_R = EG = (10 \angle -53.13°)(0.6 \angle 0°) = 6 \angle -53.13°$$
$$I_C = EB_C = (10 \angle -53.13°)(0.8 \angle 90°) = 8 \angle 36.87°$$

Kirchhoff's current law: At node *a*,

$$I - I_R - I_C = 0$$

or

$$I = I_R + I_C$$

which can also be verified (as for the *R-L* network) through vector algebra.

Phasor diagram: The phasor diagram of Fig. 15.58 indicates that **E** is in phase with the current through the resistor I_R and lags the capacitive current I_C by 90°.

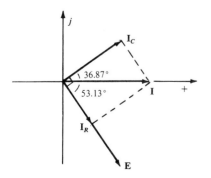

FIG. 15.58

Time domain:

$$e = \sqrt{2}(10)\sin(\omega t - 53.13°) = \mathbf{14.14\ sin(\omega t - 53.13°)}$$
$$i_R = \sqrt{2}(6)\sin(\omega t - 53.13°) = \mathbf{8.48\ sin(\omega t - 53.13°)}$$
$$i_C = \sqrt{2}(8)\sin(\omega t + 36.87°) = \mathbf{11.31\ sin(\omega t + 36.87°)}$$

A plot of all of the currents and the voltage appears in Fig. 15.59. Note that e and i_R are in phase and e lags i_C by 90°.

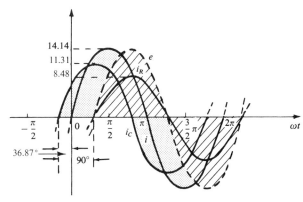

FIG. 15.59

Power:

$$P_T = EI\cos\theta = (10)(10)\cos 53.13° = (10^2)(0.6)$$
$$= \mathbf{60\ W}$$

or

$$P_T = E^2(G) = (10^2)(0.6) = \mathbf{60\ W}$$

or, finally,

$$P_T = P_R + P_C = EI_R\cos\theta_R + EI_C\cos\theta_C$$
$$= (10)(6)\cos 0° + (10)(8)\cos 90°$$
$$= \mathbf{60\ W}$$

Power factor: The power factor of the circuit is

$$F_p = \cos 53.13° = \mathbf{0.6\ leading}$$

Using Eq. (15.23), we have

$$F_p = \cos\theta = \frac{G}{Y_T} = \frac{0.6}{1.0} = \mathbf{0.6\ leading}$$

Impedance approach: The voltage **E** can also be found by first finding the total impedance of the circuit:

$$\mathbf{Z}_T = \frac{\mathbf{Z}_1\mathbf{Z}_2}{\mathbf{Z}_1 + \mathbf{Z}_2} = \frac{(1.67\ \angle 0°)(1.25\ \angle -90°)}{1.67\ \angle 0° + 1.25\ \angle -90°}$$

$$= \frac{2.09\ \angle -90°}{2.09\ \angle -36.87°} = \mathbf{1\ \angle -53.13°}$$

and then, using Ohm's law, we find

$$\mathbf{E} = \mathbf{IZ}_T = (10\ \angle 0°)(1\ \angle -53.13°) = \mathbf{10\ \angle -53.13°}$$

R-L-C (Fig. 15.60)

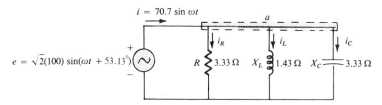

FIG. 15.60

Phasor notation: As shown in Fig. 15.61.

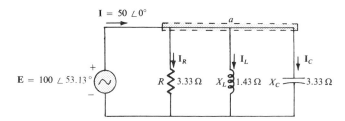

FIG. 15.61

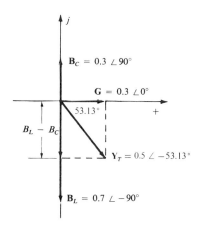

FIG. 15.62

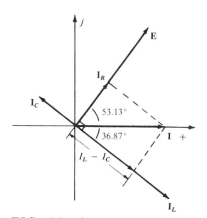

FIG. 15.63

$\mathbf{Y}_T(\mathbf{Z}_T)$:

$$\mathbf{Y}_T = \mathbf{Y}_1 + \mathbf{Y}_2 + \mathbf{Y}_3 = \mathbf{G} + \mathbf{B}_L + \mathbf{B}_C$$
$$= \frac{1}{3.33} \angle 0° + \frac{1}{1.43} \angle -90° + \frac{1}{3.33} \angle 90°$$
$$= 0.3 \angle 0° + 0.7 \angle -90° + 0.3 \angle 90°$$
$$= 0.3 - j0.7 + j0.3$$
$$\mathbf{Y}_T = 0.3 - j0.4 = \mathbf{0.5} \angle \mathbf{-53.13°}$$
$$\mathbf{Z}_T = \frac{1}{\mathbf{Y}_T} = \frac{1}{0.5 \angle -53.13°} = \mathbf{2} \angle \mathbf{53.13°}$$

Admittance diagram: As shown in Fig. 15.62.

I:

$$\mathbf{I} = \frac{\mathbf{E}}{\mathbf{Z}_T} = \mathbf{E}\mathbf{Y}_T = (100 \angle 53.13°)(0.5 \angle -53.13°) = \mathbf{50} \angle \mathbf{0°}$$

$\mathbf{I}_R$, $\mathbf{I}_L$, $\mathbf{I}_C$:

$$\mathbf{I}_R = \mathbf{E}\mathbf{G} = (100 \angle 53.13°)(0.3 \angle 0°) = \mathbf{30} \angle \mathbf{53.13°}$$
$$\mathbf{I}_L = \mathbf{E}\mathbf{B}_L = (100 \angle 53.13°)(0.7 \angle -90°) = \mathbf{70} \angle \mathbf{-36.87°}$$
$$\mathbf{I}_C = \mathbf{E}\mathbf{B}_C = (100 \angle 53.13°)(0.3 \angle +90°) = \mathbf{30} \angle \mathbf{143.13°}$$

Kirchhoff's current law: At node *a*,

$$\mathbf{I} - \mathbf{I}_R - \mathbf{I}_L - \mathbf{I}_C = 0$$

or

$$\mathbf{I} = \mathbf{I}_R + \mathbf{I}_L + \mathbf{I}_C$$

Phasor diagram: The phasor diagram of Fig. 15.63 indicates that the impressed voltage **E** is in phase with the current $\mathbf{I}_R$ through the resistor, leads the current $\mathbf{I}_L$ through the inductor by 90°, and lags the current $\mathbf{I}_C$ of the capacitor by 90°.

Time domain:

$$i = \sqrt{2}(50) \sin \omega t = \textbf{70.70 sin } \boldsymbol{\omega t}$$
$$i_R = \sqrt{2}(30) \sin(\omega t + 53.13°) = \textbf{42.42 sin}(\boldsymbol{\omega t} + \textbf{53.13°})$$
$$i_L = \sqrt{2}(70) \sin(\omega t - 36.87°) = \textbf{98.98 sin}(\boldsymbol{\omega t} - \textbf{36.87°})$$
$$i_C = \sqrt{2}(30) \sin(\omega t + 143.13°) = \textbf{42.42 sin}(\boldsymbol{\omega t} + \textbf{143.13°})$$

A plot of all of the currents and the impressed voltage appears in Fig. 15.64.

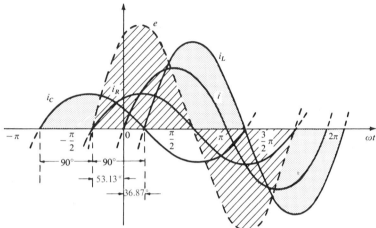

FIG. 15.64

Power: The total power in watts delivered to the circuit is

$$P_T = EI \cos \theta = (100)(50) \cos 53.13° = (5000)(0.6)$$
$$= \textbf{3000 W}$$

or

$$P_T = E^2 G = (100^2)(0.3) = \textbf{3000 W}$$

or, finally,

$$P_T = P_R + P_L + P_C$$
$$= EI_R \cos \theta_R + EI_L \cos \theta_L + EI_C \cos \theta_C$$
$$= (100)(30) \cos 0° + (100)(70) \cos 90°$$
$$+ (100)(30) \cos 90°$$
$$= 3000 + 0 + 0$$
$$= \textbf{3000 W}$$

Power factor: The power factor of the circuit is

$$F_p = \cos \theta_T = \cos 53.13° = \textbf{0.6 lagging}$$

Using Eq. (15.23), we obtain

$$F_p = \cos \theta_T = \frac{G}{Y_T} = \frac{0.3}{0.5} = \textbf{0.6 lagging}$$

Impedance approach: The input current **I** can also be determined by first finding the total impedance in the following manner:

$$\mathbf{Z}_T = \frac{\mathbf{Z}_1 \mathbf{Z}_2 \mathbf{Z}_3}{\mathbf{Z}_1 \mathbf{Z}_2 + \mathbf{Z}_2 \mathbf{Z}_3 + \mathbf{Z}_1 \mathbf{Z}_3} = \textbf{2} \angle \textbf{53.13°}$$

and, applying Ohm's law, we obtain

$$\mathbf{I} = \frac{\mathbf{E}}{\mathbf{Z}_T} = \frac{100 \angle 53.13°}{2 \angle 53.13°} = 50 \angle 0°$$

15.7 CURRENT DIVIDER RULE

The basic format for the current divider rule in ac circuits is exactly the same as that for dc circuits; that is, for two parallel branches with impedances $\mathbf{Z}_1$ and $\mathbf{Z}_2$ as shown in Fig. 15.65,

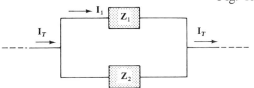

FIG. 15.65

$$\boxed{\mathbf{I}_1 = \frac{\mathbf{Z}_2\mathbf{I}_T}{\mathbf{Z}_1 + \mathbf{Z}_2} \quad \text{or} \quad \mathbf{I}_2 = \frac{\mathbf{Z}_1\mathbf{I}_T}{\mathbf{Z}_1 + \mathbf{Z}_2}} \qquad (15.24)$$

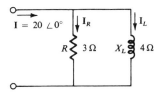

FIG. 15.66

EXAMPLE 15.15. Using the current divider rule, find the current through each impedance of Fig. 15.66.

Solution:

$$\mathbf{I}_R = \frac{\mathbf{X}_L\mathbf{I}_T}{\mathbf{R} + \mathbf{X}_L} = \frac{(4 \angle 90°)(20 \angle 0°)}{3 \angle 0° + 4 \angle 90°} = \frac{80 \angle 90°}{5 \angle 53.13°}$$
$$= 16 \angle 36.87°$$

$$\mathbf{I}_L = \frac{\mathbf{R}\mathbf{I}_T}{\mathbf{R} + \mathbf{X}_L} = \frac{(3 \angle 0°)(20 \angle 0°)}{5 \angle 53.13°} = \frac{60 \angle 0°}{5 \angle 53.13°}$$
$$= 12 \angle -53.13°$$

EXAMPLE 15.16. Using the current divider rule, find the current through each parallel branch of Fig. 15.67.

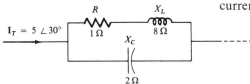

FIG. 15.67

Solution:

$$\mathbf{I}_{R-L} = \frac{\mathbf{X}_C\mathbf{I}_T}{\mathbf{X}_C + \mathbf{Z}_{R-L}} = \frac{(2 \angle -90°)(5 \angle 30°)}{-j2 + 1 + j8} = \frac{10 \angle -60°}{1 + j6}$$
$$= \frac{10 \angle -60°}{6.083 \angle 80.54°} \cong 1.644 \angle -140.54°$$

$$\mathbf{I}_C = \frac{\mathbf{Z}_{R-L}\mathbf{I}_T}{\mathbf{Z}_{R-L} + \mathbf{X}_C} = \frac{(1 + j8)(5 \angle 30°)}{6.08 \angle 80.54°}$$
$$= \frac{(8.06 \angle 82.87°)(5 \angle 30°)}{6.08 \angle 80.54°} = \frac{40.31 \angle 112.87°}{6.083 \angle 80.54°}$$
$$= 6.627 \angle 32.33°$$

15.8 EQUIVALENT CIRCUITS

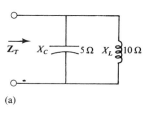

(a)

(b)

FIG. 15.68

In a series ac circuit, the total impedance of two or more elements in series is often equivalent to an impedance that can be achieved with fewer elements of different values, the elements and their values being determined by the frequency applied. This is also true for parallel circuits. For the circuit of Fig. 15.68(a),

$$\mathbf{Z}_T = \frac{\mathbf{X}_C\mathbf{X}_L}{\mathbf{X}_C + \mathbf{X}_L} = \frac{(5 \angle -90°)(10 \angle 90°)}{5 \angle -90° + 10 \angle 90°} = \frac{50 \angle 0°}{5 \angle 90°}$$
$$= 10 \angle -90°$$

The total impedance at the frequency applied is equivalent to a capacitor with a reactance of 10 Ω, as shown in Fig. 15.68(b). Always keep in mind that this equivalence is true only at the applied frequency. If the frequency changes, the reactance of each element changes, and the equivalent circuit will change—perhaps from capacitive to inductive in the above example.

Another interesting development appears if the impedance of a parallel circuit, such as the one of Fig. 15.69(a), is found in rectangular coordinates. In this case,

$$\mathbf{Z}_T = \frac{\mathbf{X}_L\mathbf{R}}{\mathbf{X}_L + \mathbf{R}} = \frac{(4 \angle 90°)(3 \angle 0°)}{4 \angle 90° + 3 \angle 0°}$$
$$= \frac{12 \angle 90°}{5 \angle 53.13°} = 2.40 \angle 36.87°$$
$$= 1.920 + j1.440$$

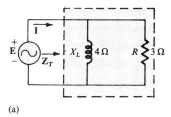

(a)

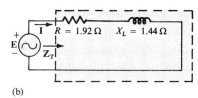

(b)

FIG. 15.69

which is the impedance of a series circuit with a resistor of 1.92 Ω and an inductive reactance of 1.44 Ω, as shown in Fig. 15.69(b).

The current **I** will be the same in each circuit of Fig. 15.68 or Fig. 15.69 if the same input voltage **E** is applied. For a parallel circuit of one resistive element and one reactive element, the series circuit with the same input impedance will always be composed of one resistive and one reactive element. The impedance of each element of the series circuit will be different from that of the parallel circuit, but the reactive elements will always be of the same type; that is, an *R-L* circuit and an *R-C* parallel circuit will have an equivalent *R-L* and *R-C* series circuit, respectively. The same is true when converting from a series to a parallel circuit. In the discussion to follow, keep in mind that *the term* equivalent *refers only to the fact that for the same applied potential, the same impedance and input current will result.*

To formulate the equivalence between the series and parallel circuits, the equivalent series circuit for a resistor and reactance in parallel can be found by determining the

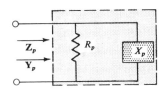

(a)

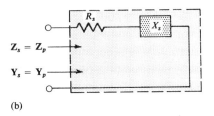

(b)

FIG. 15.70

total impedance of the circuit in rectangular form; that is, for the circuit of Fig. 15.70(a),

$$Y_p = \frac{1}{R_p} + \frac{1}{\pm jX_p}$$

and

$$Z_p = \frac{1}{Y_p} = \frac{1}{(1/R_p) \mp j(1/X_p)}$$

$$= \frac{1/R_p}{(1/R_p)^2 + (1/X_p)^2} \pm j\frac{(1/X_p)}{(1/R_p)^2 + (1/X_p)^2}$$

Multiplying the numerator and denominator of each term by $R_p^2 X_p^2$ results in

$$Z_p = \frac{R_p X_p^2}{X_p^2 + R_p^2} \pm j\frac{R_p^2 X_p}{X_p^2 + R_p^2}$$

$$= R_s \pm jX_s \qquad \text{[Fig. 15.70(b)]}$$

and

$$\boxed{R_s = \frac{R_p X_p^2}{X_p^2 + R_p^2}} \qquad\qquad \textbf{(15.25)}$$

with

$$\boxed{X_s = \frac{R_p^2 X_p}{X_p^2 + R_p^2}} \qquad\qquad \textbf{(15.26)}$$

For the network of Fig. 15.69,

$$R_s = \frac{R_p X_p^2}{X_p^2 + R_p^2} = \frac{(3)(4)^2}{4^2 + 3^2} = \frac{48}{25} = \textbf{1.920 } \Omega$$

and

$$X_s = \frac{R_p^2 X_p}{X_p^2 + R_p^2} = \frac{(3)^2(4)}{25} = \frac{36}{25} = \textbf{1.440 } \Omega$$

which agrees with the previous result.

The equivalent parallel circuit for a circuit with a resistor and reactance in series can be found by simply finding the total admittance of the system in rectangular form; that is, for the circuit of Fig. 15.70(b),

$$Z_s = R_s \pm jX_s$$

$$Y_s = \frac{1}{Z_s} = \frac{1}{R_s \pm jX_s} = \frac{R_s}{R_s^2 + X_s^2} \mp j\frac{X_s}{R_s^2 + X_s^2}$$

$$= G_p \mp jB_p = \frac{1}{R_p} \mp j\frac{1}{X_p} \qquad \text{[Fig. 15.70(a)]}$$

or

$$R_p = \frac{R_s^2 + X_s^2}{R_s} \qquad (15.27)$$

with

$$X_p = \frac{R_s^2 + X_s^2}{X_s} \qquad (15.28)$$

For the above example,

$$R_p = \frac{R_s^2 + X_s^2}{R_s} = \frac{(1.92)^2 + (1.44)^2}{1.92} = \frac{5.76}{1.92} = \mathbf{3.0\ \Omega}$$

and

$$X_p = \frac{R_s^2 + X_s^2}{X_s} = \frac{5.76}{1.44} = \mathbf{4.0\ \Omega}$$

as shown in Fig. 15.69(a).

EXAMPLE 15.17. Determine the series equivalent circuit for the network of Fig. 15.71.

Solution:

$$R_p = 8\ \text{k}\Omega \quad X_p\ (\text{resultant}) = |X_L - X_C| = |9\ \text{k}\Omega - 4\ \text{k}\Omega| \\ = 5\ \text{k}\Omega$$

and

$$R_s = \frac{R_p X_p^2}{X_p^2 + R_p^2} = \frac{(8\ \text{k}\Omega)(5\ \text{k}\Omega)^2}{(5\ \text{k}\Omega)^2 + (8\ \text{k}\Omega)^2} = \frac{200\ \text{k}\Omega}{89} = \mathbf{2.247\ \text{k}\Omega}$$

FIG. 15.71

with

$$X_s = \frac{R_p^2 X_p}{X_p^2 + R_p^2} = \frac{(8\ \text{k}\Omega)^2 5\ \text{k}\Omega}{89\ \text{k}\Omega} = \frac{320\ \text{k}\Omega}{89} \\ = \mathbf{3.596\ \text{k}\Omega} \qquad \textbf{(inductive)}$$

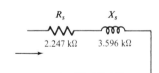

The equivalent series circuit appears in Fig. 15.72.

EXAMPLE 15.18. For the network of Fig. 15.73:

FIG. 15.72

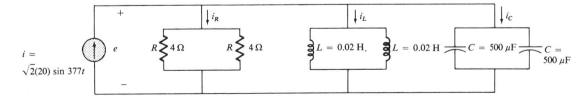

FIG. 15.73

a. Compute e, i_R, i_L, and i_C in phasor form.
b. Compute the total power factor.
c. Compute the total power delivered to the network.
d. Draw the phasor diagram.

e. Obtain the phasor sum of I_R, I_L, and I_C, and show that it equals **I**.
f. Compute the impedance of the parallel combination of X_L and X_C, and then find I_R by the current divider rule.
g. Determine the equivalent series circuit as far as the total impedance and current **I** are concerned.

Solutions:

a. Combining common elements and finding the reactance of the inductor and capacitor, we find

$$R_T = \frac{R}{2} = \frac{4}{2} = 2\ \Omega$$

$$L_T = \frac{0.02}{2} = 0.01\ \text{H}$$

$$C_T = 500\ \mu\text{F} + 500\ \mu\text{F} = 1000\ \mu\text{F}$$

$$X_L = \omega L = (377)(0.01) = 3.77\ \Omega$$

$$X_C = \frac{1}{\omega C} = \frac{1}{(377)(10^3 \times 10^{-6})} = 2.65\ \Omega$$

The network is redrawn in Fig. 15.74 with phasor notation. The total admittance is

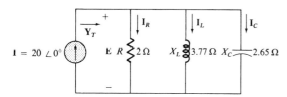

FIG. 15.74

$$\mathbf{Y}_T = \mathbf{Y}_1 + \mathbf{Y}_2 + \mathbf{Y}_3 = \mathbf{G} + \mathbf{B}_L + \mathbf{B}_C$$

$$= \frac{1}{2}\ \angle 0° + \frac{1}{3.77}\ \angle -90° + \frac{1}{2.65}\ \angle +90°$$

$$= 0.5\ \angle 0° + 0.265\ \angle -90° + 0.377\ \angle +90°$$

$$= 0.5 - j0.265 + j0.377$$

$$\mathbf{Y}_T = 0.5 + j0.112 = \mathbf{0.512}\ \angle\ \mathbf{12.63°}$$

The input voltage is

$$\mathbf{E} = \frac{\mathbf{I}}{\mathbf{Y}_T} = \frac{20\ \angle 0°}{0.512\ \angle 12.63°}$$

$$= \mathbf{39.06}\ \angle\ \mathbf{-12.63°}$$

The current through the resistor, inductor, and capacitor can be found using Ohm's law as follows:

$$\mathbf{I}_R = \mathbf{EG} = (39.06\ \angle -12.63°)(0.5\ \angle 0°)$$

$$= \mathbf{19.53}\ \angle\ \mathbf{-12.63°}$$

$$\mathbf{I}_L = \mathbf{EB}_L = (39.06\ \angle -12.63°)(0.265\ \angle -90°)$$

$$= \mathbf{10.35}\ \angle\ \mathbf{-102.63°}$$

$$\mathbf{I}_C = \mathbf{EB}_C = (39.06\ \angle -12.63°)(0.377\ \angle +90°)$$

$$= \mathbf{14.73}\ \angle\ \mathbf{+77.37°}$$

b. The total power factor is

$$F_p = \cos \theta = \frac{G}{Y_T} = \frac{0.5}{0.512} = \mathbf{0.977}$$

c. The total power in watts delivered to the circuit is

$$P_T = I_R^2 R = E^2 G = (39.06)^2(0.5)$$
$$= \mathbf{762.84 \ W}$$

d. The phasor diagram is shown in Fig. 15.75.
e. The phasor sum of $\mathbf{I}_R$, $\mathbf{I}_L$, and $\mathbf{I}_C$ is

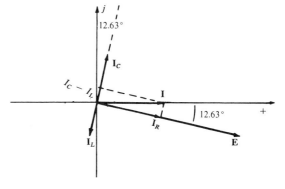

FIG. 15.75

$$\mathbf{I} = \mathbf{I}_R + \mathbf{I}_L + \mathbf{I}_C$$
$$= 19.53 \angle -12.63° + 10.35 \angle -102.63°$$
$$+ 14.73 \angle +77.37°$$
$$\mathbf{I} = 19.53 \angle -12.63° + 4.38 \angle +77.37°$$
$$I_T = \sqrt{(19.53)^2 + (4.38)^2} = 20$$

and θ_T (from phasor diagram) $= 0°$. Therefore,

$$\mathbf{I}_T = \mathbf{20 \ \angle 0°}$$

which agrees with the original input.
f. The impedance of the parallel combination of $\mathbf{X}_L$ and $\mathbf{X}_C$ is

$$\mathbf{Z}_{T_1} = \frac{\mathbf{X}_L \mathbf{X}_C}{\mathbf{X}_L + \mathbf{X}_C} = \frac{(3.77 \angle 90°)(2.65 \angle -90°)}{3.77 \angle 90° + 2.65 \angle -90°}$$
$$= \frac{10 \angle 0°}{1.12 \angle 90°}$$
$$\mathbf{Z}_{T_1} = 8.93 \angle -90°$$

The current $\mathbf{I}_R$, using the current divider rule, is

$$\mathbf{I}_R = \frac{\mathbf{Z}_{T_1} \mathbf{I}}{\mathbf{Z}_{T_1} + \mathbf{R}} = \frac{(8.93 \angle -90°)(20 \angle 0°)}{8.93 \angle -90° + 2 \angle 0°}$$
$$= \frac{178.60 \angle -90°}{9.15 \angle -77.37°} = \mathbf{19.52 \ \angle -12.63°}$$

g. $\mathbf{Z}_T = \dfrac{1}{\mathbf{Y}_T} = \dfrac{1}{0.512 \angle 12.63°} = \mathbf{1.95 \ \angle -12.63°}$

which, in rectangular form, is $1.90 - j0.427$, and

$$C = \frac{1}{\omega X_C} = \frac{1}{377(0.427)} = 0.00621 \ \text{F} = \mathbf{6210 \ \mu F}$$

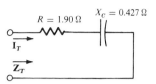

FIG. 15.76

The series circuit appears in Fig. 15.76. Since Z_{T_1} is available from part (f) above, we can apply Eqs. (15.19) and (15.20). That is, $Z_{T_1} = X_p$, and

$$R_s = \frac{R_p X_p^2}{X_p^2 + R_p^2} = \frac{(2)(8.93)^2}{(8.93)^2 + (2)^2} = \frac{159.49}{83.74} = \mathbf{1.90 \ \Omega}$$

with

$$X_s = \frac{R_p^2 X_p}{X_p^2 + R_p^2} = \frac{(2)^2(8.93)}{83.74} = \frac{35.72}{83.74} = \mathbf{0.427 \ \Omega}$$

as obtained above.

PROBLEMS

Section 15.2

1. Express the impedances of Fig. 15.77 in both polar and rectangular form.

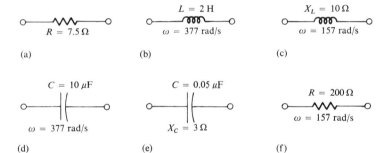

FIG. 15.77

2. Find the current i for the elements of Fig. 15.78 using phasor algebra. Sketch the waveforms for v and i on the same set of axes.

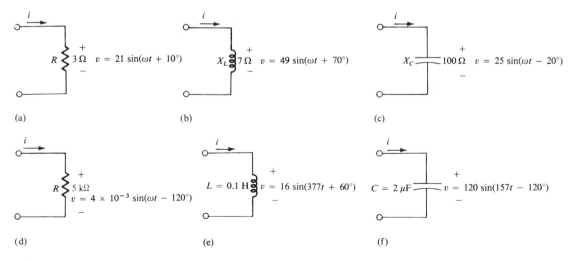

FIG. 15.78

3. Find the voltage v for the elements of Fig. 15.79 using phasor algebra. Sketch the waveforms of v and i on the same set of axes.

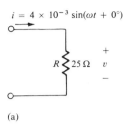
$i = 4 \times 10^{-3} \sin(\omega t + 0°)$

$R \gtrless 25 \, \Omega$ $+$ v $-$

(a)

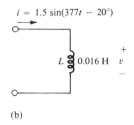
$i = 1.5 \sin(377t - 20°)$

$L \gtrless 0.016 \, H$ $+$ v $-$

(b)

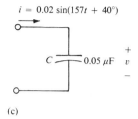
$i = 0.02 \sin(157t + 40°)$

$C \; 0.05 \, \mu F$ $+$ v $-$

(c)

Section 15.3

FIG. 15.79

4. Calculate the total impedance of the circuits of Fig. 15.80. Express your answer in rectangular and polar form, and draw the impedance diagram.

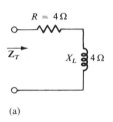
$R = 4 \, \Omega$

$\mathbf{Z}_T$ $X_L \gtrless 4 \, \Omega$

(a)

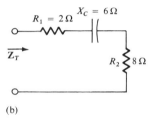
$R_1 = 2 \, \Omega$ $X_C = 6 \, \Omega$

$\mathbf{Z}_T$ $R_2 \gtrless 8 \, \Omega$

(b)

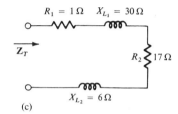
$R_1 = 1 \, \Omega$ $X_{L_1} = 30 \, \Omega$

$\mathbf{Z}_T$ $R_2 \gtrless 17 \, \Omega$

$X_{L_2} = 6 \, \Omega$

(c)

FIG. 15.80

5. Repeat Problem 4 for the circuits of Fig. 15.81.

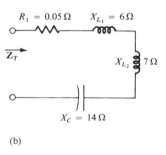
$R_1 = 3 \, \Omega$

$\mathbf{Z}_T$ $X_L \gtrless 4 \, \Omega$

$X_C = 7 \, \Omega$

(a)

$R_1 = 0.05 \, \Omega$ $X_{L_1} = 6 \, \Omega$

$\mathbf{Z}_T$ $X_{L_2} \gtrless 7 \, \Omega$

$X_C = 14 \, \Omega$

(b)

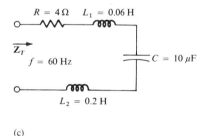
$R = 4 \, \Omega$ $L_1 = 0.06 \, H$

$\mathbf{Z}_T$ $f = 60 \, Hz$ $C = 10 \, \mu F$

$L_2 = 0.2 \, H$

(c)

FIG. 15.81

6. Find the type and impedance in ohms of the series circuit elements that must be in the closed container of Fig. 15.82 in order for the indicated voltages and currents to exist at the input terminals. (Find the simplest series circuit that will satisfy the indicated conditions.)

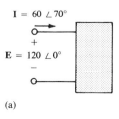
$\mathbf{I} = 60 \angle 70°$

$+$ $\mathbf{E} = 120 \angle 0°$ $-$

(a)

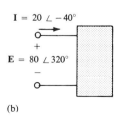
$\mathbf{I} = 20 \angle -40°$

$+$ $\mathbf{E} = 80 \angle 320°$ $-$

(b)

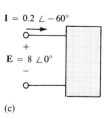
$\mathbf{I} = 0.2 \angle -60°$

$+$ $\mathbf{E} = 8 \angle 0°$ $-$

(c)

FIG. 15.82

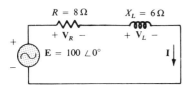

FIG. 15.83

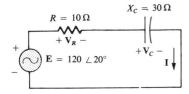

FIG. 15.84

7. For the circuit of Fig. 15.83:
   a. Find the total impedance $\mathbf{Z}_T$ in polar form.
   b. Draw the impedance diagram.
   c. Find the current $\mathbf{I}$ and the voltages $\mathbf{V}_R$ and $\mathbf{V}_L$ in phasor form.
   d. Draw the phasor diagram of the voltages $\mathbf{E}$, $\mathbf{V}_R$, and $\mathbf{V}_L$, and the current $\mathbf{I}$.
   e. Verify Kirchhoff's voltage law around the closed loop.
   f. Find the average power delivered to the circuit.
   g. Find the power factor of the circuit and indicate whether it is leading or lagging.
   h. Find the sinusoidal expressions for the voltages and current if the frequency is 60 Hz.
   i. Plot the waveforms for the voltages and current on the same set of axes.

8. Repeat Problem 7 for the circuit of Fig. 15.84, replacing $\mathbf{V}_L$ by $\mathbf{V}_C$ in parts (c) and (d).

9. For the circuit of Fig. 15.85:
   a. Find the total impedance $\mathbf{Z}_T$ in polar form.
   b. Draw the impedance diagram.
   c. Find the value of C in microfarads and L in henries.
   d. Find the current i and the voltages v_R, v_L, and v_C in phasor form.
   e. Draw the phasor diagram of the voltages $\mathbf{E}$, $\mathbf{V}_R$, $\mathbf{V}_L$, and $\mathbf{V}_C$, and the current $\mathbf{I}$.
   f. Verify Kirchhoff's voltage law around the closed loop.
   g. Find the average power delivered to the circuit.
   h. Find the power factor of the circuit and indicate whether it is leading or lagging.
   i. Find the sinusoidal expressions for the voltage and current.
   j. Plot the waveforms for the voltages and current on the same set of axes.

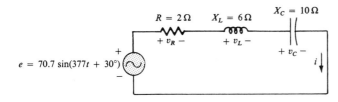

FIG. 15.85

10. Repeat Problem 9 for the circuit of Fig. 15.86.

R = 3 kΩ　　$X_L = 2$ kΩ　　$X_C = 1$ kΩ

+ v_R −　　+ v_L −　　+ v_C −　　i

$e = 6 \sin(314t - 70°)$

FIG. 15.86

Section 15.4

11. Calculate the voltages $\mathbf{V}_1$ and $\mathbf{V}_2$ for the circuit of Fig. 15.87 in phasor form using the voltage divider rule.

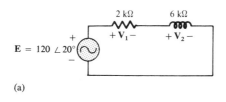

(a)

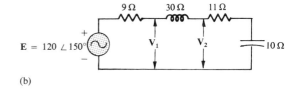

(b)

FIG. 15.87

12. Repeat Problem 11 for the circuits of Fig. 15.88.

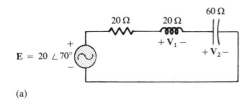

(a)

(b)

FIG. 15.88

*13. For the circuit of Fig. 15.89:
   a. Calculate i, v_R, and v_C in phasor form.
   b. Calculate the total power factor and indicate whether it is leading or lagging.
   c. Calculate the average power delivered to the circuit.
   d. Draw the impedance diagram.
   e. Draw the phasor diagram of the voltages $\mathbf{E}$, $\mathbf{V}_R$, and $\mathbf{V}_C$, and the current $\mathbf{I}$.
   f. Find the voltage across each coil using only Kirchhoff's voltage law.
   g. Find the voltages $\mathbf{V}_R$ and $\mathbf{V}_C$ using the voltage divider rule, and compare with part (a) above.
   h. Draw the equivalent series circuit of the above as far as the total impedance and the current i are concerned.

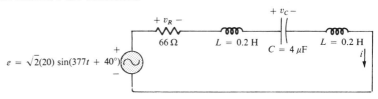

*14. Repeat Problem 13 if the capacitance is changed to 1000 μF.

FIG. 15.89

15. An electrical load has a power factor of 0.8 lagging. It dissipates 8 kW at a voltage of 200. Calculate the impedance of this load in rectangular coordinates.

*16. Find the series element or elements that must be in the enclosed container of Fig. 15.90 to satisfy the following conditions:
   a. Average power to circuit = 300 W.
   b. Circuit has a lagging power factor.

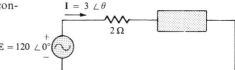

FIG. 15.90

Section 15.5

17. Find the total admittance and impedance of the circuits of Fig. 15.91. Identify the values of conductance and susceptance, and draw the admittance diagram.

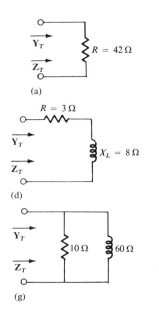
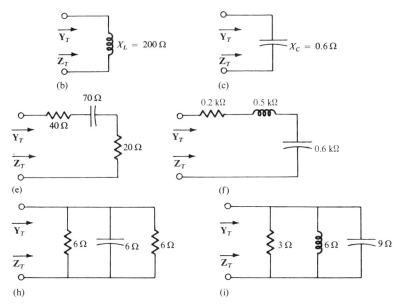

(a) (b) (c)
(d) (e) (f)
(g) (h) (i)

FIG. 15.91

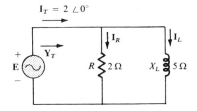

FIG. 15.92

FIG. 15.93

18. Repeat Problem 6 for the parallel circuit elements that must be in the closed container for the same voltage and current to exist at the input terminals. (Find the simplest parallel circuit that will satisfy the conditions indicated.)

Section 15.6

19. For the circuit of Fig. 15.92:
    a. Find the total admittance Y_T in polar form.
    b. Draw the admittance diagram.
    c. Find the voltage **E** and the currents I_R and I_L in phasor form.
    d. Draw the phasor diagram of the currents I_T, I_R, and I_L, and the voltage **E**.
    e. Verify Kirchhoff's current law at one node.
    f. Find the average power delivered to the circuit.
    g. Find the power factor of the circuit and indicate whether it is leading or lagging.
    h. Find the sinusoidal expressions for the currents and voltage if the frequency is 60 Hz.
    i. Plot the waveforms for the currents and voltage on the same set of axes.

20. Repeat Problem 19 for the circuit of Fig. 15.93, replacing I_L by I_C in parts (c) and (d).

21. Repeat Problem 19 for the circuit of Fig. 15.94, replacing **E** by **I**$_T$ in part (c).

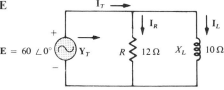

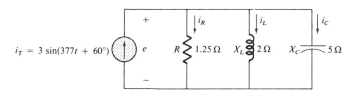

FIG. 15.94

22. For the circuit of Fig. 15.95:

$i_T = 3 \sin(377t + 60°)$

FIG. 15.95

a. Find the total admittance **Y**$_T$ in polar form.
b. Draw the admittance diagram.
c. Find the value of **C** in microfarads and **L** in henries.
d. Find the voltage e and currents i_R, i_L, and i_C in phasor form.
e. Draw the phasor diagram of the currents **I**$_T$, **I**$_R$, **I**$_L$, and **I**$_C$, and the voltage **E**.
f. Verify Kirchhoff's current law at one node.
g. Find the average power delivered to the circuit.
h. Find the power factor of the circuit and indicate whether it is leading or lagging.
i. Find the sinusoidal expressions for the currents and voltage.
j. Plot the waveforms for the currents and voltage on the same set of axes.

23. Repeat Problem 22 for the circuit of Fig. 15.96.

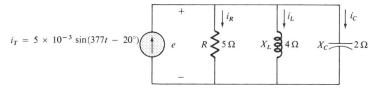

$i_T = 5 \times 10^{-3} \sin(377t - 20°)$

FIG. 15.96

24. Repeat Problem 22 for the circuit of Fig. 15.97, replacing e by i_T in part (d).

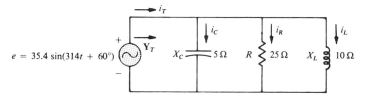

$e = 35.4 \sin(314t + 60°)$

FIG. 15.97

Section 15.7

25. Calculate the currents $\mathbf{I}_1$ and $\mathbf{I}_2$ of Fig. 15.98 in phasor form, using the current divider rule. What is the equivalent series circuit as far as the total impedance is concerned?

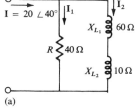

(a)

FIG. 15.98

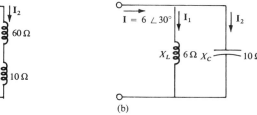

(b)

Section 15.8

26. For the series circuits of Fig. 15.99, find a parallel circuit that will have the same total impedance ($\mathbf{Z}_T$).

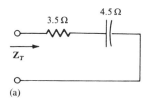

(a)

FIG. 15.99

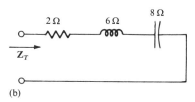

(b)

27. For the network of Fig. 15.100:
  a. Calculate e, i_R, and i_L in phasor form.
  b. Calculate the total power factor and indicate whether it is leading or lagging.
  c. Calculate the average power delivered to the circuit.
  d. Draw the admittance diagram.
  e. Draw the phasor diagram of the currents $\mathbf{I}_T$, $\mathbf{I}_R$, and $\mathbf{I}_L$, and the voltage $\mathbf{E}$.
  f. Find the current $\mathbf{I}_C$ for each capacitor using only Kirchhoff's current law.
  g. Find the series circuit of one resistive and reactive element that will have the same impedance as the original circuit.

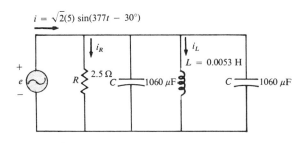

FIG. 15.100

*28. Repeat Problem 27 if the inductance is changed to 1 H.

29. Find the element or elements that must be in the closed container of Fig. 15.101 to satisfy the following conditions.

(Find the simplest parallel circuit that will satisfy the indicated conditions.)

a. Average power to the circuit = 3000 W.

b. Circuit has a lagging power factor.

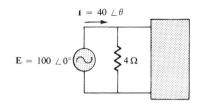

FIG. 15.101

GLOSSARY

Admittance (Y) A measure of how easily a network will "admit" the passage of current through that system. It is measured in siemens, abbreviated S.

Admittance diagram A vector display that clearly depicts the magnitude of the admittance of the conductance, capacitive susceptance, and inductive susceptance, and the magnitude and angle of the total admittance of the system.

Current divider rule A method through which the current through either of two parallel branches can be determined in an ac network without first finding the voltage across the parallel branches.

Equivalent circuits For every series ac network there is a parallel ac network (and vice versa) that will be "equivalent" in the sense that the input current and impedance are the same.

Impedance diagram A vector display that clearly depicts the magnitude of the impedance of the resistive, reactive, and capacitive components of a network, and the magnitude and angle of the total impedance of the system.

Parallel ac circuits A connection of elements in an ac network in which all of the elements have two points in common. The voltage is the same across each element.

Phasor diagram A vector display that provides at a glance the magnitude and phase relationships among the various voltages and currents of a network.

Series ac configuration A connection of elements in an ac network in which no two impedances have more than one terminal in common and the current is the same through each element.

Susceptance A measure of how "susceptible" an element is to the passage of current through it. It is measured in siemens and represented by the capital letter B.

Voltage divider rule A method through which the voltage across one element of a series of elements in an ac network can be determined without first having to find the current through the elements.

16.1 INTRODUCTION

In this chapter we shall utilize the fundamental concepts of the previous chapter to develop a technique for solving series-parallel ac networks. A brief review of Chapter 6 may be helpful before considering these networks, since the approach here will be quite similar to that undertaken earlier. The circuits to be discussed will have only one source of energy, either potential or current. Networks with two or more sources will be considered in Chapters 17 and 18, using methods previously described for dc circuits.

In general, when working with series-parallel ac networks, follow these steps:

1. Study the problem and make a brief mental sketch of the overall approach you plan to use. Doing this may result in time- and energy-saving shortcuts.
2. After the overall approach has been determined, consider each branch involved in your method independently before tying them together in series-parallel combinations. This will eliminate many of the errors that might develop due to the lack of a systematic approach.
3. When you have arrived at a solution, check to see that it is reasonable by considering the magnitudes of the energy source and the elements in the circuit. If it does not seem reasonable, either solve the network using another approach, or check over your work very carefully.

16.2 ILLUSTRATIVE EXAMPLES

EXAMPLE 16.1. For the network of Fig. 16.1, determine
a. $\mathbf{Z}_T$.
b. $\mathbf{I}_T$.
c. $\mathbf{I}_C$.
d. $\mathbf{V}_R$ and $\mathbf{V}_C$.
e. power delivered.
f. F_p of the network.

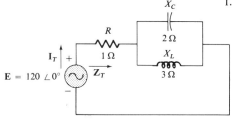

FIG. 16.1

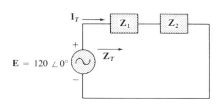

FIG. 16.2

For convenience, the network is redrawn in Fig. 16.2. It is good practice when working with complex networks to represent impedances in the manner indicated. When the unknown quantity is found in terms of these subscripted impedances, their numerical values can then be substituted to find the magnitude of the unknown. This will usually save time and prevent errors in calculations.

Solutions:

a. $\mathbf{Z}_1 = R \angle 0° = 1 \angle 0°$

$$\mathbf{Z}_2 = \mathbf{X}_C \parallel \mathbf{X}_L = \frac{\mathbf{X}_C \mathbf{X}_L}{\mathbf{X}_C + \mathbf{X}_L} = \frac{(2 \angle -90°)(3 \angle 90°)}{-j2 + j3}$$

$$= \frac{6 \angle 0°}{j1} = \frac{6 \angle 0°}{1 \angle 90°} = 6 \angle -90°$$

and

$$\mathbf{Z}_T = \mathbf{Z}_1 + \mathbf{Z}_2 = 1 - j6 = \mathbf{6.08 \angle -80.54°}$$

b. $\mathbf{I}_T = \dfrac{\mathbf{E}}{\mathbf{Z}_T} = \dfrac{120 \angle 0°}{6.08 \angle -80.5°} = \mathbf{19.74 \angle 80.54°}$

c. By the current divider rule,

$$\mathbf{I}_C = \frac{\mathbf{X}_L \mathbf{I}_T}{\mathbf{X}_L + \mathbf{X}_C} = \frac{(3 \angle 90°)(19.74 \angle 80.54°)}{1 \angle 90°}$$

$$= \frac{59.22 \angle 170.54°}{1 \angle 90°} = \mathbf{59.22 \angle 80.54°}$$

d. $\mathbf{V}_R = \mathbf{I}_T \mathbf{Z}_1 = (19.74 \angle 80.54°)(1 \angle 0°)$
$$= \mathbf{19.74 \angle 80.54°}$$

$\mathbf{V}_C = \mathbf{I}_T \mathbf{Z}_2 = (19.74 \angle 80.54°)(6 \angle -90°)$
$$= \mathbf{118.44 \angle -9.46°}$$

e. $P_{\text{del}} = I_T^2 R = (19.74)^2 \cdot 1 = \mathbf{389.67 \ W}$

f. $F_p = \cos \theta = \cos 80.54° = \mathbf{0.164 \ leading}$

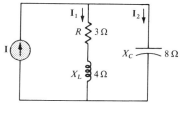

FIG. 16.3

indicating a very reactive network. That is, the closer F_p is to zero, the more reactive and less resistive is the network.

EXAMPLE 16.2. For the network of Fig. 16.3:
a. If $\mathbf{I}$ is $50 \angle 30°$, calculate $\mathbf{I}_1$ using the current divider rule.
b. Repeat part (a) for $\mathbf{I}_2$.
c. Verify Kirchhoff's current law at one node.

Solutions:
a. Redrawing the circuit as in Fig. 16.4, we have

$$\mathbf{Z}_1 = 3 + j4 = 5 \angle 53.13° \qquad \mathbf{Z}_2 = -j8 = 8 \angle -90°$$

Using the current divider rule yields

$$\mathbf{I}_1 = \frac{\mathbf{Z}_2\mathbf{I}}{\mathbf{Z}_2 + \mathbf{Z}_1} = \frac{(8 \angle -90°)(50 \angle 30°)}{(-j8) + (3 + j4)} = \frac{400 \angle -60°}{3 - j4}$$

$$= \frac{400 \angle -60°}{5 \angle -53.13°} = \mathbf{80 \angle -6.87°}$$

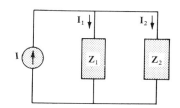

FIG. 16.4

b. $$\mathbf{I}_2 = \frac{\mathbf{Z}_1\mathbf{I}}{\mathbf{Z}_2 + \mathbf{Z}_1} = \frac{(5 \angle 53.13°)(50 \angle 30°)}{5 \angle -53.13°} = \frac{250 \angle 83.13°}{5 \angle -53.13°}$$

$$= \mathbf{50 \angle 136.26°}$$

c. $$\mathbf{I} = \mathbf{I}_1 + \mathbf{I}_2$$
$$50 \angle 30° = 80 \angle -6.87° + 50 \angle 136.26°$$
$$= (79.43 - j9.57) + (-36.12 + j34.57)$$
$$= 43.31 + j25.0$$
$$50 \angle 30° = 50 \angle 30° \qquad \text{(checks)}$$

EXAMPLE 16.3. For the network of Fig. 16.5:
a. Calculate the voltage $\mathbf{V}_C$ using the voltage divider rule.
b. Calculate the current $\mathbf{I}$.

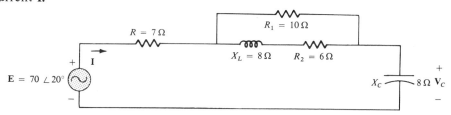

Solutions:
a. Redrawing the current as in Fig. 16.6, we have

FIG. 16.5

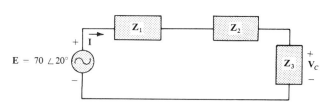

FIG. 16.6

$$\mathbf{Z}_1 = 7 \angle 0° \qquad \mathbf{Z}_3 = 8 \angle -90° = -j8$$

$$\mathbf{Z}_2 = \frac{(R_1)(R_2 + jX_L)}{R_1 + (R_2 + jX_L)} = \frac{10(6 + j8)}{10 + 6 + j8}$$

$$= \frac{10(10 \angle 53.13°)}{16 + j8} = \frac{100 \angle 53.13°}{17.89 \angle 26.57°}$$

$$\mathbf{Z}_2 = 5.59 \angle 26.56°$$

$$\mathbf{V}_C = \frac{\mathbf{Z}_3 \mathbf{E}}{\mathbf{Z}_1 + \mathbf{Z}_2 + \mathbf{Z}_3} = \frac{(8 \angle -90°)(70 \angle 20°)}{7 + 5.59 \angle 26.56° - j8}$$

$$= \frac{560 \angle -70°}{7 + (5 + j2.50) - j8} = \frac{560 \angle -70°}{12 - j5.50}$$

$$= \frac{560 \angle -70°}{13.20 \angle -24.62°} = \mathbf{42.42 \angle -45.38°}$$

b. $\mathbf{I} = \dfrac{\mathbf{V}_C}{\mathbf{X}_C} = \dfrac{42.42 \angle -45.38°}{8 \angle -90°} = \mathbf{5.30 \angle 44.62°}$

EXAMPLE 16.4. In Fig. 16.7, calculate
a. the current **I.**
b. the voltage $\mathbf{V}_{ab}$.

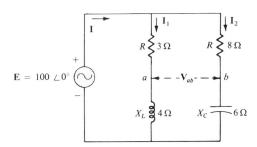

FIG. 16.7

Solutions:
a. Redrawing the circuit as in Fig. 16.8, we obtain

$$\mathbf{Z}_1 = 3 + j4 = 5 \angle 53.13°$$
$$\mathbf{Z}_2 = 8 - j6 = 10 \angle -36.87°$$

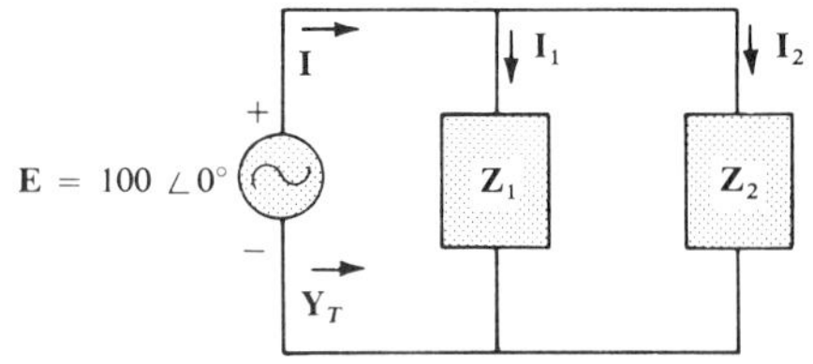

$$\mathbf{Y}_T = \mathbf{Y}_1 + \mathbf{Y}_2$$
$$= \frac{1}{\mathbf{Z}_1} + \frac{1}{\mathbf{Z}_2} = \frac{1}{5 \angle 53.13°} + \frac{1}{10 \angle -36.87°}$$
$$= 0.2 \angle -53.13° + 0.1 \angle 36.87°$$
$$= (0.12 - j0.16) + (0.08 + j0.06)$$
$$= 0.2 - j0.1 = \mathbf{0.224 \angle -26.57°}$$

FIG. 16.8

By Ohm's law,

$$\mathbf{I} = \frac{\mathbf{E}}{\mathbf{Z}_T} = \mathbf{E}\mathbf{Y}_T = (100 \angle 0°)(0.224 \angle -26.57°)$$

$$\mathbf{I} = \mathbf{22.4 \angle -26.57°}$$

Using another approach, we find the total impedance:

$$\mathbf{Z}_T = \frac{\mathbf{Z}_1 \mathbf{Z}_2}{\mathbf{Z}_1 + \mathbf{Z}_2} = \frac{(5 \angle 53.13°)(10 \angle -36.87°)}{(3 + j4) + (8 - j6)}$$

$$= \frac{50 \angle 16.26°}{11 - j2} = \frac{50 \angle 16.26°}{11.18 \angle -10.30}$$
$$= 4.47 \angle 26.30°$$

and

$$\mathbf{I} = \frac{\mathbf{E}}{\mathbf{Z}_T} = \frac{100 \angle 0°}{4.47 \angle 26.30°} = 22.37 \angle -26.30°$$

b. By Ohm's law,

$$\mathbf{I}_1 = \frac{\mathbf{E}}{\mathbf{Z}_1} = \frac{100 \angle 0°}{5 \angle 53.13°} = 20 \angle -53.13°$$

$$\mathbf{I}_2 = \frac{\mathbf{E}}{\mathbf{Z}_2} = \frac{100 \angle 0°}{10 \angle -36.87°} = 10 \angle 36.87°$$

Returning to Fig. 16.7, we have

$$\mathbf{V}_{3\Omega} = \mathbf{I}_1 \mathbf{R} = (20 \angle -53.13°)(3 \angle 0°) = 60 \angle -53.13°$$
$$\mathbf{V}_{8\Omega} = \mathbf{I}_2 \mathbf{R} = (10 \angle +36.87°)(8 \angle 0°) = 80 \angle +36.87°$$

Instead of using the three steps just shown, $\mathbf{V}_{3\Omega}$ or $\mathbf{V}_{8\Omega}$ could have been determined in one step using the voltage divider rule:

$$\mathbf{V}_{3\Omega} = \frac{(3 \angle 0°)(100 \angle 0°)}{3 \angle 0° + 4 \angle 90°} = \frac{300 \angle 0°}{5 \angle 53.13°}$$

$$= 60 \angle -53.13°$$

In order to find $\mathbf{V}_{ab}$, Kirchhoff's voltage law must be applied around the loop (Fig. 16.9) consisting of the 3-Ω and 8-Ω resistors. By Kirchhoff's voltage law,

$$-\mathbf{V}_{ab} + \mathbf{V}_{8\Omega} - \mathbf{V}_{3\Omega} = 0$$

or

$$\mathbf{V}_{ab} = \mathbf{V}_{8\Omega} - \mathbf{V}_{3\Omega}$$
$$= 80 \angle 36.87° - 60 \angle -53.13°$$
$$= (64 + j48) - (36 - j48)$$
$$= 28 + j96$$
$$\mathbf{V}_{ab} = 100 \angle 73.74°$$

EXAMPLE 16.5. For the network of Fig. 16.10, determine
a. $\mathbf{Z}'_T$ and compare it to $R_1 = 50$ kΩ.
b. $\mathbf{I}_1$ and compare it to $\mathbf{I}$, keeping in mind the results of part (a).
c. $\mathbf{V}_{\text{load}}$.

FIG. 16.9

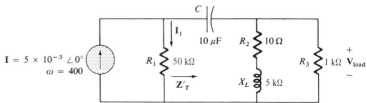

FIG. 16.10

Solutions:

a. $X_C = \dfrac{1}{\omega C} = \dfrac{1}{400(10) \times 10^{-6}} = \dfrac{10^6}{4 \times 10^3}$

$= \dfrac{10^3}{4} = 250\ \Omega = 0.25\ k\Omega$

The network is redrawn in Fig. 16.11.

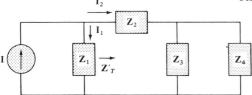

FIG. 16.11

$Z_1 = 50\ k\Omega$

$Z_2 = 0.25 \times 10^3\ \angle -90° = -j250$

$Z_3 = 10 + j5 \times 10^3 \cong j5 \times 10^3 = 5 \times 10^3\ \angle 90°$

$Z_4 = 1\ k\Omega$

and

$Z'_T = Z_2 + Z_3 \parallel Z_4$

$= -j250 + \dfrac{(5 \times 10^3\ \angle 90°)(1 \times 10^3\ \angle 0°)}{(j5 \times 10^3 + 1 \times 10^3)}$

$= -j250 + \dfrac{5 \times 10^3\ \angle 90°}{5.10\ \angle 78.69°}$

$= -j250 + 980\ \angle 11.31°$

$= -j250 + (960 + j190)$

$= 960 - j60 = \mathbf{962\ \angle -3.58°}$

$Z'_T \cong \dfrac{1}{50}$ the magnitude of $Z_1 = 50\ k\Omega$

b. $I_1 = \dfrac{Z'_T I}{Z_T + Z_1} = \dfrac{(962\ \angle -3.58°)(5 \times 10^{-3}\ \angle 0°)}{(960 - j60) + 50 \times 10^3}$

$= \dfrac{4.81\ \angle -3.58°}{\cong 50.96 \times 10^3} = 0.0944 \times 10^{-3}\ \angle -3.58°$

$\mathbf{I = 94.4 \times 10^{-6}\ \angle -3.58°}$

The magnitude of I_1 is approximately 1/50 the magnitude of I, corresponding with the ratio of the impedances above.

c. Since

$|I_1| \cong \dfrac{1}{50}\ |I|$

$I_2 \cong I = 5 \times 10^{-3}\ \angle 0°$

and

$V_{\text{load}} = I_2(Z_3 \parallel Z_4) = (5 \times 10^{-3}\ \angle 0°)(980\ \angle 11.31°)$

$= \mathbf{4.90\ \angle 11.31°}$

EXAMPLE 16.6. For the network of Fig. 16.12:
a. Compute **I**.
b. Find $\mathbf{I}_1$, $\mathbf{I}_2$, and $\mathbf{I}_3$.
c. Verify Kirchhoff's current law by showing that

$$\mathbf{I} = \mathbf{I}_1 + \mathbf{I}_2 + \mathbf{I}_3$$

d. Find the total impedance of the circuit.

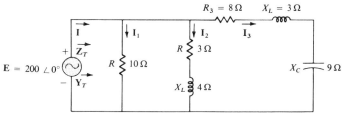

Solutions:

FIG. 16.12

a. Redrawing the circuit as in Fig. 16.13, we obtain

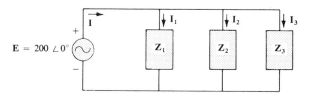

FIG. 16.13

$$\mathbf{Z}_1 = 10 \angle 0°$$
$$\mathbf{Z}_2 = 3 + j4$$
$$\mathbf{Z}_3 = 8 + j3 - j9 = 8 - j6$$

The total admittance is

$$\mathbf{Y}_T = \mathbf{Y}_1 + \mathbf{Y}_2 + \mathbf{Y}_3$$
$$= \frac{1}{\mathbf{Z}_1} + \frac{1}{\mathbf{Z}_2} + \frac{1}{\mathbf{Z}_3} = \frac{1}{10} + \frac{1}{3 + j4} + \frac{1}{8 - j6}$$

However,

$$\frac{1}{3 + j4} = \frac{3}{3^2 + 4^2} - j\frac{4}{3^2 + 4^2} = \frac{3}{25} - j\frac{4}{25}$$

and

$$\frac{1}{8 - j6} = \frac{8}{8^2 + 6^2} + j\frac{6}{8^2 + 6^2} = \frac{8}{100} + j\frac{6}{100}$$

$$\therefore \mathbf{Y}_1 = \frac{1}{10} + j0 = \frac{10}{100} + j0$$

$$\mathbf{Y}_2 = \frac{3}{25} - j\frac{4}{25} = \frac{12}{100} - j\frac{16}{100}$$

$$\mathbf{Y}_3 = \frac{8}{100} + j\frac{6}{100} = \frac{8}{100} + j\frac{6}{100}$$

$$\mathbf{Y}_T = \mathbf{Y}_1 + \mathbf{Y}_2 + \mathbf{Y}_3 = \frac{30}{100} - j\frac{10}{100}$$

$$\mathbf{I} = \mathbf{EY}_T = 200 \angle 0° \left[\frac{30}{100} - j\frac{10}{100} \right] = \mathbf{60} - \mathbf{j20}$$

b. Since the voltage is the same across parallel branches,

$$\mathbf{I}_1 = \frac{\mathbf{E}}{\mathbf{Z}_1} = \frac{200 \angle 0°}{10 \angle 0°} = \mathbf{20} \angle \mathbf{0°}$$

$$\mathbf{I}_2 = \frac{\mathbf{E}}{\mathbf{Z}_2} = \frac{200 \angle 0°}{5 \angle 53.13°} = \mathbf{40} \angle \mathbf{-53.13°}$$

$$\mathbf{I}_3 = \frac{\mathbf{E}}{\mathbf{Z}_3} = \frac{200 \angle 0°}{10 \angle -36.87°} = \mathbf{20} \angle \mathbf{+36.87°}$$

c. $\quad \mathbf{I} = \mathbf{I}_1 + \mathbf{I}_2 + \mathbf{I}_3$

$60 - j20 = 20 \angle 0° + 40 \angle -53.13° + 20 \angle +36.87°$
$= (20 + j0) + (24 - j32) + (16 + j12)$

$60 - j20 = 60 - j20 \quad$ (checks)

d. $\mathbf{Z}_T = \dfrac{1}{\mathbf{Y}_T} = \dfrac{1}{0.3 - j0.1}$

$$= \frac{0.3}{(0.3)^2 + (0.1)^2}$$

$$+ j\frac{0.1}{(0.3)^2 + (0.1)^2}$$

$$\mathbf{Z}_T = \frac{0.3}{0.1} + j\frac{0.1}{0.1} = \mathbf{3} + \mathbf{j}$$

EXAMPLE 16.7. For the network of Fig. 16.14:
a. Calculate the total impedance $\mathbf{Z}_T$.
b. Compute **I**.
c. Find the total power factor.
d. Calculate $\mathbf{I}_1$ and $\mathbf{I}_2$.
e. Find the average power delivered to the circuit.

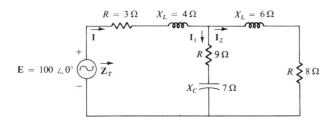

FIG. 16.14

Solutions:
a. Redrawing the circuit as in Fig. 16.15, we have

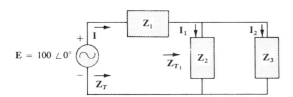

FIG. 16.15

$$\mathbf{Z}_1 = 3 + j4 = 5 \angle 53.13°$$
$$\mathbf{Z}_2 = 9 - j7 = 11.40 \angle -37.87°$$
$$\mathbf{Z}_3 = 8 + j6 = 10 \angle +36.87°$$

The total impedance is

$$\begin{aligned}
\mathbf{Z}_T &= \mathbf{Z}_1 + \mathbf{Z}_{T_1} \\
&= \mathbf{Z}_1 + \frac{\mathbf{Z}_2\mathbf{Z}_3}{\mathbf{Z}_2 + \mathbf{Z}_3} \\
&= (3 + j4) + \frac{(11.4 \angle -37.87°)(10 \angle 36.87°)}{(9 - j7) + (8 + j6)} \\
&= 3 + j4 + \frac{114 \angle -1.00°}{17.03 \angle -3.37°} \\
&= 3 + j4 + 6.69 \angle 2.37° \\
&= 3 + j4 + 6.68 + j0.28 \\
\mathbf{Z}_T &= 9.68 + j4.28 = \mathbf{10.58 \angle 23.85°}
\end{aligned}$$

b. $\mathbf{I} = \dfrac{\mathbf{E}}{\mathbf{Z}_T} = \dfrac{100 \angle 0°}{10.58 \angle 23.85°} = \mathbf{9.45 \angle -23.85°}$

c. $F_p = \cos\theta = \dfrac{R}{Z_T} = \dfrac{9.68}{10.58} = \mathbf{0.915} = \cos 23.85°$

d. $\begin{aligned}
\mathbf{I}_2 &= \frac{\mathbf{Z}_2\mathbf{I}}{\mathbf{Z}_2 + \mathbf{Z}_3} = \frac{(11.40 \angle -37.87°)(9.45 \angle -23.85°)}{(9 - j7) + (8 + j6)} \\
&= \frac{107.73 \angle -61.72°}{17 - j1} = \frac{107.73 \angle -61.72°}{17.03 \angle -3.37°}
\end{aligned}$

$$\mathbf{I}_2 = \mathbf{6.33 \angle -58.35°}$$

Applying Kirchhoff's current law yields

$$\mathbf{I} = \mathbf{I}_1 + \mathbf{I}_2$$

or

$$\begin{aligned}
\mathbf{I}_1 &= \mathbf{I} - \mathbf{I}_2 \\
&= (9.45 \angle -23.85°) - (6.33 \angle -58.35°) \\
&= (8.64 - j3.82) - (3.32 - j5.39) \\
\mathbf{I}_1 &= 5.32 + j1.57 = \mathbf{5.55 \angle 16.44°}
\end{aligned}$$

e. $\begin{aligned}
P_T &= EI\cos\theta \\
&= (100)(9.45)\cos 23.85° \\
&= (945)(0.915) \\
P_T &= \mathbf{864.68\ W}
\end{aligned}$

16.3 LADDER NETWORKS

Ladder networks were discussed in some detail in Chapter 6. This section will simply apply each method described in Section 6.3 to the general sinusoidal ac ladder network of Fig. 16.16. The current $\mathbf{I}_6$ is desired.

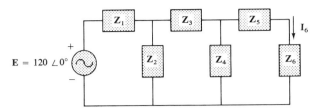

FIG. 16.16

First Method

Impedances $\mathbf{Z}_T$, $\mathbf{Z}'_T$, and $\mathbf{Z}''_T$ and currents $\mathbf{I}_1$ and $\mathbf{I}_3$ are defined in Fig. 16.17.

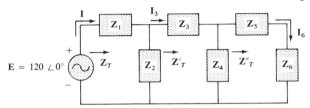

FIG. 16.17

$$\mathbf{Z}''_T = \mathbf{Z}_5 + \mathbf{Z}_6$$

and

$$\mathbf{Z}'_T = \mathbf{Z}_3 + \mathbf{Z}_4 \parallel \mathbf{Z}''_T$$

with

$$\mathbf{Z}_T = \mathbf{Z}_1 + \mathbf{Z}_2 \parallel \mathbf{Z}'_T$$

Then

$$\mathbf{I} = \frac{\mathbf{E}}{\mathbf{Z}_T}$$

and

$$\mathbf{I}_3 = \frac{\mathbf{Z}_2(\mathbf{I})}{\mathbf{Z}_2 + \mathbf{Z}'_T}$$

with

$$\mathbf{I}_6 = \frac{\mathbf{Z}_4(\mathbf{I}_3)}{\mathbf{Z}_4 + \mathbf{Z}''_T}$$

Second Method

The necessary labeling for this method appears in Fig. 16.18. Each operation is performed in the order indicated. The desired variable $\mathbf{I}_6$ is carried through each equation:

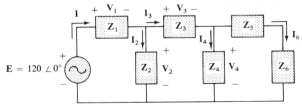

FIG. 16.18

$$I_6 = \frac{V_4}{Z_5 + Z_6}$$

and

$$V_4 = I_6(Z_5 + Z_6)$$

$$I_4 = \frac{V_4}{Z_4} = I_6\frac{(Z_5 + Z_6)}{Z_4}$$

and

$$I_3 = I_4 + I_6 \qquad \text{(in terms of } I_6 \text{ alone)}$$

so that

$$V_3 = I_3Z_3 \qquad \text{(function of } I_6 \text{)}$$

will result in

$$V_2 = V_3 + V_4 \qquad \text{(function of } I_6 \text{)}$$

and

$$I_2 = \frac{V_2}{Z_2} \qquad \text{(function of } I_6 \text{)}$$

Thus,

$$I = I_1 = I_2 + I_3 \qquad \text{(function of } I_6 \text{)}$$

and

$$V_1 = I_1Z_1 \qquad \text{(function of } I_6 \text{)}$$

Finally,

$$E = V_1 + V_2$$

where V_1 and V_2 are functions of I_6, permitting determination of I_6.

PROBLEMS

Section 16.2

1. For the series-parallel network of Fig. 16.19, determine
   a. Z_T.
   b. I.
   c. I_1.
   d. I_2 and I_3.
   e. V_L.

2. For the network of Fig. 16.20:

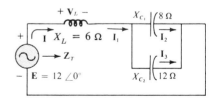

FIG. 16.19

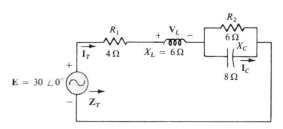

FIG. 16.20

a. Find the total impedance $\mathbf{Z}_T$.
   b. Determine the current $\mathbf{I}_T$.
   c. Calculate $\mathbf{I}_C$ using the current divider rule.
   d. Calculate $\mathbf{V}_L$ using the voltage divider rule.

3. For the network of Fig. 16.21:
   a. Find the total impedance $\mathbf{Z}_T$ and the total admittance $\mathbf{Y}_T$.
   b. Find the current $\mathbf{I}_T$.
   c. Calculate $\mathbf{I}_2$ using the current divider rule.
   d. Calculate $\mathbf{V}_C$.
   e. Calculate the average power delivered to the network.

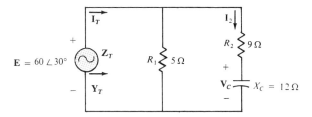

FIG. 16.21

4. For the network of Fig. 16.22:
   a. Find the total impedance $\mathbf{Z}_T$.
   b. Calculate the voltage $\mathbf{V}_2$ and the current $\mathbf{I}_L$.
   c. Find the power factor of the network.

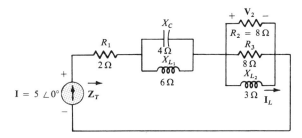

FIG. 16.22

5. For the network of Fig. 16.23:
   a. Find the current $\mathbf{I}$.
   b. Find the voltage $\mathbf{V}_C$.
   c. Find the average power delivered to the network.

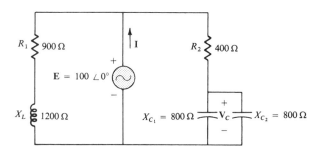

FIG. 16.23

*6. For the network of Fig. 16.24:
   a. Find the current $\mathbf{I}_1$.
   b. Calculate the voltage $\mathbf{V}_C$ using the voltage divider rule.
   c. Find the voltage $\mathbf{V}_{ab}$.

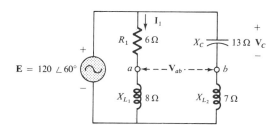

FIG. 16.24

*7. For the network of Fig. 16.25:
   a. Find the current I_1.
   b. Find the voltage V_1.
   c. Calculate the average power delivered to the network.

8. For the network of Fig. 16.26:
   a. Find the total impedance Z_T and the admittance Y_T.
   b. Find the currents I_1, I_2, and I_3.
   c. Verify Kirchhoff's current law by showing that $I_T = I_1 + I_2 + I_3$.
   d. Find the power factor of the network, and indicate whether it is leading or lagging.

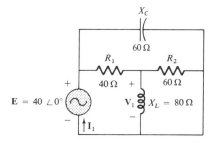

FIG. 16.25

*9. For the network of Fig. 16.27:
   a. Find the total admittance Y_T.
   b. Find the voltage V_1.
   c. Find the current I_3.
   d. Find the voltages V_2 and V_{ab}.

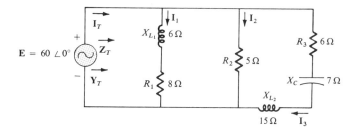

FIG. 16.26

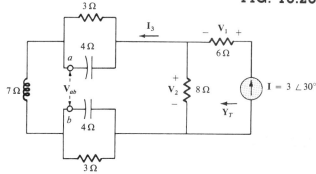

10. For the network of Fig. 16.28:
    a. Find the total impedance Z_T and the admittance Y_T.
    b. Find the current i_T in phasor form.
    c. Find the currents i_1 and i_2 in phasor form.
    d. Find the voltages v_1 and v_{ab} in phasor form.
    e. Find the average power delivered to the network.
    f. Find the power factor of the network, and indicate whether it is leading or lagging.

FIG. 16.27

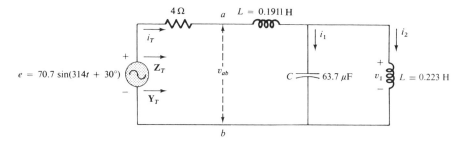

FIG. 16.28

*11. Find the current **I** in the network of Fig. 16.29.

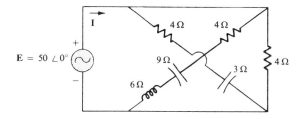

FIG. 16.29

Section 16.3

12. Find the current I_5 for the network of Fig. 16.30. Note the effect of one reactive element on the resulting calculations.

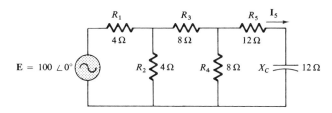

FIG. 16.30

13. Find the average power delivered to R_5 in Fig. 16.31.

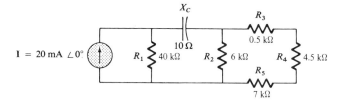

FIG. 16.31

14. Find the current I_1 for the network of Fig. 16.32.

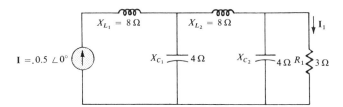

FIG. 16.32

GLOSSARY

Ladder network A repetitive combination of series and parallel branches that has the appearance of a ladder.

Series-parallel ac network A combination of series and parallel branches in the same network configuration. Each branch may contain any number of elements whose impedance is dependent on the applied frequency.

17.1 INTRODUCTION

For networks with two or more sources that are not in series or parallel, the methods described in the last two chapters cannot be applied. Rather, such methods as mesh analysis or nodal analysis must be employed. Since these methods were discussed in detail for dc circuits in Chapter 7, this chapter will consider only the variations required to apply these methods to ac circuits.

The branch-current method will not be discussed again, since it falls within the framework of mesh analysis. After completing this chapter, however, you should have no problem applying this method to ac circuits.

In addition to the methods mentioned above, the bridge network and Δ-Y, Y-Δ conversions will also be discussed for ac circuits.

Before we examine these topics, however, we must consider the subject of independent and controlled sources.

17.2 INDEPENDENT VERSUS DEPENDENT (CONTROLLED) SOURCES

In the previous chapters, each source appearing in the analysis of dc or ac networks was an *independent source*

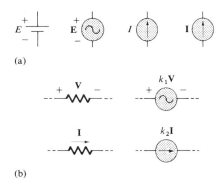

(a)

(b)

FIG. 17.1 *Sources. (a) Independent; (b) controlled or dependent.*

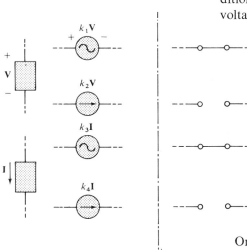

FIG. 17.2 *Conditions of V = 0 and I = 0 for a controlled source.*

such as E and I (or **E** and **I**) in Fig. 17.1. The term *independent* indicates that the magnitude of the source is *independent of the network* to which it is applied and that it displays its terminal characteristics even if completely isolated.

A *dependent* or *controlled source* is one whose magnitude is determined (or controlled) by a current or voltage of the system in which it appears. The magnitude of the voltage source $k_1\mathbf{V}$ in Fig. 17.1 is determined by the voltage **V** appearing across the resistive element and the constant k_1. The magnitude of the current source $k_2\mathbf{I}$ is determined by the current through another resistive element and the constant k_2.

Possible combinations for controlled sources are indicated in Fig. 17.2. Note that the magnitude of current sources or voltage sources can be controlled by a voltage and a current, respectively. Unlike the independent source, isolation such that **V** or **I** = 0 in Fig. 17.2 will result in the short-circuit or open-circuit equivalent as indicated in Fig. 17.2. Note that the type of representation under these conditions is controlled by whether it is a current source or a voltage source, not by the controlling agent (**V** or **I**).

One reason for our interest in controlled sources stems from electronic components such as the transistor, which employs a dependent source in its *equivalent circuit*. An equivalent circuit is a combination of elements, active (sources) and passive (R, L, C), properly chosen, that best represent the terminal characteristics and response of the device under a particular set of operating conditions. A few examples in this chapter will indicate the basic appearance of these equivalent circuits.

17.3 SOURCE CONVERSIONS

When applying the methods to be discussed, it may be necessary to convert a current source to a voltage source, or a voltage source to a current source. This can be accomplished in much the same manner as it was for dc circuits, except now we shall be dealing with phasors and impedances instead of just real numbers and resistors.

In general, the format for converting from one to the other is as shown in Fig. 17.3.

EXAMPLE 17.1. Convert the voltage source of Fig. 17.4(a) to a current source.

Solution:

$$I = \frac{E}{Z} = \frac{100 \angle 0°}{5 \angle 53.13°}$$

$$= 20 \angle -53.13° \qquad \text{Fig. 17.4(b)}$$

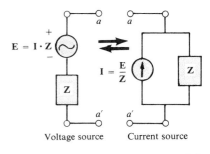

Voltage source   Current source

FIG. 17.3

Source conversion

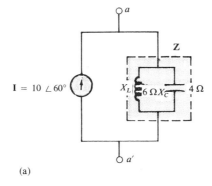

(a)   (b)

EXAMPLE 17.2. Convert the current source of Fig. 17.5(a) to a voltage source.

FIG. 17.4

(a)

(b)

Solution:

FIG. 17.5

$$Z = \frac{(4 \angle -90°)(6 \angle 90°)}{-j4 + j6} = \frac{24 \angle 0°}{2 \angle 90°}$$

$$= 12 \angle -90° \qquad \text{Fig. 17.5(b)}$$

$$E = IZ = (10 \angle 60°)(12 \angle -90°)$$

$$= 120 \angle -30° \qquad \text{Fig. 17.5(b)}$$

For dependent sources, the direct conversion of Fig. 17.3 can be applied if the controlling variable (**V** or **I** in Fig. 17.2) is not determined by a portion of the network to which the conversion is to be applied. For example, in Figs. 17.6 and 17.7, **V** and **I**, respectively, are controlled by an external portion of the network. Conversions of the other kind, where **V** and **I** are controlled by a portion of the network to be converted, will be considered in Sections 18.3 and 18.4.

EXAMPLE 17.3. Convert the voltage source of Fig. 17.6(a) to a current source.

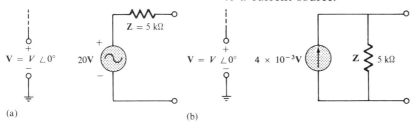

(a)   (b)

FIG. 17.6

Solution:

$$I = \frac{E}{Z} = \frac{20V \angle 0°}{5 \times 10^3 \angle 0°}$$

$$= 4 \times 10^{-3}V \angle 0° \qquad \text{Fig. 17.6(b)}$$

EXAMPLE 17.4. Convert the current source of Fig. 17.7(a) to a voltage source.

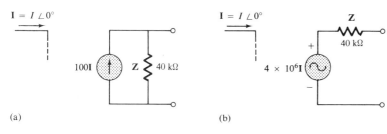

(a)   (b)

FIG. 17.7

Solution:

$$E = IZ = (100I \angle 0°)(40 \times 10^3 \angle 0°)$$

$$= 4 \times 10^6 I \angle 0° \qquad \text{Fig. 17.7(b)}$$

17.4 MESH ANALYSIS (GENERAL APPROACH)

The first method to be considered is mesh (or loop) analysis. The steps for applying this method are repeated here with changes for its use in ac circuits:

1. Assign a distinct current in the clockwise direction to each independent *closed loop* of the network.
2. Indicate the polarities within each loop for each impedance, as determined by the direction of the loop current for the closed loop.
3. Apply Kirchhoff's voltage law around each closed loop.

a. If an impedance has two or more assumed currents flowing through it, the following must be adhered to when applying Kirchhoff's voltage law: The total current through the impedance is the assumed current of the loop in which Kirchhoff's voltage law is being applied, plus the assumed currents of the other loops passing through in the same direction, minus the assumed currents passing through in the opposite direction.

b. The polarity of a voltage source is unaffected by the loop currents passing through it.

4. Solve the resulting simultaneous linear equations for the assumed loop currents using determinants.

Note that the only change in the above as compared to its appearance in Chapter 7 (dc circuits) was to replace the word *resistor* by the word *impedance*. This and the use of phasors instead of real numbers will be the only major changes in applying these methods to ac circuits.

The technique is applied as above for all networks with independent sources or networks with dependent sources where the controlling variable is not a part of the network under investigation. If the controlling variable is part of the network being examined, additional care must be taken when applying the above steps.

EXAMPLE 17.5. Using mesh analysis, find the current I_1 in Fig. 17.8.

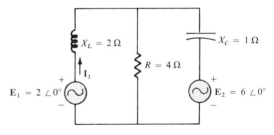

FIG. 17.8

Solution: When applying these methods to ac circuits, it is good practice to represent the resistors and reactances (or combinations thereof) by subscripted impedances. When the total solution is found in terms of these subscripted impedances, the numerical values can be substituted to find the unknown quantities.

The network is redrawn in Fig. 17.9 with subscripted impedances:

$$\mathbf{Z}_1 = +j2$$
$$\mathbf{Z}_2 = 4$$
$$\mathbf{Z}_3 = -j$$

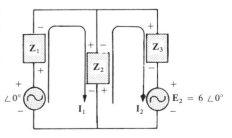

FIG. 17.9

Steps 1 and 2 are as indicated in the network.

Step 3: Kirchhoff's voltage law is applied around each closed loop:

$$+\mathbf{E}_1 - \mathbf{Z}_1\mathbf{I}_1 - \mathbf{Z}_2(\mathbf{I}_1 - \mathbf{I}_2) = 0$$
$$-\mathbf{E}_2 - \mathbf{Z}_3\mathbf{I}_2 - \mathbf{Z}_2(\mathbf{I}_2 - \mathbf{I}_1) = 0$$

which are rewritten as

$$+\mathbf{E}_1 - (\mathbf{Z}_1 + \mathbf{Z}_2)\mathbf{I}_1 + \mathbf{Z}_2\mathbf{I}_2 = 0$$
$$-\mathbf{E}_2 - (\mathbf{Z}_3 + \mathbf{Z}_2)\mathbf{I}_2 + \mathbf{Z}_2\mathbf{I}_1 = 0$$

or

$$(\mathbf{Z}_1 + \mathbf{Z}_2)\mathbf{I}_1 - \mathbf{Z}_2\mathbf{I}_2 = \mathbf{E}_1$$
$$-(\mathbf{Z}_2)\mathbf{I}_1 + (\mathbf{Z}_2 + \mathbf{Z}_3)\mathbf{I}_2 = -\mathbf{E}_2$$

Using determinants, we obtain

$$\mathbf{I}_1 = \frac{\begin{vmatrix} \mathbf{E}_1 & -\mathbf{Z}_2 \\ -\mathbf{E}_2 & \mathbf{Z}_2 + \mathbf{Z}_3 \end{vmatrix}}{\begin{vmatrix} \mathbf{Z}_1 + \mathbf{Z}_2 & -\mathbf{Z}_2 \\ -\mathbf{Z}_2 & \mathbf{Z}_2 + \mathbf{Z}_3 \end{vmatrix}}$$

$$= \frac{\mathbf{E}_1(\mathbf{Z}_2 + \mathbf{Z}_3) - \mathbf{E}_2(\mathbf{Z}_2)}{(\mathbf{Z}_1 + \mathbf{Z}_2)(\mathbf{Z}_2 + \mathbf{Z}_3) - (\mathbf{Z}_2)^2}$$

$$= \frac{(\mathbf{E}_1 - \mathbf{E}_2)\mathbf{Z}_2 + \mathbf{E}_1\mathbf{Z}_3}{\mathbf{Z}_1\mathbf{Z}_2 + \mathbf{Z}_1\mathbf{Z}_3 + \mathbf{Z}_2\mathbf{Z}_3}$$

Substituting numerical values yields

$$\mathbf{I}_1 = \frac{(2 - 6)(4) + (2)(-j)}{(+j2)(4) + (+j2)(-j) + (4)(-j)} = \frac{-16 - j2}{j8 - j^2 2 - j4}$$

$$= \frac{-16 - j2}{2 + j4} = \frac{16.1 \angle -172.87°}{4.47 \angle 63.43°}$$

$$= 3.61 \angle -236.30° \quad \text{or} \quad 3.61 \angle 123.70°$$

Therefore,

$$\mathbf{I}_1 = \mathbf{3.61} \angle \mathbf{123.70°}$$

EXAMPLE 17.6. Using mesh analysis, find the current $\mathbf{I}_2$ in Fig. 17.10.

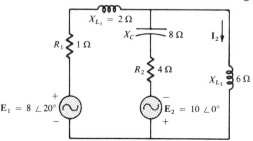

FIG. 17.10

Solution: The network is redrawn in Fig. 17.11:

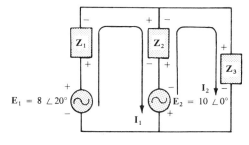

FIG. 17.11

$$\mathbf{Z}_1 = 1 + j2$$
$$\mathbf{Z}_2 = 4 - j8$$
$$\mathbf{Z}_3 = +j6$$

Note the reduction in complexity of the problem with the substitution of the subscripted impedances.

Steps 1 and 2 are as indicated in the network.

Step 3: Kirchhoff's voltage law is applied around each closed loop:

$$+\mathbf{E}_1 - \mathbf{Z}_1\mathbf{I}_1 - \mathbf{Z}_2(\mathbf{I}_1 - \mathbf{I}_2) + \mathbf{E}_2 = 0$$
$$-\mathbf{E}_2 - \mathbf{Z}_3\mathbf{I}_2 - \mathbf{Z}_2(\mathbf{I}_2 - \mathbf{I}_1) = 0$$

which are rewritten as

$$\mathbf{E}_1 + \mathbf{E}_2 - (\mathbf{Z}_1 + \mathbf{Z}_2)\mathbf{I}_1 + \mathbf{Z}_2\mathbf{I}_2 = 0$$
$$-\mathbf{E}_2 + \mathbf{Z}_2\mathbf{I}_1 - (\mathbf{Z}_2 + \mathbf{Z}_3)\mathbf{I}_2 = 0$$

or

$$(\mathbf{Z}_1 + \mathbf{Z}_2)\mathbf{I}_1 - \mathbf{Z}_2\mathbf{I}_2 = \mathbf{E}_1 + \mathbf{E}_2$$
$$-(\mathbf{Z}_2)\mathbf{I}_1 + (\mathbf{Z}_2 + \mathbf{Z}_3)\mathbf{I}_2 = -\mathbf{E}_2$$

Using determinants, we have

$$\mathbf{I}_2 = \frac{\begin{vmatrix} \mathbf{Z}_1 + \mathbf{Z}_2 & \mathbf{E}_1 + \mathbf{E}_2 \\ -\mathbf{Z}_2 & -\mathbf{E}_2 \end{vmatrix}}{\begin{vmatrix} \mathbf{Z}_1 + \mathbf{Z}_2 & -\mathbf{Z}_2 \\ -\mathbf{Z}_2 & \mathbf{Z}_2 + \mathbf{Z}_3 \end{vmatrix}}$$

$$= \frac{-(\mathbf{Z}_1 + \mathbf{Z}_2)\mathbf{E}_2 + \mathbf{Z}_2(\mathbf{E}_1 + \mathbf{E}_2)}{(\mathbf{Z}_1 + \mathbf{Z}_2)(\mathbf{Z}_2 + \mathbf{Z}_3) - \mathbf{Z}_2^2}$$

$$= \frac{-\mathbf{Z}_1\mathbf{E}_2 + \mathbf{Z}_2\mathbf{E}_1}{\mathbf{Z}_1\mathbf{Z}_2 + \mathbf{Z}_1\mathbf{Z}_3 + \mathbf{Z}_2\mathbf{Z}_3}$$

Substituting numerical values yields

$$\mathbf{I}_2 = \frac{-(1 + j2)(10 \angle 0°) + (4 - j8)(8 \angle 20°)}{(1 + j2)(4 - j8) + (1 + j2)(+j6) + (4 - j8)(+j6)}$$

$$= \frac{-(10 + j20) + (4 - j8)(7.52 + j2.74)}{20 + (j6 - 12) + (j24 + 48)}$$

$$= \frac{-(10 + j20) + (52.0 - j49.20)}{56 + j30} = \frac{+42.0 - j69.20}{56 + j30}$$

$$= \frac{80.95 \angle -58.74°}{63.53 \angle 28.18°} = 1.27 \angle -86.92°$$

Therefore,

$$\mathbf{I}_2 = 1.27 \angle -86.92°$$

EXAMPLE 17.7. Write the mesh equations for the network of Fig. 17.12. Do not solve.

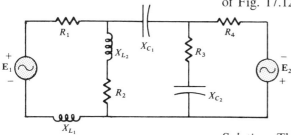

FIG. 17.12

Solution: The network is redrawn in Fig. 17.13. Again note the reduced complexity and increased clarity by use of subscripted impedances:

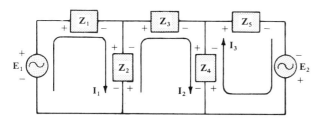

FIG. 17.13

$$\mathbf{Z}_1 = R_1 + jX_{L_1}$$
$$\mathbf{Z}_2 = R_2 + jX_{L_2}$$
$$\mathbf{Z}_3 = jX_{C_1}$$
$$\mathbf{Z}_4 = R_3 - jX_{C_2}$$
$$\mathbf{Z}_5 = R_4$$

Steps 1 and 2 are as indicated in the network.

Step 3: Kirchhoff's voltage law is applied around each loop:

$$+\mathbf{E}_1 - \mathbf{I}_1\mathbf{Z}_1 - \mathbf{Z}_2(\mathbf{I}_1 - \mathbf{I}_2) = 0$$
$$-\mathbf{Z}_2(\mathbf{I}_2 - \mathbf{I}_1) - \mathbf{Z}_3(\mathbf{I}_2) - \mathbf{Z}_4(\mathbf{I}_2 - \mathbf{I}_3) = 0$$
$$+\mathbf{E}_2 - \mathbf{Z}_4(\mathbf{I}_3 - \mathbf{I}_2) - \mathbf{Z}_5\mathbf{I}_3 = 0$$

which are rewritten as

$$\mathbf{I}_1(\mathbf{Z}_1 + \mathbf{Z}_2) \quad -\mathbf{I}_2(\mathbf{Z}_2) \quad\quad\quad\quad + 0 \quad\quad = \mathbf{E}_1$$
$$\mathbf{I}_1(\mathbf{Z}_2) \quad - \mathbf{I}_2(\mathbf{Z}_2 + \mathbf{Z}_3 + \mathbf{Z}_4) \quad + \mathbf{I}_3(\mathbf{Z}_4) \quad = 0$$
$$0 \quad\quad\quad - \mathbf{I}_2(\mathbf{Z}_4) \quad\quad + \mathbf{I}_3(\mathbf{Z}_4 + \mathbf{Z}_5) = \mathbf{E}_2$$

Substituting for subscripted impedances yields

$$\mathbf{I}_1[R_1 + R_2 + j(X_{L1} + X_{L2})] - \mathbf{I}_2(R_2 + jX_{L2}) + 0 = \mathbf{E}_1$$
$$\mathbf{I}_1(R_2 + jX_{L2}) - \mathbf{I}_2[R_2 + R_3 + j(X_{L2} - X_{C1} - X_{C2})]$$
$$+ \mathbf{I}_3(R_3 - jX_{C2}) = 0$$
$$0 - \mathbf{I}_2(R_3 - jX_{C2}) + \mathbf{I}_3(R_3 + R_4 - jX_{C2}) = \mathbf{E}_2$$

17.5 MESH ANALYSIS (FORMAT APPROACH)

We will now apply the format approach developed in Section 7.9. Again, the result will be a more rapid and accurate writing of the mesh equations.

The approach is the following:

1. Assign a loop current to each independent closed loop (as in the previous section) in a *clockwise* direction.
2. The number of required equations is equal to the number of chosen independent closed loops. Column 1 of each equation is formed by simply summing the impedance values of those impedances through which the loop current of interest passes and multiplying the result by that loop current.
3. We must now consider the mutual terms which are always subtracted from the terms in the first column. It is possible to have more than one mutual term if the loop current of interest has an element in common with more than one other loop current. Each mutual term is the product of the mutual impedance and the other loop current passing through the same element.
4. The column to the right of the equality sign is the algebraic sum of the voltage sources through which the loop current of interest passes. Positive signs are assigned to those sources of emf having a polarity such that the loop current passes from the negative to positive terminal. A negative sign is assigned to those potentials for which the reverse is true.
5. Solve resulting simultaneous equations for the desired loop currents.

Symmetry is also a property of mesh equations written for sinusoidal ac networks, as demonstrated by Example 17.7, in which the loop currents were all chosen in the clockwise direction. Networks containing controlled sources *may not* demonstrate this property.

EXAMPLE 17.8. Using the format approach, write the mesh equations for the network of Example 17.6 and compare results.

Solution:

Step 1: The network of Fig. 17.11 is redrawn as indicated in Fig. 17.14, with the loop currents chosen in the clockwise direction.

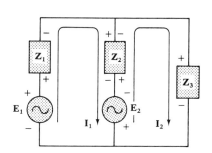

FIG. 17.14

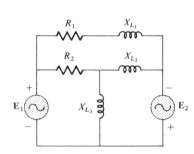

FIG. 17.15

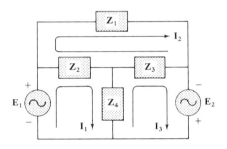

FIG. 17.16

Steps 2 to 4:

$$\mathbf{I}_1: (\mathbf{Z}_1 + \mathbf{Z}_2)\mathbf{I}_1 - (\mathbf{Z}_2)\mathbf{I}_2 = \mathbf{E}_1 + \mathbf{E}_2$$
$$\mathbf{I}_2: (\mathbf{Z}_2 + \mathbf{Z}_3)\mathbf{I}_2 - (\mathbf{Z}_2)\mathbf{I}_1 = -\mathbf{E}_2$$

or

$$(\mathbf{Z}_1 + \mathbf{Z}_2)\mathbf{I}_1 - (\mathbf{Z}_2)\mathbf{I}_2 = \mathbf{E}_1 + \mathbf{E}_2$$
$$-(\mathbf{Z}_2)\mathbf{I}_1 + (\mathbf{Z}_2 + \mathbf{Z}_3)\mathbf{I}_2 = -\mathbf{E}_2$$

Note the symmetry *about* the diagonal axis. The results are the same as those obtained in Example 17.6.

EXAMPLE 17.9. Using the format approach, write the mesh equations for the network of Fig. 17.15.

Solution: The network is redrawn as shown in Fig. 17.16, where

$$\mathbf{Z}_1 = R_1 + jX_{L_1}$$
$$\mathbf{Z}_2 = R_2$$
$$\mathbf{Z}_3 = jX_{L_2}$$
$$\mathbf{Z}_4 = jX_{L_3}$$

$$\mathbf{I}_1: \qquad (\mathbf{Z}_2 + \mathbf{Z}_4)\mathbf{I}_1 - (\mathbf{Z}_2)\mathbf{I}_2 - (\mathbf{Z}_4)\mathbf{I}_3 = \mathbf{E}_1$$
$$\mathbf{I}_2: (\mathbf{Z}_1 + \mathbf{Z}_2 + \mathbf{Z}_3)\mathbf{I}_2 - (\mathbf{Z}_2)\mathbf{I}_1 - (\mathbf{Z}_3)\mathbf{I}_3 = 0$$
$$\mathbf{I}_3: \qquad (\mathbf{Z}_3 + \mathbf{Z}_4)\mathbf{I}_3 - (\mathbf{Z}_4)\mathbf{I}_1 - (\mathbf{Z}_3)\mathbf{I}_2 = \mathbf{E}_2$$

or

$$(\mathbf{Z}_2 + \mathbf{Z}_4)\mathbf{I}_1 \qquad -(\mathbf{Z}_2)\mathbf{I}_2 \qquad -(\mathbf{Z}_4)\mathbf{I}_3 \quad = \mathbf{E}_1$$
$$-(\mathbf{Z}_2)\mathbf{I}_1 + (\mathbf{Z}_1 + \mathbf{Z}_2 + \mathbf{Z}_3)\mathbf{I}_2 \quad -(\mathbf{Z}_3)\mathbf{I}_3 \quad = 0$$
$$-(\mathbf{Z}_4)\mathbf{I}_1 \qquad -(\mathbf{Z}_3)\mathbf{I}_2 \qquad + (\mathbf{Z}_3 + \mathbf{Z}_4)\mathbf{I}_3 = \mathbf{E}_2$$

Note the symmetry *about* the diagonal axis. That is, note the location of $-\mathbf{Z}_2$, $-\mathbf{Z}_4$, and $-\mathbf{Z}_3$ off the diagonal.

EXAMPLE 17.10. The transistor network of Fig. 17.17 will appear as shown in Fig. 17.18 when the approximate "equivalent circuit" for the transistor is substituted. Note that now a current source is in evidence with a current in an isolated network as the controlling variable, and the method can be applied as before. The quantity h_{f_e} (a constant) is a characteristic of the transistor. Using mesh analysis (format approach), determine the current $\mathbf{I}_L$.

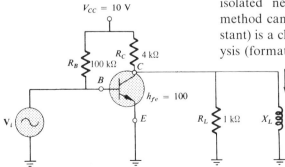

FIG. 17.17

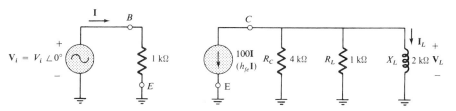

Solution: Converting the current source to a voltage source yields

FIG. 17.18

$$\mathbf{E} = \mathbf{IZ} = (100\mathbf{I})(4 \times 10^3 \angle 0°) = 4 \times 10^5\mathbf{I} \angle 0°$$

The network of Fig. 17.18 is then redrawn as shown in Fig. 17.19, where

$$\mathbf{Z}_1 = 4 \text{ k}\Omega$$
$$\mathbf{Z}_2 = 1 \text{ k}\Omega$$
$$\mathbf{Z}_3 = 2 \times 10^3 \angle 90°$$

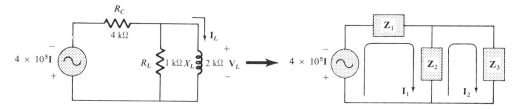

Applying the format approach, we have

FIG. 17.19

$$\mathbf{I}_1: (\mathbf{Z}_1 + \mathbf{Z}_2)\mathbf{I}_1 - (\mathbf{Z}_2)\mathbf{I}_2 = -4 \times 10^5\mathbf{I}$$
$$\mathbf{I}_2: (\mathbf{Z}_2 + \mathbf{Z}_3)\mathbf{I}_2 - (\mathbf{Z}_2)\mathbf{I}_1 = 0$$

or

$$(\mathbf{Z}_1 + \mathbf{Z}_2)\mathbf{I}_1 - (\mathbf{Z}_2)\mathbf{I}_2 = -4 \times 10^5\mathbf{I}$$
$$-\mathbf{Z}_2\mathbf{I}_1 + (\mathbf{Z}_2 + \mathbf{Z}_3)\mathbf{I}_2 = 0$$

and

$$\mathbf{I}_2 = \mathbf{I}_L = \frac{\begin{vmatrix} \mathbf{Z}_1 + \mathbf{Z}_2 & -4 \times 10^5\mathbf{I} \\ -\mathbf{Z}_2 & 0 \end{vmatrix}}{\begin{vmatrix} \mathbf{Z}_1 + \mathbf{Z}_2 & -\mathbf{Z}_2 \\ -\mathbf{Z}_2 & \mathbf{Z}_2 + \mathbf{Z}_3 \end{vmatrix}}$$

$$= \frac{-4 \times 10^5\mathbf{Z}_2\mathbf{I}}{\mathbf{Z}_1\mathbf{Z}_2 + \mathbf{Z}_2\mathbf{Z}_3 + \mathbf{Z}_1\mathbf{Z}_3}$$

Substituting values yields

$$\mathbf{I}_L = \frac{-4 \times 10^5(1 \times 10^3)\mathbf{I}}{(4 \times 10^3)(1 \times 10^3) + (1 \times 10^3)(2 \times 10^3 \angle 90° + (4 \times 10^3)(2 \times 10^3 \angle 90°}$$

$$= \frac{-4 \times 10^2\mathbf{I}}{4 + j2 + j8} = \frac{-4 \times 10^2\mathbf{I}}{4 + j10} = \frac{-4 \times 10^2\mathbf{I}}{10.77 \angle 68.2°}$$

$$\mathbf{I}_L = -37.14\mathbf{I} \angle -68.2°$$

From Fig. 17.18,

$$\mathbf{I} = \frac{\mathbf{V}_i \angle 0°}{1 \text{ k}\Omega}$$

and

$$\mathbf{I}_L = -37.14 \left(\frac{\mathbf{V}_i \angle 0°}{1 \times 10^3} \right) \angle -68.2°$$

$$= -37.14 \times 10^{-3} \mathbf{V}_i \angle -68.2°$$

and

$$\mathbf{V}_L = \mathbf{I}_2 \mathbf{Z}_3 = \mathbf{I}_L \mathbf{Z}_3$$
$$= (-37.14 \times 10^{-3} \mathbf{V}_i \angle -68.2°)(2 \times 10^3 \angle 90°)$$
$$\mathbf{V}_L = -74.28 \mathbf{V}_i \angle 21.8°$$

Note that in this case the magnitude of the amplifier gain

$$\frac{\mathbf{V}_L}{\mathbf{V}_i} \cong 74$$

17.6 NODAL ANALYSIS (FORMAT APPROACH)

Before examining the application of this method to ac circuits, the student should review the section on nodal analysis in Chapter 7, since we shall repeat only the final conclusions as they apply to ac circuits. Recall that these conclusions made the writing of the nodal equations quite direct and in a form convenient for the use of determinants. For sinusoidal ac networks, the procedure is the following:

1. Choose a reference node and assign a subscripted voltage label to the $N - 1$ remaining independent nodes of the network.
2. The number of equations required for a complete solution is equal to the number of subscripted voltages $(N - 1)$. Column 1 of each equation is formed by summing the admittances tied to the node of interest and multiplying the result by that subscripted nodal voltage.
3. The mutual terms are always subtracted from the terms of the first column. It is possible to have more than one mutual term if the nodal voltage of interest has an element in common with more than one other nodal voltage. Each mutual term is the product of the mutual admittance and the other nodal voltage tied to that admittance.
4. The column to the right of the equality sign is the algebraic sum of the current sources tied to the

node of interest. A current source is assigned a
positive sign if it supplies current to a node and a
negative sign if it draws current from the node.

5. Solve resulting simultaneous equations for the
desired nodal voltages. The comments offered for
mesh analysis regarding independent and dependent sources apply here also.

EXAMPLE 17.11. Using nodal analysis, find the voltage
across the 4-Ω resistor in Fig. 17.20.

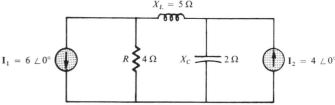

Solution: Choosing nodes (Fig. 17.21) and writing the
nodal equations, we have

FIG. 17.20

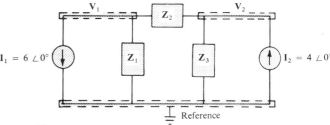

FIG. 17.21

$$\mathbf{Z}_1 = 4 \qquad \mathbf{Z}_2 = j5 \qquad \mathbf{Z}_3 = -j2$$

$$\mathbf{V}_1(\mathbf{Y}_1 + \mathbf{Y}_2) - \mathbf{V}_2(\mathbf{Y}_2) = -\mathbf{I}_1$$

$$\mathbf{V}_2(\mathbf{Y}_3 + \mathbf{Y}_2) - \mathbf{V}_1(\mathbf{Y}_2) = +\mathbf{I}_2$$

or

$$\mathbf{V}_1(\mathbf{Y}_1 + \mathbf{Y}_2) - \mathbf{V}_2(\mathbf{Y}_2) = -\mathbf{I}_1$$

$$-\mathbf{V}_1(\mathbf{Y}_2) + \mathbf{V}_2(\mathbf{Y}_3 + \mathbf{Y}_2) = +\mathbf{I}_2$$

$$\mathbf{Y}_1 = \frac{1}{\mathbf{Z}_1} \qquad \mathbf{Y}_2 = \frac{1}{\mathbf{Z}_2} \qquad \mathbf{Y}_3 = \frac{1}{\mathbf{Z}_3}$$

Using determinants yields

$$\mathbf{V}_1 = \frac{\begin{vmatrix} -\mathbf{I}_1 & -\mathbf{Y}_2 \\ +\mathbf{I}_2 & \mathbf{Y}_3 + \mathbf{Y}_2 \end{vmatrix}}{\begin{vmatrix} \mathbf{Y}_1 + \mathbf{Y}_2 & -\mathbf{Y}_2 \\ -\mathbf{Y}_2 & \mathbf{Y}_3 + \mathbf{Y}_2 \end{vmatrix}}$$

$$= \frac{-(\mathbf{Y}_3 + \mathbf{Y}_2)\mathbf{I}_1 + \mathbf{I}_2\mathbf{Y}_2}{(\mathbf{Y}_1 + \mathbf{Y}_2)(\mathbf{Y}_3 + \mathbf{Y}_2) - \mathbf{Y}_2^2}$$

$$= \frac{-(\mathbf{Y}_3 + \mathbf{Y}_2)\mathbf{I}_1 + \mathbf{I}_2\mathbf{Y}_2}{\mathbf{Y}_1\mathbf{Y}_3 + \mathbf{Y}_2\mathbf{Y}_3 + \mathbf{Y}_1\mathbf{Y}_2}$$

Substituting numerical values, we have

$$\mathbf{V}_1 = \frac{-[(1/-j2) + (1/j5)]6 \angle 0° + 4 \angle 0°(1/j5)}{(1/4)(1/-j2) + (1/j5)(1/-j2) + (1/4)(1/j5)}$$

$$= \frac{-(+j0.5 - j0.2)6 \angle 0° + 4 \angle 0°(-j0.2)}{(1/-j8) + (1/10) + (1/j20)}$$

$$= \frac{(-0.3 \angle 90°)(6 \angle 0°) + (4 \angle 0°)(0.2 \angle -90°)}{j0.125 + 0.1 - j0.05}$$

$$= \frac{-1.8 \angle 90° + 0.8 \angle -90°}{0.1 + j0.075}$$

$$= \frac{2.6 \angle -90°}{0.125 \angle 36.87°}$$

$$\mathbf{V}_1 = \mathbf{20.80} \angle -\mathbf{126.87°}$$

EXAMPLE 17.12. Write the nodal equations for the network of Fig. 17.22. In this case, a voltage source appears in the network.

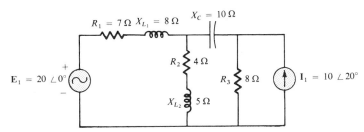

FIG. 17.22

Solution: The circuit is redrawn in Fig. 17.23.

$$\mathbf{Z}_1 = 7 + j8 \qquad \mathbf{Z}_2 = 4 + j5$$
$$\mathbf{Z}_3 = -j10 \qquad \mathbf{Z}_4 = 8$$

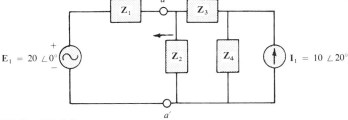

FIG. 17.23

Converting the voltage source to a current source and choosing nodes, we obtain Fig. 17.24. Note the "neat" appearance of the network using the subscripted impedances.

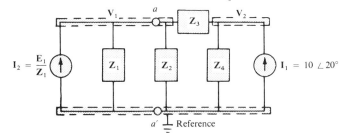

FIG. 17.24

Working directly with Fig. 17.22 would be difficult and would probably produce errors.

Write the nodal equations:

$$\mathbf{V}_1(\mathbf{Y}_1 + \mathbf{Y}_2 + \mathbf{Y}_3) - \mathbf{V}_2(\mathbf{Y}_3) = +\mathbf{I}_2$$
$$\mathbf{V}_2(\mathbf{Y}_3 + \mathbf{Y}_4) - \mathbf{V}_1(\mathbf{Y}_3) = +\mathbf{I}_1$$

$$\mathbf{Y} = \frac{1}{\mathbf{Z}_1} \qquad \mathbf{Y}_2 = \frac{1}{\mathbf{Z}_2} \qquad \mathbf{Y}_3 = \frac{1}{\mathbf{Z}_3} \qquad \mathbf{Y}_4 = \frac{1}{\mathbf{Z}_4}$$

which are rewritten as

$$\mathbf{V}_1(\mathbf{Y}_1 + \mathbf{Y}_2 + \mathbf{Y}_3) - \mathbf{V}_2(\mathbf{Y}_3) = +\mathbf{I}_2$$
$$-\mathbf{V}_1(\mathbf{Y}_3) + \mathbf{V}_2(\mathbf{Y}_3 + \mathbf{Y}_4) = +\mathbf{I}_1$$

$$\mathbf{Y}_1 = \frac{1}{7 + j8} \qquad \mathbf{Y}_2 = \frac{1}{4 + j5} \qquad \mathbf{Y}_3 = \frac{1}{-j10} \qquad \mathbf{Y}_4 = \frac{1}{8}$$

$$\mathbf{I}_2 = \frac{20 \angle 0°}{7 + j8} \qquad \mathbf{I}_1 = 10 \angle 20°$$

EXAMPLE 17.13. Write the nodal equations for the network of Fig. 17.25.

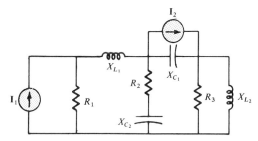

Solution: Choose nodes (Fig. 17.26) and write the nodal equations:

FIG. 17.25

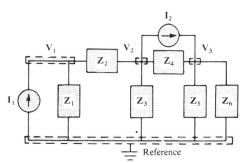

$$\mathbf{Z}_1 = R_1 \qquad \mathbf{Z}_2 = jX_{L_1} \qquad \mathbf{Z}_3 = R_2 - jX_{C_2}$$
$$\mathbf{Z}_4 = -jX_{C_1} \qquad \mathbf{Z}_5 = R_3 \qquad \mathbf{Z}_6 = jX_{L_2}$$

FIG. 17.26

$$\mathbf{V}_1(\mathbf{Y}_1 + \mathbf{Y}_2) - \mathbf{V}_2(\mathbf{Y}_2) = +\mathbf{I}_1$$
$$\mathbf{V}_2(\mathbf{Y}_2 + \mathbf{Y}_3 + \mathbf{Y}_4) - \mathbf{V}_1(\mathbf{Y}_2) - \mathbf{V}_3(\mathbf{Y}_4) = -\mathbf{I}_2$$
$$\mathbf{V}_3(\mathbf{Y}_4 + \mathbf{Y}_5 + \mathbf{Y}_6) - \mathbf{V}_2(\mathbf{Y}_4) = +\mathbf{I}_2$$

which are rewritten as

$$\mathbf{V_1}(\mathbf{Y_1} + \mathbf{Y_2}) - \mathbf{V_2}(\mathbf{Y_2}) \qquad\quad + 0 \qquad\quad = +\mathbf{I_1}$$
$$-\mathbf{V_1}(\mathbf{Y_2}) \quad + \mathbf{V_2}(\mathbf{Y_2} + \mathbf{Y_3} + \mathbf{Y_4}) - \mathbf{V_3}(\mathbf{Y_4}) = -\mathbf{I_2}$$
$$\underline{\qquad 0 \qquad\quad - \mathbf{V_2}(\mathbf{Y_4}) + \mathbf{V_3}(\mathbf{Y_4} + \mathbf{Y_5} + \mathbf{Y_6}) = +\mathbf{I_2}}$$

$$\mathbf{Y_1} = \frac{1}{R_1} \qquad \mathbf{Y_2} = \frac{1}{jX_{L_1}} \qquad \mathbf{Y_3} = \frac{1}{R_2 - jX_{C_2}}$$

$$\mathbf{Y_4} = \frac{1}{-jX_{C_1}} \qquad \mathbf{Y_5} = \frac{1}{R_3} \qquad \mathbf{Y_6} = \frac{1}{jX_{L_2}}$$

Note the symmetry about the diagonal for this example and those preceding it in this section.

EXAMPLE 17.14. Apply nodal analysis to the network of Example 17.11. That is, determine $\mathbf{V_L}$ for the network of Fig. 17.18, redrawn in Fig. 17.27 for convenience. Compare results.

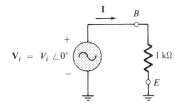
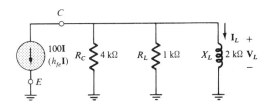

FIG. 17.27

Solution: In this case there is no need for a source conversion. The network is redrawn in Fig. 17.28 with the chosen node voltage and subscripted impedances.

Apply the format approach:

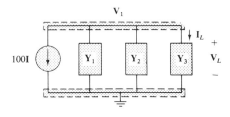

FIG. 17.28

$$\mathbf{Y_1} = \frac{1}{\mathbf{Z_1}} = \frac{1}{4 \times 10^3} = 0.25 \times 10^{-3} \angle 0° = \mathbf{G_1}$$

$$\mathbf{Y_2} = \frac{1}{\mathbf{Z_2}} = \frac{1}{1 \times 10^3} = 1.0 \times 10^{-3} \angle 0° = \mathbf{G_2}$$

$$\mathbf{Y_3} = \frac{1}{\mathbf{Z_3}} = \frac{1}{2 \times 10^3 \angle 90°} = 0.5 \times 10^{-3} \angle -90°$$
$$= -j0.5 \times 10^{-3} = \mathbf{B_L}$$

$$\mathbf{V_1}: (\mathbf{Y_1} + \mathbf{Y_2} + \mathbf{Y_3})\mathbf{V_1} = -100\mathbf{I}$$

and

$$\mathbf{V_1} = \frac{-100\mathbf{I}}{\mathbf{Y_1} + \mathbf{Y_2} + \mathbf{Y_3}}$$

$$= \frac{-100\mathbf{I}}{(0.25 \times 10^{-3}) + (1 \times 10^{-3}) + (-j0.5 \times 10^{-3})}$$

$$= \frac{-100 \times 10^3\mathbf{I}}{1.25 - j0.5} = \frac{-100 \times 10^3\mathbf{I}}{1.3463 \angle -21.80°}$$

$$= -74.28 \times 10^3\mathbf{I} \angle 21.80°$$

$$= -74.28 \times 10^3 \left(\frac{\mathbf{V_i}}{1 \text{ k}\Omega}\right) \angle 21.80°$$

$$\mathbf{V_1} = \mathbf{V_L} = -74.28\mathbf{V_i} \angle 21.80°$$

as obtained for Example 17.10.

EXAMPLE 17.15. The transistor configuration of Fig. 17.29 will result in a network very similar in appearance to Fig. 17.30 when the equivalent circuits are substituted. The quantities h_1 and h_2 are characteristic constants of the transistors. Determine V_L for the network of Fig. 17.30. The resistance values were chosen for clarity and are not typical values. In this case, one of the controlling variables is part of the network to be analyzed. Care must be exercised when applying the method.

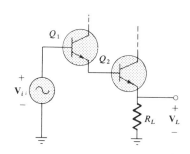

FIG. 17.29

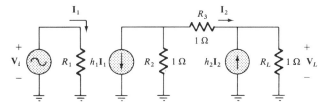

Solution: The network is redrawn in Fig. 17.31. Note that the controlling variable

FIG. 17.30

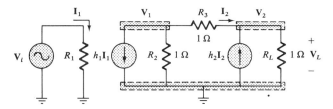

FIG. 17.31

$$I_2 = \frac{V_1 - V_2}{R_3} = \frac{V_1 - V_2}{1} = V_1 - V_2$$

and

$$I_1 = \frac{V_i}{R_1} = \frac{V_i}{1} = V_i$$

Applying the format approach, we have

$$\begin{cases} V_1: V_1(1 + 1) - (1)V_2 = -h_1I_1 \\ V_2: V_2(1 + 1) - (1)V_1 = h_2I_2 \end{cases}$$

and

$$\begin{cases} 2V_1 - V_2 = -h_1V_i \\ 2V_2 - V_1 = h_2(V_1 - V_2) = h_2V_1 - h_2V_2 \end{cases}$$

so that

$$2V_1 - V_2 = -h_1V_i$$

and

$$\underline{-V_1 - h_2V_1 + 2V_2 + h_2V_2 = 0}$$

Or

$$2V_1 - V_2 = -h_1V_i$$

with

$$\underline{-(1 + h_2)V_1 + (2 + h_2)V_2 = 0}$$

and

$$V_2 = V_L = \frac{\begin{vmatrix} 2 & -h_1 V_i \\ -(1 + h_2) & 0 \end{vmatrix}}{\begin{vmatrix} 2 & -1 \\ -(1 + h_2) & (2 + h_2) \end{vmatrix}}$$

$$= \frac{-h_1(1 + h_2)V_i}{2(2 + h_2) - (1 + h_2)}$$

$$= \frac{-h_1(1 + h_2)V_i}{4 + 2h_2 - 1 - h_2}$$

so that

$$V_L = \left[-\frac{h_1(1 + h_2)}{3 + h_2} \right] V_i$$

For

$$h_1 = h_2 = 100 \quad \text{(typical)}$$

$$V_L = \frac{-100(101)}{3 + 100} V_i \cong -98V_i$$

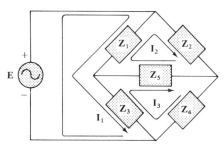

FIG. 17.32 *Maxwell bridge.*

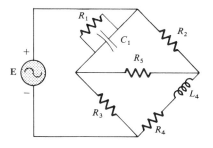

FIG. 17.33

17.7 BRIDGE NETWORKS (ac)

The basic bridge configuration was discussed in some detail in Section 7.13 for dc networks. We now continue to examine bridge networks by considering those that have reactive components and a sinusoidal ac voltage or current applied.

We will first analyze various familiar forms of the bridge network using mesh analysis and nodal analysis (the format approach). The balance conditions will be investigated throughout the section.

Apply *mesh analysis* to the network of Fig. 17.32. The network is redrawn in Fig. 17.33, where

$$Z_1 = \frac{1}{Y_1} = \frac{1}{G_1 + jB_C} = \frac{G_1}{G_1^2 + B_C^2} - j\frac{B_C}{G_1^2 + B_C^2}$$

$$Z_2 = R_2$$
$$Z_3 = R_3$$
$$Z_4 = R_4 + jX_L$$
$$Z_5 = R_5$$

Apply the format approach:

$$(\mathbf{Z}_1 + \mathbf{Z}_3)\mathbf{I}_1 \qquad - (\mathbf{Z}_1)\mathbf{I}_2 - (\mathbf{Z}_3)\mathbf{I}_3 = \mathbf{E}$$
$$(\mathbf{Z}_1 + \mathbf{Z}_2 + \mathbf{Z}_5)\mathbf{I}_2 - (\mathbf{Z}_1)\mathbf{I}_1 - (\mathbf{Z}_5)\mathbf{I}_3 = 0$$
$$\underline{(\mathbf{Z}_3 + \mathbf{Z}_4 + \mathbf{Z}_5)\mathbf{I}_3 - (\mathbf{Z}_3)\mathbf{I}_1 - (\mathbf{Z}_5)\mathbf{I}_2 = 0}$$

which are rewritten as

$$(\mathbf{Z}_1 + \mathbf{Z}_3)\mathbf{I}_1 - (\mathbf{Z}_1)\mathbf{I}_2 - (\mathbf{Z}_3)\mathbf{I}_3 = \mathbf{E}$$
$$-(\mathbf{Z}_1)\mathbf{I}_1 + (\mathbf{Z}_1 + \mathbf{Z}_2 + \mathbf{Z}_5)\mathbf{I}_2 - (\mathbf{Z}_5)\mathbf{I}_3 = 0$$
$$\underline{-(\mathbf{Z}_3)\mathbf{I}_1 - (\mathbf{Z}_5)\mathbf{I}_2 + (\mathbf{Z}_3 + \mathbf{Z}_4 + \mathbf{Z}_5)\mathbf{I}_3 = 0}$$

Note the symmetry about the diagonal of the above equations. For balance, $\mathbf{I}_{\mathbf{Z}_5} = 0$ A, and

$$\mathbf{I}_{\mathbf{Z}_5} = \mathbf{I}_2 - \mathbf{I}_3 = 0$$

From the above equations,

$$\mathbf{I}_2 = \frac{\begin{vmatrix} \mathbf{Z}_1 + \mathbf{Z}_3 & \mathbf{E} & -\mathbf{Z}_3 \\ -\mathbf{Z}_1 & 0 & -\mathbf{Z}_5 \\ -\mathbf{Z}_3 & 0 & (\mathbf{Z}_3 + \mathbf{Z}_4 + \mathbf{Z}_5) \end{vmatrix}}{\begin{vmatrix} \mathbf{Z}_1 + \mathbf{Z}_3 & -\mathbf{Z}_1 & -\mathbf{Z}_3 \\ -\mathbf{Z}_1 & \mathbf{Z}_1 + \mathbf{Z}_2 + \mathbf{Z}_5 & -\mathbf{Z}_5 \\ -\mathbf{Z}_3 & -\mathbf{Z}_5 & \mathbf{Z}_3 + \mathbf{Z}_4 + \mathbf{Z}_5 \end{vmatrix}}$$

$$= \frac{\mathbf{E}(\mathbf{Z}_1\mathbf{Z}_3 + \mathbf{Z}_1\mathbf{Z}_4 + \mathbf{Z}_1\mathbf{Z}_5 + \mathbf{Z}_3\mathbf{Z}_5)}{\Delta}$$

where Δ signifies the determinant of the denominator (or coefficients). Similarly,

$$\mathbf{I}_3 = \frac{\mathbf{E}(\mathbf{Z}_1\mathbf{Z}_3 + \mathbf{Z}_3\mathbf{Z}_2 + \mathbf{Z}_1\mathbf{Z}_5 + \mathbf{Z}_3\mathbf{Z}_5)}{\Delta}$$

and

$$\mathbf{I}_{\mathbf{Z}_5} = \mathbf{I}_2 - \mathbf{I}_3 = \frac{\mathbf{E}(\mathbf{Z}_1\mathbf{Z}_4 - \mathbf{Z}_3\mathbf{Z}_2)}{\Delta}$$

For $\mathbf{I}_{\mathbf{Z}_5} = 0$, the following must be satisfied (for a finite Δ not equal to zero):

$$\boxed{\mathbf{Z}_1\mathbf{Z}_4 = \mathbf{Z}_3\mathbf{Z}_2} \qquad \mathbf{I}_{\mathbf{Z}_5} = 0 \qquad \text{(17.1)}$$

This condition will be analyzed in greater depth later in this section.

Applying *nodal analysis* to the network of Fig. 17.34 will result in the configuration of Fig. 17.35, where

$$\mathbf{Y}_1 = \frac{1}{\mathbf{Z}_1} = \frac{1}{R_1 - jX_C} \qquad \mathbf{Y}_2 = \frac{1}{\mathbf{Z}_2} = \frac{1}{R_2}$$

$$\mathbf{Y}_3 = \frac{1}{\mathbf{Z}_3} = \frac{1}{R_3} \qquad \mathbf{Y}_4 = \frac{1}{\mathbf{Z}_4} = \frac{1}{R_4 + jX_L}$$

$$\mathbf{Y}_5 = \frac{1}{R_5}$$

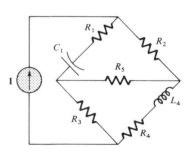

FIG. 17.34 *Hay bridge.*

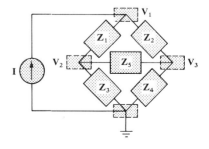

FIG. 17.35

and

$$(\mathbf{Y}_1 + \mathbf{Y}_2)\mathbf{V}_1 \quad - (\mathbf{Y}_1)\mathbf{V}_2 - (\mathbf{Y}_2)\mathbf{V}_3 = \mathbf{I}$$
$$(\mathbf{Y}_1 + \mathbf{Y}_3 + \mathbf{Y}_5)\mathbf{V}_2 - (\mathbf{Y}_1)\mathbf{V}_1 - (\mathbf{Y}_5)\mathbf{V}_3 = 0$$
$$(\mathbf{Y}_2 + \mathbf{Y}_4 + \mathbf{Y}_5)\mathbf{V}_3 - (\mathbf{Y}_2)\mathbf{V}_1 - (\mathbf{Y}_5)\mathbf{V}_2 = 0$$

which are rewritten as

$$(\mathbf{Y}_1 + \mathbf{Y}_2)\mathbf{V}_1 - (\mathbf{Y}_1)\mathbf{V}_2 \qquad - (\mathbf{Y}_2)\mathbf{V}_3 = \mathbf{I}$$
$$-(\mathbf{Y}_1)\mathbf{V}_1 \quad + (\mathbf{Y}_1 + \mathbf{Y}_3 + \mathbf{Y}_5)\mathbf{V}_2 - (\mathbf{Y}_5)\mathbf{V}_3 = 0$$
$$-(\mathbf{Y}_2)\mathbf{V}_1 \quad - (\mathbf{Y}_5)\mathbf{V}_2 + (\mathbf{Y}_2 + \mathbf{Y}_4 + \mathbf{Y}_5)\mathbf{V}_3 = 0$$

Again, note the symmetry about the diagonal axis. For balance, $\mathbf{V}_{\mathbf{Z}_5} = 0$ volts, and

$$\mathbf{V}_{\mathbf{Z}_5} = \mathbf{V}_2 - \mathbf{V}_3 = 0$$

From the above equations,

$$\mathbf{V}_2 = \frac{\begin{vmatrix} \mathbf{Y}_1 + \mathbf{Y}_2 & \mathbf{I} & -\mathbf{Y}_2 \\ -\mathbf{Y}_1 & 0 & -\mathbf{Y}_5 \\ -\mathbf{Y}_2 & 0 & (\mathbf{Y}_2 + \mathbf{Y}_4 + \mathbf{Y}_5) \end{vmatrix}}{\begin{vmatrix} \mathbf{Y}_1 + \mathbf{Y}_2 & -\mathbf{Y}_1 & -\mathbf{Y}_2 \\ -\mathbf{Y}_1 & (\mathbf{Y}_1 + \mathbf{Y}_3 + \mathbf{Y}_5) & -\mathbf{Y}_5 \\ -\mathbf{Y}_2 & -\mathbf{Y}_5 & (\mathbf{Y}_2 + \mathbf{Y}_4 + \mathbf{Y}_5) \end{vmatrix}}$$

$$= \frac{\mathbf{I}(\mathbf{Y}_1\mathbf{Y}_3 + \mathbf{Y}_1\mathbf{Y}_4 + \mathbf{Y}_1\mathbf{Y}_5 + \mathbf{Y}_3\mathbf{Y}_5)}{\Delta}$$

Similarly,

$$\mathbf{V}_3 = \frac{\mathbf{I}(\mathbf{Y}_1\mathbf{Y}_3 + \mathbf{Y}_3\mathbf{Y}_2 + \mathbf{Y}_1\mathbf{Y}_5 + \mathbf{Y}_3\mathbf{Y}_5)}{\Delta}$$

Note the similarities between the above equations and those obtained for mesh analysis. Then

$$\mathbf{V}_{\mathbf{Z}_5} = \mathbf{V}_2 - \mathbf{V}_3 = \frac{\mathbf{I}(\mathbf{Y}_1\mathbf{Y}_4 - \mathbf{Y}_3\mathbf{Y}_2)}{\Delta}$$

For $\mathbf{V}_{\mathbf{Z}_5} = 0$, the following must be satisfied for a finite Δ not equal to zero:

$$\boxed{\mathbf{Y}_1\mathbf{Y}_4 = \mathbf{Y}_3\mathbf{Y}_2} \qquad \mathbf{V}_{\mathbf{Z}_5} = 0 \qquad \textbf{(17.2)}$$

However, substituting $\mathbf{Y}_1 = 1/\mathbf{Z}_1$, $\mathbf{Y}_2 = 1/\mathbf{Z}_2$, $\mathbf{Y}_3 = 1/\mathbf{Z}_3$, and $\mathbf{Y}_4 = 1/\mathbf{Z}_4$, we have

$$\frac{1}{\mathbf{Z}_1\mathbf{Z}_4} = \frac{1}{\mathbf{Z}_3\mathbf{Z}_2}$$

or

$$\boxed{\mathbf{Z}_1\mathbf{Z}_4 = \mathbf{Z}_3\mathbf{Z}_2} \qquad \mathbf{V}_{\mathbf{Z}_5} = 0$$

corresponding with Eq. (17.1) obtained earlier.

Let us now investigate the balance criteria in more detail by considering the network of Fig. 17.36, where it is specified that $\mathbf{I}, \mathbf{V} = 0$.

Since $\mathbf{I} = 0$,

$$\boxed{\mathbf{I}_1 = \mathbf{I}_3} \qquad \text{(17.3a)}$$

and

$$\boxed{\mathbf{I}_2 = \mathbf{I}_4} \qquad \text{(17.3b)}$$

In addition, for $\mathbf{V} = 0$,

$$\boxed{\mathbf{I}_1 \mathbf{Z}_1 = \mathbf{I}_2 \mathbf{Z}_2} \qquad \text{(17.3c)}$$

and

$$\boxed{\mathbf{I}_3 \mathbf{Z}_3 = \mathbf{I}_4 \mathbf{Z}_4} \qquad \text{(17.3d)}$$

Substituting the current relations above into Eq. (17.3d), we have

$$\mathbf{I}_1 \mathbf{Z}_3 = \mathbf{I}_2 \mathbf{Z}_4$$

and

$$\mathbf{I}_2 = \frac{\mathbf{Z}_3}{\mathbf{Z}_4} \mathbf{I}_1$$

Substituting this relationship for $\mathbf{I}_2$ into Eq. (17.3c) yields

$$\mathbf{I}_1 \mathbf{Z}_1 = \left(\frac{\mathbf{Z}_3}{\mathbf{Z}_4} \mathbf{I}_1 \right) \mathbf{Z}_2$$

and

$$\mathbf{Z}_1 \mathbf{Z}_4 = \mathbf{Z}_2 \mathbf{Z}_3$$

as obtained above. Rearranging, we have

$$\boxed{\frac{\mathbf{Z}_1}{\mathbf{Z}_3} = \frac{\mathbf{Z}_2}{\mathbf{Z}_4}} \qquad \text{(17.4)}$$

corresponding with Eq. (7.7a) for dc resistive networks.

For the network of Fig. 17.34, which is referred to as a *Hay bridge* when $\mathbf{Z}_5$ is replaced by a sensitive galvanometer,

$$\mathbf{Z}_1 = R_1 + jX_C$$
$$\mathbf{Z}_2 = R_2$$
$$\mathbf{Z}_3 = R_3$$
$$\mathbf{Z}_4 = R_4 + jX_L$$

This particular network is used for measuring the resistance and inductance of coils in which the resistance is a small fraction of the reactance X_L.

Substitute into Eq. (17.4) in the following form:

$$\mathbf{Z}_2 \mathbf{Z}_3 = \mathbf{Z}_4 \mathbf{Z}_1$$
$$R_2 R_3 = (R_4 + jX_L)(R_1 - jX_C)$$

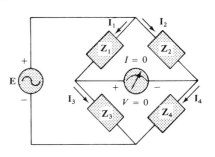

FIG. 17.36

or

$$R_2 R_3 = R_1 R_4 + j(R_1 X_L - R_4 X_C) + X_C X_L$$

so that

$$R_2 R_3 + j0 = (R_1 R_4 + X_C X_L) + j(R_1 X_L - R_4 X_C)$$

In order for the equations to be equal, *the real and imaginary parts must be equal.* Therefore, for a balanced Hay bridge,

$$R_2 R_3 = R_1 R_4 + X_C X_L \qquad \text{(17.5a)}$$

and

$$0 = R_1 X_L - R_4 X_C \qquad \text{(17.5b)}$$

or substituting

$$X_L = \omega L$$

$$X_C = \frac{1}{\omega C}$$

$$X_C X_L = \left(\frac{1}{\omega C}\right)(\omega L) = \frac{L}{C}$$

and

$$R_2 R_3 = R_1 R_4 + \frac{L}{C}$$

with

$$R_1 \omega L = \frac{R_4}{\omega C}$$

Solving for R_4 in the last equation yields

$$R_4 = \omega^2 L C R_1$$

and substituting into the previous equation, we have

$$R_2 R_3 = R_1(\omega^2 L C R_1) + \frac{L}{C}$$

Multiply through by C and factor:

$$C R_2 R_3 = L(\omega^2 C^2 R_1^2 + 1)$$

and

$$L = \frac{C R_2 R_3}{1 + \omega^2 C^2 R_1^2} \qquad \text{(17.6a)}$$

with further algebra yielding

$$R_4 = \frac{\omega^2 C^2 R_1 R_2 R_3}{1 + \omega^2 C^2 R_1^2} \qquad \text{(17.6b)}$$

Equations (17.5) and (17.6) are the balance conditions for the Hay bridge. Note that each is frequency-dependent. For different frequencies, the resistive and capacitive elements must vary for a particular coil to achieve balance. For a coil placed in the Hay bridge as shown in Fig. 17.35, the resistance and inductance of the coil can be determined by Eqs. (17.6a) and (17.6b) when balance is achieved.

The bridge of Fig. 17.32 is referred to as a *Maxwell bridge* when Z_5 is replaced by a sensitive galvanometer. This setup is used for inductance measurements when the resistance of the coil is large enough not to require a Hay bridge.

Application of Eq. (17.4) will yield the following results for the inductance and resistance of the inserted coil:

$$L = CR_2R_3 \qquad (17.7)$$

$$R_4 = \frac{R_2R_3}{R_1} \qquad (17.8)$$

The derivation of these equations is quite similar to that employed for the Hay bridge. Keep in mind that the real and imaginary parts must be equal. The derivation will appear as an exercise at the end of the chapter.

One remaining popular bridge is the *capacitance comparison bridge* of Fig. 17.37. An unknown capacitance and its associated resistance can be determined using this bridge. Application of Eq. (17.4) will yield the following results:

$$C_4 = C_3\frac{R_1}{R_2} \qquad (17.9)$$

$$R_4 = \frac{R_2R_3}{R_1} \qquad (17.10)$$

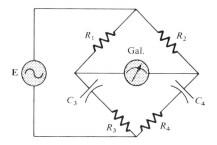

FIG. 17.37 *Capacitance comparison bridge.*

The derivation of these equations will also appear as a problem at the end of the chapter.

17.8 Δ-Y, Y-Δ CONVERSIONS

The Δ-Y, Y-Δ (or π-T, T-π as defined in Section 7.13) conversions for ac circuits will not be derived here since the development corresponds exactly with that for dc circuits. Taking the Δ-Y configuration shown in Fig. 17.38, we find the general equations for the impedances of the Y in terms of those for the Δ:

$$\mathbf{Z}_1 = \frac{\mathbf{Z}_A\mathbf{Z}_C}{\mathbf{Z}_A + \mathbf{Z}_B + \mathbf{Z}_C} \qquad (17.11)$$

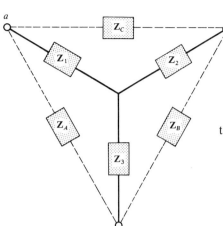

FIG. 17.38 Δ-Y configuration.

$$Z_2 = \frac{Z_B Z_C}{Z_A + Z_B + Z_C} \qquad (17.12)$$

$$Z_3 = \frac{Z_A Z_B}{Z_A + Z_B + Z_C} \qquad (17.13)$$

For the impedances of the Δ in terms of those for the Y, the equations are

$$Z_A = \frac{Z_1 Z_2 + Z_1 Z_3 + Z_2 Z_3}{Z_2} \qquad (17.14)$$

$$Z_B = \frac{Z_1 Z_2 + Z_1 Z_3 + Z_2 Z_3}{Z_1} \qquad (17.15)$$

$$Z_C = \frac{Z_1 Z_2 + Z_1 Z_3 + Z_2 Z_3}{Z_3} \qquad (17.16)$$

Note that each impedance of the Y is equal to the product of the impedances in the two ᵢlosest branches of the Δ, divided by the sum of the impedances in the Δ; and the value of each impedance of the Δ is equal to the sum of the possible product combinations of the impedances of the Y, divided by the impedances of the Y farthest from the impedance to be determined.

Drawn in different forms (Fig. 17.39), they are also referred to as the T and π configurations.

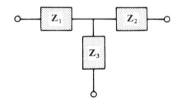

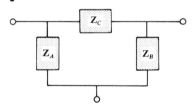

FIG. 17.39

In the study of dc networks, we found that if all of the resistors of the Δ or Y were the same, the conversion from one to the other could be accomplished using the equation

$$R_\Delta = 3R_Y \qquad \text{or} \qquad R_Y = \frac{R_\Delta}{3}$$

For ac networks,

$$Z_\Delta = 3Z_Y \qquad \text{or} \qquad Z_Y = \frac{Z_\Delta}{3} \qquad (17.17)$$

Be careful when using this simplified form. It is not sufficient for all the impedances of the Δ or Y to be of the same magnitude: *The angle associated with each must also be the same.*

EXAMPLE 17.16. Find the total impedance Z_T of the network of Fig. 17.40.

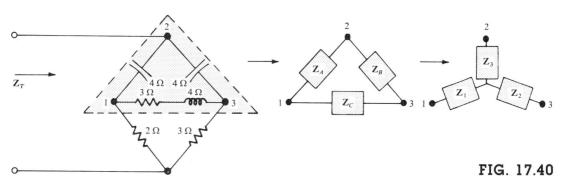

FIG. 17.40

Solution:

$$\mathbf{Z}_A = -j4 \qquad \mathbf{Z}_B = -j4 \qquad \mathbf{Z}_C = 3 + j4$$

$$\begin{aligned}
\mathbf{Z}_1 &= \frac{\mathbf{Z}_A \mathbf{Z}_C}{\mathbf{Z}_A + \mathbf{Z}_B + \mathbf{Z}_C} = \frac{(-j4)(3 + j4)}{-j4 - j4 + 3 + j4} \\
&= \frac{(4 \angle -90°)(5 \angle 53.13°)}{3 - j4} = \frac{20 \angle -36.87°}{5 \angle -53.13°} \\
&= 4 \angle 16.13° = 3.84 + j1.11
\end{aligned}$$

$$\begin{aligned}
\mathbf{Z}_2 &= \frac{\mathbf{Z}_B \mathbf{Z}_C}{\mathbf{Z}_A + \mathbf{Z}_B + \mathbf{Z}_C} = \frac{(-j4)(3 + j4)}{5 \angle -53.13°} \\
&= 4 \angle 16.13° = 3.84 + j1.11
\end{aligned}$$

Recall from the study of dc circuits that if two branches of the Y or Δ were the same, the corresponding Δ or Y, respectively, would also have two similar branches. In this example, $\mathbf{Z}_A = \mathbf{Z}_B$. Therefore, $\mathbf{Z}_1 = \mathbf{Z}_2$, and

$$\begin{aligned}
\mathbf{Z}_3 &= \frac{\mathbf{Z}_A \mathbf{Z}_B}{\mathbf{Z}_A + \mathbf{Z}_B + \mathbf{Z}_C} = \frac{(-j4)(-j4)}{5 \angle -53.13°} \\
&= \frac{16 \angle -180°}{5 \angle -53.13°} = 3.2 \angle -126.87° = -1.92 - j2.56
\end{aligned}$$

Replace the Δ by the Y (Fig. 17.41):

$$\mathbf{Z}_1 = 3.84 + j1.11 \qquad \mathbf{Z}_2 = 3.84 + j1.11$$
$$\mathbf{Z}_3 = -1.92 - j2.56 \qquad \mathbf{Z}_4 = 2$$
$$\mathbf{Z}_5 = 3$$

Impedances $\mathbf{Z}_1$ and $\mathbf{Z}_4$ are in series:

$$\begin{aligned}
\mathbf{Z}_{T_1} &= \mathbf{Z}_1 + \mathbf{Z}_4 = 3.84 + j1.11 + 2 = 5.84 + j1.11 \\
&= 5.94 \angle 10.76°
\end{aligned}$$

Impedances $\mathbf{Z}_2$ and $\mathbf{Z}_5$ are in series:

$$\begin{aligned}
\mathbf{Z}_{T_2} &= \mathbf{Z}_2 + \mathbf{Z}_5 = 3.84 + j1.11 + 3 = 6.84 + j1.11 \\
&= 6.93 \angle 9.21°
\end{aligned}$$

Impedances $\mathbf{Z}_{T_1}$ and $\mathbf{Z}_{T_2}$ are in parallel:

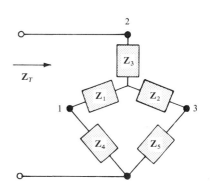

FIG. 17.41

$$\mathbf{Z}_{T_3} = \frac{\mathbf{Z}_{T_1}\mathbf{Z}_{T_2}}{\mathbf{Z}_{T_1} + \mathbf{Z}_{T_2}} = \frac{(5.94 \angle 10.76°)(6.93 \angle 9.21°)}{5.84 + j1.11 + 6.84 + j1.11}$$

$$= \frac{41.16 \angle 19.97°}{12.68 + j2.22} = \frac{41.16 \angle 19.97°}{12.87 \angle 9.93°} = 3.198 \angle 10.04°$$

$$= 3.15 + j0.56$$

Impedances $\mathbf{Z}_3$ and $\mathbf{Z}_{T_3}$ are in series. Therefore,

$$\mathbf{Z}_T = \mathbf{Z}_3 + \mathbf{Z}_{T_3} = -1.92 - j2.56 + 3.15 + j0.56$$
$$= 1.23 - j2.0 = \mathbf{2.35} \angle \mathbf{-58.41°}$$

EXAMPLE 17.17. Using both the Δ-Y and Y-Δ transformations, find the total impedance $\mathbf{Z}_T$ for the network of Fig. 17.42.

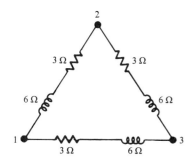

FIG. 17.42

Solutions: Using the Δ-Y transformation, we obtain Fig. 17.43. In this case, since both systems are balanced (same impedance in each branch), the center point d' of the transformed Δ will be the same as the point d of the original Y:

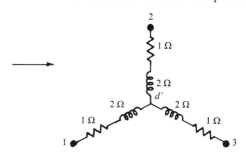

FIG. 17.43

$$\mathbf{Z}_Y = \frac{\mathbf{Z}_\Delta}{3} = \frac{3 + j6}{3} = 1 + j2$$

and (Fig. 17.44)

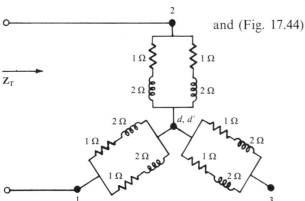

FIG. 17.44

$$\mathbf{Z}_T = 2\left(\frac{1 + j2}{2}\right) = 1 + j2$$

Using the Y-Δ transformation (Fig. 17.45), we obtain

$$\mathbf{Z}_\Delta = 3\mathbf{Z}_Y = 3(1 + j2) = 3 + j6$$

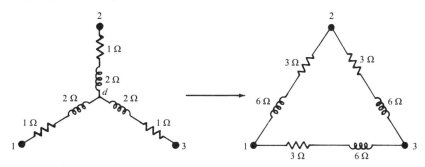

FIG. 17.45

Each resulting parallel combination in Fig. 17.46 will have the following impedance:

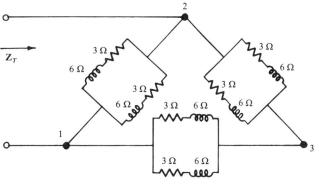

FIG. 17.46

$$\mathbf{Z}' = \frac{3 + j6}{2} = 1.5 + j3$$

and

$$\mathbf{Z}_T = \frac{\mathbf{Z}'(2\mathbf{Z}')}{\mathbf{Z}' + 2\mathbf{Z}'} = \frac{2(\mathbf{Z}')^2}{3\mathbf{Z}'} = \frac{2\mathbf{Z}'}{3}$$

$$\mathbf{Z}_T = \frac{2(1.5 + j3)}{3} = 1 + j2$$

which compares with the above result.

PROBLEMS

Section 17.2

1. Discuss, in your own words, the difference between a controlled and an independent source.

Section 17.3

2. Convert the voltage sources of Fig. 17.47 to current sources.

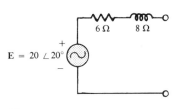

(a)

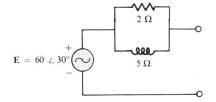

(b)

FIG. 17.47

3. Convert the current sources of Fig. 17.48 to voltage sources.

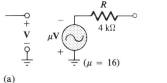

(a)

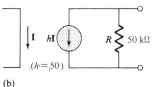

(b)

FIG. 17.48

4. Convert the voltage source of Fig. 17.49(a) to a current source and the current source of Fig. 17.49(b) to a voltage source.

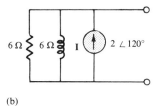

(a)

(b)

FIG. 17.49

Section 17.4

5. Write the mesh equations for the networks of Fig. 17.50. Determine the current through the resistor R_1.

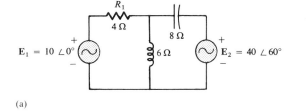

(a)

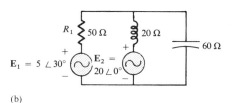

(b)

FIG. 17.50

6. Repeat Problem 5 for the networks of Fig. 17.51.

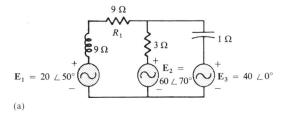

(a)

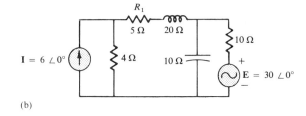

(b)

FIG. 17.51

*7. Repeat Problem 5 for the networks of Fig. 17.52.

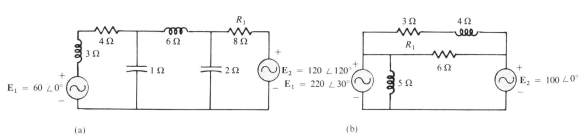

(a)

(b)

*8. Repeat Problem 5 for the networks of Fig. 17.53.

FIG. 17.52

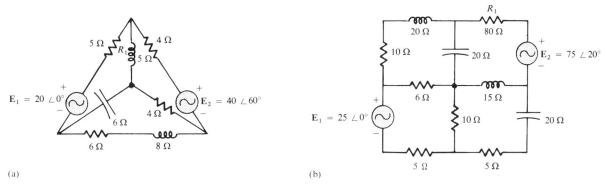

(a)

(b)

Section 17.5

FIG. 17.53

9. Using the format approach, write the mesh equations for the networks of Fig. 17.50. Is symmetry present? Determine the current through the resistor R_1.

10. Repeat Problem 9 for the networks of Fig. 17.51.

*11. Repeat Problem 9 for the networks of Fig. 17.52.

*12. Repeat Problem 9 for the networks of Fig. 17.53.

13. Using the mesh-analysis format approach, determine the current I_L (in terms of V) for the network of Fig. 17.54.

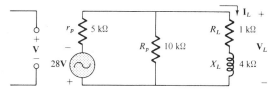

*14. Using the mesh-analysis format approach, determine the current I_L (in terms of I) for the network of Fig. 17.55.

FIG. 17.54

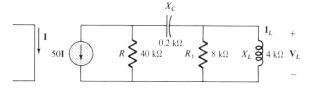

FIG. 17.55

Section 17.6

15. Using the format approach, write the nodal equations for each network of Fig. 17.56. Is symmetry present? Determine the nodal voltages.

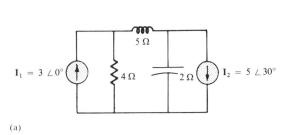

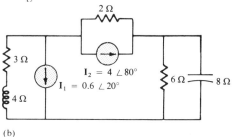

(a)   (b)

FIG. 17.56

*16. Repeat Problem 15 for the networks of Fig. 17.57.

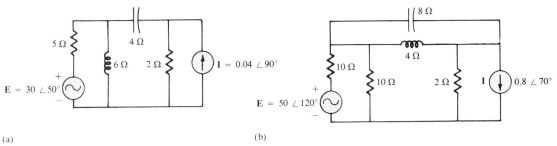

(a)   (b)

FIG. 17.57

17. Repeat Problem 15 for the networks of Fig. 17.58.

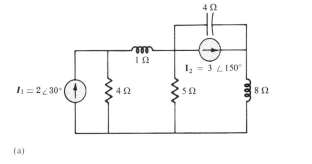

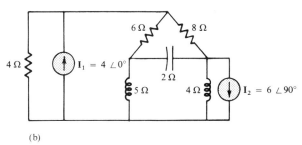

(a)   (b)

FIG. 17.58

18. Using the nodal-analysis approach, repeat Problem 13 (determine $\mathbf{V}_L$ in terms of $\mathbf{V}$).

*19. Using the nodal-analysis approach, repeat Problem 14 (determine $\mathbf{V}_L$ in terms of $\mathbf{I}$).

*20. For the network of Fig. 17.59:

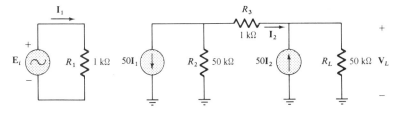

FIG. 17.59

a. Write the nodal equations.
b. Is symmetry present? If not, why?
c. Determine $\mathbf{V}_L$ in terms of $\mathbf{E}_i$.

Section 17.7

21. For the bridge network of Fig. 17.60:
    a. Is the bridge balanced?
    b. Using mesh analysis, determine the current through the capacitive reactance.
    c. Using nodal analysis, determine the voltage across the capacitive reactance.

22. Repeat Problem 21 for the bridge of Fig. 17.61.

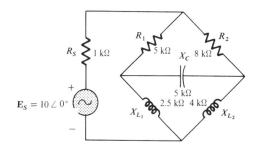

FIG. 17.60

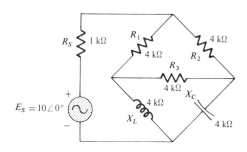

23. The Hay bridge of Fig. 17.62 is balanced. Using Eq. (17.1), determine the unknown inductance $\mathbf{L}_x$ and resistance R_x.

FIG. 17.61

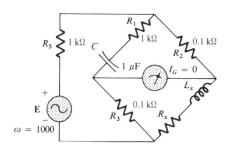

24. Determine whether the Maxwell bridge of Fig. 17.63 is balanced ($\omega = 1000$).

FIG. 17.62

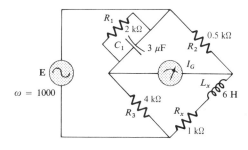

25. Derive the balance equations [Eqs. (17.9) and (17.10)] for the capacitance comparison bridge.

26. Determine the balance equations for the inductance bridge of Fig. 17.64.

FIG. 17.63

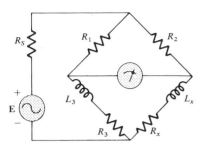

FIG. 17.64

Section 17.8

27. Using the Δ-Y or Y-Δ conversion, determine the current **I** for the networks of Fig. 17.65.

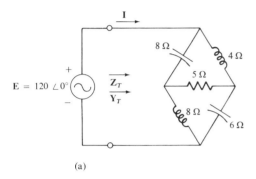

(a)

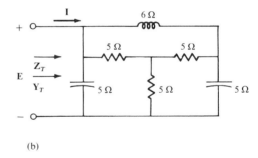

(b)

FIG. 17.65

28. Repeat Problem 27 for the networks of Fig. 17.66. ($E = 100 \angle 0°$ in each case.)

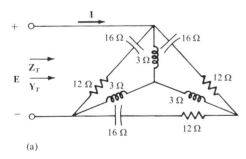

(a)

(b)

FIG. 17.66

GLOSSARY

Bridge network A network configuration having the appearance of a diamond in which no two branches are in series or parallel.

Capacitance comparison bridge A bridge configuration having a galvanometer in the bridge arm that is used to determine an unknown capacitance and associated resistance.

Delta (Δ) configuration A network configuration having the appearance of the capital Greek letter delta.

Dependent or controlled source A source whose magnitude and/or phase angle is determined (or controlled) by a current or voltage of the system in which it appears.

Hay bridge A bridge configuration used for measuring the resistance and inductance of coils in those cases where the resistance is a small fraction of the reactance of the coil.

Independent source A source whose magnitude is independent of the network to which it is applied. It displays its terminal characteristics even if completely isolated.

Maxwell bridge A bridge configuration used for inductance measurements when the resistance of the coil is large enough not to require a Hay bridge.

Mesh analysis A method through which the loop (or mesh) currents of a network can be determined. The branch currents of the network can then be determined directly from the loop currents.

Nodal analysis A method through which the node voltages of a network can be determined. The voltage across each element can then be determined through application of Kirchhoff's voltage law.

Source conversion The changing of a voltage source to a current source, or vice versa, which will result in the same terminal behavior of the source. In other words, the external network is unaware of the change in sources.

Wye (Y) configuration A network configuration having the appearance of the capital letter Y.

NETWORK THEOREMS (ac)

18

Th

18.1 INTRODUCTION

This chapter will parallel Chapter 8, which dealt with network theorems as applied to dc networks. It would be time well spent to review each theorem in Chapter 8 before beginning this chapter, as many of the comments offered there will not be repeated.

Due to the need for developing confidence in the application of the various theorems to networks with controlled (dependent) sources, some sections have been divided into two parts: independent sources and dependent sources.

Theorems to be considered in detail include the superposition theorem, Thevenin and Norton theorems and the maximum power theorem. The substitution and reciprocity theorems and Millman theorem are not discussed in detail here, since a review of Chapter 8 will enable you to apply them to sinusoidal ac networks with little difficulty.

18.2 SUPERPOSITION THEOREM

You will recall from Chapter 8 that the superposition theorem eliminated the need for solving simultaneous linear equations by considering the effects of each source independently. To consider the effects of each source, we had to remove the remaining sources. This was accomplished by setting voltage sources to zero (short-circuit representation) and current sources to zero (open-circuit representation). The current through, or voltage across, a portion of the network produced by each source was then added algebraically to find the total solution for the current or voltage.

The only variation in applying this method to ac networks with independent sources is that we will now be working with impedances and phasors instead of just resistors and real numbers.

The superposition theorem is not applicable to power effects in ac networks, since we are still dealing with a nonlinear relationship. It can be applied to networks with sources of different frequencies only if the total response for *each* frequency is found independently and the results are expanded in a nonsinusoidal expression as appearing in Chapter 23.

We will first consider networks with only independent sources. The analysis is then very similar to that for dc networks.

Independent Sources

EXAMPLE 18.1. Using the superposition theorem, find the current **I** through the 4-Ω reactance (X_{L_2}) in Fig. 18.1.

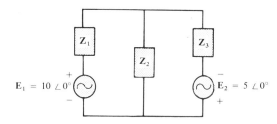

FIG. 18.1

Solution: In the redrawn circuit (Fig. 18.2),

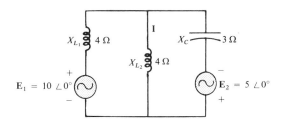

FIG. 18.2

$$\mathbf{Z}_1 = j4 \qquad \mathbf{Z}_2 = j4 \qquad \mathbf{Z}_3 = -j3$$

Considering the effects of the voltage source $\mathbf{E}_1$ (Fig. 18.3), we have

$$\mathbf{Z}_{2\|3} = \frac{\mathbf{Z}_2\mathbf{Z}_3}{\mathbf{Z}_2 + \mathbf{Z}_3} = \frac{(j4)(-j3)}{j4 - j3}$$
$$= 12 \angle -90°$$

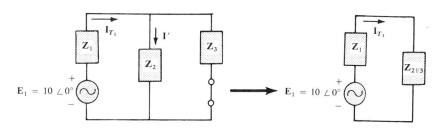

FIG. 18.3

$$\mathbf{I}_{T_1} = \frac{\mathbf{E}_1}{\mathbf{Z}_{2\|3} + \mathbf{Z}_1} = \frac{10 \angle 0°}{-j12 + j4} = \frac{10 \angle 0°}{8 \angle -90°}$$
$$= 1.25 \angle 90°$$

and

$$\mathbf{I'} = \frac{\mathbf{Z}_2 \mathbf{I}_{T_1}}{\mathbf{Z}_2 + \mathbf{Z}_3} \qquad \text{(current divider rule)}$$

$$\mathbf{I'} = \frac{(-j3)(j1.25)}{j4 - j3} = \frac{3.75}{j} = 3.75 \angle -90°$$

Considering the effects of the voltage source $\mathbf{E}_2$ (Fig. 18.4), we have

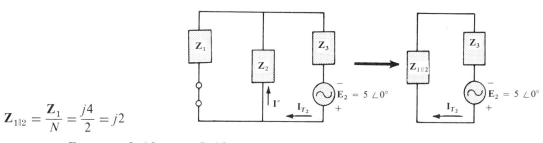

FIG. 18.4

$$\mathbf{Z}_{1\|2} = \frac{\mathbf{Z}_1}{N} = \frac{j4}{2} = j2$$

$$\mathbf{I}_{T_2} = \frac{\mathbf{E}_2}{\mathbf{Z}_{1\|2} + \mathbf{Z}_3} = \frac{5 \angle 0°}{j2 - j3} = \frac{5 \angle 0°}{1 \angle -90°} = 5 \angle 90°$$

and

$$\mathbf{I''} = \frac{\mathbf{I}_{T_2}}{2} = 2.5 \angle 90°$$

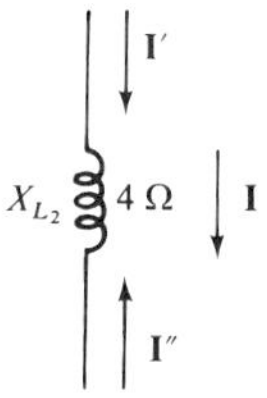

The total current through the 4-Ω reactance X_{L_2} (Fig. 18.5) is

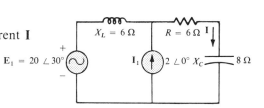

$$\mathbf{I} = \mathbf{I'} - \mathbf{I''}$$
$$= 3.75 \angle -90° - 2.50 \angle 90° = -j3.75 - j2.50$$
$$= -j6.25$$

$$\mathbf{I} = \mathbf{6.25} \angle -\mathbf{90°}$$

FIG. 18.5

EXAMPLE 18.2. Using superposition, find the current $\mathbf{I}$ through the 6-Ω resistor in Fig. 18.6.

FIG. 18.6

Solution: In the redrawn circuit (Fig. 18.7),

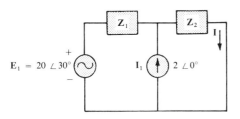

FIG. 18.7

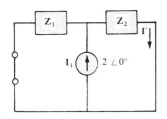

FIG. 18.8

$$\mathbf{Z}_1 = j6 \qquad \mathbf{Z}_2 = 6 - j8$$

Consider the effects of the current source (Fig. 18.8). Applying the current divider rule, we have

$$\mathbf{I}' = \frac{\mathbf{Z}_1 \mathbf{I}_1}{\mathbf{Z}_1 + \mathbf{Z}_2} = \frac{j6(2)}{j6 + 6 - j8} = \frac{j12}{6 - j2}$$
$$= \frac{12 \angle 90°}{6.32 \angle -18.43°}$$
$$\mathbf{I}' = 1.9 \angle 108.43°$$

Consider the effects of the voltage source (Fig. 18.9). Applying Ohm's law gives us

$$\mathbf{I}'' = \frac{\mathbf{E}_1}{\mathbf{Z}_T} = \frac{\mathbf{E}_1}{\mathbf{Z}_1 + \mathbf{Z}_2} = \frac{20 \angle 30°}{6.32 \angle -18.43°}$$
$$\mathbf{I}'' = 3.16 \angle 48.43°$$

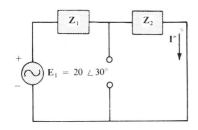

FIG. 18.9

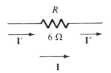

FIG. 18.10

The total current through the 6-Ω resistor (Fig. 18.10) is

$$\mathbf{I} = \mathbf{I}' + \mathbf{I}''$$
$$= 1.9 \angle 108.43° + 3.16 \angle 48.43°$$
$$= (-0.60 + j1.80) + (2.10 + j2.36)$$
$$= 1.50 + j4.16$$
$$\mathbf{I} = \mathbf{4.42} \angle \mathbf{70.2°}$$

EXAMPLE 18.3. Using superposition, find the voltage across the 6-Ω resistor in Fig. 18.6. Check the results against $\mathbf{V}_{6\Omega} = \mathbf{I} \cdot 6$, where $\mathbf{I}$ is the current found through the 6-Ω resistor in the previous example.

Solution: For the current source,

$$\mathbf{V}'_{6\Omega} = \mathbf{I}'6 = (1.9 \angle 108.43°)(6) = 11.4 \angle 108.43°$$

For the voltage source,

$$\mathbf{V}''_{6\Omega} = \mathbf{I}''(6) = (3.16 \angle 48.43°)(6) = 18.96 \angle 48.43°$$

FIG. 18.11

The total voltage across the 6-Ω resistor (Fig. 18.11) is

$$\mathbf{V}_{6\Omega} = \mathbf{V}'_{6\Omega} + \mathbf{V}''_{6\Omega}$$
$$= 11.4\ \angle\ 108.43° + 18.96\ \angle\ 48.43°$$
$$= (-3.60 + j10.82) + (12.58 + j14.18)$$
$$= 8.98 + j25.0$$
$$\mathbf{V}_{6\Omega} = \mathbf{26.5}\ \angle\ \mathbf{70.2°}$$

Checking the result, we have

$$\mathbf{V}_{6\Omega} = \mathbf{I}(6) = (4.42\ \angle\ 70.2°)(6) = \mathbf{26.5}\ \angle\ \mathbf{70.2°} \qquad \text{(checks)}$$

Dependent Sources

For dependent sources in which the controlling variable is not determined by the network to which the superposition theorem is to be applied, the application of the theorem is basically the same as for independent sources. The solution obtained will simply be in terms of the controlling variables.

EXAMPLE 18.4. Using the superposition theorem, determine the current $\mathbf{I}_2$ for the network of Fig. 18.12. The quantities μ and h are constants.

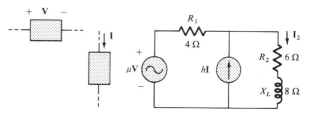

FIG. 18.12

Solution: With a portion of the system redrawn (Fig. 18.13),

$$\mathbf{Z}_1 = 4 \quad \text{and} \quad \mathbf{Z}_2 = 6 + j8$$

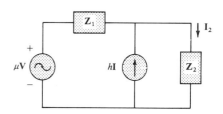

FIG. 18.13

For the voltage source (Fig. 18.14),

$$\mathbf{I}' = \frac{\mu\mathbf{V}}{\mathbf{Z}_1 + \mathbf{Z}_2} = \frac{\mu\mathbf{V}}{4 + 6 + j8} = \frac{\mu\mathbf{V}}{10 + j8}$$

$$\mathbf{I}' = \frac{\mu\mathbf{V}}{12.8\ \angle\ 38.66°} = 0.078\ \mu\mathbf{V}\ \angle\ -38.66°$$

For the current source (Fig. 18.15),

$$\mathbf{I}'' = \frac{\mathbf{Z}_1(h\mathbf{I})}{\mathbf{Z}_1 + \mathbf{Z}_2} = \frac{4(h\mathbf{I})}{12.8\ \angle\ 38.66°} = 4(0.078)h\mathbf{I}\ \angle\ -38.66°$$

$$\mathbf{I}'' = 0.312h\mathbf{I}\ \angle\ -38.66°$$

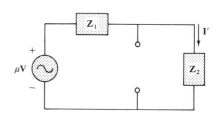

FIG. 18.14

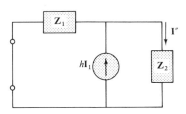

FIG. 18.15

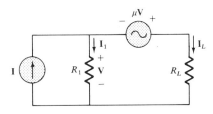

FIG. 18.16

The current

$$\mathbf{I}_2 = \mathbf{I}' + \mathbf{I}''$$
$$= 0.078 \, \mu\mathbf{V} \angle -38.66° + 0.312h\mathbf{I} \angle -38.66°$$

For $\mathbf{V} = 10 \angle 0°$, $\mathbf{I} = 20 \times 10^{-3} \angle 0°$, $\mu = 20$, $h = 100$:

$$\mathbf{I}_2 = 0.078(20)(10) \angle -38.66°$$
$$+ 0.312(100)(20 \times 10^{-3}) \angle -38.66°$$
$$= 15.60 \angle -38.66° + 0.62 \angle -38.66°$$

$$\mathbf{I}_2 = 16.22 \angle -38.66°$$

For dependent sources in which the controlling variable is determined by the network to which the theorem is to be applied, the dependent source cannot be set to zero unless the controlling variable is also zero. For networks containing dependent sources such as indicated in Example 18.4 and dependent sources of the type just introduced above, the superposition theorem is applied for each independent source and each dependent source not having a controlling variable in the portions of the network under investigation. It must be reemphasized that dependent sources are not sources of energy in the sense that if all independent sources are removed from a system, all currents and voltages must be zero.

EXAMPLE 18.5. Determine the current $\mathbf{I}_L$ through the resistor R_L of Fig. 18.16.

Solution: Note that the controlling variable $\mathbf{V}$ is determined by the network to be analyzed. From the above discussions, it is understood that the dependent source cannot be set to zero unless $\mathbf{V}$ is zero. If we set $\mathbf{I}$ to zero, the network lacks a source of emf, and $\mathbf{V} = 0$ with $\mu\mathbf{V} = 0$. The resulting $\mathbf{I}_L$ under this condition is zero. Obviously, therefore, the network must be analyzed as it appears in Fig. 18.16, with the result that neither source can be eliminated, as is normally done using the superposition theorem.

Applying Kirchhoff's voltage law, we have

$$\mathbf{V}_L = \mathbf{V} + \mu\mathbf{V} = (1 + \mu)\mathbf{V}$$

and

$$\mathbf{I}_L = \frac{(1 + \mu)\mathbf{V}}{R_L}$$

The result, however, must be found in terms of $\mathbf{I}$ since $\mathbf{V}$ and $\mu\mathbf{V}$ are only dependent variables.

Applying Kirchhoff's current law gives us

$$\mathbf{I} = \mathbf{I}_1 + \mathbf{I}_L = \frac{\mathbf{V}}{R_1} + \frac{(1 + \mu)\mathbf{V}}{R_L}$$

and

$$I = V\left(\frac{1}{R_1} + \frac{1 + \mu}{R_L}\right)$$

or

$$V = \frac{I}{(1/R_1) + [(1 + \mu)/R_L]}$$

Substituting into the above yields

$$I_L = \frac{(1 + \mu)V}{R_L} = \frac{(1 + \mu)}{R_L}\left[\frac{I}{(1/R_1) + [(1 + \mu)/R_L]}\right]$$

Therefore,

$$I_L = \frac{(1 + \mu)R_1 I}{R_L + (1 + \mu)R_1}$$

18.3 THEVENIN'S THEOREM

Thevenin's theorem, as stated for sinusoidal ac circuits, is changed only to include the term *impedance* instead of *resistance;* that is, *any two-terminal linear ac network can be replaced by an equivalent circuit consisting of a voltage source and an impedance in series* as shown in Fig. 18.17.

Since the reactances of a circuit are frequently dependent, the Thevenin circuit found for a particular network is applicable only at *one* frequency.

The steps required to apply this method to dc circuits are repeated here with changes for sinusoidal ac circuits. As before, the only change is the replacement of the term *resistance* by *impedance*. Again, dependent and independent sources will be treated separately.

Independent Sources

Step 1: Remove that portion of the network across which the Thevenin equivalent circuit is to be found.

Step 2: Mark (○, ●, and so on) the terminals of the remaining two-terminal network.

Step 3: Calculate Z_{Th} by first setting all voltage and current sources to zero (short circuit and open circuit, respectively) and then finding the resulting impedance between the two marked terminals.

Step 4: Calculate E_{Th} by first replacing the voltage and current sources and then finding the open-circuit voltage between the marked terminals.

Step 5: Draw the Thevenin equivalent circuit with the portion of the circuit previously removed replaced between the terminals of the Thevenin equivalent circuit.

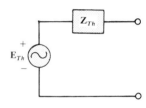

FIG. 18.17

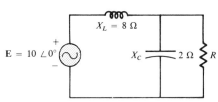

FIG. 18.18

EXAMPLE 18.6. Find the Thevenin equivalent circuit for the network external to resistor R in Fig. 18.18.

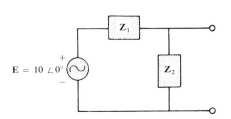

FIG. 18.19

Solution:
  Steps 1 and 2 (Fig. 18.19):

$$\mathbf{Z}_1 = j8 \quad \text{and} \quad \mathbf{Z}_2 = -j2$$

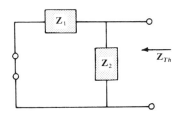

FIG. 18.20

Step 3 (Fig. 18.20):

$$\mathbf{Z}_{Th} = \frac{\mathbf{Z}_1 \mathbf{Z}_2}{\mathbf{Z}_1 + \mathbf{Z}_2} = \frac{(j8)(-j2)}{j8 - j2} = \frac{-j^2 16}{j6} = \frac{16}{6 \angle 90°}$$
$$= 2.67 \angle -90°$$

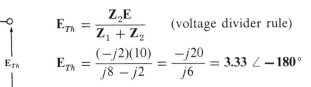

FIG. 18.21

Step 4 (Fig. 18.21):

$$\mathbf{E}_{Th} = \frac{\mathbf{Z}_2 \mathbf{E}}{\mathbf{Z}_1 + \mathbf{Z}_2} \quad \text{(voltage divider rule)}$$

$$\mathbf{E}_{Th} = \frac{(-j2)(10)}{j8 - j2} = \frac{-j20}{j6} = 3.33 \angle -180°$$

Step 5: The Thevenin equivalent circuit is shown in Fig. 18.22.

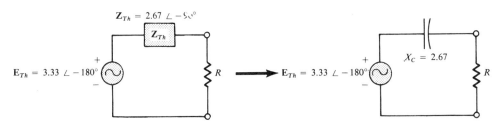

FIG. 18.22

EXAMPLE 18.7. Find the Thevenin equivalent circuit for the network external to branch a–a' in Fig. 18.23.

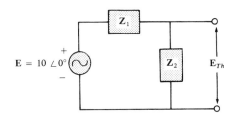

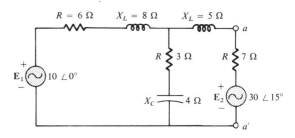

FIG. 18.23

Solution:

Steps 1 and 2 (Fig. 18.24). Note the reduced complexity with subscripted impedances:

$$\mathbf{Z}_1 = 6 + j8 \qquad \mathbf{Z}_2 = 3 - j4 \qquad \mathbf{Z}_3 = j5$$

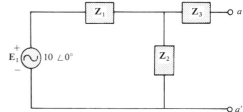

Step 3 (Fig. 18.25):

$$\mathbf{Z}_{Th} = \mathbf{Z}_3 + \frac{\mathbf{Z}_1\mathbf{Z}_2}{\mathbf{Z}_1 + \mathbf{Z}_2} = j5 + \frac{(10 \angle 53.13°)(5 \angle -53.13°)}{(6 + j8) + (3 - j4)}$$

$$= j5 + \frac{50 \angle 0°}{9 + j4} = j5 + \frac{50 \angle 0°}{9.85 \angle 23.96°}$$

$$= j5 + 5.08 \angle -23.96° = j5 + 4.64 - j2.06$$

$$\mathbf{Z}_{Th} = \mathbf{4.64 + j2.94 = 5.49 \angle 32.36°}$$

FIG. 18.24

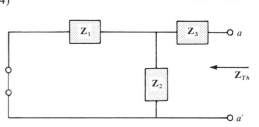

Step 4 (Fig. 18.26): Since *a–a'* is an open circuit, $\mathbf{I}_{\mathbf{Z}_3} = 0$. Then $\mathbf{E}_{Th}$ is the voltage drop across $\mathbf{Z}_2$:

$$\mathbf{E}_{Th} = \frac{\mathbf{Z}_2\mathbf{E}}{\mathbf{Z}_2 + \mathbf{Z}_1} \qquad \text{(voltage divider rule)}$$

$$= \frac{(5 \angle -53.13°)(10 \angle 0°)}{9.85 \angle 23.96°}$$

$$\mathbf{E}_{Th} = \frac{50 \angle -53.13°}{9.85 \angle 23.96°} = \mathbf{5.08 \angle -77.09°}$$

FIG. 18.25

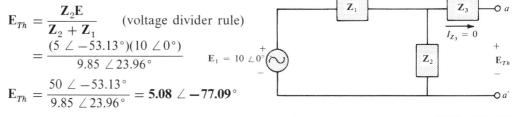

FIG. 18.26

Step 5: The Thevenin equivalent circuit is shown in Fig. 18.27.

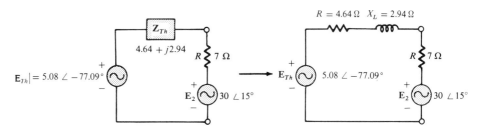

FIG. 18.27

Dependent Sources

For dependent sources with a controlling variable not in the network under investigation, the procedure indicated above can be applied. However, for dependent sources of the other type, where the controlling variable is part of the network to which the theorem is to be applied, another approach must be employed. The necessity for a different approach will be demonstrated in an example to follow. The method is not limited to dependent sources of the latter type. It can also be applied to any dc or sinusoidal ac network. However, for networks of independent sources, the method of application employed in Chapter 8 and the first portion of this section is generally more direct, with usual savings in time and errors.

The new approach to Thevenin's theorem can best be introduced at this stage in the development by considering the Thevenin equivalent circuit of Fig. 18.28(a). As indicated in Fig. 18.28(b), the open-circuit terminal voltage ($\mathbf{E}_{oc}$) of the Thevenin equivalent circuit is the Thevenin equivalent voltage. That is,

$$\boxed{\mathbf{E}_{oc} = \mathbf{E}_{Th}} \tag{18.1}$$

If the external terminals are short-circuited as in Fig. 18.28(c), the resulting short-circuit current is determined by

$$\boxed{\mathbf{I}_{sc} = \frac{\mathbf{E}_{Th}}{\mathbf{Z}_{Th}}} \tag{18.2}$$

or, rearranged,

$$\mathbf{Z}_{Th} = \frac{\mathbf{E}_{Th}}{\mathbf{I}_{sc}}$$

and

$$\boxed{\mathbf{Z}_{Th} = \frac{\mathbf{E}_{oc}}{\mathbf{I}_{sc}}} \tag{18.3}$$

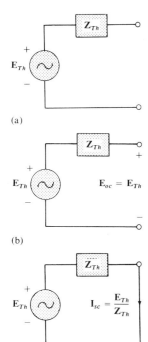

(a)

(b)

(c)

FIG. 18.28

Equations (18.1) and (18.3) indicate that for any linear bilateral dc or ac network with or without dependent sources of any type, if the open-circuit terminal voltage of a portion of a network can be determined along with the short-circuit current between the same two terminals, the Thevenin equivalent circuit is effectively known. A few examples will make the method quite clear. The advantage of the method, which was stressed earlier in this section for independent sources, should now be more obvious. The current $\mathbf{I}_{sc}$, which is necessary to find $\mathbf{Z}_{Th}$, is in general more difficult to obtain since all of the sources are present.

There is a third approach to the Thevenin equivalent circuit that is also useful from a practical viewpoint. The Thevenin voltage is found as in the two previous methods. However, the Thevenin impedance is obtained by applying a source of emf to the terminals of interest and determining the source current as indicated in Fig. 18.29. For this method, the source voltage of the original network is set to zero. The Thevenin impedance is then determined by the following equation:

$$\boxed{\mathbf{Z}_{Th} = \frac{\mathbf{E}_g}{\mathbf{I}_g}} \qquad (18.4)$$

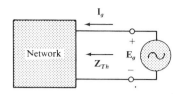

FIG. 18.29 *Determining* $\mathbf{Z}_{Th}$.

Note that for each technique $\mathbf{E}_{Th} = \mathbf{E}_{oc}$, but the Thevenin impedance is found in different ways.

The first two examples include network configurations frequently encountered in the analysis of electronic networks. They have a dependent source with an *external* controlling variable, permitting the use of any one of the three techniques described in this section. In fact, each will be applied to the network of each example to demonstrate its validity.

EXAMPLE 18.8. Using each of the three techniques described in this section, determine the Thevenin equivalent circuit for the network of Fig. 18.30.

Solution: Since for each approach the Thevenin voltage is found in exactly the same manner, it will be determined first. From Fig. 18.30, where $\mathbf{I}_{X_C} = 0$,

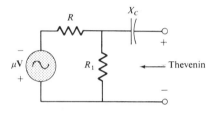

FIG. 18.30

$$\mathbf{V}_{R_1} = \mathbf{E}_{Th} = \mathbf{E}_{oc} = \underbrace{-\frac{R_1(\mu\mathbf{V})}{R_1 + R}}_{\substack{\text{due to the polarity for } \mathbf{V} \text{ and defined} \\ \text{terminal polarities}}} = -\frac{\mu R_1 \mathbf{V}}{R_1 + R}$$

The following three methods for determining the Thevenin impedance appear in the order in which they were introduced in this section:

Method 1 (Fig. 18.31):

$$\mathbf{Z}_{Th} = R \parallel R_1 - jX_C$$

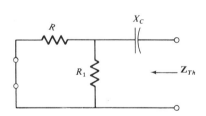

FIG. 18.31

Method 2 (Fig. 18.32): Converting the voltage source to a current source (Fig. 18.33), we have (current divider rule)

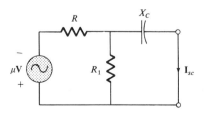

FIG. 18.32

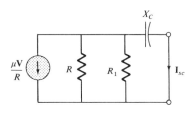

FIG. 18.33

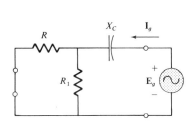

FIG. 18.34

$$\mathbf{I}_{sc} = \frac{-(R \parallel R_1)\dfrac{\mu\mathbf{V}}{R}}{(R \parallel R_1) - jX_C} = \frac{-\dfrac{RR_1}{R + R_1}\left(\dfrac{\mu\mathbf{V}}{R}\right)}{(R \parallel R_1) - jX_C}$$

$$= \frac{\dfrac{-\mu R_1 \mathbf{V}}{R + R_1}}{(R \parallel R_1) - jX_C}$$

and

$$\mathbf{Z}_{Th} = \frac{\mathbf{E}_{oc}}{\mathbf{I}_{sc}} = \frac{\dfrac{-\mu R_1 \mathbf{V}}{R_1 + R}}{\dfrac{\dfrac{-\mu R_1 \mathbf{V}}{R + R_1}}{(R \parallel R_1) - jX_C}} = \frac{1}{\dfrac{1}{(R \parallel R_1) - jX_C}}$$

$$= R \parallel R_1 - jX_C$$

Method 3 (Fig. 18.34):

$$\mathbf{I}_g = \frac{\mathbf{E}_g}{(R \parallel R_1) - jX_C}$$

and

$$\mathbf{Z}_{Th} = \frac{\mathbf{E}_g}{\mathbf{I}_g} = R \parallel R_1 - jX_C$$

In each case, the Thevenin impedance is the same. The resulting Thevenin equivalent circuit is shown in Fig. 18.35.

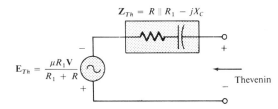

FIG. 18.35

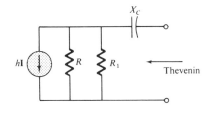

FIG. 18.36

EXAMPLE 18.9. Repeat Example 18.8 for the network of Fig. 18.36.

Solution: From Fig. 18.36, $\mathbf{E}_{Th}$ is

$$\mathbf{E}_{Th} = \mathbf{E}_{oc} = -h\mathbf{I}(R \parallel R_1) = \frac{hRR_1\mathbf{I}}{R + R_1}$$

Method 1 (Fig. 18.37):

$$\mathbf{Z}_{Th} = R \parallel R_1 - jX_C$$

Note the similarity between this solution and that obtained for the previous example.

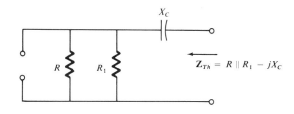

$$\mathbf{Z}_{Th} = R \parallel R_1 - jX_C$$

FIG. 18.37

Method 2 (Fig. 18.38):

$$\mathbf{I}_{sc} = \frac{-(R \parallel R_1)h\mathbf{I}}{(R \parallel R_1) - jX_C}$$

and

$$\mathbf{Z}_{Th} = \frac{\mathbf{E}_{oc}}{\mathbf{I}_{sc}} = \frac{-h\mathbf{I}(R \parallel R_1)}{\dfrac{-(R \parallel R_1)h\mathbf{I}}{(R \parallel R_1) - jX_C}} = R \parallel R_1 - jX_C$$

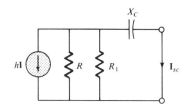

FIG. 18.38

Method 3 (Fig. 18.39):

$$\mathbf{I}_g = \frac{\mathbf{E}_g}{(R \parallel R_1) - jX_C}$$

and

$$\mathbf{Z}_{Th} = \frac{\mathbf{E}_g}{\mathbf{I}_g} = R \parallel R_1 - jX_C$$

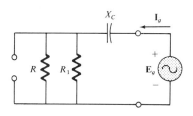

FIG. 18.39

The following example has a dependent source that will not permit the use of the method described in the beginning of this section for independent sources. All three methods will be applied, however, so that the results can be compared.

EXAMPLE 18.10. For the network of Fig. 18.40 (introduced in Example 18.5), determine the Thevenin equivalent circuit between the indicated terminals using each method described in this section. Compare your results.

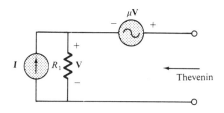

FIG. 18.40

Solution: First, using Kirchhoff's voltage law, $\mathbf{E}_{Th}$ (which is the same for each method) is written

$$\mathbf{E}_{Th} = \mathbf{V} + \mu\mathbf{V} = (1 + \mu)\mathbf{V}$$

However,

$$\mathbf{V} = \mathbf{I}R_1$$

so

$$\mathbf{E}_{Th} = (1 + \mu)\mathbf{I}R_1$$

$\mathbf{Z}_{Th}$:

Method 1 (Fig. 18.41): Since $\mathbf{I} = 0$, $\mathbf{V}$ and $\mu\mathbf{V} = 0$, and

$$\mathbf{Z}_{Th} \neq R_1 \qquad \text{(incorrect)}$$

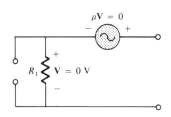

FIG. 18.41

Method 2 (Fig. 18.42): Kirchhoff's voltage law around the indicated loop gives us

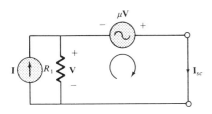

FIG. 18.42

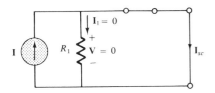

FIG. 18.43

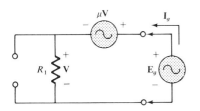

FIG. 18.44

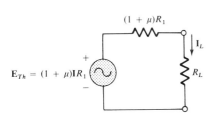

FIG. 18.45

$$V + \mu V = 0$$

and

$$V(1 + \mu) = 0$$

Since μ is a positive constant, the above equation can be satisfied only when $V = 0$. Substitution of this result into Fig. 18.42 will yield the configuration of Fig. 18.43, and

$$I_{sc} = I$$

with

$$Z_{Th} = \frac{E_{oc}}{I_{sc}} = \frac{(1 + \mu)IR_1}{I} = (1 + \mu)R_1 \qquad \text{(correct)}$$

Method 3 (Fig. 18.44):

$$E_g = V + \mu V = (1 + \mu)V$$

or

$$V = \frac{E_g}{1 + \mu}$$

and

$$I_g = \frac{V}{R_1} = \frac{E_g}{(1 + \mu)R_1}$$

and

$$Z_{Th} = \frac{E_g}{I_g} = (1 + \mu)R_1 \qquad \text{(correct)}$$

The Thevenin equivalent circuit appears in Fig. 18.45, and

$$I_L = \frac{(1 + \mu)R_1I}{R_L + (1 + \mu)R_1}$$

which compares with the result of Example 18.5.

The network of Fig. 18.46 is the basic configuration of the transistor equivalent circuit applied most frequently today. Needless to say, it is necessary to know its characteristics and be adept in its use. Note that there is a controlled voltage and current source, each controlled by variables in the configuration.

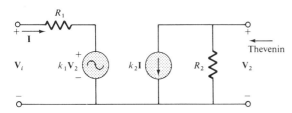

FIG. 18.46

EXAMPLE 18.11. Determine the Thevenin equivalent circuit for the indicated terminals of the network of Fig. 18.46.

Solution: Apply the second method introduced in this section.

$\mathbf{E}_{Th}$:

$$\mathbf{E}_{oc} = \mathbf{V}_2$$
$$\mathbf{I} = \frac{\mathbf{V}_i - k_1\mathbf{V}_2}{R_1} = \frac{\mathbf{V}_i - k_1\mathbf{E}_{oc}}{R_1}$$

and

$$\mathbf{E}_{oc} = -k_2 I R_2 = -k_2 R_2 \left(\frac{\mathbf{V}_i - k_1\mathbf{E}_{oc}}{R_1}\right)$$
$$= \frac{-k_2 R_2 \mathbf{V}_i}{R_1} + \frac{k_1 k_2 R_2 \mathbf{E}_{oc}}{R_1}$$

or

$$\mathbf{E}_{oc}\left(1 - \frac{k_1 k_2 R_2}{R_1}\right) = \frac{-k_2 R_2 \mathbf{V}_i}{R_1}$$

and

$$\mathbf{E}_{oc}\left(\frac{R_1 - k_1 k_2 R_2}{R_1}\right) = \frac{-k_2 R_2 \mathbf{V}_i}{R_1}$$

so

$$\boxed{\mathbf{E}_{oc} = \frac{-k_2 R_2 \mathbf{V}_i}{R_1 - k_2 k_2 R_2} = \mathbf{E}_{Th}} \qquad (18.5)$$

$\mathbf{I}_{sc}$:

For the network of Fig. 18.47, where

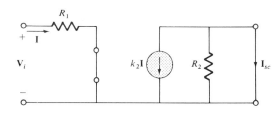

FIG. 18.47

$$\mathbf{V}_2 = 0 \qquad k_1\mathbf{V}_2 = 0 \qquad \mathbf{I} = \frac{\mathbf{V}_i}{R_1}$$

and

$$\mathbf{I}_{sc} = -k_2\mathbf{I} = \frac{-k_2\mathbf{V}_i}{R_1}$$

so

$$\mathbf{Z}_{Th} = \frac{\mathbf{E}_{oc}}{\mathbf{I}_{sc}} = \frac{\dfrac{-k_2 R_2 \mathbf{V}_i}{R_1 - k_1 k_2 R_2}}{\dfrac{-k_2\mathbf{V}_i}{R_1}} = \frac{R_1 R_2}{R_1 - k_1 k_2 R_2}$$

and

$$\boxed{\mathbf{Z}_{Th} = \dfrac{R_2}{1 - \dfrac{k_1 k_2 R_2}{R_1}}} \qquad (18.6)$$

Frequently, the approximation $k_1 \cong 0$ is applied. Then, the Thevenin voltage and impedance are

$$\boxed{\mathbf{E}_{Th} = \dfrac{-k_2 R_2 \mathbf{V}_i}{R_1}} \qquad k_1 = 0 \qquad (18.7)$$

$$\boxed{\mathbf{Z}_{Th} = \mathbf{R}_2} \qquad k_1 = 0 \qquad (18.8)$$

Apply $\mathbf{Z}_{Th} = \mathbf{E}_g / \mathbf{I}_g$ to the network of Fig. 18.48, where

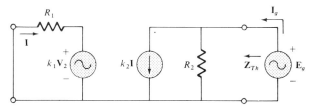

FIG. 18.48

$$\mathbf{I} = \dfrac{-k_1 \mathbf{V}_2}{R_1}$$

But

$$\mathbf{V}_2 = \mathbf{E}_g$$

so

$$\mathbf{I} = \dfrac{-k_1 \mathbf{E}_g}{R_1}$$

Applying Kirchhoff's current law, we have

$$\mathbf{I}_g = k_2 \mathbf{I} + \dfrac{\mathbf{E}_g}{R_2} = k_2 \left(-\dfrac{k_1 \mathbf{E}_g}{R_1} \right) + \dfrac{\mathbf{E}_g}{R_2}$$

$$= \mathbf{E}_g \left(\dfrac{1}{R_2} - \dfrac{k_1 k_2}{R_1} \right)$$

and

$$\dfrac{\mathbf{I}_g}{\mathbf{E}_g} = \dfrac{R_1 - k_1 k_2 R_2}{R_1 R_2}$$

or

$$\mathbf{Z}_{Th} = \dfrac{\mathbf{E}_g}{\mathbf{I}_g} = \dfrac{R_1 R_2}{R_1 - k_1 k_2 R_2}$$

as obtained above.

The last two methods presented in this section were applied only to networks in which the magnitudes of the controlled sources were dependent on a variable within the network for which the Thevenin equivalent circuit was to be obtained. Understand that both of those methods can also be applied to any dc or sinusoidal ac network containing only independent sources or dependent sources of the other kind.

18.4 NORTON'S THEOREM

The three methods described for Thevenin's theorem will each be altered to permit their use with Norton's theorem. Since the Thevenin and Norton impedances are the same for a particular network, certain portions of the discussion will be quite similar to those encountered in the previous section. We will first consider independent sources and the approach developed in Chapter 8, followed by dependent sources and the new techniques developed for Thevenin's theorem.

You will recall from Chapter 8 that Norton's theorem allows us to replace any two-terminal linear bilateral ac network by an equivalent circuit consisting of a current source and impedance, as in Fig. 18.49.

The Norton equivalent circuit, like the Thevenin equivalent circuit, is applicable at only one frequency since the reactances are frequency-dependent.

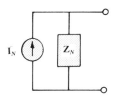

FIG. 18.49

Independent Sources

The procedure outlined below to find the Norton equivalent of a sinusoidal ac network is changed (from that in Chapter 8) in only one respect: to replace the term *resistance* with *impedance*.

Step 1: Remove that portion of the network across which the Norton equivalent circuit is to be found.

Step 2: Mark (○, ●, and so on) the terminals of the remaining two-terminal network.

Step 3: Calculate Z_N by first setting all voltage and current sources to zero (short circuit and open circuit, respectively), and then finding the resulting impedance between the two marked terminals.

Step 4: Calculate I_N by first replacing the voltage and current sources, and then finding the short-circuit current between the marked terminals.

Step 5: Draw the Norton equivalent circuit with the portion of the circuit previously removed replaced between the terminals of the Norton equivalent circuit.

The Norton and Thevenin equivalent circuits can be found from each other by using the source transformation shown in Fig. 18.50. The source transformation is applicable for any Thevenin or Norton equivalent circuit determined from a network with any combination of independent or dependent sources.

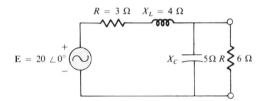

FIG. 18.50

EXAMPLE 18.12. Determine the Norton equivalent circuit for the network external to the 6-Ω resistor of Fig. 18.51.

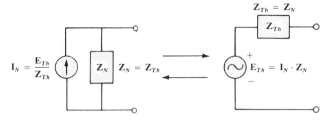

FIG. 18.51

Solution:
 Steps 1 and 2 (Fig. 18.52):

$$\mathbf{Z}_1 = 3 + j4 = 5 \angle 53.13° \qquad \mathbf{Z}_2 = -j5$$

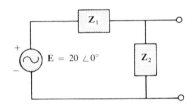

FIG. 18.52

 Step 3 is shown in Fig. 18.53.

$$\mathbf{Z}_N = \frac{\mathbf{Z}_1 \mathbf{Z}_2}{\mathbf{Z}_1 + \mathbf{Z}_2} = \frac{(5 \angle 53.13°)(5 \angle -90°)}{3 + j4 - j5} = \frac{25 \angle -36.87°}{3 - j1}$$

$$\mathbf{Z}_N = \frac{25 \angle -36.87°}{3.16 \angle -18.43°} = 7.91 \angle -18.44° = \mathbf{7.50 - j2.50}$$

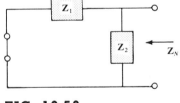

FIG. 18.53

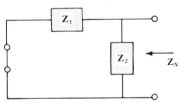

FIG. 18.54

 Step 4 (Fig. 18.54):

$$\mathbf{I}_N = \mathbf{I}_1 = \frac{\mathbf{E}}{\mathbf{Z}_1} = \frac{20 \angle 0°}{5 \angle 53.13°} = \mathbf{4 \angle -53.13°}$$

 Step 5 The Norton equivalent circuit is shown in Fig. 18.55.

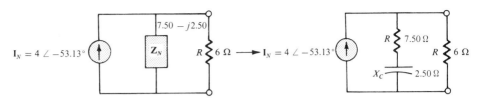

FIG. 18.55

EXAMPLE 18.13. Find the Norton equivalent circuit for the network external to the 7-Ω capacitive reactance in Fig. 18.56.

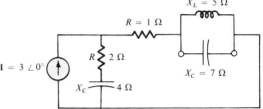

FIG. 18.56

Solution:

Steps 1 and 2 (Fig. 18.57):

$$\mathbf{Z}_1 = 2 - j4 \qquad \mathbf{Z}_2 = 1 \qquad \mathbf{Z}_3 = j5$$

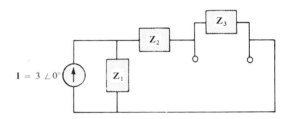

FIG. 18.57

Step 3 (Fig. 18.58):

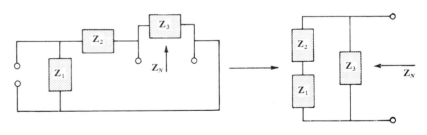

FIG. 18.58

$$\mathbf{Z}_N = \frac{\mathbf{Z}_3(\mathbf{Z}_1 + \mathbf{Z}_2)}{\mathbf{Z}_3 + (\mathbf{Z}_1 + \mathbf{Z}_2)}$$

$$\mathbf{Z}_1 + \mathbf{Z}_2 = 2 - j4 + 1 = 3 - j4 = 5 \angle -53.13°$$

$$\mathbf{Z}_N = \frac{(5 \angle 90°)(5 \angle -53.13°)}{j5 + 3 - j4} = \frac{25 \angle 36.87°}{3 + j1}$$

$$= \frac{25 \angle 36.87°}{3.16 \angle +18.43°}$$

$$= 7.91 \angle 18.44° = \mathbf{7.50 + j2.50}$$

Step 4 (Fig. 18.59):

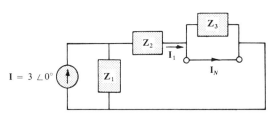

FIG. 18.59

$$\mathbf{I}_N = \mathbf{I}_1 = \frac{\mathbf{Z}_1\mathbf{I}}{\mathbf{Z}_1 + \mathbf{Z}_2} \qquad \text{(current divider rule)}$$

$$= \frac{(2 - j4)(3)}{3 - j4} = \frac{6 - j12}{5 \angle -53.13°} = \frac{13.4 \angle -63.43°}{5 \angle -53.13°}$$

$$\mathbf{I}_N = \mathbf{2.68} \angle -\mathbf{10.3}°$$

Step 5 The Norton equivalent circuit is shown in Fig. 18.60.

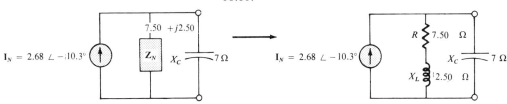

FIG. 18.60

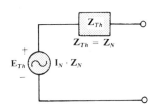

FIG. 18.61

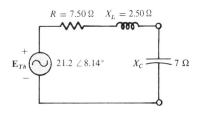

FIG. 18.62

EXAMPLE 18.14. Find the Thevenin equivalent circuit for the network external to the 7-Ω capacitive reactance in Fig. 18.56.

Solution: Using the conversion between sources (Fig. 18.61), we obtain

$$\mathbf{Z}_{Th} = \mathbf{Z}_N = \mathbf{7.50} + \mathbf{j2.50}$$

$$\mathbf{E}_{Th} = \mathbf{I}_N\mathbf{Z}_N = (2.68 \angle -10.3°)(7.91 \angle 18.44°)$$

$$\mathbf{E}_{Th} = \mathbf{21.2} \angle \mathbf{8.14}°$$

The Thevenin equivalent circuit is shown in Fig. 18.62.

Dependent Sources

As stated for Thevenin's theorem, dependent sources in which the controlling variable is not determined by the network for which the Norton equivalent circuit is to be found do not alter the procedure outlined above.

For dependent sources of the other kind, one of the following procedures must be applied. Both of these procedures can also be applied to networks with any combination of independent sources and dependent sources not controlled by the network under investigation.

The Norton equivalent circuit appears in Fig. 18.63(a). In Fig. 18.63(b), we find that

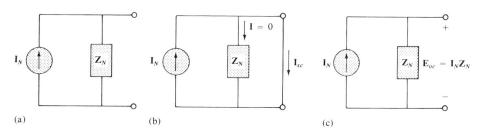

(a)   (b)   (c)

FIG. 18.63

$$\boxed{\mathbf{I}_{sc} = \mathbf{I}_N} \qquad (18.9)$$

and in Fig. 18.63(c) that

$$\mathbf{E}_{oc} = \mathbf{I}_N \mathbf{Z}_N$$

Or, rearranging, we have

$$\mathbf{Z}_N = \frac{\mathbf{E}_{oc}}{\mathbf{I}_N}$$

and

$$\boxed{\mathbf{Z}_N = \frac{\mathbf{E}_{oc}}{\mathbf{I}_{sc}}} \qquad (18.10)$$

The Norton impedance can also be determined by applying a source of emf $\mathbf{E}_g$ to the terminals of interest and finding the resulting $\mathbf{I}_g$, as shown in Fig. 18.64. All independent sources and dependent sources not controlled by a variable in the network of interest are set to zero, and

$$\boxed{\mathbf{Z}_N = \frac{\mathbf{E}_g}{\mathbf{I}_g}} \qquad (18.11)$$

For this latter approach, the Norton current is still determined by the short-circuit current.

EXAMPLE 18.15. Using each method described for dependent sources, find the Norton equivalent circuit for the network of Fig. 18.65.

Solution:
$\mathbf{I}_N$:

For each method, $\mathbf{I}_N$ is determined in the same manner. From Fig. 18.66, using Kirchhoff's current law, we have

$$0 = \mathbf{I} + h\mathbf{I} + \mathbf{I}_{sc}$$

or

$$\mathbf{I}_{sc} = -(1 + h)\mathbf{I}$$

Applying Kirchhoff's voltage law gives us

$$\mathbf{E} + \mathbf{I}R_1 - \mathbf{I}_{sc}R_2 = 0$$

and

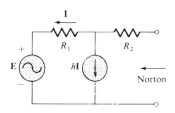

FIG. 18.64

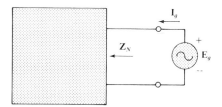

FIG. 18.65

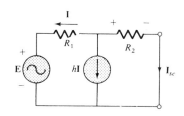

FIG. 18.66

$$IR_1 = I_{sc}R_2 - E$$

or

$$I = \frac{I_{sc}R_2 - E}{R_1}$$

so

$$I_{sc} = -(1 + h)I = -(1 + h)\left(\frac{I_{sc}R_2 - E}{R_1}\right)$$

or

$$R_1 I_{sc} = -(1 + h)I_{sc}R_2 + (1 + h)E$$

$$I_{sc}[R_1 + (1 + h)R_2] = (1 + h)E$$

$$\boldsymbol{I_{sc} = \frac{(1 + h)E}{R_1 + (1 + h)R_2} = I_N}$$

$\mathbf{Z}_N$:

Method 1: $\mathbf{E}_{oc}$ is determined from the network of Fig. 18.67. By Kirchhoff's current law,

$$0 = I + hI \quad \text{or} \quad I(h + 1) = 0$$

For h, a positive constant $\mathbf{I}$ must equal zero to satisfy the above. Therefore,

$$\mathbf{I} = 0 \quad \text{and} \quad h\mathbf{I} = 0$$

and

$$\mathbf{E}_{oc} = \mathbf{E}$$

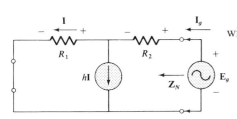

FIG. 18.67

with

$$\mathbf{Z}_N = \frac{\mathbf{E}_{oc}}{\mathbf{I}_{sc}} = \frac{\mathbf{E}}{\dfrac{(1 + h)\mathbf{E}}{R_1 + (1 + h)R_2}} = \frac{R_1 + (1 + h)R_2}{(1 + h)}$$

Method 2: Note Fig. 18.68. By Kirchhoff's current law,

$$\mathbf{I}_g = \mathbf{I} + h\mathbf{I} = (1 + h)\mathbf{I}$$

By Kirchhoff's voltage law,

$$\mathbf{E}_g - \mathbf{I}_g R_2 - \mathbf{I}R_1 = 0$$

or

$$\mathbf{I} = \frac{\mathbf{E}_g - \mathbf{I}_g R_2}{R_1}$$

Substituting, we have

$$\mathbf{I}_g = (1 + h)\mathbf{I} = (1 + h)\left(\frac{\mathbf{E}_g - \mathbf{I}_g R_2}{R_1}\right)$$

FIG. 18.68

and

$$\mathbf{I}_g R_1 = (1 + h)\mathbf{E}_g - (1 + h)\mathbf{I}_g R_2$$

so

$$\mathbf{E}_g(1 + h) = \mathbf{I}_g[R_1 + (1 + h)R_2]$$

or

$$\mathbf{Z}_N = \frac{\mathbf{E}_g}{\mathbf{I}_g} = \frac{R_1 + (1 + h)R_2}{1 + h}$$

which agrees with the above.

EXAMPLE 18.16. Find the Norton equivalent circuit for the network configuration of Fig. 18.46.

Solution: By source conversion,

$$\mathbf{I}_N = \frac{\mathbf{E}_{Th}}{\mathbf{Z}_{Th}} = \frac{\dfrac{-k_2 R_2 \mathbf{V}_i}{R_1 - k_1 k_2 R_2}}{\dfrac{R_1 R_2}{R_1 - k_1 k_2 R_2}}$$

and

$$\boxed{\mathbf{I}_N = \frac{-k_2 V_i}{R_1}} \qquad \textbf{(18.12)}$$

which is I_{sc} as determined in that example, and

$$\boxed{\mathbf{Z}_N = \mathbf{Z}_{Th} = \frac{R_2}{1 - \dfrac{k_1 k_2 R_2}{R_1}}} \qquad \textbf{(18.13)}$$

For $k_1 \cong 0$, we have

$$\boxed{\mathbf{I}_N = \frac{-k_2 V_i}{R_1}} \qquad k_1 = 0 \qquad \textbf{(18.14)}$$

$$\boxed{\mathbf{Z}_N = R_2} \qquad k_1 = 0 \qquad \textbf{(18.15)}$$

18.5 MAXIMUM POWER THEOREM

When applied to ac circuits, the maximum power theorem states that *maximum power will be delivered to a load when the load impedance is the conjugate of the Thevenin impedance across its terminals.* That is, for Fig. 18.69, for maximum power transfer to the load,

$$\boxed{Z_L = Z_{Th} \quad \text{and} \quad \theta_L = -\theta_{Th_Z}} \qquad \textbf{(18.16)}$$

or, in rectangular form,

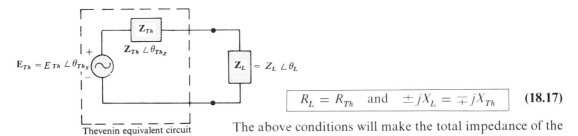

FIG. 18.69

$$R_L = R_{Th} \quad \text{and} \quad \pm jX_L = \mp jX_{Th} \qquad \textbf{(18.17)}$$

The above conditions will make the total impedance of the circuit purely resistive, as indicated in Fig. 18.70:

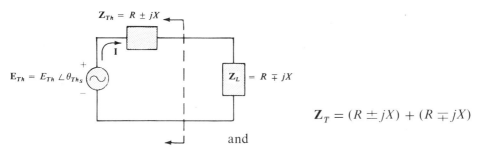

FIG. 18.70

$$\mathbf{Z}_T = (R \pm jX) + (R \mp jX)$$

and

$$\boxed{\mathbf{Z}_T = 2R} \qquad \textbf{(18.18)}$$

Since the circuit is purely resistive, the power factor of the circuit under maximum power conditions is 1. That is,

$$\boxed{F_p = 1} \qquad \text{(maximum power transfer)} \quad \textbf{(18.19)}$$

The magnitude of the current **I** of Fig. 18.70 is

$$I = \frac{E_{Th}}{Z_T} = \frac{E_{Th}}{2R}$$

The maximum power to the load is

$$P_{\max} = I^2 R = \left(\frac{E_{Th}}{2R}\right)^2 R$$

and

$$\boxed{P_{\max} = \frac{E_{Th}^2}{4R}} \qquad \textbf{(18.20)}$$

EXAMPLE 18.17. Find the load impedance in Fig. 18.71 for maximum power to the load.

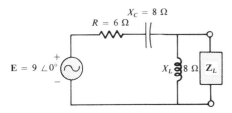
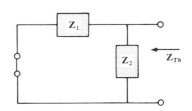

FIG. 18.71

Solution:

$$\mathbf{Z}_1 = 6 - j8 = 10 \angle -53.13°$$
$$\mathbf{Z}_2 = j8$$

$$\mathbf{Z}_{Th} = \frac{\mathbf{Z}_1\mathbf{Z}_2}{\mathbf{Z}_1 + \mathbf{Z}_2} = \frac{(10 \angle -53.13°)(8 \angle 90°)}{6 - j8 + j8} = \frac{80 \angle 36.87°}{6 \angle 0°}$$

$$\mathbf{Z}_{Th} = 13.33 \angle 36.87° = 10.66 + j8$$

and

$$\mathbf{Z}_L = 13.3 \angle -36.87° = \mathbf{10.66 - j8}$$

In order to find the maximum power, we must first find $\mathbf{E}_{Th}$ (Fig. 18.72), as follows:

$$\mathbf{E}_{Th} = \frac{\mathbf{Z}_2(\mathbf{E})}{\mathbf{Z}_2 + \mathbf{Z}_1} \quad \text{(voltage divider rule)}$$

$$\mathbf{E}_{Th} = \frac{(8 \angle 90°)(9 \angle 0°)}{j8 + 6 - j8} = \frac{72 \angle 90°}{6 \angle 0°} = \mathbf{12 \angle 90°}$$

$$P_{\max} = \frac{E_{Th}^2}{4R} = \frac{(12)^2}{4(10.66)} = \frac{144}{42.64} = \mathbf{3.38 \ W}$$

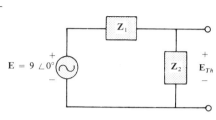

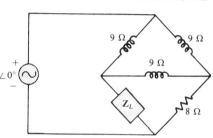

FIG. 18.72

EXAMPLE 18.18. Find the load impedance in Fig. 18.73 for maximum power to the load, and find the maximum power.

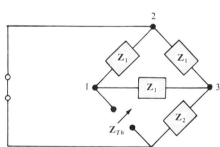

FIG. 18.73

Solution: First we must find $\mathbf{Z}_{Th}$ (Fig. 18.74).

$$\mathbf{Z}_1 = j9 \qquad \mathbf{Z}_2 = 8$$

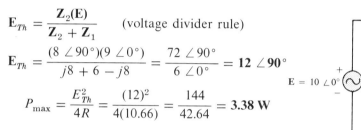

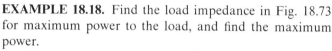

FIG. 18.74

Converting from a Δ to a Y (Fig. 18.75), we have

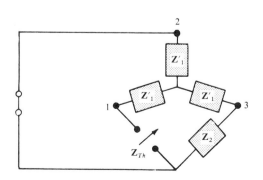

$$\mathbf{Z}'_1 = \frac{\mathbf{Z}_1}{3} = j3 \qquad \mathbf{Z}_2 = 8$$

FIG. 18.75

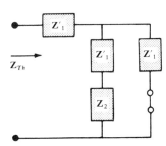

FIG. 18.76

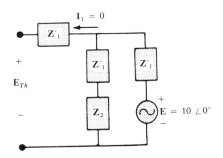

FIG. 18.77

The redrawn circuit (Fig. 18.76) shows

$$
\begin{aligned}
\mathbf{Z}_{Th} &= \mathbf{Z}'_1 + \frac{\mathbf{Z}'_1(\mathbf{Z}'_1 + \mathbf{Z}_2)}{\mathbf{Z}'_1 + (\mathbf{Z}'_1 + \mathbf{Z}_2)} \\
&= j3 + \frac{3 \angle 90°(j3 + 8)}{j6 + 8} \\
&= j3 + \frac{(3 \angle 90°)(8.54 \angle 20.56°)}{10 \angle 36.87°} \\
&= j3 + \frac{25.62 \angle 110.56°}{10 \angle 36.87°} = j3 + 2.56 \angle 73.69° \\
&= j3 + 0.72 + j2.46 \\
\mathbf{Z}_{Th} &= 0.72 + j5.46
\end{aligned}
$$

and

$$
\mathbf{Z}_L = 0.72 - j5.46
$$

For $\mathbf{E}_{Th}$, use the modified circuit of Fig. 18.77 with the voltage source replaced in its original position. Since $I_1 = 0$, $\mathbf{E}_{Th}$ is the voltage across the series impedance of $\mathbf{Z}'_1$ and $\mathbf{Z}_2$. Using the voltage divider rule gives us

$$
\begin{aligned}
\mathbf{E}_{Th} &= \frac{(\mathbf{Z}'_1 + \mathbf{Z}_2)\mathbf{E}}{\mathbf{Z}'_1 + \mathbf{Z}_2 + \mathbf{Z}'_1} = \frac{(j3 + 8)(10 \angle 0°)}{8 + j6} \\
&= \frac{(8.54 \angle 20.56°)(10 \angle 0°)}{10 \angle 36.87°} \\
\mathbf{E}_{Th} &= 8.54 \angle -16.31°
\end{aligned}
$$

and

$$
\begin{aligned}
P_{\max} &= \frac{E_{Th}^2}{4R} = \frac{(8.54)^2}{(4)(0.72)} = \frac{72.93}{2.88} \\
P_{\max} &= \mathbf{25.32 \ W}
\end{aligned}
$$

18.6 SUBSTITUTION, RECIPROCITY, AND MILLMAN'S THEOREMS

As indicated in the introduction to this chapter, the substitution and reciprocity theorems and Millman's theorem will not be considered here in detail. A careful review of Chapter 8 will enable you to apply these theorems to sinusoidal ac networks with little difficulty. A number of problems in the use of these theorems appear in the problem section.

PROBLEMS

Section 18.2

1. Using superposition, for each network of Fig. 18.78, determine the current through the inductance L_1.

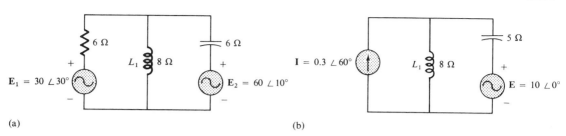

(a)   (b)

*2. Repeat Problem 1 for the networks of Fig. 18.79.   **FIG. 18.78**

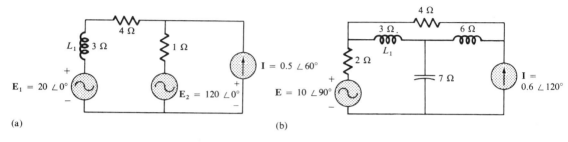

(a)   (b)

                                   FIG. 18.79

3. Using superposition, for the network of Fig. 18.80, determine the current I_L ($h = 100$).

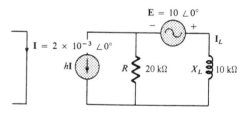

                                   FIG. 18.80

4. Using superposition, for the network of Fig. 18.81, determine the voltage V_L ($\mu = 20$).

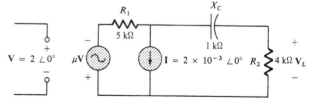

                                   FIG. 18.81

*5. Using superposition, determine the current I_L for the network of Fig. 18.82 ($\mu = 20$; $h = 100$).

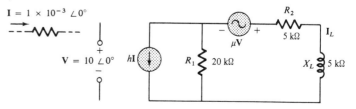

                                   FIG. 18.82

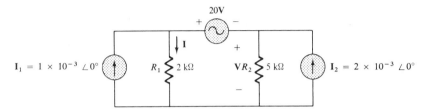

FIG. 18.83

*6. Determine V_L for the network of Fig. 18.83 ($h = 50$).

*7. Calculate the current **I** for the network of Fig. 18.84.

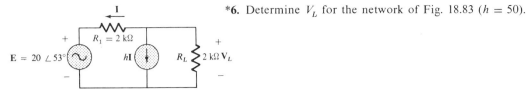

FIG. 18.84

Section 18.3

8. Find the Thevenin equivalent circuit for the portions of the networks of Fig. 18.85 external to the elements between points a and b.

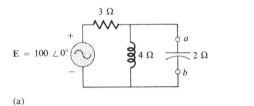

(a)

(b)

FIG. 18.85

*9. Repeat Problem 8 for the networks of Fig. 18.86.

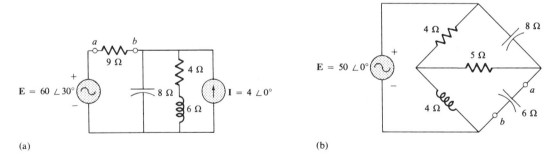

(a)

(b)

FIG. 18.86

*10. Repeat Problem 8 for the networks of Fig. 18.87.

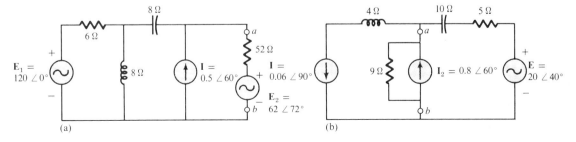

(a)

(b)

FIG. 18.87

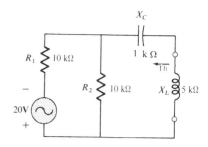
11. Determine the Thevenin equivalent circuit for the network external to the 5-kΩ inductive reactance of Fig. 18.88 (in terms of **V**).

FIG. 18.88

12. Determine the Thevenin equivalent circuit for the network external to the 4-kΩ inductive reactance of Fig. 18.89 (in terms of **I**).

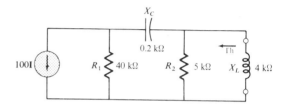

FIG. 18.89

13. Find the Thevenin equivalent circuit for the network external to the 10-kΩ inductive reactance of Fig. 18.80.

14. Determine the Thevenin equivalent circuit for the network external to the 4-kΩ resistor of Fig. 18.81.

*15. Find the Thevenin equivalent circuit for the network external to the 5-kΩ inductive reactance of Fig. 18.82.

*16. Determine the Thevenin equivalent circuit for the network external to the 2-kΩ resistor of Fig. 18.83.

*17. Find the Thevenin equivalent circuit for the network external to the resistor R_1 of Fig. 18.84.

Section 18.4

18. Find the Norton equivalent circuit for the portions of the networks of Fig. 18.85 external to the elements between points a and b.

19. Repeat Problem 18 for the networks of Fig. 18.86.

*20. Repeat Problem 18 for the networks of Fig. 18.87.

*21. Repeat Problem 18 for the networks of Fig. 18.90.

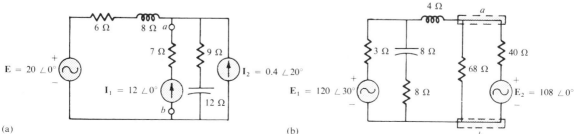

(a)   (b)

FIG. 18.90

22. Determine the Norton equivalent circuit for the network external to the 5-kΩ inductive reactance of Fig. 18.88.

23. Determine the Norton equivalent circuit for the network external to the 4-kΩ inductive reactance of Fig. 18.89.

24. Find the Norton equivalent circuit for the network external to the 4-kΩ resistor of Fig. 18.81.

*25. Find the Norton equivalent circuit for the network external to the 5-kΩ inductive reactance of Fig. 18.82.

*26. For the network of Fig. 18.91, find the Norton equivalent circuit for the network external to the 2-kΩ resistor.

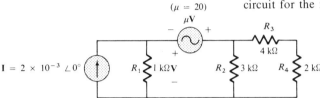

FIG. 18.91

*27. Find the Norton equivalent circuit for the network external to the I_1 current source of Fig. 18.84.

Section 18.5

28. Find the load impedance Z_L for the networks of Fig. 18.92 for maximum power to the load, and find the maximum power to the load.

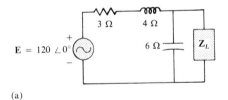

(a)

FIG. 18.92

(b)

*29. Repeat Problem 28 for the networks of Fig. 18.93.

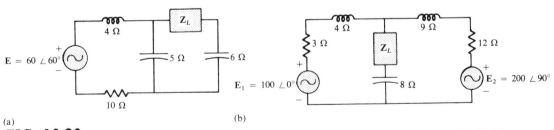

(a)

FIG. 18.93

(b)

30. Repeat Problem 28 for the network of Fig. 18.94.

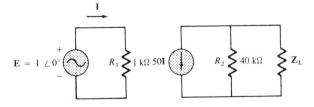

FIG. 18.94

Section 18.6

31. For the network of Fig. 18.95, determine two equivalent branches through the substitution theorem for the branch a–b.

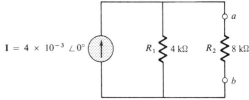

FIG. 18.95

32. **a.** For the network of Fig. 18.96(a), find the current **I**.
    b. Repeat part (a) for the network of Fig. 18.96(b).
    c. Do the results of parts (a) and (b) compare?

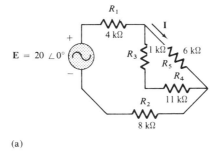

(a)

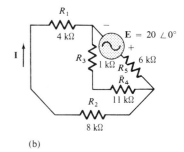

(b)

33. Using Millman's theorem, determine the current through the 4-kΩ capacitive reactance of Fig. 18.97.

FIG. 18.96

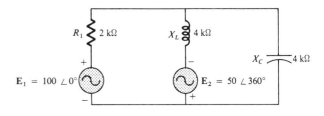

FIG. 18.97

GLOSSARY

Maximum power transfer theorem A theorem used to determine the load impedance necessary to insure maximum power to the load.

Millman's theorem A method employing voltage-to-current source conversions which will permit the determination of unknown variables in a multiloop network.

Norton's theorem A theorem that permits the reduction of any two-terminal linear ac network to one having a single current source and parallel impedance. The resulting configuration can then be employed to determine a particular current or voltage in the original network or to examine the effects of a specific portion of the network on a particular variable.

Reciprocity theorem A theorem stating that for single-source networks, the magnitude of the current in any branch of a network, due to a single voltage source anywhere else in the network, will equal the magnitude of the current through the branch in which the source was originally located if the source is placed in the branch in which the current was originally measured.

Substitution theorem A theorem stating that if the voltage across and current through any branch of an ac bilateral network are known, the branch can be replaced by any combination of elements that will maintain the same voltage across and current through the chosen branch.

Superposition theorem A method of network analysis that permits considering the effects of each source independently. The resulting current and/or voltage is the phasor sum of the currents and/or voltages developed by each source independently.

Thevenin's theorem A theorem that permits the reduction of any two-terminal linear ac network to one having a single voltage source and series impedance. The resulting configuration can then be employed to determine a particular current or voltage in the original network or to examine the effects of a specific portion of the network on a particular variable.

19.1 INTRODUCTION

The discussion of power in Chapter 13 included only the average power delivered to an ac network. We will now examine the total power equation in a slightly different form and introduce two additional forms of power: *apparent* and *reactive*.

Let us define, for the configuration of Fig. 19.1,

$$v = V_m \sin(\omega t + \theta)$$

and

$$i = I_m \sin \omega t$$

where θ is the phase angle by which v leads i, and since

$$\mathbf{Z} = \frac{V \angle \theta}{I \angle 0°} = \frac{V}{I} \angle \theta = Z \angle \theta$$

it is also the angle associated with the total impedance of the load of Fig. 19.1.

The power delivered to the load of Fig. 19.1 at any instant of time is determined by

$$p = vi$$

Substituting into the above, we have

$$p = V_m I_m \sin \omega t \sin(\omega t + \theta)$$

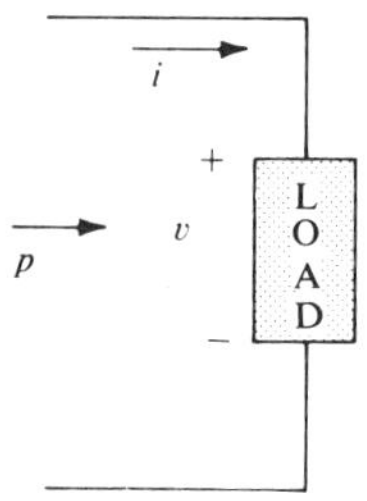

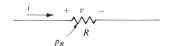

FIG. 19.1

If we now apply a number of trigonometric identities, we will find that the power equation can also be written

$$p = VI \cos\theta(1 - \cos 2\omega t) + VI \sin\theta(\sin 2\omega t)$$

(19.1)

where V and I are effective values.

It would appear initially that nothing has been gained by putting the equation in this form. However, the usefulness of the form of Eq. (19.1) will be demonstrated in the following sections. The derivation of Eq. (19.1) from the initial form will appear as an assignment at the end of the chapter.

If Eq. (19.1) is expanded to the form

$$p = \underbrace{VI \cos\theta}_{\text{average}} - \underbrace{VI \cos\theta}_{\text{peak}} \cos \underbrace{2\omega t}_{2x} + \underbrace{VI \sin\theta}_{\text{peak}} \sin \underbrace{2\omega t}_{2x}$$

there are two obvious points that can be made. First, the average power still appears as an isolated term that is time-independent. Second, both terms that follow vary at a frequency twice that of the applied voltage or current with peak values having a very similar format.

In an effort to insure completeness and order in presentation, each basic element (R, L, C) will first be treated separately.

19.2 RESISTIVE CIRCUIT

For a purely resistive circuit (such as in Fig. 19.2), v and i are in phase, and $\theta = 0°$. Substituting $\theta = 0°$ into Eq. (19.1), we obtain

$$p_R = VI \cos(0°)(1 - \cos 2\omega t) + VI \sin(0°) \sin 2\omega t$$

$$p_R = VI(1 - \cos 2\omega t) + 0$$

$$\boxed{p_R = VI - VI \cos 2\omega t}$$

(19.2)

where VI is the average or dc term and $-VI \cos 2\omega t$ is a negative cosine wave with twice the frequency of either input quantity (v or i) and a peak value of VI.

Plotting the waveform for p_R (Fig. 19.3), we see that

$$T_1 = \text{period of input quantities}$$

$$T_2 = \text{period of power curve } p_R$$

Note that in Fig. 19.3 the power curve passes through two cycles about its average value of VI for each cycle of either v or i ($T_1 = 2T_2$ or $f_2 = 2f_1$). Consider also that since the peak and average values of the power curve are the same, the curve is always above the horizontal axis. This

FIG. 19.2

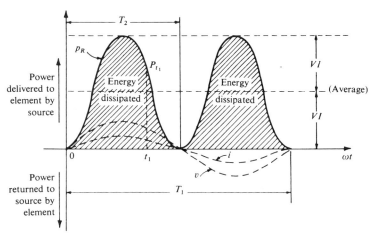

FIG. 19.3

indicates that *the total power delivered to a resistor will be dissipated*. The power returned to the source is represented by the portion of the curve below the axis, which in this case is zero. The power dissipated at any instant of time t_1 by the resistor can be found by substituting t_1 into Eq. (19.2). The average power from Eq. (19.2), or Fig. 19.3, is VI; or, as a summary,

$$P = VI = \frac{V_m I_m}{2} = I^2 R = \frac{V^2}{R} \qquad \text{(W)} \quad \textbf{(19.3)}$$

as derived in Chapter 13.

The energy dissipated by the resistor (W_R) *over one full cycle of the power curve* (Fig. 19.3) can be found using the following equation:

$$W_R = \int_0^{T_2} p_R \, dt \qquad \text{(joules, J)}$$

W_R = area under the power curve from 0 to T_2 (period of p_R) **(19.4)**

The area under the curve = (average value) × (length of curve), and

$$W_R = (VI) \times (T_2)$$

or

$$W_R = VIT_2 \qquad \text{(J)} \qquad \textbf{(19.5)}$$

or, since $T_2 = 1/f_2$, where f_2 is the frequency of the p_R curve,

$$W_R = \frac{VI}{f_2} \qquad \text{(J)} \qquad \textbf{(19.6)}$$

FIG. 19.4

19.3 APPARENT POWER

From our analysis of dc networks (and resistive elements above), it would seem *apparent* that the power delivered to the load of Fig. 19.4 is simply determined by the product of the applied voltage and current, with no concern for the components of the load. That is, $P = VI$. However, we found in Chapter 13 that the power factor ($\cos \theta$) of the load will have a pronounced effect on the power dissipated, less pronounced for more reactive loads. Although the product of the voltage and current is not always the power delivered, it is a power rating of significant usefulness in the description and analysis of sinusoidal ac networks and in the maximum rating of a number of electrical components and systems. It is called the *apparent power* and is represented symbolically by S^*. Since it is simply the product of voltage and current, its units are voltamperes, for which the abbreviation is VA. In total,

$$\boxed{S = VI} \quad \text{(VA)} \qquad \textbf{(19.7)}$$

or, since

$$V = IZ \quad \text{and} \quad I = \frac{V}{Z}$$

then

$$\boxed{S = I^2 Z} \quad \text{(VA)} \qquad \textbf{(19.8)}$$

and

$$\boxed{S = \frac{V^2}{Z}} \quad \text{(VA)} \qquad \textbf{(19.9)}$$

The average power to a system is defined by the equation

$$P = VI \cos \theta$$

However,

$$S = VI$$

Therefore,

$$\boxed{P = S \cos \theta} \quad \text{(W)} \qquad \textbf{(19.10)}$$

or the power factor of a system F_p is

$$\boxed{F_p = \cos \theta = \frac{P}{S}} \qquad \textbf{(19.11)}$$

---

*Prior to 1968, the symbol for apparent power was the more descriptive symbol P_a.

The power factor of a circuit, therefore, is the ratio of the average power to the apparent power. For a purely resistive circuit, we have

$$P = VI = S \quad \text{and} \quad F_p = \cos\theta = \frac{P}{S} = 1$$

In general, electrical equipment is rated in volt-amperes (VA) or in kilovolt-amperes (kVA), not in watts. By knowing the volt-ampere rating and the rated voltage of a device, we can readily determine the *maximum* current rating. For example, a device rated at 10 kVA at 200 V has a maximum current rating of $I = 10{,}000/200 = 50$ A when operated under rated conditions. The volt-ampere rating of a piece of equipment is equal to the wattage rating only when the F_p is 1. It is therefore a maximum power dissipation rating. This condition exists only when the total impedance of a system $Z \angle \theta$ is such that $\theta = 0°$.

The exact current demand of a device, when used under normal operating conditions, could be determined if the wattage rating and power factor were given instead of the volt-ampere rating. However, the power factor is sometimes not available, or it may vary with load.

The reason for rating electrical equipment in kVA rather than kW is obvious from the configuration of Fig. 19.5. The load has an apparent power rating of 10 kVA and a current rating of 50 A at the applied voltage, 200 V. As indicated, the current demand is above the rated value and could damage the load element, yet the reading on the wattmeter is very low since the load is highly reactive. In other words, the wattmeter reading is an indication not of the current drawn but simply of the watts dissipated. Theoretically, if the load were purely reactive, the wattmeter reading could be zero and the device burning up due to the high current demand.

19.4 INDUCTIVE CIRCUIT AND REACTIVE POWER

For a purely inductive circuit (such as that in Fig. 19.6), v leads i by 90°. Therefore, in Eq. (19.1), $\theta = 90°$. Substituting $\theta = 90°$ into Eq. (19.1) yields

$$p_L = VI\cos(90°)(1 - \cos 2\omega t) + VI\sin(90°)(\sin 2\omega t)$$
$$= 0 + VI\sin 2\omega t$$

or

$$\boxed{p_L = VI\sin 2\omega t} \qquad (19.12)$$

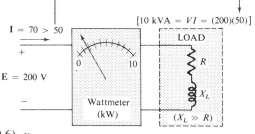

FIG. 19.5

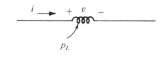

FIG. 19.6

where $VI \sin 2\omega t$ is a sine wave with twice the frequency of either input quantity (v or i) and a peak value of VI. Note the absence of an average or constant term in the equation.

Plotting the waveform for p_L (Fig. 19.7), we obtain

$$T_1 = \text{period of either input quantity}$$
$$T_2 = \text{period of } p_L \text{ curve}$$

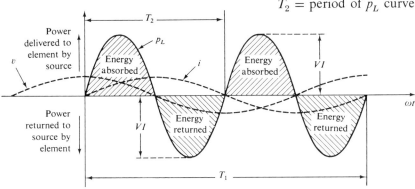

FIG. 19.7

Note that over one full cycle of p_L, the area above the horizontal axis in Fig. 19.7 is exactly equal to that below the axis. This indicates that over a full cycle of p_L, the power delivered by the source to the inductor is exactly equal to that returned to the source by the inductor. *The net flow of power to the pure inductor is therefore zero over a full cycle, and no energy is lost in the transaction.* The power absorbed or returned by the inductor at any instant of time t_1 can be found simply by substituting t_1 into Eq. (19.12). The peak value of the curve VI is defined as the *reactive power* associated with a pure inductor.

In general, the reactive power associated with any circuit is defined to be $VI \sin \theta$, a factor appearing in the second term of Eq. (19.1). Note that it is the peak value of that term of the total power equation that produces no net transfer of energy. The symbol for reactive power is Q^*, and its unit of measure is the *var* (*volt-ampere reactive*). The Q is derived from the quadrature (90°) relationship between the various powers to be discussed in detail in a later section. Therefore,

$$\boxed{Q = VI \sin \theta} \qquad \text{(var)} \qquad \textbf{(19.13)}$$

where θ is the phase angle between V and I.

For the inductor,

$$\boxed{Q_L = VI} \qquad \text{(var)} \qquad \textbf{(19.14)}$$

or, since $V = IX_L$ or $I = V/X_L$,

$$\boxed{Q_L = I^2 X_L} \qquad \text{(var)} \qquad \textbf{(19.15)}$$

---

*Prior to 1968, the symbol for reactive power was the more descriptive symbol P_q.

or

$$\boxed{Q_L = \frac{V^2}{X_L}} \qquad \text{(var)} \qquad \textbf{(19.16)}$$

the apparent power associated with an inductor $S = VI$, and the average power $P = 0$, as noted in Fig. 19.7. The power factor is therefore

$$F_p = \cos\theta = \frac{P}{S} = \frac{0}{VI} = 0$$

If the average power is zero, and the energy supplied is returned within one cycle, why is reactive power of any significance? The reason is not obvious but can be explained using the curve of Fig. 19.7. At every instant of time along the power curve that the curve is above the axis (positive), energy must be supplied to the inductor even though it will be returned during the negative portion of the cycle. This power requirement during the positive portion of the cycle requires that the generating plant provide this energy during that interval. Therefore, the effect of reactive elements such as the inductor can be to raise the power requirement of the generating plant even though the reactive power is not dissipated but simply "borrowed." The increased power demand during these intervals is a cost factor that must be passed on to the industrial consumer. In fact, most larger users of electrical energy pay for the apparent power demand rather than the watts dissipated since the voltamperes used are sensitive to the reactive power requirement (see Section 19.6). In other words, the closer the power factor of an industrial outfit is to 1, the more efficient is its operation, since it is limiting its use of "borrowed" power.

The energy stored by the inductor during the positive portion of the cycle (Fig. 19.7) is equal to that returned during the negative portion and can be determined using the integral

$$W_L = \int_{T_2/2}^{T_2} p_L \, dt \qquad \text{(J)}$$

$$= \text{area under power curve from } T_2/2 \text{ to } T_2$$

Recall from Chapter 13 that the average value of the positive portion of a sinusoid equals 2 (peak value/π), and that the area under any curve = (average value) × (length of the curve):

$$W_L = \left(\frac{2VI}{\pi}\right) \times \left(\frac{T_2}{2}\right)$$

or

$$\boxed{W_L = \frac{VIT_2}{\pi}} \qquad \text{(J)} \qquad \textbf{(19.17)}$$

or, since $T_2 = 1/f_2$, where f_2 is the frequency of the p_L curve, we have

$$\boxed{W_L = \frac{VI}{\pi f_2}} \qquad \text{(J)} \qquad \textbf{(19.18)}$$

Since the frequency f_2 of the power curve is twice that of the input quantity, if we substitute the frequency f_1 of the input voltage or current, Eq. (19.18) becomes

$$W_L = \frac{VI}{\pi(2f_1)}$$

or

$$\boxed{W_L = \frac{VI}{\omega_1}} \qquad \text{(J)} \qquad \textbf{(19.19)}$$

but

$$V = IX_L = I\omega_1 L$$

so

$$W_L = \frac{(I\omega_1 L)I}{\omega_1}$$

or

$$\boxed{W_L = LI^2} \qquad \text{(J)} \qquad \textbf{(19.20)}$$

19.5 CAPACITIVE CIRCUIT

FIG. 19.8

For a purely capacitive circuit (such as in Fig. 19.8), i leads v by 90°. Therefore, in Eq. (19.1), $\theta = -90°$. Substituting $\theta = -90°$ into Eq. (19.1), we obtain

$$p_C = VI\cos(-90°)(1 - \cos 2\omega t) \\ + VI\sin(-90°)(\sin 2\omega t)$$

$$p_C = 0 - VI\sin 2\omega t$$

or

$$\boxed{p_C = -VI\sin 2\omega t} \qquad \textbf{(19.21)}$$

where $-VI\sin 2\omega t$ is a negative sine wave with twice the frequency of either input (v or i) and a peak value of VI. Again, note the absence of an average or constant term.

Plotting the waveform for p_C (Fig. 19.9) gives us

$$T_1 = \text{period of either input quantity}$$
$$T_2 = \text{period of } p_C \text{ curve}$$

Note that the same situation exists here for the p_C curve as existed for the p_L curve. The power delivered by the source

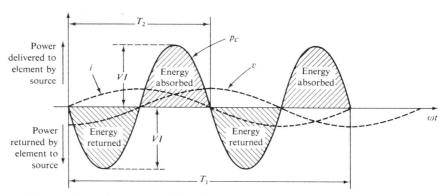

FIG. 19.9

to the capacitor is exactly equal to that returned to the source by the capacitor over one full cycle. *The net flow of power to the pure capacitor is therefore zero over a full cycle, and no energy is lost in the transaction.* The power absorbed or returned by the capacitor at any instant of time t_1 can be found by substituting t_1 into Eq. (19.21).

The reactive power associated with the capacitor is equal to the peak value of the p_C curve, as follows:

$$\boxed{Q_C = VI} \qquad \text{(var)} \qquad \textbf{(19.22)}$$

but, since $V = IX_C$ and $I = V/X_C$, the reactive power to the capacitor can also be written

$$\boxed{Q_C = I^2 X_C} \qquad \text{(var)} \qquad \textbf{(19.23)}$$

and

$$\boxed{Q_C = \frac{V^2}{X_C}} \qquad \text{(var)} \qquad \textbf{(19.24)}$$

The apparent power associated with the capacitor is

$$\boxed{S = VI} \qquad \text{(VA)} \qquad \textbf{(19.25)}$$

and the average power is $P = 0$, as noted from Eq. (19.21) or Fig. 19.9. The power factor is therefore

$$F_p = \cos\theta = \frac{P}{S} = \frac{0}{VI} = 0$$

The energy stored by the capacitor during the positive portion of the cycle (Fig. 19.9) is equal to that returned during the negative portion and can be determined using the integral

$$W_C = \int_0^{T_2/2} p_C \, dt \qquad \text{(J)}$$

$$= \text{area under the power curve from 0 to } T_2/2$$

Proceeding in a manner similar to that used for the inductor, we can show that

$$W_C = \frac{VIT_2}{\pi} \quad (J) \qquad \textbf{(19.26)}$$

or, since $T_2 = 1/f_2$, where f_2 is the frequency of the p_C curve,

$$W_C = \frac{VI}{\pi f_2} \quad (J) \qquad \textbf{(19.27)}$$

In terms of the frequency f_1 of the input quantities v and i,

$$W_C = \frac{VI}{\omega_1} \quad (J) \qquad \textbf{(19.28)}$$

and

$$W_C = CV^2 \quad (J) \qquad \textbf{(19.29)}$$

19.6 THE POWER TRIANGLE

The three quantities—average power, apparent power, and reactive power—can be related graphically by a power triangle, as indicated in Fig. 19.10(a) for an R-L circuit. The power triangle for an R-C circuit is shown in Fig. 19.10(b).

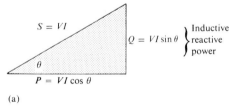

(a)   (b)

FIG. 19.10

If a network should have both capacitive and inductive elements, the reactive component of the power triangle will be determined by the *difference* between the reactive power delivered to each. If $Q_L > Q_C$, the resultant power triangle will be similar to Fig. 19.10(a). If $Q_C > Q_L$, the resultant power triangle will be similar to Fig. 19.10(b).

That the total reactive power is the difference between that of the inductive and capacitive elements can be demonstrated by considering Eqs. (19.12) and (19.21). Using these equations, the reactive power delivered to each reactive element has been plotted for a series L-C circuit on the same set of axes in Fig. 19.11. The reactive elements were chosen such that $X_L > X_C$. Note that the power curve for each is exactly 180° out of phase. The curve for the resultant reactive power is therefore determined by the algebraic resultant of the two at each instant of time. Since the reac-

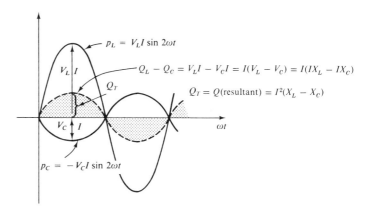

FIG. 19.11

tive power is defined as the peak value, the reactive component of the power triangle is as indicated in the figure: $I^2(X_L - X_C)$.

An additional verification can be derived by first considering the impedance diagram of a series *R-L-C* circuit (Fig. 19.12). If we multiply each radius vector by the current squared (I^2), we obtain the results shown in Fig. 19.13, which is the power triangle for a predominantly inductive circuit.

Since the reactive power and average power are always angled 90° to each other, the three powers are related by the Pythagorean theorem; that is,

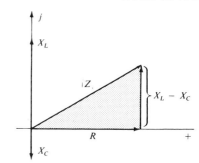

FIG. 19.12

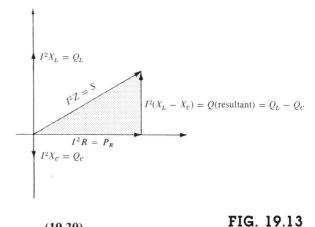

FIG. 19.13

$$\boxed{S^2 = P^2 + Q^2} \qquad \text{(19.30)}$$

Therefore, the third power can always be found if the other two are known.

19.7 THE TOTAL *P*, *Q*, AND *S*

The total number of watts, vars, and volt-amperes and the power factor of any system can be found using the following procedure:

1. Find the total number of watts and vars for each branch of the circuit.
2. The total number of watts of the system is then the sum of the average power delivered to each branch.
3. The total reactive power is the difference between the reactive power of the inductive loads and that of the capacitive loads.
4. The total apparent power is $\sqrt{P_T^2 + Q_T^2}$.
5. The total power factor is P_T/S_T.

There are two important points in the above tabulation. First, the total apparent power must be determined from the total average and reactive powers and *cannot* be determined from the apparent powers of each branch. Second, and more important, it is *not necessary* to consider the series-parallel arrangement of branches. In other words, the total real, reactive, or apparent power is independent of whether the loads are in series, parallel, or series-parallel. The following examples will demonstrate the relative ease with which all the quantities of interest can be found.

EXAMPLE 19.1. Find the total number of watts, vars, and volt-amperes and the power factor F_p of the network in Fig. 19.14. Draw the power triangle and find the current in phasor form.

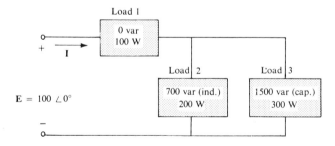

FIG. 19.14

Solution: Use a table:

Load	W	var	VA
1	100	0	100
2	200	700 (ind.)	$\sqrt{200^2 + 700^2} = 728.0$
3	300	1500 (cap.)	$\sqrt{300^2 + 1500^2} = 1529.71$
	$P_T = 600$ Total watts dissipated	$Q_T = 800$ **(cap.)** Resultant vars of circuit	$S_T = \sqrt{600^2 + 800^2}$ $= 1000$ (Note that $S_T \neq$ sum of each branch: $1000 \neq 100 + 728 + 1529.71$)

$$F_p = \frac{P_T}{S_T} = \frac{600}{1000} = 0.6 \text{ leading (cap.)}$$

The power triangle is shown in Fig. 19.15.

Since $S_T = VI = 1000$, $I = 1000/100 = 10$ A; and since θ of $\cos\theta = F_p$ is the angle between the input voltage and current,

$$\mathbf{I} = 10 \angle +53.13°$$

The plus sign is associated with the phase angle since the circuit is predominantly capacitive.

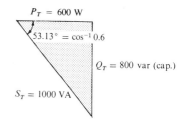

FIG. 19.15

EXAMPLE 19.2.

a. Find the total number of watts, var, and volt-amperes and the power factor F_p for the network of Fig. 19.16.
b. Sketch the power triangle.
c. Find the energy dissipated by the resistor over one full cycle of the input voltage if the frequency of the input quantities is 60 Hz.
d. Find the energy stored in, or returned by, the capacitor or inductor over 1/2 cycle of the power curve for each if the frequency of the input quantities is 60 Hz.

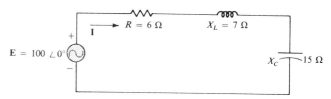

Solutions:

a. $\mathbf{I} = \dfrac{\mathbf{E}}{\mathbf{Z}_T} = \dfrac{100 \angle 0°}{6 + j7 - j15} = \dfrac{100 \angle 0°}{10 \angle -53.13°} = 10 \angle 53.13°$

FIG. 19.16

$\mathbf{V}_R = \mathbf{I}R = (10 \angle 53.13°)(6 \angle 0°) = 60 \angle 53.13°$

$\mathbf{V}_L = \mathbf{I}X_L = (10 \angle 53.13°)(7 \angle 90°) = 70 \angle 143.13°$

$\mathbf{V}_C = \mathbf{I}X_C = (10 \angle 53.13°)(15 \angle -90°) = 150 \angle -36.87°$

$P_T = EI \cos\theta = (100)(10) \cos 53.13° = \mathbf{600\ W}$

$\quad = I^2R = (10^2)(6) = \mathbf{600\ W}$

$\quad = \dfrac{V_R^2}{R} = \dfrac{60^2}{6} = \mathbf{600\ W}$

$S_T = EI = (100)(10) = \mathbf{1000\ VA}$

$\quad = I^2Z_T = (10^2)(10) = \mathbf{1000\ VA}$

$\quad = \dfrac{E^2}{Z_T} = \dfrac{100^2}{10} = \mathbf{1000\ VA}$

$Q_T = EI \sin\theta = (100)(10) \sin 53.13° = \mathbf{800\ var}$

$\quad = Q_C - Q_L$

$\quad = I^2(X_C - X_L) = (100)(15 - 7) = \mathbf{800\ var}$

$\quad = \dfrac{V_C^2}{X_C} - \dfrac{V_L^2}{X_L} = \dfrac{150^2}{15} - \dfrac{70^2}{7} = 1500 - 700$

$\quad = \mathbf{800\ var}$

$$F_p = \frac{P_T}{S_T} = \frac{600}{1000} = \textbf{0.6 leading (cap.)}$$

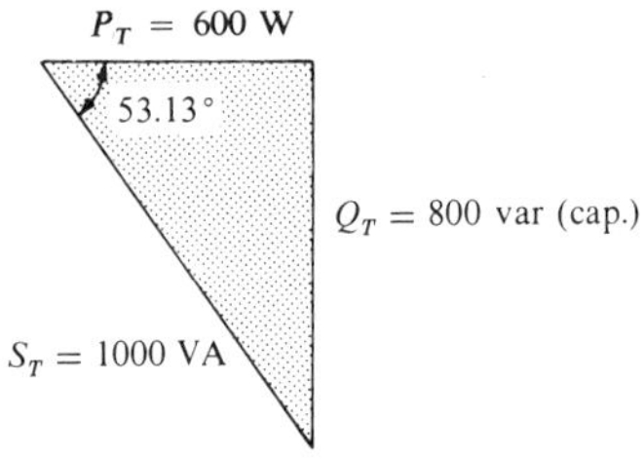

b. The power triangle is as shown in Fig. 19.17.

c. $W_R = 2\left(\dfrac{V_R I}{f_2}\right) = 2\left(\dfrac{V_R I}{2f_1}\right) = \dfrac{V_R I}{f_1} = \dfrac{(60)(10)}{60} = \textbf{10 J}$

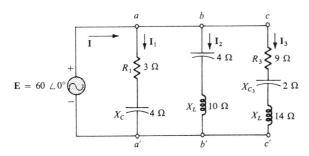

FIG. 19.17

d. $W_L = \dfrac{V_L I}{2\pi f_1} = \dfrac{(70)(10)}{(6.28)(60)} = \dfrac{700}{377} = \textbf{1.86 J}$

$W_C = \dfrac{V_C I}{2\pi f_1} = \dfrac{(150)(10)}{377} = \dfrac{1500}{377} = \textbf{3.98 J}$

EXAMPLE 19.3. For the network of Fig. 19.18:
a. Find the average power, apparent power, reactive power, and F_p for each branch.
b. Find the total number of watts, var, and volt-amperes and the power factor F_p of the circuit. Sketch the power triangle.
c. Find the current **I**.

FIG. 19.18

Solutions:
a. For branch *a-a'*:

$$\mathbf{I_1} = \frac{\mathbf{E}}{\mathbf{Z}_{a\text{-}a'}} = \frac{60\angle 0°}{3 - j4} = \frac{60\angle 0°}{5\angle -53.13°} = 12\angle +53.13°$$

$$P = I_1^2 R = (12^2)(3) = (144)(3) = \textbf{432 W}$$

$$S = EI_1 = (60)(12) = \textbf{720 VA}$$

$$Q = I_1^2 X_C = (12^2)(4) = \textbf{576 var}$$

$$F_p = \frac{P}{S} = \frac{432}{720} = \textbf{0.6 leading (cap.)}$$

For branch *b-b'*:

$$\mathbf{I_2} = \frac{\mathbf{E}}{\mathbf{Z}_{b\text{-}b'}} = \frac{60\angle 0°}{j10 - j4} = \frac{60\angle 0°}{6\angle 90°} = 10\angle -90°$$

$$P = I_2^2 R = I_2^2(0) = \textbf{0 W}$$

$$S = EI_2 = (60)(10) = \textbf{600 VA}$$

$$Q_T = I_2^2(X_L - X_C) = (10^2)(6) = \textbf{600 var}$$

$$F_p = \frac{P}{S} = \frac{0}{600} = \textbf{0}$$

For branch *c-c'*:

$$I_3 = \frac{E}{Z_{c\text{-}c'}} = \frac{60 \angle 0°}{9 - j2 + j14} = \frac{60 \angle 0°}{9 + j12} = \frac{60 \angle 0°}{15 \angle 53.13°}$$

$$= 4 \angle -53.13°$$

$$P = I_3^2 R = (4^2)(9) = \textbf{144 W}$$

$$S = EI_3 = (60)(4) = \textbf{240 VA}$$

$$Q = I_3^2(X_L - X_C) = (4^2)(12) = \textbf{192 var}$$

$$F_p = \cos \theta = \frac{P}{S} = \frac{144}{240} = \textbf{0.6 lagging (ind.)}$$

b. The total system can now be represented as shown in Fig. 19.19. Using a table,

$$S_T = \sqrt{(576)^2 + (216)^2}$$

$$S_T = \textbf{615.17 VA}$$

Branch	W	var
a-a'	432	576 (cap.)
b-b'	0	600 (ind.)
c-c'	144	192 (ind.)
	$P_T = \textbf{576 W}$	$Q_T = \textbf{216 (ind.)}$

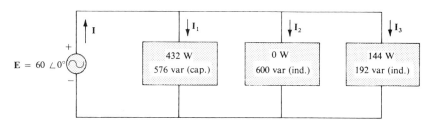

$$F_{p(T)} = \cos \theta = \frac{P_T}{S_T} = \frac{576}{615.17} = \textbf{0.9363 lagging (ind.)}$$

$$\theta = \cos^{-1} 0.9363 = 20.56°$$

The power triangle is shown in Fig. 19.20.

FIG. 19.19

c. $S_T = EI = 615.17$ **VA**

Therefore,

$$I = \frac{615}{60} = 10.25 \text{ A}$$

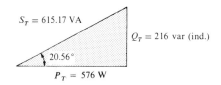

The circuit is inductive. Therefore, **I** lags **E** by 20°, and

I = **10.25** $\angle$ **−20.56°**

FIG. 19.20

EXAMPLE 19.4. An electrical device is rated 5 kVA, 100 V at a 0.6 power-factor lag. What is the impedance of the device in rectangular coordinates?

Solution:

$$S_T = EI = 5000 \ VA$$

Therefore,

$$I = \frac{5000}{100} = 50 \text{ A}$$

For $F_p = 0.6$, we have

$$\theta = \cos^{-1}0.6 = 53.13°$$

Since the power factor is lagging, the circuit is predominantly inductive, and **I** lags **E**. Or, for $\mathbf{E} = 100 \angle 0°$,

$$\mathbf{I} = 50 \angle -53.13°$$

However,

$$\mathbf{Z}_T = \frac{\mathbf{E}}{\mathbf{I}} = \frac{100 \angle 0°}{50 \angle -53.13°} = 2 \angle 53.13° = \mathbf{1.2 + j1.6}$$

which is the impedance of the circuit of Fig. 19.21.

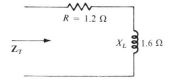

FIG. 19.21

19.8 EFFECTIVE RESISTANCE

The resistance of a conductor as determined by the equation $R = \rho(l/A)$ is often called the *dc, ohmic,* or *geometric* resistance. It is a constant quantity determined only by the material used and its physical dimensions. In ac circuits, the actual resistance of a conductor (called the *effective* resistance) differs from the dc resistance because of the varying currents and voltages which introduce effects not present in dc circuits.

Some of these effects include radiation losses, skin effect, eddy currents, and hysteresis losses. The first two effects apply to any network, while the latter two are concerned with the additional losses introduced by the presence of ferromagnetic materials in a changing magnetic field.

The effective resistance of an ac circuit cannot be measured by the ratio V/I, since this is now the impedance of a circuit which may have both resistance and reactance. The effective resistance can be found, however, by using the power equation $P = I^2R$, where

$$R_{\text{eff}} = \frac{P}{I^2} \qquad (19.31)$$

A wattmeter and an ammeter are therefore necessary for measuring the effective resistance of an ac circuit.

Let us now examine the various losses in greater detail. The radiation loss is the loss of energy in the form of electromagnetic waves during the transfer of energy from one element to another. This loss in energy requires that the input power be larger to establish the same current I, causing R to increase as determined by Eq. (19.31). At a frequency of 60 Hz, the effects of radiation losses can be completely ignored. However, at radio frequencies, this is an important effect and may in fact become the main effect in an electromagnetic device such as an antenna.

The explanation of skin effect requires the use of some basic concepts previously described. It will be remembered from Chapter 10 that a magnetic field exists around every current-carrying conductor (Fig. 19.22). Since the amount of charge flowing in ac circuits changes with time, the magnetic field surrounding the moving charge (current) also changes. Recall also that a wire placed in a changing magnetic field will have an emf developed across its terminals as determined by Faraday's law $e = N(d\phi/dt)$. The higher the frequency of the changing flux as determined by an alternating current, the greater the induced emf e will be.

For a conductor carrying alternating current, the changing magnetic field surrounding the wire links the wire itself, thus developing within the wire an emf that opposes the original flow of current. These effects are more pronounced at the center of the conductor than at the surface since the center is linked by the changing flux inside the wire as well as that outside the wire. As the frequency of the applied signal increases, the flux linking the wire will change at a greater rate. An increase in frequency will therefore increase the counter-emf at the center of the wire to the point where the current will, for all practical purposes, flow on the surface of the conductor. At 60 Hz, the effects of skin effect are just about noticeable. However, at radio frequencies the skin effect is so pronounced that large conductors are frequently made hollow since the center part is relatively ineffective. The skin effect, therefore, reduces the effective area through which the current can flow, and causes the resistance of the conductor, given by the equation $R\uparrow = \rho(l/A\downarrow)$, to increase.

As mentioned earlier, hysteresis and eddy current losses will appear when a ferromagnetic material is placed in the region of a changing magnetic field. To describe eddy current losses in greater detail, we will consider the effects of an alternating current passing through a coil that is wrapped around a ferromagnetic core. As the alternating current passes through the coil, it will develop a changing magnetic flux ϕ linking both the coil and the core which will develop an emf within the core as determined by Faraday's law. This induced emf and the geometric resistance of the core $R_c = \rho(l/A)$ cause currents to be developed within the core, $i_{core} = (e_{ind}/R_c)$, called *eddy currents*. They flow in the circular paths, as shown in Fig. 19.23, changing direction with the applied ac potential.

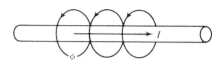

FIG. 19.22

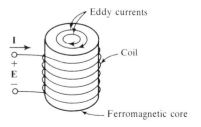

FIG. 19.23

The eddy current losses are determined by

$$P_{\text{eddy}} = i_{\text{eddy}}^2 R_{\text{core}}$$

The magnitude of these losses is determined primarily by the type of core used. If the core is nonferromagnetic—and has a high resistivity like wood or air—the eddy current losses can be neglected. In terms of the frequency of the applied signal and the magnetic field strength produced, the eddy current loss is proportional to the square of the frequency times the square of the magnetic field strength:

$$P_{\text{eddy}} \propto f^2 B^2$$

Eddy current losses can be reduced if the core is constructed of thin, laminated sheets of ferromagnetic material insulated from one another and aligned parallel to the magnetic flux. Such construction reduces the magnitude of the eddy currents by placing more resistance in their path.

Hysteresis losses were described in Section 10.8. You will recall that in terms of the frequency of the applied signal and the magnetic field strength produced, the hysteresis loss is proportional to the frequency to the first power times the magnetic field strength to the nth power:

$$P_{\text{hys}} \propto f^1 B^n$$

where n can vary from 1.4 to 2.6, depending on the material under consideration.

Hysteresis losses can be effectively reduced by the injection of small amounts of silicon into the magnetic core, constituting some 2% or 3% of the total composition of the core. This must be done carefully, however, as too much silicon makes the core brittle and difficult to machine into the shape desired.

EXAMPLE 19.5.

a. An air-core coil is connected to a 120-V, 60-Hz source as shown in Fig. 19.24. The current is found to be 5 A, and a wattmeter reading of 75 W is observed. Find the effective resistance and the inductance of the coil.

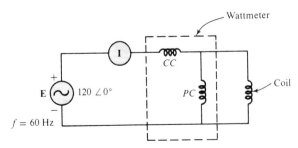

FIG. 19.24

b. A brass core is then inserted in the coil, and the ammeter reads 4 A and the wattmeter reads 80 W. Calculate

the effective resistance of the core. To what do you attri-
bute the increase in value over that of part (a)?

c. If a solid iron core is inserted into the coil, the current is
found to be 2 A, and the wattmeter reads 52 W. Calculate the resistance and the inductance of the coil. Compare these values to those of part (a), and account for
the changes.

Solutions:

a. $R = \dfrac{P}{I^2} = \dfrac{75}{5^2} = 3 \, \Omega$

$Z_T = \dfrac{E}{I} = \dfrac{120}{5} = 24 \, \Omega$

$X_L = \sqrt{Z_T^2 - R^2} = \sqrt{24^2 - 3^2} = 23.81 \, \Omega$

and $\quad X_L = 2\pi fL$

or $\quad L = \dfrac{X_L}{2\pi f} = \dfrac{23.81}{377} = \textbf{63.16 mH}$

b. $R = \dfrac{P}{I^2} = \dfrac{80}{4^2} = \dfrac{80}{16} = 5 \, \Omega$

The brass core has less reluctance than the air core.
Therefore a greater magnetic flux density B will be cre-
ated in it. Since $P_{eddy} \propto f^2 B^2$, and $P_{hys} \propto f^1 B^n$, as the
flux density increases, the core losses and the effective
resistance increase.

c. $R = \dfrac{P}{I^2} = \dfrac{52}{2^2} = \dfrac{52}{4} = 13 \, \Omega$

$Z_T = \dfrac{E}{I} = \dfrac{120}{2} = 60 \, \Omega$

$X_L = \sqrt{Z_T^2 - R^2} = \sqrt{60^2 - 13^2} = 58.57 \, \Omega$

$L = \dfrac{X_L}{2\pi f} = \dfrac{58.57}{377} = \textbf{155.36 mH}$

The iron core has less reluctance than the air or brass
cores. Therefore a greater magnetic flux density B will be
developed in the core. Again, since $P_{eddy} \propto f^2 B^2$ and P_{hys}
$\propto f^1 B^n$, the increased flux density will cause the core losses
and the effective resistance to increase.

Since the inductance L is related to the change in flux by
the equation $L = N(d\phi/di)$, the inductance will be greater
for the iron core because the changing flux linking the core
will increase.

PROBLEMS

Section 19.7

1. For the network of Fig. 19.25:

a. Find the average power delivered to each element.
b. Find the reactive power for each element.
c. Find the apparent power for each element.
d. Find the total number of watts, vars, and volt-amperes and the power factor F_p of the circuit.
e. Sketch the power triangle.
f. Find the energy dissipated by the resistor over one full cycle of the input voltage.
g. Find the energy stored or returned by the capacitor and the inductor over 1/2 cycle of the power curve for each.

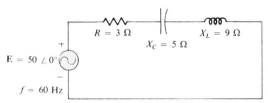

$R = 3\ \Omega$ $X_C = 5\ \Omega$ $X_L = 9\ \Omega$

$E = 50 \angle 0°$

$f = 60$ Hz

FIG. 19.25

2. For the system of Fig. 19.26:
   a. Find the total number of watts, vars, and volt-amperes and the power factor F_p.
   b. Draw the power triangle.
   c. Find the current I_T.

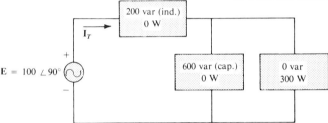

200 var (ind.)
0 W

I_T

$E = 100 \angle 90°$

600 var (cap.)
0 W

0 var
300 W

FIG. 19.26

3. Repeat Problem 2 for the system of Fig. 19.27.

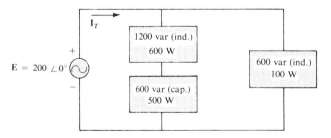

I_T

1200 var (ind.)
600 W

$E = 200 \angle 0°$

600 var (cap.)
500 W

600 var (ind.)
100 W

FIG. 19.27

4. Repeat Problem 2 for the system of Fig. 19.28.

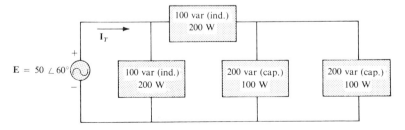

100 var (ind.)
200 W

I_T

$E = 50 \angle 60°$

100 var (ind.)
200 W

200 var (cap.)
100 W

200 var (cap.)
100 W

FIG. 19.28

5. For the circuit of Fig. 19.29:
  a. Find the average, reactive, and apparent power for the 5-Ω resistor.
  b. Repeat part (a) for the 10-Ω inductive reactance.
  c. Find the total number of watts, vars, and volt-amperes and the power factor F_p.
  d. Find the current $\mathbf{I}_T$.

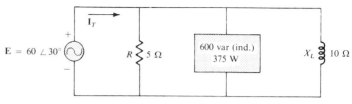

6. Repeat Problem 1 for the circuit of Fig. 19.30.

FIG. 19.29

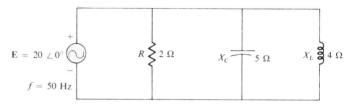

7. Repeat Problem 1 for the circuit of Fig. 19.31.

FIG. 19.30

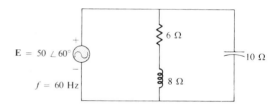

FIG. 19.31

*8. Repeat Problem 1 for the circuit of Fig. 19.32.

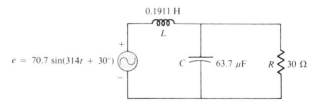

9. An electrical system is rated 10 kVA, 200 V at a 0.5 leading power factor.
  a. What is the impedance of the system in rectangular coordinates?
  b. Find the average power delivered to the system.

FIG. 19.32

10. Repeat Problem 9 for an electrical system rated 5 kVA, 120 V at a 0.8 lagging power factor.

*11. For the system of Fig. 19.33:
  a. Find the total number of watts, vars, and volt-amperes and F_p.
  b. Find the current $\mathbf{I}_T$.
  c. Draw the power triangle.

d. Find the type elements and their impedance in ohms within each electrical box. (Assume that all elements are in series within the boxes.)

e. Verify that the results of part (b) are correct by finding the current I_T using only the input voltage **E** and the results of part (d). Compare the value of I_T with that obtained for part (a).

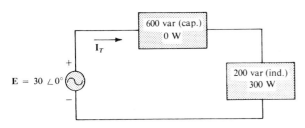

FIG. 19.33

*12. For the system of Fig. 19.34:
   a. Find the total number of watts, vars, and volt-amperes and F_p.
   b. Find the current **I**.
   c. Find the type elements and their impedance in each box. (Assume that the elements within each box are in series.)

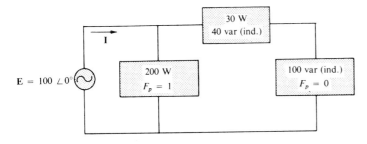

FIG. 19.34

13. For the circuit of Fig. 19.35:
   a. Find the total number of watts, vars, and volt-amperes and F_p.
   b. Find the voltage **E**.
   c. Find the type elements and their impedance in each box. (Assume that the elements within each box are in series.)

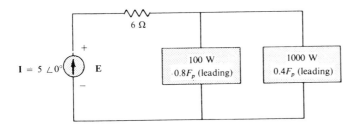

`5

*14. Repeat Problem 11 for the system of Fig. 19.36.

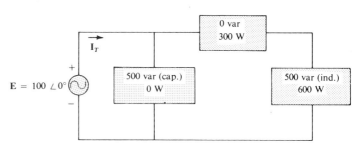

FIG. 19.36

15. **a.** A wattmeter is connected with its current coil as shown in Fig. 19.37 and the potential coil across points *f-g*. What does the wattmeter read?
   b. Repeat part (a) with the potential coil (*PC*) across *a-b*, *b-c*, *a-c*, *a-d*, *c-d*, *d-e*, and *f-e*.

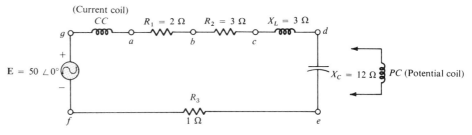

FIG. 19.37

Section 19.8

16. **a.** An air-core coil is connected to a 200-V, 60-Hz source. The current is found to be 4 A, and a wattmeter reading of 80 W is observed. Find the effective resistance and the inductance of the coil.
   b. A brass core is inserted in the coil. The ammeter reads 3 A, and the wattmeter reads 90 W. Calculate the effective resistance of the core. Explain the increase over the value of part (a).
   c. If a solid iron core is inserted into the coil, the current is found to be 2 A, and the wattmeter reads 60 W. Calculate the resistance and inductance of the coil. Compare these values to the values of part (a), and account for the changes.

17. **a.** The inductance of an air core coil is 0.08 H, and the effective resistance is 4 Ω when a 60-V 50-Hz source is connected across the coil. Find the current passing through the coil and the reading of a wattmeter across the coil.
   b. If a brass core is inserted in the coil, the effective resistance increases to 7 Ω, and the wattmeter reads 30 W. Find the current passing through the coil and the inductance of the coil.
   c. If a solid iron core is inserted in the coil the effective resistance of the coil increases to 10 Ω, and the current decreases to 1.7 A. Find the wattmeter reading and the inductance of the coil.

GLOSSARY

Apparent power The power delivered to a load without consideration of the effects of a power factor angle of the load. It is determined solely by the product of the terminal voltage and current of the load.

Average (real) power The delivered power that is dissipated in the form of heat by a network or system.

Eddy currents Small, circular currents that flow in a paramagnetic core, causing an increase in the power losses and the effective resistance of the material.

Effective resistance The resistance value that includes: the effects of radiation losses, skin effect, eddy currents, and hysteresis losses.

Hysteresis losses Losses in a magnetic material introduced by changes in the direction of the magnetic flux within the material.

Radiation losses The loss of energy in the form of electromagnetic waves during the transfer of energy from one element to another.

Reactive power The power associated with reactive elements that provides a measure of the energy associated with setting up the magnetic and electric fields of inductive and capacitive elements, respectively.

Skin effect At high frequencies, a counter-emf builds up at the center of a conductor resulting in an increased flow near the surface (skin) of the conductor and a great reduction near the center. As a result, resistance increases as determined by the basic equation for resistance in terms of the geometric shape of the conductor.

20.1 INTRODUCTION

The purpose of this chapter is to introduce the very important resonant (or tuned) circuit, which is fundamental to the operation of a wide variety of electrical and electronic systems in use today. The resonant circuit is a combination of R, L, and C elements having a frequency-response characteristic as shown in Fig. 20.1. Note in the figure that the response is a maximum for the frequency f_r, decreasing to the right and left of this frequency. In other words, the resonant circuit selects a range of frequencies for which the response will be near or equal to the maximum. The frequencies to the far left or right are, for all practical purposes, nullified with respect to their effect on the system's response. The radio or television receiver has a response curve of the type indicated in Fig. 20.1 for each broadcast station. When the receiver is set (or tuned) to a particular station, it is set on or near the frequency f_r of Fig. 20.1. Stations transmitting at frequencies to the far right or left of this resonant frequency are not carried through with significant power to affect the program of interest. The tuning process (setting the dial to f_r) as described above is the reason for the terminology *tuned circuit*. When the response is a maximum, the circuit is said to be in a state of *resonance,* with f_r as the *resonant frequency*.

The concept of resonance is not limited to electrical or electronic systems. If mechanical impulses are applied to a

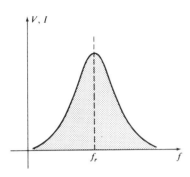

FIG. 20.1 *Resonance curve.*

mechanical system at the proper frequency, the system will enter a state of resonance in which sustained vibrations of very large amplitude will develop. The frequency at which this occurs is called the *natural frequency* of the system. The classic example of this effect was the Tacoma Narrows Bridge built in 1940 over Puget Sound. It had a suspended span of 2800 feet. Four months after the Tacoma Narrows Bridge was completed, a 42 mi/h pulsating gale set the bridge into oscillations at its natural frequency. The amplitude of the oscillations increased to the point where the main span broke up and fell into the water below. It has since been replaced by the new Tacoma Narrows Bridge, completed in 1950.

The resonant electrical circuit *must* have both inductance and capacitance. In addition, resistance will always be present due either to the lack of ideal elements or to the control offered on the shape of the resonance curve. When resonance occurs due to the application of the proper frequency (f_r), the energy absorbed at any instant by one reactive element is exactly equal to that released by another reactive element within the system. In other words, energy pulsates from one reactive element to the other. Therefore, once the system has reached a state of resonance, it requires no further reactive power since it is self-sustaining. The total apparent power is then simply equal to the average power dissipated by the resistive elements. The *average power absorbed by the system will also be a maximum at resonance,* just as the transfer of energy to the mechanical system above was a maximum at the natural frequency.

There are two types of resonant circuits: series and parallel. Each will be considered in some detail in this chapter.

SERIES RESONANCE

20.2 SERIES RESONANT CIRCUIT

The basic circuit configuration for the series resonant circuit appears in Fig. 20.2. The resistance R_l is the internal

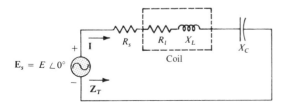

FIG. 20.2 *Series resonant circuit.*

resistance of the coil. The resistance R_s is the source resistance and any other resistance added in series to affect the shape of the resonance curve.

Defining $R = R_s + R_l$, the total impedance of this network is determined by

$$\mathbf{Z}_T = R + jX_L - jX_C = R + j(X_L - X_C)$$

Series resonance will occur when

$$\boxed{X_L = X_C} \tag{20.1}$$

The total impedance at resonance is then simply

$$\boxed{\mathbf{Z}_{T_s} = R} \tag{20.2}$$

since the reactive components will drop out of the equation for $\mathbf{Z}_T$ above. The subscript s will be employed to indicate series resonant conditions.

The resonant frequency can be determined in terms of the inductance and capacitance by examining the defining equation for resonance [Eq. (20.1)]:

$$X_L = X_C$$

Substituting yields

$$\omega L = \frac{1}{\omega C} \quad \text{and} \quad \omega^2 = \frac{1}{LC}$$

and

$$\boxed{\omega_s = \frac{1}{\sqrt{LC}}} \tag{20.3}$$

or

$$\boxed{f_s = \frac{1}{2\pi \sqrt{LC}}} \quad \begin{pmatrix} f \text{ in Hz} \\ L \text{ in H} \\ C \text{ in F} \end{pmatrix} \tag{20.4}$$

The current through the circuit at resonance is

$$\mathbf{I} = \frac{E \angle 0°}{R \angle 0°} = \frac{E}{R} \angle 0°$$

which you will note is the maximum current for the circuit of Fig. 20.2 for an applied voltage **E.** Consider also that *the input voltage and current are in phase at resonance.*

Since the current is the same through the capacitor and inductor, the voltage across each is equal in magnitude but 180° out of phase at resonance:

$$\left. \begin{aligned} \mathbf{V}_L &= \mathbf{IX}_L = (I \angle 0°)(X_L \angle 90°) = IX_L \angle 90° \\ \mathbf{V}_C &= \mathbf{IX}_C = (I \angle 0°)(X_C \angle -90°) = IX_C \angle -90° \end{aligned} \right\} \begin{aligned} &180° \\ &\text{out of} \\ &\text{phase} \end{aligned}$$

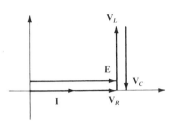

FIG. 20.3 *Phasor diagram for the series resonant circuit at resonance.*

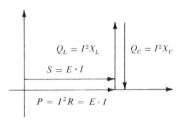

FIG. 20.4 *Power triangle for the series resonant circuit at resonance.*

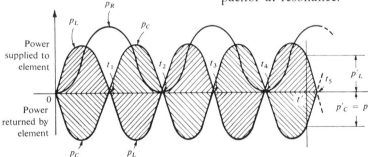

FIG. 20.5 *Power curves at resonance for the series resonant circuit.*

and, since $X_L = X_C$,

$$\boxed{V_{L_S} = V_{C_S}} \qquad (20.5)$$

Figure 20.3, a phasor diagram of the voltages and current, clearly indicates that the voltage across the resistor at resonance is the input voltage.

The average power to the resistor at resonance is equal to I^2R, and the reactive power to the capacitor and inductor are I^2X_C and I^2X_L, respectively.

The power triangle (Fig. 20.4) at resonance shows that the total apparent power is equal to the average power dissipated by the resistor, since $Q_L = Q_C$. The power factor of the circuit at resonance is

$$F_p = \cos \theta = \frac{P}{S} = 1$$

Plotting the power curves of each element on the same set of axes (Fig. 20.5), we note that even though the total reactive power at any instant is equal to zero ($t = t'$), energy is still being absorbed and released by the inductor and capacitor at resonance.

A closer examination reveals that the energy absorbed by the inductor from time 0 to t_1 is the same as the energy being released by the capacitor from 0 to t_1. The reverse occurs from t_1 to t_2, and so on. Therefore, the total apparent power continues to be equal to the average power, even though the inductor and capacitor are absorbing and releasing energy. This condition occurs only at resonance. The slightest change in frequency introduces a reactive component into the power triangle which will increase the apparent power of the system above the average power dissipation, and resonance will no longer exist.

20.3 THE QUALITY FACTOR (*Q*)

The *quality factor Q* of a series resonant circuit is defined as the ratio of the reactive power of either the inductor or the capacitor to the average power of the resistor at resonance; that is,

$$Q_s = \frac{\text{reactive power}}{\text{average power}} \qquad (20.6)$$

The quality factor is also an indication of how much energy is placed in storage (and transferred from one reactive element to the other) as compared to that dissipated. The lower the level of dissipation, the larger the Q_s factor and the more intense the region of resonance.

Substituting for an inductive reactance in Eq. (20.6) at resonance gives us

$$Q_s = \frac{I^2 X_L}{I^2 R}$$

and

$$Q_s = \frac{X_L}{R} = \frac{\omega_s L}{R} \qquad (20.7)$$

or, for the capacitive reactance,

$$Q_s = \frac{I^2 X_C}{I^2 R}$$

and

$$Q_s = \frac{X_C}{R} = \frac{1}{\omega_s C R} \qquad (20.8)$$

If the resistance R is just the resistance of the coil (R_l), we can speak of the Q of the coil, where

$$Q_{\text{coil}} = Q = \frac{X_L}{R_l} \qquad (R = R_l) \qquad (20.9)$$

Since the quality factor of a coil is typically the information provided by manufacturers of inductors, it is given the symbol Q without an associated subscript. It would appear from Eq. (20.9) that Q will increase linearly with frequency. That is, if the frequency doubles, then Q will also increase by a factor of two. This is approximately true for the low range to the mid-range of frequencies such as shown for the coil of Fig. 20.6. Unfortunately, however, as the frequency increases, the effective resistance of the coil will also increase and the resulting Q will decrease. In addition, the capacitive effects between the windings will increase, reducing the net inductance of the coil, further reducing the Q of the coil. For this reason, the Q of a coil must be specified at a particular frequency (usually at the maximum). For wide frequency applications, a plot of Q vs. frequency is often provided. The maximum Q for most commercially available coils approaches 100.

(a)

Courtesy of United Transformer Corp.

FIG. 20.6 *Q vs. frequency for a TRW/UTC 10-mH coil.*

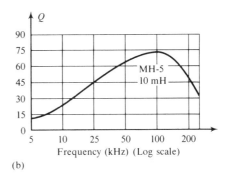

(b)

If we substitute into Eq. (20.7) the fact that

$$\omega_s = 2\pi f_s \quad \text{and} \quad f_s = \frac{1}{2\pi \sqrt{LC}}$$

then

$$Q_s = \frac{\omega_s L}{R} = \frac{2\pi f_s L}{R} = \frac{2\pi}{R}\left(\frac{1}{2\pi \sqrt{LC}}\right) L$$

$$Q_s = \frac{L}{R}\left(\frac{1}{\sqrt{LC}}\right) = \left(\frac{\sqrt{L}}{\sqrt{L}}\right)\frac{L}{R\sqrt{LC}}$$

and

$$\boxed{Q_s = \frac{1}{R}\sqrt{\frac{L}{C}}} \qquad (20.10)$$

For series resonant circuits that are used in communication systems, Q_s is usually greater than one. By applying the voltage divider rule to the circuit of Fig. 20.2, we obtain

$$V_L = \frac{X_L E}{Z_T} = \frac{X_L E}{R} \qquad \text{(at resonance)}$$

and

$$\boxed{V_{L_s} = Q_s E} \qquad (20.11)$$

or

$$V_C = \frac{X_C E}{Z_T} = \frac{X_C E}{R}$$

and

$$\boxed{V_{C_s} = Q_s E} \qquad (20.12)$$

Since Q_s is usually greater than one, the voltage across the capacitor or inductor of a series resonant circuit is usually greater than the input voltage. In fact, in many cases, the Q_s is so high that careful design and handling (including adequate insulation) are mandatory with respect to the voltage across the capacitor and inductor.

In the circuit of Fig. 20.7, for example, which is in the state of resonance,

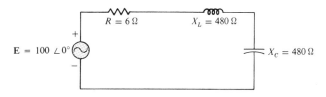

$$Q_s = \frac{X_L}{R} = \frac{480}{6} = 80$$

FIG. 20.7 *High Q series resonant circuit.*

and

$$V_L = V_C = Q_s E = (80)(100) = 8000 \text{ V}$$

which is certainly a potential to be handled with great care.

20.4 Z_T VS. FREQUENCY

The total impedance of the series *R-L-C* circuit of Fig. 20.2 at any frequency is determined by

$$\mathbf{Z}_T = R + jX_L - jX_C \quad \text{or} \quad \mathbf{Z}_T = R + j(X_L - X_C)$$

The magnitude of the impedance $\mathbf{Z}_T$ is

$$Z_T = \sqrt{R^2 + (X_L - X_C)^2}$$

The total-impedance-vs.-frequency curve for the series resonant circuit of Fig. 20.2 can be found by applying the impedance-vs.-frequency-curve for each element of the equation just derived, written in the following form:

$$\boxed{Z_T(f) = \sqrt{[R(f)]^2 + [X_L(f) - X_C(f)]^2}} \quad \textbf{(20.13)}$$

where $Z_T(f)$ "means" the total impedance as a *function* of frequency. Ideally, the resistance R does not change with frequency, so its curve is a straight horizontal line with a magnitude R above the frequency axis (Fig. 20.8). The curve for the inductance, as determined by the reactance equation, is a straight line intersecting the origin with a slope equal to the inductance of the coil. The mathematical expression for any straight line in a two-dimensional plane is given by

$$y = mx + b$$

Thus, for the coil,

$$X_L = 2\pi f L + 0 = \underbrace{(2\pi L)}_{m}\underbrace{(f)}_{x} + \underbrace{0}_{b}$$
$$\underset{y \;=}{\downarrow} \qquad\qquad \downarrow \quad \downarrow \quad \downarrow$$

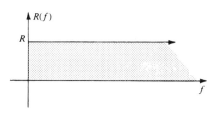

FIG. 20.8 *Resistance vs. frequency.*

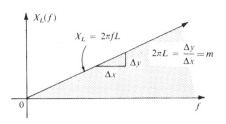

FIG. 20.9 *Inductive reactance vs. frequency.*

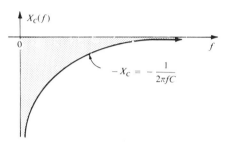

FIG. 20.10 *Capacitive reactance vs. frequency.*

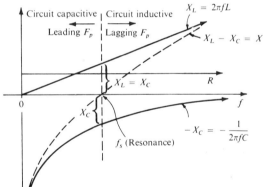

FIG. 20.11

(where $2\pi L$ is the slope), producing the results shown in Fig. 20.9.

For the capacitor,

$$X_C = \frac{1}{2\pi fC} \quad \text{or} \quad X_C f = \frac{1}{2\pi C}$$

which becomes $yx = k$, the equation for a hyperbola, where

$$y \text{ (variable)} = X_C$$
$$x \text{ (variable)} = f$$
$$k \text{ (constant)} = \frac{1}{2\pi C}$$

Since the equation for the total impedance Z_T includes a term $-X_C(f)$, the curve plotted in Fig. 20.10 is that of $-1/2\pi fC$. Plotting all three quantities on the same set of axes gives the results shown in Fig. 20.11. Note first that resonance occurs at f where $X_L = X_C$. Since the capacitive reactance of the circuit is greater to the left of f than the inductive reactance, the circuit is predominantly *capacitive* to the left of f_s. To the right, the opposite is true, and the circuit is predominantly *inductive*. This also means that a leading power factor exists to the left of f_s, and a lagging power factor to the right. (Note the curve of Fig. 20.13.)

Applying

$$Z_T(f) = \sqrt{[R(f)]^2 + [X_L(f) - X_C(f)]^2}$$
$$= \sqrt{[R(f)]^2 + [X(f)]^2}$$

to the curves of Fig. 20.11 where $X(f)$ is as shown, we obtain the curve for $Z_T(f)$ as shown in Fig. 20.12. The minimum impedance occurs at the resonant frequency and is equal to the resistance R. Note that the curve is not symmetrical about the resonant frequency (especially at higher values of Z_T).

At very low frequencies, the circuit is almost purely capacitive, and the current leads the applied voltage by 90°. At very high frequencies, the circuit is almost purely induc-

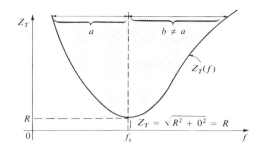

FIG. 20.12 Z_T vs. frequency for the series resonant circuit.

tive, and the current lags the voltage by 90°. The applied voltage and resulting current are in phase only at resonance as indicated by the provided phase plot of Fig. 20.13.

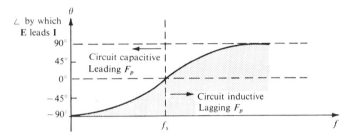

FIG. 20.13 Phase plot for the series resonant circuit.

20.5 SELECTIVITY

If we now plot the magnitude of the current $I = E/Z_T$ vs. frequency for a *fixed* applied voltage E, we obtain the curve shown in Fig. 20.14, which rises from zero to a maximum value of E/R (where Z_T is a minimum) and then drops toward zero (as Z_T increases) at a slower rate than it rose to its peak value. The curve is actually the inverse of the impedance-vs.-frequency curve. Since the Z_T curve is not absolutely symmetrical about the resonant frequency, the curve of the current vs. frequency has the same property.

There is a definite range of frequencies at which the current is near its maximum value and the impedance is at a minimum. Those frequencies corresponding to 0.707 of the

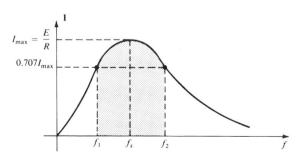

FIG. 20.14 I vs. f for the series resonant circuit.

maximum current are called the *band frequencies, cutoff frequencies,* or *half-power frequencies.* They are indicated by f_1 and f_2 in Fig. 20.14. The range of frequencies between the two is referred to as the *bandwidth* (abbreviated BW) of the resonant circuit.

Half-power frequencies are those frequencies at which the power delivered is one-half that delivered at the resonant frequency; that is,

$$\boxed{P_{\text{HPF}} = \frac{1}{2}P_{\text{max}}} \qquad (20.14)$$

The above condition is derived

$$P_{\text{max}} = I_{\text{max}}^2 R$$

and

$$P_{\text{HPF}} = I^2 R = (0.707 I_{\text{max}})^2 R$$
$$= 0.5 I_{\text{max}}^2 R = \frac{1}{2}P_{\text{max}}$$

Since the resonant circuit is adjusted to select a band of frequencies, the curve of Fig. 20.14 is called the *selectivity curve.* The term is derived from the fact that one must be selective in choosing the frequency to insure that it is in the bandwidth. The smaller the bandwidth, the higher the selectivity. The shape of the curve, as shown in Fig. 20.15, depends on each element of the series *R-L-C* circuit. If the resistance is made smaller with a fixed inductance and capacitance, the bandwidth decreases and the selectivity increases. Similarly, if the ratio L/C increases with fixed resistance, the bandwidth again decreases with an increase in selectivity.

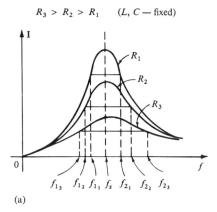

(a)

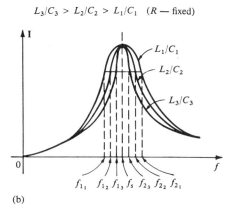

(b)

FIG. 20.15 *Effect of R, L, and C on the selectivity curve for the series resonant circuit.*

In terms of Q_s, if R is larger for the same X_L, then Q_s is less, as determined by the equation $Q_s = \omega_s L/R$. *A small Q_s, therefore, is associated with a resonant curve with a large*

*oandwidth and a small selectivity, while a large Q_s indicates
the opposite.*

For circuits where $Q_s \geq 10$, a widely accepted approximation is that the resonant frequency bisects the bandwidth and
that the resonant curve is symmetrical about the resonant
frequency. These conditions are shown in Fig. 20.16, indicating that the cutoff frequencies are then equidistant from
the resonant frequency.

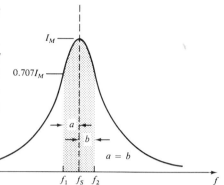

FIG. 20.16 *Approximate series
resonance curve for $Q_s \geq 10$.*

For any Q_s, the preceding is not true. The cutoff frequencies f_1 and f_2 can be found for the general case (any Q_s) by
first employing the fact that a drop in current to 0.707 of its
resonant value corresponds to an increase in impedance
equal to $1/0.707 = \sqrt{2}$ times the resonant value, which
is R.

Substituting $\sqrt{2}R$ into the equation for the magnitude of
$\mathbf{Z}_T$, we find that

$$\mathbf{Z}_T = \sqrt{R^2 + (X_L - X_C)^2}$$

becomes

$$\sqrt{2}R = \sqrt{R^2 + (X_L - X_C)^2}$$

or, squaring both sides, that

$$2R^2 = R^2 + (X_L - X_C)^2$$

and

$$R^2 = (X_L - X_C)^2$$

Taking the square root of both sides gives us

$$R = X_L - X_C$$

Let us first consider the case where $X_L > X_C$ which re-
lates to f_2 or ω_2. Substituting $\omega_2 L$ for X_L and $1/\omega_2 C$ for X_C
and bringing both quantities to the left of the equal sign,
we have

$$R - \omega_2 L + \frac{1}{\omega_2 C} = 0 \quad \text{or} \quad R\omega_2 - \omega_2^2 L + \frac{1}{C} = 0$$

which can be written

$$\omega_2^2 - \frac{R}{L}\omega_2 - \frac{1}{LC} = 0$$

Solving the quadratic, we have

$$\omega_2 = \frac{-(-R/L) \pm \sqrt{[-(R/L)]^2 - [-(4/LC)]}}{2}$$

and

$$\omega_2 = + \frac{R}{2L} \pm \frac{1}{2} \sqrt{\frac{R^2}{L^2} + \frac{4}{LC}}$$

so, finally,

$$\boxed{\omega_2 = \frac{R}{2L} + \frac{1}{2} \sqrt{\left(\frac{R}{L}\right)^2 + \frac{4}{LC}}} \qquad \text{(rad/s)} \quad \textbf{(20.15)}$$

The negative sign in front of the second factor was dropped because $(1/2)\sqrt{R^2/L^2 + 4/LC}$ is always greater than $R/(2L)$. If it were not dropped, there would be a negative solution for the radian frequency ω. Since $\omega_2 = 2\pi f_2$,

$$\boxed{f_2 = \frac{1}{2\pi} \left[\frac{R}{2L} + \frac{1}{2} \sqrt{\left(\frac{R}{L}\right)^2 + \frac{4}{LC}} \right]} \qquad \text{(Hz)} \quad \textbf{(20.16)}$$

If we repeat the same procedure for $X_C > X_L$ which relates to ω_1 or f_1 such that $Z_T = \sqrt{R^2 + (X_C - X_L)^2}$, the solution for $\omega = \omega_1$ becomes

$$\boxed{\omega_1 = -\frac{R}{2L} + \frac{1}{2} \sqrt{\left(\frac{R}{L}\right)^2 + \frac{4}{LC}}} \qquad \text{(rad/s)} \quad \textbf{(20.17)}$$

or, since $\omega_1 = 2\pi f_1$, it can be written

$$\boxed{f_1 = \frac{1}{2\pi} \left[-\frac{R}{2L} + \frac{1}{2} \sqrt{\left(\frac{R}{L}\right)^2 + \frac{4}{LC}} \right]} \qquad \text{(Hz)}$$

$$\textbf{(20.18)}$$

The bandwidth (BW) is

$$\text{BW} = f_2 - f_1$$

$$\text{BW} = \left[\frac{R}{4\pi L} + \frac{1}{4\pi} \sqrt{\left(\frac{R}{L}\right)^2 + \frac{4}{LC}} \right]$$
$$- \left[-\frac{R}{4\pi L} + \frac{1}{4\pi} \sqrt{\left(\frac{R}{L}\right)^2 + \frac{4}{LC}} \right]$$

and

$$\boxed{\text{BW} = f_2 - f_1 = \frac{R}{2\pi L}} \qquad \textbf{(20.19)}$$

Substituting $R/L = \omega_s/Q_s$ from $Q_s = \omega_s L/R$ and $1/2\pi = f_s/\omega_s$ from $\omega_s = 2\pi f_s$ gives us

$$\text{BW} = \frac{R}{2\pi L} = \left(\frac{f_s}{\omega_s}\right)\left(\frac{\omega_s}{Q_s}\right)$$

or

$$\boxed{\text{BW} = \frac{f_s}{Q_s}} \qquad \textbf{(20.20)}$$

which is a very convenient form, since it relates the bandwidth to the Q_s of the circuit. As mentioned earlier, Eq. (20.20) verifies that the larger the Q_s, the smaller the bandwidth, and vice versa.

Written in a slightly different form, Eq. (20.20) becomes

$$\boxed{\frac{f_2 - f_1}{f_s} = \frac{1}{Q_s}} \qquad \textbf{(20.21)}$$

The ratio $(f_2 - f_1)/f_s$ is sometimes called the *fractional bandwidth*.

20.6 V_R, V_L, AND V_C

Plotting the magnitude (effective value) of the voltages V_R, V_L, and V_C and the current $\mathbf{I}$ versus frequency for the series resonant circuit on the same set of axes, we obtain the curves shown in Fig. 20.17. Note that the V_R curve has the same shape as the I curve and a peak value equal to the magnitude of the input voltage E. The V_C curve builds up slowly at first from a value equal to the input voltage, since the reactance of the capacitor is infinite (open circuit) at zero frequency and the reactance of the inductor is zero (short circuit) at this frequency. As the frequency increases, $1/(\omega C)$ of the equation

$$V_C = IX_C = I\frac{1}{\omega C}$$

becomes smaller, but I increases at a rate faster than that at which $1/(\omega C)$ drops. Therefore, V_C rises and will continue to rise, due to the quickly rising current, until the frequency nears resonance. As it approaches the resonant condition, the rate of change of I decreases. When this occurs, the factor $1/(\omega C)$, which decreased as the frequency rose, will overcome the rate of change of I, and V_C will start to drop. The peak value will occur at a frequency just before resonance. After resonance, both V_C and I drop in magnitude, and V_C approaches zero.

The higher the Q_s of the circuit, the closer $f_{C_{max}}$ will be to f_s, and the closer $V_{C_{max}}$ will be to $Q_s E$. For circuits with $Q_s \geq 10$, $f_{C_{max}} \cong f_s$ and $V_{C_{max}} \cong Q_s E$.

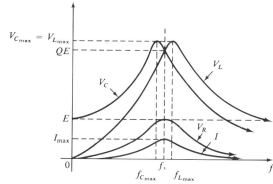

FIG. 20.17 V_R, V_L, V_C, and I vs. *frequency for a series resonant circuit.*

The curve for V_L increases steadily from zero to the resonant frequency, since both quantities ωL and I of the equation $V_L = IX_L = \omega LI$ increase over this frequency range. At resonance, I has reached its maximum value, but ωL is still rising. Therefore, V_L will reach its maximum value after resonance. After reaching its peak value, the voltage V_L will drop toward E, since the drop in I will overcome the rise in ωL. It approaches E because X_L will eventually be infinite, and X_C will be zero.

As Q_s of the circuit increases, the frequency $f_{L_{max}}$ drops toward f_s, and $V_{L_{max}}$ approaches $Q_s E$. For circuits with $Q_s \geq 10$, $f_{L_{max}} \cong f_s$, and $V_{L_{max}} \cong Q_s E$.

The V_L curve has a greater magnitude than the V_C curve for any frequency above resonance, and the V_C curve has a greater magnitude than the V_L curve for any frequency below resonance. This again verifies that the series R-L-C circuit is predominantly capacitive from zero to the resonant frequency and predominantly inductive for any frequency above resonance.

For the condition $Q_s \geq 10$, the curves of Fig. 20.17 will appear as shown in Fig. 20.18. Note that they each peak (on an approximate basis) at the resonant frequency and have a similar shape.

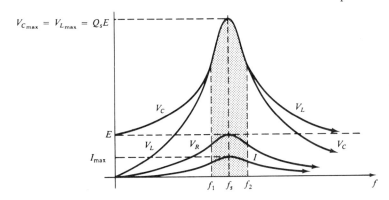

FIG. 20.18 *Approximate curves for V_R, V_L, V_C, and I for a series resonant circuit where $Q_s \geq 10$.*

20.7 EXAMPLES (SERIES RESONANCE)

EXAMPLE 20.1. For the resonant circuit shown in Fig. 20.19, find i, v_R, v_L, and v_C in phasor form. What is the Q_s of the circuit? If the resonant frequency is 5000 Hz, find the

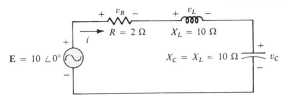

FIG. 20.19

bandwidth. What is the power dissipated in the circuit at the half-power frequencies?

Solution:

$$\mathbf{Z}_{T_s} = R = 2\,\Omega$$

$$\mathbf{I} = \frac{\mathbf{E}}{\mathbf{Z}_{T_s}} = \frac{10\,\angle 0°}{2\,\angle 0°} = \mathbf{5\,\angle 0°}$$

$$\mathbf{V}_R = \mathbf{E} = 10\,\angle 0°$$

$$\mathbf{V}_L = \mathbf{IX}_L = (5\,\angle 0°)(10\,\angle 90°) = \mathbf{50\,\angle 90°}$$

$$\mathbf{V}_C = \mathbf{IX}_C = (5\,\angle 0°)(10\,\angle -90°) = \mathbf{50\,\angle -90°}$$

$$Q_s = \frac{X_L}{R} = \frac{10}{2} = \mathbf{5}$$

$$\mathrm{BW} = f_2 - f_1 = \frac{f_s}{Q_s} = \frac{5000}{5} = \mathbf{1000\ Hz}$$

$$P_{\mathrm{HPF}} = \frac{1}{2}P_{\max} = \frac{1}{2}I^2_{\max}R = \left(\frac{1}{2}\right)(5)^2(2) = \mathbf{25.0\ W}$$

EXAMPLE 20.2. The bandwidth of a series resonant circuit is 400 Hz. If the resonant frequency is 4000 Hz, what is the value of Q_s? If $R = 10\,\Omega$, what is the value of X_L at resonance? Find the inductance L and capacitance C of the circuit.

Solution:

$$\mathrm{BW} \doteq \frac{f_s}{Q_s} \quad \text{or} \quad Q_s = \frac{f_s}{\mathrm{BW}} = \frac{4000}{400} = \mathbf{10}$$

$$Q_s = \frac{X_L}{R} \quad \text{or} \quad X_L = Q_s R = (10)(10) = \mathbf{100\,\Omega}$$

$$X_L = 2\pi f_s L \quad \text{or} \quad L = \frac{X_L}{2\pi f_s} = \frac{100}{(6.28)(4000)} = \mathbf{3.98\ mH}$$

$$X_C = \frac{1}{2\pi f_s C}$$

or

$$C = \frac{1}{2\pi f_s X_C} = \frac{1}{(6.28)(4000)(100)} = \mathbf{0.398\ \mu F}$$

EXAMPLE 20.3. A series R-L-C circuit has a series resonant frequency of 12,000 Hz. If $R = 5\,\Omega$ and X_L at resonance is 300 Ω, find the bandwidth. Find the cutoff frequencies.

Solution:

$$Q_s = \frac{X_L}{R} = \frac{300}{5} = \mathbf{60}$$

$$\mathrm{BW} = \frac{f_s}{Q_s} = \frac{12,000}{60} = \mathbf{200\ Hz}$$

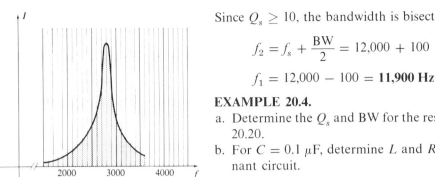

FIG. 20.20

Since $Q_s \geq 10$, the bandwidth is bisected by f_s. Therefore,

$$f_2 = f_s + \frac{BW}{2} = 12{,}000 + 100 = \textbf{12,100 Hz}$$

$$f_1 = 12{,}000 - 100 = \textbf{11,900 Hz}$$

EXAMPLE 20.4.
a. Determine the Q_s and BW for the response curve of Fig. 20.20.
b. For $C = 0.1 \ \mu F$, determine L and R for the series resonant circuit.

Solutions:
a. The resonant frequency is 2800 Hz. At 0.707 times the peak value, the BW = **200 Hz,** and

$$Q_s = \frac{f_s}{BW} = \frac{2800}{200} = \textbf{14}$$

b. $f_s = \dfrac{1}{2\pi \sqrt{LC}}$

or

$$\begin{aligned}
L &= \frac{1}{4\pi^2 f_s^2 C} \\
&= \frac{1}{4(\pi)^2 (2.8 \times 10^3)^2 (0.1 \times 10^{-6})} \\
&= \frac{1}{30.951} = \textbf{32.31 mH}
\end{aligned}$$

$$Q_s = \frac{X_L}{R} \quad \text{or} \quad R = \frac{X_L}{Q_s} = \frac{(17.58 \times 10^3)(32.31 \times 10^{-3})}{14}$$

$$= \textbf{40.572} \ \Omega$$

20.8 REACTANCE CHART

The reactance chart appearing in Fig. 20.21 is very useful in the analysis and design of series or parallel resonant circuits. The left vertical scale is the reactance in ohms of the inductor or capacitor having the values indicated on the other axis. The frequency applied, or to be determined, appears on the bottom horizontal scale. Its use is best demonstrated through a few examples.

EXAMPLE 20.5. Using the reactance chart, repeat Example 20.2.

Solution: As before, $Q_s = 10$, and $X_L = 100 \ \Omega$. If we find the intersection of $100 \ \Omega$ on the left vertical axis and 4000 Hz on the bottom scale and follow the line going to the top right corner, we find an inductance level of 0.004 or 4 mH, as determined before. These operations appear on Fig.

20.21. For the capacitive value, we progress to the bottom right corner and read $0.4\,\mu F$, as obtained before.

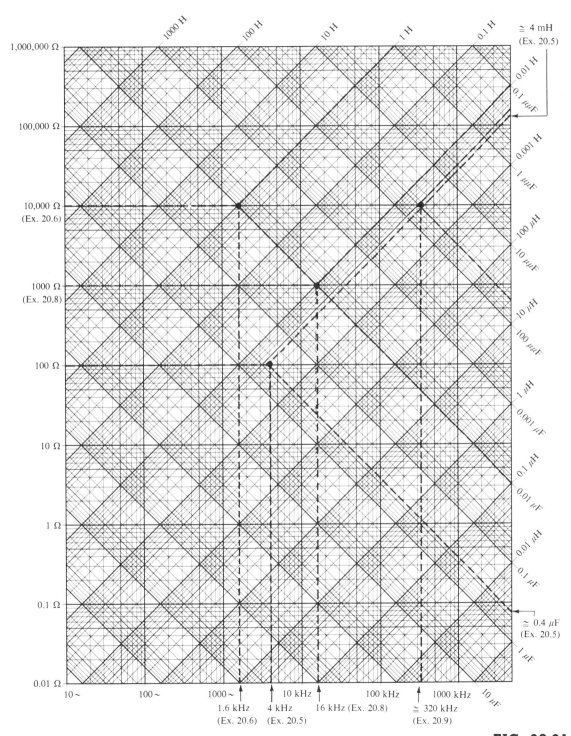

FIG. 20.21

EXAMPLE 20.6. Given $L = 1$ H and $C = 0.01$ μF, determine the series resonant frequency and the inductive and capacitive reactances.

Solution: The value 1 H appears on the top horizontal scale, and 0.01 μF to the bottom right. Their intersection is shown in Fig. 20.21. If we drop a vertical line down to the frequency axis, we find a resonant frequency of approximately 1600 Hz. The accuracy of this reading is certainly a limitation of using the chart. The reactance of each element is determined by drawing a horizontal line to the left-hand axis to find $X_L = 10{,}000$ Ω.

PARALLEL RESONANCE

20.9 PARALLEL RESONANT CIRCUIT

The parallel resonant circuit has the basic configuration of Fig. 20.22. This circuit is often called the *tank circuit* due to the storage of energy by the inductor and capacitor. A transfer of energy similar to that discussed for the series circuit also occurs in the parallel resonant circuit. In the ideal case (no radiation losses, and so on), the capacitor absorbs energy during one half-cycle of the power curves at the same rate at which it is released by the inductor. During the next half-cycle of the power curves, the inductor absorbs energy at the same rate at which the capacitor releases it. The total reactive power at resonance is therefore zero, and the total power factor is one.

Since tank circuits are frequently used with devices such as the transistor, which is essentially a constant-current source device, a current source will be used to supply the input to the parallel resonant circuits in the following analysis, as shown in Fig. 20.22. In addition, our analysis of parallel resonant circuits will be well served if we first replace the series *R-L* branch by an equivalent parallel combination as shown in Fig. 20.23.

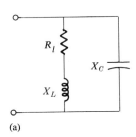

(a)

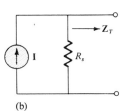

(b)

FIG. 20.22

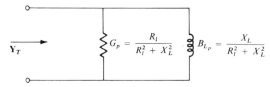

FIG. 20.23

In Section 15.8, we found that every series combination of a resistor and a reactive element has a parallel circuit equivalent. For the *R-L* branch in Fig. 20.22, the parallel circuit equivalent (Fig. 20.23) is derived below, following the procedure of Section 15.8.

$$\mathbf{Z}_{R\text{-}L} = R_l + jX_L \qquad \mathbf{Y}_{R\text{-}L} = \frac{1}{Z_{R\text{-}L}} = \frac{1}{R_l + jX_L}$$

$$\mathbf{Y}_{R\text{-}L} = \frac{R_l}{R_l^2 + X_L^2} - j\frac{X_L}{R_l^2 + X_L^2}$$

Replacing the series $R\text{-}L$ combination of Fig. 20.22 by the parallel combination just derived, we obtain the configuration shown in Fig. 20.24 with the current source applied.

For resonance,

$$X_{L_p} = X_C$$

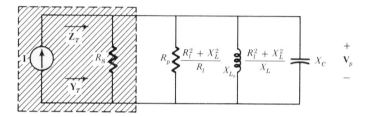

Source

FIG. 20.24

Substituting from above gives us

$$\frac{R_l^2 + X_L^2}{X_L} = X_C$$

Thus,

$$X_C X_L = R_l^2 + X_L^2 \quad \text{or} \quad X_L^2 = X_C X_L - R_l^2$$

and, since

$$X_C X_L = \frac{1}{\omega C}\omega L = \frac{L}{C}$$

then

$$X_L^2 = \frac{L}{C} - R_l^2 \quad \text{or} \quad X_L = \sqrt{\frac{L}{C} - R_l^2}$$

Therefore,

$$f_p = \frac{1}{2\pi L}\sqrt{\frac{L}{C} - R_l^2} = \frac{1}{2\pi L}\sqrt{\frac{1 - R_l^2(C/L)}{C/L}}$$

$$= \frac{1}{2\pi L\sqrt{C/L}}\sqrt{1 - \frac{R_l^2 C}{L}}$$

and

$$\boxed{f_p = \frac{1}{2\pi\sqrt{LC}}\sqrt{1 - \frac{R_l^2 C}{L}}} \qquad \textbf{(20.22)}$$

or

$$\boxed{f_p = f_s\sqrt{1 - R_l^2\frac{C}{L}}} \qquad \textbf{(20.23)}$$

where f_p is the resonant frequency of a parallel resonant circuit and f_s is the resonant frequency of a series resonant circuit of the same reactive elements. Note that unlike a series resonant circuit, the resonant frequency of a parallel resonant circuit is dependent on the resistance R_l. Note also, however, the absence of the source resistance R_s in Eqs. (20.22) and (20.23).

20.10 SELECTIVITY CURVE FOR PARALLEL RESONANT CIRCUITS

The magnitude-vs.-frequency curve of the impedance $\mathbf{Z}_T$ in Fig. 20.24 is similar to the inverse of that obtained for the series resonant circuit as shown in Fig. 20.25. $\mathbf{Z}_T$ is the total impedance of the parallel resonant circuit as defined by Fig. 20.24.

The value of Z_{T_p} (the total impedance at resonance) can be found by summing the admittances of the parallel branches and applying the resonance condition $X_{L_p} = X_C$. That is, for the capacitor, the total admittance is

$$\mathbf{Y}_C = \mathbf{B}_C = \frac{1}{X_C} \angle 90° = 0 + j\frac{1}{X_C}$$

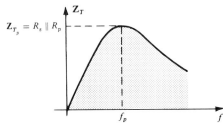

FIG. 20.25 Z_T *vs.* f *for the parallel resonant circuit.*

and for the inductive branch, it is

$$\mathbf{Y}_{L_p} = \mathbf{B}_{L_p} = \frac{1}{X_{L_p}} \angle -90° = 0 - j\frac{1}{X_{L_p}}$$

The total admittance is

$$\mathbf{Y}_T = \mathbf{G}_s + \mathbf{G}_p + \mathbf{B}_{L_p} + \mathbf{B}_C$$

$$\mathbf{Y}_T = \frac{1}{R_s} + \frac{1}{R_p} - j\frac{1}{X_{L_p}} + j\frac{1}{X_C}$$

At resonance, $X_{L_p} = X_C$, and

$$\mathbf{Y}_{T_p} = \frac{1}{R_s} + \frac{1}{R_p}$$

or

$$\mathbf{Z}_{T_p} = \frac{1}{\mathbf{Y}_{T_p}} = \boxed{R_s \parallel R_p} \qquad \textbf{(20.24)}$$

as noted by Fig. 20.24 when the reactive elements are removed.

The impedance at resonance is therefore resistive and a maximum value. It can be determined in terms of the reactive elements if we use the fact that at resonance

$$\frac{X_L}{R_l^2 + X_L^2} = \frac{1}{X_C}$$

or

$$R_l^2 + X_L^2 = X_L X_C$$

Substituting into

$$R_p = \frac{R_l^2 + X_L^2}{R_l}$$

we have

$$R_p = \frac{X_L X_C}{R_l}$$

or

$$R_p = \frac{(\omega L)(1/\omega C)}{R_l}$$

and

$$R_p = \frac{L}{R_l C}$$

and

$$\boxed{Z_{T_p} = R_s \parallel R_p = R_s \parallel \frac{L}{R_l C}} \qquad \text{(20.25)}$$

Since the current I is constant (current source) for any value of Z_T, the voltage across the parallel circuit will have the same shape as the total impedance Z_T, as shown in Fig. 20.26.

For the parallel circuit under discussion, the selectivity curve is usually considered to be that of the voltage V_C across the capacitor. The reason for this interest in V_C is derived from electronic considerations not discussed in this text that often place the capacitor at the input to another stage of a network.

Since the voltage across parallel elements is the same, then

$$\boxed{\mathbf{V}_C = \mathbf{V}_p = \mathbf{I}\mathbf{Z}_T} \qquad \text{(20.26)}$$

The resonant value of $\mathbf{V}_C$ is therefore determined by the value of Z_{T_p} and the magnitude of the current source $\mathbf{I}$.

For series resonance, we defined $Q = X_L/R_l$. The Q of a coil can have a pronounced effect on the parallel resonance equations. Continuing with the notation Q for this ratio, let us reexamine the parallel equivalent resistance R_p and inductance X_{L_p}.

R_p:

$$R_p = \frac{R_l^2 + X_L^2}{R_l} = R_l + \frac{X_L^2 \; (R_l)}{R_l \; (R_l)}$$

$$\boxed{R_p = R_l + Q^2 R_l} \qquad \text{(20.27)}$$

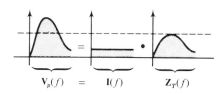

$$V_p(f) \quad = \quad \mathbf{I}(f) \quad \quad Z_T(f)$$

FIG. 20.26

X_{L_p}:

$$X_{L_p} = \frac{R_l^2 + X_L^2}{X_L} = \frac{R_l^2(X_L)}{X_L(X_L)} + X_L$$

$$\boxed{X_{L_p} = \frac{X_L}{Q^2} + X_L} \qquad \textbf{(20.28)}$$

For $Q \geq 10$,

$$Q^2R \gg R \quad \text{and} \quad X_L \gg \frac{X_L}{Q^2}$$

leaving

$$\boxed{R_p \cong Q^2 R_l} \qquad \textbf{(20.29)}$$

and

$$\boxed{X_{L_p} \cong X_L} \qquad \textbf{(20.30)}$$

resulting in the neater and quite useful approximate equivalent circuit of Fig. 20.27.

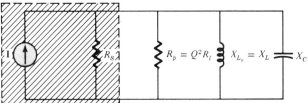

Source

FIG. 20.27 *Approximate equivalent circuit for $Q \geq 10$.*

From Fig. 20.27, it is now quite obvious that when $Q \geq 10$, resonance is defined by

$$\boxed{X_L = X_C} \qquad (Q \geq 10) \qquad \textbf{(20.31)}$$

and recalling a similar circumstance for series resonant circuits, we know that

$$\boxed{f_p = \frac{1}{2\pi\sqrt{LC}}} \qquad (Q \geq 10) \qquad \textbf{(20.32)}$$

The quality factor of the parallel resonant circuit is determined by the ratio of the reactive power to the real power. That is,

$$Q_p = \frac{V_p^2/X_{L_p}}{V_p^2/R}$$

where $R = R_s \parallel R_p$, and V_p is the voltage across the parallel branches. The result is

$$\boxed{Q_p = \frac{R}{X_{L_p}}} \qquad \textbf{(20.33)}$$

or, since $X_{L_p} = X_C$ at resonance,

$$\boxed{Q_p = \frac{R}{X_C}} \qquad \textbf{(20.34)}$$

Note the use of the subscript p to denote the quality factor of a parallel resonant circuit. For $Q \geq 10$,

$$Q_p = \frac{R}{X_L} = \frac{R}{\omega_p L} \qquad \text{(20.35)}$$

and

$$Q_p = \frac{R}{X_C} = \omega_p CR \qquad \text{(20.36)}$$

For a situation where R_s is considered to be sufficiently large and can be ignored,

$$R = R_s \parallel R_p = R_p$$

and

$$Q_p = \frac{R_p}{X_{L_p}} = \frac{(R_l^2 + X_L^2)/R_l}{(R_l^2 + X_L^2)/X_L}$$

so

$$Q_p = \frac{X_L}{R_l} = Q \qquad (R_s = \infty\ \Omega) \qquad \text{(20.37)}$$

The bandwidth of the parallel resonant is related to the resonant frequency and Q_p in the same manner as for series resonant circuits. That is,

$$\text{BW} = f_2 - f_1 = \frac{f_p}{Q_p} \qquad \text{(20.38)}$$

The effect of R_l, L, and C on the shape of the parallel resonance curve, as shown in Fig. 20.28 for the input impedance, is quite similar to their effect on the series resonance curve. Whether or not R_l is zero, the parallel resonant circuit will frequently appear in a network schematic as shown in Fig. 20.28.

At resonance, since $Z_{T_p} = L/R_l C$, an increase in R_l or a decrease in the ratio L/C will result in a decrease in the

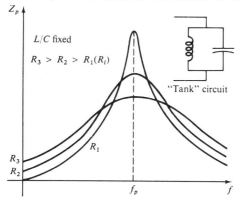

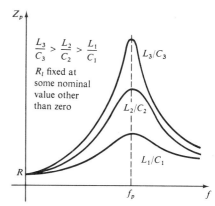

FIG. 20.28 *Effect of L, C, and R_l on the parallel resonance curve.*

resonant impedance, with a corresponding increase in the current. The bandwidth of the resonance curves is given by Eq. (20.34). For increasing R_l or decreasing L (or L/C for constant C), the bandwidth will increase as shown in Fig. 20.28.

At low frequencies, the capacitive reactance is quite high, and the inductive reactance is low. Since the elements are in parallel, the total impedance at low frequencies will therefore be inductive. At high frequencies, the reverse is true, and the network is capacitive. At resonance, the network appears resistive. These facts lead to the phase plot of Fig. 20.29. Note that it is the inverse of that appearing for the series resonant circuit in that at low frequencies the series resonant circuit was capacitive and at high frequencies inductive.

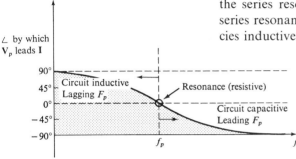

FIG. 20.29 *Phase plot for the parallel resonant circuit.*

Equations (20.31) and (20.32) were derived under the condition that $Q \geq 10$. If we examine the total impedance at resonance, we find

$$Z_{T_p} = R_s \parallel R_p$$

or

$$Z_{T_p} = R_s \parallel \frac{R_l^2 + X_L^2}{R_l} = R_s \parallel (R_l + Q^2 R_l)$$

For $Q \geq 10$,

$$\boxed{Z_{T_p} = R_s \parallel Q^2 R_l} \qquad (Q \geq 10) \qquad \textbf{(20.39)}$$

as defined by Fig. 20.27.

The resonant frequency of a parallel resonant circuit can be found in terms of $Q = X_L/R_l$ if the following manipulations are made on the factor within the square root:

$$1 - \frac{R_l^2 C}{L} = 1 - \frac{R_l^2 \omega C}{\omega L} = 1 - \frac{R_l^2}{X_L X_C} = 1 - \frac{R_l^2}{R_l^2 + X_L^2}$$

$$= \frac{R_l^2 + X_L^2 - R_l^2}{R_l^2 + X_L^2} = \frac{X_L^2}{R_l^2 + X_L^2}$$

However,

$$X_L = QR_l \quad \text{and} \quad \frac{X_L^2}{R_l^2 + X_L^2} = \frac{Q^2 R_l^2}{R_l^2 + Q^2 R_l^2} = \frac{Q^2}{1 + Q^2}$$

so Eq. (20.22) becomes

$$\boxed{f_p = \frac{1}{2\pi\sqrt{LC}}\sqrt{\frac{Q^2}{1+Q^2}}} \qquad \text{(for any } Q\text{)} \quad \textbf{(20.40)}$$

For $Q \geq 10$, $1 + Q^2 \cong Q^2$ and $Q^2/(1 + Q^2) \cong 1$, resulting in

$$\boxed{f_p = \frac{1}{2\pi\sqrt{LC}}} \qquad (Q \geq 10)$$

as obtained earlier.

You will recall that for the series resonant circuit, $V_L = V_C = QE$ at the resonant condition. A similar result can be obtained for the parallel resonant circuit if we carefully examine the network of Fig. 20.30. The current $\mathbf{I}_T$ is not the source current (due to R_s) but the current entering the tank circuit. Of course, for the condition $R_s \cong \infty \ \Omega$ (open circuit), which often occurs, $\mathbf{I}_T$ is equal to the source current $\mathbf{I}$.

At resonance, $\mathbf{Z}_{T_p} = Q^2 R_l$, as determined earlier and through Ohm's law, $\mathbf{V} = \mathbf{I}_T \cdot Q^2 R_l$.

The current $\mathbf{I}_L$ is

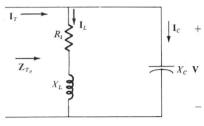

FIG. 20.30

$$\mathbf{I}_L = \frac{\mathbf{V}}{\mathbf{Z}_L} = \frac{\mathbf{I}_T Q^2 R_l}{R_l + jX_L}$$

Dividing by R_l in the numerator and denominator gives us

$$\mathbf{I}_L = \frac{\mathbf{I}_T Q^2}{1 + j\dfrac{X_L}{R_l}} = \frac{\mathbf{I}_T Q^2}{1 + jQ}$$

The magnitude of $\mathbf{I}_L$ is given by

$$I_L = \frac{I_T Q^2}{\sqrt{1 + Q^2}}$$

which for $Q \geq 10$ becomes

$$I_L = \frac{I_T Q^2}{Q}$$

and

$$\boxed{I_L = QI_T} \qquad \textbf{(20.41)}$$

In a parallel resonant circuit, therefore, the magnitude of the current through the inductive branch is Q times the current entering the tank circuit (at resonance only). Furthermore, since

$$\mathbf{I}_C = \frac{\mathbf{V}}{-jX_C} = \frac{\mathbf{I}_T Q^2 R_l}{-jX_C}$$

if we divide by R_l, we get

$$\mathbf{I}_C = \frac{\mathbf{I}_T Q^2}{-j\dfrac{X_C}{R_l}}$$

However, at resonance, for $Q \geq 10$,

$$X_L = X_C \quad \text{and} \quad \frac{X_C}{R_l} = \frac{X_L}{R_l} = Q$$

and

$$\mathbf{I}_C = \frac{\mathbf{I}_T Q^2}{-jQ}$$

The magnitude is

$$I_C = \frac{I_T Q^2}{Q}$$

and

$$\boxed{I_C = QI_T} \qquad \text{(at resonance)} \qquad \textbf{(20.42)}$$

20.11 SUMMARY TABLE AND REACTANCE CHART

Table 20.1 is included because $Q \geq 10$ has such a pronounced effect on the basic equations for parallel resonance.

For $Q \geq 10$, we can use the reactance chart of Fig. 20.21 again, since

TABLE 20.1 *Parallel resonant circuit.*

	Any Q	$Q \geq 10$	$R_s = \infty\ \Omega$ $(Q \geq 10)$
Resonance	$\dfrac{R_l^2 + X_L^2}{X_L} = X_C$	$X_L = X_C$	$X_L = X_C$
f_p	$\dfrac{1}{2\pi\sqrt{LC}}\sqrt{1 - \dfrac{R_l^2 C}{L}},\ \dfrac{1}{2\pi\sqrt{LC}}\sqrt{\dfrac{Q^2}{1 + Q^2}}$	$\dfrac{1}{2\pi\sqrt{LC}}$	$\dfrac{1}{2\pi\sqrt{LC}}$
Z_{Tp}	$R_s \parallel \dfrac{R_l^2 + X_L^2}{R_l},\ R_s \parallel \dfrac{L}{R_l C}$	$R_s \parallel Q^2 R_l$	$Q^2 R_l$
Q_p	$\dfrac{R}{X_{Lp}}, \dfrac{R}{X_C},\quad \text{or}\quad \omega_p C R \quad (R = R_s \parallel R_p)$	$\dfrac{R}{\omega_p L},\ \omega_p C R$	Q
BW	$\dfrac{f_p}{Q_p}$	$I_L = I_C = Q_p I_T$ $\dfrac{f_p}{Q_p}$	$I_L = I_C = QI_T$ $\dfrac{f_p}{Q}$

$$X_L = X_C \quad \text{and} \quad f_p = f_s = \frac{1}{2\pi\sqrt{LC}}$$

Its use will be demonstrated in Example 20.8.

20.12 EXAMPLES (PARALLEL RESONANCE)

EXAMPLE 20.7. For the network of Fig. 20.31, determine
a. Q.
b. R_p.
c. Z_{T_p}.
d. C at resonance.
e. Q_p.
f. BW.

Solutions:

a. $Q = \dfrac{X_L}{R_l} = \dfrac{2\pi f_p L}{R_l} = \dfrac{(6.28)(0.04 \times 10^6)(10^{-3})}{10}$

$= \mathbf{25.12}$

b. $Q \geq 10$

Therefore,

$R_p = Q^2 R_l = (25.12)^2 10 = \mathbf{6.31\ k\Omega}$

c. $Z_{T_p} = R_s \parallel R_p = 40\ k\Omega \parallel 6.31\ k\Omega = \mathbf{5.45\ k\Omega}$

d. $Q \geq 10$

Therefore,

$f_p = \dfrac{1}{2\pi\sqrt{LC}}$ and

$C = \dfrac{1}{L(f2\pi)^2} = \dfrac{1}{(10^{-3})(0.04 \times 10^6 \cdot 2\pi)^2} = \mathbf{0.0159\ F}$

e. $Q \geq 10$

Therefore,

$Q_p = \dfrac{R}{\omega_p L} = \dfrac{5.45\ k\Omega}{(6.28)(0.04 \times 10^6)(10^{-3})} = \mathbf{21.71}$

f. BW $= \dfrac{f_p}{Q_p} = \dfrac{0.04 \times 10^6}{21.71} = \mathbf{1.84\ kHz}$

EXAMPLE 20.8. For the network of Fig. 20.32, determine the resonant frequency and X_L using the proper equation and the reactance chart of Fig. 20.21. Compare your results.

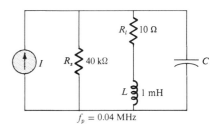

$f_p = 0.04$ MHz

FIG. 20.31

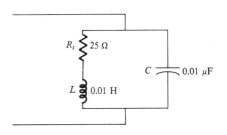

FIG. 20.32

Solution: From the reactance chart, we find that $f_p \cong$ 16 kHz and $X_L = 1000\ \Omega$. These data indicate that $Q = X_L/R_l = 1000/25 = 40$, which is sufficiently large compared with $Q \geq 10$ for us to assume that $X_L = X_C$ and use the chart. In other words, if we went to the complete equation for f_p in Table 20.1, the results would be essentially the same as those obtained using the condition $X_L = X_C$. For values of Q close to 10, however, it would be necessary to turn to the complete equations in Table 20.1 and discontinue use of the chart.

Using the equations, we find

$$f_p = \frac{1}{2\pi\sqrt{LC}} = \frac{1}{6.28\ \sqrt{(0.01)(0.01\times 10^{-6})}}$$

$$= \frac{1}{6.28\sqrt{10^{-10}}} = \frac{1}{6.28\times 10^{-5}}$$

$$= \frac{10^5}{6.28} = 15,924\ \text{Hz} = \textbf{15.924 kHz}$$

and

$$X_L = 2\pi f_p L = (6.28)(15.924\times 10^3)(0.01) = \textbf{1000.03}\ \Omega$$

The results are certainly very close in magnitude and validate the use of the reactance chart.

EXAMPLE 20.9. The equivalent network for the transistor configuration of Fig. 20.33 is shown in Fig. 20.34. Determine

a. f_p (using the reactance chart).
b. Q.
c. Q_p.
d. BW.
e. V_p at resonance.
f. Sketch the curve of V_C versus frequency.

Solutions:

a. From the chart, $f_p \cong \textbf{320 kHz}$

b. $Q = \dfrac{X_L}{R_l} = \dfrac{2\pi f_p L}{R_l} = \dfrac{(6.28)(320\times 10^3)(5\times 10^{-3})}{50}$

$$= \frac{10,048}{50} = \textbf{200.96}$$

c. $Q_p = \dfrac{R}{\omega_p L} = \dfrac{R}{X_L}$

$R = R_s \parallel R_p = R_s \parallel Q^2 R_l = 50\ \text{k}\Omega \parallel (200.96)^2 \cdot 50$
$\quad = 50\ \text{k}\Omega \parallel 2{,}019.2\ \text{k}\Omega = 48.79\ \text{k}\Omega$

Therefore,

$$Q_p = \frac{48.79\ \text{k}\Omega}{10,048} = \textbf{4.86}$$

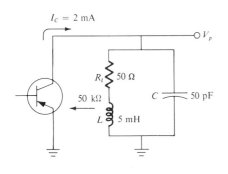

$I_C = 2$ mA

R_l 50 Ω

50 kΩ

C 50 pF

L 5 mH

V_p

FIG. 20.33

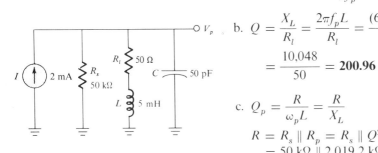

I 2 mA

R_s 50 kΩ

R_l 50 Ω

C 50 pF

L 5 mH

V_p

FIG. 20.34

d. $\text{BW} = \dfrac{f_p}{Q_p} = \dfrac{320 \times 10^3}{4.86} = \textbf{65.84 kHz}$

e. At resonance, $Z_{T_p} = R_s \parallel R_p = R = 48.79 \text{ k}\Omega$

and

$V_p = IZ_{T_p} = (2 \times 10^{-3})(48.79 \times 10^3) = \textbf{97.58 V}$

f. See Fig. 20.35.

EXAMPLE 20.10. Design a parallel resonant circuit to have the response curve of Fig. 20.36 using a 1-mH, 10-Ω inductor, with $R_s = 40$ kΩ.

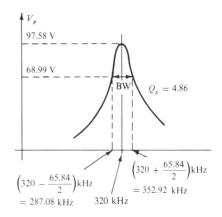

$\left(320 - \dfrac{65.84}{2}\right)\text{kHz}$
$= 287.08 \text{ kHz}$ 320 kHz $\left(320 + \dfrac{65.84}{2}\right)\text{kHz}$
$= 352.92$ kHz

FIG. 20.35

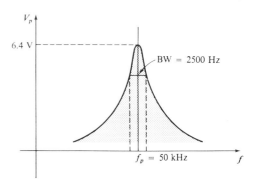

FIG. 20.36

Solution:

$$\text{BW} = \frac{f_p}{Q_p}$$

Therefore,

$$Q_p = \frac{f_p}{\text{BW}} = \frac{50,000}{2500} = 20$$

$$X_L = 2\pi f_p L = (6.28)(50 \times 10^3)(10^{-3}) = 314.0 \ \Omega$$

and

$$Q = \frac{X_L}{R_l} = \frac{314}{10} = 31.4$$

$$R_p = Q^2 R_l = (31.4)^2 10 = 9859.60 \ \Omega$$

$$Q_p = \frac{R}{X_L} = \frac{R_s \parallel 9859.60}{314} = 20$$

$$\frac{(R_s)(9859.6)}{R_s + 9859.6} = 6280$$

$$9859.60 R_s = 6280 R_s + 61.92 \times 10^6$$

and

$$3579.6 R_s = 61.92 \times 10^6$$
$$R_s = 17.298 \text{ k}\Omega$$

However, the source resistance was given as 40 kΩ. We must therefore add a parallel resistor that will reduce the 40 kΩ to approximately 17.298 kΩ:

$$\frac{(40 \text{ k}\Omega)(R')}{40 \text{ k}\Omega + R'} = 17.298 \text{ k}\Omega$$

and

$$(40 \text{ k}\Omega)(R') = (17.298 \text{ k}\Omega)(R') + 691.92 \times 10^6$$

or

$$22.70R' = 691.92 \times 10^3$$

and

$$R' = \mathbf{30.481 \text{ k}\Omega}$$

At resonance, $X_L = X_C$, and

$$X_C = \frac{1}{\omega_p C}$$

or

$$314 = \frac{1}{(6.28)(50 \times 10^3)(C)} = \frac{1}{314 \times 10^3 C}$$

or

$$C = \frac{1}{(314 \times 10^3)(314)} = \frac{10^{-6}}{98.596}$$

or, finally,

$$C \cong \mathbf{0.01 \ \mu F}$$

The network appears in Fig. 20.37.

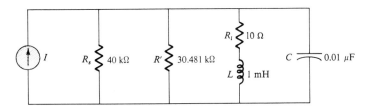

FIG. 20.37

20.13 FILTERS

Filters are networks that either pass or stop the passage of a specific range of frequencies. A proper treatment of the subject would require a lengthy chapter in itself. In this chapter, we will simply introduce the subject by examining the very basic *pass-band, band-stop,* and *double-tuned* filters. The terminology applied indicates their basic function.

Pass-Band Filter

The basic construction of the pass-band filter is usually one of the three indicated in Fig. 20.38. It is the purpose of each of these networks to insure that $\mathbf{V}_L$ is a maximum for a particular frequency range of $\mathbf{E}$.

For the network of Fig. 20.38(a), we will assume any load (R_L) connected to the output terminals is much greater

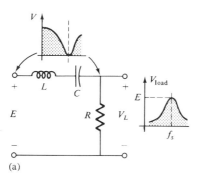

(a)

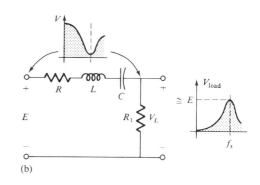

(b)

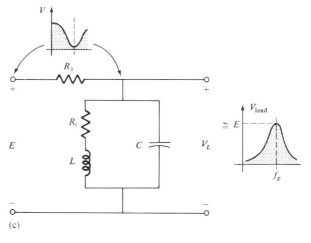

(c)

FIG. 20.38

than R; that is, $R_L \gg R$, to permit the approximation $R_L \parallel R \cong R$. The Q_s of the network is then

$$Q_s = \frac{X_L}{R} \qquad (20.43)$$

Applying the voltage divider rule, we get

$$\mathbf{V}_L = \frac{R\mathbf{E}}{R + j(X_L - X_C)}$$

At resonance, $X_L = X_C$, and

$$\mathbf{V}_L = \frac{R\mathbf{E}}{R} = \mathbf{E}$$

At frequencies to the right and left of resonance, the impedance of the L-C combination increases over that of R, resulting in a decreasing voltage drop across R. The resulting response curve is shown in Fig. 20.38(a). For cases where R_L is not sufficiently greater than R, replace R in all equations above by $R \parallel R_L$.

For the network of Fig. 20.38(b), R_1 is chosen many times greater than R of the series resonant circuit, and $R_L \gg R_1$. At resonance,

$$V_L = \frac{R_1 E}{R_1 + R + j(X_L - X_C)}$$

or

$$V_L = \frac{R_1 E}{R_1 + R} \cong \frac{R_1 E}{R_1} \quad (\text{since } R_1 \gg R)$$

$$= E$$

and

$$\boxed{Q_{s_{(\text{unloaded})}} = \frac{X_L}{R + R_1}} \tag{20.44}$$

For R_L not $\gg R_1$,

$$\boxed{Q_{s_{(\text{loaded})}} = \frac{X_L}{R + R_1 \parallel R_L}} \tag{20.45}$$

The parallel resonant circuit is employed in the network of Fig. 20.38(c). For this configuration, $V_L = E$ at resonance due to the high impedance of the tank circuit compared with R_1. For the lower and higher frequency range, the impedance of the parallel resonant circuit is considerably less than that of R_1, producing the response curves shown in Fig. 20.38(c).

For $R_L \gg Q^2 R_l$ (and $Q \geq 10$) as shown in Fig. 20.39,

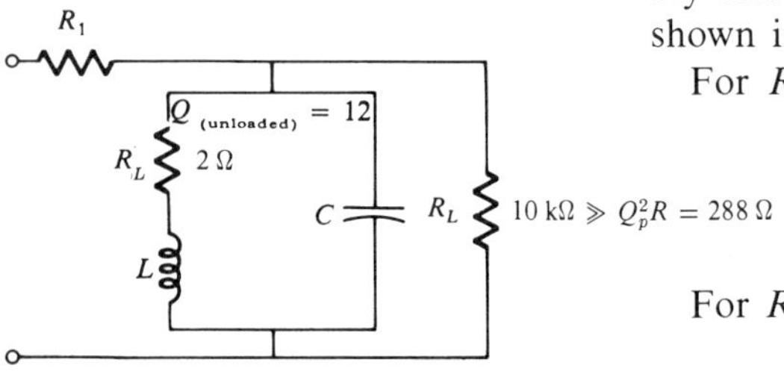

$$\boxed{Q_{p_{(\text{loaded})}} = Q = \frac{X_L}{R_l}} \tag{20.46}$$

For R_L not $\gg Q^2 R$ (but $Q \geq 10$),

$$Q_{p_{(\text{loaded})}} = \frac{R_p}{X_{L_p}}$$

and

$$\boxed{Q_{p_{(\text{loaded})}} = \frac{R_L \parallel Q^2 R_l}{X_L}} \tag{20.47}$$

R_1 does not affect the quality factor, since near resonance the tank impedance is sufficiently greater than R_1 that it can be ignored.

EXAMPLE 20.11. Sketch the response curve for V_L for the pass-band filter of Fig. 20.40.

Solution:

$$Q = \frac{200}{5} = 40 = Q_p \quad (R_s \cong \infty \ \Omega)$$

Since

$$Q = 40 \geq 10$$

FIG. 20.39

R_1

$Q_{(\text{unloaded})} = 12$

$R_L \ 2\,\Omega$

C $R_L \ 10 \text{ k}\Omega \gg Q_p^2 R = 288\,\Omega$

L

then

$$Z_{T_p} = Q^2 R_l = (40)^2 5 = 8 \text{ k}\Omega$$

and

$$V_L = \frac{(8 \text{ k}\Omega)E}{8 \text{ k}\Omega + 0.5 \text{ k}\Omega} = 0.94E \quad \text{(at resonance)}$$

$$\text{BW} = \frac{f_p}{Q_p} = \frac{100{,}000}{40} = 2500 \text{ Hz}$$

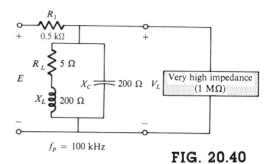

$f_p = 100 \text{ kHz}$

FIG. 20.40

Therefore,

$$f_1 = 100 \text{ kHz} - 1.25 \text{ kHz} = 98.75 \text{ kHz}$$
$$f_2 = 100 \text{ kHz} + 1.25 \text{ kHz} = 101.25 \text{ kHz}$$

At the half-power frequencies,

$$Z_{T_{\text{HPF}}} = 0.707 Z_{T_p}$$
$$= 0.707(8 \text{ k}\Omega)$$
$$Z_{T_{\text{HPF}}} = 5.66 \text{ k}\Omega$$

and

$$V_L = \frac{(5.66 \text{ k}\Omega)E}{5.66 \text{ k}\Omega + 0.5 \text{ k}\Omega} = 0.92E$$

The sketch of V_L appears in Fig. 20.41.
At $f_p = 10$ kHz,

$$X_L = 20 \,\Omega \quad \text{and} \quad X_C = 2 \text{ k}\Omega$$

so the impedance of the tank circuit is given by

$$\mathbf{Z}_T = (R_l + jX_L) \parallel X_C \cong R_l + jX_L$$
$$= 5 + j20 = 20.62 \angle 75.96°$$

and the magnitude of V_L is

$$|V_L| = \frac{Z_T E}{Z_T + R_1} = \frac{(20.62)E}{5 + j20 + 0.5 \text{ k}\Omega} = \frac{20.62E}{505 + j20}$$

$$= \frac{20.62E}{505.4} \cong 0.041E$$

or 4.1% of E, as compared with 94% and 92% at the resonant and half-power frequencies, respectively.
For $R_L = 50$ kΩ, we find

$$Q_{p_{(\text{loaded})}} = \frac{R_L \parallel Q^2 R_l}{X_L}$$

$$= \frac{50 \text{ k}\Omega \parallel 8 \text{ k}\Omega}{200} = \frac{6897}{200} = 34.49$$

as compared with 40 unloaded.

FIG. 20.41

Band-Stop Filter

A close examination of the provided curves of Fig. 20.38 should reveal the basic configuration of the band-stop filter. For the circuit of Fig. 20.38(a), for instance, all frequencies have a high relative response across the L-C combination except at the resonance frequency. In other words, for $V_{L\text{-}C}$, all frequencies pass well except at or near resonance. For a band-stop characteristic, therefore, the output voltage V_L should be taken across the L-C combination.

For each configuration of Fig. 20.38, the band-stop output (V_L) is taken across the elements indicated in Fig. 20.42.

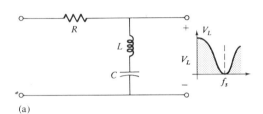

(a)

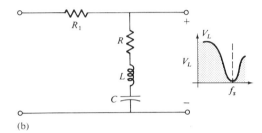

(b)

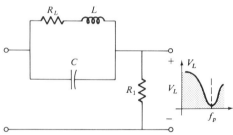

(c)

FIG. 20.42

For the network of Fig. 20.42(a), since R_L is in parallel with the series L-C combination, it will not affect the Q_s of the network because at $Z_{L\text{-}C} = 0\ \Omega$, or near resonance, $R_L \gg Z_{L\text{-}C}$ and the parallel effect of R_L can be ignored. Therefore,

$$Q_{s_{\text{(loaded or unloaded)}}} = \frac{X_L}{R} \qquad \textbf{(20.48)}$$

At resonance, the series resonant circuit (R-L-C) of Fig. 20.42(b) has a total impedance of $R\ \Omega$. For $R_L \gg R$ (which is normally the case), R_L can again be ignored when Q_s is determined. However, R_1 is in series with R, and

$$Q_{s_{\text{(loaded or unloaded)}}} = \frac{X_L}{R + R_1} \qquad \textbf{(20.49)}$$

For the parallel resonant network of Fig. 20.42(c), the parallel combination of R_L and R_1 will result in an impedance less than R_1, which in series with the high impedance of the tank circuit at resonance can be ignored. Therefore, R_L normally will not affect the quality factor of the resonant circuit. Therefore,

$$Q_{p_{\text{(loaded or unloaded)}}} = Q = \frac{X_L}{R_l} \qquad (20.50)$$

The band-reject and band-pass characteristics of miniature filters manufactured by TRW/UTC inductive products appear in Fig. 20.43 with a photograph of a typical unit. The -3 db point in the band-pass characteristics is the 0.707 level described in this chapter. [Decibels (dB) and their application will be examined in your electronics courses.] Note that the band-pass characteristics are a universal set since the resonant frequency is undefined and the horizontal axis is the ratio f/f_r. When examining the curves, remember that the vertical scale is a measure of the attenuation of the input signal. That is, at 0 dB the output is equal to the input, while at -30 dB the output is significantly less than the input. Since each of these units can be used for the pass-band or stop-band function, data about each application are provided. Note that the pass band is inserted by the center frequency for each unit and the type number of the unit reflects the center frequency. The stop band is implying essentially zero response with the -35 dB criteria. Note that the center frequency does not bisect the stop band since it is so near the bottom of the resonance curve.

Double-Tuned Filter

There are some network configurations that display both a pass-band and a band-stop characteristic, such as shown in Fig. 20.44. For the network of Fig. 20.44(a), the parallel resonant circuit will establish the band stop by resonating at the frequency not permitted to establish a V_L. The greater part of the applied voltage **E** will appear across this parallel resonant circuit at this frequency due to its very high impedance compared with R_L.

For the pass band, the parallel resonant circuit is designed to be capacitive (inductive if L_s is replaced by C_s). The inductance L_s is chosen to cancel the effects of the resulting capacitance reactance at the resonant pass-band frequency of the tank circuit, thereby acting as a series resonant circuit. The applied voltage **E** will then appear across R_L at this frequency.

For the network of Fig. 20.44(b), the series resonant circuit will still determine the pass band, acting as a very low impedance across the parallel inductor at resonance, establishing $V_L = E$. At the desired band-stop resonant frequency, the series resonant circuit is capacitive. The induct-

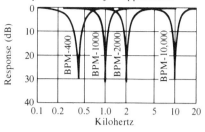

BAND REJECT
BPM units are designed for both band-pass and band-reject applications.

BAND PASS
Typical normalized BPM response

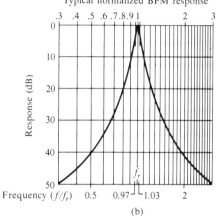

(b)

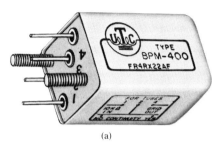

(a)

Courtesy of United Transformer Corp.

Type No.	Center Frequency (Hz)	Pass Band (Less than 2 db) (Hz)	Stop Band (More than 35 db)	
			Below (Hz)	Above (Hz)
BPM 400	400	388–412	200	800
BPM 440	440	427–453	220	880
BPM 500	500	485–515	250	1000
BPM 600	600	582–618	300	1200
BPM 800	800	776–824	400	1600
BPM 1000	1000	970–1030	500	2000
BPM 1200	1200	1164–1236	600	2400
BPM 1500	1500	1455–1545	750	3000
BPM 1600	1600	1552–1648	800	3200
BPM 2000	2000	1940–2060	1000	4000
BPM 2500	2500	2425–2575	1250	5000
BPM 3000	3000	2910–3090	1500	6000
BPM 3200	3200	3104–3296	1600	6400
BPM 4000	4000	3880–4120	2000	8000
BPM 4800	4800	4656–4944	2400	9600
BPM 5000	5000	4850–5150	2500	10000
BPM 6000	6000	5820–6180	3000	12000
BPM 8000	8000	7760–8240	4000	16000
BPM 10000	10000	9700–10300	5000	20000
BPM 20000	20000	19400–20600	10000	40000

FIG. 20.43 *Band-pass and band-reject characteristics for BPM TRW/UTC filters.*

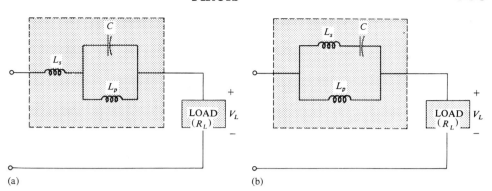

(a)   (b)

ance L_p is chosen to establish parallel resonance at the resonant band-stop frequency. The high impedance of the parallel resonant circuit will result in a very low load voltage V_L.

For rejected frequencies below the pass band, the networks should appear as shown in Fig. 20.44. For the reverse situation, L_s in Fig. 20.44(a) and L_p in Fig. 20.44(b) are replaced by capacitors.

FIG. 20.44 *Double-tuned networks.*

EXAMPLE 20.12. For the network of Fig. 20.44(b), determine L_s and L_p for a capacitance C of 500 pF if a frequency of 200 kHz is to be rejected and a frequency of 600 kHz accepted.

Solution: For series resonance, we have

$$f_s = \frac{1}{2\pi\sqrt{LC}}$$

and

$$L_s = \frac{1}{4\pi^2 f_s^2 C} = \frac{1}{4(3.14)^2(600\times10^3)^2 500\times10^{-12}}$$
$$= \mathbf{141\ \mu H}$$

At 200 kHz,

$$X_L = \omega L = 2\pi f_s L = (6.28)(200\times10^3)(141)\times10^{-6}$$
$$= 177.2\ \Omega$$

$$X_C = \frac{1}{\omega C} = \frac{1}{(6.28)(200\times10^3)(500\times10^{-12})}$$
$$= 1.59\ k\Omega$$

For the series elements,

$$j(X_L - X_C) = j(177.1 - 1592) = -j(1414.8)\ \text{capacitive}$$

At parallel resonance ($Q \geq 10$ assumed),

$$X_L = X_C$$

and

$$L_p = \frac{X_C}{\omega} = \frac{1412.9}{(6.28)(200\times10^3)} = \mathbf{1.126\ mH}$$

PROBLEMS

Sections 20.1 through 20.8

1. Find the resonant ω_s and f_s for the series circuit with the following parameters:
   a. $R = 10\,\Omega$, $L = 1\,H$, $C = 16\,\mu F$
   b. $R = 300\,\Omega$, $L = 0.5\,H$, $C = 0.16\,\mu F$
   c. $R = 20\,\Omega$, $L = 0.28\,mH$, $C = 7.46\,\mu F$

2. **a.** Using the reactance chart, repeat Problem 1.
   b. Using the chart, find a combination of inductance and capacitance that will resonate at 10 kHz. Determine the reactance of X_L and X_C at this frequency.
   c. Using the chart, determine the elements of a series resonant circuit with $X_L = X_C = 2\,k\Omega$ and $f_s = 100$ kHz, with $Q_s = 50$.

3. For the series circuit of Fig. 20.45:
   a. Find the value of X_C for resonance.
   b. Find the current **I** and the voltages $\mathbf{V}_R$, $\mathbf{V}_L$, and $\mathbf{V}_C$ in phasor form at resonance.
   c. Draw the phasor diagram of the voltages and current.
   d. Sketch the power triangle for the circuit at resonance.
   e. Find the Q_s of the circuit.

4. Repeat Problem 3 for the circuit of Fig. 20.46.

5. For the circuit of Fig. 20.47:
   a. Find the value of L in millihenries if the resonant frequency is 1800 Hz.
   b. Repeat Problem 3, parts (b)–(e).
   c. Calculate the cutoff frequencies.
   d. Find the bandwidth of the series resonant circuit.

6. **a.** Find the bandwidth of a series resonant circuit having a resonant frequency of 6000 Hz and a Q_s of 15.
   b. Find the cutoff frequencies.
   c. If the resistance of the circuit at resonance is $3\,\Omega$, what are the values of X_L and X_C in ohms?
   d. What is the power dissipated at the half-power frequencies if the maximum current flowing through the circuit is 0.5 A?

7. **a.** A series circuit has a resonant frequency of 10 kHz. The resistance of the circuit is $5\,\Omega$, and X_C at resonance is $200\,\Omega$. Find the bandwidth.
   b. Find the cutoff frequencies.
   c. Find Q_s.
   d. If the input voltage is $30\,\angle\,0°$, find the voltage across the coil and capacitor.
   e. Find the power dissipated at resonance.

8. **a.** The bandwidth of a series resonant circuit is 200 Hz. If the resonant frequency is 2000 Hz, what is the value of Q_s for the circuit?
   b. If $R = 2\,\Omega$, what is the value of X_L at resonance?
   c. Find the value of L and C at resonance.
   d. Find the cutoff frequencies.

FIG. 20.45

FIG. 20.46

FIG. 20.47

9. The cutoff frequencies of a series resonant circuit are 5400 Hz and 6000 Hz:
   a. Find the bandwidth of the circuit.
   b. If Q_s is 9.5, find the resonant frequency of the circuit.
   c. If the resistance of the circuit is 2 Ω, find the value of X_L and X_C at resonance.
   d. Find the value of L and C at resonance.

*10. Design a series resonant circuit with an input voltage 5 ∠0° to have the following specifications:
   a. A peak current of 500 mA at resonance.
   b. A bandwidth of 120 Hz.
   c. A resonant frequency of 8400 Hz.
   Find the value of L and C and the cutoff frequencies.

*11. Design a series resonant circuit to have a bandwidth of 400 Hz using a coil with a Q of 20 and a resistance of 2 Ω. Find the value of L and C and the cutoff frequencies.

Sections 20.9 through 20.12

12. For the circuit of Fig. 20.48:
   a. Find the value of X_C for resonance.
   b. Find the total impedance Z_T at resonance.
   c. Find the currents I_L and I_C at resonance.
   d. If the resonant frequency is 20,000 Hz, find the value of L and C at resonance.
   e. Find Q_p and BW.

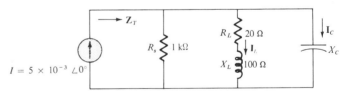

13. Repeat Problem 12 for the circuit of Fig. 20.49.

FIG. 20.48

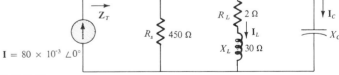

14. For the circuit of Fig. 20.50:
   a. Find the resonant frequency.
   b. Find the value of X_L and X_C at resonance.
   c. Is the coil a high-Q or low-Q coil at resonance?
   d. Find the impedance Z_{T_p} at resonance.
   e. Find the currents I_L and I_C at resonance.
   f. Calculate Q_p and BW.

FIG. 20.49

FIG. 20.50

15. Repeat Problem 14 for the circuit of Fig. 20.51.

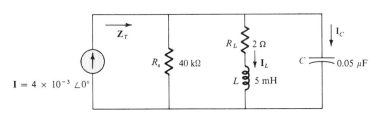

FIG. 20.51

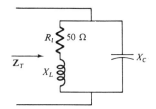

FIG. 20.52

16. It is desired that the impedance Z_T of the circuit of Fig. 20.52 be a resistor of 50 kΩ at resonance.
  a. Find the value of X_L.
  b. Compute X_C.
  c. Find the resonant frequency if $L = 16$ mH.
  d. Find the value of C in microfarads.

17. For the network of Fig. 20.53:
  a. Find f_p.
  b. Calculate V_C at resonance.
  c. Determine the power absorbed at resonance.
  d. Find BW.

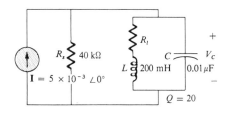

FIG. 20.53

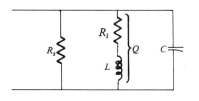

FIG. 20.54

*18.** For the network of Fig. 20.54, the following are specified:

$$f_p = 100 \text{ kHz}$$
$$\text{BW} = 2500 \text{ Hz}$$
$$L = 2 \text{ mH}$$
$$Q = 80$$

Find R_s and C.

*19.** For the network of Fig. 20.55:
  a. Find the value of X_L for resonance.
  b. Find Q.
  c. Find the resonant frequency if the bandwidth is 1000 Hz.
  d. Find the maximum value of the voltage V_C.
  e. Sketch the curve of V_C vs. frequency. Indicate its peak value, resonant frequency, and band frequencies.

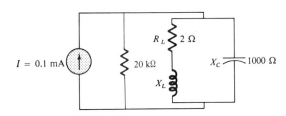

FIG. 20.55

*20. Repeat Problem 19 for the network of Fig. 20.56.

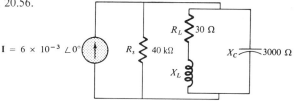

FIG. 20.56

*21. Design the network of Fig. 20.57 to have the following characteristics:
a. Bandwidth of 500 Hz.
b. $Q_p = 30$.
c. $V_{C_{max}} = 1.8$ V.

Section 20.13

22. For the pass-band filter of Fig. 20.58:
a. Determine Q_s.
b. Find the cutoff frequencies.
c. Sketch the frequency characteristics.
d. Find $Q_{s(loaded)}$ if a load of 200 Ω is applied.
e. Indicate on the curve of part (c) the change in the frequency characteristics with the load applied.

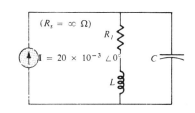

FIG. 20.57

23. For the pass-band filter of Fig. 20.59:
a. Determine Q_p ($R_L = \infty$ Ω, open circuit).
b. Sketch the frequency characteristics.
c. Find $Q_{p(loaded)}$ for $R_L = 100$ kΩ, and indicate the effect of R_L on the characteristics of part (b).
d. Repeat part (c) for $R_L = 20$ kΩ.

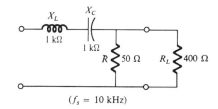

FIG. 20.58

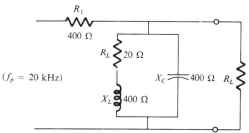

24. For the band-stop filter of Fig. 20.60:
a. Determine Q_s.
b. Find the BW and half-power frequencies.
c. Sketch the frequency characteristics.
d. What is the effect on the curve of part (c) if a load of 2 kΩ is applied?

25. The network of Problem 23 is used as a band-stop filter:
a. Sketch the network when used as a band-stop filter.
b. Sketch the band-stop characteristics.
c. What is the effect on the band-stop characteristics if the 100-kΩ and 20-kΩ loads are applied?

26. a. For the network of Fig. 20.44(a), if $L_p = 400$ μH ($Q > 10$), $L_s = 60$ μH, and $C = 120$ pF, determine the rejected and accepted frequencies.
b. Sketch the response curve for part (a).

FIG. 20.59

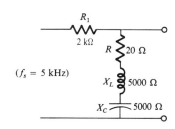

FIG. 20.60

27. a. For the network of Fig. 20.44(b), if the rejected frequency is 30 kHz and the accepted 100 kHz, determine the values of L_s and L_p ($Q \geq 10$) for a capacitance of 200 pF.
   b. Sketch the response curve for part (a).

GLOSSARY

Band (cutoff, half-power, corner, −3dB) frequencies Frequencies that define the points on the resonance curve that are 0.707 of the peak current or voltage value. In addition, they define the frequencies at which the power transfer to the resonant circuit will be half the maximum power level (−3dB).

Band-stop filter A network designed to reject (not pass) signals within a particular range of frequencies.

Bandwidth The range of frequencies between the band, cutoff, or half-power frequencies.

Double-tuned filter A network having both a pass-band and a band-stop region.

Filter Networks designed to either pass or reject the transfer of signals at certain frequencies to a load.

Pass-band filter A network designed to pass signals within a particular range of frequencies.

Quality factor (Q) A ratio that provides an immediate indication of the sharpness of the peak of a resonant curve. The higher the Q, the sharper the peak and the more quickly it drops off to the right and left of the resonant frequency.

Reactance chart A graph that is very useful in the analysis and design of series or parallel resonant circuits.

Resonance A condition established by the application of a particular frequency (the resonant frequency) to a series or parallel R-L-C network. The transfer of power to the system is a maximum, and, for frequencies above and below, the power transfer drops off to significantly lower levels.

Selectivity A characteristic of resonant networks directly related to the bandwidth of the resonant system. High selectivity is associated with small bandwidths (low Q's), and low selectivity with larger bandwidths (low Q's).

21.1 INTRODUCTION

The system described in Chapter 13, which developed only one sinusoidal voltage for each rotation of the rotor, is referred to as a *single-phase* ac generator. If the number of coils on the rotor is increased in a specified manner, the result is a *polyphase generator*, which develops more than one ac phase voltage per rotation of the rotor. In this chapter, the three-phase system will be discussed in detail since it is the most frequently used for transmitting power.

In general, the three-phase system is more economical for transmitting power at a fixed power loss than the single-phase system. This economy is due primarily to the reduction of the I^2R losses of the transmission lines. A reduction in these losses permits the use of smaller conductors, which in turn reduces the weight of copper required.

The three-phase system is used in almost all commercial electric generators. This does not mean that single-phase and two-phase generating systems are obsolete. Most small emergency generators, such as the gasoline type, are one-phase generating systems.

One of the more common applications of the two-phase system is in servomechanisms, which are self-correcting control systems capable of detecting and adjusting their own operation. Servomechanisms are used in ships and air-craft to keep them on course automatically, or, in simpler devices such as a thermostatic circuit, to regulate heat output. In most cases, however, where single-phase and two-phase inputs are required, they are supplied by one and two phases of a three-phase generating system rather than generated independently.

The number of phase voltages that can be produced by a polyphase generator is not limited to three. Any number of phases can be obtained by spacing the windings for each phase at the proper angular position around the rotor. Some electrical systems operate more efficiently if more than three phases are used. One such system involves the process of rectification, which is used to convert alternating current to direct current. The greater the number of phases, the more ideal the dc output of the system.

21.2 THE THREE-PHASE GENERATOR

The three-phase generator has three coils placed 120° apart on the rotor (armature), as shown in Fig. 21.1(a). Since the three coils have an equal number of turns, and each coil rotates with the same angular velocity, the emf induced in each coil will have the same peak value, shape, and frequency. For each coil, the direction of increasing induced emf, as determined by the right-hand rule, is shown in Fig. 21.1(a). The induced voltage $e_{A'A}$ rises toward its positive maximum while the induced voltage $e_{B'B}$ drops toward its negative maximum since $e_{BB'} = -e_{B'B}$. The induced voltage $e_{C'C}$ has reached its positive maximum and now decreases toward zero with rotation.

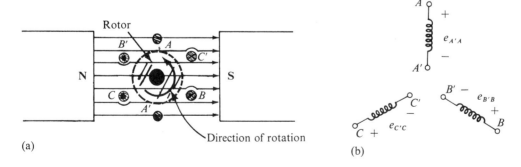

(a)

(b)

FIG. 21.1 *Three-phase generator.*

Plotting the induced voltages $e_{A'A}$, $e_{B'B}$, and $e_{C'C}$ on the same set of axes will give the results shown in Fig. 21.2. *At any particular instant of time, the algebraic sum of the three phase voltages is zero.* This is shown at $\omega t = 0$ in Fig. 21.2.

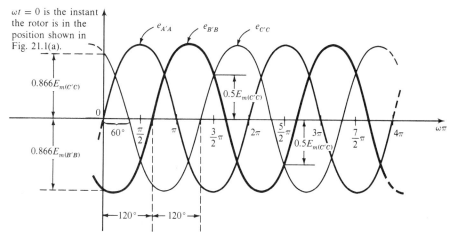

FIG. 21.2

Note that when one induced voltage is zero, the other two are 86.6% of their positive or negative maximums. In addition, when any two are equal in magnitude and sign (at $0.5E_m$), the remaining induced voltage has the opposite polarity and a peak value.

The sinusoidal expression for each of the induced voltages is

$$e_{A'A} = E_{m(A'A)} \sin \omega t$$
$$e_{B'B} = E_{m(B'B)} \sin(\omega t - 120°)$$
$$e_{C'C} = E_{m(C'C)} \sin(\omega t - 240°) = E_{m(C'C)} \sin(\omega t + 120°)$$

(21.1)

The phasor diagram of the induced voltages is shown in Fig. 21.3, where

$$E_{A'A} = 0.707E_{m(A'A)}$$
$$E_{B'B} = 0.707E_{m(B'B)}$$
$$E_{C'C} = 0.707E_{m(C'C)}$$

and

$$\mathbf{E}_{A'A} = E_{A'A} \angle 0°$$
$$\mathbf{E}_{B'B} = E_{B'B} \angle -120°$$
$$\mathbf{E}_{C'C} = E_{C'C} \angle +120°$$

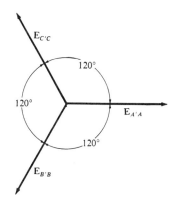

FIG. 21.3

By rearranging the phasors as shown in Fig. 21.4 and applying a law of vectors which states that *the vector sum of any number of vectors drawn such that the "head" of each is connected to the "tail" of the next, and that the "tail" of the first vector is connected to the "head" of the last is zero,* we can conclude that the phasor sum of the phase voltages in a three-phase system is zero. That is,

$$\Sigma(\mathbf{E}_{A'A} + \mathbf{E}_{B'B} + \mathbf{E}_{C'C}) = 0 \qquad \textbf{(21.2)}$$

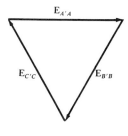

FIG. 21.4

The symbolic representation of the three-phase system is shown in Fig. 21.1(b). Note the correspondence between the representative coils and their lettered terminals as compared to the actual coil positions on the rotor and their lettered terminals.

21.3 THE Y-CONNECTED GENERATOR

If the three terminals A', B', and C' of Fig. 21.1(b) are connected together, the generator is referred to as a *Y-connected, three-phase generator* (Fig. 21.5). As indicated in Fig. 21.5, the Y is inverted for ease of notation and for clarity. The point at which all the terminals are connected is called the *neutral point*. If a conductor is not attached from this point to the load, the system is called a *Y-connected, three-phase, three-wire generator*. If the neutral is connected, the system is a *Y-connected, three-phase, four-wire generator*. The function of the neutral will be discussed in detail when we consider the load circuit.

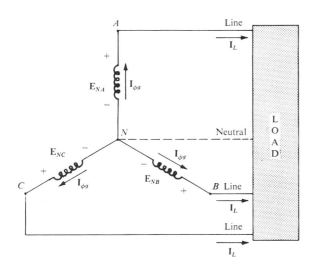

FIG. 21.5 *Y-connected generator.*

Since A', B', and C' now correspond to the neutral point N, the phase voltages can be defined to be $\mathbf{E}_{NA}$, $\mathbf{E}_{NB}$, and $\mathbf{E}_{NC}$.

The three conductors connected from A, B, and C to the load are called *lines*. For the Y-connected system it should be obvious from Fig. 21.5 that the line current equals the phase current for each phase; that is,

$$\boxed{\mathbf{I}_L = \mathbf{I}_{\phi g}} \tag{21.3}$$

The voltage across the lines is called the *line voltage*. On the phasor diagram (Fig. 21.6) it is the phasor drawn from the end of one phase to another in the counterclockwise direction. Note again that the direction of the phasor is from the first to the second subscript.

Applying Kirchhoff's voltage law around the indicated loop of Fig. 21.6, we obtain

$$\mathbf{E}_{BA} - \mathbf{E}_{NA} + \mathbf{E}_{NB} = 0$$

or

$$\mathbf{E}_{BA} = \mathbf{E}_{NA} - \mathbf{E}_{NB} = \mathbf{E}_{NA} + \mathbf{E}_{BN}$$

The phasor diagram is redrawn to find $\mathbf{E}_{BA}$ as shown in Fig. 21.7. Since each phase voltage when reversed ($\mathbf{E}_{BN}$) will bisect the other two, $\alpha = 60°$. The angle $\beta = 30°$, since a line drawn from opposite ends of a rhombus will divide in half both the angle of origin and the opposite angle. Lines drawn between opposite corners of a rhombus will also bisect each other at right angles.

The length

$$x = E_{NA} \cos 30° = \frac{\sqrt{3}}{2} E_{NA}$$

and, since

$$E_{BA} = 2x = (2)\frac{\sqrt{3}}{2} E_{NA} = \sqrt{3} E_{NA}$$

and noting from the phasor diagram that θ of $\mathbf{E}_{BA} = 30°$, then

$$\mathbf{E}_{BA} = E_{BA} \angle 30° = \sqrt{3} E_{NA} \angle 30°$$

or, in words, the magnitude of the line voltage of a Y-connected generator is $\sqrt{3}$ times the phase voltage:

$$\boxed{E_L = \sqrt{3} E_\phi} \tag{21.4}$$

In addition, the phase angle between a line voltage and the nearest phase voltage is 30°.

In sinusoidal notation,

$$e_{BA} = \sqrt{2} E_{BA} \sin(\omega t + 30°)$$

Repeating the same procedure for the other line voltages gives

$$e_{AC} = \sqrt{2} E_{AC} \sin(\omega t + 150°)$$

and

$$e_{CB} = \sqrt{2} E_{CB} \sin(\omega t + 270°)$$

The phasor diagram of the line and phase voltages is shown in Fig. 21.8. If the phasors representing the line voltages in Fig. 21.8(a) are rearranged slightly, they will form a closed loop [Fig. 21.8(b)]. Therefore, we can conclude that the sum of the line voltages is also zero; that is,

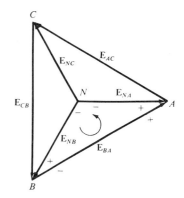

FIG. 21.6 *Line and phase voltages of the Y-connected three-phase generator.*

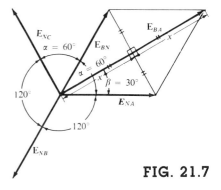

FIG. 21.7

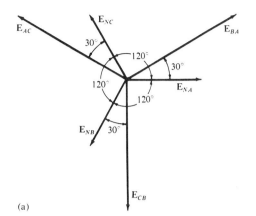

(a)

FIG. 21.8

$$\Sigma(\mathbf{E}_{BA} + \mathbf{E}_{AC} + \mathbf{E}_{CB}) = 0 \qquad \textbf{(21.5)}$$

(b)

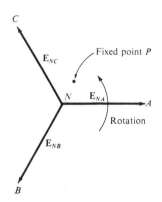

FIG. 21.9

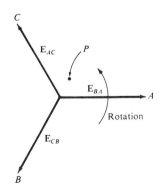

FIG. 21.10

21.4 PHASE SEQUENCE (Y-CONNECTED GENERATOR)

The phase sequence is the order in which the phasors representing the phase voltages pass through a fixed point on the phasor diagram if the phasors are rotated in a counterclockwise direction. For example, in Fig. 21.9 the phase sequence is *ABC*. However, since the fixed point can be chosen anywhere on the phasor diagram, the sequence can also be written as *BCA* or *CAB*. The phase sequence is quite important in the three-phase distribution of power. In a three-phase motor, for example, if two phase voltages are interchanged, the sequence will change and the direction of rotation of the motor will be reversed. Other effects will be described when we consider the loaded three-phase system.

The phase sequence can also be described in terms of the line voltages. Drawing the line voltages on a phasor diagram in Fig. 21.10, we are able to determine the phase sequence by again rotating the phasors in the counterclockwise direction. In this case, however, the sequence can be determined by noting the order of the passing first or second subscripts. In the system of Fig. 21.10, for example, the phase sequence of the first subscripts passing point *P* is *ABC*, and of the second subscripts *BCA*. But we know that *BCA* is equivalent to *ABC*, so the sequence is the same for each. Note that the phase sequence is the same as that of the phase voltages described in Fig. 21.9.

If the sequence is given, the phasor diagram can be drawn by simply picking a reference voltage, placing it on the reference axis, and then drawing the other voltages at the proper angular position. For a sequence of *ACB*, for example, we might choose E_{BA} to be the reference [Fig. 21.11(a)] if we wanted the phasor diagram of the line voltages, or E_{NA} for the phase voltages [Fig. 21.11(b)]. For the sequence indicated, the phasor diagrams would be as in Fig. 21.11. In phasor notation,

$$\text{Line voltages} \begin{cases} \mathbf{E}_{BA} = E_{BA} \angle 0° \quad \text{(reference)} \\ \mathbf{E}_{AC} = E_{AC} \angle -120° \\ \mathbf{E}_{CB} = E_{CB} \angle +120° \end{cases}$$

$$\text{Phase voltages} \begin{cases} \mathbf{E}_{NA} = E_{NA} \angle 0° \quad \text{(reference)} \\ \mathbf{E}_{NC} = E_{NC} \angle -120° \\ \mathbf{E}_{NB} = E_{NB} \angle +120° \end{cases}$$

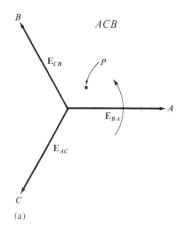

(a)

21.5 THE Y-CONNECTED GENERATOR WITH A Y-CONNECTED LOAD

Loads connected to three-phase supplies are of two types: the Y and the Δ.

If a Y-connected load is connected to a Y-connected generator, the system is symbolically represented by Y-Y. The physical setup of such a system is shown in Fig. 21.12.

If the load is balanced, the neutral can be removed without affecting the circuit in any manner. That is, if

$$\mathbf{Z}_1 = \mathbf{Z}_2 = \mathbf{Z}_3$$

then I_N will be zero. (This will be demonstrated in Example 21.1.) Note that in order to have a balanced load, the phase angle must also be the same for each impedance—a condition that was not necessary in dc circuits when we considered balanced systems.

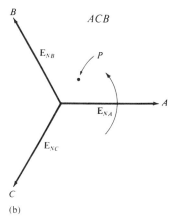

(b)

FIG. 21.11

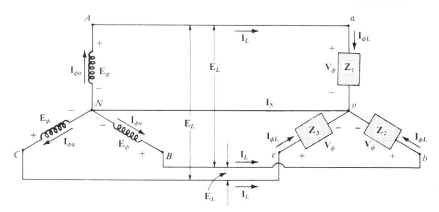

FIG. 21.12 *Y-connected generator with a Y-connected load.*

In practice, if a factory, for example, had only balanced three-phase loads, the absence of the neutral would have no effect, since ideally the system would always be balanced. The cost would therefore be less since the number of required conductors would be reduced. However, lighting and most other electrical equipment will use only one of the phase voltages, and even if the loading is designed to be balanced (as it should be), there will never be perfect continuous balancing, since lights and other electrical equipment will be turned on and off, upsetting the balanced condition. The neutral is therefore necessary to carry the resulting current away from the load and back to the Y-connected generator. This will be demonstrated when we consider unbalanced Y-connected systems.

We shall now examine the *four-wire, Y-Y connected system*. The current passing through each phase of the generator is the same as its corresponding line current, which, in turn, for a Y-connected load is equal to the current in the phase of the load to which it is attached:

$$\mathbf{I}_{\phi g} = \mathbf{I}_L = \mathbf{I}_{\phi L} \tag{21.6}$$

For a balanced or unbalanced load, since the generator and load have a common neutral point, then

$$\mathbf{V}_\phi = \mathbf{E}_\phi \tag{21.7}$$

In addition, since $\mathbf{I}_{\phi L} = \mathbf{V}_\phi / \mathbf{Z}_\phi$, the magnitude of the current in each phase will be equal for a balanced load and unequal for an unbalanced load. You will recall that for the Y-connected generator, the magnitude of the line voltage is equal to $\sqrt{3}$ times the phase voltage. This same relationship can be applied to a balanced or unbalanced four-wire Y-connected load:

$$E_L = \sqrt{3} V_\phi \tag{21.8}$$

For a voltage drop across a load element, the first subscript refers to that terminal through which the current enters the load element, and the second subscript to the terminal from which the current leaves. In other words, the first subscript is, by definition, positive with respect to the second for a voltage drop. Note Fig. 21.13, in which the standard double subscripts for a source of emf and a voltage drop are indicated.

EXAMPLE 21.1. The phase sequence of the Y-connected generator in Fig. 21.13 is *ABC*.

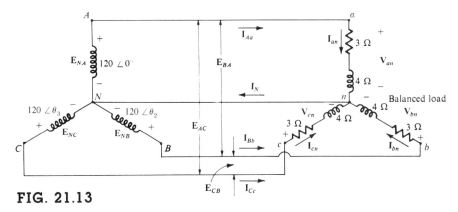

FIG. 21.13

a. Find the phase angles θ_2 and θ_3.
b. Find the magnitude of the line voltages.
c. Find the line currents.
d. Verify that since the load is balanced, $\mathbf{I}_N = 0$.

Solutions:
a. For an *ABC* phase sequence,

$$\theta_2 = -\mathbf{120°} \quad \text{and} \quad \theta_3 = +\mathbf{120°}$$

b. $E_L = \sqrt{3}E_\phi = 1.73(120) = 208$ V. Therefore,

$$E_{BA} = E_{CB} = E_{AC} = \mathbf{208 \ V}$$

c. $\mathbf{V}_\phi = \mathbf{E}_\phi$. Therefore,

$$\mathbf{V}_{an} = \mathbf{E}_{NA}, \quad \mathbf{V}_{bn} = \mathbf{E}_{NB}, \quad \mathbf{V}_{cn} = \mathbf{E}_{NC}$$

$$\mathbf{I}_{\phi L} = \mathbf{I}_{an} = \frac{\mathbf{V}_{an}}{\mathbf{Z}_{an}} = \frac{120 \ \angle 0°}{3 + j4} = \frac{120 \ \angle 0°}{5 \ \angle 53.13°}$$
$$= 24 \ \angle -53.13°$$

$$\mathbf{I}_{bn} = \frac{\mathbf{V}_{bn}}{\mathbf{Z}_{bn}} = \frac{120 \ \angle -120°}{5 \ \angle 53.13°} = 24 \ \angle -173.13°$$

$$\mathbf{I}_{cn} = \frac{\mathbf{V}_{cn}}{\mathbf{Z}_{cn}} = \frac{120 \ \angle +120°}{5 \ \angle 53.13°} = 24 \ \angle 66.87°$$

and since $\mathbf{I}_L = \mathbf{I}_{\phi L}$,

$$\mathbf{I}_{Aa} = \mathbf{I}_{an} = \mathbf{24 \ \angle -53.13°}$$
$$\mathbf{I}_{Bb} = \mathbf{I}_{bn} = \mathbf{24 \ \angle -173.13°}$$
$$\mathbf{I}_{Cc} = \mathbf{I}_{cn} = \mathbf{24 \ \angle 66.87°}$$

d. Applying Kirchhoff's current law, we have

$$\mathbf{I}_N = \mathbf{I}_{Aa} + \mathbf{I}_{Bb} + \mathbf{I}_{Cc}$$

In rectangular form,

$$
\begin{aligned}
\mathbf{I}_{Aa} &= 24 \ \angle -53.13° &&= & 14.40 &- j19.20 \\
\mathbf{I}_{Bb} &= 24 \ \angle -173.13° &&= & -23.83 &- \ j2.87 \\
\mathbf{I}_{Cc} &= 24 \ \angle 66.87° &&= & 9.43 &+ j22.07 \\
\Sigma\mathbf{I}_{Aa} + \mathbf{I}_{Bb} + \mathbf{I}_{Cc} &&&= & 0 &+ j0
\end{aligned}
$$

and $\mathbf{I}_N$ is in fact equal to **0.**

21.6 THE Y-Δ SYSTEM

There is no neutral connection for the Y-Δ system of Fig. 21.14. Any variation in the impedance of a phase which produces an unbalanced system will simply vary the line and phase currents of the system.

For a balanced load,

$$\boxed{\mathbf{Z}_1 = \mathbf{Z}_2 = \mathbf{Z}_3} \tag{21.9}$$

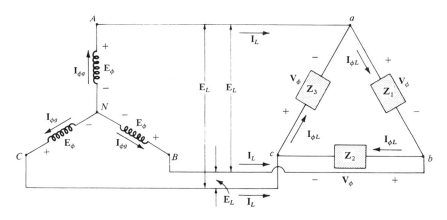

FIG. 21.14 *Y-connected generator with a Δ-connected load.*

The voltage across each phase of the load is equal to the line voltage of the generator for a balanced or unbalanced load:

$$\boxed{\mathbf{V}_\phi = \mathbf{E}_L} \qquad (21.10)$$

The relationship between the line currents and phase currents of a balanced Δ load can be found using an approach very similar to that used in Section 21.3 to find the relationship between the line voltages and phase voltages of a Y-connected generator. For this case, however, Kirchhoff's current law is employed instead of Kirchhoff's voltage law.

The results obtained are

$$\boxed{I_L = \sqrt{3}I_\phi} \qquad (21.11)$$

and the phase angle between a line current and the nearest phase current is 30°. A more detailed discussion of this relationship between the line and phase currents of a Δ-connected system can be found in Section 21.7.

For a balanced load, the line currents will be equal in magnitude and the phase currents will be equal in magnitude.

EXAMPLE 21.2. For the three-phase system of Fig. 21.15:
a. Find the phase angles θ_2 and θ_3.
b. Find the current in each phase of the load.
c. Find the magnitude of the line currents.

Solutions:
a. For an *ABC* sequence,

$$\theta_2 = -\mathbf{120°} \quad \text{and} \quad \theta_3 = +\mathbf{120°}$$

b. $\mathbf{V}_\phi = \mathbf{E}_L$. Therefore,

$$\mathbf{V}_{ab} = \mathbf{E}_{BA}, \quad \mathbf{V}_{ca} = \mathbf{E}_{AC}, \quad \text{and} \quad \mathbf{V}_{bc} = \mathbf{E}_{CB}$$

The phase currents are

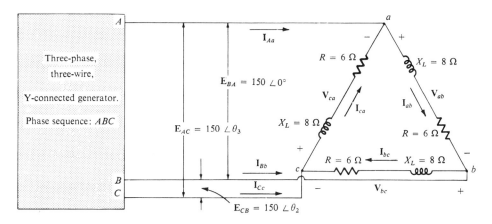

FIG. 21.15

$$\mathbf{I}_{ab} = \frac{\mathbf{V}_{ab}}{\mathbf{Z}_{ab}} = \frac{150 \angle 0°}{6 + j8} = \frac{150 \angle 0°}{10 \angle 53.13°} = \mathbf{15} \angle -\mathbf{53.13°}$$

$$\mathbf{I}_{bc} = \frac{\mathbf{V}_{bc}}{\mathbf{Z}_{bc}} = \frac{150 \angle -120°}{10 \angle 53.13°} = \mathbf{15} \angle -\mathbf{173.13°}$$

$$\mathbf{I}_{ca} = \frac{\mathbf{V}_{ca}}{\mathbf{Z}_{ca}} = \frac{150 \angle +120°}{10 \angle 53.13°} = \mathbf{15} \angle \mathbf{66.87°}$$

c. $I_L = \sqrt{3}I_\phi = \sqrt{3}(15) = \mathbf{25.95 \ A}$. Therefore,

$$I_{Aa} = I_{Bb} = I_{Cc} = \mathbf{25.95 \ A}$$

21.7 THE Δ-CONNECTED GENERATOR

If we rearrange the coils of the generator in Fig. 21.16(a) as shown in Fig. 21.16(b), with A connected to C', B to A', and C to B', the system is referred to as a *three-phase, three-wire, Δ-connected ac generator*. In this system, the phase and line voltages are equivalent and equal to the voltage induced across each coil of the generator. That is,

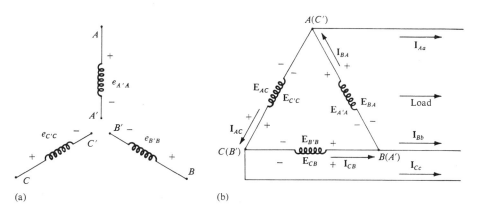

(a)   (b)

FIG. 21.16 *Δ-connected generator.*

$$\left.\begin{array}{ll}\mathbf{E}_{BA} = \mathbf{E}_{A'A} & \text{and} \quad e_{A'A} = \sqrt{2}E_{A'A}\sin\omega t \\ \mathbf{E}_{CB} = \mathbf{E}_{B'B} & \text{and} \quad e_{B'B} = \sqrt{2}E_{B'B}\sin(\omega t - 120°) \\ \mathbf{E}_{AC} = \mathbf{E}_{C'C} & \text{and} \quad e_{C'C} = \sqrt{2}E_{C'C}\sin(\omega t + 120°) \end{array}\right\} \begin{array}{l}\text{Phase} \\ \text{sequence} \\ ABC\end{array}$$

or

$$\boxed{\mathbf{E}_L = \mathbf{E}_{\phi g}} \tag{21.12}$$

Note that only one voltage (magnitude) is available instead of the two available in the Y-connected system.

Unlike the line current for the Y-connected generator, the line current for the Δ-connected system is not equal to the phase current. The relationship between the two can be found by applying Kirchhoff's current law at one of the nodes and solving for the line current in terms of the phase currents. That is, at node A,

$$\mathbf{I}_{BA} = \mathbf{I}_{Aa} + \mathbf{I}_{AC}$$

or

$$\mathbf{I}_{Aa} = \mathbf{I}_{BA} - \mathbf{I}_{AC} = \mathbf{I}_{BA} + \mathbf{I}_{CA}$$

The phasor diagram is shown in Fig. 21.17 for a balanced load.

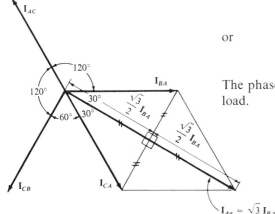

FIG. 21.17

Using the same procedure to find the line current as was used to find the line voltage of a Y-connected generator produces the following general results:

$$\boxed{I_L = \sqrt{3}I_{\phi g}} \tag{21.13}$$

The phase angle between a line current and the nearest phase current is 30°. The phasor diagram of the currents is shown in Fig. 21.18.

It can be shown in the same manner employed for the voltages of a Y-connected generator that the phasor sum of the line currents or phase currents for Δ-connected systems with balanced loads is zero.

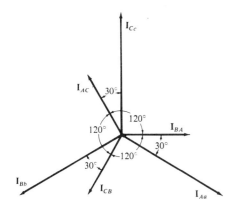

FIG. 21.18

21.8 PHASE SEQUENCE (Δ-CONNECTED GENERATOR)

Even though the line and phase voltages of a Δ-connected system are the same, it is standard practice to describe the phase sequence in terms of the line voltages. The method used is the same as that described for the line voltages of the Y-connected generator. For example, the phasor diagram of the line voltages for a phase sequence ABC is shown in Fig. 21.19. In drawing such a diagram, one must take care to have the sequence of the first and second subscripts the same.

In phasor notation,

$$\mathbf{E}_{BA} = E_{BA} \angle 0°$$
$$\mathbf{E}_{CB} = E_{CB} \angle -120°$$
$$\mathbf{E}_{AC} = E_{AC} \angle 120°$$

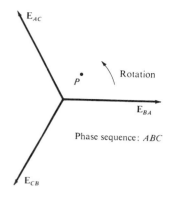

Phase sequence: ABC

FIG. 21.19

21.9 THE Δ-Δ, Δ-Y THREE-PHASE SYSTEMS

The basic equations necessary to analyze either of the two systems (Δ-Δ, Δ-Y) have been presented at least once in this chapter. We will therefore proceed directly to two descriptive examples, one with a Δ-connected load and one with a Y-connected load.

EXAMPLE 21.3. For the Δ-Δ system shown in Fig. 21.20:
a. Find the phase angles θ_2 and θ_3 for the specified phase sequence.
b. Find the current in each phase of the load.
c. Find the magnitude of the line currents.

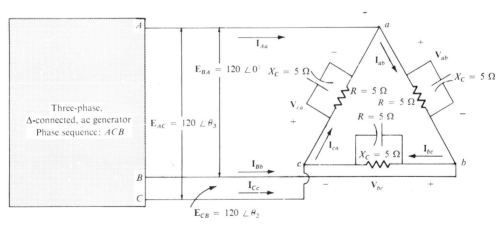

FIG. 21.20 Δ-Δ *system.*

Solutions:
a. For an *ACB* phase sequence,

$$\theta_2 = \mathbf{120°} \quad \text{and} \quad \theta_3 = \mathbf{-120°}$$

b. $\mathbf{V}_\phi = \mathbf{E}_L$. Therefore,

$$\mathbf{V}_{ab} = \mathbf{E}_{BA}, \quad \mathbf{V}_{ca} = \mathbf{E}_{AC}, \quad \text{and} \quad \mathbf{V}_{bc} = \mathbf{E}_{CB}$$

The phase currents are

$$\mathbf{I}_{ab} = \frac{\mathbf{V}_{ab}}{\mathbf{Z}_{ab}} = \frac{120 \angle 0°}{\dfrac{(5 \angle 0°)(5 \angle -90°)}{5 - j5}} = \frac{120 \angle 0°}{\dfrac{25 \angle -90°}{7.07 \angle -45°}}$$

$$= \frac{120 \angle 0°}{3.54 \angle -45°} = \mathbf{33.9 \angle 45°}$$

$$\mathbf{I}_{bc} = \frac{\mathbf{V}_{bc}}{\mathbf{Z}_{bc}} = \frac{120 \angle 120°}{3.54 \angle -45°} = \mathbf{33.9 \angle 165°}$$

$$\mathbf{I}_{ca} = \frac{\mathbf{V}_{ca}}{\mathbf{Z}_{ca}} = \frac{120 \angle -120°}{3.54 \angle -45°} = \mathbf{33.9 \angle -75°}$$

c. $I_L = \sqrt{3}I_\phi = (1.73)(34) = 58.82$ A. Therefore,

$$I_{Aa} = I_{Bb} = I_{Cc} = \mathbf{58.82\ A}$$

EXAMPLE 21.4. For the Δ-Y system shown in Fig. 21.21:
a. Find the voltage across each phase of the load.
b. Find the magnitude of the line voltages.

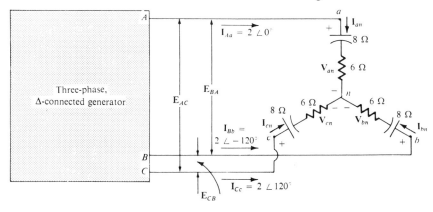

FIG. 21.21 Δ-Y system.

Solutions:
a. $\mathbf{I}_{\phi L} = \mathbf{I}_L$. Therefore,

$$\mathbf{I}_{an} = \mathbf{I}_{Aa} = 2 \angle 0°$$
$$\mathbf{I}_{bn} = \mathbf{I}_{Bb} = 2 \angle -120°$$
$$\mathbf{I}_{cn} = \mathbf{I}_{Cc} = 2 \angle 120°$$

The phase voltages are

$$\mathbf{V}_{an} = \mathbf{I}_{an}\mathbf{Z}_{an} = (2 \angle 0°)(10 \angle -53.13°) = \mathbf{20 \angle -53.13°}$$
$$\mathbf{V}_{bn} = \mathbf{I}_{bn}\mathbf{Z}_{bn} = (2 \angle -120°)(10 \angle -53.13°)$$
$$= \mathbf{20 \angle -173.13°}$$
$$\mathbf{V}_{cn} = \mathbf{I}_{cn}\mathbf{Z}_{cn} = (2 \angle 120°)(10 \angle -53.13°) = \mathbf{20 \angle 66.87°}$$

b. $E_L = \sqrt{3}V_\phi = (1.73)(20) = 34.6$ V or

$$E_{BA} = E_{CB} = E_{AC} = \mathbf{34.6\ V}$$

21.10 POWER

Y-Connected Balanced Load (Fig. 21.22)

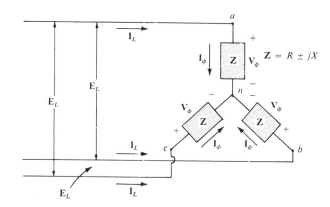

Average power The average power delivered to each phase can be determined by any one of Eqs. (21.14), (21.15), and (21.16).

FIG. 21.22

$$\boxed{P_\phi = V_\phi I_\phi \cos \theta_{I_\phi}^{V_\phi} = I_\phi^2 R_\phi = \frac{V_R^2}{R_\phi}} \qquad \text{(W)} \quad \textbf{(21.14)}$$

where $\theta_{I_\phi}^{V_\phi}$ indicates that θ is the phase angle between V_ϕ and I_ϕ. The total power to the balanced load is

$$\boxed{P_T = 3P_\phi} \qquad \text{(W)} \qquad \textbf{(21.15)}$$

or, since

$$V_\phi = \frac{E_L}{\sqrt{3}} \quad \text{and} \quad I_\phi = I_L$$

$$P_T = 3\frac{E_L}{\sqrt{3}}I_L \cos \theta_{I_\phi}^{V_\phi}$$

But

$$\frac{3}{\sqrt{3}}(1) = \left(\frac{3}{\sqrt{3}}\right)\left(\frac{\sqrt{3}}{\sqrt{3}}\right) = \frac{3\sqrt{3}}{3} = \sqrt{3}$$

Therefore,

$$\boxed{P_T = \sqrt{3}E_L I_L \cos \theta_{I_\phi}^{V_\phi} = 3I_L^2 R_\phi} \qquad \text{(W)} \quad \textbf{(21.16)}$$

Reactive power The reactive power of each phase is

$$\boxed{Q_\phi = V_\phi I_\phi \sin \theta_{I_\phi}^{V_\phi} = I_\phi^2 X_\phi = \frac{V_X^2}{X_\phi}} \qquad \text{(var)} \quad \textbf{(21.17)}$$

The total reactive power of the load is

$$\boxed{Q_T = 3Q_\phi} \qquad \text{(var)} \qquad \textbf{(21.18)}$$

or, proceeding in the same manner as above, we have

$$\boxed{Q_T = \sqrt{3}E_L I_L \sin \theta_{I_\phi}^{V_\phi} = 3I_L^2 X_\phi} \qquad \text{(var)} \quad \textbf{(21.19)}$$

Apparent power The apparent power of each phase is

$$S_\phi = V_\phi I_\phi \qquad \text{(VA)} \qquad \textbf{(21.20)}$$

The total apparent power of the load is

$$S_T = 3S_\phi \qquad \text{(VA)} \qquad \textbf{(21.21)}$$

or, as before,

$$S_T = \sqrt{3} E_L I_L \qquad \text{(VA)} \qquad \textbf{(21.22)}$$

Power factor The power factor of the system is given by

$$F_p = \frac{P_T}{S_T} = \cos\theta \text{ (leading or lagging)} \qquad \textbf{(21.23)}$$

EXAMPLE 21.5. See Fig. 21.23.

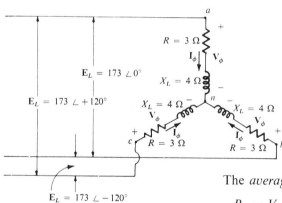

FIG. 21.23

$$Z_\phi = 3 + j4 = 5 \angle 53.13°$$

$$V_\phi = \frac{V_L}{\sqrt{3}} = \frac{173}{1.73} = 100 \text{ V}$$

$$I_\phi = \frac{V_\phi}{Z_\phi} = \frac{100}{5} = 20 \text{ A}$$

The *average power* is

$$P_\phi = V_\phi I_\phi \cos\theta_{I_\phi}^{V_\phi} = (100)(20)\cos 53.13° = (2000)(0.6)$$
$$= \textbf{1200 W}$$

$$P_\phi = I_\phi^2 R_\phi = (20^2)(3) = (400)(3) = \textbf{1200 W}$$

$$P_\phi = \frac{V_R^2}{R_\phi} = \frac{60^2}{3} = \frac{3600}{3} = \textbf{1200 W}$$

$$P_T = 3P_\phi = (3)(1200) = \textbf{3600 W}$$

or

$$P_T = \sqrt{3} E_L I_L \cos\theta_{I_\phi}^{V_\phi} = (1.73)(173)(20)(0.6)$$
$$= \textbf{3600 W}$$

The *reactive power* is

$$Q_\phi = V_\phi I_\phi \sin\theta_{I_\phi}^{V_\phi} = (100)(20)\sin 53.13° = 2000(0.8)$$
$$= \textbf{1600 var}$$

or

$$Q_\phi = I_\phi^2 X_\phi = (20^2)(4) = (400)(4) = \textbf{1600 var}$$

$$Q_T = 3Q_\phi = (3)(1600) = \textbf{4800 var}$$

or

$$Q_T = \sqrt{3}E_L I_L \sin\theta_{I_\phi}^{V_\phi} = (1.73)(173)(20)(0.8) = \textbf{4800 var}$$

The *apparent power* is

$$S_\phi = V_\phi I_\phi = (100)(20) = \textbf{2000 VA}$$
$$S_T = 3S_\phi = (3)(2000) = \textbf{6000 VA}$$

or

$$S_T = \sqrt{3}E_L I_L = (1.73)(173)(20) = \textbf{6000 VA}$$

The *power factor* is

$$F_p = \frac{P_T}{S_T} = \frac{3600}{6000} = \textbf{0.6 lagging}$$

Δ-Connected Balanced Load (Fig. 21.24)

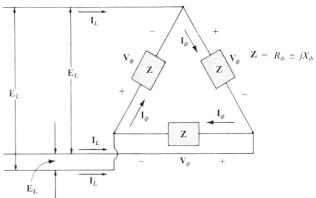

FIG. 21.24

Average power

$$\boxed{P_\phi = V_\phi I_\phi \cos\theta_{I_\phi}^{V_\phi} = I_\phi^2 R_\phi = \frac{V_R^2}{R_\phi}} \quad \text{(W)} \quad \textbf{(21.24)}$$

$$\boxed{P_T = 3P_\phi} \quad \text{(W)} \quad \textbf{(21.25)}$$

Reactive power

$$\boxed{Q_\phi = V_\phi I_\phi \sin\theta_{I_\phi}^{V_\phi} = I_\phi^2 X_\phi = \frac{V_X^2}{X_\phi}} \quad \text{(var)} \quad \textbf{(21.26)}$$

$$\boxed{Q_T = 3Q_\phi} \quad \text{(var)} \quad \textbf{(21.27)}$$

Apparent power

$$\boxed{S_\phi = V_\phi I_\phi} \quad \text{(VA)} \quad \textbf{(21.28)}$$

$$\boxed{S_T = 3S_\phi = \sqrt{3}E_L I_L} \quad \text{(VA)} \quad \textbf{(21.29)}$$

Power factor

$$\boxed{F_p = \frac{P_T}{S_T}} \quad \textbf{(21.30)}$$

EXAMPLE 21.6. Determine the total watts, vars, and volt-amperes for the network of Fig. 21.25. In addition, calculate the total power factor of the load.

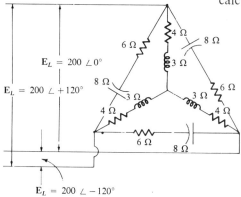

$\mathbf{E}_L = 200 \angle 0°$

$\mathbf{E}_L = 200 \angle +120°$

$\mathbf{E}_L = 200 \angle -120°$

FIG. 21.25

Solution: Consider the Δ and Y separately. For the Δ,

$$\mathbf{Z} = 6 - j8 = 10 \angle -53.13°$$

$$I_\phi = \frac{200}{10} = 20 \text{ A}$$

$$P_T = 3I_\phi^2 R_\phi = (3)(20^2)(6) = \textbf{7200 W}$$

$$Q_T = 3I_\phi^2 X_\phi = (3)(20^2)(8) = \textbf{9600 var (cap.)}$$

$$S_T = 3V_\phi I_\phi = (3)(200)(20) = \textbf{12,000 VA}$$

For the Y,

$$\mathbf{Z} = 4 + j3 = 5 \angle 36.87°$$

$$I_\phi = \frac{200/\sqrt{3}}{5} = \frac{116}{5} = 23.12 \text{ A}$$

$$P_T = 3I_\phi^2 R_\phi = (3)(23.12^2)(4) = \textbf{6414.41 W}$$

$$Q_T = 3I_\phi^2 X_\phi = (3)(23.12^2)(3) = \textbf{4810.81 var (ind.)}$$

$$S_T = 3V_\phi I_\phi = (3)(116)(23.12) = \textbf{8045.76 VA}$$

$$P_T = P_{T_\Delta} + P_{T_Y} = 7200 + 6414.41 = \textbf{13,614.41 W}$$

$$Q_T = Q_{T_\Delta} - Q_{T_Y} = 9600 \text{ (cap.)} - 4810.81 \text{ (ind.)}$$
$$= \textbf{4789.19 var (cap.)}$$

$$S_T = \sqrt{P_T^2 + Q_T^2} = \sqrt{(13,614.41)^2 + (4,789.19)^2}$$
$$= \textbf{14,432.2 VA}$$

$$F_p = \frac{P_T}{S_T} = \frac{13,614.41}{14,432.20} = \textbf{0.943 leading}$$

21.11 THE THREE-WATTMETER METHOD

The power delivered to a balanced or an unbalanced four-wire, Y-connected load can be found using three watt-meters in the manner shown in Fig. 21.26. Each wattmeter measures the power delivered to each phase. The potential

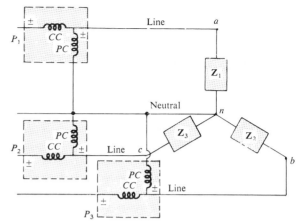

FIG. 21.26

coil of each wattmeter is connected parallel with the load, while the current coil is in series with the load. The total average power of the system can be found by summing the three wattmeter readings; that is,

$$P_{T_Y} = P_1 + P_2 + P_3 \qquad \textbf{(21.31)}$$

For the load (balanced or unbalanced), the wattmeters are connected as shown in Fig. 21.27. The total power is again the sum of the three wattmeter readings:

$$P_{T_\Delta} = P_1 + P_2 + P_3 \qquad \textbf{(21.32)}$$

If in either of the cases just described the load is balanced, the power delivered to each phase will be the same. The total power is then just three times any one wattmeter reading.

21.12 THE TWO-WATTMETER METHOD

The power delivered to a three-phase, three-wire, Y- or Δ-connected balanced or unbalanced load can be found using only two wattmeters if the proper connection is used and if the wattmeter readings are properly interpreted. The basic connections are shown in Fig. 21.28. One end of each potential coil is connected to the same line. The current coils are then placed in the remaining lines.

The connection shown in Fig. 21.29 will also satisfy the requirements. A third hookup is also possible, but this is left to the reader as an exercise.

The total power delivered to the load is the algebraic sum of the two wattmeter readings. For a *balanced* load, we will now consider two methods of determining whether the total power is the sum or the difference of the two watt-meter readings. The first method to be described requires

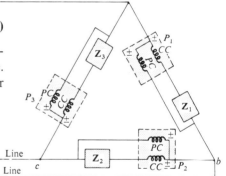

FIG. 21.27

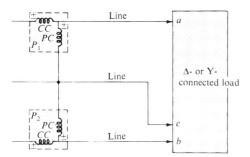

FIG. 21.28

that we know or be able to find the power factor (leading or lagging) of any one phase of the load. When this informa-

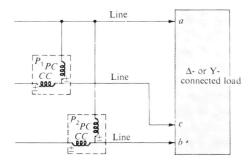

FIG. 21.29

tion has been obtained, it can be applied directly to the curve of Fig. 21.30.

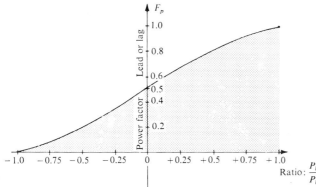

FIG. 21.30

The curve in Fig. 21.30 is a plot of the power factor of the load (phase) vs. the ratio P_l/P_h where P_l and P_h are the magnitudes of the lower- and higher-reading wattmeters, respectively. Note that for a power factor (leading or lagging) greater than 0.5, the ratio has a positive value. This indicates that both wattmeters are reading positive and the total power is the sum of the two wattmeter readings; that is, $P_T = P_l + P_h$. For a power factor less than 0.5 (leading or lagging), the ratio has a negative value. This indicates that the smaller-reading wattmeter is reading negative and the total power is the difference of the two wattmeter readings; that is, $P_T = P_h - P_l$.

A closer examination will reveal that when the power factor is 1 ($\cos 0° = 1$) corresponding to a purely resistive load, $P_l/P_h = 1$ or $P_l = P_h$, and both wattmeters will have the same wattage indication. At a power factor equal to 0 ($\cos 90° = 0$) corresponding to a purely reactive load, $P_l/P_h = -1$ or $P_l = -P_h$, and both wattmeters will again have the same wattage indication but with opposite signs. The transition from a negative to a positive ratio occurs when the power factor of the load is 0.5 or $\theta = \cos^{-1} 0.5 = 60°$. At this power factor, $P_l/P_h = 0$, so that $P_l = 0$, while P_h will read the total power delivered to the load.

The second method for determining whether the total power is the sum or difference of the two wattmeter readings involves a simple laboratory test. In order to apply the test, both wattmeters must first have an up-scale deflection. If one of the wattmeters has a below-zero indication, an up-scale deflection can be obtained by simply reversing the leads of the current coil of the wattmeter. To perform the test, first remove the lead of the potential coil of the *low-reading* wattmeter from the line that has no current coil in it. Take this lead and touch it to the line that has the current coil of the *high-reading* wattmeter in it. If the pointer of the low-reading wattmeter deflects upward, the two wattmeter readings should be added. If the pointer deflects downward, below zero watts, the wattage reading of the low-reading wattmeter should be subtracted from that of the high-reading wattmeter.

For a *balanced system,* since

$$P_T = P_h \pm P_l = \sqrt{3}E_L I_L \cos \theta_{I_\phi}^{V_\phi}$$

the power factor of the load (phase) can be found from the wattmeter readings and the magnitude of the line voltage and current:

$$\boxed{F_p = \cos \theta_{I_\phi}^{V_\phi} = \frac{P_h \pm P_l}{\sqrt{3}E_L I_L}} \qquad \textbf{(21.33)}$$

21.13 UNBALANCED THREE-PHASE, FOUR-WIRE, Y-CONNECTED LOAD

For the three-phase, four-wire, Y-connected unbalanced load in Fig. 21.31,

$$\mathbf{Z}_1 \neq \mathbf{Z}_2 \neq \mathbf{Z}_3$$

Since the neutral is a common point between the load and source, no matter what the impedance of each phase of the

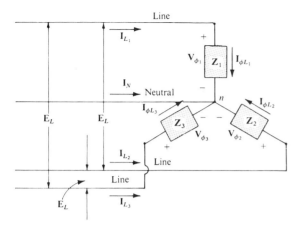

FIG. 21.31 *Unbalanced Y-connected load.*

load, the voltage across each phase is the phase voltage of the generator:

$$\boxed{\mathbf{V}_\phi = \mathbf{E}_\phi} \qquad (21.34)$$

The phase currents, therefore, can be determined by Ohm's law:

$$\boxed{\mathbf{I}_{\phi_1} = \frac{\mathbf{V}_{\phi_1}}{\mathbf{Z}_1} = \frac{\mathbf{E}_{\phi_1}}{\mathbf{Z}_1}, \text{ and so on}} \qquad (21.35)$$

The current in the neutral for any unbalanced system can then be found by applying Kirchhoff's current law at the common point N':

$$\boxed{\mathbf{I}_N = \mathbf{I}_{\phi_1} + \mathbf{I}_{\phi_2} + \mathbf{I}_{\phi_3} = \mathbf{I}_{L_1} + \mathbf{I}_{L_2} + \mathbf{I}_{L_3}} \qquad (21.36)$$

21.14 UNBALANCED THREE-PHASE, THREE-WIRE, Y-CONNECTED LOAD

For the system shown in Fig. 21.32, the required equations can be derived by first applying Kirchhoff's voltage law around each closed loop to produce

$$\mathbf{E}_{BA} - \mathbf{V}_{an} + \mathbf{V}_{bn} = 0$$
$$\mathbf{E}_{CB} - \mathbf{V}_{bn} + \mathbf{V}_{cn} = 0$$
$$\mathbf{E}_{AC} - \mathbf{V}_{cn} + \mathbf{V}_{an} = 0$$

Substituting, we have

$$\mathbf{V}_{an} = \mathbf{I}_{an}\mathbf{Z}_1 \qquad \mathbf{V}_{bn} = \mathbf{I}_{bn}\mathbf{Z}_2 \qquad \mathbf{V}_{cn} = \mathbf{I}_{cn}\mathbf{Z}_3$$

$$\boxed{\begin{aligned} \mathbf{E}_{BA} &= \mathbf{I}_{an}\mathbf{Z}_1 - \mathbf{I}_{bn}\mathbf{Z}_2 \\ \mathbf{E}_{CB} &= \mathbf{I}_{bn}\mathbf{Z}_2 - \mathbf{I}_{cn}\mathbf{Z}_3 \\ \mathbf{E}_{AC} &= \mathbf{I}_{cn}\mathbf{Z}_3 - \mathbf{I}_{an}\mathbf{Z}_1 \end{aligned}}$$

(21.37a)
(21.37b)
(21.37c)

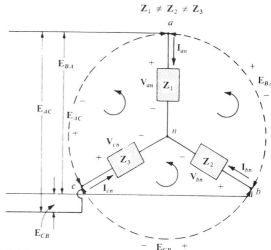

FIG. 21.32

Applying Kirchhoff's current law at node N' gives us

$$\mathbf{I}_{an} + \mathbf{I}_{bn} + \mathbf{I}_{cn} = 0 \quad \text{and} \quad \mathbf{I}_{bn} = -\mathbf{I}_{an} - \mathbf{I}_{cn}$$

Substituting for I_{bn} in Eqs. (21.37a) and (21.37b) yields

$$\mathbf{E}_{BA} = \mathbf{I}_{an}\mathbf{Z}_1 - [-(\mathbf{I}_{an} + \mathbf{I}_{cn})]\mathbf{Z}_2$$
$$\mathbf{E}_{CB} = -(\mathbf{I}_{an} + \mathbf{I}_{cn})\mathbf{Z}_2 - \mathbf{I}_{cn}\mathbf{Z}_3$$

which are rewritten as

$$\mathbf{E}_{BA} = \mathbf{I}_{an}(\mathbf{Z}_1 + \mathbf{Z}_2) + \mathbf{I}_{cn}\mathbf{Z}_2$$
$$\mathbf{E}_{CB} = \mathbf{I}_{an}(-\mathbf{Z}_2) + \mathbf{I}_{cn}[-(\mathbf{Z}_2 + \mathbf{Z}_3)]$$

Using determinants, we have

$$\mathbf{I}_{an} = \frac{\begin{vmatrix} \mathbf{E}_{BA} & \mathbf{Z}_2 \\ \mathbf{E}_{CB} & -(\mathbf{Z}_2 + \mathbf{Z}_3) \end{vmatrix}}{\begin{vmatrix} \mathbf{Z}_1 + \mathbf{Z}_2 & \mathbf{Z}_2 \\ -\mathbf{Z}_2 & -(\mathbf{Z}_2 + \mathbf{Z}_3) \end{vmatrix}}$$

$$= \frac{-(\mathbf{Z}_2 + \mathbf{Z}_3)\mathbf{E}_{BA} - \mathbf{E}_{CB}\mathbf{Z}_2}{-\mathbf{Z}_1\mathbf{Z}_2 - \mathbf{Z}_1\mathbf{Z}_3 - \mathbf{Z}_2\mathbf{Z}_3 - \mathbf{Z}_2^2 + \mathbf{Z}_2^2}$$

$$\mathbf{I}_{an} = \frac{-\mathbf{Z}_2(\mathbf{E}_{BA} + \mathbf{E}_{CB}) - \mathbf{Z}_3\mathbf{E}_{BA}}{-\mathbf{Z}_1\mathbf{Z}_2 - \mathbf{Z}_1\mathbf{Z}_3 - \mathbf{Z}_2\mathbf{Z}_3}$$

Apply Kirchhoff's voltage law to the line voltages:

$$\mathbf{E}_{BA} + \mathbf{E}_{AC} + \mathbf{E}_{CB} = 0 \quad \text{or} \quad \mathbf{E}_{BA} + \mathbf{E}_{CB} = -\mathbf{E}_{AC}$$

Substitute for $\mathbf{E}_{BA} + \mathbf{E}_{BC}$ in the above equation for $\mathbf{I}_{an}$:

$$\mathbf{I}_{an} = \frac{-\mathbf{Z}_2(-\mathbf{E}_{AC}) - \mathbf{Z}_3\mathbf{E}_{BA}}{-\mathbf{Z}_1\mathbf{Z}_2 - \mathbf{Z}_1\mathbf{Z}_3 - \mathbf{Z}_2\mathbf{Z}_3}$$

and

$$\boxed{\mathbf{I}_{an} = \frac{\mathbf{E}_{BA}\mathbf{Z}_3 - \mathbf{E}_{AC}\mathbf{Z}_2}{\mathbf{Z}_1\mathbf{Z}_2 + \mathbf{Z}_1\mathbf{Z}_3 + \mathbf{Z}_2\mathbf{Z}_3}} \qquad \text{(21.38)}$$

In the same manner, it can be shown that

$$\mathbf{I}_{cn} = \frac{\mathbf{E}_{AC}\mathbf{Z}_2 - \mathbf{E}_{CB}\mathbf{Z}_1}{\mathbf{Z}_1\mathbf{Z}_2 + \mathbf{Z}_1\mathbf{Z}_3 + \mathbf{Z}_2\mathbf{Z}_3} \qquad (21.39)$$

Substituting Eq. (21.39) for $\mathbf{I}_{cn}$ in the right-hand side of Eq. (21.37b), we obtain

$$\mathbf{I}_{bn} = \frac{\mathbf{E}_{CB}\mathbf{Z}_1 - \mathbf{E}_{BA}\mathbf{Z}_3}{\mathbf{Z}_1\mathbf{Z}_2 + \mathbf{Z}_1\mathbf{Z}_3 + \mathbf{Z}_2\mathbf{Z}_3} \qquad (21.40)$$

EXAMPLE 21.7. A *phase sequence indicator* (Fig. 21.33) is an instrument that can determine the phase sequence of a polyphase circuit. The numbers 1-2-3 correspond with the terminals *A-B-C* described in this chapter.

A network that will perform the same function as the indicator of Fig. 21.33 appears in Fig. 21.34. As noted, the applied phase sequence is *ABC*. The bulb corresponding to this phase sequence burns more brightly than the bulb indicating the *ACB* sequence because a greater current is passing through the *ABC* bulb. Calculating the phase currents will show if this situation does in fact exist:

Courtesy of General Electric Co.

FIG. 21.33 *Phase-sequence indicator.*

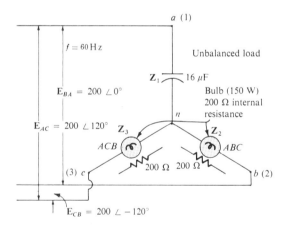

FIG. 21.34

$$X_C = \frac{1}{\omega C} = \frac{1}{(377)(16 \times 10^{-6})} = 166 \ \Omega$$

By Eq. (21.39),

$$\mathbf{I}_{cn} = \frac{\mathbf{E}_{AC}\mathbf{Z}_2 - \mathbf{E}_{CB}\mathbf{Z}_1}{\mathbf{Z}_1\mathbf{Z}_2 + \mathbf{Z}_1\mathbf{Z}_3 + \mathbf{Z}_2\mathbf{Z}_3}$$

$$= \frac{(200 \angle 120°)(200 \angle 0°) - (200 \angle -120°)(166 \angle -90°)}{(166 \angle -90°)(200 \angle 0°) + (166 \angle -90°)(200 \angle 0°)}{+ (200 \angle 0°)(200 \angle 0°)}$$

$$\mathbf{I}_{cn} = \frac{40,000 \angle 120° + 33,200 \angle -30°}{33,200 \angle -90° + 33,200 \angle -90° + 40,000 \angle 0°}$$

Dividing the numerator and denominator by 1000 and converting to the rectangular domain yields

$$I_{cn} = \frac{(-20 + j34.64) + (28.75 - j16.60)}{40 - j66.4}$$

$$= \frac{8.75 + j18.04}{77.52 \angle -58.93°} = \frac{20.05 \angle 64.13°}{77.52 \angle -58.93°}$$

$$I_{cn} = 0.259 \angle 123.06°$$

By Eq. (21.40),

$$I_{bn} = \frac{E_{CB}Z_1 - E_{BA}Z_3}{Z_1Z_2 + Z_1Z_3 + Z_2Z_3}$$

$$= \frac{(200 \angle -120°)(166 \angle -90°) - (200 \angle 0°)(200 \angle 0°)}{77.5 \times 10^3 \angle -59.3°}$$

$$I_{bn} = \frac{33,200 \angle -210° - 40,000 \angle 0°}{77.5 \times 10^3 \angle -59.3°}$$

Dividing by 1000 and converting to the rectangular domain yields

$$I_{bn} = \frac{-28.75 + j16.60 - 40.0}{77.52 \angle -58.93°}$$

$$= \frac{-68.75 + j16.60}{77.52 \angle -58.93°}$$

$$= \frac{70.73 \angle 166.43°}{77.52 \angle -58.93°}$$

$$I_{bn} = 0.91 \angle 225.36°$$

and $I_{bn} > I_{cn}$. Therefore, the bulb indicating an *ABC* sequence will burn more brightly due to the greater current. If the phase sequence were *ACB*, the reverse would be true.

PROBLEMS

Section 21.5

1. A balanced Y load having a 10-Ω resistance in each leg is connected to a three-phase, four-wire, Y-connected generator having a line voltage of 208 V. Calculate the magnitude of the
   a. phase voltage of the generator.
   b. phase voltage of the load.
   c. phase current of the load.
   d. line current.

2. Repeat Problem 1 if each phase impedance is changed to a 12-Ω resistor in series with a 16-Ω capacitive reactance.

3. Repeat Problem 1 if each phase impedance is changed to a 10-Ω resistor in parallel with a 10-Ω capacitive reactance.

4. The phase sequence for the Y-Y system of Fig. 21.35 is *ABC*.
   a. Find the angles θ_2 and θ_3 for the specified phase sequence.

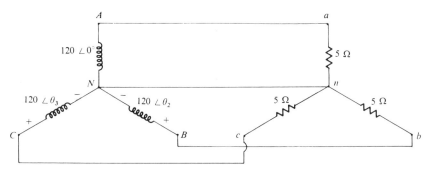

FIG. 21.35

b. Find the voltage across each phase impedance in phasor form.

c. Find the current through each phase impedance in phasor form.

d. Draw the phasor diagram of the currents found in part (c) and show that their phasor sum is 0.

e. Find the magnitude of the line currents.

f. Find the magnitude of the line voltages.

5. Repeat Problem 4 if the phase impedances are changed to a 9-Ω resistor in series with a 12-Ω inductive reactance.

6. Repeat Problem 4 if the phase impedances are changed to a 6-Ω resistance in parallel with an 8-Ω capacitive reactance.

7. For the system of Fig. 21.36, find the magnitude of the unknown voltages and currents.

*8. Compute the magnitude of the voltage E_{BA} for the balanced three-phase system of Fig. 21.37.

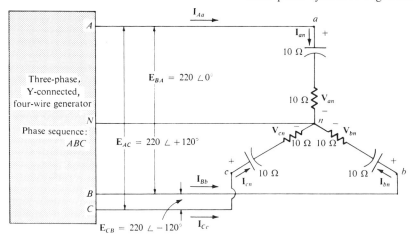

FIG. 21.36

Section 21.6

9. A balanced Δ load having a 20-Ω resistance in each leg is connected to a three-phase, three-wire, Y-connected generator having a line voltage of 208 V. Calculate the magnitude of the

a. phase voltage of the generator.

b. phase voltage of the load.

c. phase current of the load.

d. line current.

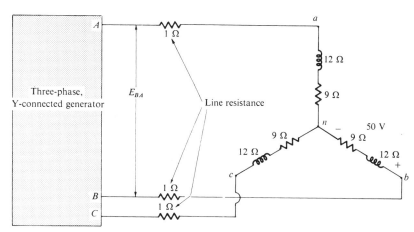

FIG. 21.37

10. Repeat Problem 9 if each phase impedance is changed to a 7-Ω resistor in series with a 14-Ω inductive reactance.

11. Repeat Problem 9 if each phase impedance is changed to an 8-Ω resistance in parallel with an 8-Ω capacitive reactance.

12. The phase sequence for the Y-Δ system of Fig. 21.38 is *ABC*.
   a. Find the angles θ_2 and θ_3 for the specified phase sequence.
   b. Find the voltage across each phase impedance in phasor form.
   c. Draw the phasor diagram of the voltages found in part (b) and show that their sum is 0 around the closed loop of the Δ load.
   d. Find the current through each phase impedance in phasor form.
   e. Find the magnitude of the line currents.
   f. Find the magnitude of the generator phase voltages.

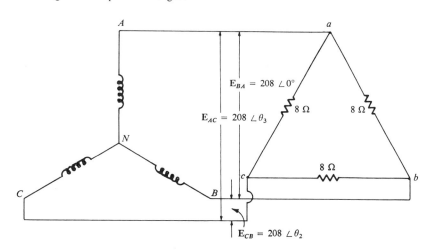

FIG. 21.38

13. Repeat Problem 12 if the phase impedances are changed to a 100-Ω resistor in series with a capacitive reactance of 100 Ω.

14. Repeat Problem 13 if the phase impedances are changed to a 3-Ω resistor in parallel with an inductive reactance of 4 Ω.

15. For the system of Fig. 21.39, find the magnitude of the unknown voltages and currents.

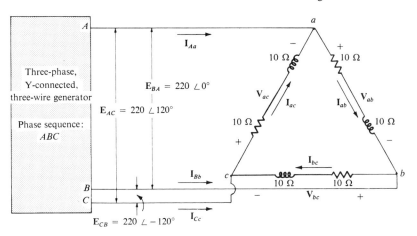

FIG. 21.39

Section 21.9

16. A balanced Y load having a 6-Ω resistance in each leg is
connected to a three-phase, Δ-connected generator having a
line voltage of 208 V. Calculate the magnitude of the
a. phase voltage of the generator.
b. phase voltage of the load.
c. phase current of the load.
d. line current.

17. Repeat Problem 16 if each phase impedance is changed to a
9-Ω resistor in series with a 9-Ω inductive reactance.

18. Repeat Problem 16 if each phase impedance is changed to a
15-Ω resistor in parallel with a 20-Ω capacitive reactance.

*19. For the system of Fig. 21.40, find the magnitude of the unknown voltages and currents.

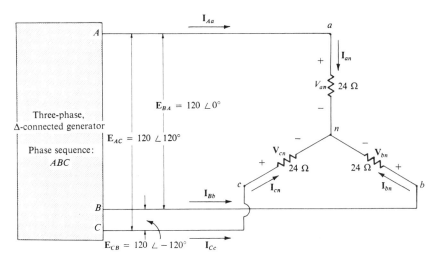

FIG. 21.40

20. Repeat Problem 19 if each phase impedance is changed to a
10-Ω resistor in series with a 20-Ω inductive reactance.

21. Repeat Problem 19 if each phase impedance is changed to a 20-Ω resistor in parallel with a 15-Ω capacitive reactance.

22. A balanced Δ load having a 50-Ω resistance in each leg is connected to a three-phase, Δ-connected generator having a line voltage of 440 V. Calculate the magnitude of the
    a. phase voltage of the generator.
    b. phase voltage of the load.
    c. phase current of the load.
    d. line current.

23. Repeat Problem 22 if each phase impedance is changed to a 12-Ω resistor in series with a 9-Ω capacitive reactance.

24. Repeat Problem 22 if each phase impedance is changed to a 6-Ω resistor in parallel with a 6-Ω inductive reactance.

25. The phase sequence for the Δ-Δ system of Fig. 21.41 is ABC.
    a. Find the angles θ_2 and θ_3 for the specified phase sequence.
    b. Find the voltage across each phase impedance in phasor form.
    c. Draw the phasor diagram of the voltages found in part (b) and show that their phasor sum is 0 around the closed loop of the Δ load.
    d. Find the current through each phase impedance in phasor form.
    e. Find the magnitude of the line currents.

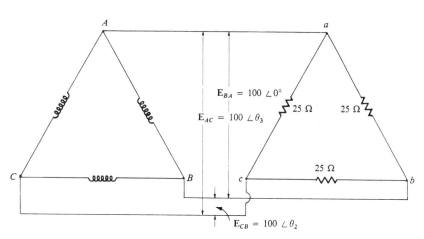

FIG. 21.41

26. Repeat Problem 25 if each phase impedance is changed to a 12-Ω resistor in series with a 16-Ω inductive reactance.

27. Repeat Problem 25 if each phase impedance is changed to a 20-Ω resistor in parallel with a 20-Ω capacitive reactance.

Section 21.10

28. Find the total watts, vars, volt-amperes, and F_p of the three-phase system of Problem 2.

29. Find the total watts, vars, volt-amperes, and F_p of the three-phase system of Problem 4.

30. Find the total watts, vars, volt-amperes, and F_p of the three-phase system of Problem 7.

31. Find the total watts, vars, volt-amperes, and F_p of the three-phase system of Problem 11.

32. Find the total watts, vars, volt-amperes, and F_p of the three-phase system of Problem 13.

33. Find the total watts, vars, volt-amperes, and F_p of the three-phase system of Problem 15.

34. Find the total watts, vars, volt-amperes, and F_p of the three-phase system of Problem 18.

35. Find the total watts, vars, volt-amperes, and F_p of the three-phase system of Problem 20.

36. Find the total watts, vars, volt-amperes, and F_p of the three-phase system of Problem 24.

37. Find the total watts, vars, volt-amperes, and F_p of the three-phase system of Problem 26.

38. A balanced three-phase, Δ-connected load has a line voltage of 200 and a total power consumption of 4800 W at a lagging power factor of 0.8. Find the impedance of each phase in rectangular coordinates.

39. A balanced three-phase, Y-connected load has a line voltage of 208 and a total power consumption of 1200 W at a leading power factor of 0.6. Find the impedance of each phase in rectangular coordinates.

*40. Find the total watts, vars, volt-amperes, and F_p of the system of Fig. 21.42.

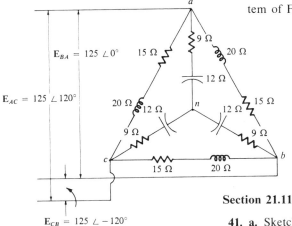

$\mathbf{E}_{BA} = 125 \angle 0°$

$\mathbf{E}_{AC} = 125 \angle 120°$

$\mathbf{E}_{CB} = 125 \angle -120°$

FIG. 21.42

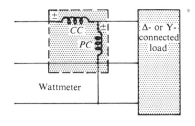

FIG. 21.43

Section 21.11

41. a. Sketch the connections required to measure the total watts delivered to the load of Fig. 21.36 using three wattmeters.
 b. Determine the total wattage dissipation and the reading of each wattmeter.

42. Repeat Problem 41 for the network of Fig. 21.38.

Section 21.12

43. a. For the three-wire system of Fig. 21.43, properly connect a second wattmeter so that the two will measure the total power delivered to the load.
 b. If one wattmeter has a reading of 200 W and the other a reading of 85 W, what is the total dissipation in watts if the total power factor is 0.8 (leading)?

c. Repeat part (b) if the total power factor is 0.2 (lagging) and $P_l = 100$ W.

44. Sketch three different ways two wattmeters can be connected to measure the total power delivered to the load of Problem 15.

Section 21.13

*45. For the system of Fig. 21.44, find
   a. the magnitude of the voltage across each phase of the load.
   b. the magnitude of the current through each phase of the load.
   c. the total watts, vars, volt-amperes, and F_p of the system.

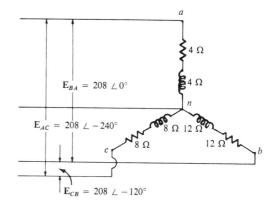

Section 21.14

*46. For the three-phase, three-wire system of Fig. 21.45, find the magnitude of the current through each phase of the load and the total watts, vars, volt-amperes, and F_p of the load.

FIG. 21.44

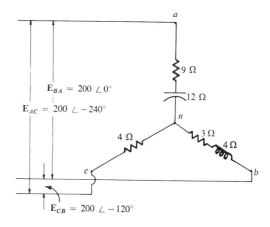

FIG. 21.45

GLOSSARY

Delta (Δ)-connected generator A three-phase generator having the three phases connected in the shape of the capital Greek letter delta (Δ).

Line current The current that flows from the generator to the load of a single-phase or polyphase system.

Line voltage The emf that exists between the lines of a single-phase or polyphase system.

Neutral connection The connection between the generator and the load that, under balanced conditions, will have zero current associated with it.

Phase current The current that flows through each phase of a single-phase or polyphase generator load.

Phase sequence The order in which the generated sinusoidal voltages of a polyphase generator will affect the load to which they are applied.

Phase voltage The emf that appears between the line and neutral of a Y-connected generator and from line to line in a Δ-connected generator.

Polyphase ac generator An electromechanical source of ac power that generates more than one sinusoidal voltage per rotation of the rotor. The frequency generated is determined by the speed of rotation and the number of poles of the rotor.

Single-phase ac generator An electromechanical source of ac power that generates a single sinusoidal voltage having a frequency determined by the speed of rotation and the number of poles of the rotor.

Three-wattmeter method A method for determining the total power delivered to a three-phase load using three wattmeters.

Two-wattmeter method A method for determining the total power delivered to a Δ- or Y-connected three-phase load using only two wattmeters and considering the power factor of the load.

Unbalanced polyphase load A load not having the same impedance in each phase.

WYE (Y)-connected generator A three-phase source of ac power in which the three phases are connected in the shape of the letter Y.

ac METERS

22

22.1 INTRODUCTION

Both the d'Arsonval- and the electrodynamometer-type movements introduced in Chapter 12 can be employed in the design of ac meters. In fact, meters using electrodynamometer-type movements can measure both dc and ac levels without a change in internal circuitry. The d'Arsonval movement, however, does require the addition of a rectifying network which will be described shortly.

In addition to the basic ac ammeter, voltmeter, and wattmeter, the frequency meter, frequency counter, vector voltmeter, Amp-Clamp®, impedance bridge, and oscilloscope will be introduced.

22.2 RECTIFIER-TYPE ac METERS USING THE d'ARSONVAL MOVEMENT

If current flows through a d'Arsonval movement in the direction dictated by the polarities of its external terminals, the pointer will indicate an up-scale deflection. However, if the current is reversed, the torque developed will reverse, causing the pointer to have a below-zero indication. If an alternating current with a frequency of 60 Hz were passed through the movement, the pointer would try to follow the alternating current but would manage only to vibrate about the zero position, producing no meaningful indication.

For the d'Arsonval movement to measure ac quantities, the current first must be converted to one of a steady nature, such as direct current. One method of accomplishing this is rectification, which is well described in most introductory electronic texts. A basic rectifying circuit is shown in Fig. 22.1.

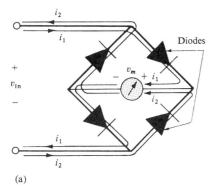

(a)

FIG. 22.1

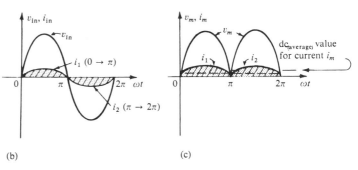

(b)   (c)

The circuit is constructed of four diodes and a d'Arsonval movement in a bridge configuration. The diode is a circuit element that will permit current to flow through it in only one direction. Ideally, it is a short circuit to current flowing through it in the designated direction and an open circuit to current flowing in the opposite direction. For conventional current flow, the diode is a short when the current has the same direction as the arrow in the circuit symbol for the element. In Fig. 22.1, the negative portion of the input current will be "flipped over," so the current always flows in one direction through the movement. The paths of current flow are shown for the positive and negative portions of the sinusoidal input. The meter, however, will read the *average* value of this pulsating waveform. The deflection must therefore be calibrated to indicate the rms value on the face of the movement. If the input quantity to be measured is not sinusoidal, the meter will not indicate the correct effective value since it has been calibrated for that specific signal. Figure 22.2 shows a typical ac voltmeter using a d'Arsonval movement.

22.3 ac METERS USING THE ELECTRODYNAMOMETER MOVEMENT

The electrodynamometer movement can measure both ac and dc quantities without a change in internal circuitry. An electrodynamometer can in fact read the effective value of any periodic or nonperiodic waveform, because a reversal

Courtesy of Simpson Electric Co.

FIG. 22.2 *ac voltmeter (d'Arsonval movement).*

in current direction reverses the fields of both the stationary and the movable coils, so the deflection of the pointer is always up-scale. A typical voltmeter using this movement is shown in Fig. 22.3.

22.4 VOM

As mentioned in Chapter 12, the VOM (volt-ohm-milliammeter) can measure both ac and dc voltages and some currents. Figure 12.13 shows the internal circuitry of a 260 VOM; it illustrates how the rectifier circuit is incorporated in the meter to permit the d'Arsonval movement to indicate the effective value of a sinusoidal input voltage. The bridge circuit is a modified form of the one discussed in Section 22.2.

Courtesy of Weston Instruments, Inc.

FIG. 22.3 *ac voltmeter (electrodynamometer movement).*

22.5 METER INDICATIONS

The indication of an ac or dc voltmeter (or ammeter) when placed in a dc or ac circuit, respectively, must be clearly understood. In Figs. 22.4 and 22.5, two types of dc voltmeters have been placed across a resistor having a 10-V (effective value) ac voltage drop across it. The deflection of each is discussed.

In Fig. 22.4, the d'Arsonval movement cannot follow the 60-Hz oscillations, resulting in a zero volt indication, while in Fig. 22.5, the electrodynamometer movement will indicate the effective value of the waveform.

In Figs. 22.6 through 22.8, four types of ac voltmeters have been placed across a resistor having a 10-V dc drop across it. The deflection of each is again discussed.

In Fig. 22.6, the d'Arsonval movement responds to the average value of the full-wave rectified signal across the movement, which is $2V_m/\pi$. In other words, the $\mathrm{dc}_{(average)}$ value $= 2V_m/\pi$. The required effective value indication is

$$V_{\mathrm{eff}} = 0.707 V_m = 0.707 \left(\frac{\pi}{2}\right)(\mathrm{dc}_{(average)})$$

from above, and

| Meter indication = 1.11 ($\mathrm{dc}_{(average)}$) | (full-wave) |

(22.1)

The scale of the meter is calibrated to indicate 1.11 times the average value of the voltage across the movement.

For Fig. 22.6,

$$\text{Meter indications} = 1.11(10) = 11.1 \text{ V}$$

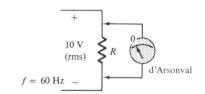

FIG. 22.4

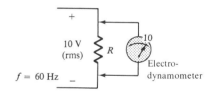

FIG. 22.5

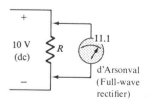

FIG. 22.6

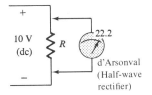

FIG. 22.7

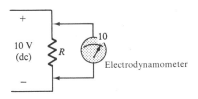

FIG. 22.8

For the ac meter of Fig. 22.7, the calibrating factor is 2.22, since the average value for a half-wave rectified signal is one-half that of the full-wave rectified signal or V_m/π.

$$\boxed{\text{Meter indication} = 2.22 \ (\text{dc}_{(\text{average})})} \qquad \text{(half-wave)}$$

(22.2)

For Fig. 22.7,

$$\text{Meter indications} = 2.22(10) = 22.2 \text{ V}$$

The electrodynamometer movement in the meter of Fig. 22.8 will also read the effective value of any waveform.

22.6 SINGLE-PHASE WATTMETER (ELECTRODYNAMOMETER TYPE)

Since the wattmeter uses the electrodynamometer-type movement, the same instrument can be used to measure the power in a dc or an ac circuit. It can, in fact, be used to measure the wattage of any circuit with periodic or nonperiodic inputs. It is hooked up in the circuit in the same manner as described for dc circuits. The correction factor is still $(V_{PC})^2/R_{PC}$, where V_{PC} is now the effective value of the voltage across the potential coil. The correction factor is taken into account in the design of some wattmeters, thus eliminating the need for adjusting the meter reading. When using this meter, the operator must take care not to exceed the current, voltage, or wattage rating. The product of the voltage and current ratings may or may not equal the wattage rating. In the high-power-factor wattmeter, the product of the voltage and current ratings is usually equal to the wattage rating, or at least 80% of it. For a low-power-factor wattmeter, the product of the current and voltage ratings is much greater than the wattage rating. For obvious reasons, the low-power-factor meter is used only in circuits with low power factors (total impedance highly reactive). Typical ratings for high-power-factor (HPF) and low-power-factor (LPF) meters are shown in Table 22.1. Meters of both high and low power factors have an accuracy of 0.5% to 1% of full scale.

22.7 FREQUENCY METER

Of the many frequency meters available, some of the more common include the vibrating-reed type, the moving-coil

TABLE 22.1

Meter	Current Ratings	Voltage Ratings	Wattage Ratings
HPF	2.5 A	150 V	1500/750/375
	5.0 A	300 V	
LPF	2.5 A	150 V	300/150/75
	5.0 A	300 V	

type, and the resonant-circuit type. We shall discuss briefly the vibrating-reed type, which is the simplest in design (Fig. 22.9).

Courtesy of James G. Biddle Co.

FIG. 22.9 *Frequency meter.*

The vibrating-reed type of frequency meter consists basically of a number of reeds with weighted ends mounted in a row on the same support. Above each reed on the face of the meter is a number corresponding to the frequency of the current passing through the instrument. Each reed is designed to have a natural frequency of vibration. When a current with an unknown frequency is applied, it flows through a coil of an electromagnet. The strength of the coil will then vary with the frequency of the input current and cause the different reeds to vibrate. Since the strength of the coil will alternate between a maximum and minimum twice during each cycle of the input current, a magnet is required to cancel the pulse in one direction. The reed that has a natural frequency equal to that of the input current will vibrate with maximum displacement. For an input current with a frequency of 60 Hz, the vibrating reeds will appear as in Fig. 22.10.

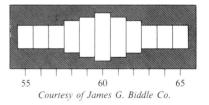

Courtesy of James G. Biddle Co.

FIG. 22.10

FIG. 22.11 *Frequency counter.*

22.8 FREQUENCY COUNTERS

The frequency counter of Fig. 22.11 provides a digital readout of the frequency or period of waveforms having a frequency range from 5 Hz to 80 MHz. In a period average mode it can average the cycle time over 10, 100, or 1000 cycles. It has an input impedance of 1 MΩ and an internal rechargeable battery for portability. Note the high degree of accuracy available from the six-digit display.

22.9 VECTOR VOLTMETER

In the radio frequency (R-F) spectrum of 1 to 1,000 MHz, the Hewlett-Packard vector voltmeter of Fig. 22.12 will provide an indication of both the magnitude and the phase of a voltage measurement. It has two input channels to permit a display of the phase difference on a zero-center meter with end-scale ranges of $\pm 180°$, $\pm 60°$, $\pm 18°$, and $\pm 6°$. The $\pm 6°$ scale provides a 0.1° resolution. The areas of application in this frequency range require voltage measurements in the range from 100 μV to 1 V rms. For this meter, the maximum ac input is 2 V peak, and the dc input, ± 50 V.

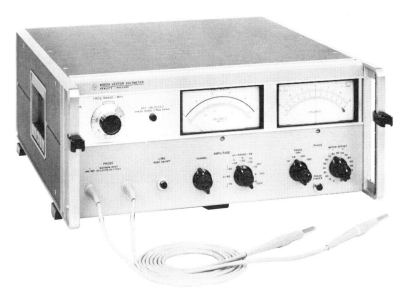

FIG. 22.12 *Vector voltmeter.*

22.10 THE AMP-CLAMP®

The Amp-Clamp® is an instrument that can measure alternating current in the ampere range without having to open the circuit. The loop in Fig. 22.13 is opened by squeezing the "trigger"; then it is placed around the current-carrying conductor. Through transformer action, the level of current in rms units will appear on the appropriate scale. The accuracy of this instrument is ±3% of full scale at 60 Hz, and its scales have maximum values ranging from 6 A to 300 A. The addition of two leads as indicated in the figure permit its use as both a voltmeter and an ohmmeter.

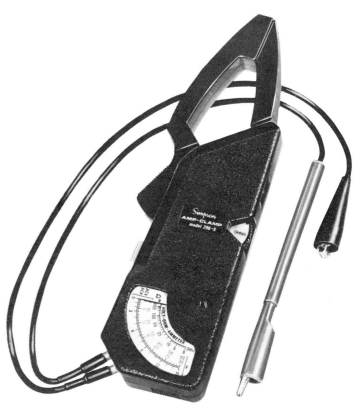

Courtesy of Simpson Instruments, Inc.

FIG. 22.13 *Amp-Clamp*®

22.11 IMPEDANCE BRIDGE

The Hewlett-Packard universal bridge appearing in Fig. 22.14 can be used to make precision measurements of L, C, and R in the range of 0.1 μH to 1111 H, 0.1 pF to 1111 μF, and 0.1 mΩ to 1.111 MΩ, respectively. It can also be used to determine the loss parameters and Q level of the reactive elements. The measurement frequency range is 50 Hz to 10 kHz through the use of an external oscillator, and 1 kHz with the internal oscillator. After the unknown is connected to the bottom left-hand set of terminals, a sequence of steps will result in a balanced condition on the top-center movement. The parameter values can then be read directly off the analog and digital scales.

Courtesy of Hewlett Packard Co.

FIG. 22.14 *Universal bridge.*

22.12 OSCILLOSCOPE

One of the most versatile and important instruments in the electronics industry is the *oscilloscope*. It provides a display of the waveform on a cathode-ray tube to permit the detection of irregularities and the determination quantities such as magnitude, frequency, period, dc component, and so on. The unit of Fig. 22.15 is particularly interesting for two reasons: It is portable (working off internal batteries), and it is very small and lightweight. It weighs only 3.5 pounds and is approximately $3'' \times 5'' \times 10''$ in size. It has an input impedance of 1 MΩ and a time base that can be set for 5 μs to 500 ms per horizontal division. The vertical scale can be set to sensitivities extending from 1 mV to 50 V per division. This oscilloscope can also display two signals (dual trace) at the same time for magnitude and phase comparisons.

PROBLEMS

Section 22.5

1. Determine the indication of each of the meters of Table 22.2 if placed across 125-V dc line.

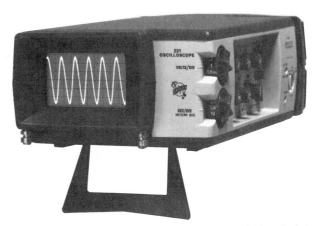

Courtesy of Tektronix, Inc.

FIG. 22.15 *Miniscope.*

TABLE 22.2

Meter	Meter Indication
a. dc voltmeter (d'Arsonval movement)	
b. ac voltmeter (Half-wave rectifier type)	
c. ac voltmeter (Full-wave rectifier type)	
d. ac voltmeter (Electrodynamometer)	

2. Determine the indication of each of the meters of Table 22.3 if placed across a 110-V ac line.

TABLE 22.3

Meter	Meter Indication
a. ac voltmeter (Half-wave rectifier type)	
b. ac voltmeter (Full-wave rectifier type)	
c. dc voltmeter (d'Arsonval movement)	
d. dc voltmeter (Electrodynamometer)	

GLOSSARY

Amp-Clamp® A clamp-type instrument that will permit noninvasive current measurements and that can be used as a conventional voltmeter or ohmmeter.

Electrodynamometer meters Instruments that can measure both ac and dc quantities without a change in internal circuitry.

Frequency counter An instrument that will provide a digital display of the frequency or period of a periodic time-varying signal.

Frequency meter An instrument that employs a vibrating-reed, moving-coil, or resonant-type circuit and that can measure the frequency of an alternating signal.

Impedance bridge An instrument used to measure the parameter values of an inductor, capacitor, and resistor.

Oscilloscope An instrument that will display, through the use of a CRT, the characteristics of a time-varying signal.

Rectifier-type ac meter An instrument calibrated to indicate the effective value of a current or voltage through the use of a rectifier network and d'Arsonval-type movement.

Single-phase wattmeter An instrument employing an electrodynamometer-type movement that can measure the power to a dc or an ac single-phase network.

Vector voltmeter An instrument that will indicate both magnitude and phase of a potential difference.

VOM A multimeter with the capability to measure resistance and both ac and dc levels of current and voltage.

23.1 INTRODUCTION

Prior to this chapter, the sine wave described in Chapter 13 has been the only form of alternating quantity that we have used in our analysis. Any waveform that differs from the basic description of the sinusoidal waveform is referred to as *nonsinusoidal*. The most obvious is the dc voltage or current; others are voice patterns, such as those in Fig. 23.1, which were recorded as a person held continuous "A" and "O" sounds. Note that the waveforms are almost periodic, but that a variation in tone causes a slight change in the waveform. Voice patterns are unique to people in much the same manner as fingerprints.

The output of many electrical and electronic devices will be a nonsinusoidal quantity if a sinusoidal input is applied. For example, the nonlinear characteristics of a diode are shown in Fig. 23.2(a). If a sinusoidal voltage is applied across the diode, the current passing through the diode will have the nonsinusoidal wave shape shown in Fig. 23.2(b). Note, however, that the nonsinusoidal output is also periodic: It has the same period as the input voltage. A periodic nonsinusoidal output will result for any nonlinear system that has no residual effects associated with it if a sinusoidal quantity is applied at the input.

In this chapter, we shall discuss one method of obtaining the response of a system to a periodic nonsinusoidal input.

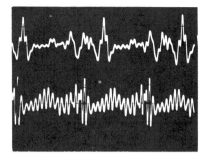

FIG. 23.1 *Voice patterns.*

23.2 FOURIER SERIES

Fourier series refers to a series of terms, developed in 1826 by Baron Jean Fourier, that can be used to represent a nonsinusoidal waveform. In the analysis of these waveforms, we solve for each term in the Fourier series:

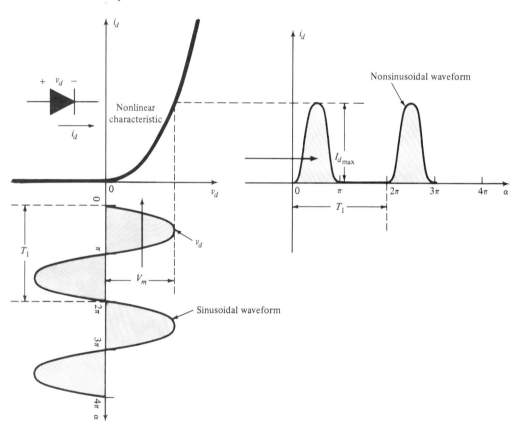

FIG. 23.2

$$f(\alpha) = \underbrace{A_0}_{\substack{\text{dc or} \\ \text{average value}}} + \underbrace{A_1 \cos \alpha + A_2 \cos 2\alpha + A_3 \cos 3\alpha + \cdots + A_n \sin n\alpha}_{\text{cosine terms}}$$
$$+ \underbrace{B_1 \sin \alpha + B_2 \sin 2\alpha + B_3 \sin 3\alpha + \cdots + B_n \sin n\alpha}_{\text{sine terms}} \quad \textbf{(23.1)}$$

Depending on the waveform, a large number of these terms may be required to approximate the waveform closely for the purpose of circuit analysis.

The Fourier series has three basic parts. The first is the dc term A_0, which is the average value of the waveform over one full cycle. The second is a series of cosine terms. There are no restrictions on the values or relative values of the amplitudes of these cosine terms, but each will have a frequency that is an integer multiple of the frequency of the first cosine term of the series. The third part is a series of sine terms. There are again *no* restrictions on the values or relative values of the amplitudes of these sine terms, but each will have a frequency that is an integer multiple of the frequency of the first sine term of the series. For a particular waveform, it is quite possible that all of the sine *or* cosine terms may be zero. Characteristics of this type can be determined by simply examining the nonsinusoidal waveform and its position on the horizontal axis.

If a waveform is symmetric about the vertical axis, it is called an *even* function or is said to have *axis symmetry* [Fig. 23.3(a)]. For all even functions, the $B_{1 \to n}$ constants will all be zero, and the function can be completely described by just the *dc and cosine terms*. Note that the cosine wave itself is also symmetrical about the vertical axis (ordinate) [Fig. 23.3(b)].

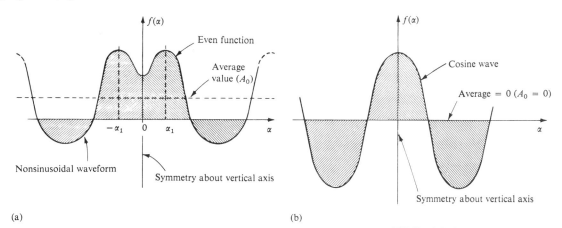

(a)

(b)

FIG. 23.3 *Axis symmetry.*

For both waveforms of Fig. 23.3, the following mathematical relationship is true:

$$f(\alpha) = f(-\alpha) \qquad \text{(even function)} \qquad \textbf{(23.2)}$$

In words, it states that the magnitude of the function is the same at $+\alpha$ as at $-\alpha$ [α_1 in Fig. 23.3(a)].

If the waveform is such that its value for $+\alpha$ is the negative of that for $-\alpha$ [Fig. 23.4(a)], it is called an *odd* function or is said to have *point symmetry* (about any point of intersection on the horizontal axis), and all of the constants $A_{1 \to n}$ will be zero. The function can then be represented by the *dc and sine terms*. Note that the sine wave itself is an odd function [Fig. 23.4(b)].

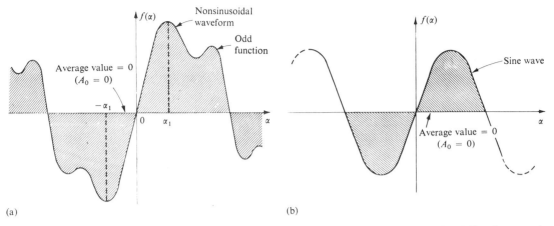

(a)

(b)

FIG. 23.4 *Point symmetry.*

For both waveforms of Fig. 23.4, the following mathematical relationship is true:

$$f(\alpha) = -f(-\alpha) \qquad \text{(odd function)} \qquad \textbf{(23.3)}$$

In words, it states that the magnitude of the function at $+\alpha$ is equal to the negative of the magnitude at $-\alpha$ [α_1 in Fig. 23.4(a)].

The first term of the sine and cosine series is called the *fundamental component*. It represents the minimum frequency term required to represent a particular waveform, and it also has the same frequency as the waveform being represented. A fundamental term, therefore, must be present in any Fourier series representation. The other terms with higher-order frequencies (integer multiples of the fundamental) are called the *harmonic terms*. A term that has a frequency equal to twice the fundamental is the second harmonic; three times, the third harmonic; and so on.

If the waveform is such that

$$f(t) = f\left(\frac{T}{2} + t\right) \qquad \textbf{(23.4)}$$

the odd harmonics of the series of cosine and sine terms are zero. Fig. 23.5 is an example of this type of function.

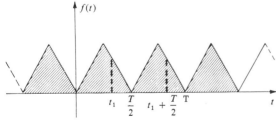

FIG. 23.5

Equation (23.4) states that the function repeats itself after each $T/2$ time interval (t_1 in Fig. 23.5). The waveform, however, will also repeat itself after each period T. In general, therefore, for a function of this type, if the period

T of the waveform is chosen to be twice that of the minimum period ($T/2$), the odd harmonics will all be zero.

If the waveform is such that

$$f(t) = -f\left(\frac{T}{2} + t\right) \qquad (23.5)$$

the waveform is said to have *half-wave* or *mirror symmetry* and *the even harmonics of the series of cosine and sine terms will be zero*. Fig. 23.6 is an example of this type of function.

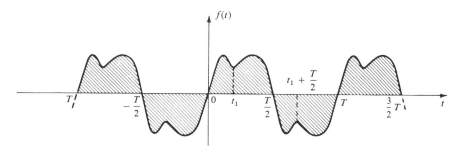

Equation (23.5) states that the waveform encompassed in one time interval $T/2$ will repeat itself in the next $T/2$ time interval, but in the negative sense (t_1 in Fig. 23.6). For example, the waveform of Fig. 23.6 from zero to $T/2$ will repeat itself in the time interval $T/2$ to T, but below the horizontal axis.

FIG. 23.6 *Mirror symmetry.*

An instrument for measuring harmonic frequencies and corresponding amplitudes is the wave analyzer shown in Fig. 23.7.

Courtesy of Hewlett Packard Co.

The constants A_0, $A_{1 \to n}$, $B_{1 \to n}$ can be determined by using the following integral formulas:

FIG. 23.7 *Wave analyzer.*

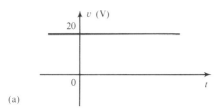

(a)

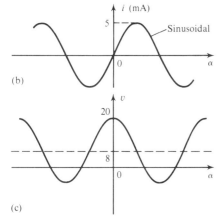

(b)

(c)

FIG. 23.8

$$A_0 = \frac{1}{T} \int_0^T f(t)\, dt = \frac{1}{2\pi} \int_0^{2\pi} f(\alpha)\, d\alpha \qquad \textbf{(23.6)}$$

$$A_n = \frac{2}{T} \int_0^T f(t) \cos n\omega t\, dt = \frac{1}{\pi} \int_0^{2\pi} f(\alpha) \cos n\alpha\, d\alpha$$

(23.7)

$$B_n = \frac{2}{T} \int_0^T f(t) \sin n\omega t\, dt = \frac{1}{\pi} \int_0^{2\pi} f(\alpha) \sin n\alpha\, d\alpha$$

(23.8)

These equations have been presented for recognition purposes only; they will not be used in the following analysis.

The following examples will demonstrate the use of the equations and concepts introduced thus far in this chapter.

EXAMPLE 23.1. Write the Fourier series expansion for the waveforms of Fig. 23.8.

Solution:
a. $A_0 = 20$, $A_{1 \to n} = 0$, $B_{1 \to n} = 0$
   $v = 20$
b. $A_0 = 0$, $A_{1 \to n} = 0$, $B_1 = 5 \times 10^{-3}$, $B_{2 \to n} = 0$
   $i = 5 \times 10^{-3} \sin \alpha$
c. $A_0 = 8$, $A_1 = 12$, $A_{2 \to n} = 0$, $B_{1 \to n} = 0$
   $v = 8 + 12 \cos \alpha$

EXAMPLE 23.2. Sketch the following Fourier series expansion:

$$v = 2 + 1 \cos \alpha + 2 \sin \alpha$$

Solution: Note Fig. 23.9.

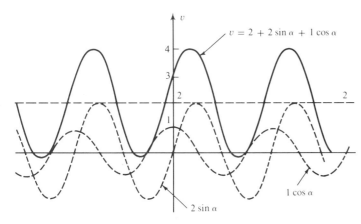

FIG. 23.9

The solution could be obtained graphically by first plotting all of the functions and considering a sufficient number of points on the horizontal axis; or phasor algebra could be employed as follows:

$$1 \cos \alpha + 2 \sin \alpha = 1 \angle 90° + 2 \angle 0° = j1 + 2$$
$$= 2 + j1 = 2.236 \angle 26.57°$$
$$= 2.236 \sin(\alpha + 26.57°)$$

and

$$v = 2 + 2.236 \sin(\alpha + 26.57°)$$

which is simply the sine wave portion riding on a dc level of 2 volts. That is, its positive maximum is $2 + 2.236 = 4.236$ V, and its minimum is $2 - 2.236 = -0.236$ V.

EXAMPLE 23.3. Sketch the following Fourier series expansion:

$$i = \sin \alpha + \sin 2\alpha$$

Solution: See Fig. 23.10. Note that in this case the sum of the two sinusoidal waveforms of different frequencies is *not* a sine wave. Recall that complex algebra can be applied only to waveforms having the *same* frequency. In this case the solution is obtained graphically.

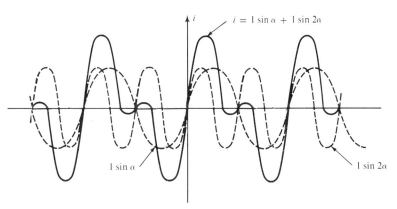

FIG. 23.10

As a further example in the use of the Fourier series approach, consider the square wave shown in Fig. 23.11. The average value is zero, so $A_0 = 0$. It is an odd function, so all the constants $A_{1 \to n}$ equal zero; only sine terms will be present in the series expansion. Since the waveform satisfies the criteria for $f(t) = -f(T/2 + t)$, the even harmonics will also be zero.

The expression obtained after evaluating the various coefficients from Eq. (23.8) is

$$v = \frac{4}{\pi} V_m \left[\sin \omega t + \frac{1}{3} \sin 3\omega t + \frac{1}{5} \sin 5\omega t \right.$$
$$\left. + \frac{1}{7} \sin 7\omega t + \cdots + \frac{1}{n} \sin n\omega t \right] \quad \textbf{(23.9)}$$

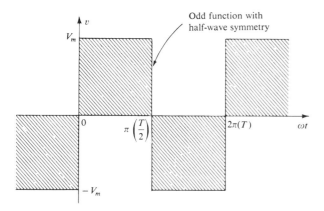

FIG. 23.11 *Square wave.*

Note that the fundamental does indeed have the frequency of the square wave. If we add the fundamental and third harmonics, we obtain the results shown in Fig. 23.12.

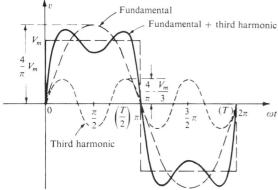

FIG. 23.12

Even with only the first two terms, the wave shape is beginning to look like a square wave. If we add the next two terms (Fig. 23.13), the width of the pulse increases, and the number of peaks increases.

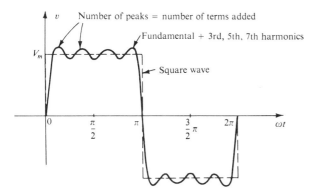

FIG. 23.13

As we continue to add terms, the series will better approximate the square wave. Note, however, that the amplitude of each succeeding term diminishes to the point at which it will be negligible compared with those of the first few terms. A good approximation would be to assume that

the waveform is composed of the harmonics up to and including the ninth. Any higher harmonics would be less
than one-tenth the fundamental. If the waveform just described were shifted above or below the horizontal axis, the
Fourier series would be altered only by a change in the dc
term. Figure 23.14(a), for example, is the sum of Figs.
23.14(b) and (c). The Fourier series for the complete waveform is therefore

$$v_T = V_m + \text{Eq. (23.9)}$$

$$= V_m + \frac{4}{\pi} V_m \left(\sin \omega t + \frac{1}{3} \sin 3\omega t \right.$$

$$\left. + \frac{1}{5} \sin 5\omega t + \frac{1}{7} \sin 7\omega t + \cdots \right)$$

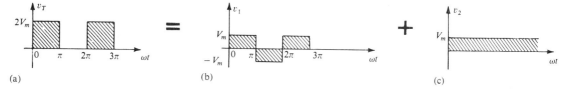

(a)   (b)   (c)

The equation for the pulsating waveform of Fig. 23.15(b)
is

FIG. 23.14

$$\boxed{\begin{array}{l} v = 0.318 V_m + 0.500 V_m \sin \alpha - 0.212 V_m \\ \cos 2\alpha - 0.0424 V_m \cos 4\alpha - \cdots \end{array}} \quad \textbf{(23.10)}$$

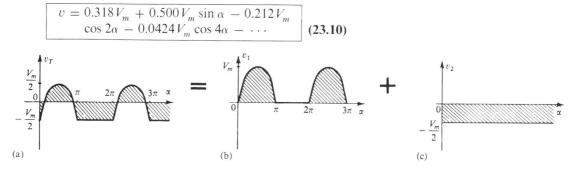

(a)   (b)   (c)

FIG. 23.15

The waveform in Fig. 23.15(a) is the sum of the two in Figs.
23.15(b) and (c). The Fourier series for the complete waveform is therefore

$$v_T = -\frac{V_m}{2} + \text{Eq. (23.10)}$$

$$= [\underbrace{-0.500 + 0.318}_{-0.182}]V_m + 0.500 V_m \sin \alpha$$

$$- 0.212 V_m \cos 2\alpha - 0.0424 V_m \cos 4\alpha + \cdots$$

If either waveform were shifted to the right or left, the
phase shift would be subtracted or added, respectively,
from the sine and cosine terms. The dc term would not
change with a shift to the right or left.

If the half-wave rectified signal is shifted 90° to the left,
as in Fig. 23.16, the Fourier series becomes

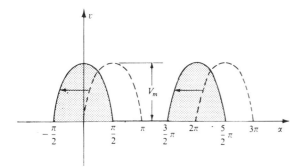

FIG. 23.16

$$v_o = 0.318 V_m + 0.500 V_m \underbrace{\sin(\alpha + 90°)}_{\cos \alpha} - 0.212 V_m \cos 2(\alpha + 90°) - 0.0424 V_m \cos 4(\alpha + 90°) + \cdots$$

$$= 0.318 V_m + 0.500 V_m \cos \alpha - 0.212 V_m \cos(2\alpha + 180°) - 0.0424 V_m \cos(4\alpha + 360°) + \cdots$$

$$= 0.318 V_m + 0.500 V_m \cos \alpha + 0.212 V_m \cos 2\alpha - 0.0424 V_m \cos 4\alpha + \cdots$$

If the waveform is shifted 90° to the right, the Fourier series becomes

$$v_o = 0.318 V_m + 0.500 V_m \sin(\alpha - 90°) - 0.212 V_m \cos 2(\alpha - 90°) - 0.0424 V_m \cos 4(\alpha - 90°) + \cdots$$

$$= 0.318 V_m - 0.500 V_m \cos \alpha + 0.212 V_m \cos 2\alpha - 0.0424 V_m \cos 4\alpha + \cdots$$

23.3 CIRCUIT RESPONSE TO A NONSINUSOIDAL INPUT

The Fourier series representation of a nonsinusoidal input can be applied to a linear network using the principle of superposition. Recall that this theorem allowed us to consider the effects of each source of a circuit independently. If we replace the nonsinusoidal input by the terms of the Fourier series deemed necessary for practical considerations, we can use superposition to find the response of the network to each term (Fig. 23.17).

$$e = A_0 + A_1 \cos \alpha + \cdots + A_n \cos n\alpha + \cdots$$
$$+ B_1 \sin \alpha + \cdots + B_n \sin n\alpha + \cdots$$

FIG. 23.17

The total response of the system is then the algebraic sum of the values obtained for each term. The major change between using this theorem for nonsinusoidal circuits and using it for the circuits previously described is that the frequency will be different for each term in the nonsinusoidal application. Therefore, the reactances

$$X_L = 2\pi f L \quad \text{and} \quad X_C = \frac{1}{2\pi f C}$$

will change for each term of the input voltage or current.

In Chapter 13, we found that the effective value of any waveform was given by

$$\sqrt{\frac{1}{T}\int_0^T [f(t)]^2\, dt}$$

If we apply this equation to the following Fourier series:

$$v(\alpha) = V_o + V_{m_1}\cos\alpha + \cdots + V_{m_n}\cos n\alpha$$
$$+ V'_{m_1}\sin\alpha + \cdots + V'_{m_n}\sin n\alpha$$

$$V_{\text{eff}} = \sqrt{V_o^2 + \frac{V_{m_1}^2 + \cdots + V_{m_n}^2 + V'^2_{m_1} + \cdots + V'^2_{m_n}}{2}}$$

(23.11)

However, since

$$\frac{V_{m_1}}{2} = \left(\frac{V_{m_1}}{\sqrt{2}}\right)\left(\frac{V_{m_1}}{\sqrt{2}}\right) = (V_{1_{\text{eff}}})(V_{1_{\text{eff}}}) = V_{1_{\text{eff}}}^2$$

and

$$V_{\text{eff}} = \sqrt{\begin{array}{c} V_o^2 + V_{1_{\text{eff}}}^2 + \cdots + V_{n_{\text{eff}}}^2 \\ + V'^2_{1_{\text{eff}}} + \cdots + V'^2_{n_{\text{eff}}} \end{array}}$$

(23.12)

Similarly, for

$$i(\alpha) = I_o + I_{m_1}\cos\alpha + \cdots + I_{m_n}\cos n\alpha$$
$$+ I'_{m_1}\sin\alpha + \cdots + I'_{m_n}\sin n\alpha$$

we have

$$I_{\text{eff}} = \sqrt{I_o^2 + \frac{I_{m_1}^2 + \cdots + I_{m_n}^2 + I'^2_{m_1} + \cdots + I'^2_{m_n}}{2}}$$

(23.13)

and

$$I_{\text{eff}} = \sqrt{I_o^2 + I_{1_{\text{eff}}}^2 + \cdots + I_{n_{\text{eff}}}^2 + I'^2_{1_{\text{eff}}} + \cdots + I'^2_{n_{\text{eff}}}}$$

(23.14)

The total power delivered is the sum of that delivered by the corresponding terms of the voltage and current. In the following equations, all voltages and currents are effective values:

$$P_T = V_0 I_0 + V_1 I_1 \cos\theta_1 + \cdots + V_n I_n \cos\theta_n + \cdots$$

(23.15)

$$P_T = I_0^2 R + I_1^2 R + \cdots + I_n^2 R + \cdots \quad \text{(23.16)}$$

or

$$P_T = I_{\text{eff}}^2 R \qquad \text{(23.17)}$$

with I_{eff} as defined by Eq. (23.13), and, similarly,

$$P_T = \frac{V_{\text{eff}}^2}{R} \qquad \text{(23.18)}$$

with V_{eff} as defined by Eq. (23.11).

EXAMPLE 23.4. The input to the circuit of Fig. 23.18 is the following:

$$e = 12 + 10\sin 2t$$

a. Find the current i and the voltages v_R and v_C.
b. Find the effective values of i, v_R, and v_C.
c. Find the power delivered to the circuit.

Solutions:
a. Redraw the original circuit as shown in Fig. 23.19. Then apply superposition:

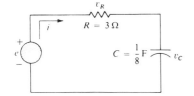

FIG. 23.18

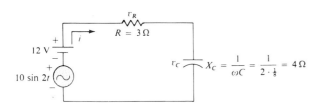

FIG. 23.19

1. *For the 12-V dc supply portion of the input, $I = 0$* since the capacitor is an open circuit to dc when v_C has reached its final (steady-state) value.

$$V_R = IR = 0 \text{ V}$$

and

$$V_C = 12 \text{ V}$$

2. *For the ac supply,*

$$\mathbf{Z} = 3 - j4 = 5 \angle -53.13°$$

and

$$\mathbf{I} = \frac{\mathbf{E}}{\mathbf{Z}} = \frac{\frac{10}{\sqrt{2}} \angle 0°}{5 \angle -53.13°} = \frac{2}{\sqrt{2}} \angle +53.13°$$

$$\mathbf{V}_R = \mathbf{I}R = \left(\frac{2}{\sqrt{2}} \angle +53.13°\right)(3 \angle 0°) =$$

$$\frac{6}{\sqrt{2}} \angle +53.13°$$

and

$$\mathbf{V}_C = \mathbf{I}\mathbf{X}_C = \left(\frac{2}{\sqrt{2}} \angle +53.13°\right)(4 \angle -90°)$$

$$= \frac{8}{\sqrt{2}} \angle -36.87°$$

In the time domain,

$$i = 0 + 2\sin(2t + 53.13°)$$

Note that even though the dc term was present in the expression for the input voltage, the dc term for the current in this circuit is zero:

$$v_R = 0 + 6\sin(2t + 53.13°)$$

and

$$v_C = 12 + 8\sin(2t - 36.87°)$$

b. Eq. (23.14): $I_{\text{eff}} = \sqrt{0^2 + \frac{2^2}{2}} = \sqrt{2} = \mathbf{1.414\ A}$

Eq. (23.12): $V_{R_{\text{eff}}} = \sqrt{0^2 + \frac{6^2}{2}} = \sqrt{18} = \mathbf{4.243\ V}$

Eq. (23.12): $V_{C_{\text{eff}}} = \sqrt{12^2 + \frac{8^2}{2}} = \sqrt{176} = \mathbf{13.267\ V}$

c. $P = I_{\text{eff}}^2 R = (\sqrt{2})^2 \cdot 3 = \mathbf{6\ W}$

EXAMPLE 23.5. Find the response of the circuit of Fig. 23.20 to the input shown.

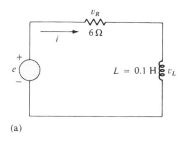

(a)

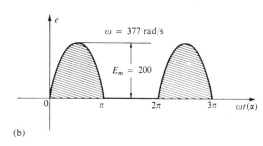

(b)

FIG. 23.20

$e = 0.318E_m + 0.500E_m \sin \omega t - 0.212E_m \cos 2\omega t$
$\quad - 0.0424E_m \cos 4\omega t + \cdots$

Solution: For discussion purposes, only the first three terms will be used to represent *e*. Converting the cosine terms to sine terms and substituting for E_m gives us

$$e = 63.60 + 100.0 \sin \omega t - 42.40 \sin(2\omega t + 90°)$$

Using phasor notation, the original circuit becomes as shown in Fig. 23.21. Apply superposition:

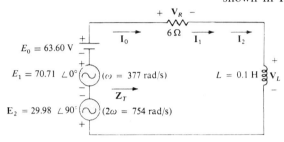

$E_0 = 63.60$ V

$E_1 = 70.71 \angle 0°$ ($\omega = 377$ rad/s)

$\mathbf{Z}_T$

$E_2 = 29.98 \angle 90°$ ($2\omega = 754$ rad/s)

$+ \; V_R \; -$

$6\,\Omega$ I_1 I_2

I_0

$L = 0.1$ H V_L

FIG. 23.21

For the dc term ($E_o = 63.6$ V),

$$X_L = 0 \quad \text{(short for dc)}$$
$$\mathbf{Z}_T = 6 \angle 0° = \mathbf{R}$$
$$I_o = \frac{E_o}{R} = \frac{63.6}{6} = 10.60 \text{ A}$$
$$V_{R_o} = I_o R = E_o = 63.60 \text{ V}$$
$$V_{L_o} = 0$$

The average power is

$$P_o = I_o^2 R = (10.60)^2(6) = 675 \text{ W}$$

For the fundamental term ($\mathbf{E}_1 = 70.71 \angle 0°$, $\omega = 377$),

$$X_{L_1} = \omega L = (377)(0.1) = 37.7 \; \Omega$$
$$\mathbf{Z}_{T_1} = 6 + j37.7 = 38.17 \angle 80.96°$$
$$\mathbf{I}_1 = \frac{\mathbf{E}_1}{\mathbf{Z}_{T_1}} = \frac{70.71 \angle 0°}{38.17 \angle 80.96°} = 1.85 \angle -80.96°$$
$$\begin{aligned}\mathbf{V}_{R_1} = \mathbf{I}_1 \mathbf{R} &= (1.85 \angle -80.96°)(6 \angle 0°) \\ &= 11.10 \angle -80.96°\end{aligned}$$
$$\begin{aligned}\mathbf{V}_{L_1} = \mathbf{I}_1 \mathbf{X}_L &= (1.85 \angle -80.96°)(37.7 \angle 90°) \\ &= 69.75 \angle 9.04°\end{aligned}$$

The average power is

$$P_1 = I_1^2 R = (1.85)^2(6) = 20.54 \text{ W}$$

For the second harmonic ($\mathbf{E}_2 = 30 \angle 90°$, $\omega = 754$),

$$X_{L_2} = \omega L = (754)(0.1) = 75.4 \; \Omega$$
$$\mathbf{Z}_{T_2} = 6 + j75.4 = 75.64 \angle 85.45°$$
$$\mathbf{I}_2 = \frac{\mathbf{E}_2}{\mathbf{Z}_{T_2}} = \frac{30 \angle -90°}{75.64 \angle 85.45} = 0.397 \angle -174.45°$$

The phase angle of $\mathbf{E}_2$ was changed to $-90°$ to give it the same polarity as the input voltages $\mathbf{E}_0$ and $\mathbf{E}_1$.

$$V_{R_2} = I_2 R = (0.397 \angle -174.45°)(6 \angle 0°)$$
$$= 2.38 \angle -174.45°$$
$$V_{L_2} = I_2 X_{L_2} = (0.397 \angle -174.45°)(75.4 \angle 90°)$$
$$= 29.9 \angle -84.45°$$

The average power is

$$P_2 = I_2^2 R = (0.397)^2(6) = 0.946 \text{ W}$$

The Fourier series expansion for i is

$$i = \mathbf{10.6 + \sqrt{2}(1.85) \sin(377t - 80.96°)}$$
$$\mathbf{+ \sqrt{2}(0.397) \sin(754t - 174.45°)}$$

and

$$I_{\text{eff}} = \sqrt{10.6^2 + 1.85^2 + 0.397^2} = \mathbf{10.77 \text{ A}}$$

The Fourier series expansion for v_R is

$$v_R = \mathbf{63.6 + \sqrt{2}(11.10) \sin(377t - 80.96°)}$$
$$\mathbf{+ \sqrt{2}(2.38) \sin(754t - 174.45°)}$$

and

$$V_{R_{\text{eff}}} = \sqrt{63.6^2 + 11.10^2 + 2.38^2} = \mathbf{64.61 \text{ V}}$$

The Fourier series expansion for v_L is

$$v_L = \mathbf{\sqrt{2}(69.75) \sin(377t + 9.04°)}$$
$$\mathbf{+ \sqrt{2}(29.93) \sin(754t - 84.45°)}$$

and

$$V_{L_{\text{eff}}} = \sqrt{69.75^2 + 29.93^2}$$
$$V_{L_{\text{eff}}} = \mathbf{75.90 \text{ V}}$$

The total average power is

$$P_T = I_{\text{eff}}^2 R = (10.77)^2(6) = \mathbf{695.96 \text{ W}} = P_0 + P_1 + P_2$$

23.4 ADDITION AND SUBTRACTION OF NONSINUSOIDAL WAVEFORMS

The Fourier series expression for the waveform resulting from the addition or subtraction of two nonsinusoidal waveforms can be found using phasor algebra if the terms having the same frequency are considered separately.

For example, the sum of the following two nonsinusoidal waveforms is found using this method:

$$v_1 = 30 + 20 \sin 20t + \cdots + 5 \sin(60t + 30°)$$
$$v_2 = 60 + 30 \sin 20t + 20 \sin 40t + 10 \cos 60t$$

1. dc terms:

$$V_{T_o} = 30 + 60 = 90$$

2. $\omega = 20$:

$$V_{T_{1(max)}} = 30 + 20 = 50$$

and

$$V_{T_1} = 50 \sin 20t$$

3. $\omega = 40$:

$$V_{T_2} = 20 \sin 40t$$

4. $\omega = 60$:

$$5 \sin(60t + 30°) = (0.707)(5) \angle 30° = 3.54 \angle 30°$$
$$10 \cos 60t = 10 \sin(60t + 90°) \Rightarrow (0.707)(10) \angle 90°$$
$$= 7.07 \angle 90°$$
$$\mathbf{V}_{T_3} = 3.54 \angle 30° + 7.07 \angle 90°$$
$$= 3.07 + j1.77 + j7.07 = 3.07 + j8.84$$
$$\mathbf{V}_{T_3} = 9.36 \angle 70.85°$$

and

$$\mathbf{V}_{T_3} = 13.24 \sin(60t + 70.85°)$$

and

$$v_1 + v_2 = v_T = 90 + 50 \sin 20t + 20 \sin 40t$$
$$+ 13.24 \sin(60t + 70.85°)$$

PROBLEMS

Section 23.2

1. For the waveforms of Fig. 23.22, determine whether the following will be present in the Fourier series representation:
   a. dc term
   b. cosine terms
   c. sine terms
   d. even-ordered harmonics
   e. odd-ordered harmonics

2. If the Fourier series for the waveform of Fig. 23.23(a) is

$$i = \frac{2I_m}{\pi}\left(1 + \frac{2}{3}\cos 2\omega t - \frac{2}{15}\cos 4\omega t + \frac{2}{35}\cos 6\omega t \ldots\right)$$

   find the Fourier series representation for waveforms (b), (c), and (d).

3. Sketch the following nonsinusoidal waveforms with $\alpha = \omega t$ as the abscissa:
   a. $v = -4 + 2\sin\alpha$
   b. $v = \cos\alpha + \sin\alpha$
   c. $i = 2 - 2\cos\alpha$

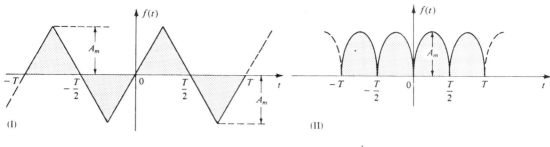

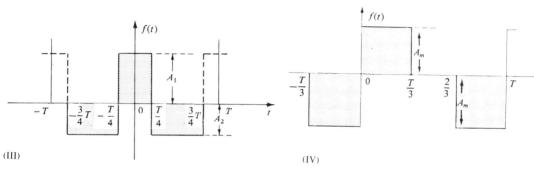

d. $i = 3 \sin \alpha - 6 \sin 2\alpha$

e. $v = 2 \cos 2\alpha + \sin 2\alpha$

FIG. 23.22

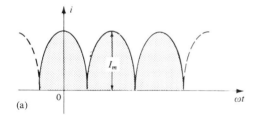

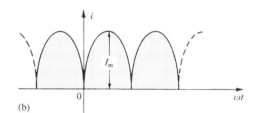

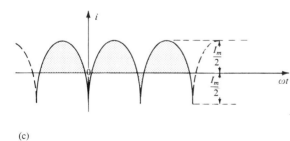

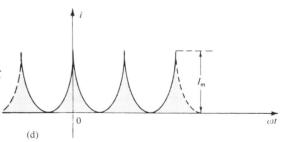

4. Sketch the following nonsinusoidal waveforms with ωt at the abscissa:

a. $i = 50 \sin \omega t + 25 \sin 3\omega t$

b. $i = 50 \sin \alpha - 25 \sin 3\alpha$

c. $i = 4 + 3 \sin \omega t + 2 \sin 2\omega t - 1 \sin 3\omega t$

FIG. 23.23

Section 23.3

5. Find the average and effective values of the following nonsinusoidal waves:

a. $v = 100 + 50 \sin \omega t + 25 \sin 2\omega t$

b. $i = 3 + 2 \sin(\omega t - 53°) + 0.8 \sin(2\omega t - 70°)$

6. Find the average and effective values of the following non-sinusoidal waves:
   a. $v = 20 \sin \omega t + 15 \sin 2\omega t - 10 \sin 3\omega t$
   b. $i = 6 \sin(\omega t + 20°) + 2 \sin(2\omega t + 30°)$
   $\qquad - 1 \sin(3 \omega t + 60°)$

7. Find the total average power to a circuit whose voltage and current are as indicated in Problem 5.

8. Find the total average power to a circuit whose voltage and current are as indicated in Problem 6.

9. The Fourier series representation for the input voltage to the circuit of Fig. 23.24 is

$$e = 18 + 30 \sin 400t$$

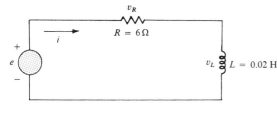

FIG. 23.24

a. Find the nonsinusoidal expression for the current i.
b. Calculate the effective value of the current.
c. Find the expression for the voltage across the resistor.
d. Calculate the effective value of the voltage across the resistor.
e. Find the expression for the voltage across the reactive element.
f. Calculate the effective value of the voltage across the reactive element.
g. Find the average power delivered to the resistor.

10. Repeat Problem 9 for $e = 24 + 30 \sin 400t + 10 \sin 800t$.

11. Repeat Problem 9 for the following input voltage:

$$e = -60 + 20 \sin 300t - 10 \sin 600t$$

12. Repeat Problem 9 for the circuit of Fig. 23.25.

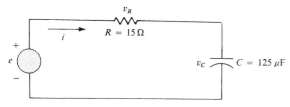

FIG. 23.25

*13. The input voltage to the circuit of Fig. 23.26 is a full-wave rectified signal having the following Fourier series expansion:

$$e = \frac{200}{\pi}\left(1 + \frac{2}{3}\cos 2\omega t - \frac{2}{15}\cos 4\omega t + \frac{2}{53}\cos 6\omega t \dots\right)$$

where $\omega = 377$.
a. Find the Fourier series expression for the voltage v_o using only the first three terms of the expression.
b. Find the effective value of v_o.
c. Find the average power delivered to the 1-kΩ resistor.

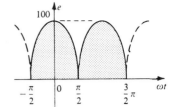

*14. Find the Fourier series expression for the voltage v_o of Fig. 23.27.

Section 23.4

15. Perform the indicated operations on the following nonsinusoidal waveforms:

   a. $[60 + 70 \sin \omega t + 20 \sin(2\omega t + 90°) + 10 \sin(3\omega t + 60°)] + [20 + 30 \sin \omega t - 20 \cos 2\omega t + 5 \cos 3\omega t]$

   b. $[20 + 60 \sin \alpha + 10 \sin(2\alpha - 180°) + 5 \cos(3\alpha + 90°)] - [5 - 10 \sin \alpha + 4 \sin(3\alpha - 30°)]$

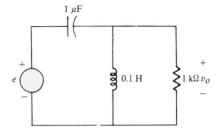

FIG. 23.26

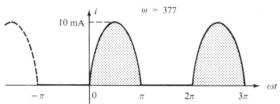

16. Find the nonsinusoidal expression for the current i_T (Fig. 23.28).

$$i_2 = 10 + 30 \sin 20t - 0.5 \sin(40t + 90°)$$
$$i_1 = 20 + 4 \sin(20t + 90°) + 0.5 \sin(40t + 30°)$$

17. Find the nonsinusoidal expression for the voltage e of the diagram of Fig. 23.29.

$$v_1 = 20 - 200 \sin 600t + 100 \cos 1200t + 75 \sin 1800t$$
$$v_2 = -10 + 150 \sin(600t + 30°) + 50 \sin(1800t + 60°)$$

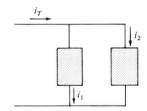

FIG. 23.27

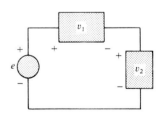

FIG. 23.28

GLOSSARY

Axis symmetry A sinusoidal or nonsinusoidal function that has symmetry about the vertical axis.

Even harmonics The terms of the Fourier series expansion that have frequencies that are even multiples of the fundamental component.

Fourier series A series of terms, developed in 1826 by Baron Jean Fourier, that can be used to represent a nonsinusoidal function.

Fundamental component The minimum frequency term required to represent a particular waveform in the Fourier series expansion.

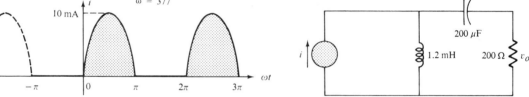

FIG. 23.29

Half-wave (mirror) symmetry   A sinusoidal or nonsinusoidal function that satisfies the relationship $f(t) = -f\left(\dfrac{T}{2} + t\right)$.

Harmonic   The terms of the Fourier series expansion that have frequencies that are integer multiples of the fundamental component.

Nonsinusoidal waveform   Any waveform that differs from the fundamental sinusoidal function.

Odd harmonics   The terms of the Fourier series expansion that have frequencies that are odd multiples of the fundamental component.

Point symmetry   A sinusoidal or nonsinusoidal function that satisfies the relationship $f(\alpha) = -f(-\alpha)$.

24.1 INTRODUCTION

Chapter 11 discussed the *self-inductance* of a coil. We shall now examine the *mutual inductance* that exists between coils of the same or different dimensions. Mutual inductance is a phenomenon basic to the operation of the transformer, an electrical device used today in almost every field of electrical engineering. This device plays an integral part in power distribution systems and can be found in many electronic circuits and measuring instruments. In this chapter, we will discuss three of the basic applications of a transformer. These are its ability to build up or step down the voltage or current, to act as an impedance matching device, and to isolate (no physical connection) one portion of a circuit from another. In addition, we will introduce the dot convention and consider the transformer equivalent circuit. The chapter will conclude with a word about the effect of mutual inductance on the mesh equations of a network.

24.2 MUTUAL INDUCTANCE

The transformer is constructed of two coils placed so that the changing flux developed by one will link the other as shown in Fig. 24.1. This will result in an induced voltage across each coil. To distinguish between the coils, we will apply the transformer convention that the coil to which the source is applied is called the *primary,* and the coil to which the load is applied is called the *secondary.* For the primary of the transformer of Fig. 24.1, an application of Faraday's law [Eq. (11.1)] will result in

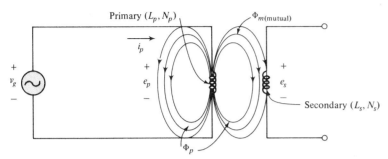

Primary (L_p, N_p)

$\Phi_{m(mutual)}$

i_p

$+$ e_p $-$

$+$ e_s $-$

Secondary (L_s, N_s)

Φ_p

v_g

$+$ $-$

FIG. 24.1

$$e_p = N_p \frac{d\phi_p}{dt}$$

or, from Eq. (11.4),

$$e_p = L_p \frac{di_p}{dt}$$

The magnitude of the voltage induced across the secondary e_s is determined by

$$\boxed{e_s = N_s \frac{d\phi_m}{dt}} \qquad \textbf{(24.1a)}$$

where ϕ_m is the portion of the primary flux ϕ_p that links the secondary.

If all of the flux linking the primary links the secondary, then

$$\phi_p = \phi_m$$

and

$$\boxed{e_s = N_s \frac{d\phi_p}{dt}} \qquad \textbf{(24.1b)}$$

The *coefficient of coupling* between two coils is determined by

$$\boxed{k \text{ (coefficient of coupling)} = \frac{\phi_m}{\phi_p}} \qquad \textbf{(24.2)}$$

Since the maximum changing flux that can link the secondary is ϕ_p, the coefficient of coupling between two coils can never be greater than one. The coefficient of coupling between various coils is indicated in Fig. 24.2. Note that for the iron core, $k \cong 1$, while for the air core, k is considerably less. Those coils with low coefficients of coupling are said to be *loosely coupled.*

For the secondary, we have

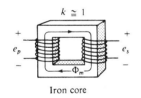

$k \cong 1$

$+$ e_p $-$

$+$ e_s $-$

Φ_m

Iron core

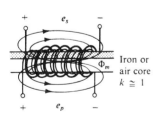

$+$ e_s $-$

Φ_m

Iron or air core

$k \cong 1$

$+$ e_p $-$

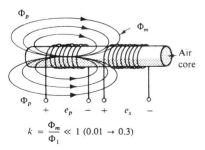

Φ_p

Φ_m

Air core

Φ_p

$+$ e_p $-$ $+$ e_s $-$

$k = \dfrac{\Phi_m}{\Phi_1} \ll 1 \ (0.01 \rightarrow 0.3)$

FIG. 24.2

$$e_s = N_s \frac{d\phi_m}{dt} = N_s \frac{dk\phi_p}{dt}$$

and

$$\boxed{e_s = kN_s \frac{d\phi_p}{dt}} \qquad \text{(24.3)}$$

The mutual inductance between the two coils of Fig. 24.1 is determined by

$$\boxed{M = N_s \frac{d\phi_m}{di_p}} \qquad \text{(henries, H)} \qquad \text{(24.4a)}$$

or

$$\boxed{M = N_p \frac{d\phi_m}{di_s}} \qquad \text{(H)} \qquad \text{(24.4b)}$$

Note in the above equations that the symbol for mutual inductance is the capital letter M, and that its unit of measurement, like that of self-inductance, is the *henry*. In words, Eqs. (24.4a) and (24.4b) state that the mutual inductance between two coils is proportional to the instantaneous change in flux linking one coil due to an instantaneous change in current through the other coil.

In terms of the inductance of each coil and the coefficient of coupling, the mutual inductance is determined by

$$\boxed{M = k\sqrt{L_p L_s}} \qquad \text{(H)} \qquad \text{(24.5)}$$

The greater the coefficient of coupling (greater flux linkages), or the greater the inductance of either coil, the higher the mutual inductance between the coils. Relate this fact to the configurations of Fig. 24.2.

The secondary voltage e_s can also be found in terms of the mutual inductance if we rewrite Eq. (24.1) as

$$e_s = N_s \left(\frac{d\phi_m}{di_p} \right) \left(\frac{di_p}{dt} \right)$$

and, since $M = N_s(d\phi_m/di_p)$, it can also be written

$$\boxed{e_s = M \frac{di_p}{dt}} \qquad \text{(24.6a)}$$

Similarly,

$$\boxed{e_p = M \frac{di_s}{dt}} \qquad \text{(24.6b)}$$

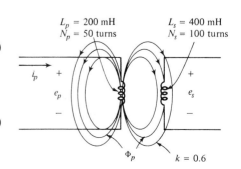

EXAMPLE 24.1. For the transformer in Fig. 24.3:
a. Find the mutual inductance M.

FIG. 24.3

b. Find the induced voltage e_p if the flux changes at the rate of 450 mWb/s.
c. Find the induced voltage e_s for the same rate of change indicated in part (b).
d. Find the induced voltages e_p and e_s if the current i_p changes at the rate of 2 A/s.

Solutions:

a. $M = k \sqrt{L_p L_s} = 0.6 \sqrt{(200 \times 10^{-3})(400 \times 10^{-3})}$
$\qquad = 0.6 \sqrt{8 \times 10^{-2}} = (0.6)(2.828 \times 10^{-1}) = \mathbf{169.7 \ mH}$

b. $e_p = N_p \dfrac{d\phi_p}{dt} = (50)(450 \times 10^{-3}) = \mathbf{22.5 \ V}$

c. $e_s = k N_s \dfrac{d\phi_p}{dt} = (0.6)(100)(450 \times 10^{-3}) = \mathbf{27 \ V}$

d. $e_p = L_p \dfrac{di_p}{dt} = (200 \times 10^{-3})(2) = \mathbf{400 \ mV}$

$\qquad = M \dfrac{di_p}{dt} = (170 \times 10^{-3})(2) = \mathbf{340 \ mV}$

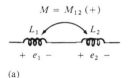

(a)

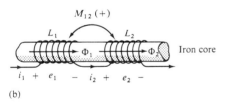

(b)

FIG. 24.4 *Mutually coupled coils connected in series.*

24.3 SERIES CONNECTION OF MUTUALLY COUPLED COILS

In Chapter 11, we found that the total inductance of series isolated coils was determined simply by the sum of the inductances. For two coils that are connected in series but also share the same flux linkages, such as those in Fig. 24.4(a), a mutual term is introduced that will alter the total inductance of the series combination. The physical picture of how the coils are connected is indicated in Fig. 24.4(b). An iron core is included, although the equations to be developed are for any two mutually coupled coils with any value of coefficient of coupling k. When referring to the voltage induced across the inductance L_1 (or L_2) due to the change in flux linkages of the inductance L_2 (or L_1, respectively), the mutual inductance is represented by M_{12}. This type of subscript notation is particularly important when there are two or more mutual terms.

Due to the presence of the mutual term, the induced voltage e_1 is composed of that due to the self-inductance L_1 and that due to the mutual inductance M_{12}. That is,

$$e_1 = L_1 \frac{di_1}{dt} + M_{12} \frac{di_2}{dt}$$

However, since $i_1 = i_2 = i$,

$$e_1 = L_1 \frac{di}{dt} + M_{12} \frac{di}{dt}$$

$$\boxed{e_1 = (L_1 + M_{12}) \frac{di}{dt}} \qquad (24.7a)$$

and, similarly,

$$\boxed{e_2 = (L_2 + M_{12}) \frac{di}{dt}} \qquad (24.7b)$$

For the series connection, the total induced voltage across the series coils, represented by e_T, is

$$e_T = e_1 + e_2 = (L_1 + M_{12}) \frac{di}{dt} + (L_2 + M_{12}) \frac{di}{dt}$$

or

$$e_T = (L_1 + L_2 + M_{12} + M_{12}) \frac{di}{dt}$$

and the total inductance is

$$\boxed{L_{T(+)} = L_1 + L_2 + 2M_{12}} \qquad (H) \qquad (24.8a)$$

The subscript $(+)$ was included to indicate that the mutual terms have a positive sign. If the coils were wound such as shown in Fig. 24.5, where ϕ_1 and ϕ_2 are in opposition, the induced voltage due to the mutual term would oppose that due to the self-inductance, and the total inductance would be determined by

$$\boxed{L_{T(-)} = L_1 + L_2 - 2M_{12}} \qquad (H) \qquad (24.8b)$$

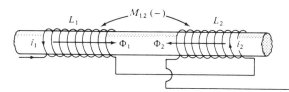

FIG. 24.5 *Mutually coupled coils connected in series with negative mutual inductance.*

Through Eqs. (24.8a) and (24.8b), the mutual inductance can be determined by

$$\boxed{M = \frac{1}{4}(L_{T(+)} - L_{T(-)})} \qquad (H) \qquad (24.9)$$

Equation (24.9) is very effective in determining the mutual inductance between two coils. It states that the mutual inductance is equal to one-quarter the difference between the total inductance with positive mutual inductance and with negative mutual inductance.

From the preceding, it should be clear that the mutual inductance will directly affect the magnitude of the voltage

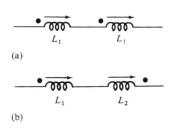

(a)

(b)

FIG. 24.6 *Dot convention for the series coils of (a) Fig. 24.4 and (b) Fig. 24.5.*

induced across a coil, since it will determine the net inductance of the coil. Further examination reveals that the sign of the mutual turn for each coil of a coupled pair is the same. For $L_{T(+)}$ they were both positive, and for $L_{T(-)}$ they were both negative. On a network schematic where it is inconvenient to indicate the windings and the flux path, a system of dots is employed which will determine whether the mutual terms are to be positive or negative. The dot convention is shown in Fig. 24.6 for the series coils of Figs. 24.4 and 24.5.

If the current through *each* of the mutually coupled coils is going away from (or toward) the dot as it *passes through the coil*, the mutual term will be positive, as shown for the case in Fig. 24.6(a). If the arrow indicating current direction through the coil is leaving the dot for one coil and entering for the other, the mutual term is negative.

A few possibilities for mutually coupled transformer coils are indicated in Fig. 24.7(a). The sign of M is indicated for each. When determining the sign, be sure to examine the current direction within the coil itself. In Fig. 24.7(b), one direction was indicated outside for one coil and through for the other. It initially might appear that the sign should be positive since both currents enter the dot, but the current *through* coil 1 is leaving the dot; hence a negative sign is in order.

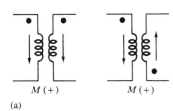

(a)

FIG. 24.7

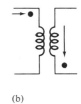

(b)

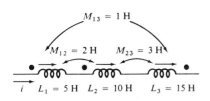

FIG. 24.8

EXAMPLE 24.2. Find the total inductance of the series coils of Fig. 24.8.

Solution:

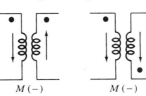

Coil 1: $L_1 + M_{12} - M_{13}$

Coil 2: $L_2 + M_{12} - M_{23}$

Coil 3: $L_3 - M_{23} - M_{13}$

and

$$L_T = (L_1 + M_{12} - M_{13}) + (L_2 + M_{12} - M_{23}) + (L_3 - M_{23} - M_{13})$$
$$L_T = L_1 + L_2 + L_3 + 2M_{12} - 2M_{23} - 2M_{13}$$

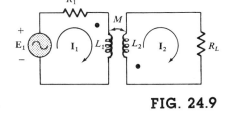

FIG. 24.9

Substituting values, we find

$$L_T = 5 + 10 + 15 + 2(2) - 2(3) - 2(1)$$
$$L_T = 34 - 8 = \mathbf{26\ H}$$

EXAMPLE 24.3. Write the mesh equations for the transformer network in Fig. 24.9.

Solution: For each coil, the mutual term is positive, and the sign of M in $\mathbf{X}_m = \omega M \angle 90°$ is positive, as determined by the direction of $\mathbf{I}_1$ and $\mathbf{I}_2$. Thus,

$$\mathbf{E}_1 - \mathbf{I}_1\mathbf{R}_1 - \mathbf{I}_1\mathbf{X}_{L_1} - \mathbf{I}_2\mathbf{X}_m = 0$$

or

$$\mathbf{E}_1 - \mathbf{I}_1(\mathbf{R}_1 + \mathbf{X}_{L_1}) - \mathbf{I}_2\mathbf{X}_m = 0$$

For the other loop,

$$-\mathbf{I}_2\mathbf{X}_{L_2} - \mathbf{I}_1\mathbf{X}_m - \mathbf{I}_2\mathbf{R}_L = 0$$

or

$$\mathbf{I}_2(\mathbf{X}_{L_2} + \mathbf{R}_L) + \mathbf{I}_1\mathbf{X}_m = 0$$

24.4 THE IRON-CORE TRANSFORMER

An iron-core transformer under loaded conditions is shown in Fig. 24.10. The iron core will serve to increase the coefficient of coupling between the coils by increasing the mutual flux ϕ_m. Recall from Chapter 11 that magnetic flux lines will always take the path of least reluctance, which in this case is the iron core.

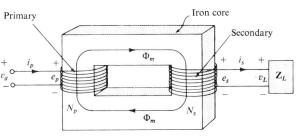

FIG. 24.10 *Iron-core transformer.*

We will assume in the analyses to follow in this chapter that all of the flux linking coil 1 will link coil 2. In other words, the coefficient of coupling is its maximum value, one, and $\phi_m = \phi_1$. In addition, we will first analyze the transformer from an ideal viewpoint. That is, we will neglect such losses as the geometric or dc resistance of the coils, the leakage reactance due to the flux linking either coil that forms no part of ϕ_m, and the hysteresis and eddy current losses. This is not to convey the impression, however, that we will be far from the actual operation of a transformer. Most transformers manufactured today can be considered almost ideal. The equations we will develop under ideal conditions will be, in general, a first approximation to the actual response that will never be off by more than a few percent. The losses will be considered in greater detail in Section 24.6.

When the current i_1 through the primary circuit of the iron-core transformer is a maximum, the flux ϕ_m linking both coils is also a maximum. In fact, the magnitude of the flux is directly proportional to the current through the primary windings. Therefore, the two are in phase, and for sinusoidal inputs, the magnitude of the flux will vary as a sinusoid also. That is, if

$$i_p = \sqrt{2} I_p \sin \omega t$$

then

$$\phi_m = \Phi_{m(\max)} \sin \omega t$$

The induced voltage across the primary due to a sinusoidal input can be determined by Faraday's law:

$$e_p = N_p \frac{d\phi_p}{dt} = N_p \frac{d\phi_m}{dt}$$

Substituting for ϕ_m gives us

$$e_p = N_p \frac{d}{dt} (\Phi_{m(\max)} \sin \omega t)$$

and differentiating, we obtain

$$e_p = \omega N_p \Phi_{m(\max)} \cos \omega t$$

or

$$e_p = \omega N_p \Phi_{m(\max)} \sin(\omega t + 90°)$$

indicating that the induced voltage e_p leads the current through the primary coil by 90°.

The effective value of e_p is

$$E_p = \frac{\omega N_p \Phi_{m(\max)}}{\sqrt{2}} = \frac{2\pi f N_p \Phi_{m(\max)}}{\sqrt{2}}$$

and

$$\boxed{E_p = 4.44 f N_p \Phi_{m(\max)}} \qquad \textbf{(24.10a)}$$

which is an equation for the effective value of the voltage across the primary coil in terms of the frequency of the input current or voltage, the number of turns of the primary, and the maximum value of the magnetic flux linking the primary.

For the case under discussion, where the flux linking the secondary equals that of the primary, if we repeat the procedure just described for the induced voltage across the secondary, we get

$$E_s = 4.44fN_s\Phi_{m(\max)} \qquad \text{(24.10b)}$$

Dividing Eq. (24.10a) by Eq. (24.10b), as follows:

$$\frac{E_p}{E_s} = \frac{4.44fN_p\Phi_{m(\max)}}{4.44fN_s\Phi_{m(\max)}}$$

we obtain

$$\frac{E_p}{E_s} = \frac{N_p}{N_s} \qquad \text{(24.11)}$$

Note that the ratio of the magnitudes of the induced voltages is the same as the ratio of the corresponding turns.

If we consider that

$$e_p = N_p\frac{d\phi_m}{dt}$$

and

$$e_s = N_s\frac{d\phi_m}{dt}$$

and divide one by the other; that is,

$$\frac{e_p}{e_s} = \frac{N_p(d\phi_m/dt)}{N_s(d\phi_m/dt)}$$

then

$$\frac{e_p}{e_s} = \frac{N_p}{N_s}$$

The *instantaneous* values of e_1 and e_2 are therefore related by a constant determined by the turns ratio. Since their instantaneous magnitudes are related by a constant, the induced voltages are in phase, and Eq. (24.11) can be changed to include phasor notation; that is

$$\frac{\mathbf{E}_p}{\mathbf{E}_s} = \frac{N_p}{N_s} \qquad \text{(24.12)}$$

or, since $\mathbf{V}_g = \mathbf{E}_1$ and $\mathbf{V}_L = \mathbf{E}_2$ for the ideal situation,

$$\frac{\mathbf{V}_g}{\mathbf{V}_L} = \frac{N_p}{N_s} \qquad \text{(24.13)}$$

The ratio N_p/N_s, usually represented by the lower-case letter a, is referred to as the *transformation ratio:*

$$a = \frac{N_p}{N_s} \qquad \text{(24.14)}$$

If $a < 1$, the transformer is called a *step-up transformer,* since the voltage $E_s > E_p$:

$$\frac{E_p}{E_s} = \frac{N_p}{N_s} = a$$

so

$$E_s = \frac{E_p}{a}$$

and, if $a < 1$,

$$E_s > E_p$$

If $a > 1$, the transformer is called a *step-down transformer,* since $E_s < E_p$; that is,

$$E_p = aE_s$$

and, if $a > 1$, then

$$E_p > E_s$$

EXAMPLE 24.4. For the iron-core transformer of Fig. 24.11:

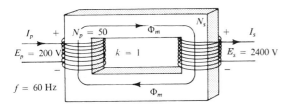

FIG. 24.11

a. Find the maximum flux $\Phi_{m(\max)}$.
b. Find the number of turns N_s.

Solutions:

a. $E_p = 4.44 N_p f \Phi_{m(\max)}$

Therefore,

$$\Phi_{m(\max)} = \frac{E_p}{4.44 N_p f} = \frac{200}{(4.44)(50)(60)}$$

and

$$\Phi_{m(\max)} = \textbf{15.02 mWb}$$

b. $\dfrac{E_p}{E_s} = \dfrac{N_p}{N_s}$

Therefore,

$$N_s = \frac{N_p E_s}{E_p} = \frac{(50)(2400)}{200}$$

$$N_s = \textbf{600 turns}$$

The induced voltage across the secondary of the transformer of Fig. 24.10 will establish a current i_s through the load Z_L and the secondary windings. This current and the turns N_s will develop an mmf $N_s i_s$ that would not be present under no-load conditions, since $i_s = 0$ and $N_s i_s = 0$. Under loaded or unloaded conditions, however, the net ampere-turns on the core produced by both the primary and the secondary must remain unchanged for the same flux ϕ_m to be established in the core. The flux ϕ_m must remain the same to have the same induced voltage across the primary to balance the voltage impressed across the primary. In order to counteract the mmf of the secondary, which is tending to change ϕ_m, an additional current must flow in the primary. This current is called the *load component of the primary current* and is represented by the notation i'_p.

For the balanced or equilibrium condition,

$$N_p i'_p = N_s i_s$$

The total current in the primary under loaded conditions is

$$i_p = i'_p + i_{\phi_m}$$

where i_{ϕ_m} is the current in the primary necessary to establish the flux ϕ_m. For most practical applications, $i'_p > i_{\phi_m}$. For our analysis, we will assume $i_p \cong i'_p$, so

$$N_p i_p = N_s i_s$$

Since the instantaneous values of i_p and i_s are related by the turns ratio, the phasor quantities $\mathbf{I}_p$ and $\mathbf{I}_s$ are also related by the same ratio:

$$N_p \mathbf{I}_p = N_s \mathbf{I}_s$$

or

$$\boxed{\frac{\mathbf{I}_p}{\mathbf{I}_s} = \frac{N_s}{N_p}} \qquad \textbf{(24.15)}$$

Keep in mind that Eq. (24.15) holds true only if we neglect the effects of i_{ϕ_m}. Otherwise, the magnitudes of $\mathbf{I}_p$ and $\mathbf{I}_s$ are not related by the turns ratio, and $\mathbf{I}_p$ and $\mathbf{I}_s$ are not in phase.

For the step-up transformer, $a < 1$, and the current in the secondary, $I_s = aI_p$, is less in magnitude than that in the primary. For a step-down transformer, the reverse is true.

24.5 REFLECTED IMPEDANCE AND POWER

In the previous sections, we found that

$$\frac{\mathbf{V}_g}{\mathbf{V}_L} = \frac{N_p}{N_s} \quad \text{and} \quad \frac{\mathbf{I}_p}{\mathbf{I}_s} = \frac{N_s}{N_p}$$

Dividing one by the other, we have

$$\frac{\mathbf{V}_g/\mathbf{V}_L = N_p/N_s = a}{\mathbf{I}_p/\mathbf{I}_s \;\; = N_s/N_p = 1/a}$$

or

$$\frac{\mathbf{V}_g/\mathbf{I}_p}{\mathbf{V}_L/\mathbf{I}_s} = a^2$$

However, since

$$\mathbf{Z}_p = \frac{\mathbf{V}_g}{\mathbf{I}_p} \quad \text{and} \quad \mathbf{Z}_L = \frac{\mathbf{V}_L}{\mathbf{I}_s}$$

then

$$\boxed{\mathbf{Z}_p = a^2\mathbf{Z}_L} \qquad\qquad (24.16)$$

which in words states that the impedance of the primary circuit of an ideal transformer is the transformation ratio squared times the impedance of the load. If a transformer is used, therefore, an impedance can be made to appear larger or smaller at the primary by placing it in the secondary of a step-down ($a > 1$) or step-up ($a < 1$) transformer, respectively. Note that if the load is capacitive or inductive, the reflected impedance will also be capacitive or inductive.

For the ideal iron-core transformer,

$$\frac{E_p}{E_s} = a = \frac{I_s}{I_p}$$

or

$$\boxed{E_pI_p = E_sI_s} \qquad\qquad (24.17)$$

and

$$\boxed{P_{\text{in}} = P_{\text{out}}} \qquad \text{(ideal case)} \qquad (24.18)$$

EXAMPLE 24.5. For the iron-core transformer of Fig. 24.12:

a. Find the magnitude of the current in the primary and the impressed voltage across the primary.

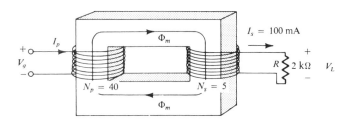

FIG. 24.12

b. Find the input resistance of the transformer.

Solutions:

a. $\dfrac{I_p}{I_s} = \dfrac{N_s}{N_p}$

or

$I_p = \dfrac{N_s}{N_p} I_s = \left(\dfrac{5}{40}\right)(0.1)$

and

$I_p = \textbf{12.5 mA}$

$V_L = I_s Z_L = (0.1)(2 \times 10^3)$

$V_L = 200 \text{ V}$

and

$\dfrac{V_g}{V_L} = \dfrac{N_p}{N_s}$

or

$V_g = \dfrac{N_p}{N_s} V_L = \left(\dfrac{40}{5}\right)(200)$

and

$V_g = \textbf{1600 V}$

b. $Z_p = a^2 Z_L \qquad a = \dfrac{N_p}{N_s} = 8$

$Z_p = (8)^2(2 \times 10^3)$

$Z_p = R_p = \textbf{128 k}\Omega$

EXAMPLE 24.6. For the speaker in Fig. 24.13 to receive maximum power from the circuit, the internal resistance of the speaker should be 540 Ω. If a transformer is used, the speaker resistance of 15 Ω can be made to appear 540 Ω at the primary. Find the transformation ratio required and the number of turns in the primary if the secondary winding has 40 turns.

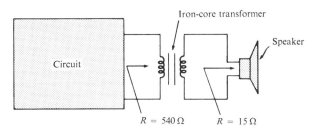

FIG. 24.13

Solution:

$$Z_p = a^2 Z_L$$

or

$$a = \sqrt{\frac{Z_p}{Z_L}} = \sqrt{\frac{540}{15}} = \sqrt{36} = 6$$

Therefore,

$$a = 6 = \frac{N_p}{N_s}$$

or

$$N_p = 6N_s$$
$$N_p = 6(40)$$

and

$$N_p = \textbf{240 turns}$$

EXAMPLE 24.7. For the residential supply appearing in Fig. 24.14, determine (assuming a totally resistive load)

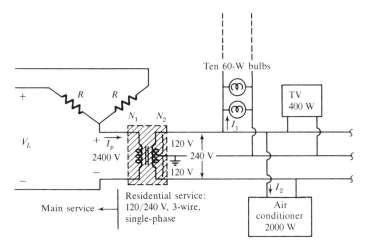

FIG. 24.14

a. the value of R to insure a balanced load.
b. the magnitude of I_1 and I_2.
c. the line voltage V_L.
d. the total power delivered.
e. the turns ratio N_1/N_2.

Solutions:
a. Total power:

$$P_T = 600 + 400 + 2000 = 3000 \text{ W}$$
$$P_{in} = P_{out}$$
$$V_p I_p = V_s I_s = 3000 \text{ W (purely resistive load)}$$
$$2400 I_p = 3000$$
$$I_p = 1.25 \text{ A}$$
$$R = \frac{V}{I} = \frac{2400}{1.25} = \textbf{1920 } \boldsymbol{\Omega}$$

b. $P = (10)(60) = 600 \text{ W} = VI_1 = 120I_1$

and $I_1 = \textbf{5 A}$

$P = 2000 \text{ W} = VI_2 = 240I_2$

and $I_2 = \textbf{8}\tfrac{1}{3} \textbf{A}$

c. $V_L = \sqrt{3}V_\phi = 1.73(2400) = \textbf{4152 V}$

d. $P_T = 3P_\phi = 3(3000) = \textbf{9 kW}$

e. $\dfrac{N_1}{N_2} = \dfrac{V_p}{V_s} = \dfrac{2400}{240} = \textbf{10}$

24.6 EQUIVALENT CIRCUIT (IRON-CORE TRANSFORMER)

For the nonideal or practical iron-core transformer, the equivalent circuit appears as in Fig. 24.15. As indicated,

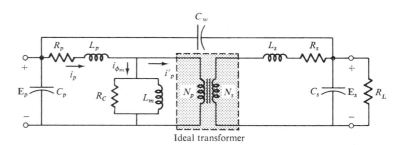

Ideal transformer

FIG. 24.15 *Equivalent circuit for the practical iron-core transformer.*

part of this equivalent circuit is an ideal transformer. The remaining elements of Fig. 24.15 are those elements that contribute to the nonideal characteristics of the device. The resistances R_p and R_s are simply the dc or geometric resistance of the primary and secondary coils, respectively. For the primary and secondary coils of a transformer, there is a definite amount of flux that links each coil that does not pass through the core. This situation is shown in Fig. 24.16. This *leakage* flux, serving as a definite loss to the system since it employs an amount of input energy to be established but serves no useful purpose, is represented by an inductance L_p in the primary circuit and an inductance L_s in the secondary. A coil is employed due to the 90° phase shift between v and i (or ϕ), with v leading i.

The resistance R_C represents the hysteresis and eddy current losses (core losses) within the core due to an ac flux through the core. The inductance L_m (magnetizing inductance) is the inductance associated with the magnetization of the core, that is, the establishing of the flux ϕ_m in the core. The capacitances C_p and C_s are the lumped capacitances of the primary and secondary circuits, respectively,

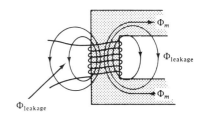

FIG. 24.16

and C_w is the equivalent lumped capacitances between the windings of the transformer.

Since i'_p is normally considerably larger than i_{ϕ_m}, we will ignore i_{ϕ_m} for the moment (set it equal to 0), resulting in the absence of R_C and L_m in the reduced equivalent circuit of Fig. 24.17. In addition, the capacitances C_p, C_w, and C_s do not appear in the equivalent circuit of Fig. 24.17, since their reactance in the present frequency range of interest will not appreciably affect the transfer characteristics of the transformer.

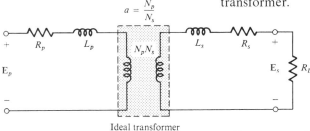

FIG. 24.17 *Reduced equivalent circuit for the nonideal iron-core transformer.*

If we now reflect the secondary circuit through the ideal transformer, as shown in Fig. 24.18(a), we will have the load and generator voltage in the same physical circuit. The total resistance and inductive reactance are determined by

$$R_{\text{equivalent}} = R_e = R_p + a^2 R_s \qquad \textbf{(24.19a)}$$

and

$$X_{\text{equivalent}} = X_e = X_p + a^2 X_s \qquad \textbf{(24.19b)}$$

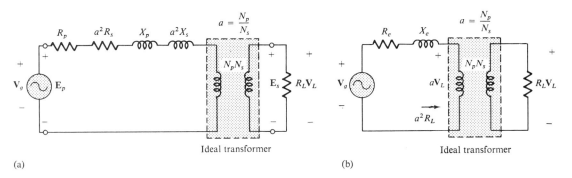

(a)

(b)

FIG. 24.18

which result in the useful equivalent circuit of Fig. 24.18(b). The load voltage can be obtained directly from the circuit of Fig. 24.18(b) through the voltage divider rule:

$$a\mathbf{V}_L = \frac{a^2 R_L \mathbf{V}_g}{(R_e + a^2 R_L) + jX_e} \qquad \textbf{(24.20)}$$

In a different light, the generator voltage necessary to establish a particular load voltage can also be determined through Eq. (24.20). The voltages across the elements of Fig. 24.18(b) have the phasor relationship indicated in Fig.

24.19(a). Note that the current is the reference phasor for drawing the phasor diagram. That is, the voltages across the resistive elements are *in phase* with the current phasor, while the voltage across the equivalent inductance leads the current by 90°. The primary voltage, by Kirchhoff's voltage law, is then the phasor sum of these voltages, as indicated in Fig. 24.19(a). For an inductive load, the phasor diagram appears in Fig. 24.19(b). Note that $a\mathbf{V}_L$ leads $\mathbf{I}$ by the power factor angle of the load. The remainder of the diagram is then similar to that for a resistive load. (The phasor diagram for a capacitive load will be left to the reader as an exercise.)

The effect of R_e and X_e on the magnitude of $\mathbf{V}_g$ for a particular $\mathbf{V}_L$ is obvious from Eq. (24.20) or Fig. 24.19. For increased values, R_e or X_e, an increase in $\mathbf{V}_g$ is required for the same load voltage. For R_e and $X_e = 0$, V_L and V_g are related by the turns ratio.

EXAMPLE 24.8. For a transformer having the equivalent circuit of Fig. 24.20, determine

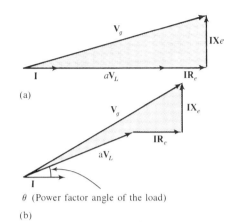

(a)

(b)

θ (Power factor angle of the load)

FIG. 24.19 *Phasor diagram for the iron-core transformer with (a) unity power factor load (resistive) and (b) lagging power factor load (inductive).*

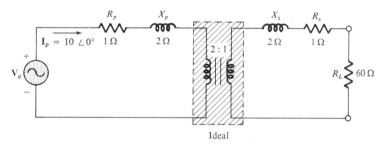

a. R_e and X_e.
b. V_g.

Solutions:
a. $R_e = R_p + a^2 R_s = 1 + (2)^2 1 = 5\ \Omega$
   $X_e = X_p + a^2 X_s = 2 + (2)^2 2 = 10\ \Omega$

b. The transformed equivalent circuit appears in Fig. 24.21.

$a V_L = (I_p)(a^2 R_L) = 2400\ \text{V}$

Thus,

$V_L = \dfrac{2400}{a} = \dfrac{2400}{2} = 1200\ \text{V}$

and

$\mathbf{V}_g = \mathbf{I}_p (R_e + a^2 R_L + jX_e)$
$= 10(5 + 240 + j10) = 10(245 + j10)$
$\mathbf{V}_g = 2450 + j100$

and

$\mathbf{V}_g = \mathbf{2452.04\ V}$

For R_e and $X_e = 0$, $V_g = aV_L = 2400$. Therefore, it is necessary to increase the generator voltage by 50 V (due to R_e and X_e) to obtain the same load voltage.

FIG. 24.20

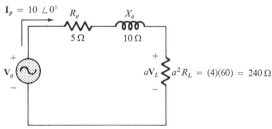

FIG. 24.21

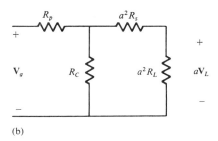

(a)

(b)

FIG. 24.22 (a) Low-frequency reflected equivalent circuit and (b) mid-frequency reflected circuit.

24.7 FREQUENCY CONSIDERATIONS

For certain frequency ranges, the effect of some parameters in the equivalent circuit of the iron-core transformer of Fig. 24.15 can be neglected. Since it is convenient to consider a low-, mid-, and high-frequency region, the equivalent circuits for each will now be introduced and briefly examined.

For the low-frequency region, the reactance ($2\pi fL$) of the primary and secondary leakage reactances can be ignored, and the reflected equivalent circuit will appear as shown in Fig. 24.22(a). The magnetizing inductance must be included, since it appears in parallel with the secondary reflected circuit. As the frequency approaches zero, the reactance of the magnetizing inductance will reduce in magnitude, causing a reduction in the voltage across the secondary circuit. For $f = 0$ Hz, L_m is ideally a short circuit, and $V_L = 0$. As the frequency increases, the reactance of L_m will eventually be sufficiently large compared with the reflected secondary impedance to be neglected. The mid-frequency reflected equivalent circuit will then appear as shown in Fig. 24.22(b). Note the absence of reactive elements, resulting in an *in-phase* relationship between load and generator voltages.

For higher frequencies, the capacitive elements and primary and secondary leakage reactances must be considered, as shown in Fig. 24.23. For discussion purposes, the effect of C_w and C_s appear as a lumped capacitor C in the reflected network of Fig. 24.23; C_p does not appear since the effect of C will predominate. As the frequency of interest increases, the capacitive reactance ($X_C = 1/2\pi fC$) will decrease to the point that it will have a shorting effect across the secondary circuit of the transformer, causing V_L to decrease in magnitude.

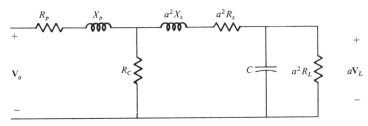

FIG. 24.23 High-frequency reflected equivalent circuit.

A typical transformer-frequency response curve appears in Fig. 24.24. For the low- and high-frequency regions, the primary element responsible for the drop-off is indicated. The peaking that occurs in the high-frequency region is due to the series resonant circuit of Fig. 24.23(b), composed of R_e, $X_e(L)$, and C. In the peaking region, the series resonant circuit is in, or near, its resonant or tuned state.

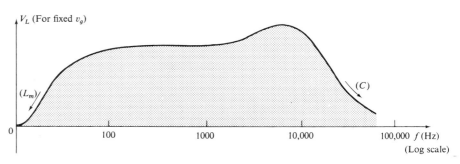

FIG. 24.24 *Transformer-frequency response curve.*

The network discussed in some detail earlier in this chapter was for the high mid-frequency region.

24.8 AIR-CORE TRANSFORMER

As the name implies, the air-core transformer does not have a ferromagnetic core to link the primary and secondary coils. Rather, the coils are placed sufficiently close to have a mutual inductance that will establish the desired transformer action. In Fig. 24.25, current direction and polarities have been defined for the air-core transformer. Note the presence of a mutual inductance term M, which will be positive in this case, as determined by the dot convention.

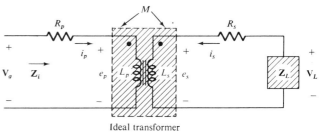

FIG. 24.25 *Air-core transformer equivalent circuit.*

From past analysis in this chapter, we now know that

$$e_p = L_p \frac{di_p}{dt} + M \frac{di_s}{dt} \qquad (24.21)$$

for the primary circuit.

We found in Chapter 11 that for the pure inductor, with no mutual inductance present, the mathematical relationship

$$v_1 = L_1 \frac{di_1}{dt}$$

resulted in the following useful form of the voltage across an inductor:

$$\mathbf{V}_1 = \mathbf{I}_1 \mathbf{X}_{L_1} \quad \text{where} \quad \mathbf{X}_{L_1} = \omega L_1 \angle 90° = j\omega L_1$$

Similarly, it can be shown, for a mutual inductance, that

$$v_1 = M\frac{di_2}{dt}$$

will result in

$$\boxed{\mathbf{V}_1 = \mathbf{I}_2\mathbf{X}_m \quad \text{where} \quad \mathbf{X}_m = \omega M \angle 90° = j\omega M}$$

(24.22)

Equation (24.21) can then be written (using phasor notation)

$$\boxed{\mathbf{E}_p = \mathbf{I}_p\mathbf{X}_{L_p} + \mathbf{I}_s\mathbf{X}_m}$$

(24.23)

and

$$\mathbf{V}_g = \mathbf{I}_p\mathbf{R}_p + \mathbf{I}_p\mathbf{X}_{L_p} + \mathbf{I}_s\mathbf{X}_m$$

or

$$\boxed{\mathbf{V}_g = \mathbf{I}_p(\mathbf{R}_p + \mathbf{X}_{L_p}) + \mathbf{I}_s\mathbf{X}_m}$$

(24.24)

For the secondary circuit,

$$\boxed{\mathbf{E}_s = \mathbf{I}_s\mathbf{X}_{L_s} + \mathbf{I}_p\mathbf{X}_m}$$

(24.25)

and

$$\mathbf{V}_L = \mathbf{I}_s\mathbf{R}_s + \mathbf{I}_s\mathbf{X}_{L_s} + \mathbf{I}_p\mathbf{X}_m$$

or

$$\boxed{\mathbf{V}_L = \mathbf{I}_s(\mathbf{R}_s + \mathbf{X}_{L_s}) + \mathbf{I}_p\mathbf{X}_m}$$

(24.26)

Substituting gives us

$$\mathbf{V}_L = -\mathbf{I}_s\mathbf{Z}_L$$

and

$$0 = \mathbf{I}_s(\mathbf{R}_s + \mathbf{X}_{L_s} + \mathbf{Z}_L) + \mathbf{I}_p\mathbf{X}_m$$

Solving for $\mathbf{I}_s$, we have

$$\mathbf{I}_s = \frac{-\mathbf{I}_p\mathbf{X}_m}{\mathbf{R}_s + \mathbf{X}_{L_s} + \mathbf{Z}_L}$$

and substituting into Eq. (24.24), we obtain

$$\mathbf{V}_g = \mathbf{I}_p(\mathbf{R}_p + \mathbf{X}_{L_p}) + \left(\frac{-\mathbf{I}_p\mathbf{X}_m}{\mathbf{R}_s + \mathbf{X}_{L_s} + \mathbf{Z}_L}\right)\mathbf{X}_m$$

Thus, the input impedance is

$$\mathbf{Z}_i = \frac{\mathbf{V}_g}{\mathbf{I}_p} = \mathbf{R}_p + \mathbf{X}_{L_p} - \frac{\mathbf{X}_m^2}{\mathbf{R}_s + \mathbf{X}_{L_s} + \mathbf{Z}_L}$$

or, defining,

$$\mathbf{Z}_p = \mathbf{R}_p + \mathbf{X}_{L_p} \quad \text{and} \quad \mathbf{Z}_s = \mathbf{R}_s + \mathbf{X}_{L_s}$$

and, substituting, we have

$$\mathbf{X}_m = j\omega M$$

$$\mathbf{Z}_i = \mathbf{Z}_p - \frac{(+j\omega M)^2}{\mathbf{Z}_s + \mathbf{Z}_L}$$

and

$$\boxed{\mathbf{Z}_i = \mathbf{Z}_p + \frac{(\omega M)^2}{\mathbf{Z}_s + \mathbf{Z}_L}} \qquad (24.27)$$

The term $(\omega M)^2/(\mathbf{Z}_s + \mathbf{Z}_L)$ is called the *coupled imped-ance*. Note that it is independent of the sign of M. Consider also that since $(\omega M)^2$ is a constant with $0°$ phase angle, if $\mathbf{Z}_2$ is resistive, the resulting coupled impedance term will appear capacitive, due to division of $(\mathbf{R}_L + \mathbf{Z}_s)$ into $(\omega M)^2$. This resulting capacitive reactance will oppose the series primary inductance L_p, causing a reduction in $\mathbf{Z}_i$. Includ-ing the effect of the mutual term, the input impedance to the network will appear as shown in Fig. 24.26.

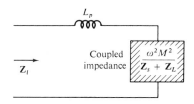

FIG. 24.26 *Input characteristics for the air-core transformer.*

EXAMPLE 24.9. Determine the input impedance to the air-core transformer in Fig. 24.27.

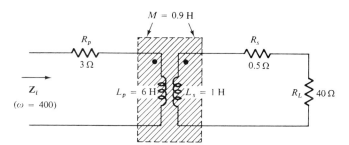

FIG. 24.27

Solution:

$$\mathbf{Z}_i = \mathbf{Z}_p + \frac{(\omega M)^2}{\mathbf{Z}_s + \mathbf{Z}_L}$$

$$= R_p + jX_{L_p} + \frac{(\omega M)^2}{R_s + jX_{L_s} + R_L}$$

$$= 3 + j2.4 \times 10^3 + \frac{(360)^2}{0.5 + j400 + 40}$$

$$\cong j2.4 \times 10^3 + \frac{129.6 \times 10^3}{40.5 + j400}$$

$$= j2.4 \times 10^3 + \frac{129.6 \times 10^3}{402.05 \angle 84.22°}$$

$$= j2.4 \times 10^3 + 322.4 \angle -84.22° = j2.4 \times 10^3$$
$$+ \underbrace{(0.0325 \times 10^3 - j0.3208 \times 10^3)}_{\text{capacitive}}$$

$$= 0.0325 \times 10^3 + j(2.40 \times 10^3 - 0.3208 \times 10^3)$$

$$\mathbf{Z}_i = \mathbf{32.5} + \mathbf{j2079} = R_i + jX_{L_i} = \mathbf{2079.25} \angle \mathbf{89.10°}$$

24.9 THE TRANSFORMER AS AN ISOLATION DEVICE

The transformer is frequently used to isolate one portion of an electrical system from another. By *isolation,* we mean the absence of any direct physical connection. As a first example of its use as an isolation device, consider the measurement of line voltages on the order of 40,000 V (Fig. 24.28).

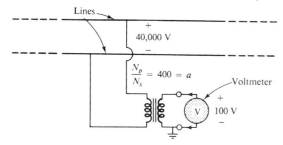

FIG. 24.28

To apply a voltmeter across 40,000 V would obviously be a dangerous task due to the possibility of physical contact with the lines when making the necessary connections. By including a transformer in the transmission system as original equipment, one can bring the potential down to a safe level for measurement purposes and can determine the line voltage using the turns ratio. Therefore, the transformer will serve both to isolate and to step down the voltage.

As a second example, consider the application of the voltage v_x to the vertical input of the oscilloscope (a measuring instrument) in Fig. 24.29.

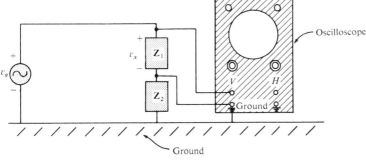

FIG. 24.29

If the connections are made as shown and the generator and oscilloscope have a common ground, the impedance $\mathbf{Z}_2$ has been effectively shorted out of the circuit by the ground connection of the oscilloscope. The input voltage to the oscilloscope will therefore be meaningless as far as the voltage v_x is concerned. In addition, if $\mathbf{Z}_2$ is the current-limiting impedance in the circuit, the current in the circuit may rise to a level that will cause severe damage to the circuit. If a transformer is used as shown in Fig. 24.30, this problem will be eliminated, and the input voltage to the oscilloscope will be v_x.

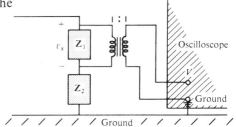

FIG. 24.30

24.10 NAMEPLATE DATA

A typical power transformer rating might be the following:

$$5 \text{ kVA}, \quad 2000/100 \text{ V}, \quad 60 \text{ Hz}$$

The 2000 V or the 100 V can be either the primary or the secondary voltage; that is, if 2000 V is the primary voltage, then 100 V is the secondary voltage, and vice versa. The 5 kVA is the apparent power $(S = VI)$ rating of the transformer. If the secondary voltage is 100 V, then the maximum load current is

$$I_L = \frac{S}{V_L} = \frac{5000}{100} = 50 \text{ A}$$

and if the secondary voltage is 2000 V, then the maximum load current is

$$I_L = \frac{S}{V_L} = \frac{5000}{2000} = 2.5 \text{ A}$$

The transformer is rated in terms of the apparent power rather than the average power for the reason demonstrated by the circuit of Fig. 24.31.

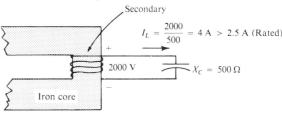

FIG. 24.31

Since the current through the load is greater than that determined by the apparent power rating, the transformer may be permanently damaged. Note, however, that since the load is purely capacitive, the average power to the load is zero. The wattage rating would therefore be meaningless regarding the ability of this load to damage the transformer.

The transformation ratio of the transformer under discussion can be either of two values. If the secondary voltage is 2000 V, the transformation ratio is $a = N_p/N_s = V_g/V_L =$

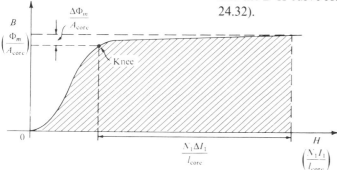

$100/2000 = 1/20$, and the transformer is a step-up transformer. If the secondary voltage is 100 V, the transformation ratio is $a = N_p/N_s = V_g/V_L = (2000/100) = 20$, and the transformer is a step-down transformer.

The rated primary current can be determined simply by applying Eq. (24.15):

$$I_1 = \frac{I_2}{a}$$

which is equal to $[2.5/(1/20)] = 50$ A if the secondary voltage is 2000 V, and $(50/20) = 2.5$ A if the secondary voltage is 100 V.

To explain the necessity for including the frequency in the nameplate data, consider Eq. (24.10a):

$$E_p = 4.44 f_p N_p \Phi_{m(max)}$$

and the *B-H* curve for the iron core of the transformer (Fig. 24.32).

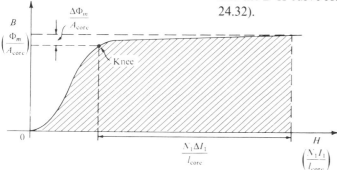

FIG. 24.32

The point of operation on the *B-H* curve for most transformers is at the knee of the curve. If the frequency of the applied signal should drop, and N_p and E_p remain the same, then $\Phi_{m(max)}$ must increase in magnitude, as determined by Eq. (24.10a):

$$\Phi_{m(max)}\uparrow = \frac{E_p}{4.44 f_p\downarrow N_p}$$

Note on the *B-H* curve that this increase in $\Phi_{m(max)}$ will cause a very high current to flow in the primary, resulting in possible damage of the transformer.

24.11 TYPES OF TRANSFORMERS

Transformers are available in many different shapes and sizes. Some of the more common types include the power transformer, audio transformer, I-F (intermediate-frequency) transformer, and R-F (radio-frequency) transformer. Each is designed to fulfill a particular requirement in a specific area of application. The symbols for some of the basic types of transformers are shown in Fig. 24.33.

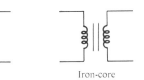

FIG. 24.33

The method of construction varies from one transformer to another. Two of the many different ways in which the primary and secondary coils can be wound around an iron core are shown in Fig. 24.34. In either case, the core is made of laminated sheets of ferromagnetic material separated by an insulator to reduce the eddy-current losses. The sheets themselves will also contain a small percentage of silicon to reduce the hysteresis losses.

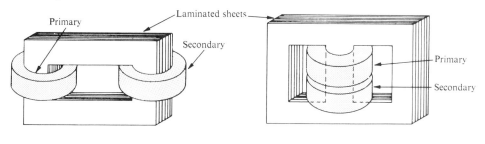

(a) Core type.   (b) Shell type.

FIG. 24.34

A shell-type power transformer with its schematic representation is shown in Fig. 24.35.

(a) Shell-type power transformer.
Courtesy of United Transformer Co.

(b) Schematic representation.

FIG. 24.35

The *autotransformer* [Fig. 24.36(b)] is a type of power transformer which, instead of employing the two-circuit principle (complete isolation between coils), has one winding common to both the input and output circuits. The induced voltages are related to the turns ratio in the same manner as that described for the two-circuit transformer. If the proper connection is used, a two-circuit power transformer can be used as an autotransformer. The advantage of using it as an autotransformer is that a larger kVA can be transformed. This can be demonstrated by the two-circuit filament transformer shown in Fig. 24.36(a). It is shown in Fig. 24.36(b) as an autotransformer.

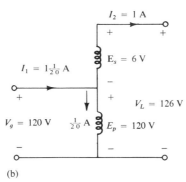

(a)

(b)

FIG. 24.36 *Autotransformer.*

For the two-circuit transformer, note that $S = (1/20) \cdot (120) = 6$ VA, while for the autotransformer, $S = (1\frac{1}{20}) \cdot (120) = 126$ VA, which is many times that of the two-circuit transformer. Note also that the current and voltage of each coil are the same as those for the two-circuit configuration.

The disadvantage of the autotransformer is obvious: loss of the isolation between the primary and secondary circuits.

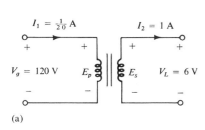

(a) I-F transformer.

(b) R-F transformer.

FIG. 24.37

The R-F and I-F transformers are used extensively in radio and television transmitters and receivers. Their construction is somewhat different from the two just described. Typical R-F and I-F transformers are shown in Fig. 24.37. Both are available with or without a shield. The types shown here both have an air core. Permeability-tuned I-F transformers, which permit the changing of ϕ_m, and thereby k, by moving a ferromagnetic core within the primary and secondary coils of the transformer, are available.

A dual-in-line pulse transformer package appears in Fig. 24.38 for use with integrated circuits and printed circuit board applications. Note the availability of four isolated transformers and the appearance of the dot convention. The data for one such unit include a 2:1 primary-to-secondary turns ratio; a leakage inductance of 0.50 μH; a coupling capacitance of 7 pF; a primary dc resistance of 0.19 Ω; and a secondary resistance of 0.13 Ω.

Courtesy of Bourns®, Inc.

FIG. 24.38 *Dual-in-line pulse transformer package.*

A hybrid compatible transformer appears in Fig. 24.39. This is one approach to providing a preassembled transformer with the standard hybrid integrated circuit board to which other elements can be applied.

Courtesy of Burr-Brown Research Corp., Inc.

FIG. 24.39 *Hybrid compatible transformer.*

A pulse transformer designed to generate or transfer a pulse of electrical energy appears in Fig. 24.40. Units of this type can be as small as 1/4 inch in diameter.

Courtesy of TRW/UTC Transformers

FIG. 24.40 *Pulse transformer.*

A miniature radio transformer designed for direct mounting on a printed circuit board appears in Fig. 24.41, with a variable transformer that can vary the coefficient of coupling between the windings.

(a) Miniature radio transformer. Designed for plug-in printed circuit applications.
Courtesy of Microtran Company, Inc.

FIG. 24.41

(b) Variable transformer.
Courtesy of Basler Electric Co.

24.12 TAPPED AND MULTIPLE-LOAD TRANSFORMERS

For the center-tapped (primary) transformer of Fig. 24.42, where the voltage from the center top to either outside lead is defined as $E_p/2$, the relationship between E_p and E_s is

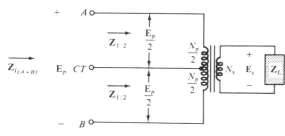

FIG. 24.42 *Ideal transformer with a center-tapped primary.*

$$\boxed{\frac{\mathbf{E}_p}{\mathbf{E}_s} = \frac{N_p}{N_s}} \qquad (24.28)$$

For each half-section of the primary,

$$\mathbf{Z}_{1/2} = \left(\frac{N_p/2}{N_s}\right)^2 \mathbf{Z}_L$$

$$\mathbf{Z}_{1/2} = \frac{1}{4}\left(\frac{N_p}{N_s}\right)^2 \mathbf{Z}_L$$

and

$$\boxed{\mathbf{Z}_{1/2} = \frac{1}{4}\mathbf{Z}_i} \qquad (24.29)$$

as indicated in Fig. 24.42.

For the multiple-load transformer of Fig. 24.43, the following equations apply:

$$\boxed{\frac{\mathbf{E}_1}{\mathbf{E}_2} = \frac{N_1}{N_2}, \qquad \frac{\mathbf{E}_1}{\mathbf{E}_3} = \frac{N_1}{N_3}, \qquad \frac{\mathbf{E}_2}{\mathbf{E}_3} = \frac{N_2}{N_3}}$$

(24.30)

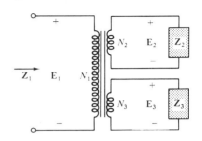

FIG. 24.43 *Ideal transformer with multiple loads.*

The total input impedance can be determined by first noting that for the ideal transformer, the power delivered to the primary is equal to the power dissipated by the load; that is,

$$P_1 = P_{L_2} + P_{L_3}$$

and for $Z_1 = R_1$, $Z_2 = R_2$ and $Z_3 = R_3$, or

$$\frac{E_1^2}{R_1} = \frac{E_2^2}{R_2} + \frac{E_3^2}{R_3}$$

or, since

$$E_2 = \frac{N_2}{N_1}E_1 \quad \text{and} \quad E_3 = \frac{N_3}{N_1}E_1$$

then

$$\frac{E_1^2}{R_1} = \frac{[(N_2/N_1)E_1]^2}{R_2} + \frac{[(N_3/N_1)E_1]^2}{R_3}$$

and

$$\frac{E_1^2}{R_1} = \frac{E_1^2}{(N_1/N_2)^2 R_2} + \frac{E_1^2}{(N_1/N_3)^2 R_3}$$

Thus,

$$\frac{1}{R_1} = \frac{1}{(N_1/N_2)^2 R_2} + \frac{1}{(N_1/N_3)^2 R_3} \qquad (24.31)$$

indicating that the load impedances are reflected in parallel.

For the configuration of Fig. 24.44 with E_2 and E_3 defined as shown, Eqs. (24.30) and (24.31) are applicable.

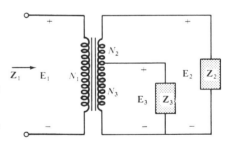

FIG. 24.44 *Ideal transformer with tapped secondary and multiple loads.*

24.13 NETWORKS WITH MAGNETICALLY COUPLED COILS

For multiloop networks with magnetically coupled coils, the mesh-analysis approach is most frequently applied. A firm understanding of the dot convention discussed earlier should make the writing of the equations quite direct and free of errors. Before writing the equations for any particular loop, first determine whether the mutual term is positive or negative, keeping in mind that it will have the same sign as that for the other magnetically coupled coil. For the two-loop network of Fig. 24.45, for example, the mutual term has a positive sign since the current through each coil leaves the dot. For the primary loop,

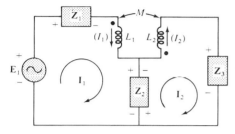

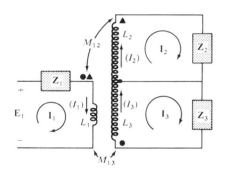

FIG. 24.45

$$E_1 - I_1Z_1 - I_1Z_{L_1} - I_2Z_m - Z_2(I_1 - I_2) = 0$$

where M of $\mathbf{Z}_m = \omega M \angle 90°$ is positive and

$$I_1(Z_1 + Z_{L_1} + Z_2) - I_2(Z_2 - Z_m) = E_1$$

Note in the above that the mutual impedance was treated as if it were an additional inductance in series with the inductance L_1 having a sign determined by the dot convention and the voltage across which is determined by the current in the magnetically coupled loop.

For the secondary loop,

$$-Z_2(I_2 - I_1) - I_2Z_{L_2} - I_1Z_m - I_2Z_3 = 0$$

or

$$I_2(Z_2 + Z_{L_2} + Z_3) - I_1(Z_2 - Z_m) = 0$$

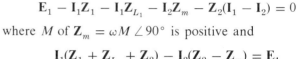

FIG. 24.46

For the network of Fig. 24.46, we find a mutual term between L_1 and L_2 and L_1 and L_3 labeled M_{12} and M_{13}, respectively.

For the coils with the dots (L_1 and L_3), since each current through the coils leaves the dot, M_{13} is positive for the chosen direction of I_1 and I_3. However, since the current I_1 leaves the dot through L_1, and I_2 enters the dot through coil L_2, M_{12} is negative. Consequently, for the input circuit,

$$E_1 - I_1Z_1 - I_1Z_{L_1} - I_2(-Z_{m_{12}}) - I_3Z_{m_{13}} = 0$$

or

$$E_1 - I_1(Z_1 + Z_{L_1}) + I_2Z_{m_{12}} - I_3Z_{m_{13}} = 0$$

For loop 2,

$$-I_2Z_2 - I_2Z_{L_2} - I_1(-Z_{m_{12}}) = 0$$
$$-I_1Z_{m_{12}} + I_2(Z_2 + Z_{L_2}) = 0$$

and for loop 3,

$$-\mathbf{I}_3\mathbf{Z}_3 - \mathbf{I}_3\mathbf{Z}_{L_3} - \mathbf{I}_1\mathbf{Z}_{m_{13}} = 0$$

or

$$\mathbf{I}_1\mathbf{Z}_{m_{13}} + \mathbf{I}_3(\mathbf{Z}_3 + \mathbf{Z}_{L_3}) = 0$$

In determinant form,

$$
\begin{array}{lll}
\mathbf{I}_1(\mathbf{Z}_1 + \mathbf{Z}_{L_1}) - \mathbf{I}_2\mathbf{Z}_{m_{12}} & + \mathbf{I}_3\mathbf{Z}_{m_{13}} & = \mathbf{E}_1 \\
-\mathbf{I}_1\mathbf{Z}_{m_{12}} & + \mathbf{I}_2(\mathbf{Z}_2 + \mathbf{Z}_{L_2}) + 0 & = 0 \\
\mathbf{I}_1\mathbf{Z}_{m_{13}} & + 0 & + \mathbf{I}_3(\mathbf{Z}_3 + \mathbf{Z}_{13}) = 0
\end{array}
$$

PROBLEMS

Section 24.2

1. For the air-core transformer of Fig. 24.47:
   a. Find the value of L_s if the mutual inductance M is equal to 80 mH.
   b. Find the induced voltages e_p and e_s if the flux linking the primary coil changes at the rate of 0.08 Wb/s.
   c. Find the induced voltages e_p and e_s if the current i_p changes at the rate of 0.3 A/ms.

2. a. Repeat Problem 1 if k is changed to 1.
   b. Repeat Problem 1 if k is changed to 0.2.
   c. Compare the results of parts (a) and (b).

3. Repeat Problem 1 for $k = 0.9$, $N_p = 300$ turns, and $N_s = 25$ turns.

Section 24.3

4. Determine the total inductance of the series coils of Fig. 24.48.

5. Determine the total inductance of the series coils of Fig. 24.49.

6. Determine the total inductance of the series coils of Fig. 24.50.

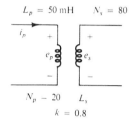

$L_p = 50\,\text{mH}$ $N_s = 80$

$N_p - 20$ L_s
$k = 0.8$

FIG. 24.47

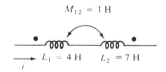

$M_{12} = 1\,\text{H}$

$L_1 = 4\,\text{H}$ $L_2 = 7\,\text{H}$

FIG. 24.48

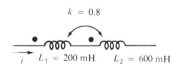

$k = 0.8$

$L_1 = 200\,\text{mH}$ $L_2 = 600\,\text{mH}$

FIG. 24.49

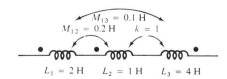

$M_{13} = 0.1\,\text{H}$
$M_{12} = 0.2\,\text{H}$ $k = 1$

$L_1 = 2\,\text{H}$ $L_2 = 1\,\text{H}$ $L_3 = 4\,\text{H}$

FIG. 24.50

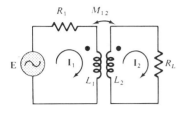

FIG. 24.51

7. Write the mesh equations for the network of Fig. 24.51.

Section 24.4

8. For the iron-core transformer ($k = 1$) of Fig. 24.52:
   a. Find the magnitude of the induced voltage E_s.
   b. Find the maximum flux $\Phi_{m(max)}$.

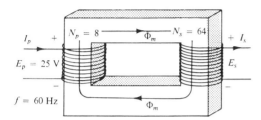

FIG. 24.52

9. Repeat Problem 8 for $N_p = 240$ and $N_s = 30$.

10. Find the input voltage of an iron-core transformer with a secondary voltage of 240 and $N_p = 60$ with $N_s = 720$.

11. If the maximum flux passing through the core of Problem 8 is 12.5 mWb, find the frequency of the input voltage.

Section 24.5

12. For the iron-core transformer of Fig. 24.53:
   a. Find the magnitude of the current I_L and the voltage V_L if $a = 1/5$, $I_p = 2$ A, and $Z_L = 2$-Ω resistor.
   b. Find the input resistance for the data specified in part (a).

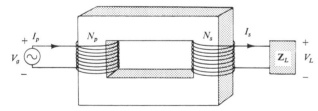

FIG. 24.53

13. Find the input impedance for the iron-core transformer of Fig. 24.53 if $a = 2$, $I_p = 4$ A, and $V_g = 1600$ V.

14. Find the voltage V_g and the current I_p if the input impedance of the iron-core transformer of Fig. 24.53 is 4 Ω and $V_L = 1200$ V with $a = 1/4$.

15. If $V_L = 240$ V, $Z_L = 20$-Ω resistor, $I_p = 0.05$ A, and $N_s = 50$, find the number of turns in the primary circuit of the iron-core transformer of Fig. 24.53.

16. **a.** If $N_p = 400$, $N_s = 1200$, and $V_g = 100$ V, find the magnitude of I_p for the iron-core transformer of Fig. 24.53 if $Z_L = 9 + j12$.
   b. Find the magnitude of the voltage V_L and the current I_L for the conditions of part (a).

17. **a.** For the circuit of Fig. 24.54, find the transformation ratio required to deliver maximum power to the speaker.
   b. Find the maximum power delivered to the speaker.

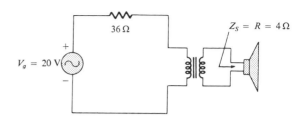

FIG. 24.54

Section 24.6

18. For the transformer of Fig. 24.55, determine
    a. the equivalent resistance R_e.
    b. the equivalent reactance X_e.
    c. the equivalent circuit reflected to the primary.
    d. the primary current for $V_g = 50 \angle 0°$.
    e. the load voltage V_L.
    f. the phasor diagram of the reflected primary circuit.
    g. the new load voltage if we assume the transformer to be ideal with a 4 : 1 turns ratio. Compare the result with that of part (e).

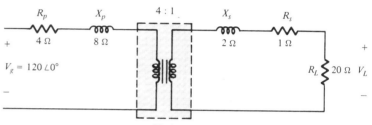

19. For the transformer of Fig. 24.55, if the resistive load is replaced by an inductive reactance of 20 Ω:
    a. Determine the total reflected primary impedance.
    b. Calculate the primary current.
    c. Determine the voltage across R_e, X_e, and the reflected load.
    d. Draw the phasor diagram.

FIG. 24.55

20. Repeat Problem 19 for a capacitive load having a reactance of 20 Ω.

Section 24.7

21. Discuss in your own words the frequency characteristics of the transformer. Employ the applicable equivalent circuit and frequency characteristics appearing in this chapter.

Section 24.8

22. Determine the input impedance to the air-core transformer of Fig. 24.56. Sketch the reflected primary network.

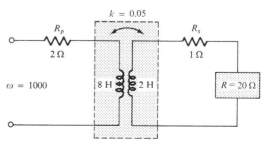

FIG. 24.56

23. An ideal transformer is rated 10 kVA, 2400/120 V, 60 Hz.
    a. Find the transformation ratio if the 120 V is the secondary voltage.
    b. Find the current rating of the secondary if the 120 V is the secondary voltage.
    c. Find the current rating of the primary if the 120 V is the secondary voltage.
    d. Repeat parts (a) through (c) if the 2400 V is the secondary voltage.

Section 24.11

24. Determine the primary and secondary voltages and currents for the autotransformer of Fig. 24.57.

Section 24.12

25. For the center-tapped transformer of Fig. 24.42 where $N_p = 100$, $N_s = 25$, $\mathbf{Z}_L = \mathbf{R}_L = 5 \angle 0°$, $\mathbf{E}_p = 100 \angle 0°$, determine
    a. the load voltage and current.
    b. the impedance $\mathbf{Z}_i$.
    c. the impedance $\mathbf{Z}_{1/2}$.

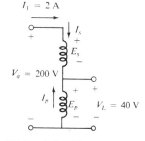

FIG. 24.57

26. For the multiple-load transformer of Fig. 24.43 where $N_1 = 90$, $N_2 = 15$, $N_3 = 45$, $\mathbf{Z}_2 = \mathbf{R}_2 = 8 \angle 0°$, $\mathbf{Z}_3 = \mathbf{R}_3 = 5 \underline{/0°}$, $\mathbf{E}_1 = 60 \angle 0°$:
    a. Determine the load voltages and currents.
    b. Calculate $\mathbf{Z}_1$.

27. For the multiple-load transformer of Fig. 24.44 where $N_1 = 120$, $N_2 = 40$, $N_3 = 30$, $\mathbf{Z}_2 = \mathbf{R}_2 = 12 \angle 0°$, $\mathbf{Z}_3 = \mathbf{R}_3 = 10 \underline{/0°}$, $\mathbf{E}_1 = 120 \angle 60°$:
    a. Determine the load voltages and currents.
    b. Calculate $\mathbf{Z}_1$.

Section 24.13

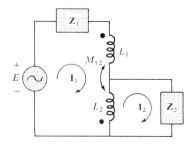

FIG. 24.58

28. Write the mesh equations for the network of Fig. 24.58.

29. Write the mesh equations for the network of Fig. 24.59.

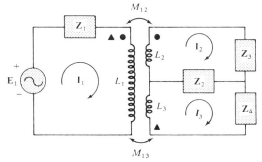

FIG. 24.59

GLOSSARY

Auto transformer   A transformer with one winding common to both the primary and the secondary circuits. A loss in isolation is balanced with the increase in its kVA rating.

Coefficient of coupling (k)   A measure of the magnetic coupling of two coils that ranges from a minimum of 0 to a maximum of 1.

Dot convention   A technique for labeling the effect of the mutual inductance on a net inductance of a network or system.

Leakage flux   The flux linking the coil that does not pass through the ferromagnetic path of the magnetic circuit.

Loosely coupled   A term applied to two coils that have a low coefficient of coupling.

Multiple-load transformers   Transformers having more than a single load connected to the secondary winding.

Mutual inductance   The inductance that exists between magnetically coupled coils of the same or different dimensions.

Nameplate data   Information such as the kVA rating, voltage transformation ratio, and frequency of application that is of primary importance in choosing the proper transformer for a particular application.

Primary   The coil or winding to which the source of energy is normally applied.

Reflected impedance   The impedance appearing at the primary of a transformer due to a load connected to the secondary. Its magnitude is controlled directly by the transformation ratio.

Secondary   The coil or winding to which the load is normally applied.

Step-down transformer   A transformer whose secondary voltage is less than its primary voltage. The transformation ratio a is greater than one.

Step-up transformer   A transformer whose secondary voltage is greater than its primary voltage. The magnitude of the transformation ratio a is less than one.

Tapped transformer   A transformer having an additional connection between the terminals of the primary or secondary windings.

Transformation ratio (a)   The ratio of primary to secondary turns of a transformer.

25.1 INTRODUCTION

In the broad spectrum of courses to follow, there will be an increasing need for the ability to *model* both devices and systems and employ these models in the analysis and synthesis of combined and enlarged systems.

This chapter will be limited to the configuration most frequently subject to the modeling technique: the two-port network shown in Fig. 25.1.

Note that in Fig. 25.1 there are two ports of entry or interest, each having a pair of terminals. For some devices, the two-port configuration of Fig. 25.1 may appear as shown in Fig. 25.2(a). The block diagram of Fig. 25.2(a) simply indicates that terminals 1' and 2' are in common—a particular case of the general two-port network. A single-port and a multiport network appear in Fig. 25.2(b). The former has been analyzed throughout the text, while the characteristics of the latter will be left for a more advanced course.

The primary purpose of this chapter is to develop a set of equations (and, subsequently, networks) that will allow us to model the device or system appearing within the black-box structure of Fig. 25.1. That is, we will be able to establish a network that will display the same terminal characteristics as those of the original system, device, and so on. In Fig. 25.3, for example, a transistor appears between the four external terminals. Through the analysis to follow, we

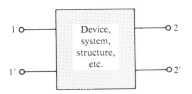

FIG. 25.1 *Two-port configuration.*

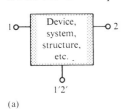

(a)

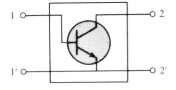

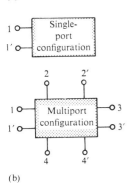

(b)

FIG. 25.2

will find a combination of network elements that will allow us to replace the transistor by a network that will behave very much like the original device for a specific set of operating conditions. Methods such as mesh and nodal analysis can then be applied to determine any unknown quantities. The models, when reduced to their simplest forms as determined by the operating conditons, can also provide very quick estimates of network behavior without a lengthy mathematical derivation. In other words, someone well-versed in the use of models can analyze the operation of large, complex systems in short order. The results may be only approximate in most cases, but typically this quick return for a minimum of effort is often worthwhile.

The following analysis may initially appear very mathematical and devoid of any practical meaning. However, you will note that the parameters are determined by the ratio of electrical quantities under very specific network conditions. The resulting quantities also have a terminology that will be applied very frequently in the electronics course to follow. The analysis of this chapter will be limited to linear (fixed-value) systems with bilateral elements. Three sets of parameters will be developed for the two-port configuration, referred to as the *impedance* (**z**), *admittance* (**y**), and *hybrid* (**h**) parameters. A table will be provided at the end of the chapter relating the three sets of parameters.

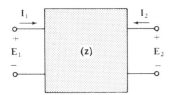

FIG. 25.3

FIG. 25.4 *Two-port impedance parameter configuration.*

25.2 IMPEDANCE (z) PARAMETERS

For the two-port configuration of Fig. 25.4, four variables are specified. For most situations, if any two are specified, the remaining two variables can be determined. The four variables can be related by the following equations:

$$E_1 = z_{11}I_1 + z_{12}I_2 \qquad \text{(25.1a)}$$

$$E_2 = z_{21}I_1 + z_{22}I_2 \qquad \text{(25.1b)}$$

The *impedance parameters* z_{11}, z_{12}, and z_{22} are measured in ohms.

To model the black box, each impedance parameter must be determined. They are determined by setting a particular variable to zero.

z_{11}

For z_{11}, if I_2 is set to zero, as shown in Fig. 25.5, Eq. (25.1a) becomes

$$E_1 = z_{11}I_1 + z_{12}(0)$$

$$\boxed{z_{11} = \frac{E_1}{I_1}}\Big|_{I_2 = 0} \quad \text{(ohms, } \Omega\text{)} \quad \textbf{(25.2)}$$

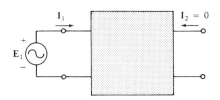

FIG. 25.5 z_{11} *determination.*

Equation (25.2) dictates that with I_2 set to zero, the impedance parameter is determined by the resulting ratio of E_1 to I_1. Since E_1 and I_1 are both input quantities, with I_2 set to zero, the parameter z_{11} is formally referred to in the following manner:

$z_{11} = $ *open-circuit, input-impedance parameter*

z_{12}

For z_{12}, I_1 is set to zero, and Eq. (25.1a) results in

$$\boxed{z_{12} = \frac{E_1}{I_2}}\Big|_{I_1 = 0} \quad (\Omega) \quad \textbf{(25.3)}$$

For most systems where input and output quantities are to be compared, the ratio of interest is usually that of the output over the input quantity. In this case, the *reverse* is true, resulting in the following:

$z_{12} = $ *open-circuit, reverse-transfer impedance parameter*

The term *transfer* is included to indicate that z_{12} will relate an input and output quantity (for the condition $I_1 = 0$). The network configuration for determining z_{12} is shown in Fig. 25.6.

For an applied source E_2, the ratio E_1/I_2 will determine z_{12} with I_1 set to zero.

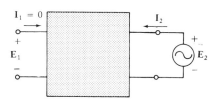

FIG. 25.6 z_{12} *determination.*

z_{21}

To determine z_{21}, set I_2 to zero and find the ratio E_2/I_1 as determined by Eq. (25.1b). That is,

$$\boxed{z_{21} = \frac{E_2}{I_1}}\Big|_{I_2 = 0} \quad (\Omega) \quad \textbf{(25.4)}$$

In this case, input and output quantities are again the determining variables, requiring the term *transfer* in the nomenclature. However, the ratio is that of an output to

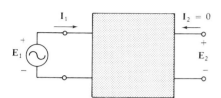

FIG. 25.7 z_{21} *determination.*

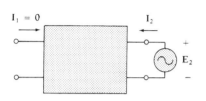

FIG. 25.8 z_{22} *determination.*

an input quantity, so the descriptive term *forward* is applied, and

z_{21} = *open-circuit, forward-transfer impedance parameter*

The determining network is shown in Fig. 25.7. For an applied voltage E_1, it is determined by the ratio E_2/I_1 with I_2 set to zero.

z_{22}

The remaining parameter, z_{22}, is determined by

$$z_{22} = \frac{E_2}{I_2} \bigg|_{I_1 = 0} \quad (\Omega) \quad \textbf{(25.5)}$$

as derived from Eq. (25.1b) with I_1 set to zero. Since it is the ratio of the output voltage to the output current with I_1 set to zero, it has the terminology

z_{22} = *open-circuit, output-impedance parameter*

The required network is shown in Fig. 25.8. For an applied voltage E_2, it is determined by the resulting ratio E_2/I_2 with $I_1 = 0$.

EXAMPLE 25.1 Determine the impedance (**z**) parameters for the T network of Fig. 25.9.

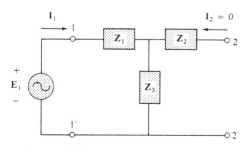

FIG. 25.9 *T configuration.*

Solution: For z_{11}, the network will appear as shown in Fig. 25.10.

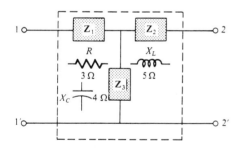

FIG. 25.10

$$I_1 = \frac{E_1}{Z_1 + Z_3}$$

$$z_{11} = \frac{E_1}{I_1} \bigg|_{I_2 = 0} \quad \boxed{= Z_1 + Z_3} \quad \textbf{(25.6)}$$

For z_{12}, the network will appear as shown in Fig. 25.11:

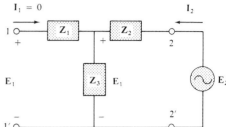

$$E_1 = I_2 Z_3$$

Thus,

$$z_{12} = \frac{E_1}{I_2}\bigg|_{I_1 = 0} = \frac{I_2 Z_3}{I_2} \boxed{= Z_3} \qquad (25.7)$$

FIG. 25.11

For z_{21}, the required network appears in Fig. 25.12:

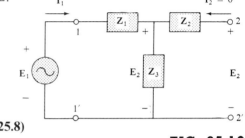

$$E_2 = I_1 Z_3$$

and

$$z_{21} = \frac{E_2}{I_1}\bigg|_{I_2 = 0} = \frac{I_1 Z_3}{I_1} \boxed{= Z_3} \qquad (25.8)$$

FIG. 25.12

For z_{22}, the determining configuration is shown in Fig. 25.13:

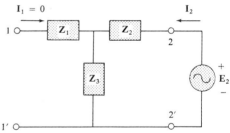

$$I_2 = \frac{E_2}{Z_2 + Z_3}$$

and

$$z_{22} = \frac{E_2}{I_2}\bigg|_{I_1 = 0} = \frac{I_2(Z_2 + Z_3)}{I_2} \boxed{= Z_2 + Z_3} \quad (25.9)$$

FIG. 25.13

Note that for the T configuration, $z_{12} = z_{21}$. For $Z_1 = 3 \angle 0°$, $Z_2 = 5 \angle 90°$, and $Z_3 = 4 \angle -90°$, we have

$z_{11} = Z_1 + Z_3 = 3 - j4$

$z_{12} = z_{21} = Z_3 = 4 \angle -90° = -j4$

$z_{22} = Z_2 + Z_3 = 5 \angle 90° + 4 \angle -90° = 1 \angle 90° = j1$

For a set of impedance parameters, the terminal (external) behavior of the device or network within the black box of Fig. 25.1 is determined. An *equivalent circuit* for the black box can be developed using the impedance parameters and Eqs. (25.1a) and (25.1b). Two possibilities for the impedance parameters appear in Fig. 25.14.

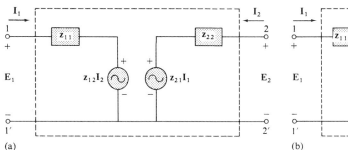

(a)

(b)

FIG. 25.14 *Two-port, **z**-parameter equivalent networks.*

Applying Kirchhoff's voltage law to the input and output loops of the network of Fig. 25.14(a) results in

$$\mathbf{E}_1 - \mathbf{z}_{11}\mathbf{I}_1 - \mathbf{z}_{12}\mathbf{I}_2 = 0$$

and

$$\mathbf{E}_2 - \mathbf{z}_{22}\mathbf{I}_2 - \mathbf{z}_{21}\mathbf{I}_1 = 0$$

which, when rearranged, become

$$\mathbf{E}_1 = \mathbf{z}_{11}\mathbf{I}_1 + \mathbf{z}_{12}\mathbf{I}_2$$
$$\mathbf{E}_2 = \mathbf{z}_{21}\mathbf{I}_1 + \mathbf{z}_{22}\mathbf{I}_2$$

corresponding exactly with Eqs. (25.1a) and (25.1b).

For the network of Fig. 25.14(b),

$$\mathbf{E}_1 - \mathbf{I}_1(\mathbf{z}_{11} - \mathbf{z}_{12}) - \mathbf{z}_{12}(\mathbf{I}_1 + \mathbf{I}_2) = 0$$
$$\mathbf{E}_2 - \mathbf{I}_1(\mathbf{z}_{21} - \mathbf{z}_{12}) - \mathbf{I}_2(\mathbf{z}_{22} - \mathbf{z}_{12}) - \mathbf{z}_{12}(\mathbf{I}_1 + \mathbf{I}_2) = 0$$

which, when rearranged, are

$$\mathbf{E}_1 = \mathbf{I}_1(\mathbf{z}_{11} - \mathbf{z}_{12} + \mathbf{z}_{12}) + \mathbf{I}_2\mathbf{z}_{12}$$
$$\mathbf{E}_2 = \mathbf{I}_1(\mathbf{z}_{21} - \mathbf{z}_{12} + \mathbf{z}_{12}) + \mathbf{I}_2(\mathbf{z}_{22} - \mathbf{z}_{12} + \mathbf{z}_{12})$$

and

$$\mathbf{E}_1 = \mathbf{z}_{11}\mathbf{I}_1 + \mathbf{z}_{12}\mathbf{I}_2$$
$$\mathbf{E}_2 = \mathbf{z}_{21}\mathbf{I}_1 + \mathbf{z}_{22}\mathbf{I}_2$$

Note in each network the necessity for a current-controlled voltage source, that is, a voltage source the magnitude of which is determined by a particular current of the network.

The usefulness of the impedance parameters and the resulting equivalent networks can best be described by considering the system of Fig. 25.15(a), which contains a device (or system) for which the impedance parameters have been determined. As shown in Fig. 25.15(b), the equivalent network for the device (or system) can then be substituted, and methods such as mesh analysis,

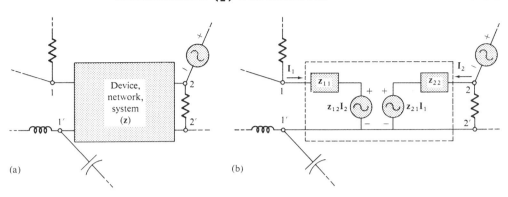

FIG. 25.15

nodal analysis, and so on, can be employed to determine required unknown quantities. The device itself can then be replaced by an equivalent circuit and the desired solutions obtained more directly and with less effort than is required using only the characteristics of the device.

EXAMPLE 25.2. Draw the equivalent circuit in the form shown in Fig. 25.14(a) using the impedance parameters determined in Example 25.1.

Solution: The circuit appears in Fig. 25.16.

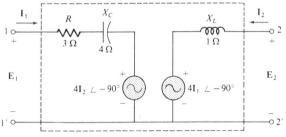

FIG. 25.16

25.3 ADMITTANCE (y) PARAMETERS

The equations relating the four terminal variables of Fig. 25.1 can also be written in the following form:

$$\boxed{I_1 = y_{11}E_1 + y_{12}E_2} \qquad \text{(25.10a)}$$

$$\boxed{I_2 = y_{21}E_1 + y_{22}E_2} \qquad \text{(25.10b)}$$

Note that in this case each term of each equation has the units of current as compared with voltage for each term of Eqs. (25.1a) and (25.1b). In addition, the unit of each coefficient is siemens, compared with the ohm for the impedance parameters.

The impedance parameters were determined by setting a particular current to zero through an open-circuit condition. For the admittance parameters of Eqs. (25.10a) and (25.10b), a voltage is set to zero through a short-circuit condition.

The terminology applied to each of the admittance parameters follows directly from the descriptive terms applied to each of the impedance parameters. The equations for each are determined directly from Eqs. (25.10a) and (25.10b) by setting a particular voltage to zero.

Y₁₁

$$\boxed{y_{11} = \frac{I_1}{E_1}}\Bigg|_{E_2 = 0} \qquad \text{(siemens, S)} \qquad \textbf{(25.11)}$$

$y_{11} =$ *short-circuit, input-admittance parameter*

The determining network appears in Fig. 25.17.

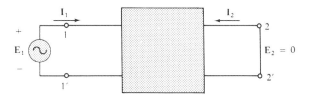

FIG. 25.17 y_{11} *determination.*   **Y₁₂**

$$\boxed{y_{12} = \frac{I_1}{E_2}}\Bigg|_{E_1 = 0} \qquad \text{(S)} \qquad \textbf{(25.12)}$$

$y_{12} =$ *short-circuit, reverse-transfer admittance parameter*

The network for determining y_{12} appears in Fig. 25.18.

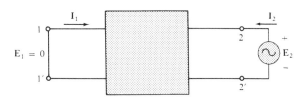

FIG. 25.18 y_{12} *determination.*   **Y₂₁**

$$\boxed{y_{21} = \frac{I_2}{E_1}}\Bigg|_{E_2 = 0} \qquad \text{(S)} \qquad \textbf{(25.13)}$$

$y_{21} =$ *short-circuit, forward-transfer admittance parameter*

The network for determining y_{21} appears in Fig. 25.19.

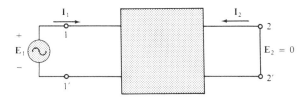

FIG. 25.19 y_{21} *determination.*

$\mathbf{Y}_{22}$

$$\boxed{y_{22} = \frac{I_2}{E_2}}\bigg|_{E_1 = 0} \quad (S) \qquad (25.14)$$

y_{22} = *short-circuit, output-admittance parameter*

The required network appears in Fig. 25.20.

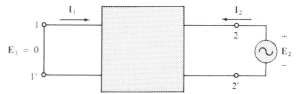

EXAMPLE 25.3. Determine the admittance parameters for the π network of Fig. 25.21.

FIG. 25.20 y_{22} *determination.*

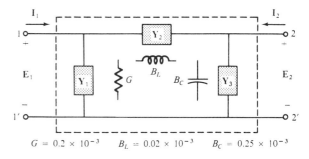

FIG. 25.21 π *network.*

Solution: The network for y_{11} will appear as shown in Fig. 25.22:

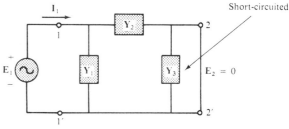

$$I_1 = E_1 Y_T = E_1(Y_1 + Y_2)$$

and

$$y_{11} = \frac{I_1}{E_1}\bigg|_{E_2 = 0} \quad \boxed{= Y_1 + Y_2}. \qquad (25.15)$$

FIG. 25.22

The determining network for y_{12} appears in Fig. 25.23. Y_1 is short-circuited; so $I_{Y_2} = I_1$, and

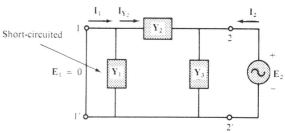

FIG. 25.23

$$I_{Y_2} = I_1 = -E_2Y_2$$

The minus sign results because the defined direction of I_1 in Fig. 25.23 is opposite to the actual flow direction due to the applied source E_2; and

$$y_{12} = \left.\frac{I_1}{E_2}\right|_{E_1 = 0} \qquad \boxed{= -Y_2} \qquad (25.16)$$

The network employed for y_{21} appears in Fig. 25.24. In this case, y_3 is short-circuited, resulting in

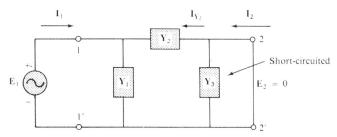

FIG. 25.24

$$I_{Y_2} = I_2$$

and

$$I_{Y_2} = I_2 = -E_1Y_2$$

Thus,

$$y_{21} = \left.\frac{I_2}{E_1}\right|_{E_2 = 0} \qquad \boxed{= -Y_2} \qquad (25.17)$$

Note that for the π configuration, $y_{12} = y_{21}$, which was expected, since the impedance parameters for the T network were such that $z_{12} = z_{21}$. A T network can be converted directly to a π network using the Y-Δ transformation.

The determining network for y_{22} appears in Fig. 25.25:

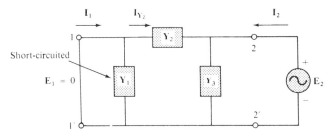

FIG. 25.25

$$Y_T = Y_2 + Y_3 \quad \text{and} \quad I_2 = E_2(Y_2 + Y_3)$$

Thus,

$$y_{22} = \left.\frac{I_2}{E_2}\right|_{E_1 = 0} \qquad \boxed{= Y_2 + Y_3} \qquad (25.18)$$

Substituting values, we have

$$\mathbf{Y}_1 = 0.2 \times 10^{-3} \angle 0°$$
$$\mathbf{Y}_2 = 0.02 \times 10^{-3} \angle -90°$$
$$\mathbf{Y}_3 = 0.25 \times 10^{-3} \angle 90°$$

$$\mathbf{y}_{11} = \mathbf{Y}_1 + \mathbf{Y}_2$$
$$= \mathbf{0.2 \times 10^{-3} - j0.02 \times 10^{-3} \text{ (inductive)}}$$

$$\mathbf{y}_{12} = \mathbf{y}_{21} = -\mathbf{Y}_2 = -(-j0.02 \times 10^{-3})$$
$$= \mathbf{j0.02 \times 10^{-3} \text{ (capacitive)}}$$

$$\mathbf{y}_{22} = \mathbf{Y}_2 + \mathbf{Y}_3 = -j0.02 \times 10^{-3} + j0.25 \times 10^{-3}$$
$$= \mathbf{j0.23 \times 10^{-3} \text{ (capacitive)}}$$

Note the similarities between the results for $\mathbf{y}_{11}$ and $\mathbf{y}_{22}$ for the π network compared with $\mathbf{z}_{11}$ and $\mathbf{z}_{22}$ for the T network.

Two networks satisfying the terminal relationships of Eqs. (25.10a) and (25.10b) are shown in Fig. 25.26. Note the use of parallel branches, since each term of Eqs. (25.10a) and (25.10b) has the units of current, and the most direct route to the equivalent circuit is an application of Kirchhoff's current law in reverse. That is, find the network that satisfies Kirchhoff's current law relationship. For the impedance parameters, each term had the units of volts, so Kirchhoff's voltage law was applied in reverse to determine the series combination of elements in the equivalent circuit of Fig. 25.14(a).

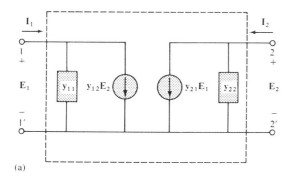

(a)

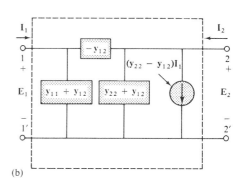

(b)

FIG. 25.26 *Two-port, y-parameter equivalent networks.*

Applying Kirchhoff's current law to the network of Fig. 25.26(b), we have

$$\text{Node } a: \quad \mathbf{I}_1 \overset{\text{entering}}{=} \overset{\text{leaving}}{\mathbf{y}_{11}E_1 - \mathbf{y}_{12}E_2}$$
$$\text{Node } b: \quad \mathbf{I}_2 = \mathbf{y}_{22}E_2 - \mathbf{y}_{21}E_1$$

which, when rearranged, are Eqs. (25.10a) and (25.10b).

25.4 HYBRID (h) PARAMETERS

The hybrid (**h**) parameters are employed extensively in the analysis of transistor networks. The term *hybrid* is derived from the fact that the parameters have a mixture of units (a hybrid set) rather than a single unit of measurement such as ohms or siemens for the **z** and **y** parameters, respectively. The defining hybrid equations have a mixture of current *and* voltage variables on one side, as follows:

$$\boxed{E_1 = h_{11}I_1 + h_{12}E_2} \qquad \textbf{(25.19a)}$$

$$\boxed{I_2 = h_{21}I_1 + h_{22}E_2} \qquad \textbf{(25.19b)}$$

To determine the hybrid parameters, it will be necessary to establish both the short-circuit and the open-circuit condition, depending on the parameter desired.

h_{11}

$$\boxed{h_{11} = \frac{E_1}{I_1}}\Bigg|_{E_2 = 0} \qquad \text{(ohms, } \Omega) \qquad \textbf{(25.20)}$$

h_{11} = *short-circuit, **i**nput-impedance parameter*

The determining network is shown in Fig. 25.27.

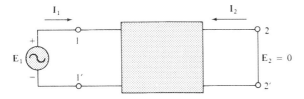

FIG. 25.27 h_{11} *determination.*

h_{12}

$$\boxed{h_{12} = \frac{E_1}{E_2}}\Bigg|_{I_1 = 0} \qquad \text{(dimensionless)} \qquad \textbf{(25.21)}$$

h_{12} = *open-circuit, reverse-transfer voltage ratio parameter*

The network employed in determining h_{12} is shown in Fig. 25.28.

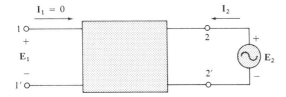

FIG. 25.28 h_{12} *determination.*

h$_{21}$

$$h_{21} = \frac{I_2}{I_1}\Bigg|_{E_2 = 0} \quad \text{(dimensionless)} \quad \textbf{(25.22)}$$

h_{21} = *short-circuit, forward-transfer current ratio*

The determining network appears in Fig. 25.29.

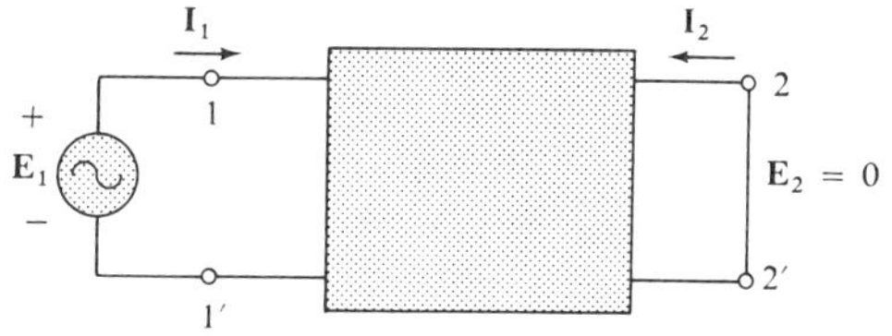

h$_{22}$

$$h_{22} = \frac{I_2}{E_2}\Bigg|_{I_1 = 0} \quad \text{(siemens, S)} \quad \textbf{(25.23)}$$

h_{22} = *open-circuit, output admittance parameter*

The network employed to determine h_{22} is shown in Fig. 25.30.

The subscript notation for the hybrid parameters is reduced to the following for most applications. The letter chosen is that letter appearing in boldface in the preceding description of each parameter.

$$\mathbf{h}_{11} = \mathbf{h}_i \qquad \mathbf{h}_{12} = \mathbf{h}_r \qquad \mathbf{h}_{21} = \mathbf{h}_f \qquad \mathbf{h}_{22} = \mathbf{h}_o$$

The hybrid equivalent circuit appears in Fig. 25.31. Since the unit of measurement for each term of Eq. (25.19a) has the units of volts, Kirchhoff's voltage law was applied in reverse to obtain the series input circuit indicated. The unit of measurement of each turn of Eq. (25.19b) has the units of current, resulting in the parallel elements of the output circuit as obtained by applying Kirchhoff's current law in reverse.

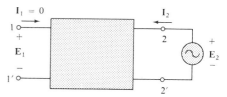

FIG. 25.29 h_{21} *determination.*

FIG. 25.30 h_{22} *determination.*

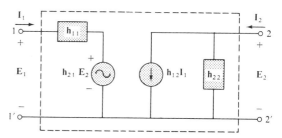

FIG. 25.31 *Two-port, hybrid-parameter equivalent network.*

Note that the input circuit has a voltage-controlled voltage source whose controlling voltage is the output terminal voltage, while the output circuit has a current-controlled current source whose controlling current is the current of the input circuit.

EXAMPLE 25.4. For the hybrid equivalent circuit of Fig. 25.32:
a. Determine the current ratio (gain) I_2/I_1.
b. Determine the voltage ratio (gain) E_2/E_1.

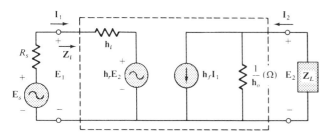

FIG. 25.32

Solutions:

a. Using the current divider rule, we have

$$I_2 = \frac{(1/h_o)h_f I_1}{(1/h_o) + Z_L} = \frac{h_f I_1}{1 + h_o Z_L}$$

and

$$\boxed{A_i = \frac{I_2}{I_1} = \frac{h_f}{1 + h_o Z_L}} \qquad \text{(25.24)}$$

b. Applying Kirchhoff's voltage law to the input circuit gives us

$$E_1 - h_i I_1 - h_r E_2 = 0$$

and

$$I_1 = \frac{E_1 - h_r E_2}{h_i}$$

Apply Kirchhoff's current law to the output circuit:

$$I_2 = h_f I_1 + h_o E_2$$

However,

$$I_2 = -\frac{E_2}{Z_L}$$

so

$$-\frac{E_2}{Z_L} = h_f I_1 + h_o E_2$$

Substituting for I_1 gives us

$$-\frac{E_2}{Z_L} = h_f\left(\frac{E_1 - h_r E_2}{h_i}\right) + h_o E_2$$

or

$$h_i E_2 = -h_f Z_L E_1 + h_r h_f Z_L E_2 - h_i h_o Z_L E_2$$

and

$$E_2(h_i - h_r h_f Z_L + h_i h_o Z_L) = -h_f Z_L E_1$$

with the result that

$$A_v = \frac{E_2}{E_1} = \frac{-h_f Z_L}{h_i(1 + h_o Z_L) - h_r h_f Z_L} \qquad (25.25)$$

EXAMPLE 25.5. For a particular transistor, $h_i = 1$ kΩ, $h_r = 4 \times 10^{-4}$, $h_f = 50$, and $h_o = 25 \times 10^{-6}$ S. Determine the current and the voltage gain if Z_L is a 2-kΩ resistive load.

Solutions:

a. $A_i = \dfrac{h_f}{1 + h_o Z_L} = \dfrac{50}{1 + (25 \times 10^{-6})(2 \times 10^3)}$

$A_i = \dfrac{50}{1 + 50 \times 10^{-3}} = \dfrac{50}{1.050} = \mathbf{47.62}$

b. $A_v = \dfrac{-h_f Z_L}{h_i(1 + h_o Z_L) - h_r h_f Z_L}$

$\quad = \dfrac{-(50)(2 \times 10^3)}{(1 \times 10^3)(1.050) - (4 \times 10^{-4})(50)(2 \times 10^3)}$

$A_v = \dfrac{-100 \times 10^3}{1.050 \times 10^3 - 0.04 \times 10^3} = -\dfrac{100}{1.01} = \mathbf{-99}$

The minus sign simply indicates a phase shift of 180° between E_2 and E_1 for the defined polarities in Fig. 25.32.

25.5 INPUT AND OUTPUT IMPEDANCES

The input and output impedances will now be determined for the hybrid equivalent circuit and a **z** parameter equivalent circuit. The input impedance can always be determined by the ratio of the input voltage to the input current with or without a load applied. The output impedance is always determined with the source voltage or current set to zero. We found in the previous section that for the hybrid equivalent circuit of Fig. 25.32,

$$E_1 = h_i I_1 + h_r E_2$$
$$E_2 = -I_2 Z_L$$

and

$$\frac{I_2}{I_1} = \frac{h_f}{1 + h_o Z_L}$$

By substituting for I_2 in the second equation (using the relationship of the last equation), we have

$$E_2 = -\left(\frac{h_f I_1}{1 + h_o Z_L}\right) Z_L$$

so the first equation becomes

$$E_1 = h_i I_1 + h_r \left(- \frac{h_f I_1 Z_L}{1 + h_o Z_L}\right)$$

and

$$E_1 = I_1 \left(h_i - \frac{h_r h_f Z_L}{1 + h_o Z_L}\right)$$

Thus,

$$Z_i = \frac{E_1}{I_1} = h_i - \frac{h_r h_f Z_L}{1 + h_o Z_L} \qquad (25.26)$$

For the output impedance, we will set the source voltage to zero but preserve its internal resistance R_s as shown in Fig. 25.33.

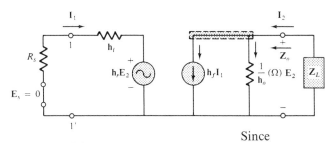

FIG. 25.33

Since

$$E_s = 0$$

then

$$I_1 = -\frac{h_r E_2}{h_i + R_s}$$

From the output circuit,

$$I_2 = h_f I_1 + h_o E_2$$

or

$$I_2 = h_f \left(-\frac{h_r E_2}{h_i + R_s}\right) + h_o E_2$$

and

$$I_2 = \left(-\frac{h_r h_f}{h_i + R_s} + h_o\right) E_2$$

Thus,

$$Z_o = \frac{E_2}{I_2} = \frac{1}{h_o - \dfrac{h_r h_f}{h_i + R_s}} \qquad (25.27)$$

EXAMPLE 25.6. Determine $\mathbf{Z}_i$ and $\mathbf{Z}_o$ for the transistor having the parameters of Example 25.5 if $R_s = 1$ kΩ.

Solution:

$$\mathbf{Z}_i = \mathbf{h}_i - \frac{\mathbf{h}_r \mathbf{h}_f \mathbf{Z}_L}{1 + \mathbf{h}_o \mathbf{Z}_L} = 1 \times 10^3 - \frac{0.04 \times 10^3}{1.050}$$

$$= 1 \times 10^3 - 0.0381 \times 10^3 = \mathbf{961.9 \ \Omega}$$

$$\mathbf{Z}_o = \frac{1}{\mathbf{h}_o - \dfrac{\mathbf{h}_r \mathbf{h}_f}{\mathbf{h}_i + R_s}} = \frac{1}{25 \times 10^{-6} - \dfrac{(4 \times 10^{-4})(50)}{1 \text{ k}\Omega + 1 \text{ k}\Omega}}$$

$$= \frac{1}{25 \times 10^{-6} - \dfrac{200 \times 10^{-4}}{2 \times 10^3}}$$

$$= \frac{1}{25 \times 10^{-6} - 10 \times 10^{-6}}$$

$$\mathbf{Z}_o = \frac{1}{15 \times 10^{-6}} = \mathbf{66.67 \ k\Omega}$$

For the **z** parameter equivalent circuit of Fig. 25.34,

$$\mathbf{I}_2 = \frac{-\mathbf{z}_{21}\mathbf{I}_1}{\mathbf{z}_{22} + \mathbf{Z}_L}$$

and

$$\mathbf{I}_1 = \frac{\mathbf{E}_1 - \mathbf{z}_{12}\mathbf{I}_2}{\mathbf{z}_{11}}$$

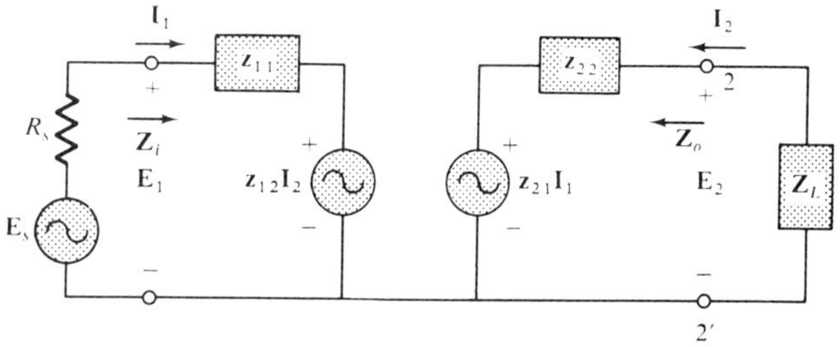

or

$$\mathbf{E}_1 = \mathbf{z}_{11}\mathbf{I}_1 + \mathbf{z}_{12}\mathbf{I}_2 = \mathbf{z}_{11}\mathbf{I}_1 + \mathbf{z}_{12}\left(\frac{-\mathbf{z}_{21}\mathbf{I}_1}{\mathbf{z}_{22} + \mathbf{Z}_L}\right)$$

and

$$\boxed{\mathbf{Z}_i = \frac{\mathbf{E}_1}{\mathbf{I}_1} = \mathbf{z}_{11} - \frac{\mathbf{z}_{12}\mathbf{z}_{21}}{\mathbf{z}_{22} + \mathbf{Z}_L}} \qquad (25.28)$$

For the output impedance, $\mathbf{E}_s = 0$, and

$$\mathbf{I}_1 = -\frac{\mathbf{z}_{12}\mathbf{I}_2}{R_s + \mathbf{z}_{11}} \quad \text{and} \quad \mathbf{I}_2 = \frac{\mathbf{E}_2 - \mathbf{z}_{21}\mathbf{I}_1}{\mathbf{z}_{22}}$$

or

$$\mathbf{E}_2 = \mathbf{z}_{22}\mathbf{I}_2 + \mathbf{z}_{21}\mathbf{I}_1 = \mathbf{z}_{22}\mathbf{I}_2 + \mathbf{z}_{21}\left(-\frac{\mathbf{z}_{12}\mathbf{I}_2}{R_s + \mathbf{z}_{11}}\right)$$

and

$$\mathbf{E}_2 = \mathbf{z}_{22}\mathbf{I}_2 - \frac{\mathbf{z}_{12}\mathbf{z}_{21}\mathbf{I}_2}{R_s + \mathbf{z}_{11}}$$

Thus,

FIG. 25.34

$$Z_o = \frac{E_2}{I_2} = z_{22} - \frac{z_{12}z_{21}}{R_s + z_{11}} \qquad (25.29)$$

25.6 CONVERSION BETWEEN PARAMETERS

The equations relating the **z** and **y** parameters can be determined directly from Eqs. (25.1) and (25.10). For Eqs. (25.10a) and (25.10b),

$$I_1 = y_{11}E_1 + y_{12}E_2$$
$$I_2 = y_{21}E_1 + y_{22}E_2$$

The use of determinants will result in

$$E_1 = \frac{\begin{vmatrix} I_1 & y_{12} \\ I_2 & y_{22} \end{vmatrix}}{\begin{vmatrix} y_{11} & y_{12} \\ y_{21} & y_{22} \end{vmatrix}} = \frac{y_{22}I_1 - y_{12}I_2}{y_{11}y_{22} - y_{12}y_{21}}$$

Substituting the notation

$$\Delta_y = y_{11}y_{22} - y_{12}y_{21}$$

we have

$$E_1 = \frac{y_{22}}{\Delta_y}I_1 - \frac{y_{12}}{\Delta_y}I_2$$

which, when related to Eq. (25.1a),

$$E_1 = z_{11}I_1 + z_{12}I_2$$

indicates that

$$z_{11} = \frac{y_{22}}{\Delta_y} \quad \text{and} \quad z_{12} = \frac{y_{12}}{\Delta_y}$$

and, similarly,

$$z_{21} = \frac{y_{21}}{\Delta_y} \quad \text{and} \quad z_{22} = \frac{y_{11}}{\Delta_y}$$

For the conversion of **z** parameters to the admittance domain, determinants are applied to Eqs. (25.1a) and (25.1b). The impedance parameters can be found in terms of the hybrid parameters by first forming the determinant for I_1 from the hybrid equations:

$$E_1 = h_{11}I_1 + h_{12}E_2$$
$$I_2 = h_{21}I_1 + h_{22}E_2$$

and

$$I_1 = \frac{\begin{vmatrix} E_1 & h_{12} \\ I_2 & h_{22} \end{vmatrix}}{\begin{vmatrix} h_{11} & h_{12} \\ h_{21} & h_{22} \end{vmatrix}} = \frac{h_{22}}{\Delta_h} E_1 - \frac{h_{12}}{\Delta_h} I_2$$

and

$$\frac{h_{22}}{\Delta_h} E_1 = I_1 - \frac{h_{12}}{\Delta_h} I_2$$

or

$$E_1 = \frac{\Delta_h I_1}{h_{22}} - \frac{h_{12}}{h_{22}} I_2$$

which, when related to the impedance parameter equation,

$$E_1 = z_{11} I_1 + z_{12} I_2$$

indicates that

$$z_{11} = \frac{\Delta_h}{h_{22}} \quad \text{and} \quad z_{12} = -\frac{h_{12}}{h_{22}}$$

The remaining conversions are left as an exercise. A complete table of conversions appears in Table 25.1.

TABLE 25.1 *Conversions between* **z**, **y**, *and* **h** *parameters.*

FROM → [z] TO ↓	[z]		[y]		[h]	
[z]	z_{11}	z_{12}	$\dfrac{y_{22}}{\Delta_y}$	$\dfrac{-y_{12}}{\Delta_y}$	$\dfrac{\Delta_h}{h_{22}}$	$\dfrac{h_{12}}{h_{22}}$
	z_{21}	z_{22}	$\dfrac{-y_{21}}{\Delta_y}$	$\dfrac{y_{11}}{\Delta_y}$	$\dfrac{-h_{21}}{h_{22}}$	$\dfrac{1}{h_{22}}$
[y]	$\dfrac{z_{22}}{\Delta_z}$	$\dfrac{-z_{12}}{\Delta_z}$	y_{11}	y_{12}	$\dfrac{1}{h_{11}}$	$\dfrac{-h_{12}}{h_{11}}$
	$\dfrac{-z_{21}}{\Delta_z}$	$\dfrac{z_{11}}{\Delta_z}$	y_{21}	y_{22}	$\dfrac{h_{21}}{h_{11}}$	$\dfrac{\Delta_h}{h_{11}}$
[h]	$\dfrac{\Delta_z}{z_{22}}$	$\dfrac{z_{12}}{z_{22}}$	$\dfrac{1}{y_{11}}$	$\dfrac{-y_{12}}{y_{11}}$	h_{11}	h_{12}
	$\dfrac{-z_{21}}{z_{22}}$	$\dfrac{1}{z_{22}}$	$\dfrac{y_{21}}{y_{11}}$	$\dfrac{\Delta_y}{y_{11}}$	h_{21}	h_{22}

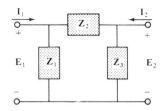

FIG. 25.35

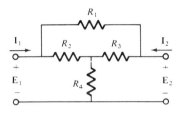

FIG. 25.36

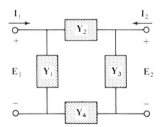

FIG. 25.37

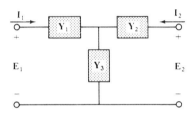

FIG. 25.38

PROBLEMS

Section 25.2

1. **a.** Determine the impedance (**z**) parameters for the π network of Fig. 25.35.
   b. Sketch the **z** parameter equivalent circuit (using either form of Fig. 25.14).

2. **a.** Determine the impedance (**z**) parameters for the network of Fig. 25.36.
   b. Sketch the **z** parameter equivalent circuit (using either form of Fig. 25.14).

Section 25.3

3. **a.** Determine the admittance (**y**) parameters for the T network of Fig. 25.37.
   b. Sketch the **y** parameter equivalent circuit (using either form of Fig. 25.26).

4. **a.** Determine the admittance (**y**) parameters for the network of Fig. 25.38.
   b. Sketch the **y** parameter equivalent circuit (using either form of Fig. 25.26).

Section 25.4

5. **a.** Determine the **h** parameters for the network of Fig. 25.35.
   b. Sketch the hybrid equivalent circuit.

6. **a.** Determine the **h** parameters for the network of Fig. 25.36.
   b. Sketch the hybrid equivalent circuit.

7. **a.** Determine the **h** parameters for the network of Fig. 25.37.
   b. Sketch the hybrid equivalent circuit.

8. **a.** Determine the **h** parameters for the network of Fig. 25.38.
   b. Sketch the hybrid equivalent circuit.

9. **a.** For the hybrid equivalent circuit of Fig. 25.39, determine
   a. the current gain $A_i = I_2/I_1$.
   b. the voltage gain $A_v = E_2/E_1$.

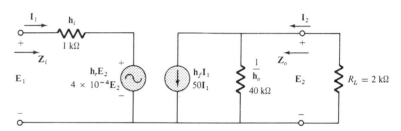

FIG. 25.39

Section 25.5

10. For the hybrid equivalent circuit of Fig. 25.39, determine
    a. the input impedance
    b. the output impedance

11. Determine the input and output impedances for the **z**-parameter equivalent circuit of Fig. 25.40.

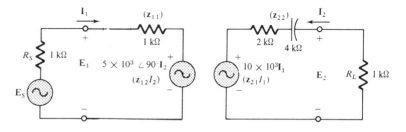

12. Determine the expression for the input and output impedance of the **y**-parameter equivalent circuit.

FIG. 25.40

Section 25.6

13. Determine the **h** parameters for the following **z** parameters:

$$z_{11} = 4 \text{ k}\Omega$$
$$z_{12} = 2 \text{ k}\Omega$$
$$z_{21} = 3 \text{ k}\Omega$$
$$z_{22} = 4 \text{ k}\Omega$$

14. a. Determine the **z** parameters for the following **h** parameters:

$$h_{11} = 1 \text{ k}\Omega$$
$$h_{12} = 2 \times 10^{-4}$$
$$h_{21} = 100$$
$$h_{22} = 20 \times 10^{-6} \text{ S}$$

b. Determine the **y** parameters for the hybrid parameters indicated in part(a).

GLOSSARY

Admittance (y) parameters A set of parameters, having the units of siemens, that can be used to establish two-port equivalent network for a system.

Hybrid (h) parameters A set of mixed parameters (ohms, siemens, some unitless) that can be used to establish a two-port equivalent network for a system.

Impedance (z) parameters A set of parameters, having the units of ohms, that can be used to establish a two-port equivalent network for a system.

Input impedance The impedance appearing at the input terminals of a system.

Output impedance The impedance appearing at the output terminals of a system with the energizing source set to zero.

Single-port network A network having a single set of access terminals.

Two-port network A network having two pairs of access terminals.

COMPUTER ANALYSIS

26

26.1 INTRODUCTION

The impact of computers on our everyday lives has become increasingly obvious in recent years. The concept of the *personal computer* has now been widely accepted, as evidenced by the increasing number of manufacturers offering units for use in the home, in small businesses, and so on. The belief that computers had to be huge, unwieldly, expensive units available only to large companies or utilities has given way to a realization that there are small, portable units that are of significantly less cost and that permit a wide range of applications in support of individual needs. Computer languages have also been developed that permit the use of a computer after a relatively few hours of training. Once the advertising media have removed the fear element associated with computers, it is reasonable to expect that the growth of the use of the computer in the next decade will match that of television during its introductory years.

The primary purpose of this chapter is to introduce the use of the computer in the analysis of some very simple electrical configurations. Although the unit appearing in this chapter may not be available at your location, the similarities between computer systems are so enormous that the presentation here can be easily translated to the computer facilities available. The language employed is the easiest to learn of those commonly available today and is used more than any other at educational institutions.

If you continue to apply the computer to its fullest advantage in the solution of engineering-related problems, there will come a time when you will have to turn to a more structured language to properly analyze or design the systems of interest. However, be assured that this chapter and the resulting use of computers using this language will result in a breadth of knowledge that will have a significant carry-over to the more advanced programming techniques.

26.2 TYPES OF COMPUTERS

In general, computers can be placed in one of three categories: *large-scale, mini,* and *micro.* Typically, a large-scale unit with its high-speed printers, input/output terminals, card sorters, and so on, will require the space of a single room of about 1000 square feet—a vast improvement over the first systems that employed vacuum tubes, needed in excess of 15,000 square feet, and weighed well over 20 tons.

The *large-scale* unit is normally designed to handle a wide variety of problems, and it has the ability to handle lengthy programs that deal with a significant amount of data. The *minicomputer* is usually oriented toward a particular area of interest but does have the potential for a significant level of expandability if interfacing with peripheral equipment is required. A number of such computers are built into a desk-type unit and, with the associated equipment, may occupy the corner of a room (usually less than 100 square feet). The *microcomputer* is designed to satisfy a particular need (or a broad range of problems of lesser difficulty) and has limited expandability. It is the type of unit that might appear in the home or in the office of a small business. Since the keyboard, the heart of the computer, is normally no larger than a typewriter, the microcomputer is quite portable. It normally has an associated video display, but this attachment is usually no larger than a small TV screen. The entire unit can easily fit on the top of an average desk and weighs no more than 20 to 30 pounds.

To some degree, the type of computer can also be defined by the available *memory.* The memory component stores the programs, data, and processing control of the computer. For large-scale units, the memory capability is typically several million *bytes* of information, a byte being a variable-length string of bits of information used to represent alphanumeric characters or control functions. Most computers today use either 8-bit or 16-bit bytes. The latter is applied more frequently to large-scale units, and the former to minicomputers and microcomputers. The minicomputer may have a capacity of a few million bytes, while the microcomputer typically has 32 to 64 kilobytes, with levels approaching one million for some newer units. Although there should naturally be some concern about the memory capability of the computer you choose, the nature of the problem will define the appropriate computer, and the manufacturer is usually astute enough to insure that the required memory is available. Of course, as your skills with the use of the computer increase, your concern about the available memory will probably also increase.

There is no question that the cost of a particular system will further define its family. Today, costs for most microcomputers range from $1000 to $10,000; for minicomputers, from $10,000 to $100,000; and for large-scale computers, from $100,000 to several million dollars. There are, of course, systems available today that do not fit into these defined limits, but the above can be used as an expected range of costs for such units.

The internal construction details of the wide variety of computers available today are very similar. They all employ LSI (*Large-Scale Integrated* circuits) or VLSI (*Very-Large-Scale Integrated* circuits) that result in nanosecond operating times for particular operations. The larger the computer, the larger the number of these units and the peripheral circuitry. The memory component of the computer employs RAMs (*Random-Access Memory*) and ROMs (*Read-Only Memory*). The latter has a set of factory-installed programs that define the sequential operating procedure of the computer. It cannot be accessed to be changed but can *only* be *read* by the computer to define procedures. The content of the RAM chip can be changed to store different programs and data. The RAMs and ROMs define the main internal memory of the computer. There are also external memory units, such as magnetic types and discs, that can be attached to store or provide data. The *microprocessor* chip (LSI or VLSI) introduced above is essentially the *brains* of the computer and defines how the program stored in memory will be implemented.

26.3 COMPUTER LANGUAGES

The popular languages of the day are the result of a need defined by a set of relatively similar problems. The FORTRAN and ALGOL-60 (the European equivalent of FORTRAN) languages are applied most effectively to problems of a technical or mathematical nature. COBOL (*Common Business-Oriented Language*) is typically applied to business problems. BASIC (*Beginning All-Purpose Symbolic Instruction Code*) is a general-purpose language that uses abbreviated statements of the English language to make it the easiest of the popular languages to learn. It can typically be applied after a few hours of instruction, while FORTRAN normally requires some formal course work. Languages such as PASCAL and PL/1 (a melting pot of ALGOL, FORTRAN, and COBOL) are more structured than the other languages introduced and require serious study or course work for proficiency of application. Keep in mind, however, that there are a multitude of similarities between languages and their application. Proficiency in any one language will always assist in development of a working knowledge of another. Always be aware that the FORTRAN language (or any other language) specified for one machine may not be exactly the same as the FORTRAN language specified for another. There may be a number of modifications in the language necessary to insure peak operating efficiency as the computer performs a sequence of commands directly related to a particular type of problem. However, be assured that normally the variations are not major and will not impede to any great measure your ability to use that same language in a different computer model.

The frequent use of the word *similar* in the past few paragraphs and sections is the primary reason a chapter devoted to computers appears in this text. If the languages and computer structure were totally different from one unit to another, the content of this chapter would be of little value unless you had the same computer described in this text. Fortunately, the similarities are very strong, and, in addition, the content of this chapter is directed more toward demonstrating how a computer can be applied at this early stage in your development rather than toward examining the details of computer programming.

26.4 PERSONAL COMPUTERS: THE HP-85A

Access to the first computers was very limited. It was not until computer design was improved by the introduction of semiconductor components that the number of users of a large-scale computer could be significantly increased. Time sharing, the sharing of a computer through the use of a number of terminals on or off location, had a significant impact on availability. In recent years with the introduction of minicomputers and microcomputers, individuals are beginning to purchase their own computers (commonly referred to as *personal computers*) to meet their computational needs. A number of very nicely designed units are now available that have a very reasonable price tag, sufficient memory to attack ·a variety of problems, excellent portability, and relative ease of operation. A few commercially available microcomputers include the Apple Computer Company's Apple II, the Radio Shack TRS-80, and the Heath Kit HII, all of which employ the BASIC language. The HP-85A microcomputer employed in this text (Fig. 26.1) was chosen because of its industrial and scientific orientation and the *software support,* which will appear in this chapter. *Software* refers to the written program, and *software support* refers to manufacturer-supplied prerecorded programs of particular tasks. In contrast, *hardware* refers to the physical components that constitute the working elements of the computer and to the design that insures that the operation of the computer is the most efficient possible.

Courtesy of Hewlett Packard Co.

The software support has become an increasingly important part of the sales campaign for computers. In fact, recent indications are that the computer itself may drop in cost while the software increases in cost.

Note the relatively small size of the HP-85A and the fact that it has a video screen and a print-out capability. Considering that a large keyboard (Fig. 26.2) is often desirable,

FIG. 26.1 *HP-85A personal computer.*

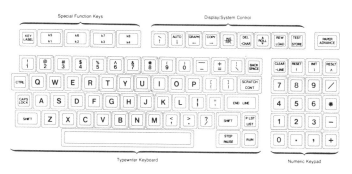

Courtesy of Hewlett Packard Co.

FIG. 26.2 *HP-85 keyboard.*

Courtesy of Hewlett Packard Co.

FIG. 26.3 *HP-85 magnetic tape unit.*

one must wonder what improvements could be made to a system in the future. As indicated on the front right of the machine in Fig. 26.1, there is a slot in which magnetic tapes, such as that shown in Fig. 26.3, can be inserted to store data and programs or to introduce software packages provided by the manufacturer. One can expand the memory of this machine from 16,000 bytes to 32,000 bytes by connecting a 16,000-byte memory module to the back panel. As indicated in Fig. 26.4, there are a wide variety of peripherals that can be attached for increased versatility. The unit weighs only 21 pounds (completely portable) and uses a language referred to as *Extended BASIC*. The term *Extended* indicates simply that the unit has features beyond those of the standard BASIC language.

Courtesy of Hewlett Packard Co.

FIG. 26.4 *HP-85 peripheral equipment selection.*

Input is provided to the HP-85A computer through the keyboard or magnetic tape. Other methods employed throughout the range of computers include magnetic discs, punch cards, and paper tape. Output for the HP-85A is obtained from the display, printer, or magnetic tape. Some computers employ magnetic discs, punch cards, or paper tape.

It would be impossible to cover all the details of computer operation in a single chapter and still introduce the programs that relate to the content of this text. However, sufficient introduction has now been provided so that now we can examine a few single programs. In addition, you will develop a sense for how a computer can be helpful in the practice of engineering and how it can become a "friendly" associate if properly approached and appreciated. As you progress through this course, spend a few hours examining some relatively simple programs, and you will develop a firm foundation for the more advanced computer work to follow. When discussing your efforts with individuals well-versed in the use of the computer, you will find that there is a pattern of logical progression that will develop through the writing and testing of a multitude and variety of problems.

26.5 INTRODUCTORY PROGRAMS

The programs to be introduced in this chapter appear in an order similar to the sequence in which the various equations appear in the text. The content assumes no prior knowledge of computer programming, although a large percentage of the readers probably have had some earlier exposure. It is strongly suggested that, if time permits, you attempt the same operations and programs on the computer facilities available at your location. The programs are introductory in level, and adaptation should require a minimum level of effort. The fact that computers such as the HP-85A can be put to actual use after a very brief period of study is a major selling feature of such systems.

For most *students* there is an initial period of real concern (fear?) that develops when they are placed in front of this modern-day miracle machine, the computer. Initially, therefore, since most students are quite familiar with present day calculators, the first application of the computer will be in the calculator mode. The fact that the results are relatively easy to obtain and are easy to check will, it is hoped, develop an initial confidence level and communication link with the system.

For each calculation of Fig. 26.5, refer to the keyboard of Fig. 26.2 to determine which keys are utilized. Be aware that when the shift key is pressed, the symbol appearing at

```
5+8-3
10
```
$5 + 8 - 3 = 10$

```
34.5*17.9
617.55
```
$34.5 \times 17.9 = 617.55$

```
80*4/(8-1.5)
49.2307692308
```
$80 \times 4/(8 - 1.5) \cong 49.231$

```
3^4-5.6/(1.05*200)
80.9733333333
```
$3^4 - 5.6/(1.05 \times 200) \cong 80.973$

```
SQR(45.67)
6.75795827155
```
$\sqrt{45.67} \cong 6.758$

the top of a key results. Also, unlike that on typical typewriters, the computer shift key results in lower-case letters rather than upper-case letters. Program statements are typically in upper-case letters. In Fig. 26.5, the operations performed appear to the right of the printout. Note the strong similarity between the standard mathematical notation and that used by the computer. The extra set of parentheses, as in the division process, simply specifies that the operation

FIG. 26.5 *HP-85 computer in the calculator mode.*

within the parentheses is performed before the other operations on the same line. After a few minutes of practice, you will find yourself as comfortable using the computer (with its video-checking capability) as you are using the calculator.

Let us now consider the relatively simple equation for the total resistance of two parallel resistors. The program in BASIC appears in Fig. 26.6. First note that each step is

```
AUTO
10 DISP"TWO PARALLEL RESISTORS"⎫    Display: TWO PARALLEL RESISTORS
20 DISP"RESISTOR VALUES"       ⎬              RESISTOR VALUES?
30 INPUT R1,R2                 ⎭    Provide values of R₁ and R₂ in ohms.
40 T=R1*R2/(R1+R2)                  Rᴛ = R₁ × R₂/(R₁ + R₂)
50 DISP"TOTAL RESISTANCE=";T        Display: TOTAL RESISTANCE = Rᴛ
60 END                              End of program


TWO PARALLEL RESISTORS
RESISTOR VALUES
?
3,4
TOTAL RESISTANCE= 1.71429571429     Answer in ohms
```

Line 10: `DISP"TWO PARALLEL RESISTORS"` → Display: TWO PARALLEL RESISTORS

$R_T = R_1 \times R_2/(R_1 + R_2)$

Display: TOTAL RESISTANCE $= R_T$

FIG. 26.6 *Program to determine the total resistance of two parallel resistors.*

numbered and the steps appear in a numerical sequence that identifies the order in which the program will be executed. The program also permits access to any line if changes are to be made and permits adding steps between those that appear, such as 31, 32, and so on. The AUTO appearing at the top of the page simply specifies that the numbering will be done AUTOmatically as the program is written. The *command* DISP comes from the word DISPlay and tells the computer to display the indicated phrase on the screen when the program is run. The equation for the total resistance is exactly as it would appear in the written form. The last DISP statement displays the result, and the END statement tells the computer that the program has been completed.

When the RUN button on the computer is pressed, the first two lines and question mark will appear as shown below the program. The values of R_1 and R_2 are inserted in the order defined in the program. The total resistance then appears in the following line at virtually the same instant the input data are entered into the computer. You can run the entire program again for different values of R_1 and R_2 by simply pressing the RUN key.

Questions regarding the above program may exist due to the brevity of the description. Unfortunately, such will be the case throughout this chapter due to the limited space available. The primary purpose here is to simply introduce the format and the logical progression employed in writing a program.

Before continuing, it is important to realize that a program will run properly only if every character in every line is absolutely correct. Fortunately, systems such as the HP-85A will let you know if you have made an error in writing a command statement when you try to enter the statement in the computer. It does so when you press END LINE to progress to the next statement. If the statement is correct, the next number will appear. Otherwise, an indication of the error will appear on the video display. However, if a negative sign is inserted rather than a division sign, the computer will not correct your error and will perform the operation indicated. Be aware, therefore, that *the computer cannot think for itself;* it can only perform operations of logic as defined in the entered program.

A second program appears in Fig. 26.7. It will provide the area in circular mils of a conductor and the total resistance of the conductor. The input data are exactly the same as those appearing in Example 3.1 of the text. Note the similarities in appearance of this program and the program just described. These similarities demonstrate the fact that once you can properly input one type of problem into the computer, you can analyze a wide variety of similar problems.

```
AUTO
10 DISP "RESISTANCE OF A COPPER
CONDUCTOR"
20 DISP "DIAMETER IN INCHES"
30 INPUT D
40 DISP "LENGTH IN FEET"
50 INPUT L
60 A=(D*1000)^2
70 DISP "AREA IN CM=";A
80 R=10.37*L/A
90 DISP "RESISTANCE=";R
100 END
```

Display: RESISTANCE OF A COPPER CONDUCTOR
DIAMETER IN INCHES?
Provide diameter in inches.

Display: LENGTH IN FEET?
Provide length in feet.
$A_{CM} = (\text{diam.} \times 1000)^2$
Display: AREA IN CM $= A_{CM}$
$R = 10.37 \times l/A_{CM}$
Display: RESISTANCE $= R$
End of program

```
RESISTANCE OF A COPPER CONDUCTOR
DIAMETER IN INCHES
?
.020
LENGTH IN FEET
?
100
AREA IN CM= 400
RESISTANCE= 2.5925
```

Answer in ohms

```
RESISTANCE OF A COPPER CONDUCTOR
DIAMETER IN INCHES
?
.0237
LENGTH IN FEET
?
389.7
AREA IN CM= 561.69
RESISTANCE= 7.19469636276
```

Answer in ohms

As indicated, once the program is run, the necessary data are requested and the results are printed as shown. To test your knowledge of the program steps, assume for the moment that you are the control element of the computer, reading the steps as they are entered. Taking each step independently, determine whether you can follow each step through the program to obtain the desired result.

FIG. 26.7 *Program to determine the resistance of a copper conductor.*

It was mentioned earlier that you can run the same program again simply by pressing the RUN key. We did so for this example to demonstrate that results for a variety of input data can be obtained very quickly once the program is properly inserted into the memory—a time-saving feature. A nonprogrammable calculator would have to repeat each step for each new set of input data.

The next program (appearing Fig. 26.8) is an application of the voltage divider rule that results in the value of the desired voltage V for a range of values from 1 Ω to the value of R_1, with increments of 1 Ω. That is, the voltage V will be determined for a value of R_2 equal to 1, 2, 3, . . . , 7 and 8 Ω since $R_1 = 8\ \Omega$. It provides the opportunity to demonstrate a LOOP routine (steps 40 through 70). Command number 40 states that for values of R_2 from 1 Ω to R_1 (in 1-Ω intervals), steps 50 through 70 are to be performed. During the first passage, $R_2 = 1\ \Omega$, V is determined and displayed, and then statement 70 specifies that the loop should be repeated with the next value of R_2 (specifically, 2 Ω). The printout displays the series of results for $R_1 = 8\ \Omega$. As before, the results begin to appear at almost the same instant the input data E and R_1 are introduced. Consider the enormous amount of time saved by not having to repeat the same calculation eight times. Also consider the feeling of assurance that results from knowing that if one

`AUTO`	
`10 DISP "VOLTAGE DIVIDER RULE"` `20 DISP "DC SUPPLY,R1 VALUE"`	Display: VOLTAGE DIVIDER RULE dc·SUPPLY, R_1 VALUE?
`30 INPUT E,R1`	Provide E and R_1 in order requested.
`40 FOR R2=1 TO R1`	For R_2 equal to every value from 1 to R_1 (in 1-ohm increments)
`50 V=R2*E/(R2+R1)`	Calculate $V = R_2 \times E/(R_2 + R_1)$.
`60 DISP V`	Display the results for V for each value of R_2.
`70 NEXT R2`	Repeat calculation for next value of R_2.
`80 END`	End of program

`VOLTAGE DIVIDER RULE`
`DC SUPPLY,R1 VALUE`
`?`
`36,8`	

`4`	Value of V for R_2 = 1 ohm
`7.2`	Value of V for R_2 = 2 ohms
`9.81818181818`	Value of V for R_2 = 3 ohms
`12`	Value of V for R_2 = 4 ohms
`13.8461538462`	Value of V for R_2 = 5 ohms
`15.4285714286`	Value of V for R_2 = 6 ohms
`16.8`	Value of V for R_2 = 7 ohms
`18`	Value of V for R_2 = 8 ohms

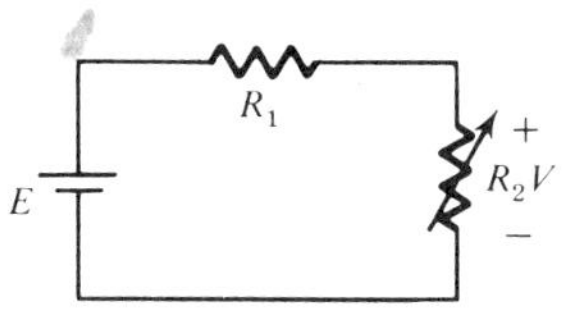

FIG. 26.8 *Employing a LOOP routine in the application of the voltage divider rule.*

calculation is correct, the computer will insure that the remaining calculations are correct. You can check the results by considering the case of $R_2 = 8 \ \Omega$ and solving for V:

$$V = \frac{R_2 \cdot E}{(R_1 + R_2)} = \frac{8 \cdot 36}{(8 + 8)}$$

$$= \frac{288}{16} = 18 \text{ V}$$

Solving third-order determinants using the method appearing in Section 7.6 is a long and tedious process, a fact that is quite obvious from Example 7.10. The program for solving third-order determinants using the equations appearing in this section is provided in Fig. 26.9. The writing

```
AUTO
10 DISP "3RD ORDER DETERMINANT"
20 DISP "COEFFICIENTS A1,A2,A3"
30 INPUT A1,A2,A3
40 DISP "COEFFICIENTS B1,B2,B3"
50 INPUT B1,B2,B3
60 DISP "COEFFICIENTS C1,C2,C3"
70 INPUT C1,C2,C3
80 DISP "VALUES D1,D2,D3"
90 INPUT D1,D2,D3
100 D=A1*B2*C3+B1*C2*A3+C1*A2*B3
-A3*B2*C1-B3*C2*A1-C3*A2*B1
110 DISP "D=";D
120 N1=D1*B2*C3+B1*C2*D3+C1*D2*B
3-D3*B2*C1-B3*C2*D1-C3*D2*B1
130 X=N1/D
140 DISP "X=";X
150 N2=A1*D2*C3+D1*C2*A3+C1*A2*D
3-A3*D2*C1-D3*C2*A1-C3*A2*D1
160 Y=N2/D
170 DISP "Y=";Y
180 N3=A1*B2*D3+B1*D2*A3+D1*A2*B
3-A3*B2*D1-B3*D2*A1-D3*A2*B1
190 Z=N3/D
200 DISP "Z=";Z
210 END
```

FIG. 26.9 *Program to determine the complete solution of three simultaneous equations.*

of the program must be done very carefully. One incorrect sign will result in a totally incorrect answer. However, recall the concern of using the method of Section 7.6 when it came to inserting values and making the long series of calculations. The advantage of the computer program is therefore very obvious. Once the program is correctly inserted into the computer and *stored* for further use, it is necessary only to input the correct data—the algebraic calculations will be performed without error. The term *store* simply means assigning the program a "name" and storing it on a separate magnetic tape unit. It can then be "loaded" back into computer memory whenever required for the calculation. The statements for this program are

STORE "PRGM 1"

LOAD "PRGM 1"

```
3RD ORDER DETERMINANT
COEFFICIENTS A1,A2,A3
?
1,0,1
COEFFICIENTS B1,B2,B3
?
0,3,2
COEFFICIENTS C1,C2,C3
?
-2,1,3
VALUES D1,D2,D3
?
-1,2,0
D= 13
X=-1.15384615385
Y= .692307692308
Z=-7.69230769231E-2
```

FIG. 26.10 *Applying the program of Fig. 26.9 to the set of equations appearing in Example 7.10 of the text.*

```
3RD ORDER DETERMINANT
COEFFICIENTS A1,A2,A3
?
1,2,-2
COEFFICIENTS B1,B2,B3
?
0,3,0
COEFFICIENTS C1,C2,C3
?
2,-4,-6
VALUES D1,D2,D3
?
1,0,3
D=-6
X= 6
Y=-7.33333333333
Z=-2.5

Z=-7.69230769231E-2
```

FIG. 26.11 *Applying the program of Fig. 26.9 to the set of equations appearing in Problem 17.5(b) of the text.*

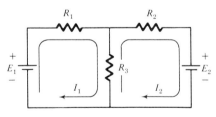

FIG. 26.12 *Two-loop dc network.*

The equations appearing in the program can be obtained directly from the text. Once the RUN key is pressed, the display appearing in Fig. 26.10 will appear, asking for the required coefficients. Once these coefficients are inserted, the next series of coefficients is requested, and so on. The results will appear immediately after the last set of data is provided. The results appearing in Fig. 26.10 are for Example 7.10 of the text.

As indicated above, the program can be stored and recalled from the magnetic tape storage when required. Such was the case for the data appearing in Fig. 26.11. They are the coefficients of the simultaneous equations of Problem 7.15(b).

The general solution for the two-loop network of Fig. 26.12 was obtained using *mesh* analysis as indicated below:

$$I_1(R_1 + R_3) - I_2 R_3 = E_1$$
$$I_2(R_2 + R_3) - I_1 R_3 = -E_2$$

or

$$(R_1 + R_3)I_1 - R_3 I_2 = E_1$$
$$-R_3 I_1 + (R_2 + R_3)I_2 = -E_2$$

$$\frac{I_1}{D} = \frac{\begin{vmatrix} E_1 & -R_3 \\ -E_2 & (R_2 + R_3) \end{vmatrix}}{\begin{vmatrix} (R_1 + R_3) & -R_3 \\ -R_3 & (R_2 + R_3) \end{vmatrix}} = \frac{E_1(R_2 + R_3) - E_2 R_3}{(R_1 + R_3)(R_2 + R_3) - R_3^2}$$

$$I_2 = \frac{\begin{vmatrix} (R_1 + R_3) & E_1 \\ -R_3 & -E_2 \end{vmatrix}}{D} = \frac{-E_2(R_1 + R_3) + E_1 R_3}{D}$$

The equations above can be applied to any two-loop network of the configuration indicated in Fig. 26.12. R_1, R_2, and R_3 (for the network of Fig. 7.19 of the text) can represent the equivalent resistance of the indicated branches. The dc supplies can be reversed through the use of a negative sign in the input data.

The program required to calculate both I_1 and I_2 appears in Fig. 26.13. Note that D was calculated as an independent quantity since it appears in each equation. When the RUN key is pressed, the first three lines below the program will appear. Once the values of R_1, R_2, and R_3 are inserted, the program will ask for E_1 and E_2. The results will then appear almost instantaneously after the dc supply values are entered.

```
AUTO
10 DISP "DC TWO LOOP NETWORK"
20 DISP "RESISTORS R1,R2,R3"
30 INPUT R1,R2,R3
40 DISP "DC SOURCES E1,E2"
50 INPUT E1,E2

60 D=(R1+R3)*(R2+R3)-R3^2
70 I1=(E1*(R2+R3)-E2*R3)/D
80 DISP "I1=";I1
90 I2=(-E2*(R1+R3)+E1*R3)/D

100 DISP "I2=";I2
110 END
```

$D = (R_1 + R_3)(R_2 + R_3) - R_3^2$
$I_1 = (E_1(R_2 + R_3) - E_2R_3)/D$
Display: I_1 = solution
$I_2 = (-E_2(R_1 + R_3) + E_1R_3)/D$

Display: I_2 = Solution

(a)

```
DC TWO LOOP NETWORK
RESISTORS R1,R2,R3
?
2,1,4
DC SOURCES E1,E2
?
2,6
I1=-1
I2=-2
```

Answers in amperes

(b)

A second run appears in Fig. 26.14 for the network of Fig. 7.25. In this case the value of R_1 is the sum of the 8- and 6-Ω series resistors. Note the level of accuracy obtained from the computer. Certainly an answer to three or four significant places is sufficient as indicated in the text material.

Equation (20.16) and (20.18) are repeated here for convenience:

$$f_2 = \frac{1}{2\pi}\left[\frac{R}{2L} + \frac{1}{2}\sqrt{\left(\frac{R}{L}\right)^2 + \frac{4}{LC}}\right]$$

$$f_1 = \frac{1}{2\pi}\left[-\frac{R}{2L} + \frac{1}{2}\sqrt{\left(\frac{R}{L}\right)^2 + \frac{4}{LC}}\right]$$

A brief review of these equations clearly indicates that it would be a time-consuming, dull, and in some ways difficult task to solve for f_1 and f_2 for different values of R, L, or C. However, once the equations are properly placed in computer memory, as shown in Fig. 26.15, the values of f_1, f_2, and BW can be calculated for a range of values with very little difficulty. Note in the program that particular parts of the equations are calculated separately due to the similarities between the equations for f_1 and f_2.

The results indicated in Fig. 26.16 are for values of L and C equal to 5 mH and 1 μF, respectively, with R equal to 5, 10, 15, 20, and 25 Ω. Note how the bandwidth has increased linearly with values of R as determined by Eq. (20.19) for fixed values of L. That is, note that twice the value of R will result in twice the bandwidth, and so on.

FIG. 26.13 (a) Program designed to calculate the loop currents of the network of Fig. 26.12 and (b) solution for the network of Fig. 7.19.

```
DC TWO LOOP NETWORK
RESISTORS R1,R2,R3
?
14,7,2
DC SOURCES E1,E2
?
4,9
I1= .128571428571
I2=-.971428571429
```

FIG. 26.14 Printout resulting from applying the program of Fig. 26.13 to the network of Fig. 7.25.

```
AUTO
10 DISP "BAND PASS FREQ"
20 DISP "INDUCTOR VALUE"
30 INPUT L
40 DISP "CAPACITOR VALUE"
50 INPUT C
60 FOR R=5 TO 25 STEP 5
70 X=.5*SQR((R/L)^2+4/(L*C))
80 Y=R/(2*L)
90 Z=1/2*PI

100 F2=Z*(Y+X)
110 DISP "F2=";F2
120 F1=Z*(-Y+X)
130 DISP "F1=";F1
140 B=F2-F1
150 DISP" BANDWIDTH =";B
160 NEXT R
170 END
```

Calculate for values of R from 5 to 25 ohms (5-ohm increments).
$X = \frac{1}{2} \sqrt{(R/L)^2 + 4/LC}$
$Y = R/2L$
$Z = 1/2\pi$
$f_2 = Z \times (Y + X)$
Display: $f_2 =$ solution
$f_1 = Z \times (-Y + X)$
Display: $f_1 =$ solution
$BW = f_2 - f_1$
Display BANDWIDTH = solution
Repeat for next value of R_1

FIG. 26.15 *Program to determine the cutoff frequencies and bandwidth of a series resonant circuit.*

```
BAND PASS FREQ
INDUCTOR VALUE
?
5.E-3
CAPACITOR VALUE
?
1.E-6
F2= 23013.6925274
F1= 21442.8962006
  BANDWIDTH = 1570.7963268
F2= 23840.6778074
F1= 20699.0851538
  BANDWIDTH = 3141.5926536
F2= 24695.2157879
F1= 19982.8268075
  BANDWIDTH = 4712.3889804
F2= 25577.0517409
F1= 19293.8664337
  BANDWIDTH = 6283.1853072
F2= 26485.8355777
F1= 18631.8539437
  BANDWIDTH = 7853.981634
```

FIG. 26.16 *Printout for the program of Fig. 26.15 for a series RLC network with $L = 5\,mH$, $C = 1\,\mu F$, and $R = 5, 10, 15, 20,$ and $25\,\Omega$.*

26.6 PLOTTING ROUTINES

The HP-85A is a very versatile unit in that it has both a visual display and a recording capability. This combination is extremely useful when plots or curves are desired using the results of a repeated series of calculations.

The plotting of curves naturally requires an understanding of how to set up the scales of the display, how to choose the labels to appear, and, of course, how to define the variables to be plotted. The program of Fig. 26.17 is designed to calculate the power to the resistor R of the network appearing in Fig. 26.17 for values of R (external or load resistance) spanning $R = 1, 2, 3, \ldots, 50\,\Omega$. The text has revealed that maximum power will be delivered to R when $R = R_1$ (internal resistance) $= 10\,\Omega$. The plot will confirm whether this is, in fact, the case and also whether the maximum power is $P_{max} = E^2/4R_1 = (100)^2/4(10) = 250$ W.

The initial steps of the program clear the display (GCLEAR) and request the component values of the network. Statement 70 defines the minimum and maximum values of the x axis ($0 \rightarrow 300$ for a maximum power less than 300 watts), while statements 80 and 90 define the spacing between tick marks. In other words, there are 50 divisions on the horizontal axis and $300/50 = 6$ divisions on the vertical axis.

Command statements 100 through 140 represent a loop sequence of calculating the power for values of R between 1 and 50 (increments of 1) and DRAWing the graph between successive values of R and P. Statements 150 and 160 simply label the graph by moving to a position defined by $X = 10$ and $Y = 100$ and writing P vs. R. Command statements 170 through 240 label specified tick marks on the graph. For both the y and x axes, there is a loop routine that defines the interval and the notation to appear.

Pressing the RUN key after the program is properly inserted will result in the title of the plot and the request for component values as appearing in Fig. 26.18. The moment

```
AUTO
10 DISP"MAXIMUM POWER TRANSFER"
20 GCLEAR
30 DISP"DC SUPPLY VOLTAGE"
40 INPUT E
50 DISP"INTERNAL RESISTANCE"
60 INPUT R1
70 SCALE 0,50,0,300
80 XAXIS 0,1
90 YAXIS 0,50
100 FOR R=1 TO 50
110 I=E/(R1+R)
120 P=I^2*R
130 DRAW R,P
140 NEXT R

150 MOVE 10,100
160 LABEL "P VS R"

170 FOR Y=50 TO 300 STEP 50
180 MOVE 1,Y
190 LABEL VAL$(Y)

200 NEXT Y
210 FOR X=5 TO 50 STEP 5
220 MOVE X,1
230 LABEL VAL$(X)
240 NEXT X
250 END
```

Defining R_{min}, R_{max}, P_{min}, P_{max}
Defining increment of R (1-ohm)
Defining increment of P (50-watt)
Calculate for values of R from 1 to 50 ohms.
$I = E/(R_1 + R)$
$P = I^2 R$
Draw the curve to the resulting value of R and resulting value of P.

Move to this position on the graph: $R = 10$, $P = 100$.
Display: P vs. R to label graph

For each 50-watt increment of P
Move to $R = 1$ and succeeding values of P.
Label the vertical axis in 50-watt increments.
Repeat for each 50-watt increment.
Statements 210 through 240 label 5-ohm increments of R.

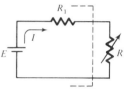

the value of internal resistance is inserted, the horizontal and vertical axes will be drawn, followed by the actual plot.

FIG. 26.17 *Program for plotting the power to R for various values of R.*

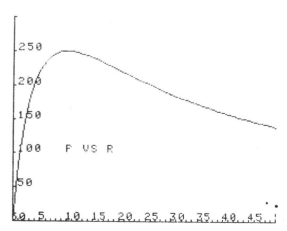

FIG. 26.18 *Resulting printout for the program of Fig. 26.17 for E = 100 V and $R_1 = 10\,\Omega$.*

Note that it does peak at $R = R_1 = 10\ \Omega$, and the maximum value is 250 watts. It is interesting to note that the maximum power curve drops off much more slowly to the right of the peak value than to the left. This result suggests that if the value of R representing maximum power cannot be achieved, higher values should be considered rather than lower values. Even at $R = 50\ \Omega\ (R_1 \times 5)$, the power to R has not dropped to the power resulting at $R_1 = 2\ \Omega$ $(R_1/5)$.

At first exposure, numerous attempts at obtaining the curve of Fig. 26.18 will be required before the desired results are obtained. Be assured, however, that once you have developed a curve such as just examined, you will be able to approach a number of different plots with confidence. Consider the similarities that appear in the plotting routine of Fig. 26.19, developed to display the charging curve of a capacitor. As in the previous example, you must have some idea of the maximum voltage and time value to properly define the horizontal and vertical axes. In addition, note the expanded use of the exponential notation for the component values. The E simply refers to a power of 10. For example, $5.E - 2 = 5 \times 10^{-2}$ and $2.E4 = 2 \times 10^{+4}$.

```
AUTO
10 GCLEAR
20 DISP "CAP CHARGING CURVE"
30 DISP "APPLIED DC VOLTAGE"
40 INPUT E
50 DISP "RESISTOR VALUE"
60 INPUT R
70 DISP "CAPACITOR VALUE"
80 INPUT C
90 SCALE 0,1.1,0,11
100 XAXIS 0, 1
110 YAXIS 0,1
120 FOR T=5.E-2 TO 1 STEP 5.E-2
130 V=E*(1-EXP(-T/(R*C)))
140 DRAW T,V
150 NEXT T
160 FOR Y=1 TO 11 STEP 1
170 MOVE .01,Y
180 LABEL VAL$(Y)
190 NEXT Y
200 FOR X=0 TO 1.1 STEP .1
210 MOVE X,.2
220 LABEL VAL$(X)
230 NEXT X
240 END
```

For $t = 0.05$ s to 1.0 s (increments of 0.05 s)
$v_C(t) = E(1 - e^{-t/RC})$

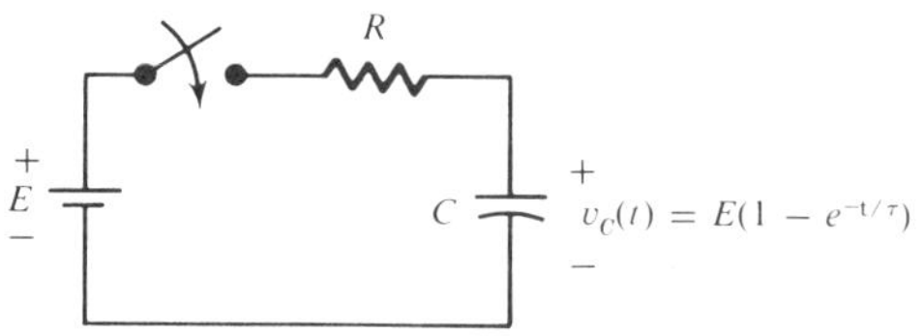

FIG. 26.19 *Program for plotting the voltage $v_C(t)$ as a function of time.*

Since the program is stored in computer memory, it can be recalled by simply pressing the RUN key after the plot has been obtained. The program will again request the new values of the components and plot the new curve.

The plot of Fig. 26.20 has a steady-state value of 8 volts and a time constant of $\tau = RC = (2 \times 10^4)(1 \times 10^{-5}) = 0.2$ second. Note that after $5\tau = 1$ second, the curve has indeed reached the steady-state level (for all practical purposes). In Fig. 26.21(a) the dc supply was reduced to 2.5 volts and the capacitor reduced to 0.5×10^{-5} F, resulting in $\tau = (2 \times 10^4)(0.5 \times 10^{-5}) = 0.1$ second, or one-half the time constant of the curve of Fig. 26.20. Note the earlier rise to the steady-state value and the reduced dc level. In Fig. 26.21(b) the only change from the input data for Fig. 20.21(a) was to increase the dc supply voltage to 10 volts. The steady-state value is still attained at about 0.5 second, but the final dc level is now 10 volts.

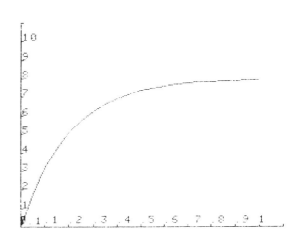

```
CAP CHARGING CURVE
APPLIED DC VOLTAGE
?
8
RESISTOR VALUE
?
2.E4
CAPACITOR VALUE
?
1.E-5
```

FIG. 26.20 *Printout for the program of Fig. 26.19 for $E = 8$ V, $R = 20$ kΩ, and $C = 10$ μF.*

26.7 THE HP-85 CIRCUIT ANALYSIS PAC®

It was noted in the introduction to this chapter that the software support for a computer is frequently the deciding element when a purchase decision is to be made. Realizing the importance of the software support, the Hewlett-Packard Corporation developed a selection of software packages which are described in the initial advertising of the desk-top HP-85A Personal Computer. Some of the support

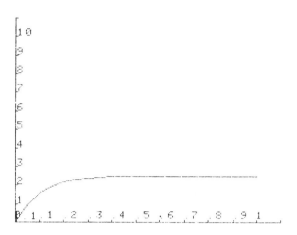

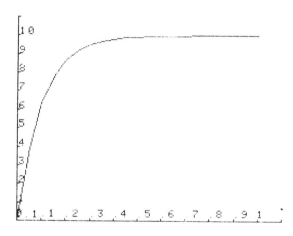

FIG. 26.21 *Printout for the program of Fig. 26.19 for (a) E = 2.5 V, R = 20 kΩ, and C = 5 μF; and (b) E = 10 V, R = 20 kΩ, and C = 5 μF.*

is described in Fig. 26.22, which is from the advertising literature for the HP-85.

Each support unit contains a magnetic tape unit in which the program is stored and a manual to introduce the use of the package (Fig. 26.23). The additional storage bins are for data and supporting programs that are also stored on similar-sized magnetic tape units.

The support manuals are very carefully written and tested. Every step is carefully defined with statements that permit the analysis of networks with up to 20 nodes and 45 components (using an optional 16,000-byte memory unit).

The use of the PAC is best described by considering the network of Fig. 26.24, which appears as Fig. 7.19 in the text. First, the program on the magnetic tape must be LOADed into the computer memory. This is accomplished by the LOAD command as indicated in Fig. 26.25(a).

BASIC Training (00085-13002) Provides a valuable learning aid for novice users. Contains a thorough HP-85 BASIC language tutorial. Demonstrates the HP-85's capabilities with graphics emphasis. Includes keyboard learning aid.

Standard* (00085-13001) Moving Average, Annuities and Compound Amounts with Amortization, Polynomial Solutions, Simultaneous Equations, Calculus and Roots of f(x), Curve Fitting, Auto Function Plot, Auto Data Plot, Histogram Generator, Arithmetic Teacher, Calendar Functions, Biorythms, Timer, Music Composer, Ski Game.

General Statistics (00085-13003) One Sample Analysis, Paired Sample Analysis, Test Statistics, Distributions, Multiple Linear Regression.

Finance (00085-13004) Compound Interest and Loan Amortization, Discounted Cash Flow Analysis, Depreciation, Simple Interest and Interest Conventions, Bonds, Notes, Breakeven Analysis.

Waveform Analysis (00085-13035) Enables you to perform fast Fourier transform on time domain data and inverse transform on frequency domain data. Lets you compute the correlation function and power spectrum. Also provides dual data block entry for cross correlation, cross power spectrum, and convolution. Results can be plotted or printed.

Math (00085-13005) Simultaneous Equations, Solution to f(x) = 0 on an Interval, Integration with Equally-Spaced and Unequally-Spaced Data Points, Ordinary Differential Equations, Chebyshev Polynomial, Fourier Series for Equally-Spaced and Unequally-Spaced Points, Fast Fourier Transform, Hyperbolic Functions, Complex Operations, Triangle Solutions.

Circuit Analysis (00085-13006) Determines steady-state AC behavior of electrical networks consisting of resistors, capacitors, inductors, voltage-controlled current sources and independent current sources.

Games (00085-13010) Blackjack, Slot Machine, Poker, Solitaire, Reversi, Gomoku, Cribbage, Wari, Sea Skirmish, Blockade, Race Track, Lander, Race, Hangman, Hunt the Wumpus, King, Nim, Maze Generator, Life, Birthday Plot.

Linear Programming (00085-13011) Solves a wide variety of optimization problems using a modified simplex method that incorporates variable bounds. Example applications include chemical blending, feed mix, production scheduling, investment portfolio selection and market media selection.

Text Editing (00085-13034) Provides variable input and output of text files and enables you to specify tabs and indents as well as add, delete, replace, renumber, and move lines. Also enables you to do matched string search and replace.

Courtesy of the Hewlett-Packard Corporation.

*The Standard Applications Pac is included as a standard accessory with HP-85.

FIG. 26.22 *HP-85 Applications Pacs.*

Courtesy of Hewlett Packard Co.

Courtesy of the Hewlett-Packard Corporation.

FIG. 26.23 *Applications Pac*®.

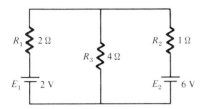

FIG. 26.24

When the RUN key is pressed, the display will include the information appearing in Fig. 26.25(b). It defines the order

```
LOAD "CAP"
```
(a)

```
Some hints for CAP:
 All interaction is initiated by
 pressing the desired special
 function keys.
 Load your circuit first.
 Select frequencies and output
 quantities before calculation.
 Plotting is done after calcula-
 tion.
 Tables are output automatically
 during calculation.
```

```
--------------------------------
PRT CIR  STORE  RECALL   PURGE
NEW CIR   ADD   DELETE  OUT SEL
```
(b)

FIG. 26.25 (*a*) *Loading the Circuit Analysis Application Pac® and (b) introductory comments offered by the Pac.*

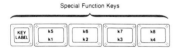

Courtesy of Hewlett Packard Co.

FIG. 26.26 *HP-85 Special Function Keys.*

of events that you must follow in order to obtain the desired results. The first statement defines a set of *special* keys which appear in Fig. 26.26. The positions of those keys correspond directly with the positions of all of the statements appearing in Fig. 26.25(b). That is, K1 corresponds to NEW CIR, Key K5 (obtained by simultaneously depressing the shift key) corresponds to PRT CIR, and so on.

Since we will be inserting a NEW CIRcuit, we should press K1. The result is the output of Fig. 26.27. Again the positions of the letters correspond to those of the keys of Fig. 26.26. Since we have resistive (*R*) elements, we again press K1, and the request for data appears as shown in Fig.

```
--------------------------------
         VCIS     IS
R         L        C        COMPLETE
```

FIG. 26.27 *Computer request for network components.*

```
R NODES: FROM, TO?
```

FIG. 26.28 *Computer request for nodes connected to resistive element.*

26.28. Before we input these data, we must first redraw the network of Fig. 26.24 as shown in Fig. 26.29.

The HP program requires that only current sources be present. Therefore, both sources are converted as indicated in Fig. 26.29 using the conversion equation. The nodes are then identified as indicated, with the 0 always referring to ground potential or the reference voltage for the network. Quite obviously, there is only one node voltage to be determined for this network. The last step is to assign a current

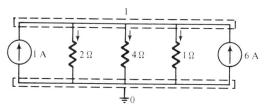

FIG. 26.29 *The network of Fig. 26.24 following the conversion of the voltage sources to current sources.*

direction to each branch as also indicated in the figure. If the result is the opposite direction, a negative sign will simply appear. Since the branch current through the 2-Ω resistor passes FROM node 1 TO node 0 and has a value of 2 Ω, the data are supplied as shown. The instant the 2 Ω is provided, the table will appear again and we will again choose *R* for the other two resistors. The result is shown in the displays appearing in Fig. 26.30.

After the resistor values are inserted, the table display will appear again as shown in Fig. 26.31, but now the key K7 will be pressed to input the independent sources. Note in the same figure a request for magnitude and phase. We made this request to avoid making a different set of statements for ac networks. For dc networks we simply indicate 0° phase.

```
R NODES: FROM, TO?
1,0
VALUE?
2

R NODES: FROM, TO?
1,0
VALUE?
4

R NODES: FROM, TO?
1,0
VALUE?
1
```

FIG. 26.30 *Inserting the resistor values for the network of Fig. 26.29.*

```
------------------------------------
        VCIS      IS
R        L         C            COMPLETE

IS NODES: FROM, TO?
0,1
AMPLITUDE?
1
PHASE?
0


IS NODES: FROM, TO?
0,1
AMPLITUDE?
6
PHASE?
0
```

Once the current sources have been inserted, the table will reappear as shown in Fig. 26.32, but key K4 will be pressed to indicate COMPLETE. The computer will then automatically display the entire network but with the independent current sources appearing first as shown in Fig. 26.32. Incidentally, the term VCIS in the above tables is an abbreviation for *Voltage-Controlled Current (I) Sources* such as appear in electronic systems.

Note at the bottom of Fig. 26.32 a new table through which the next step of the program is to be defined. We will press key K4 to SELect the type of OUTput we would pre-

FIG. 26.31 *Inserting the magnitude and phase for the independent sources of the network of Fig. 26.29.*

```
---------------------------------
        VCIS    IS
R       L       C       COMPLETE


CIRCUIT DESCRIPTION
    1 IS     0 TO 1   6 AMPS
             0 DEG
    2 IS     0 TO 1   1 AMPS
             0 DEG
    3 R      1 TO 0   2 Ω
    4 R      1 TO 0   4 Ω
    5 R      1 TO 0   1 Ω



---------------------------------
PRT CIR   STORE   RECALL   PURGE
NEW CIR   ADD     DELETE  OUT SEL
```

FIG. 26.32 *Complete circuit description for the network of Fig. 26.29.*

fer. The table choice of Fig. 26.33 will then appear, which gives us the choice of the CIRCUIT again, requesting a PLOT using a dB or NOT dB scale, choosing the

```
-----------------------------------
             dB      FREQS
CIRCUIT    TABLE    PLOT    CALC
```

FIG. 26.33 *Computer request for the next operation to be performed.*

FREQuencies of interest, performing the CALCulation, or requesting the TABLE from which the desired output quantities can be requested. Since we have not picked our output quantities, we press key K2, which results in the TABLE of Fig. 26.34. The magnitude of the node voltage 1

```
---------------------------------
        BP        /
NV      BV      BI      CONT


NODE?
1
```

FIG. 26.34 *Table request for the results desired.*

will permit determining any quantity in the network of Fig. 26.34. We will therefore press key K1 for NV. The display appears in Fig. 26.34 with our request for node voltage 1. The table will then reappear as shown in Fig. 26.35, but we will now press key K4 for COMPLETE since we are presently not interested in additional nodes or branch currents (BI), branch voltages (BV), branch powers (BP), or ratios (/).

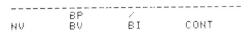

```
---------------------------------
        BP        /
NV      BV      BI      CONT
```

FIG. 26.35 *Table request for additional unknowns or continue.*

The table appearing in Fig. 26.36 will then appear, and pressing key K7 will indicate that we are ready to specify frequencies. You certainly have the right to question this need to specify frequencies for a dc system, but remember that the program is geared toward ac networks, and dc networks are only a special case of frequency-dependent systems.

The request for data as shown in Fig. 26.37 will then appear. Note that for dc systems the minimum and maximum frequencies were chosen to be the same (it could have been any frequency level) and the INCrement between frequencies is only one. This is simply a defined sequence of steps for dc systems and should not raise any concerns. The phrase within the brackets requests a negative sign in front of the increment if a log scale is to be employed.

When the data of Fig. 26.37 are inserted, the table of Fig. 26.36 will reappear. You should then depress the CALC key to request execution of the required CALCulations. The result will appear as shown in Fig. 26.38. The $F = 60$ is of no interest, but we now know that the nodal voltage V_1 is 4 volts. The current through the 4-Ω resistor is then $I = V_1/4 = 4/4 = 1$ A. The current through the resistor R_1 in Fig. 26.24 is $I = (E - V_1)/R_1 = (2 - 4)/2 = -2/2 = -1$ A to indicate that in actuality it has the opposite direction.

The steps just described are for a rather simple network. Thus using the program may appear to be a long and tedious process for a large network. However, the next program will reveal that once you pass through the steps a few times, the sequence becomes easy to follow, and complex networks can be attacked with almost no added difficulty. For example, consider the network of Fig. 26.39, which appears as Fig. 7.38 in Example 7.23 in the text. This network is certainly a magnitude or two more difficult than the network just analyzed.

```
- - - - - - - - - - - - - - - - - - - - - - - -
        dB          FREQS
CIRCUIT TABLE      PLOT       CALC
```

FIG. 26.36 *Table request for FREQuencies.*

```
MIN FREQ?
60
MAX FREQ?
60
INC [(-) FOR LOG]?
1
```

FIG. 26.37 *Inserting the frequency data.*

```
F= 60
NV 1
   4    0
```

FIG. 26.38 *The results for node voltage 1 for the network of Fig. 26.29.*

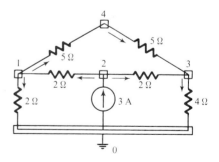

FIG. 26.39 *Network appearing in Example 7.23 (Fig. 7.38) of the text.*

Note in Fig. 26.40 that the circuit description of Fig. 26.39 is not that much longer or more difficult to insert than the one appearing in Fig. 26.32. In this case a number of additional calculations were requested, such as the branch voltage across the 5-Ω resistor between nodes 3 and 4. Branches are defined by the number appearing to the left of each element in the circuit description. In addition, the branch current through the same resistor was requested.

```
CIRCUIT DESCRIPTION
   1 IS    0 TO 3   3 AMPS
           0 DEG
   2 R     1 TO 0    2 Ω
   3 R     2 TO 1    2 Ω
   4 R     2 TO 3    2 Ω
   5 R     3 TO 0    4 Ω
   6 R     1 TO 4    5 Ω
   7 R     4 TO 3    5 Ω


  F= 60
  NV3
    6.58064516131    0
  BV5
    6.58064516131    0
  BI5
    1.64516129033    0
  BP1
    19.7419354839    0
  BP2
    3.67117585854    0
  BP3
    1.87304890739    0
  BP4
    1.87304890739    0
  BP5
    10.8262226848    0
  BP6
    .749219562957    0
  BP7
    .749219562957    0


  3.67117585854+1.87304890739+1.87
  304890739+10.8262226848+.7492195
  62957+.749219562957
     19.7419354841
```

FIG. 26.40 *Results for the network of Fig. 26.39.*

Note the BI5 = BV5/5. Finally, the power associated with each branch was also requested. For the current source it is the power supplied; for the resistors it is the power dissipated. As indicated in the bottom of Fig. 26.40, the sum of power dissipated by all of the resistive elements (BP2 through BP7) equals the power supplied by the current source (BP1). The slight difference is due primarily to the rounding off that occurs at the last place for each branch power.

The above results appeared a few seconds after the input data were provided. Consider the time required to perform the same calculations using the method presented in the text.

An ac network of some complexity appears in Fig. 26.41. It is, in fact, Problem 10 in Chapter 16. Note that the nodes

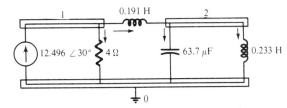

FIG. 26.41 *Network appearing as Problem 10 of Chapter 16 in the text.*

have been assigned a numerical value, and the current directions have been chosen. The value of the current source was determined using the conversion technique. The frequency to apply was determined from $f = \omega/2\pi = 314/6.28 = 50$ Hz. The circuit description, the display for the frequency selection, and the results appear in Fig. 26.42.

The node voltages and currents now include a magnitude and an angle. Further, the power to a branch is determined by $P = VI\cos\theta$, where V and I are branch values and θ is the phase angle between them. For branch 3, which is a pure inductor, the real (average) power should be zero. Substituting $90°$ in $\cos\theta$ will result in $P = 0$ as required. The results for the various currents and voltages will compare very favorably with the text results. Note that i_T in the text is the same as branch current B13.

The CAP program also has a plotting routine that is very easy to use. When the appropriate tables appear, PLOT is chosen, followed by the identification of which quantities (NV, BI, BV, and so on) are to be plotted. The program will do all of the vertical and horizontal scaling unless you prefer to have control. In the two examples to follow, control was left to the computer.

The first few plots are the result of the parallel resonant circuit of Fig. 26.43 (Problem 15 in Chapter 20). Note again that the nodes have been chosen, along with the direction

```
CIRCUIT DESCRIPTION
  1 IS    0 TO 1   12.496 AMPS
           30 DEG
  2 R     1 TO 0    4 Ω
  3 L     1 TO 2    1.91 E-1 H
  4 L     3 TO 0    2.33 E-1 H
  5 C     2 TO 0    6.37 E-5 F

MIN FREQ?
50
MAX FREQ?
50
INC [<-> FOR LOG]?
1


F= 50
NV2
  80.6898426795     27.6497966275
NV1
  49.9419558343     27.6497966275
BI4
  1.10233367552    -62.3503033625
BI5
  1.61476071017    117.649796638
BI3
  .512427034666    117.649796638
BP3
  15.7560484784     90
```

FIG. 26.42 *Results for the network of Fig. 26.41.*

of the branch currents. The description appears in Fig. 26.44.

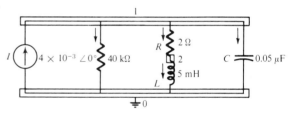

FIG. 26.43 *Network appearing as Problem 15 in Chapter 20.*

```
CIRCUIT DESCRIPTION
  1 IS    0 TO 1   .004 AMPS
           0 DEG
  2 R     1 TO 0    40000 Ω
  3 R     1 TO 2    2 Ω
  4 L     2 TO 0    5 E-3 H
  5 C     1 TO 0    5 E-8 F
```

FIG. 26.44 *Circuit description for the network of Fig. 26.43.*

The plotting capability of the HP-85 can be used to determine the resonant frequency. As indicated by the plot of Fig. 26.45, a log scale was chosen for the horizontal to increase the range of frequencies that could be examined for node voltage 1 (NV1). The minimum frequency chosen during the request cycle was 100 Hz, and the maximum 10^6 Hz. An increment of 100 Hz was chosen and the minus sign included (-100) to indicate a log scale. The results indicate that the maximum voltage is about 65 V and the resonant frequency around 10,000 Hz. Since the region of

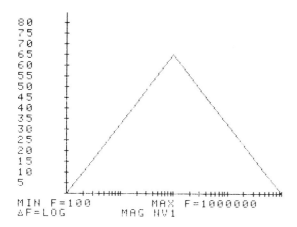

FIG. 26.45 *Magnitude of NV1 vs. frequency for the network of Fig. 26.43.*

maximum response had been identified, a smaller portion of the frequency range was chosen for the plot of Fig. 26.46. Note also that the horizontal axis is not a log scale,

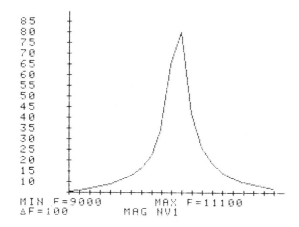

FIG. 26.46 *Magnitude of NV1 vs. frequency for a reduced frequency range for the network of Fig. 26.43.*

and so the more familiar curves result. It is interesting also to note that the maximum value has also increased. This increase is due to the fact that an increased number of readings were taken near resonance. That is, a frequency closer to the resonant value was tested with this set of test values. A more defined curve can then be obtained, as shown in Fig. 26.47, which clearly identifies the resonant frequency, the maximum voltage, and the bandwidth. Therefore, through a simple plotting routine, the computer has provided a wealth of material about the resonant characteristics of the network with a minimum of effort.

The last network to be analyzed appears in Fig. 26.48 with its identified nodes and assumed current directions. A plot of the magnitude and phase of node voltage three will

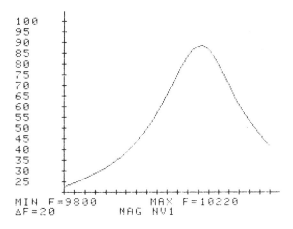

FIG. 26.47 *Magnitude of NV1 vs. frequency for the region near the resonant value.*

be developed through a sequence of plots. The circuit description appears in Fig. 26.49.

At low frequencies the reactance of the 0.08-μF capacitor is sufficiently large to be ignored. The reactance of the

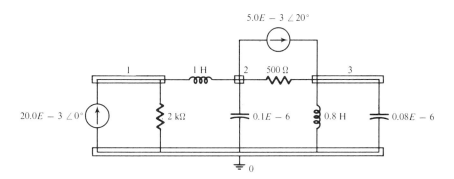

FIG. 26.48 *Complex ac network.*

0.8-H coil is very low, resulting in a shorting effect between node 3 and ground. Of course, as the frequency increases, the reactance of the coil will be increasing and node voltage NV3 will probably increase in magnitude. This is demonstrated very nicely in Fig. 26.50. Note also that the phase plot reveals a heavy inductive (90°) effect at low frequencies and a lowering of this effect as the other elements of the network begin to balance the inductive nature of the branch. Note the frequency range of the first plot. The second set of curves will begin where this set leaves off.

Interestingly enough, the results of Fig. 26.51 reveal a resonant effect due to the parallel *L-C* combination at a frequency of about 550 Hz. The resonant frequency will not be exactly the value determined solely by the parallel pair (629.4 Hz) due to the other elements of the network, but it is very close. Note that the magnitude scale starts where the first set of curves ends. The phase plot also indicates a typical resonance characteristic. The increasing ef-

```
CIRCUIT DESCRIPTION
  1 IS    2 TO 3    005 AMPS
          20 DEG
  2 IS    0 TO 1    02 AMPS
          0 DEG
  3 R     1 TO 0    2000 Ω
  4 R     2 TO 3    500 Ω
  5 L     1 TO 2    10 E-1 H
  6 L     3 TO 0    8 E-1 H
  7 C     2 TO 0    10 E-8 F
  8 C     3 TO 0    8 E-8 F
```

FIG. 26.49 *Circuit description for the network of Fig. 26.48.*

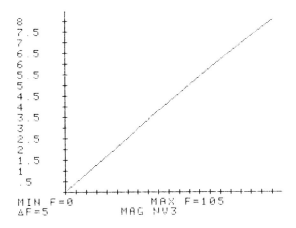

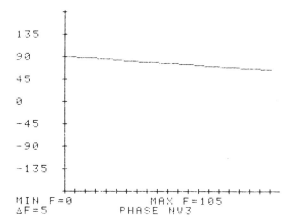

FIG. 26.50 *Magnitude and phase of the voltage NV3 for the network of Fig. 26.48 for the low-frequency range.*

fect of the parallel capacitor has resulted in a reduced level of NV3 and an increasing leading power factor.

The last set of curves appearing in Fig. 26.52 reveals a continuing decline in the magnitude until the inductive reactance of the network increases sufficiently to balance the network and introduce a lagging power factor for a brief period of time. Eventually, the capacitive reactance between node 3 and ground will approach a short-circuit state, and the magnitude and phase will approach zero and $-90°$, respectively, at very high frequencies.

A number of unanswered questions have probably arisen during this chapter. However, if the chapter has resulted in at least an inquiry about the local computer facilities and perhaps an introduction to the machine, then the content has served its purpose well. Any questions regarding the content of this chapter or the HP-85 unit can be directed to Charles Merrill Publishing Company, who will forward them to the author.

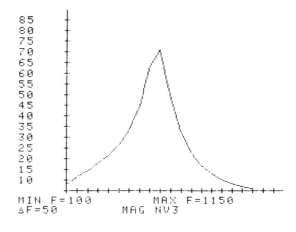

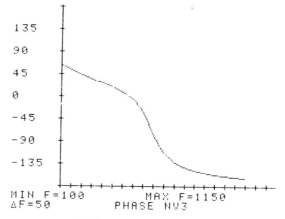

FIG. 26.51 *Magnitude and phase of NV3 for a higher-frequency range.*

GLOSSARY

ALGOL-60 Essentially the European version of the FORTRAN language.

BASIC An abbreviation for *B*eginning *A*ll-Purpose *S*ymbolic *I*nstruction *C*ode which employs abbreviated statements of the English language.

Byte A variable-length string of bits used to represent alphanumeric characters or control functions.

Card sorter A machine designed to separate punch cards into specific categories with a definite order.

COBOL An abbreviation for *CO*mmon *B*usiness-*O*riented *L*anguage, typically applied to business problems.

Command statement A set of characters or symbols that specify a particular operation of the computer.

Computer A machine designed to perform mathematical operations, compile data, and sort information in a very short period of time utilizing modern semiconductor technology.

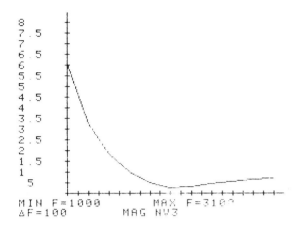

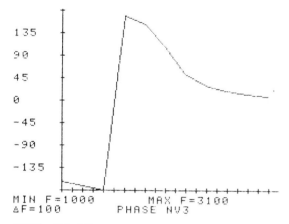

FIG. 26.52 *Continuation of the magnitude and phase plots of Fig. 26.51.*

Computer language The characters, symbols, or statements that provide the user with a channel of communication with the computer.

Data Information to be acted on by the program or output resulting from a series of computer operations.

FORTRAN A computer language designed primarily for mathematically, scientifically, and technically oriented programs.

Hardware The physical components of the computer and peripheral equipment that constitute the working elements of the system.

Large-scale computers Computers with a large memory capacity and a range of peripheral equipment that permits the analysis of a wide variety of problems of a significant nature.

LOAD command A computer statement that searches for a particular program in temporary storage on a magnetic tape, disc, and so on, and loads it into active memory.

LOOP routine A series of statements that specifies a repeated series of steps with a continually changing set of data points.

LSI Abbreviation for *Large-Scale Integration*.

Magnetic disc A flat disc on which programs, data, and so on, can be stored or retrieved using electromagnetic principles.

Magnetic tape A cassette of magnetic tape on which programs, data, and so on, can be stored and retrieved using electro-magnetic principles.

Memory The capacity of a computer to store programs, data, and internal processing controls.

Microcomputer A computer designed to satisfy a particular need or a broad range of problems of lesser difficulty with a limited expandability and memory capacity.

Microprocessor A semiconductor chip (LSI or VLSI) containing a sufficient number of electronic components to perform and control the operation of the computer.

Minicomputer A computer normally oriented toward a particular area of application with a potential for a significant level of expandability.

PASCAL A structured language requiring advanced study for proper application to the variety of complex problems that it was designed to investigate.

Peripherals Add-on components such as high-speed printers, plotters, card sorters, and so on, which can significantly expand the versatility of the central processor unit.

Personal computer A computer designed to meet the range of requirements (cost, memory, software support, and so on) associated with personal use.

PL/1 A computer language commonly referred to as a melting pot of ALGOL, FORTRAN, and COBOL.

PLOTTING routines A defined series of steps that permits the plotting of graphs, curves, and so on, on a visual display or printout.

Printer A peripheral or integral component of a computer that can provide a copy of the program, data, and results obtained.

Program A sequence of command statements, written in the computer language, that direct the operations of the computer.

Punch card A card with a pattern of holes that will define a command statement or data point.

Punched paper tape A long, narrow paper tape with a pattern of holes that will define a series of command statements or data points.

RAM An abbreviation for the *Random-Access-Memory* units which can be accessed and whose content can be changed.

ROM An abbreviation for *Read-Only Memory* units which can be read only by the computer to define procedure and which cannot be used to temporarily store data or programs.

Software All aspects of computer communications that appear in the printed form, such as the programs, data, and so on.

STORE command A computer statement that labels a program for future retrieval and stores it on a magnetic tape or similar memory system.

Terminal A computer input/output channel that can be in close proximity to or remote from the processing system.

Video display A peripheral or integral component of a computer that will provide a visual display of the program, data, and results obtained.

VLSI Abbreviation for *Very-Large-Scale Integration*.

APPENDIX A

CONVERSION FACTORS

To Convert from	To	Multiply by
Btus	Calorie grams	251.996
	Ergs	1.054×10^{10}
	Foot-pounds	777.649
	Hp-hours	0.000393
	Joules	1054.35
	Kilowatthours	0.000293
	Watt-seconds	1054.35
Centimeters	Angstrom units	1×10^8
	Feet	0.0328
	Inches	0.3937
	Meters	0.01
	Miles (statute)	6.214×10^{-6}
	Millimeters	10
Circular mils	Square centimeters	5.067×10^{-6}
	Square inches	7.854×10^{-7}
Cubic inches	Cubic centimeters	16.387
	Gallons (U.S. liquid)	0.00433
Cubic meters	Cubic feet	35.315
Days	Hours	24
	Minutes	1440
	Seconds	86,400
Dynes	Gallons (U.S. liquid)	264.172
	Newtons	0.00001
	Pounds	2.248×10^{-6}
Electronvolts	Ergs	1.60209×10^{-12}
Ergs	Dyne-centimeters	1.0
	Electronvolts	6.242×10^{11}
	Foot-pounds	7.376×10^{-8}
	Joules	1×10^{-7}
	Kilowatthours	2.777×10^{-14}
Feet	Centimeters	30.48
	Meters	0.3048
Foot-candles	Lumens/square foot	1.0
	Lumens/square meter	10.764
Foot-pounds	Dyne-centimeter	1.3558×10^7
	Ergs	1.3558×10^7
	Horsepower-hours	5.050×10^{-7}
	Joules	1.3558
	Newton-meters	1.3558

Gallon (U.S. liquid)	Cubic inches	231
	Liters	3.785
	Ounces	128
	Pints	8
Gausses	Maxwells/square centimeter	1.0
	Lines/square centimeter	1.0
	Lines/square inch	6.4516
Gilberts	Ampere-turns	0.7958
Grams	Dynes	980.665
	Ounces	0.0352
	Pounds	0.0022
Horsepower	Btus/hour	2547.16
	Ergs/second	7.46×10^9
	Foot-pounds/second	550.221
	Joules/second	746
	Watts	746
Hours	Seconds	3600
Inches	Angstrom units	2.54×10^8
	Centimeters	2.54
	Feet	12
	Meters	0.0254
Joules	Btus	0.000948
	Ergs	1×10^7
	Foot-pounds	0.7376
	Horsepower-hours	3.725×10^{-7}
	Kilowatthours	2.777×10^{-7}
	Watt-seconds	1.0
Kilograms	Dynes	980,665
	Ounces	35.2
	Pounds	2.2
Lines	Maxwells	1.0
Lines/square centimeter	Gausses	1.0
Lines/square inch	Gausses	0.1550
	Webers/square inch	1×10^{-8}
Liter	Cubic centimeters	1000.028
	Cubic inches	61.025
	Gallons (U.S. liquid)	0.2642
	Ounces (U.S. liquid)	33.815
	Quarts (U.S. liquid)	1.0567
Lumens	Candle power (spher.)	0.0796
Lumens/square centimeter	Lamberts	1.0
Lumens/square foot	Foot-candles	1.0
Maxwells	Lines	1.0
	Webers	1×10^{-8}

Meters	Angstrom units	1×10^{10}
	Centimeters	100
	Feet	3.2808
	Inches	39.370
	Miles (statute)	0.000621
Miles (statute)	Feet	5280
	Kilometers	1.609
	Meters	1609.344
Miles/hour	Kilometers/hour	1.609344
Newton-meters	Dyne-centimeters	1×10^{7}
	Kilogram meters	0.10197
Oerstads	Ampere-turns/inch	2.0212
	Ampere-turns/meter	79.577
	Gilberts/centimeter	1.0
Quarts (U.S. liquid)	Cubic centimeters	946.353
	Cubic inches	57.75
	Gallons (U.S. liquid)	0.25
	Liters	0.9463
	Pints (U.S. liquid)	2.
	Ounces (U.S. liquid)	32
Radians	Degrees	57.2958
Slugs	Kilograms	14.5939
	Pounds	32.1740
Watts	Btus/hour	3.4144
	Ergs/second	1×10^{7}
	Horsepower	0.00134
	Joules/second	1.0
Webers	Lines	1×10^{8}
	Maxwells	1×10^{8}
Years	Days	365
	Hours	8760
	Minutes	525,600
	Seconds	3.1536×10^{7}

APPENDIX B

THIRD-ORDER DETERMINANTS

Consider the three following simultaneous equations:

$$a_1 x + b_1 y + c_1 z = d_1$$
$$a_2 x + b_2 y + c_2 z = d_2$$
$$a_3 x + b_3 y + c_3 z = d_3$$

The determinant configuration for x, y, and z can be found in a manner similar to that for two simultaneous equations. That is, to solve for x, obtain the determinant in the numerator by replacing the first column by the elements on the right of the equal sign, and the denominator is the determinant of the coefficients of the variables (the same approach applies to y and z). Again, the denominator is the same for each variable. Therefore,

$$x = \frac{\begin{vmatrix} d_1 & b_1 & c_1 \\ d_2 & b_2 & c_2 \\ d_3 & b_3 & c_3 \end{vmatrix}}{D = \begin{vmatrix} a_1 & b_1 & c_1 \\ a_2 & b_2 & c_2 \\ a_3 & b_3 & c_3 \end{vmatrix}} \qquad y = \frac{\begin{vmatrix} a_1 & d_1 & c_1 \\ a_2 & d_2 & c_2 \\ a_3 & d_3 & c_3 \end{vmatrix}}{D = \begin{vmatrix} a_1 & b_1 & c_1 \\ a_2 & b_2 & c_2 \\ a_3 & b_3 & c_3 \end{vmatrix}} \qquad z = \frac{\begin{vmatrix} a_1 & b_1 & d_1 \\ a_2 & b_2 & d_2 \\ a_3 & b_3 & d_3 \end{vmatrix}}{D = \begin{vmatrix} a_1 & b_1 & c_1 \\ a_2 & b_2 & c_2 \\ a_3 & b_3 & c_3 \end{vmatrix}}$$

There is more than one expanded format for the third-order determinant. Each, however, will give the same result. One expansion of the determinant (D) is the following:

$$D = \begin{vmatrix} a_1 & b_1 & c_1 \\ a_2 & b_2 & c_2 \\ a_3 & b_3 & c_3 \end{vmatrix} = a_1 \left(+ \begin{vmatrix} b_2 & c_2 \\ b_3 & c_3 \end{vmatrix} \right) + b_1 \left(- \begin{vmatrix} a_2 & c_2 \\ a_3 & c_3 \end{vmatrix} \right) + c_1 \left(+ \begin{vmatrix} a_2 & b_2 \\ a_3 & b_3 \end{vmatrix} \right)$$

This expansion was obtained by multiplying the elements of the first row of D by their corresponding cofactors. It is not a requirement that the first row be used as the multiplying factors. In fact, any *row* or *column* (not diagonals) may be used to expand a third-order determinant.

The sign of each cofactor is dictated by the position of the multiplying factors (a_1, b_1, and c_1 in this case) as in the following standard format:

$$\begin{vmatrix} + & \rightarrow & - & + \\ \downarrow & & & \\ - & & + & - \\ + & & - & + \end{vmatrix}$$

Note that the proper sign for each element can be obtained by simply assigning the upper left element a positive sign and then changing sign as you move horizontally or vertically to the neighboring position.

For the determinant D, the elements would have the following signs:

$$\begin{vmatrix} a_1^{(+)} & b_1^{(-)} & c_1^{(+)} \\ a_2^{(-)} & b_2^{(+)} & c_2^{(-)} \\ a_3^{(+)} & b_3^{(-)} & c_3^{(+)} \end{vmatrix}$$

The minors associated with each multiplying factor are obtained by covering up the row and column in which the multiplying factor is located and writing a second-order determinant to include the remaining elements in the same relative positions that they have in the third-order determinant.

Consider the cofactors associated with a_1 and b_1 in the expansion of D. The sign is positive for a_1 and negative for b_1 as determined by the standard format. Following the procedure outlined above, we can find the minors of a_1 and b_1 as follows:

$$a_{1(\text{minor})} = \begin{vmatrix} a_1 & b_1 & c_1 \\ a_2 & b_2 & c_2 \\ a_3 & b_3 & c_3 \end{vmatrix} = \begin{vmatrix} b_2 & c_2 \\ b_3 & c_3 \end{vmatrix}$$

$$b_{1(\text{minor})} = \begin{vmatrix} a_1 & b_1 & c_1 \\ a_2 & b_2 & c_2 \\ a_3 & b_3 & c_3 \end{vmatrix} = \begin{vmatrix} a_2 & c_2 \\ a_3 & c_3 \end{vmatrix}$$

It was pointed out that any row or column may be used to expand the third-order determinant, and the same result will still be obtained. Using the first column of D, we obtain the expansion

$$D = \begin{vmatrix} a_1 & b_1 & c_1 \\ a_2 & b_2 & c_2 \\ a_3 & b_3 & c_3 \end{vmatrix} = a_1 \left(+ \begin{vmatrix} b_2 & c_2 \\ b_3 & c_3 \end{vmatrix} \right) + a_2 \left(- \begin{vmatrix} b_1 & c_1 \\ b_3 & c_3 \end{vmatrix} \right) + a_3 \left(+ \begin{vmatrix} b_1 & c_1 \\ b_2 & c_2 \end{vmatrix} \right)$$

The proper choice of row or column can often effectively reduce the amount of work required to expand the third-order determinant. For example, in the following determinants, the first column and third row, respectively, would reduce the number of cofactors in the expansion.

$$D = \begin{vmatrix} 2 & 3 & -2 \\ 0 & 4 & 5 \\ 0 & 6 & 7 \end{vmatrix} = 2 \left(+ \begin{vmatrix} 4 & 5 \\ 6 & 7 \end{vmatrix} \right) + 0 + 0 = 2(28 - 30)$$

$$= -4$$

$$D = \begin{vmatrix} 1 & 4 & 7 \\ 2 & 6 & 8 \\ 2 & 0 & 3 \end{vmatrix} = 2 \left(+ \begin{vmatrix} 4 & 7 \\ 6 & 8 \end{vmatrix} \right) + 0 + 3 \left(+ \begin{vmatrix} 1 & 4 \\ 2 & 6 \end{vmatrix} \right)$$

$$= 2(32 - 42) + 3(6 - 8) = 2(-10) + 3(-2)$$

$$= -26$$

EXAMPLES. Expand the following third-order determinants:

a. $D = \begin{vmatrix} 1 & 2 & 3 \\ 3 & 2 & 1 \\ 2 & 1 & 3 \end{vmatrix} = 1 \left(+ \begin{vmatrix} 2 & 1 \\ 1 & 3 \end{vmatrix} \right) + 3 \left(- \begin{vmatrix} 2 & 3 \\ 1 & 3 \end{vmatrix} \right) + 2 \left(+ \begin{vmatrix} 2 & 3 \\ 2 & 1 \end{vmatrix} \right)$

$$= 1[6 - 1] + 3[-(6 - 3)] + 2[2 - 6]$$

$$= 5 + 3(-3) + 2(-4)$$

$$= 5 - 9 - 8$$

$$= -12$$

b. $D = \begin{vmatrix} 0 & 4 & 6 \\ 2 & 0 & 5 \\ 8 & 4 & 0 \end{vmatrix} = 0 + 2 \left(- \begin{vmatrix} 4 & 6 \\ 4 & 0 \end{vmatrix} \right) + 8 \left(+ \begin{vmatrix} 4 & 6 \\ 0 & 5 \end{vmatrix} \right)$

$$= 0 + 2[-(0 - 24)] + 8[(20 - 0)]$$

$$= 0 + 2(24) + 8(20)$$

$$= 48 + 160$$

$$= 208$$

APPENDIX C

COLOR CODING OF MOLDED MICA CAPACITORS (PICOFARADS)

RETMA *and standard* MIL *specifications*

Color	Sig- nificant Figure	Decimal Multiplier	Tolerance $\pm\%$	Class	Temp. Coeff. PPM/°C Not More than	Cap. Drift Not More than
Black	0	1	20	A	± 1000	$\pm (5\% + 1\text{ pF})$
Brown	1	10	—	B	± 500	$\pm (3\% + 1\text{ pF})$
Red	2	100	2	C	± 200	$\pm (0.5\% + 0.5\text{ pF})$
Orange	3	1000	3	D	± 100	$\pm (0.3\% + 0.1\text{ pF})$
Yellow	4	10,000	—	E	$+100 - 20$	$\pm (0.1\% + 0.1\text{ pF})$
Green	5	—	5	—	—	—
Blue	6	—	—	—	—	—
Violet	7	—	—	—	—	—
Gray	8	—	—	I	$+150 - 50$	$\pm (0.03\% + 0.2\text{ pF})$
White	9	—	—	J	$+100 - 50$	$\pm (0.2\% + 0.2\text{ pF})$
Gold	—	0.1	—	—	—	—
Silver	—	0.01	10	—	—	—

Courtesy of Sprague Electric Co.
NOTE: If both rows of dots are not on one face, rotate capacitor about axis of its leads to read second row on side or rear.

Present RETMA 6-Dot Color Code   Present MIL Color Code

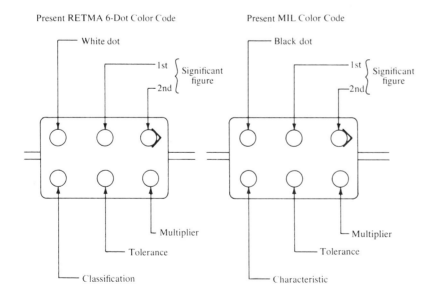

APPENDIX D

COLOR CODING OF MOLDED TUBULAR CAPACITORS (PICOFARADS)

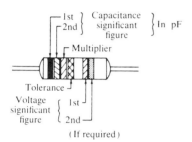

Color	Significant Figure	Decimal Multiplier	Tolerance ±%
Black	0	1	20
Brown	1	10	—
Red	2	100	—
Orange	3	1000	30
Yellow	4	10,000	40
Green	5	10^5	5
Blue	6	10^6	—
Violet	7	—	—
Gray	8	—	—
White	9	—	10

Courtesy of Sprague Electric Co.
NOTE: Voltage rating is identified by a single-digit number for ratings up to 900 volts; a two-digit number above 900 volts. Two zeros follow the voltage figure.

FIG. D.1

APPENDIX E

COMPLEX NUMBER CONVERSION

The Calculator

In recent years, the hand-held calculator has introduced a procedure for complex number conversions that has resulted in increased accuracy and a tremendous saving in time and energy. Depending on the type of calculator, conversions can be performed by using the equations appearing in Fig. E-1 or by taking advantage of an internally stored program that will perform the operation in a minimum number of steps. To be useful in complex number conversions, the calculator should be able to perform the following operations (in addition to addition, multiplication, subtraction, and division): $\sqrt{}$, $()^2$, $\tan\theta$, $\sin\theta$, $\cos\theta$, $\tan^{-1}$, $\sin^{-1}$, $\cos^{-1}$.

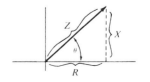

$$Z = \sqrt{R^2 + X^2}$$
$$\theta = \tan^{-1}\frac{X}{R}$$

FIG. E.1

Rectangular to polar form The required equations as determined by Fig. E-1 are

$$Z = \sqrt{R^2 + X^2} \tag{E.1}$$

$$\theta = \tan^{-1}\frac{X}{R} \tag{E.2}$$

For Eq. (E.1), the real and imaginary parts are first squared and then added together. Then the square root of the resulting number is found.

For Eq. (E.2), first the ratio X/R is determined and then the $\tan^{-1}$ key depressed. On many calculators, the $\tan^{-1}$ operation will require first pressing the function button.

Polar to rectangular form The required equations as determined by Fig. E-1 are

$$X = Z \sin \theta \tag{E.3}$$

$$R = Z \cos \theta \tag{E.4}$$

For the imaginary part, $\sin \theta$ is first determined and then multiplied by Z. For the real part, $\cos \theta$ is first determined and then multiplied by Z.

APPENDIX F

THE GREEK ALPHABET

Name	Capital	Lowercase	Used to Designate
Alpha	A	α	Area, angles, coefficients
Beta	B	β	Angles, coefficients, flux density
Gamma	Γ	γ	Specific gravity, conductivity
Delta	Δ	δ	Density, variation
Epsilon	E	ϵ	Base of natural logarithms
Zeta	Z	ζ	Coefficients, coordinates, impedance
Eta	H	η	Efficiency, hysteresis coefficient
Theta	Θ	θ	Phase angle, temperature
Iota	I	ι	
Kappa	K	κ	Dielectric constant, susceptibility
Lambda	Λ	λ	Wavelength
Mu	M	μ	Amplification factor, micro, permeability
Nu	N	ν	Reluctivity
Xi	Ξ	ξ	
Omicron	O	o	
Pi	Π	π	3.1416
Rho	P	ρ	Resistivity
Sigma	Σ	σ	Summation
Tau	T	τ	Time constant
Upsilon	Υ	υ	
Phi	Φ	ϕ	Angles, magnetic flux
Chi	X	χ	
Psi	Ψ	ψ	Dielectric flux, phase difference
Omega	Ω	ω	Ohms, angular velocity

APPENDIX G

MAGNETIC PARAMETER CONVERSIONS

	SI (MKS)	CGS	English
Φ	webers (Wb)	maxwells	lines
	1 Wb	$= 10^8$ maxwells	$= 10^8$ lines
B	Wb/m^2	gauss (maxwells/cm^2)	lines/in.^{2}
	1 Wb/m^2	$= 10^4$ gauss	$= 6.452 \times 10^4$ lines/in.^{2}
A	1 m^2	$= 10^4$ cm^2	$= 1550$ in.^{2}
μ_0	$4\pi \times 10^{-7}$ Wb/A-m	$= 1$ gauss/oersted	$= 3.20$ lines/A-m
$\mathscr{F}$	NI (ampere-turns, At)	$0.4\pi NI$ (gilberts)	NI (At)
	1 At	$= 1.257$ gilberts	1 gilbert $= 0.7958$ At
H	NI/l (At/m)	$0.4\pi\, NI/l$ (oersteds)	NI/l (At/in.)
	1 At/m	$= 1.26 \times 10^{-2}$ oersted	$= 2.54 \times 10^{-2}$ At/in.
H_g	$7.97 \times 10^5\, B_g$ (At/m)	B_g (oersteds)	$0.313\, B_g$ (At/in.)

APPENDIX H

EXPONENTIAL FUNCTIONS

x	e^x	e^{-x}	x	e^x	e^{-x}	x	e^x	e^{-x}	x	e^x	e^{-x}
0.00	1.0000	1.000000	0.80	2.2255	0.449329	1.60	4.9530	0.201897	2.40	11.023	0.090718
0.01	1.0101	0.990050	0.81	2.2479	.444858	1.61	5.0028	.199888	2.41	11.134	.089815
0.02	1.0202	.980199	0.82	2.2705	.440432	1.62	5.0531	.197899	2.42	11.246	.088922
0.03	1.0305	.970446	0.83	2.2933	.436049	1.63	5.1039	.195930	2.43	11.359	.088037
0.04	1.0408	.960789	0.84	2.3164	.431711	1.64	5.1552	.193980	2.44	11.473	.087161
0.05	1.0513	0.951229	0.85	2.3396	0.427415	1.65	5.2070	0.192050	2.45	11.588	0.086294
0.06	1.0618	.941765	0.86	2.3632	.423162	1.66	5.2593	.190139	2.46	11.705	.085435
0.07	1.0725	.932394	0.87	2.3869	.418952	1.67	5.3122	.188247	2.47	11.822	.084585
0.08	1.0833	.923116	0.88	2.4109	.414783	1.68	5.3656	.186374	2.48	11.941	.083743
0.09	1.0942	.913931	0.89	2.4351	.410656	1.69	5.4195	.184520	2.49	12.061	.082910
0.10	1.1052	0.904837	0.90	2.4596	0.406570	1.70	5.4739	0.182684	2.50	12.182	0.082085
0.11	1.1163	.895834	0.91	2.4843	.402524	1.71	5.5290	.180866	2.51	12.305	.081268
0.12	1.1275	.886920	0.92	2.5093	.398519	1.72	5.5845	.179066	2.52	12.429	.080460
0.13	1.1388	.878095	0.93	2.5345	.394554	1.73	5.6407	.177284	2.53	12.554	.079659
0.14	1.1503	.869358	0.94	2.5600	.390628	1.74	5.6973	.175520	2.54	12.680	.078866
0.15	1.1618	0.860708	0.95	2.5857	0.386741	1.75	5.7546	0.173774	2.55	12.807	0.078082
0.16	1.1735	.852144	0.96	2.6117	.382893	1.76	5.8124	.172045	2.56	12.936	.077305
0.17	1.1853	.843665	0.97	2.6379	.379083	1.77	5.8709	.170333	2.57	13.066	.076536
0.18	1.1972	.835270	0.98	2.6645	.375311	1.78	5.9299	.168638	2.58	13.197	.075774
0.19	1.2092	.826959	0.99	2.6912	.371577	1.79	5.9895	.166960	2.59	13.330	.075020
0.20	1.2214	0.818731	1.00	2.7183	0.367879	1.80	6.0496	0.165299	2.60	13.464	0.074274
0.21	1.2337	.810584	1.01	2.7456	.364219	1.81	6.1104	.163654	2.61	13.599	.073535
0.22	1.2461	.802519	1.02	2.7732	.360595	1.82	6.1719	.162026	2.62	13.736	.072803
0.23	1.2586	.794534	1.03	2.8011	.357007	1.83	6.2339	.160414	2.63	13.874	.072078
0.24	1.2712	.786628	1.04	2.8292	.353455	1.84	6.2965	.158817	2.64	14.013	.071361

x	e^x	e^{-x}	x	e^x	e^{-x}	x	e^x	e^{-x}	x	e^x	e^{-x}
0.25	1.2840	0.778801	1.05	2.8577	0.349938	1.85	6.3598	0.157237	2.65	14.154	0.070651
0.26	1.2969	.771052	1.06	2.8864	.346456	1.86	6.4237	.155673	2.66	14.296	.069948
0.27	1.3100	.763379	1.07	2.9154	.343009	1.87	6.4883	.154124	2.67	14.440	.069252
0.28	1.3231	.755784	1.08	2.9447	.339596	1.88	6.5535	.152590	2.68	14.585	.068563
0.29	1.3364	.748264	1.09	2.9743	.336216	1.89	6.6194	.151072	2.69	14.732	.067881
0.30	1.3499	0.740818	1.10	3.0042	0.332871	1.90	6.6859	0.149569	2.70	14.880	0.067206
0.31	1.3634	.733447	1.11	3.0344	.329559	1.91	6.7531	.148080	2.71	15.029	.066537
0.32	1.3771	.726149	1.12	3.0649	.326280	1.92	6.8210	.146607	2.72	15.180	.065875
0.33	1.3910	.718924	1.13	3.0957	.323033	1.93	6.8895	.145148	2.73	15.333	.065219
0.34	1.4049	.711770	1.14	3.1268	.319819	1.94	6.9588	.143704	2.74	15.487	.064570
0.35	1.4191	0.704688	1.15	3.1582	0.316637	1.95	7.0287	0.142274	2.75	15.643	0.063928
0.36	1.4333	.697676	1.16	3.1899	.313486	1.96	7.0993	.140858	2.76	15.800	.063292
0.37	1.4477	.690734	1.17	3.2220	.310367	1.97	7.1707	.139457	2.77	15.959	.062662
0.38	1.4623	.683861	1.18	3.2544	.307279	1.98	7.2427	.138069	2.78	16.119	.062039
0.39	1.4770	.677057	1.19	3.2871	.304221	1.99	7.3155	.136695	2.79	16.281	.061421
0.40	1.4918	0.670320	1.20	3.3201	0.301194	2.00	7.3891	0.135335	2.80	16.445	0.060810
0.41	1.5068	.663650	1.21	3.3535	.298197	2.01	7.4633	.133989	2.81	16.610	.060205
0.42	1.5220	.657047	1.22	3.3872	.295230	2.02	7.5383	.132655	2.82	16.777	.059606
0.43	1.5373	.650509	1.23	3.4212	.292293	2.03	7.6141	.131336	2.83	16.945	.059013
0.44	1.5527	.644036	1.24	3.4556	.289384	2.04	7.6906	.130029	2.84	17.116	.058426
0.45	1.5683	0.637628	1.25	3.4903	0.286505	2.05	7.7679	0.128735	2.85	17.288	0.057844
0.46	1.5841	.631284	1.26	3.5254	.283654	2.06	7.8460	.127454	2.86	17.462	.057269
0.47	1.6000	.625002	1.27	3.5609	.280832	2.07	7.9248	.126186	2.87	17.637	.056699
0.48	1.6161	.618783	1.28	3.5966	.278037	2.08	8.0045	.124930	2.88	17.814	.056135
0.49	1.6323	.612626	1.29	3.6328	.275271	2.09	8.0849	.123687	2.89	17.993	.055576
0.50	1.6487	0.606531	1.30	3.6693	0.272532	2.10	8.1662	0.122456	2.90	18.174	0.055023
0.51	1.6653	.600496	1.31	3.7062	.269820	2.11	8.2482	.121238	2.91	18.357	.054476
0.52	1.6820	.594521	1.32	3.7434	.267135	2.12	8.3311	.120032	2.92	18.541	.053934
0.53	1.6989	.588605	1.33	3.7810	.264477	2.13	8.4149	.118837	2.93	18.728	.053397
0.54	1.7160	.582748	1.34	3.8190	.261846	2.14	8.4994	.117655	2.94	18.916	.052866
0.55	1.7333	0.576950	1.35	3.8574	0.259240	2.15	8.5849	0.116484	2.95	19.106	0.052340
0.56	1.7507	.571209	1.36	3.8962	.256661	2.16	8.6711	.115325	2.96	19.298	.051819
0.57	1.7683	.565525	1.37	3.9354	.254107	2.17	8.7583	.114178	2.97	19.492	.051303
0.58	1.7860	.559898	1.38	3.9749	.251579	2.18	8.8463	.113042	2.98	19.688	.050793
0.59	1.8040	.554327	1.39	4.0149	.249075	2.19	8.9352	.111917	2.99	19.886	.050287
0.60	1.8221	0.548812	1.40	4.0552	0.246597	2.20	9.0250	0.110803	3.00	20.086	0.049787
0.61	1.8404	.543351	1.41	4.0960	.244143	2.21	9.1157	.109701	3.01	20.287	.049292
0.62	1.8589	.537944	1.42	4.1371	.241714	2.22	9.2073	.108609	3.02	20.491	.048801
0.63	1.8776	.532592	1.43	4.1787	.239309	2.23	9.2999	.107528	3.03	20.697	.048316
0.64	1.8965	.527292	1.44	4.2207	.236928	2.24	9.3933	.106459	3.04	20.905	.047835
0.65	1.9155	0.522046	1.45	4.2631	0.234570	2.25	9.4877	0.105399	3.05	21.115	0.047359
0.66	1.9348	.516851	1.46	4.3060	.232236	2.26	9.5831	.104350	3.06	21.328	.046888
0.67	1.9542	.511709	1.47	4.3492	.229925	2.27	9.6794	.103312	3.07	21.542	.046421
0.68	1.9739	.506617	1.48	4.3929	.227638	2.28	9.7767	.102284	3.08	21.758	.045959
0.69	1.9937	.501576	1.49	4.4371	.225373	2.29	9.8749	.101266	3.09	21.977	.045502
0.70	2.0138	0.496585	1.50	4.4817	0.223130	2.30	9.9742	0.100259	3.10	22.198	0.045049
0.71	2.0340	.491644	1.51	4.5267	.220910	2.31	10.074	.099261	3.11	22.421	.044601
0.72	2.0544	.486752	1.52	4.5722	.218712	2.32	10.176	.098274	3.12	22.646	.044157
0.73	2.0751	.481909	1.53	4.6182	.216536	2.33	10.278	.097296	3.13	22.874	.043718
0.74	2.0959	.477114	1.54	4.6646	.214381	2.34	10.381	.096328	3.14	23.104	.043283
0.75	2.1170	0.472367	1.55	4.7115	0.212248	2.35	10.486	0.095369	3.15	23.336	0.042852
0.76	2.1383	.467666	1.56	4.7588	.210136	2.36	10.591	.094420	3.16	23.571	.042426
0.77	2.1598	.463013	1.57	4.8066	.208045	2.37	10.697	.093481	3.17	23.807	.042004
0.78	2.1815	.458406	1.58	4.8550	.205975	2.38	10.805	.092551	3.18	24.047	.041586
0.79	2.2034	.453845	1.59	4.9037	.203926	2.39	10.913	.091630	3.19	24.288	.041172
3.20	24.533	0.040762	4.00	54.598	0.018316	4.80	121.51	0.008230	6.00	403.43	0.0024788
3.21	24.779	.040357	4.01	55.147	.018133	4.81	122.73	.008148	6.05	424.11	.0023579
3.22	25.028	.039955	4.02	55.701	.017953	4.82	123.97	.008067	6.10	445.86	.0022429
3.23	25.280	.039557	4.03	56.261	.017774	4.83	125.21	.007987	6.15	468.72	.0021335
3.24	25.534	.039164	4.04	56.826	.017597	4.84	125.47	.007907	6.20	492.75	.0020294
3.25	25.790	0.038774	4.05	57.397	0.017422	4.85	127.74	0.007828	6.25	518.01	0.0019305
3.26	26.050	.038388	4.06	57.974	.017249	4.86	129.02	.007750	6.30	544.57	.0018363
3.27	26.311	.038006	4.07	58.557	.017077	4.87	130.32	.007673	6.35	572.49	.0017467
3.28	26.576	.037628	4.08	59.145	.016907	4.88	131.63	.007597	6.40	601.85	.0016616
3.29	26.843	.037254	4.09	59.740	.016739	4.89	132.95	.007521	6.45	632.70	.0015805
3.30	27.113	0.036883	4.10	60.340	0.016573	4.90	134.29	0.007447	6.50	665.14	0.0015034
3.31	27.385	.036516	4.11	60.947	.016408	4.91	135.64	.007372	6.55	699.24	.0014301
3.32	27.660	.036153	4.12	61.559	.016245	4.92	137.00	.007299	6.60	735.10	.0013604
3.33	27.938	.035793	4.13	62.178	.016083	4.93	138.38	.007227	6.65	772.78	.0012940
3.34	28.219	.035437	4.14	62.803	.015923	4.94	139.77	.007155	6.70	812.41	.0012309
3.35	28.503	0.035084	4.15	63.434	0.015764	4.95	141.17	0.007083	6.75	854.06	0.0011709
3.36	28.789	.034735	4.16	64.072	.015608	4.96	142.59	.007013	6.80	897.85	.0011138
3.37	29.079	.034390	4.17	64.715	.015452	4.97	144.03	.006943	6.85	943.88	.0010595
3.38	29.371	.034047	4.18	65.366	.015299	4.98	145.47	.006874	6.90	992.27	.0010078
3.39	29.666	.033709	4.19	66.023	.015146	4.99	146.94	.006806	6.95	1043.1	.0009586
3.40	29.964	0.033373	4.20	66.686	0.014996	5.00	148.41	0.006738	7.00	1096.6	0.0009119
3.41	30.265	.033041	4.21	67.357	.014846	5.01	149.90	.006671	7.05	1152.9	.0008674
3.42	30.569	.032712	4.22	68.033	.014699	5.02	151.41	.006605	7.10	1212.0	.0008251
3.43	30.877	.032387	4.23	68.717	.014552	5.03	152.93	.006539	7.15	1274.1	.0007849
3.44	31.187	.032065	4.24	69.408	.014408	5.04	154.47	.006474	7.20	1339.4	.0007466

x	e^x	e^{-x}
3.45	31.500	0.031746
3.46	31.817	.031430
3.47	32.137	.031117
3.48	32.460	.030807
3.49	32.786	.030501
3.50	33.115	0.030197
3.51	33.448	.029897
3.52	33.784	.029599
3.53	34.124	.029305
3.54	34.467	.029013
3.55	34.813	0.028725
3.56	35.163	.028439
3.57	35.517	.028156
3.58	35.874	.027876
3.59	36.234	.027598
3.60	36.598	0.027324
3.61	36.966	.027052
3.62	37.338	.026783
3.63	37.713	.026516
3.64	38.092	.026252
3.65	38.475	0.025991
3.66	38.861	.025733
3.67	39.252	.025476
3.68	39.646	.025223
3.69	40.045	.024972
3.70	40.447	0.024724
3.71	40.854	.024478
3.72	41.264	.024234
3.73	41.679	.023993
3.74	42.098	.023754
3.75	42.521	0.023518
3.76	42.948	.023284
3.77	43.380	.023052
3.78	43.816	.022823
3.79	44.256	.022596
3.80	44.701	0.022371
3.81	45.150	.022148
3.82	45.604	.021928
3.83	46.063	.021710
3.84	46.525	.021494
3.85	46.993	0.021280
3.86	47.465	.021068
3.87	47.942	.020858
3.88	48.424	.020651
3.89	48.911	.020445
3.90	49.402	0.020242
3.91	49.899	.020041
3.92	50.400	.019841
3.93	50.907	.019644
3.94	51.419	.019448
3.95	51.935	1.019255
3.96	52.457	.019063
3.97	52.985	.018873
3.98	53.617	.018686
3.99	54.055	.018500

x	e^x	e^{-x}
4.25	70.105	0.014264
4.26	70.810	.014122
4.27	71.522	.013982
4.28	72.240	.013843
4.29	72.966	.013705
4.30	73.700	0.013569
4.31	74.440	.013434
4.32	75.189	.013300
4.33	75.944	.013108
4.34	76.708	.013037
4.35	77.478	0.012907
4.36	78.257	.012778
4.37	79.044	.012651
4.38	79.838	.012525
4.39	80.640	.012401
4.40	81.451	0.012277
4.41	82.269	.012155
4.42	83.096	.012034
4.43	83.931	.011914
4.44	84.775	.011796
4.45	85.627	0.011679
4.46	86.488	.011562
4.47	87.357	.011447
4.48	88.235	.011333
4.49	89.121	.011221
4.50	90.017	0.011109
4.51	90.922	.010998
4.52	91.836	.010889
4.53	92.759	.010781
4.54	93.691	.010673
4.55	94.632	0.010567
4.56	95.583	.010462
4.57	96.544	.010358
4.58	97.514	.010255
4.59	98.494	.010153
4.60	99.484	0.010052
4.61	100.48	.009952
4.62	101.49	.009853
4.63	102.51	.009755
4.64	103.54	.009658
4.65	104.58	0.009562
4.66	105.64	.009466
4.67	106.70	.009372
4.68	107.77	.009279
4.69	108.85	.009187
4.70	109.95	0.009095
4.71	111.05	.009005
4.72	112.17	.008915
4.73	113.30	.008826
4.74	114.43	.008739
4.75	115.58	0.008652
4.76	116.75	.008566
4.77	117.92	.008480
4.78	119.10	.008396
4.79	120.30	.008312

x	e^x	e^{-x}
5.05	156.02	0.006409
5.06	157.59	.006346
5.07	159.17	.006282
5.08	160.77	.006220
5.09	162.39	.006158
5.10	164.02	0.006097
5.11	165.67	.006036
5.12	167.34	.005976
5.13	169.02	.005917
5.14	170.72	.005858
5.15	172.43	0.005799
5.16	174.16	.005742
5.17	175.91	.005685
5.18	177.68	.005628
5.19	179.47	.005572
5.20	181.27	0.005517
5.21	183.09	.005462
5.22	184.93	.005407
5.23	186.79	.005354
5.24	188.67	.005300
5.25	190.57	0.005248
5.26	192.48	.005195
5.27	194.42	.005144
5.28	196.37	.005092
5.29	198.34	.005042
5.30	200.34	0.004992
5.31	202.35	.004942
5.32	204.38	.004893
5.33	206.44	.004844
5.34	208.51	.004796
5.35	210.61	0.004748
5.36	212.72	.004701
5.37	214.86	.004654
5.38	217.02	.004608
5.39	219.20	.004562
5.40	221.41	0.004517
5.41	223.63	.004472
5.42	225.88	.004427
5.43	228.15	.004383
5.44	230.44	.004339
5.45	232.76	0.004296
5.46	235.10	.004254
5.47	237.46	.004211
5.48	239.85	.004169
5.49	242.26	.004128
5.50	244.69	0.0040868
5.55	257.24	.0038875
5.60	270.43	.0036979
5.65	284.29	.0035175
5.70	298.87	.0033460
5.75	314.19	0.0031828
5.80	330.30	.0030276
5.85	347.23	.0028799
5.90	365.04	.0027394
5.95	383.75	.0026058

x	e^x	e^{-x}
7.25	1408.1	0.0007102
7.30	1480.3	.0006755
7.35	1556.2	.0006426
7.40	1636.0	.0006113
7.45	1719.9	.0005814
7.50	1808.0	0.0005531
7.55	1900.7	.0005261
7.60	1998.2	.0005005
7.65	2100.6	.0004760
7.70	2208.3	.0004528
7.75	2321.6	0.0004307
7.80	2440.6	.0004097
7.85	2565.7	.0003898
7.90	2697.3	.0003707
7.95	2835.6	.0003527
8.00	2981.0	0.0003355
8.05	3133.8	.0003191
8.10	3294.5	.0003035
8.15	3463.4	.0002887
8.20	3641.0	.0002747
8.25	3827.6	0.0002613
8.30	4023.9	.0002485
8.35	4230.2	.0002364
8.40	4447.1	.0002249
8.45	4675.1	.0002139
8.50	4914.8	0.0002035
8.55	5166.8	.0001935
8.60	5431.7	.0001841
8.65	5710.1	.0001751
8.70	6002.9	.0001666
8.75	6310.7	0.0001585
8.80	6634.2	.0001507
8.85	6974.4	.0001434
8.90	7332.0	.0001364
8.95	7707.9	.0001297
9.00	8103.1	0.0001234
9.05	8518.5	.0001174
9.10	8955.3	.0001117
9.15	9414.4	.0001062
9.20	9897.1	.0001010
9.25	10405	0.0000961
9.30	10938	.0000914
9.35	11499	.0000870
9.40	12088	.0000827
9.45	12708	.0000787
9.50	13360	0.0000749
9.55	14045	.0000712
9.60	14765	.0000677
9.65	15522	.0000644
9.70	16318	.0000613
9.75	17154	0.0000583
9.80	18034	.0000555
9.85	18958	.0000527
9.90	19930	.0000502
9.95	20952	0.0000477
10.00	22026	0.0000454

APPENDIX I

NATURAL TRIGONOMETRIC FUNCTIONS FOR ANGLES IN DEGREES AND DECIMALS

Deg.	Sin	Tan	Cot	Cos	Deg.	Deg.	Sin	Tan	Cot	Cos	Deg.
0.0	.00000	.00000	∞	1.0000	90.0	4.0	.06976	.06993	14.301	0.9976	86.0
.1	.00175	.00175	573.0	1.0000	89.9	.1	.07150	.07168	13.951	.9974	85.9
.2	.00349	.00349	286.5	1.0000	.8	.2	.07324	.07344	13.617	.9973	.8
.3	.00524	.00524	191.0	1.0000	.7	.3	.07498	.07519	13.300	.9972	.7
.4	.00698	.00698	143.24	1.0000	.6	.4	.07672	.07695	12.996	.9971	.6
.5	.00873	.00873	114.59	1.0000	.5	.5	.07846	.07870	12.706	.9969	.5
.6	.01047	.01047	95.49	0.9999	.4	.6	.08020	.08046	12.429	.9968	.4
.7	.01222	.01222	81.85	.9999	.3	.7	.08194	.08221	12.163	.9966	.3
.8	.01396	.01396	71.62	.9999	.2	.8	.08368	.08397	11.909	.9965	.2
.9	.01571	.01571	63.66	.9999	89.1	.9	.08542	.08573	11.664	.9963	85.1
1.0	.01745	.01746	57.29	0.9998	89.0	5.0	.08716	.08749	11.430	0.9962	85.0
.1	.01920	.01920	52.08	.9998	88.9	.1	.08889	.08925	11.205	.9960	84.9
.2	.02094	.02095	47.74	.9998	.8	.2	.09063	.09101	10.988	.9959	.8
.3	.02269	.02269	44.07	.9997	.7	.3	.09237	.09277	10.780	.9957	.7
.4	.02443	.02444	40.92	.9997	.6	.4	.09411	.09453	10.579	.9956	.6
.5	.02618	.02619	38.19	.9997	.5	.5	.09585	.09629	10.385	.9954	.5
.6	.02792	.02793	35.80	.9996	.4	.6	.09758	.09805	10.199	.9952	.4
.7	.02967	.02968	33.69	.9996	.3	.7	.09932	.09981	10.019	.9951	.3
.8	.03141	.03143	31.82	.9995	.2	.8	.10106	.10158	9.845	.9949	.2
.9	.03316	.03317	30.14	.9995	88.1	.9	.10279	.10334	9.677	.9947	84.1
2.0	.03490	.03492	28.64	0.9994	88.0	6.0	.10453	.10510	9.514	0.9945	84.0
.1	.03664	.03667	27.27	.9993	87.9	.1	.10626	.10687	9.357	.9943	83.9
.2	.03839	.03842	26.03	.9993	.8	.2	.10800	.10863	9.205	.9942	.8
.3	.04013	.04016	24.90	.9992	.7	.3	.10973	.11040	9.058	.9940	.7
.4	.04188	.04191	23.86	.9991	.6	.4	.11147	.11217	8.915	.9938	.6
.5	.04362	.04366	22.90	.9990	.5	.5	.11320	.11394	8.777	.9936	.5
.6	.04536	.04541	22.02	.9990	.4	.6	.11494	.11570	8.643	.9934	.4
.7	.04711	.04716	21.20	.9989	.3	.7	.11667	.11747	8.513	.9932	.3
.8	.04885	.04891	20.45	.9988	.2	.8	.11840	.11924	8.386	.9930	.2
.9	.05059	.05066	19.74	.9987	87.1	.9	.12014	.12101	8.264	.9928	83.1
3.0	.05234	.05241	19.081	0.9986	87.0	7.0	.12187	.12278	8.144	0.9925	83.0
.1	.05408	.05416	18.464	.9985	86.9	.1	.12360	.12456	8.028	.9923	82.9
.2	.05582	.05591	17.886	.9984	.8	.2	.12533	.12633	7.916	.9921	.8
.3	.05756	.05766	17.343	.9983	.7	.3	.12706	.12810	7.806	.9919	.7
.4	.05931	.05941	16.832	.9982	.6	.4	.12880	.12988	7.700	.9917	.6
.5	.06105	.06116	16.350	.9981	.5	.5	.13053	.13165	7.596	.9914	.5
.6	.06279	.06291	15.895	.9980	.4	.6	.13226	.13343	7.495	.9912	.4
.7	.06453	.06467	15.464	.9979	.3	.7	.13399	.13521	7.396	.9910	.3
.8	.06627	.06642	15.056	.9978	.2	.8	.13572	.13698	7.300	.9907	.2
.9	.06802	.06817	14.669	.9977	86.1	.9	.13744	.13876	7.207	.9905	82.1
Deg.	Cos	Cot	Tan	Sin	Deg.	Deg.	Cos	Cot	Tan	Sin	Deg.

Deg.	Sin	Tan	Cot	Cos	Deg.	Deg.	Sin	Tan	Cot	Cos	Deg.
8.0	.13917	.14054	7.115	0.9903	82.0	13.0	.2250	.2309	4.331	0.9744	77.0
.1	.14090	.14232	7.026	.9900	81.9	.1	.2267	.2327	4.297	.9740	76.9
.2	.14263	.14410	6.940	.9898	.8	.2	.2284	.2345	4.264	.9736	.8
.3	.14436	.14588	6.855	.9895	.7	.3	.2300	.2364	4.230	.9732	.7
.4	.14608	.14767	6.772	.9893	.6	.4	.2317	.2382	4.198	.9728	.6
.5	.14781	.14945	6.691	.9890	.5	.5	.2334	.2401	4.165	.9724	.5
.6	.14954	.15124	6.612	.9888	.4	.6	.2351	.2419	4.134	.9720	.4
.7	.15126	.15302	6.535	.9885	.3	.7	.2368	.2438	4.102	.9715	.3
.8	.15299	.15481	6.460	.9882	.2	.8	.2385	.2456	4.071	.9711	.2
.9	.15471	.15660	6.386	.9880	81.1	.9	.2402	.2475	4.041	.9707	76.1
9.0	.15643	.15838	6.314	0.9877	81.0	14.0	.2419	.2493	4.011	0.9703	76.0
.1	.15816	.16017	6.243	.9874	80.9	.1	.2436	.2512	3.981	.9699	75.9
.2	.15988	.16196	6.174	.9871	.8	.2	.2453	.2530	3.952	.9694	.8
.3	.16160	.16376	6.107	.9869	.7	.3	.2470	.2549	3.923	.9690	.7
.4	.16333	.16555	6.041	.9866	.6	.4	.2487	.2568	3.895	.9686	.6
.5	.16505	.16734	5.976	.9863	.5	.5	.2504	.2586	3.867	.9681	.5
.6	.16677	.16914	5.912	.9860	.4	.6	.2521	.2605	3.839	.9677	.4
.7	.16849	.17093	5.850	.9857	.3	.7	.2538	.2623	3.812	.9673	.3
.8	.17021	.17273	5.789	.9854	.2	.8	.2554	.2642	3.785	.9668	.2
.9	.17193	.17453	5.730	.9851	80.1	.9	.2571	.2661	3.758	.9664	75.1
10.0	.1736	.1763	5.671	0.9848	80.0	15.0	0.2588	0.2679	3.732	0.9659	75.0
.1	.1754	.1781	5.614	.9845	79.9	.1	.2605	.2698	3.706	.9655	74.9
.2	.1771	.1799	5.558	.9842	.8	.2	.2622	.2717	3.681	.9650	.8
.3	.1788	.1817	5.503	.9839	.7	.3	.2639	.2736	3.655	.9646	.7
.4	.1805	.1835	5.449	.9836	.6	.4	.2656	.2754	3.630	.9641	.6
.5	.1822	.1853	5.396	.9833	.5	.5	.2672	.2773	3.606	.9636	.5
.6	.1840	.1871	5.343	.9829	.4	.6	.2689	.2792	3.582	.9632	.4
.7	.1857	.1890	5.292	.9826	.3	.7	.2706	.2811	3.558	.9627	.3
.8	.1874	.1908	5.242	.9823	.2	.8	.2723	.2830	3.534	.9622	.2
.9	.1891	.1926	5.193	.9820	79.1	.9	.2740	.2849	3.511	.9617	74.1
11.0	.1908	.1944	5.145	0.9816	79.0	16.0	0.2756	0.2867	3.487	0.9613	74.0
.1	.1925	.1962	5.097	.9813	78.9	.1	.2773	.2886	3.465	.9608	73.9
.2	.1942	.1980	5.050	.9810	.8	.2	.2790	.2905	3.442	.9603	.8
.3	.1959	.1998	5.005	.9806	.7	.3	.2807	.2924	3.420	.9598	.7
.4	.1977	.2016	4.959	.9803	.6	.4	.2823	.2943	3.398	.9593	.6
.5	.1994	.2035	4.915	.9799	.5	.5	.2840	.2962	3.376	.9588	.5
.6	.2011	.2053	4.872	.9796	.4	.6	.2857	.2981	3.354	.9583	.4
.7	.2028	.2071	4.829	.9792	.3	.7	.2874	.3000	3.333	.9578	.3
.8	.2045	.2089	4.787	.9789	.2	.8	.2890	.3019	3.312	.9573	.2
.9	.2062	.2107	4.745	.9785	78.1	.9	.2907	.3038	3.291	.9568	73.1
12.0	.2079	.2126	4.705	0.9781	78.0	17.0	0.2924	0.3057	3.271	0.9563	73.0
.1	.2096	.2144	4.665	.9778	77.9	.1	.2940	.3076	3.251	.9558	72.9
.2	.2113	.2162	4.625	.9774	.8	.2	.2957	.3096	3.230	.9553	.8
.3	.2130	.2180	4.586	.9770	.7	.3	.2974	.3115	3.211	.9548	.7
.4	.2147	.2199	4.548	.9767	.6	.4	.2990	.3134	3.191	.9542	.6
.5	.2164	.2217	4.511	.9763	.5	.5	.3007	.3153	3.172	.9537	.5
.6	.2181	.2235	4.474	.9759	.4	.6	.3024	.3172	3.152	.9532	.4
.7	.2198	.2254	4.437	.9755	.3	.7	.3040	.3191	3.133	.9527	.3
.8	.2215	.2272	4.402	.9751	.2	.8	.3057	.3211	3.115	.9521	.2
.9	.2233	.2290	4.366	.9748	77.1	.9	.3074	.3230	3.096	.9516	72.1
Deg.	Cos	Cot	Tan	Sin	Deg.	Deg.	Cos	Cot	Tan	Sin	Deg.

Deg.	Sin	Tan	Cot	Cos	Deg.	Deg.	Sin	Tan	Cot	Cos	Deg.
18.0	0.3090	0.3249	3.078	0.9511	72.0	23.0	0.3907	0.4245	2.356	0.9205	67.0
.1	.3107	.3269	3.060	.9505	71.9	.1	.3923	.4265	2.344	.9198	66.9
.2	.3123	.3288	3.042	.9500	.8	.2	.3939	.4286	2.333	.9191	.8
.3	.3140	.3307	3.024	.9494	.7	.3	.3955	.4307	2.322	.9184	.7
.4	.3156	.3327	3.006	.9489	.6	.4	.3971	.4327	2.311	.9178	.6
.5	.3173	.3346	2.989	.9483	.5	.5	.3987	.4348	2.300	.9171	.5
.6	.3190	.3365	2.971	.9478	.4	.6	.4003	.4369	2.289	.9164	.4
.7	.3206	.3385	2.954	.9472	.3	.7	.4019	.4390	2.278	.9157	.3
.8	.3223	.3404	2.937	.9466	.2	.8	.4035	.4411	2.267	.9150	.2
.9	.3239	.3424	2.921	.9461	71.1	.9	.4051	.4431	2.257	.9143	66.1
19.0	0.3256	0.3443	2.904	0.9455	71.0	24.0	0.4067	0.4452	2.246	0.9135	66.0
.1	.3272	.3463	2.888	.9449	70.9	.1	.4083	.4473	2.236	.9128	65.9
.2	.3289	.3482	2.872	.9444	.8	.2	.4099	.4494	2.225	.9121	.8
.3	.3305	.3502	2.856	.9438	.7	.3	.4115	.4515	2.215	.9114	.7
.4	.3322	.3522	2.840	.9432	.6	.4	.4131	.4536	2.204	.9107	.6
.5	.3338	.3541	2.824	.9426	.5	.5	.4147	.4557	2.194	.9100	.5
.6	.3355	.3561	2.808	.9421	.4	.6	.4163	.4578	2.184	.9092	.4
.7	.3371	.3581	2.793	.9415	.3	.7	.4179	.4599	2.174	.9085	.3
.8	.3387	.3600	2.778	.9409	.2	.8	.4195	.4621	2.164	.9078	.2
.9	.3404	.3620	2.762	.9403	70.1	.9	.4210	.4642	2.154	.9070	65.1
20.0	0.3420	0.3640	2.747	0.9397	70.0	25.0	0.4226	0.4663	2.145	0.9063	65.0
.1	.3437	.3659	2.733	.9391	69.9	.1	.4242	.4684	2.135	.9056	64.9
.2	.3453	.3679	2.718	.9385	.8	.2	.4258	.4706	2.125	.9048	.8
.3	.3469	.3699	2.703	.9379	.7	.3	.4274	.4727	2.116	.9041	.7
.4	.3486	.3719	2.689	.9373	.6	.4	.4289	.4748	2.106	.9033	.6
.5	.3502	.3739	2.675	.9367	.5	.5	.4305	.4770	2.097	.9026	.5
.6	.3518	.3759	2.660	.9361	.4	.6	.4321	.4791	2.087	.9018	.4
.7	.3535	.3779	2.646	.9354	.3	.7	.4337	.4813	2.078	.9011	.3
.8	.3551	.3799	2.633	.9348	.2	.8	.4352	.4834	2.069	.9003	.2
.9	.3567	.3819	2.619	.9342	69.1	.9	.4368	.4856	2.059	.8996	64.1
21.0	0.3584	0.3839	2.605	0.9336	69.0	26.0	0.4384	0.4877	2.050	0.8988	64.0
.1	.3600	.3859	2.592	.9330	68.9	.1	.4399	.4899	2.041	.8980	63.9
.2	.3616	.3879	2.578	.9323	.8	.2	.4415	.4921	2.032	.8973	.8
.3	.3633	.3899	2.565	.9317	.7	.3	.4431	.4942	2.023	.8965	.7
.4	.3649	.3919	2.552	.9311	.6	.4	.4446	.4964	2.014	.8957	.6
.5	.3665	.3939	2.539	.9304	.5	.5	.4462	.4986	2.006	.8949	.5
.6	.3681	.3959	2.526	.9298	.4	.6	.4478	.5008	1.997	.8942	.4
.7	.3697	.3979	2.513	.9291	.3	.7	.4493	.5029	1.988	.8934	.3
.8	.3714	.4000	2.500	.9285	.2	.8	.4509	.5051	1.980	.8926	.2
.9	.3730	.4020	2.488	.9278	68.1	.9	.4524	.5073	1.971	.8918	63.1
22.0	0.3746	0.4040	2.475	0.9272	68.0	27.0	0.4540	0.5095	1.963	0.8910	63.0
.1	.3762	.4061	2.463	.9265	67.9	.1	.4555	.5117	1.954	.8902	62.9
.2	.3778	.4081	2.450	.9259	.8	.2	.4571	.5139	1.946	.8894	.8
.3	.3795	.4101	2.438	.9252	.7	.3	.4586	.5161	1.937	.8886	.7
.4	.3811	.4122	2.426	.9245	.6	.4	.4602	.5184	1.929	.8878	.6
.5	.3827	.4142	2.414	.9239	.5	.5	.4617	.5206	1.921	.8870	.5
.6	.3843	.4163	2.402	.9232	.4	.6	.4633	.5228	1.913	.8862	.4
.7	.3859	.4183	2.391	.9225	.3	.7	.4648	.5250	1.905	.8854	.3
.8	.3875	.4204	2.379	.9219	.2	.8	.4664	.5272	1.897	.8846	.2
.9	.3891	.4224	2.367	.9212	67.1	.9	.4679	.5295	1.889	.8838	62.1
Deg.	Cos	Cot	Tan	Sin	Deg.	Deg.	Cos	Cot	Tan	Sin	Deg.

Deg.	Sin	Tan	Cot	Cos	Deg.	Deg.	Sin	Tan	Cot	Cos	Deg.
28.0	0.4695	0.5317	1.881	0.8829	62.0	33.0	0.5446	0.6494	1.5399	0.8387	57.0
.1	.4710	.5340	1.873	.8821	61.9	.1	.5461	.6519	1.5340	.8377	56.9
.2	.4726	.5362	1.865	.8813	.8	.2	.5476	.6544	1.5282	.8368	.8
.3	.4741	.5384	1.857	..8805	.7	.3	.5490	.6569	1.5224	.8358	.7
.4	.4756	.5407	1.849	.8796	.6	.4	.5505	.6594	1.5166	.8348	.6
.5	.4772	.5430	1.842	.8788	.5	.5	.5519	.6619	1.5108	.8339	.5
.6	.4787	.5452	1.834	.8780	.4	.6	.5534	.6644	1.5051	.8329	.4
.7	.4802	.5475	1.827	.8771	.3	.7	.5548	.6669	1.4994	.8320	.3
.8	.4818	.5498	1.819	.8763	.2	.8	.5563	.6694	1.4938	.8310	.2
.9	.4833	.5520	1.811	.8755	61.1	.9	.5577 .	.6720	1.4882	.8300	56.1
29.0	0.4848	0.5543	1.804	0.8746	61.0	34.0	0.5592	0.6745	1.4826	0.8290	56.0
.1	.4863	.5566	1.797	.8738	60.9	.1	.5606	.6771	1.4770	.8281	55.9
.2	.4879	.5589	1.789	.8729	.8	.2	.5621	.6796	1.4715	.8271	.8
.3	.4894	.5612	1.782	.8721	.7	.3	.5635	.6822	1.4659	.8261	.7
.4	.4909	.5635	1.775	.8712	.6	.4	.5650	.6847	1.4605	.8251	.6
.5	.4924	.5658	1.767	.8704	.5	.5	.5664	.6873	1.4550	.8241	.5
.6	.4939	.5681	1.760	.8695	.4	.6	.5678	.6899	1.4496	.8231	.4
.7	.4955	.5704	1.753	.8686	.3	.7	.5693	.6924	1.4442	.8221	.3
.8	.4970	.5727	1.746	.8678	.2	.8	.5707	.6950	1.4388	.8211	.2
.9	.4985	.5750	1.739	.8669	60.1	.9	.5721	.6976	1.4335	.8202	55.1
30.0	0.5000	0.5774	1.7321	0.8660	60.0	35.0	0.5736	0.7002	1.4281	0.8192	55.0
.1	.5015	.5797	1.7251	.8652	59.9	.1	.5750	.7028	1.4229	.8181	54.9
.2	.5030	.5820	1.7182	.8643	.8	.2	.5764	.7054	1.4176	.8171	.8
.3	.5045	.5844	1.7113	.8634	.7	.3	.5779	.7080	1.4124	.8161	.7
.4	.5060	.5867	1.7045	.8625	.6	.4	.5793	.7107	1.4071	.8151	.6
.5	.5075	.5890	1.6977	.8616	.5	.5	.5807	.7133	1.4019	.8141	.5
.6	.5090	.5914	1.6909	.8607	.4	.6	.5821	.7159	1.3968	.8131	.4
.7	.5105	.5938	1.6842	.8599	.3	.7	.5835	.7186	1.3916	.8121	.3
.8	.5120	.5961	1.6775	.8590	.2	.8	.5850	.7212	1.3865	.8111	.2
.9	.5135	.5985	1.6709	.8581	59.1	.9	.5864	.7239	1.3814	.8100	54.1
31.0	0.5150	0.6009	1.6643	0.8572	59.0	36.0	0.5878	0.7265	1.3764	0.8090	54.0
.1	.5165	.6032	1.6577	.8563	58.9	.1	.5892	.7292	1.3713	.8080	53.9
.2	.5180	.6056	1.6512	.8554	.8	.2	.5906	.7319	1.3663	.8070	.8
.3	.5195	.6080	1.6447	.8545	.7	.3	.5920	.7346	1.3613	.8059	.7
.4	.5210	.6104	1.6383	.8536	.6	.4	.5934	.7373	1.3564	.8049	.6
.5	.5225	.6128	1.6319	.8526	.5	.5	.5948	.7400	1.3514	.8039	.5
.6	.5240	.6152	1.6255	.8517	.4	.6	.5962	.7427	1.3465	.8028	.4
.7	.5255	.6176	1.6191	.8508	.3	.7	.5976	.7454	1.3416	.8018	.3
.8	.5270	.6200	1.6128	.8499	.2	.8	.5990	.7481	1.3367	.8007	.2
.9	.5284	.6224	1.6066	.8490	58.1	.9	.6004	.7608	1.3319	.7997	53.1
32.0	0.5299	0.6249	1.6003	0.8480	58.0	37.0	0.6018	0.7536	1.3270	0.7986	53.0
.1	.5314	.6273	1.5941	.8471	57.9	.1	.6032	.7563	1.3222	.7976	52.9
.2	.5329	.6297	1.5880	.8462	.8	.2	.6046	.7590	1.3175	.7965	.8
.3	.5344	.6322	1.5818	.8453	.7	.3	.6060	.7618	1.3127	.7955	.7
.4	.5358	.6346	1.5757	.8443	.6	.4	.6074	.7646	1.3079	.7944	.6
.5	.5373	.6371	1.5697	.8434	.5	.5	.6088	.7673	1.3032	.7934	.5
.6	.5388	.6395	1.5637	.8425	.4	.6	.6101	.7701	1.2985	.7923	.4
.7	.5402	.6420	1.5577	.8415	.3	.7	.6115	.7729	1.2938	.7912	.3
.8	.5417	.6445	1.5517	.8406	.2	.8	.6129	.7757	1.2892	.7902	.2
.9	.5432	.6469	1.5458	.8396	57.1	.9	.6143	.7785	1.2846	.7891	52.1
Deg.	Cos	Cot	Tan	Sin	Deg.	Deg.	Cos	Cot	Tan	Sin	Deg.

Deg.	Sin	Tan	Cot	Cos	Deg.		Deg.	Sin	Tan	Cot	Cos	Deg.
38.0	0.6157	0.7813	1.2799	0.7880	52.0		42.0	0.6691	0.9004	1.1106	0.7431	48.0
.1	.6170	.7841	1.2753	.7869	51.9		.1	.6704	.9036	1.1067	.7420	47.9
.2	.6184	.7869	1.2708	.7859	.8		.2	.6717	.9067	1.1028	.7408	.8
.3	.6198	.7898	1.2662	.7848	.7		.3	.6730	.9099	1.0990	.7396	.7
.4	.6211	.7926	1.2617	.7837	.6		.4	.6743	.9131	1.0951	.7385	.6
.5	.6225	.7954	1.2572	.7826	.5		.5	.6756	.9163	1.0913	.7373	.5
.6	.6239	.7983	1.2527	.7815	.4		.6	.6769	.9195	1.0875	.7361	.4
.7	.6252	.8012	1.2482	.7804	.3		.7	.6782	.9228	1.0837	.7349	.3
.8	.6266	.8040	1.2437	.7793	.2		.8	.6794	.9260	1.0799	.7337	.2
.9	.6280	.8069	1.2393	.7782	51.1		.9	.6807	.9293	1.0761	.7325	47.1
39.0	0.6293	0.8098	1.2349	0.7771	51.0		43.0	0.6820	0.9325	1.0724	0.7314	47.0
.1	.6307	.8127	1.2305	.7760	50.9		.1	.6833	.9358	1.0686	.7302	46.9
.2	.6320	.8156	1.2261	.7749	.8		.2	.6845	.9391	1.0649	.7290	.8
.3	.6334	.8185	1.2218	.7738	.7		.3	.6858	.9424	1.0612	.7278	.7
.4	.6347	.8214	1.2174	.7727	.6		.4	.6871	.9457	1.0575	.7266	.6
.5	.6361	.8243	1.2131	.7716	.5		.5	.6884	.9490	1.0538	.7254	.5
.6	.6374	.8273	1.2088	.7705	.4		.6	.6896	.9523	1.0501	.7242	.4
.7	.6388	.8302	1.2045	.7694	.3		.7	.6909	.9556	1.0464	.7230	.3
.8	.6401	.8332	1.2002	.7683	.2		.8	.6921	.9590	1.0428	.7218	.2
.9	.6414	.8361	1.1960	.7672	50.1		.9	.6934	.9623	1.0392	.7206	46.1
40.0	0.6428	0.8391	1.1918	0.7660	50.0		44.0	0.6947	0.9657	1.0355	0.7193	46.0
.1	.6441	.8421	1.1875	.7649	49.9		.1	.6959	.9691	1.0319	.7181	45.9
.2	.6455	.8451	1.1833	.7638	.8		.2	.6972	.9725	1.0283	.7169	.8
.3	.6468	.8481	1.1792	.7627	.7		.3	.6984	.9759	1.0247	.7157	.7
.4	.6481	.8511	1.1750	.7615	.6		.4	.6997	.9793	1.0212	.7145	.6
							.5	.7009	.9827	1.0176	.7133	.5
40.5	0.6494	0.8541	1.1708	0.7604	49.5		.6	.7022	.9861	1.0141	.7120	.4
.6	.6508	.8571	1.1667	.7593	.4		.7	.7034	.9896	1.0105	.7108	.3
.7	.6521	.8601	1.1626	.7581	.3		.8	.7046	.9930	1.0070	.7096	.2
.8	.6534	.8632	1.1585	.7570	.2		.9	.7059	.9965	1.0035	.7083	45.1
.9	.6547	.8662	1.1544	.7559	49.1		45.0	0.7071	1.0000	1.0000	0.7071	45.0
41.0	0.6561	0.8693	1.1504	0.7547	49.0		**Deg.**	**Cos**	**Cot**	**Tan**	**Sin**	**Deg.**
.1	.6574	.8724	1.1463	.7536	48.9							
.2	.6587	.8754	1.1423	.7524	.8							
.3	.6600	.8785	1.1383	.7513	.7							
.4	.6613	.8816	1.1343	.7501	.6							
.5	.6626	.8847	1.1303	.7490	.5							
.6	.6639	.8878	1.1263	.7478	.4							
.7	.6652	.8910	1.1224	.7466	.3							
.8	.6665	.8941	1.1184	.7455	.2							
.9	.6678	.8972	1.1145	.7443	48.1							
Deg.	**Cos**	**Cot**	**Tan**	**Sin**	**Deg.**							

ANSWERS TO ODD-NUMBERED EXERCISES

Chapter 1

1. **a.** 10^4 **b.** 10^{-4} **c.** 10^3 **d.** 10^6
   e. 10^{-7} **f.** 10^{-5}
3. **a.** 10^4 **b.** 10 **c.** 10^9 **d.** 10^{-2}
   e. 10 **f.** 10^{31}
5. **a.** 10^6 **b.** 10^{-2} **c.** 10^{32} **d.** 10^{-63}
7. **a.** 900 **b.** 2×10^5 **c.** 9×10^{12}
   d. $\frac{1}{7} \times 10^{-6}$ **e.** 24×10^{10}
   f. 8×10^{20} **g.** $28{,}224 \times 10^{-14}$
9. **a.** 10^5 pF **b.** $467 \, \Omega$
   c. 63.9×10^{-3} H
   d. 69×10^{-5} km
   e. 11.52×10^6 ms **f.** $16 \, \mu$H
   g. 60×10^{-4} m^2
11. **a.** Length $\neq 10$ ft $\cdot$ 2 s or
    72 in. $\cdot$ 6 cm/s
    b. Time $\neq$ 40 h/200 min
    or 50 s $\cdot$ 2 min
13. 3.408×10^{-2} min $= 2.045$ s
15. 2.6823 m/s
17. **a.** 4.74×10^{-3} Btu
    b. 7.098×10^{-4} m^3
    c. 1.2096×10^5 s
    d. 22.113×10^3 pints

Chapter 2

3. **a.** 111.197×10^{-6} N
   b. 288×10^4 N
   c. 114.195×10^6 N
5. 13 A
7. 2400 C
9. 2.3 s
11. 374.52×10^{18} electrons
13. 1.194 A > 1 A (yes)
15. **a.** 9 V **b.** 3 V **c.** 12 V
17. 6 C
19. 2.67 V
21. 5 A
23. 25 h
25. 0.773 h

Chapter 3

1. **a.** 500 mils **b.** 10 mils
   c. 4 mils **d.** 1000 mils
   e. 600 mils **f.** 3.937 mils
3. **a.** 0.04 in. **b.** 0.03 in.
   c. 0.24495 in. **d.** 0.025 in.
   e. 0.00314 in. **f.** 0.009 in.
5. $51.33 \, \Omega$
7. **a.** 470.86 ft **b.** 493.21 ft
9. **a.** 1244.40 cm **b.** larger
   c. smaller

11. $47 \Rightarrow$ nickel
13. $2.409 \, \Omega$
15. $3.334 \, \Omega$
17. **a.** $1.842 \times 10^{-3} \, \Omega$ ($0\,°$C),
    $2.628 \times 10^{-3} \, \Omega$ ($100\,°$C)
    b. $7.85 \times 10^{-5} \, \Omega/10\,°$C
19. **a.** $40.29\,°$C **b.** $-195.61\,°$C
21. $39.371 \, \Omega$
23. **a.** $0.567 \, \Omega$ ($\#11$),
    $1.136 (\#14)$
    b. $\#14 : \#11 \cong 2 : 1$
    c. $\#14 : \#11 \cong 1 : 2$
25. **a.** $\#2$ **b.** $\#0$
29. $10fc \rightarrow 3 \, \text{k}\Omega$, $100fc \rightarrow 0.4 \, \text{k}\Omega$
31. **a.** $51{,}300 \, \Omega - 56{,}700 \, \Omega$
    b. $198 \, \Omega - 242 \, \Omega$
    c. $8 \, \Omega - 12 \, \Omega$
33. **a.** 0.1566 S **b.** 0.0955 S
    c. 0.0219 S

Chapter 4

1. 15 V
3. $4 \, \text{k}\Omega$
5. 72 nV
7. $54.55 \, \Omega$
9. $1.68 \, \text{kV}$
11. 6 A
13. 1 W
15. **a.** $57{,}600$ J **b.** 16×10^{-3} kWh
17. 2 s
19. $196 \, \mu$W
21. 4 A
23. 795.99 V
25. 0.833 A, $144.058 \, \Omega$
27. 70.71 mA, $1.414 \, \text{kV}$
29. 80%
31. 84.77%
33. **a.** 1657.78 W **b.** 15.07 A
    c. 19.38 A
35. 38.40 J
37. $\eta_1 = 40\%$, $\eta_2 = 80\%$
39. **a.** 1350 J
    b. Energy doubles,
    Power the same
41. 6.67 h
43. \$2.35
45. 49.28¢

Chapter 5

1. **a.** $20 \, \Omega$, 3 A **b.** $14 \, \text{k}\Omega$, 10 mA
   c. $110 \, \Omega$, 318.2 mA
   d. $10 \, \text{k}\Omega$, 12 mA
3. **a.** 16 V **b.** 1.56 V

5. **a.** $6 \, \text{k}\Omega$, 20 mA, $V_1 = 60$ V,
   $V_2 = 20$ V, $V_3 = 40$ V
   c. $P_1 = 1.2$ W, $P_2 = 0.4$ W,
   $P_3 = 0.8$ W, $P_{\text{del.}} = 2.4$ W
   d. $R_1 = 2$ W, $R_2 = \frac{1}{2}$ W,
   $R_3 = 1$ W
7. **a.** 60 V, 2 A, $30 \, \Omega$
   b. 4 A, $V_1 = 16$ V, $V_2 = 24$ V,
   $E = 48$ V
   c. $21 \, \Omega$, $V_1 = 2$ V, $V_2 = 1$ V,
   $V_3 = 21$ V, $E = 24$ V
   d. 2 A, $R_1 = 2 \, \Omega$, $R_2 = 13 \, \Omega$,
   $E = 32$ V
9. **a.** $V_{ab} = 66.67$ V **b.** $V_{ab} = 8$ V
   $\substack{+-}$ $\substack{-+}$
   c. $V_{ab} = 20$ V **d.** $V_{ab} = 0.18$ V
   $\substack{+-}$ $\substack{+-}$
11. **a.** 0.25 S, $4 \, \Omega$, 30 A
    b. 1×10^{-3} S, $1 \, \text{k}\Omega$, 40 mA
    c. 0.2 S, $5 \, \Omega$, 12 A
    d. $\frac{3}{4}$ S, $\frac{4}{3} \, \Omega$, 15 A
    e. 0.100525 S, $9.948 \, \Omega$
    201.05 mA
    f. 1.333 S, $0.75 \, \Omega$, 53.33 A
13. **a.** $18 \, \Omega$ **b.** $R_1 = R_2 = 24 \, \Omega$
15. **a.** 1.167 S, $0.8571 \, \Omega$
    b. $I_T = 1.050$ A, $I_1 = 0.30$ A,
    $I_2 = 0.15$ A, $I_3 = 0.60$ A
    d. $P_1 = 0.27$ W,
    $P_2 = 0.135$ W,
    $P_3 = 0.54$ W,
    $P_{\text{del.}} = 0.945$ W
    e. $R_1 = \frac{1}{2}$ W, $R_2 = \frac{1}{2}$ W,
    $R_3 = 1$ W
17. **a.** $R_1 = 5 \, \Omega$, $I_2 = 1$ A,
    $R_2 = 10 \, \Omega$
    b. $E = 12$ V, $I_2 = 1.333$ A,
    $I_3 = 1$ A, $R_3 = 12 \, \Omega$,
    $I = 4.333$ A
    c. $I_1 = 220$ mA, $I_3 = 55$ mA,
    $I_2 = 725$ mA,
    $R = 303.45 \, \Omega$
    d. $V_1 = 30$ V, $E = 30$ V,
    $I_1 = 1$ A, $I_2 = 0.5$ A,
    $R_2 = 60 \, \Omega = R_3$
19. **a.** bulbs : $I = 5$ A,
    washer : $I = \frac{10}{3}$ A,
    TV : $I = 3$ A
    b. $11\frac{1}{3}$ A, no **c.** $10.59 \, \Omega$
    d. $P_{\text{del.}} = 1360$ W
21. **a.** $I = 4$ A, $I_1 = 3$ A
    b. $I_3 = I = 6 \, \mu$A, $I_2 = 2 \, \mu$A,
    $I_1 = 2 \, \mu$A, $R = 9 \, \Omega$
23. $2 \, \Omega$
25. $100 \, \Omega$
27. 1.471%

Chapter 6

1. a. yes (KCL) **b.** 3 A
   c. yes (KCL) **d.** 4 V
   e. $2\,\Omega$ **f.** 5 A
   g. $P_{del.} = 50$ W, $P_1 = 12$ W,
   $P_2 = 18$ W
3. a. $I = I_1 = 1.4$ A
   b. $V_1 = 2.8$ V, $V_2 = 7.0$ V,
   $V_3 = 4.2$ V
   c. $I_3 = I_4 = 0.7$ A
   d. 9.8 W **e.** 10.8 V
5. a. 4 A
   b. $I_2 = 1.333$ A,
   $I_3 = 0.6665$ A
   c. $V_A = 8$ V, $V_B = 4$ V
7. a. $I = 16$ mA, $I_6 = 2$ mA
   b. $V_1 = 28$ V, $V_5 = 7.2$ V
   c. 8.64 mW
9. a. $V_G = 3.2$ V, $V_S = 4.95$ V
   b. 4.55 V **c.** 2.8 V
11. a. 2 A
   b. $I_1 = I_3 = \frac{2}{3}$ A, $I_8 = \frac{2}{9}$ A
   c. 1.037 W
13. a. $I_2 = 1\frac{2}{3}$ A, $I_3 = 3\frac{1}{3}$ A,
   $I_8 = 0$ A
   b. $V_4 = 10$ V, $V_8 = 0$ V
15. a. 24 A **b.** 8 A
   c. $V_3 = 48$ V, $V_5 = 24$ V,
   $V_7 = 16$ V
   d. $P(R_7) = 128$ W,
   $P_{del.} = 5760$ W
17. a. 12 A **b.** 0.5 A
   c. 0.5 A **d.** 6 A

Chapter 7

1. $V_{ab} = 28$ V
         $+-$
3. a. $V_{ab} = 45$ V
         $+-$
   b. $V_{cd} = 5.33$ V
         $-+$
5. a. $I = 3$ A, $\mathcal{R}_P = 6\,\Omega$
   b. $I = 3.5$ A, $\mathcal{R}_P = 2\,\Omega$
7. a. 8 A
   b. $E = 48$ V, $R_S = 4\,\Omega$,
   $I = 8$ A
9. 2.4 A, 9.6 V
11. a. $I_2 = 0.5107$ A, $I_5 = 2$ A
   b. 3.575 V **c.** 782.4 mW
13. a. 29 **b.** 18
15. a. $x = 1.705$, $y = 2.591$
   $z = 0.546$
   b. $x = 6$, $y = -7.33$
   $z = -2.5$

17. a. I_1 $(cw) = I_{50} = 0.1613$ A,
   I_2 $(ccw) = I_{30} = 0.9355$ A,
   I_3 $(down) = I_{20} = 1.097$ A,
   $V_{ab} = 1.935$ V
   $+-$
   b. I_1 $(ccw) = I_{20v} = 7.286$ A,
   I_2 $(cw) = I_{12v} = -0.714$ A,
   I_3 $(\mu p) = I_4 = 6.572$ A,
   $V_{ab} = 70$ V
   $-+$
19. a. I_1 $(cw) = I_4 = -\frac{1}{7}$ A,
   I_2 $(ccw) = I_2 = \frac{5}{7}$ A,
   $V_{ab} = 2$ V
   $-+$
   b. I_1 $(ccw) = I_4 = 3.0625$ A,
   I_2 $(cw) = I_{12} = 0.1875$ A,
   $V_{ab} = 22$ V
   $-+$
21. a. I_1 $(cw) = I_2 = 1.871$ A,
   I_2 $(ccw) = I_5 = 8.548$ A,
   $V_{ab} = 22.74$ V
   $-+$
   b. I_1 $(cw) = I_4 = 1.274$ A,
   I_2 $(cw) = I_8 = 0.260$ A,
   $V_{ab} = 0.904$ V
   $-+$
23. a. I_1 $(cw) = I_2 = 1.231$ A,
   I_2 $(cw) = I_7 = -0.4957$ A,
   I_3 $(cw) = I_3 = -0.6638$ A,
   b. I_1 $(cw) = I_{R_5} = -0.2385$ A,
   I_2 $(cw) = I_{10} = -0.5169$ A,
   I_3 $(cw) = I_7 = -1.278$ A
25. a. I_1 $(cw) = I_4 = -\frac{1}{7}$ A,
   I_2 $(cw) = I_2 = -\frac{5}{7}$ A
   b. I_1 $(cw) = I_4 = -3.0625$ A,
   I_2 $(cw) = I_{12} = 0.1875$ A
27. a. I_1 $(cw) = I_2 = 1.871$ A,
   I_2 $(cw) = I_5 = -8.548$ A
   b. I_1 $(cw) = I_4 = 1.274$ A,
   I_2 $(cw) = I_8 = 0.26$ A
29. a. I_1 $(cw) = I_2 = 1.231$ A,
   I_2 $(cw) = I_7 = -0.4957$ A,
   I_3 $(cw) = I_3 = -0.6638$ A
   b. I_1 $(cw) = I_{R_5} = -0.2385$ A,
   I_2 $(cw) = I_{10} = -0.5169$ A,
   I_3 $(cw) = I_7 = -1.268$ A
31. a. V_1 (left node) $= 8.077$ V,
   V_2 (right node) $= 9.385$ V
   b. V_1 (left node) $= 4.8$ V,
   V_2 (right node) $= 6.4$ V
33. a. V_1 (left node) $= -2.462$ V
   V_2 (right node) $= 1.009$ V
   b. V_1 (top left node)
   $= 8.877$ V
   V_2 (top right node)
   $= 9.831$ V
   V_3 (center node)
   $= -3.005$ V

35. a. V_1 (far left node)
   $= -5.311$ V
   V_2 (center node)
   $= -0.6219$ V
   V_3 (far right node)
   $= 3.751$ V
   b. V_1 (top left node)
   $= -6.917$ V,
   V_2 (top right node)
   $= 12$ V,
   V_3 (bottom left node)
   $= 2.3$ V
37. b. $V_{R_3} = 0.1967$ V **c.** no
39. b. $V_{R_3} = 0$ V **c.** balanced
41. a. $I_{10\,V} = 3.33$ mA
   b. $I_{20\,V} = 0.8235$ A
43. a. 133.33 mA **b.** 7 A

Chapter 8

1. a. $I_{R_1} = \frac{5}{6}$ A, $I_{R_2} = 0$ A,
   $I_{R_3} = \frac{5}{6}$ A
   b. $E_1 : P = 5.333$ W,
   $E_2 : P = 0.333$ W
   c. $P_1 = 8.333$ W **d.** no
3. a. 0.143 A **b.** 0.1916 A
5. a. $R_{Th} = 6\,\Omega$, $E_{Th} = 6$ V
   b. $2\,\Omega : I = 0.75$ A,
   $30\,\Omega : I = 0.1667$ A,
   $100\,\Omega : I = 0.0566$ A
7. (I) **a.** $R_{Th} = 2\,\Omega$, $E_{Th} = 84$ V
   b. $2\,\Omega : 882$ W,
   $100\,\Omega : 67.82$ W
   (II) **a.** $R_{Th} = 2\,\Omega$, $E_{Th} = 8$ V
   b. $2\,\Omega : 8$ W,
   $100\,\Omega : 0.6152$ W
9. (I) $R_{Th} = 45\,\Omega$, $E_{Th} = -5$ V
   (II) $R_{Th} = 13.5\,\Omega$,
   $E_{Th} = 207$ V
11. (I) **a.** $R_N = 14\,\Omega$,
   $I_N = 2.571$ A
   (II) **a.** $R_N = 7.5\,\Omega$,
   $I_N = 1.333$ A
13. (I) **a.** $R_N = 13\,\Omega$,
   $I_N = 1.539$ A
   (II) **a.** $R_N = 2\,\Omega$,
   $I_N = 30$ A
15. (I) **a.** $R_N = 9.333\,\Omega$,
   $I_N = 4.429$ A
   (II) **a.** $R_N = 1.8462\,\Omega$,
   $I_N = 50$ A
17. a. (I) $R = 14\,\Omega$
   (II) $R = 7.5\,\Omega$
   b. (I) 23.14 W,
   (II) 3.332 W

19. a. (I) $R = 13\,\Omega$,
(II) $R = 2\,\Omega$
b. (I) 7.698 W,
(II) 450 W
21. $R_1 = 0\,\Omega$
23. $V_L = 18.34$ V, $I_L = 6.113$ A
25. $V_L = 89.5$ V, $I_L = 0.448$ A
27. $I_L = 2.2$ mA, $V_L = 13.2$ V
29. Voltage source = 9.75 V or
Current source = 0.5 mA
31. a. 0.5 mA **b.** 0.5 mA **c.** yes
33. a. 4 V **b.** 4 V **c.** yes

Chapter 9

1. 9×10^3 N/C
3. 70 μF
5. 50 V/m
7. 8×10^3 V/m
9. 937.5 pF
11. 5 (mica)
13. a. 10^6 V/m **b.** 4.96 μC
c. 0.0248 μF
15. 29,035 V
17. a. $v_C = 20(1 - e^{-t/1 \times 10^{-3}})$
b. 1τ : 12.64 V, 3τ : 19.02 V,
5τ : 19.86 V
c. $i_C = 50 \times 10^{-6} e^{-t/1 \times 10^{-3}}$
$v_R = 20e^{-t/1 \times 10^{-3}}$
d. i_C : $1\tau = 31.6\,\mu$A,
$3\tau = 47.55\,\mu$A,
$5\tau = 49.65\,\mu$A,
v_R : $1\tau = 7.36$ V,
$3\tau = 0.98$ V,
$5\tau = 0.14$ V
19. a. $v_C = 50(1 - e^{-t/0.46})$
b. $i_C = 2.174 \times 10^{-3}e^{-t/0.46}$,
$v_{R_2} = 43.48e^{-t/0.46}$
c. $v_C = 33.14$ V,
$i_C = 0.733$ mA,
$v_{R_2} = 14.66$ V
d. $v_C = 33.14e^{-t/0.4}$,
$i_C = -1.657 \times 10^{-3}\,e^{-t/0.4}$,
$v_{R_2} = -33.14e^{-t/0.4}$
21. a. $v_C = 60(1 - e^{-t/0.6})$
b. $i_C = 120 \times 10^{-6}e^{-t/0.6}$,
$v_{R_1} = 24e^{-t/0.6}$
c. $v_C = 60e^{-t/0.6}$,
$i_C = -120 \times 10^{-6}e^{-t/0.6}$,
$v_{R_1} = -24e^{-t/0.6}$
23. a. $v_C = 66.67(1 - e^{-t/0.08})$,
$i_C = 16.67 \times 10^{-3}e^{-t/0.08}$
25. $0 \to 2\,\mu$s : 90 mA,
$2 \to 4\,\mu$s : -180 mA,

$4 \to 6\,\mu$s : 90 mA,
$6 \to 8\,\mu$s : 60 mA,
$8 \to 10\,\mu$s : -30 mA,
$10 \to 15\,\mu$s : 0 mA,
$15 \to 16\,\mu$s : -60 mA
27. a. 6 μF (right-hand branch);
$Q = 60\,\mu$C,
$V = 10$ V : 12 μF;
$Q = 40\,\mu$C,
$V = 3.33$ V : 6 μF;
$Q = 40\,\mu$C,
$V = 6.67$ V
b. 1200 pF; $Q = 9600$ pC,
$V = 8$ V : 200 pF;
$Q = 3200$ pC,
$V = 16$ V : 400 pF;
$Q = 6400$ pC,
$V = 16$ V : 600 pF;
$Q = 9600$ pC, $V = 16$ V
29. 8640×10^{-12} J
31. a. 5 J **b.** 0.1 C **c.** 200 A
d. 10,000 W **e.** 10 s

Chapter 10

1. CGS : 5×10^4 maxwells,
8 gauss-English : 5×10^4
lines, 51.616 lines/in.^{2}
3. a. 4×10^{-3} T
5. 95.24 rels (At/Wb)
7. 2624.67 At/m
9. a. $\mu_{dc} = 7.62 \times 10^{-3}$ Wb/Am
b. $\mu_\Delta = 3.08 \times 10^{-3}$ Wb/Am
c. $\mu_{av} = 10 \times 10^{-3}$ Wb/Am
11. 2.133 A
13. 2.76 A
15. a. 187.53 At
b. μ (sheet steel)
$= 28.9 \times 10^{-4}$ Wb/Am,
μ (cast steel)
$= 9.5 \times 10^{-4}$ Wb/Am
17. 27.27 A
19. a. 2.55 A **b.** 3.58 N
21. 1.45 A
23. 0.592×10^{-4} Wb
$= 5920$ lines

Chapter 11

1. 4.25 V
3. 14 turns
5. 15.65 μH
7. a. 2.5 V **b.** 0.3 V **c.** 200 V
9. 0–3 ms = 0 V,
3–8 ms = 1.6 V,

8–13ms $= -1.6$ V,
13–14 ms = 0 V,
14–15 ms = 8 V,
15–16 ms $= -8$ V,
16–17 ms = 0 V
11. a. $i_L = 2 \times 10^{-6}$
$\times (1 - e^{-t/12.5 \times 10^{-6}})$
b. $1\tau = 1.264\,\mu$A,
$2\tau = 1.730\,\mu$A,
$3\tau = 1.902\,\mu$A,
$4\tau = 1.962\,\mu$A,
$5\tau = 1.986\,\mu$A
c. $v_L = 40 \times 10^{-3}$
$\times e^{-t/12.5 \times 10^{-6}}$,
$v_R = 40 \times 10^{-3}$
$\times (1 - e^{-t/12.5 \times 10^{-6}})$
d. v_L; $1\tau = 14.72$ mV,
$2\tau = 5.4$ mV,
$3\tau = 1.96$ mV,
$4\tau = 0.76$ mV,
$5\tau = 0.28$ mV : v_R;
$1\tau = 25.28$ mV,
$2\tau = 34.60$ mV,
$3\tau = 38.04$ mV,
$4\tau = 39.24$ mV,
$5\tau = 39.72$ mV
13. a. $i_L = 8 \times 10^{-3}$
$\times (1 - e^{-t/0.5 \times 10^{-3}})$
b. $v_L = 32e^{-t/0.5 \times 10^{-3}}$,
$v_{R_1} = 32$
$\times (1 - e^{-t/0.5 \times 10^{-3}})$
c. $i_L = 8 \times 10^{-3}e^{-t/31.3 \times 10^{-6}}$,
$v_{R_1} = 32e^{-t/31.3 \times 10^{-6}}$,
$v_{R_2} = 480e^{-t/31.3 \times 10^{-6}}$,
$v_L = 512e^{-t/31.3 \times 10^{-6}}$
15. a. $i_L = 3(1 - e^{-t/0.5})$
b. $v_L = 18e^{-t/0.5}$
c. $i_L = 174 \times 10^{-3}e^{-t}$,
$v_L = v_{R_1} = 0.522e^{-t}$
17. a. $i_L = 2.75e^{-t/5.25 \times 10^{-3}}$
b. $v_L = 22e^{-t/5.25 \times 10^{-3}}$
19. $I_1 = 4$ mA, $V_1 = 16$ V,
$V_2 = 0$ V
21. $I_1 = 2$ A, $V_1 = 12$ V
23. a. $W_{2\,H} = 16\,\mu$J,
$W_{3\,H} = 24\,\mu$J
b. $W_{3\,mH} = 6$ nJ,
$W_{6\,mH} = 27$ nJ
25. $W_{0.5\,H} = 1$ J,
$W_{4\,H} = 3.56$ J

Chapter 12

1. a. 1 mA **b.** $R_{shunt} = 5$ mΩ
3. a. $R_s = 300$ kΩ

b. $\Omega/v = 20{,}000$
5. $0.05 \, \mu A$
9. 3 A
11. b. $\cong 50 \, \mu A$ **c.** $25 \, \mu A$
13. b. $\cong 61$ mA
15. a. $V_1 = 20$ V, $V_2 = 10$ V
  b-c. $R_{int} = 1 \, M\Omega$,
    $V_m = 19.99$ V vs 20 V
  d. $\cong 20$ V
17. b. $P_{actual} = 57.6$ W,
  $P_{10 \, k\Omega} = 230.4$ mV
  (above actual)

Chapter 13

1. a. 1.2 s **b.** 2 **c.** 0.833 Hz
  d. 2 V
3. a. 40 ms **b.** 28.57 ns
  c. $18.18 \, \mu s$ **d.** 1 s
5. 0.3 ms
7. 142.86 ms
9. a. 45° **b.** 30° **c.** 150°
  d. 210° **e.** 240° **f.** 99°
11. a. 314 rad/s **b.** 3768 rad/s
  c. 0.628 rad/s
  d. 25.12×10^{-3} rad/s
13. a. 20, 60.03 Hz
  b. 5, 120.06 Hz
  c. 10^6, 1592.36 Hz
  d. 0.0001, 150 Hz
  e. 7.6, 6.94 Hz
  f. $\frac{1}{42}$, 1 Hz
15. 0.628 ms
17. 20 ms
21. a. in phase
  b. i leads v by 90°
  c. v leads i by 190°
23. a. $v = 0.01 \sin$
    $\times (157t - 110°)$
  b. $v = 2 \times 10^{-3} \sin$
    $\times (62.8t + 135°)$
25. a. $V_{av} = 2.5$ V
  b. $I_{av} = -4.78$ mA
27. a. 14.14 V **b.** 5 V
  c. 4.24 mA **d.** 1.244 A
29. 1.43 V
31. $V_{av} = 0$ V, $V_{eff} = 10$ V
33. a. $i = 30 \sin 377t$
  b. $i = 6 \sin (377t + 20°)$
  c. $i = 8 \sin (\omega t + 100°)$
  d. $i = 16 \sin (\omega t + 220°)$
35. a. 0 **b.** $314 \, \Omega$ **c.** $753.6 \, \Omega$
  d. $25.13 \, k\Omega$ **e.** $1.256 \, M\Omega$
37. a. 0.796 Hz **b.** 60.03 Hz
  c. 250 Hz **d.** 3.87 Hz

39. a. $v = 90 \sin (30t + 90°)$
  b. $v = 226.2 \times 10^3$
    $\sin (377t + 90°)$
  c. $v = 200 \times 10^{-6}$
    $\sin (400t + 110°)$
  d. $v = 8 \sin (20t - 70°)$
41. a. $i = 0.125 \sin (60t - 90°)$
  b. $i = 80 \times 10^{-3} \sin (t - 86°)$
  c. $i = 480 \sin (0.05t + 140°)$
  d. $i = 0.1194 \times 10^{-3} \sin 377t$
43. a. $10.62 \, \mu F$ **b.** $9.28 \, \mu F$
  c. $636.94 \, \mu F$
45. a. $i = 40 \sin (\omega t + 90°)$
  b. $i = 0.16 \sin (\omega t + 110°)$
  c. $i = 3.2 \sin (\omega t + 190°)$
  d. $i = 28 \sin (\omega t - 50°)$
47. a. $v = 500 \sin (\omega t - 90°)$
  b. $v = 400 \sin (\omega t - 30°)$
  c. $v = 60 \sin (\omega t + 60°)$
  d. $v = 30 \sin (\omega t + 10°)$
49. a. $L = 132.63$ mH
  b. $C = 147.36 \, \mu F$
  c. $R = 7 \, \Omega$
51. a. 0 W **b.** 0 W **c.** 7.875 W
53. a. 389.7 W, 0.866
  b. 25 W, 0.5
  c. 64.9 W, 0.866
  d. 2.598 W, 0.866
55. 100 W : 0.333, 0 W : 0,
  300 W : 1
57. a. $i = 10 \sin (377t + 20°)$
  b. 150 W **c.** $\cong 0.1$ s
59. a. $e = 1200 \sin (377t - 110°)$
  b. $6.63 \, \mu F$ **c.** 0 W

Chapter 14

1. a. $5.0 \, \underline{/36.87°}$ **b.** $2.83 \, \underline{/45°}$
  c. $16.38 \, \underline{/77.66°}$
  d. $657.65 \, \underline{/81.25°}$
  e. $1250 \, \underline{/36.87°}$
  f. $0.00658 \, \underline{/81.25°}$
  g. $11.78 \, \underline{/-49.82°}$
  h. $6.72 \, \underline{/143.47°}$
  i. $61.85 \, \underline{/-104.04°}$
  j. $101.53 \, \underline{/-39.81°}$
  k. $4326.66 \, \underline{/123.69°}$
  l. $25.495 \times 10^{-3} \, \underline{/-78.69°}$
3. a. $15.033 \, \underline{/86.19°}$
  b. $60.208 \, \underline{/4.76°}$
  c. $0.30 \, \underline{/88.09°}$
  d. $2504.50 \, \underline{/-86.57°}$
  e. $86.182 \, \underline{/93.73°}$
  f. $38.694 \, \underline{/-94.0°}$

5. a. $11.80 + j6.82$
  b. $151.90 + j49.90$
  c. $4.72 \times 10^{-6} + j75.10$
  d. $5.20 + j1.60$
  e. $209.30 + j311.0$
  f. $-21.20 + j12.0$
  g. $14.30 + j7.05$
  h. $95.698 + j22.768$
7. q. $6.0 \, \underline{/-50.0°}$
  r. $0.000120 \, \underline{/140.0°}$
  s. $109.0 \, \underline{/-230.0°}$
  t. $76.471 \, \underline{/-80.0°}$
  u. $4.0 \, \underline{/0°}$
  v. $0.645 \, \underline{/-15.975°}$
  w. $4.21 \times 10^{-3} \, \underline{/161.10°}$
  x. $18.191 \, \underline{/-50.91°}$
9. a. $100.0 \, \underline{/30.0°}$
  b. $0.250 \, \underline{/-40.0°}$
  c. $70.71 \, \underline{/-90.0°}$
  d. $29.69 \, \underline{/0.0°}$
  e. $4.242 \times 10^{-6} \, \underline{/90.0°}$
  f. $2.546 \times 10^{-6} \, \underline{/70°}$
11. $v_a = 41.769$
    $\sin (377t + 29.43°)$
13. $v_c = 140.940 \sin (\omega t - 58.37°)$

Chapter 15

1. a. $\mathbf{R} = 7.5 \, \underline{/0°} = 7.5$
  b. $\mathbf{X}_L = 754 \, \underline{/90°} = j754$
  c. $\mathbf{X}_L = 10 \, \underline{/90°} = +j10$
  d. $\mathbf{X}_C = 265.25 \, \underline{/-90°}$
    $= -j265.25$
  e. $\mathbf{X}_C = 3 \, \underline{/-90°} = -j3$
  f. $\mathbf{R} = 200 \, \underline{/0°} = 200$
3. a. $v = 100 \times 10^{-3} \sin \omega t$
  b. $v = 9.045 \sin (377t + 70°)$
  c. $v = 2{,}545 \sin (157t - 50°)$
5. a. $\mathbf{Z}_T = 3 - j3 = 4.24 \, \underline{/-45°}$
  b. $\mathbf{Z}_T = 0.05 - j1$
    $= 1.001 \, \underline{/-87.14°}$
  c. $\mathbf{Z}_T = 4 - j167.42$
    $= 167.47 \, \underline{/-88.63°}$
7. a. $\mathbf{Z}_T = 10 \, \underline{/36.87°}$
  c. $\mathbf{I} = 10 \, \underline{/-36.87°}$,
    $\mathbf{V}_R = 80 \, \underline{/-36.87°}$,
    $\mathbf{V}_L = 60 \, \underline{/53.13°}$
  f. 800 W **g.** $F_p = 0.8$ lagging
9. a. $\mathbf{Z}_T = 4.47 \, \underline{/-63.43°}$
  c. $C = 265.25 \, \mu F$,
    $L = 15.92$ mH
  d. $\mathbf{I} = 11.19 \, \underline{/93.43°}$,
    $\mathbf{V}_R = 22.38 \, \underline{/93.43°}$,
    $V_L = 67.14 \, \underline{/183.43°}$,
    $V_C = 111.90 \, \underline{/3.43°}$

g. 250.43 W

h. $F_p = 0.447$ leading

i. $i = 15.82$
$$\sin (377t + 93.43°),$$
$$v_R = 31.65$$
$$\sin (377t + 93.43°)$$
$$v_L = 94.94$$
$$\sin (377t + 183.43°),$$
$$v_C = 158.23$$
$$\sin (377t + 3.43°)$$

11. a. $V_1 = 37.97 \,\underline{/-51.57°}$,
$V_2 = 113.92 \,\underline{/38.43°}$

b. $V_1 = 55.71 \,\underline{/26.8°}$,
$V_2 = 12.53 \,\underline{/-63.20°}$

13. a. $I = 38.72 \times 10^{-3} \,\underline{/122.66°}$,
$V_R = 2.56 \,\underline{/122.66°}$,
$V_C = 25.68 \,\underline{/32.66°}$

b. $F_p = 0.128$ leading

c. 98.95 mW

f. $2.92 \,\underline{/212.66°}$

g. $V_R = 2.56 \,\underline{/122.66°}$,
$V_C = 25.68 \,\underline{/32.66°}$

h. $Z_T = 66 - j512.33$

15. $Z_T = 3.2 + j2.4$

17. a. $Z = 42 \,\underline{/0°}$, $Y = 0.0238$

b. $Z = 200 \,\underline{/90°}$,
$Y = 5 \times 10^{-3} \,\underline{/-90°}$

c. $Z = 0.6 \,\underline{/-90°}$,
$Y = 1.667 \,\underline{/90°}$

d. $Z = 8.544 \,\underline{/69.44°}$,
$Y = 0.117 \,\underline{/-69.44°}$

e. $Z = 92.195 \,\underline{/-49.40°}$,
$Y = 0.0109 \,\underline{/49.40°}$

f. $Z = 223.61 \,\underline{/-26.57°}$,
$Y = 4.47 \times 10^{-3} \,\underline{/26.57°}$

g. $Z_T = 9.86 \,\underline{/9.46°}$,
$Y_T = 0.1014 \,\underline{/-9.46°}$

h. $Z_T = 2.68 \,\underline{/-26.57°}$,
$Y_T = 0.373 \,\underline{/26.57°}$

i. $Y_T = 0.3379 \,\underline{/-9.47°}$,
$Z_T = 2.959 \,\underline{/9.47°}$

19. a. $Y_T = 0.539 \,\underline{/-21.8°}$

c. $E = 3.71 \,\underline{/21.8°}$,
$I_R = 1.855 \,\underline{/21.8°}$,
$I_L = 0.742 \,\underline{/-68.2°}$

f. 6.88 W

g. $F_p = 0.928$ lagging

h. $e = 5.25$
$$\sin (377t + 21.8°),$$
$$i_R = 2.62$$
$$\sin (377t + 21.8°),$$
$$i_L = 1.049$$
$$\sin (377t - 68.2°),$$
$$i_T = 2.828 \sin 377t$$

21. a. $Y_T = 0.13 \,\underline{/-50.31°}$

c. $I_T = 7.8 \,\underline{/-50.31°}$,
$I_R = 5 \,\underline{/0°}$, $I_L = 6 \,\underline{/-90°}$

f. 300 W

g. $F_p = 0.638$ lagging

h. $e = 84.84 \sin 377t$,
$i_R = 7.07 \sin 377t$,
$i_L = 8.484 \sin (377t - 90°)$,
$i_T = 11.03$
$$\sin (377t - 50.31°)$$

23. a. $Y_T = 0.32 \,\underline{/51.24°}$

c. $C = 1326 \,\mu F$,
$L = 10.61$ mH
$E = 11.05$
$$\times 10^{-3} \,\underline{/-71.34°},$$
$I_R = 2.21$
$$\times 10^{-3} \,\underline{/-71.34°},$$
$I_L = 2.76$
$$\times 10^{-3} \,\underline{/-161.34°},$$
$I_C = 5.53 \times 10^{-3} \,\underline{/18.66°}$

g. 24.42 μW

h. $F_p = 0.625$ leading

i. $e = 15.62 \times 10^{-3}$
$$\sin (377t - 71.34°),$$
$i_R = 3.12 \times 10^{-3}$
$$\sin (377t - 71.34°),$$
$i_L = 3.90 \times 10^{-3}$
$$\sin (377t - 161.34°),$$
$i_C = 7.82 \times 10^{-3}$
$$\sin (377t + 18.66°)$$

25. a. $I_1 = 17.37 \,\underline{/69.74°}$,
$I_2 = 9.92 \,\underline{/-20.26°}$,
$Z_T = 30.16 + j17.23$

b. $I_1 = 15 \,\underline{/30°}$,
$I_2 = 9 \,\underline{/210°}$,
$Z_T = 15 \,\underline{/90°}$

27. a. $E = 10 \,\underline{/-66.87°}$,
$I_R = 4 \,\underline{/-66.87°}$,
$I_L = 5 \,\underline{/-156.87°}$

b. $F_p = 0.8$ leading

c. 40 W **f.** $4 \,\underline{/23.13°}$

g. $Z_T = 1.6 - j1.2$

29. $Y = G_T - jB_L = 0.3 - j0.265$
$= > R = 20 \,\Omega$, $X_L = 3.77 \,\Omega$

Chapter 16

1. a. $Z_T = 1.2 \,\underline{/90°}$

b. $I = 10 \,\underline{/-90°}$

c. $I_1 = 10 \,\underline{/-90°}$

d. $I_2 = 6 \,\underline{/-90°}$,
$I_3 = 4 \,\underline{/-90°}$

e. $V_L = 60 \,\underline{/0°}$

3. a. $Z_T = 4.07 \,\underline{/-12.53°}$,
$Y_T = 0.246 \,\underline{/12.53°}$

b. $I_T = 14.74 \,\underline{/42.53°}$

c. $I_2 = 3.997 \,\underline{/83.13°}$

d. $V_C = 47.96 \,\underline{/-6.87°}$

e. 863.17 W

5. a. $I = 179.89 \times 10^{-3} \,\underline{/23.48°}$

b. $V_C = 70.71 \,\underline{/-45°}$

c. 16.5 W

7. a. $I_1 = 0.674 \,\underline{/48.13°}$

b. $V_1 = 22.99 \,\underline{/16.70°}$

c. 17.99 W

9. a. $Y_T = 0.107 \,\underline{/-10.31°}$

b. $V_1 = 18 \,\underline{/30°}$

c. $I_3 = 1.91 \,\underline{/10.81°}$

d. $V_2 = 10.77 \,\underline{/57.81°}$
$V_{ab} = 13.37 \,\underline{/100.81°}$

11. $I = 30.86 \,\underline{/22.87°}$

13. 257.07 mW

Chapter 17

3. a. $E = 8.245 \,\underline{/-15.96°}$,
$Z_s = 4 - j16$

b. $E = 8.485 \,\underline{/165°}$,
$Z_s = 3 + j3$

5. a. $5.15 \,\underline{/24.5°}$

b. $0.442 \,\underline{/143.48°}$

7. a. $13.07 \,\underline{/-33.71°}$

b. $28.49 \,\underline{/-3.13°}$

9. a. $5.15 \,\underline{/24.5°}$

b. $0.442 \,\underline{/143.48°}$

11. a. $13.07 \,\underline{/-33.71°}$

b. $28.49 \,\underline{/-3.13°}$

13. $I_L = 3.165 \times 10^{-3} V \,\underline{/137.29°}$

15. a. V_1 (left-hand node)
$$= 14.68 \,\underline{/68.89°},$$
V_2 (right-hand node)
$$= 12.97 \,\underline{/155.88°}$$

b. V_1 (left-hand node)
$$= 5.12 \,\underline{/-79.36°},$$
V_2 (right-hand node)
$$= 2.71 \,\underline{/39.96°}$$

17. a. V_1 (far left node)
$$= 5.74 \,\underline{/122.76°},$$
V_2 (center node)
$$= 4.04 \,\underline{/145.03°},$$
V_3 (far right node)
$$= 25.94 \,\underline{/78.07°}$$

b. V_1 (top left node)
$$= 15.13 \,\underline{/1.29°},$$
V_2 (left center node)
$$= 17.24 \,\underline{/3.73°},$$
V_3 (right center node)
$$= 10.59 \,\underline{/-0.11°}$$

19. $V_L = 171.63I \,\underline{/-120.70°}$

21. a. yes **b.** $I_C = 0$ A

c. $V_C = 0$ V
23. $L_x = 5$ mH, $R_x = 5\ \Omega$
27. a. $I = 12.02\ /-38.61°$
  b. $I = 7.02\ /20.56°$

Chapter 18

1. a. $9.18\ /-25.9°$
  b. $3.77\ /-93.8°$
3. $178.55 \times 10^{-3}\ /-26.57°$
5. $70.61 \times 10^{-3}\ /-11.31°$
7. $2.944 \times 10^{-3}\ /0°$
9. a. $Z_{Th} = 12.9\ /-7.12°$
    $E_{Th} = 36.41\ /-88.8°$
  b. $Z_{Th} = 6.32\ /-33.46°$,
    $E_{Th} = 60.68\ /30.21°$
11. $Z_{Th} = 5.1 \times 10^3\ /-11.31°$
    $E_{Th} = 10$ V
13. $Z_{Th} = 20 \times 10^3\ /0°$,
    $E_{Th} = -3990\ /0°$
15. $Z_{Th} = 25 \times 10^3\ /0°$,
    $E_{Th} = -1800\ /0°$
17. $Z_{Th} = 105 \times 10^3\ /0°$,
    $E_{Th} = 315\ /0°$
19. a. $Z_N = 12.9\ /-7.12°$,
    $I_N = 2.82\ /95.94°$
  b. $Z_N = 6.32\ /-33.46°$,
    $I_N = 9.6\ /63.73°$
21. a. $Z_N = 9.66\ /14.93°$,
    $I_N = 2.15\ /-42.87°$
  b. $Z_N = 4.37\ /55.67°$,
    $I_N = 22.83\ /-34.65°$
23. $Z_N = 4.44 \times 10^3\ /-0.031°$,
    $I_N = 100I\ /0.286°$
25. $Z_N = 25 \times 10^3\ /0°$,
    $I_N = 72 \times 10^{-3}\ /0°$
27. $Z_N = 1.9625 \times 10^3\ /0°$,
    $I_N = 2 \times 10^{-3}\ /0°$
29. a. $Z_L = 11.035\ /77.03°$,
    $P_M = 90$ W
  b. $Z_L = 5.71\ /64.30°$,
    $P_M = 618.33$ W
31. $I_{ab} = \frac{4}{3} \times 10^{-3}\ /0°$ (current
      source of this value),
    $V_{ab} = 10.67\ /0°$ (voltage
      source of this value)
33. $25.77 \times 10^{-3}\ /104.04°$

Chapter 19

1. a. $R: 300$ W, $L: 0$ W, $C: 0$ W
  b. $R: 0$ var, $L: 900$ var,
    $C: 500$ var
  c. $R: 300$ VA, $L: 900$ VA,
    $C: 500$ VA

d. $P_T = 300$ W,
    $Q_T = 400$ var (L),
    $S_T = 500$ VA,
    $F_p = 0.6$ lagging
  f. $W_R = 5$ J
  g. $W_C = 1.327$ J,
    $W_L = 2.389$ J
3. a. $P_T = 1200$ W,
    $Q_T = 1200$ var (L),
    $S_T = 1697$ VA,
    $F_p = 0.7071$ lagging
  c. $I = 8.485\ /-45°$
5. a. 720 W, 0 var, 720 VA
  b. 0 W, 360 var, 360 VA
  c. $P_T = 1095$ W,
    $Q_T = 960$ var (L),
    $S_T = 1456.24$ VA,
    $F_p = 0.752$ lagging
  d. $I = 24.271\ /-11.24°$
7. a. $R: 150$ W, $L: 0$ W
  b. $R: 0$ var, $L: 200$ var,
    $C: 250$ var
  c. $R: 150$ VA, $L: 200$ VA,
    $C: 250$ VA
  d. $P_T = 150$ W,
    $Q_T = 50$ var (C),
    $S_T = 158.11$ VA,
    $F_p = 0.949$ leading
  f. $W_R = 2.5$ J
  g. $W_L = 0.531$ J,
    $W_C = 0.663$ J
9. a. $Z_T = 2 - j3.464$
  b. 5000 W
11. a. $P_T = 300$ W,
    $Q_T = 400$ var (C),
    $S_T = 500$ VA,
    $F_p = 0.6$ leading
  b. $I_T = 16.67\ /53.13°$
  d. Box (600 var (C)): $R = 0$,
    $L = 0$, $X_C = 2.159\ \Omega$,
    Box (200 var (L)): $C = 0$,
    $R = 1.0796\ \Omega$,
    $X_L = 0.7197\ \Omega : Z_T$
    $= 1.0796 - j1.4393$
13. a. $P_T = 1250$ W,
    $Q_T = 2366.26$ var (C),
    $S_T = 2676.14$ VA,
    $F_p = 0.4671$ leading
  b. $E = 535.23\ /-62.15°$
  c. Box (100 W): R
      $= 1743.38\ \Omega$,
    $X_C = 1307.53\ \Omega$,
    Box (1000 W): $R = 43.59\ \Omega$,
    $X_C = 99.88\ \Omega$

15. a. 128.14 W
  b. $a\text{-}b$: 42.69 W,
    $b\text{-}c$: 64.03 W,
    $a\text{-}c$: 106.72 W,
    $a\text{-}d$: 106.72 W,
    $c\text{-}d$: 0 W, $d\text{-}e$: 0 W,
    $f\text{-}e$: 21.34 W
17. a. $I = 2.358$ A,
    $P = 22.24$ W
  b. $I = 2.070$ A,
    $L = 89.6$ mH
  c. $P = 28.9$ W,
    $L = 107.77$ mH

Chapter 20

1. a. $\omega_s = 250$ rad/s,
    $f_s = 39.81$ Hz
  b. $\omega_s = 3535.53$ rad/s,
    $f_s = 562.98$ Hz
  c. $\omega_s = 21,880$ rad/s,
    $f_s = 3484.08$ Hz
3. a. $30\ \Omega$
  b. $I = 10\ /0°$, $V_R = 50\ /0°$,
    $V_L = 300\ /90°$,
    $V_C = 300\ /-90°$
  d. $P_T = 500$ W,
    $Q_L = 3$ kvar,
    $Q_C = 3$ kvar
  e. 6
5. a. 3.91 mH
  b. $I = 2.5\ /0°$, $V_R = 10\ /0°$,
    $V_L = 110.5\ /90°$,
    $V_C = 110.5\ /-90°$,
    $Q_S = 11.05$
  c. $f_2 = 1883.92$ Hz,
    $f_1 = 1721.1$ Hz
  d. $BW = 162.82$ Hz
7. a. 250 Hz
  b. $f_2 = 10,125$ Hz,
    $f_1 = 9875$ Hz
  c. 40
  d. $V_L = 1200\ /90°$,
    $V_C = 1200\ /-90°$
  e. 180 W
9. a. 600 Hz **b.** 5700 Hz
  c. $X_L = X_C = 19\ \Omega$
  d. $L = 0.531$ mH,
    $C = 1.47\ \mu F$
11. $X_L = X_C = 40\ \Omega$,
    $f_s = 8 \times 10^3$ Hz,
    $L = 0.796$ mH,
    $C = 0.5\ \mu F$,
    $f_2 = 8200$ Hz,
    $f_1 = 7800$ Hz

13. a. 30 Ω **b.** 225.5 Ω

  c. $\mathbf{I}_C \cong 0.6 \; /90°$,
  $\mathbf{I}_L \cong 0.6 \; /-86.19°$

  d. $L = 238.85 \; \mu H$,
  $C = 265 \; nF$

  e. $Q_p = 7.52$,
  $BW = 2659.6 \; Hz$

15. a. 10,071 Hz

  b. $X_L = X_C = 316.23 \; \Omega$

  c. 158.12 (high) **d.** 22.22 kΩ

  e. $\mathbf{I}_C = 0.281 \; /90°$,
  $\mathbf{I}_L = 0.281 \; /-89.64°$

  f. $Q_p = 70.27$,
  $BW = 143.32 \; Hz$

17. a. 3560.7 Hz

  b. $\mathbf{V}_C = 138.2 \; /0°$

  c. 691 mW **d.** 576.17 Hz

19. a. 1000 Ω **b.** 500

  c. 19.23 kHz **d.** 1.923 V

21. $X_L = X_C = 3 \; \Omega$,
  $f_p = 15 \; kHz$, $L = 31.85 \; \mu H$,
  $C = 3.54 \; \mu F$, $R_l = 0.1 \; \Omega$

23. a. 20

  b. $BW = 1000 \; Hz$,
  $f_1 = 19,500 \; Hz$,
  $f_2 = 20,500 \; Hz$

  c. $Q_{p(loaded)} = 18.52$,
  $BW = 1079.91 \; Hz$,
  $f_2 = 20,539.96 \; Hz$,
  $f_1 = 19,460.04 \; Hz$

  d. $Q_{p(loaded)} = 14.29$,
  $BW = 1399.58 \; Hz$,
  $f_2 = 20,699.79 \; Hz$,
  $f_1 = 19,300.21 \; Hz$

25. b. complement of problem 23

  c. 100 kΩ, 20 kΩ ≫ 400 Ω
  ∴ Q_p unaffected

27. a. $L_s = 12.68 \; mH$,
  $L_p = 128.19 \; mH$

Chapter 21

1. a. 120.1 V **b.** 120.1 V

  c. 12.01 A **d.** 12.01 A

3. a. 120.1 V **b.** 120.1 V

  c. 16.98 A **d.** 16.98 A

5. a. $\theta_2 = -120°$, $\theta_3 = +120°$

  b. $\mathbf{V}_{an} = 120 \; /0°$,
  $\mathbf{V}_{bn} = 120 \; /-120°$,
  $\mathbf{V}_{cn} = 120 \; /120°$

  c. $\mathbf{I}_{an} = 8 \; /-53.13°$,
  $\mathbf{I}_{bn} = 8 \; /-173.13°$,
  $\mathbf{I}_{cn} = 8 \; /66.87°$

  e. 8 A **f.** 207.84 V

7. $V_{an} = V_{bn} = V_{cn} = 127.0 \; V$,
  $I_{an} = I_{bn} = I_{cn} = 8.98 \; A$,
  $I_{Aa} = I_{Bb} = I_{Cc} = 8.98 \; A$

9. a. 120.1 V **b.** 208 V

  c. 10.4 A **d.** 18.0 A

11. a. 120.1 V **b.** 208 V

  c. 36.75 A **d.** 63.651 A

13. a. $\theta_2 = -120°$, $\theta_3 = +120°$

  b. $\mathbf{V}_{ab} = 208 \; /0°$,
  $\mathbf{V}_{bc} = 208 \; /-120°$,
  $\mathbf{V}_{ac} = 208 \; /120°$

  d. $\mathbf{I}_{ab} = 1.471 \; /45°$,
  $\mathbf{I}_{bc} = 1.471 \; /-75°$,
  $\mathbf{I}_{ca} = 1.471 \; /165°$

  e. 2.548 A **f.** 120.1 V

15. $V_{ab} = V_{bc} = V_{ac} = 220 \; V$,
  $I_{ab} = I_{bc} = I_{ac} = 15.56 \; A$,
  $I_{Aa} = I_{Bb} = I_{Cc} = 26.95 \; A$

17. a. 208 V **b.** 120.1 V

  c. 9.43 A **d.** 9.43 A

19. $V_{an} = V_{bn} = V_{cn} = 69.28 \; V$,
  $I_{an} = I_{bn} = I_{cn} = 2.89 \; A$,
  $I_{Aa} = I_{Bb} = I_{Cc} = 2.89 \; A$

21. $V_{an} = V_{bn} = V_{cn} = 69.28 \; V$,
  $I_{an} = I_{bn} = I_{cn} = 5.77 \; A$,
  $I_{Aa} = I_{Bb} = I_{Cc} = 5.77 \; A$

23. a. 440 V **b.** 440 V

  c. 29.33 A **d.** 50.8 A

25. a. $\theta_2 = -120°$, $\theta_3 = +120°$

  b. $\mathbf{V}_{ba} = 100 \; /0°$,
  $\mathbf{V}_{cb} = 100 \; /-120°$,
  $\mathbf{V}_{ac} = 100 \; /120°$

  d. $\mathbf{I}_{ab} = 4 \; /0°$,
  $\mathbf{I}_{bc} = 4 \; /-120°$,
  $\mathbf{I}_{ca} = 4 \; /120°$

  e. $I_{Aa} = I_{Bb} = I_{Cc} = 6.928 \; A$

27. a. $\theta_2 = -120°$, $\theta_3 = 120°$

  b. $\mathbf{V}_{ba} = 100 \; /0°$,
  $\mathbf{V}_{cb} = 100 \; /-120°$,
  $\mathbf{V}_{ac} = 100 \; /120°$

  d. $\mathbf{I}_{ab} = 7.072 \; /45°$,
  $\mathbf{I}_{bc} = 7.072 \; /-75°$,
  $\mathbf{I}_{ca} = 7.072 \; /165°$

  e. $I_{Aa} = I_{Bb} = I_{Cc} = 12.25 \; A$

29. 8640 W, 0 var, 8640 A,
  $F_p = 1$

31. 16.224 kW, 16.224 kvar (Cap),
  22.944 kVA,
  $F_p = 0.7071$ leading

33. 7.263 kW, 7.263 kvar (Ind),
  10.272 kVA,
  $F_p = 0.7071$ lagging

35. 287.93 W, 575.86 var (Ind),
  643.83 VA,
  $F_p = 0.4472$ lagging

37. 900 W, 1200 var (Ind), 1500 VA,
  0.6 lagging

39. $\mathbf{Z} = 12.98 - j17.31$

41. b. $P_T = 2419.2 \; W$,
  P (each) = 806.4 W

43. b. 285 W

  c. $P_h = 200 \; W$, $P_T = 100 \; W$

45. a. $\mathbf{V}_{an} = 120.1 \; /0°$,
  $\mathbf{V}_{bn} = 120.1 \; /-120°$,
  $\mathbf{V}_{cn} = 120.1 \; /120°$

  b. $\mathbf{I}_{an} = 21.22 \; /-45°$,
  $\mathbf{I}_{bn} = 7.077 \; /-165°$,
  $\mathbf{I}_{cn} = 10.62 \; /75°$

  c. 3304.44 W,
  3304.44 var (L),
  4673.18 VA,
  $F_p = 0.7071$ lagging

Chapter 22

1. a. 125 V **b.** 277.5 V

  c. 138.75 V **d.** 125 V

Chapter 23

1. (I) **a.** no **b.** no **c.** yes
  d. no **e.** yes

  (II) **a.** yes **b.** yes **c.** no
  d. yes **e.** yes

  (III) **a.** yes **b.** yes **c.** no
  d. yes **e.** yes

  (IV) **a.** no **b.** no **c.** yes
  d. yes **e.** yes

5. a. $V_{av} = 100 \; V$,
  $V_{eff} = 107.53 \; V$

  b. $I_{av} = 3 \; A$, $I_{eff} = 3.36 \; A$

7. 333.52 W

9. a. $i = 3 + 3$
  $\sin (400t - 53.13°)$

  b. 3.67 A

  c. $v_R = 18 + 18$
  $\sin (400t - 53.13°)$

  d. 22.05 V

  e. $v_L = 0 + 24$
  $\sin (400t + 36.87°)$

  f. 16.97 V **g.** 80.81 W

11. a. $i = -10 + 2.36$
  $\sin (300t - 45°) - 0.75$
  $\sin (600t - 63.43°)$

  b. 10.15 A

  c. $v_R = -60 + 14.17$
  $\sin (300t - 45°) - 4.5$
  $\sin (600t - 63.43°)$

  d. 60.91 V

e. $v_L = 0 + 14.17$
$\sin(300t + 45°) - 8.99$
$\sin(600t + 26.57°)$
f. 11.87 V   g. 618.14 W

13. a. $v_0 = 2.54$
$\sin(754t - 94.57°)$
$- 2.45 \sin(1508t - 101.1°)$
b. 2.495 V   c. 6.225 mW

15. a. $80 + 100 \sin \omega t$
$+ 0 + 14.55$
$\sin(3\omega t + 69.9°)$
b. $15 + 70 \sin \alpha$
$+ 10 \sin(2\alpha - 180°)$
$+ 8.69 \sin(3\alpha + 166.7°)$

17. $e = 10 + 102.66$
$\sin(600t + 133.07°)$
$+ 100 \sin(1200t + 90°)$
$+ 108.97 \sin(1800t + 23.41°)$

Chapter 24

1. a. 0.2 H
b. $e_p = 1.6$ V, $e_s = 5.12$ V
c. $e_p = 15$ V, $e_s = 24$ V

3. a. 158.02 mH
b. $e_p = 24$ V, $e_s = 1.8$ V
c. $e_p = 15$ V, $e_s = 24$ V

5. 1.354 H

7. $I_1(R_1 + X_{L_1}) + I_2(X_m) \qquad = E_1$
$I_1(X_m) \qquad + I_2(X_{L_2} + R_L) = 0$
$\overline{x_m = -\omega M \; \underline{/90°}}$

9. a. 3.125 V
b. 391.02×10^{-6} Wb

11. 56.31 Hz

13. 400 Ω

15. 12,000t

17. a. $a = 3$ b. 2.78 W

19. a. $360.56 \; \underline{/86.82°}$
b. $332.82 \times 10^{-3} \; \underline{/-86.82°}$
c. $V_{R_e} = 6.656 \; \underline{/-86.82°}$
$V_{X_e} = 13.313 \; \underline{/3.18°}$,
$V_{X_L} = 106.50 \; \underline{/3.18°}$

23. a. 20   b. 83.33 A   c. 4.167 A
d. $a = \frac{1}{20}$, $I_s = 4.167$ A,
$I_p = 83.33$ A

25. a. $V_L = 25 \; \underline{/0°}$, $I_L = 5 \; \underline{/0°}$
b. $80 \; \underline{/0°}$ c. $20 \; \underline{/0°}$

27. a. $E_2 = 40 \; \underline{/60°}$,
$I_2 = 3.33 \; \underline{/60°}$,
$E_3 = 30 \; \underline{/60°}$,
$I_3 = 3 \; \underline{/60°}$
b. $64.478 \; \underline{/0°}$

29. $[Z_1 + X_{L_1}]I_1 - \qquad\qquad Z_{M_{12}}I_2 + \qquad\qquad Z_{M_{13}}I_3 = E_1$
$Z_{M_{12}}I_1 - [Z_2 + Z_3 + X_{L_2}]I_2 + \qquad\qquad Z_2I_3 = 0$
$Z_{M_{13}}I_1 - \qquad\qquad Z_2I_2 + [Z_2 + Z_4 + X_{L_3}]I_3 = 0$

Chapter 25

1. a. $Z_{11} = \dfrac{Z_1Z_2 + Z_1Z_3}{Z_1 + Z_2 + Z_3}$,
$Z_{12} = Z_{21} = \dfrac{Z_1Z_3}{Z_1 + Z_2 + Z_3}$,
$Z_{22} = \dfrac{Z_1Z_3 + Z_2Z_3}{Z_1 + Z_2 + Z_3}$

3. a. $y_{11} = \dfrac{Y_1Y_2 + Y_1Y_3}{Y_1 + Y_2 + Y_3}$,
$y_{12} = y_{21} = \dfrac{-Y_1Y_2}{Y_1 + Y_2 + Y_3}$,
$y_{22} = \dfrac{Y_1Y_2 + Y_2Y_3}{Y_1 + Y_2 + Y_3}$

5. a. $h_{11} = \dfrac{Z_1Z_2}{Z_1 + Z_2}$,
$h_{21} = \dfrac{-Z_1}{Z_1 + Z_2}$,
$h_{12} = \dfrac{Z_1}{Z_1 + Z_2}$,
$h_{22} = \dfrac{Z_1 + Z_2 + Z_3}{Z_1Z_3 + Z_2Z_3}$

7. a. $h_{11} = \dfrac{Y_1 + Y_2 + Y_3}{Y_1Y_2 + Y_1Y_3}$,
$h_{21} = \dfrac{-Y_2}{Y_2 + Y_3}$,
$h_{12} = \dfrac{Y_2}{Y_2 + Y_3}$,
$h_{22} = \dfrac{Y_2Y_3}{Y_2 + Y_3}$

9. a. 47.62   b. -99

11. $Z_i = 9219.5 \; \underline{/-139.40°}$,
$Z_0 = 29.07 \times 10^3 \; \underline{/-86.05°}$

13. $h_{11} = 2.5$ kΩ, $h_{12} = 0.5$,
$h_{21} = -0.75$,
$h_{22} = 0.25 \times 10^{-3}$ S